Synthetic Aperture Radar Image Processing Algorithms for Nonlinear Oceanic Turbulence and Front Modeling

Synthetic Aperture Radar Image Processing Algorithms for Nonlinear Oceanic Turbulence and Front Modeling

Maged Marghany
Innovative Information Sdn. Bhd., Safanad Information Technology,
Kuala Lumpur, Malaysia

ELSEVIER

Elsevier
Radarweg 29, PO Box 211, 1000 AE Amsterdam, Netherlands
125 London Wall, London EC2Y 5AS, United Kingdom
50 Hampshire Street, 5th Floor, Cambridge, MA 02139, United States

ISBN: 978-0-443-19155-8

For Information on all Elsevier publications
visit our website at https://www.elsevier.com/books-and-journals

Publisher: Candice G. Janco
Acquisitions Editor: Maria Elekidou
Editorial Project Manager: Aleksandra Packowska
Production Project Manager: Rashmi Manoharan
Cover Designer: Mark Rogers

Typeset by MPS Limited, Chennai, India
Transferred to Digital Printing in 2024

Dedication

The book is dedicated to my mother Faridah and Aleksandra Packowska and Richard Feynman and Michio Kaku taught me that a real distinguished professor invents novelty not exploiting his students and colleagues to award the highest H-index by false publications.

Richard Feynman and Michio Kaku also instilled in me the understanding that fake professors are those who collaborate with extensive networks without any contribution, particularly in publishing, to conceal the lack of authenticity in their positions.

Contents

Preface

This book was accomplished when I was ranked among the top 2% of scientists in a global record compiled four times by Stanford University. This book explores the inclusive implementation of synthetic aperture radar (SAR) for turbulence and front dynamics. In this sense, the book is devoted to demonstrating the capabilities of SAR in nonlinear ocean dynamic studies without integration with optical remote sensing sensors. In the majority of developed countries, researchers do not believe in SAR imaging mechanism capabilities in ocean dynamic explorations without referring to optical remote sensing sensors due to the absence of spectral signatures in SAR data. Consequently, the literature involves repeated and similar image processing and modeling tools in ocean dynamic explorations either using remote sensing techniques or conventional models over several decades. In other words, all published research work for instance in the South China Sea territory just paraphrases previous work done in several decades using open-source models without understanding the mathematical and numerical analysis behind these tools. This sort of constraint does not allow for further developments in the field of ocean dynamic models and remote sensing image processing. Thus numerous remote sensing oceanography works in a developed country are ill-defined.

Since the initiation of satellite oceanography applications, there has been no precise algorithm development to state the complete image of the nonlinearity of ocean turbulence and ocean front in SAR data. This book, consequently, is assigned to provide novel algorithms in modeling and simulation of nonlinear turbulence flow and ocean front features in SAR data. The majority of the advanced algorithms explored in this book are based on quantum computing. In this view, this book is filling a gap between advanced quantum computing algorithms, microwave remote sensing, and nonlinearity turbulence and front dynamic features with a new generation of microwave remote sensing technology such as Sentinel-1, and TerraSAR-X satellite data, which are implemented to deliver a wide understanding of the mechanism of radar in imaging and modeling the nonlinearity of ocean turbulence and ocean front flows.

Chapter 1 provides an in-depth examination of the concept of nonlinearity, focusing on the significant gap that separates radar satellite images from the intricate dynamics of oceanic turbulence. The elucidation and resolution of this gap are presented through the lens of nonlinearity concepts, utilizing mathematical philosophical foundations, especially in the context of nonlinear differential equations, to precisely define nonlinearity. Therefore Chapter 2, which is a key chapter, typically discusses the complex relationship between turbulence and front zone formation, as oceanic fronts form in specific areas of extreme turbulence and convection. Consequently, Chapter 3 presents a novel perspective on integrating quantum mechanics to elucidate the complexities of SAR. Distinguishing itself from previous works, this chapter avoids an exhaustive exploration of radar theory and signal processing, as these topics have been thoroughly addressed elsewhere. Instead, the focus is on introducing a pioneering concept of quantization to comprehend the mechanisms involved in radar imaging of sea turbulence. Therefore this chapter is the keystone to bridge the gap between nonlinearity between microwave remote sensing, quantum computing algorithms, and turbulence dynamic flows.

Thus Chapter 4 unveils a groundbreaking discovery—the quantized Marghany turbulence boundary in the South China Sea. This finding aligns with established Western boundary theories by Sverdrup, Stommel, Ekman, and Munk. However, these theories fall short in tracking or simulating turbulence flow under wind stress curl. In a recent chapter, the author introduces the "quantized Marghany turbulence boundary" theory and an accompanying algorithm, "quantized Marghany turbulence spectra energy," designed to reconstruct and track this boundary in SAR data across diverse wind monsoon periods at a large scale. In this sense, the "quantized Marghany turbulence boundary" is a groundbreaking theory introduced in the previous chapter, outlining a new turbulence boundary in the South China Sea. This boundary is formed by the entanglement and superposition of features within the same time and topology zone. The turbulence features within the quantized Marghany turbulence boundary showcase nonlinear characteristics, highlighting its nature as a nonlinear turbulence boundary. Chapter 5 delves into the discussion of the nonlinearity of internal waves within this boundary, specifically at the western boundary of Peninsular Malaysia.

Hence, Chapter 6 extends the early theory of the quantized Marghany's front, initially discovered and detailed in the author's earlier work "Nonlinear Ocean Dynamics." It explores the manifestation of a shallow coastal front on the east coast of the Malaysian Peninsula within the context of nonlinear ocean dynamics. In 1994, the author identified the phenomenon of "quantized bound front states," triggered by slight variations in water masses, eventually leading to the Marghany front's identification.

Utilizing advanced quantum image processing tools, the chapter demonstrates the automated detection of the quantized Marghany bound front state (QMBFS) in a single SAR dataset. Additionally, the team introduces groundbreaking quantum image fusion technology to track thermal fronts using multi-SAR and MODIS data.

Therefore Chapter 7 serves as a bridge between earlier discussions, presenting a novel method for detecting QMBFSs within the quantized Marghany turbulence boundary in multi-SAR images. This approach leverages texture algorithms, including dissimilarity, entropy, and homogeneity. The chapter explores the impact of speckles on the topology of turbulence occurrence imaging in SAR images. Notably, homogeneity outperforms other texture algorithms in identifying quantized Marghany turbulence boundaries. Meanwhile, the quantum entropy algorithm excels in detecting various QMBFSs, particularly in regions abundant with filaments such as Sumatra, Bali, and the coastal waters of the Netherlands.

Chapter 8 introduces two innovative theories: (1) a newly developed quantum algorithm, quantum_superposition_MDD, based on quantum superposition and Marghany Distance Decomposition. This algorithm enhances the visualization of spiral flows, providing sharper patterns, and (2) the Theory of Quantized Spiral Turbulence, a novel framework shedding light on the occurrence and characteristics of spiral patterns in turbulent fluid dynamics. This theory is particularly applicable to coastal regions and internal wave zones, addressing a significant gap in Malaysian coastal waters' spiral flow studies.

It is noteworthy that this work contributes to pioneering research in spiral flow studies within Malaysian coastal waters. The focus on fundamental research has been a characteristic of Malaysian oceanography researchers for decades, given the absence of ocean dynamic studies. Their reliance on in situ data collection and open-source software has shaped their research approach over the years.

Chapter 9 unveils a pioneering method for the automatic detection of oceanic fronts and turbulence structures. This approach combines a multifractal dimension algorithm with a wavelet-based multifractal technique. The method derives multifractal measures by analyzing structure functions, with a specific focus on the Lagrangian structure function. The insights gained from this chapter notably advance our understanding of ocean fronts and turbulence structures, particularly in the context of analyzing multi-SAR data.

Chapter 10, hence, presents a groundbreaking algorithm designed to automatically identify rain cells causing turbulence on the sea surface using SAR data. This chapter acquaints with the innovative concept of understanding the interactions between rain cells and the sea surface, crucial for identifying rain cell imprints in SAR data. Coined as the "Quantized Marghany Rain Interaction Coefficient (QMRIC)," this concept establishes a parameter to quantify the quantum behavior exhibited by rain cells during their interaction with the sea surface. The QMRIC serves as a metric to measure the level of uncertainty or potential entanglement resulting from raindrop impacts.

Chapter 11 introduces an inventive method for automatically detecting upwelling from SAR images. This addresses challenges posed by the limited thermal bands in SAR compared to optical data. The proposed approach harnesses the quantum cellular automaton, a pioneering technique, to identify upwelling zones. It is amazing that Chapter 11 validates the existence of the quantized Marghany turbulence boundary and the QMBFS, explored in detail in Chapters 4 and 5.

Chapter 12, subsequently, introduces a revolutionary method for automatically identifying eddy and vorticity flows in SAR data. Leveraging platforms such as ALOS PALSAR and Sentinel HH polarization data, the quantum multiobjective evolutionary algorithm is developed and evaluated. The effectiveness of this approach is demonstrated using SAR imagery captured over the Farasan Islands in the Red Sea and the Mozambique Channel. The method involves the immediate transformation of SAR backscatter and texture information into qubit chromosomes.

Finally, Chapter 13 presents a pioneering method for the automated detection of shear-wave chaotic turbulence flow spectra and associated dynamic features in SAR data, specifically sequences from TerraSAR-X. This method employs quantum hologram interferometry based on 4D phase unwrapping, transforming 2D TerraSAR-X data into 4D images using a quality map derived from the 4D quantum walk search. The resulting 4D images unveil the Lyapunov spectrum, a crucial indicator of the intricate nature of shear-wave chaotic flows and supporting evidence for quantum entanglement theory. The outcomes underscore the significant impact of 4D shear-wave chaotic flows in generating phenomena such as eddies, vorticity, and quantized Marghany turbulence boundary layers.

I would like to express my gratitude to the editorial project manager, Aleksandra Packowska, who consistently provided me with the opportunity to publish not only this book but also previous ones. Their unwavering commitment has been instrumental in shaping this book into a valuable repository of novel knowledge. Without her dedication, this achievement would not have been possible.

Distinguished Prof. Dr. Maged Marghany

Chapter 1

Nonlinearity: fundamental concepts and understanding

1.1 What is the magic of nonlinearity?

The precise definition of nonlinearity is not restricted to a specific one. Numerous definitions have been delivered from different views of fields of science and engineering. For instance, from the mathematical of point view, nonlinearity represents the attribute of a function that articulates different system correlations that are not in direct form. However, in reliable electricity, the nonlinearity is sufficiently expressed as the considerable divergence of a recorded—outturn electric current or signal affiliation with one of the straightforward proportionalities. Put differently, nonlinearity indicates the performance of a circuit, principally an amplifier, in which the yield signal power does not fluctuate in direct proportion to the input signal intensity (Roberts et al., 2006; Snieder, 1998). Therefore, in a nonlinear circuit, the output-to-input amplitude ratio (correspondingly termed the gain) relies on the input signal strength. Consequently, what is nonlinearity of physics? In physics and alternative sciences, nonlinearity is contrary to linearity, in which a system does not fulfill the principle of superposition, which suggests that the yield is not straightforwardly proportional to the input variables or causes.

Consequently, in science and mathematics, nonlinearity is a complex in which the change of the output is not proportional to the change of the input. Nonlinear problems, therefore, realistically are of considerable interest to aware engineers, modern biologists, famed physicists, notable mathematicians, and many other scientists because most systems are inherently nonlinear in complex nature. In this comprehensive view, the nonlinearity of dynamical systems, portraying fluctuations in variable quantity throughout considerable time, possibly will seem erratic, chaotic, or counter-spontaneous, analogizing with considerable elementary linear systems (Aarseth, 1994; Barnett & Serletis, 2000; Berger, 1977; Chen et al., 2010).

Subsequently, in statistics, nonlinearity identifies the circumstances allying two impartial systems or variables, whether a nonstraight line or direct correlation between independent parameters and dependent ones. In other distinct words, both necessary changes in specific input and desired output (Fig. 1.1) in the nonlinear system over appropriate time are not correlated or proportional directly to each other.

Naturally, the performance of nonlinearity is pronounced in mathematics by a nonlinear mathematical problem. The significant question realistically is: what are remarkably the nonlinear system equations? It is an array of synchronized identifications wherein the unpredicted (or the predicted functions in the instance of differential equations) seem as the variable quantity of a polynomial of degree, for instance, which is sophisticated than one or the claim of a function, which is not expressed as a polynomial of degree 1. In this understanding, the nonlinearity mathematical correspondence solutions cannot be in the form of the linear amalgamation of unknown functions or parameters. Consequently, the nonlinearity can be addressed, nonetheless of whether identified linear operations exist in the equations. In precise, a differential equation is identified as linear in the circumstances its unknown variables or functions and its derivatives are

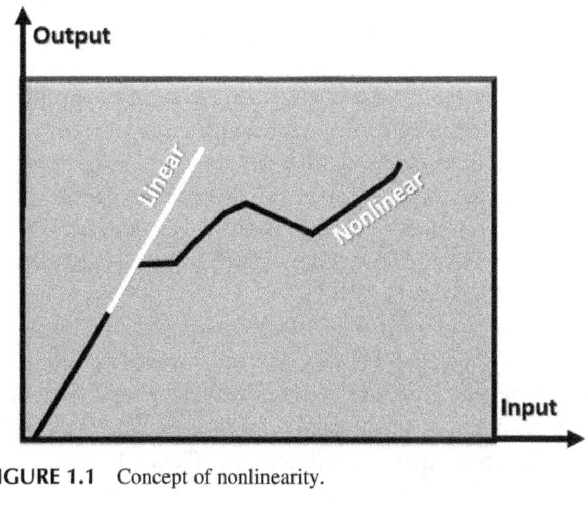

FIGURE 1.1 Concept of nonlinearity.

Synthetic Aperture Radar Image Processing Algorithms for Nonlinear Oceanic Turbulence and Front Modeling. DOI: https://doi.org/10.1016/B978-0-443-19155-8.00001-5

linear, although other nonlinear variables have existed (Aarseth, 1994; Coppens et al., 1965; Cortina, 1993; Emancipator & Kroll, 1993; Fletcher & Gent, 1954).

Nonlinearity could be a universal matter once inspecting reason and repercussion correlations. Such specific cases necessitate convoluted modeling, and postulation analysis to tentatively suggest the needed clarifications of nonlinear consequences. Nonlinearity, therefore, deprived of clarification, perhaps precedes chaotic and unreliable consequences. For specific instances, in sensitive devices and systems that exploit digital transition, and correspondingly in frequency modulation wireless transmitters, nonlinearity is perhaps tolerated. Therefore, these signals are either full-on or full-off; the amplitude waveforms are not equivalent, and eventually, analog distortion is not accurately fixed. In this pessimistic view, nonlinear circuits commonly trigger deliberate distortion in applications, for instance, hi-fi audio and amplitude-modulation (AM) wireless communication. Generally, both cause and effect cannot be simply equated indirectly in which the system consequences are often unpredicted and chaotically correlated with dependent and independent variables (Coppens et al., 1965; Haber, 1985; Liu & Li, 2015).

1.2 Nonlinearity and dynamics

Conventional science obeys the laws of linearity. Newton's prevailing theories of fundamental physics were originally premised on the consistent basis of a static and steady universe. Convenient time and considerable space adequately maintained sequential abstractions; they represent genuinely independent measures without any starting or finish, providing mostly a framework for which events occurred. The deterministic and absolute system of specific information that merely constituted the chaotic nature of the chaotic universe and was grounded on Newtonian mechanics and Euclidean geometry began precisely to unravel in the late-19th century. Before Einstein's relativity, Hendricks Lorentz genuinely thought he was close to explaining the ultimate mysteries of universal physics.

The issues and problems with classical science nonlinearity suggested by these early scientists, first proposed by Maxwell in electromagnetic and thermodynamics and later by Plank in electrodynamics, prove to be the foundations for a history of nonlinear science, dynamics, and complexity (Aarseth, 1994; Barnett & Serletis, 2000; Fletcher & Gent, 1954; Ghil, 2019; Haber, 1985; Kroll et al., 2000; Liu & Li, 2015).

As the assumptions in electrodynamics set by Hans Oersted and Michel Faraday were demolished, scientists who believed structures were universal and absolute began to disclose the dynamic qualities of the atom. By 1915, Einstein had formulated the General Theory of Relativity, based on Plank's work from the turn of the century. Space and time are intertwined and relative to the observer, as revealed by Relativity. Furthermore, by adding a fourth dimension to traditional science's spatial dimensions, space-time becomes a dynamic non-Euclidean geometry rather than being flat and constant. Consequently, quantum physicists such as Bohr, Heisenberg, Schrödinger, Dirac, and others faced questions akin to those posed by Greek philosophers who pondered atomic structure and scratched their chins over the nature of knowledge as they questioned classical assumptions. As a consequence of Heisenberg's measuring difficulties, the natural world's complex and dynamic patterns were revealed.

Whenever Einstein's theoretical approach cannot explain the linear determination of world boundaries in Minkowski's space-time matrix mechanics, physics must provide explanations. Given the dynamic topography, it is indeed questionable if the Euclidean metric can be tweaked to resolve the inconsistencies between relativity and quantum theory, or if nonlinear quantum physics through string theory is required. Nonlinear mathematics has remained consistent with recent advancements in physics, and a new study paradigm is already evolving.

The uncertainties of linearity led the scientist to create the first mathematical model of nonlinear processes, the Lorenz Attractor, which became basic to dynamical systems and a symbol for "chaos theory" throughout the 1960s. The 1960s' culture and nonlinear fads regularly caused unorthodox scientists of an uncommon science to be dismissed. Nonlinearity gained respectability once Lorenz's work on paper was proven by an experiment utilizing a water wheel test to generate the fluid dynamical qualities (not unlike an experiment proposed by Newton). Therefore, Benoit Mandelbrot revealed theoretical geometries of nature, which can be accomplished by classical Euclidean dimensions in time and space when battling with conventional statistics while working with economic data. While nonlinear dynamics struggled to develop much of applied science, with challenges even for physics, but success in areas such as fluid dynamics and light, the emerging mathematical discipline of theoretical dynamics thrived. However, according to Karl Popper, nonlinear science and dynamics essentially depicted a universe of purely speculative mathematical forms. Despite the challenges, Thomas Kuhn's history and philosophy of science aided the early abstract mathematicians' work as biologists and ecologists became interested in nonlinearity (Aarseth, 1994; Barnett & Serletis, 2000; Snieder, 1998; Tsay, 1986).

Nowadays, scientists who have to use nonlinear techniques frequently argue that linear methods are just indefensible. Nonetheless, nonlinear science, originally defined by Penrose, does not fully bridge the "Platonic" realm of ideal mathematical forms in theory with empirical fact in the physical universe.

The most compelling reason for the advancement of nonlinear dynamic systems is that if linear techniques have proven problematic in the scientific and hard sciences, then linear methods cannot claim infallibility in the different field of sciences. Physical science advancements at the outset of the 20th century showed an immeasurably more dynamic and complex universe than classical scientists had previously recognized. Mathematicians advanced dynamical models, but applied fields encounter challenges, which are similar to measurement issues in physics and the physical sciences. The difficulties of nonlinear approaches coexist with the constraints of linearity (Barnett & Serletis, 2000; Berger, 1977; Chen et al., 2010; Coppens et al., 1965; Cortina, 1993).

One thing to remember is that nonlinear techniques have constraints, whereas linear models have advantages and risks. Continued research techniques are critical to creating applicable nonlinear approaches with changing assumptions in a rapidly evolving world, as a consequence of a modification in perceptions, as evaluated through science philosophy in the formation of technological principles. Needless to say, nonlinear approaches provide guidance for overcoming barriers with the logic of linear assumptions. They cannot, however, be undermined. Computational theory, current mathematics, and the hard sciences do not take the place of induction, qualitative judgment, and practical reasoning. Continuing work on these themes intends to build mixed-methods approaches in which qualitative research and legal logic contribute to the construction of a framework for technological ethics. As explored further in the next sections and chapters, this conception is crucial to bridging the gap between the nonlinearity of the ocean dynamic system and radar signal interaction.

1.3 Mathematical definition of nonlinearity

There are various sorts of models in the realm of mathematics that scientists, economists, statisticians, and other professionals are using to forecast, analyze, and explain the universe around them. These equations integrate variables in a manner that allows one to influence or forecast the output of another. Linear equations are the most common method of analysis in elementary mathematics, but nonlinear equations dominate in advanced math and science. The significant question is: what is nonlinear and what is an example? A nonlinear equation is one in which the maximum degree of a term is 2 or more than 2. For instance, $4x^2 + 3x + 1 = 0$ and $5x + 4y = 7$ are nonlinear equations since first equation has the maximum degree of 2 and the second equation comprises variables x and y. Needless to say, a nonlinear equation has the generic form $ax^2 + by^2 = c$, where a, b, and c are constants, and a^0, x, and y are variables. On the contrary, the general form of a linear equation is $ax + b = c$, where a, b, c are constants, and a^0 and x and y are variables (Barnett & Serletis, 2000; Berger, 1977; Chen et al., 2010; Cortina, 1993).

The easiest form of a linear equation is $y = a + bx$, where a and b represent constants in an equation and two variables are present. This is the foundation of the linear equation. Whenever this linear equation is plotted on graph paper, it produces a straight line that passes through the origin. It has a constant slope value along the straight line that passes through the origin (Fig. 1.2). In contrast, while discussing linear and nonlinear equations, it is crucial to understand that linear equations will have no exponents, while nonlinear equations will have exponents raised to powers greater than 1.

In other words, the degree of a nonlinear equation must be at least 2 or greater. The curvature of the graph rises as the degree of the equation grows (Fig. 1.3). According to this viewpoint, the superposition principle does not apply to systems with nonlinear equations. As a result, as discussed earlier in Section 1.1, the input and output of a nonlinear system are not directly related. In this understanding, a nonlinear function, as its term

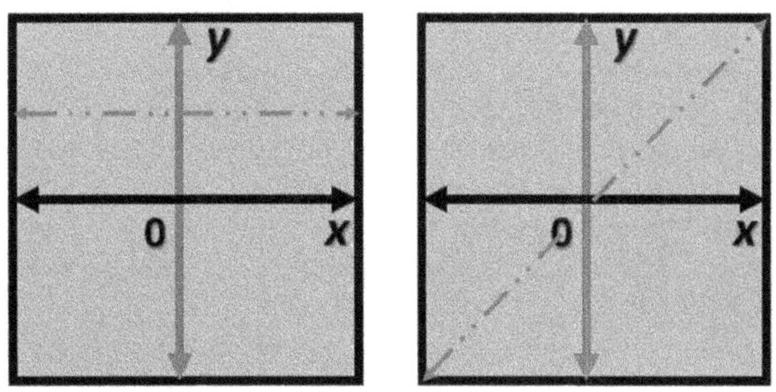

FIGURE 1.2 Examples of linear equations.

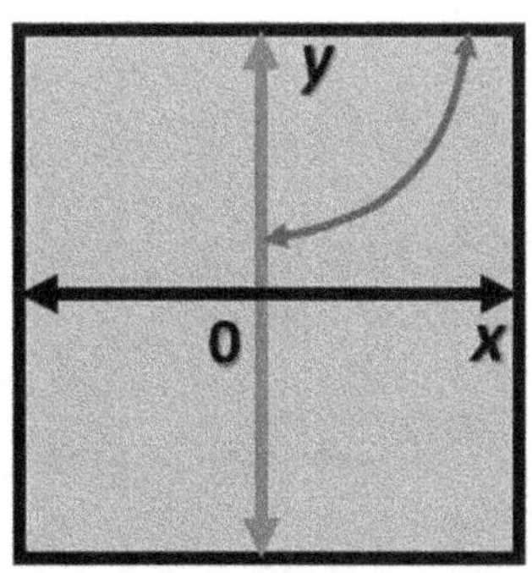

FIGURE 1.3 Nonlinear curve growth.

recommends, is a function that is NOT linear. In other words, its graph can be anything other than a line, i.e., the graph of a nonlinear function is NOT a line. Therefore, nonlinear functions have graphs that are not straight lines. It can have any curve other than a straight line as its graph. For instance, if the initial ocean current speed is 0.1 ms^{-1} and increases twice every 12 hours, the function $f(x) = 0.1\ (2)^x$ can be used to simulate the current speed fluctuation in a given periodic range, where x is the periodic range, and $f(x)$ is the periodic fluctuations in current speeds (Berger, 1977; Chen et al., 2010; Snieder, 1998).

1.4 Sort of nonlinear functions

Consequently, quadratic functions, cubic functions, and polynomial functions are forms of nonlinear functions. In this view, quadratic functions are being used to acquire values for various parameters in several disciplines of engineering and research. A parabola is used to illustrate them graphically. The curve's direction is determined by the coefficient of the highest degree. The term "quadratic" is derived from the term "quad," which means "square." A quadratic function is a "polynomial function of degree 2 (Fig. 1.4)," in other words. Quadratic functions are employed in a variety of situations. Did you know that a rocket's route is represented by the equation of a quadratic function when it is launched (Fig. 1.5)? A quadratic function has the usual form $f(x) = ax^2 + bx + c$, where a, b, and c are all real values with $a \neq 0$ (Aarseth, 1994; Barnett & Serletis, 2000; Berger, 1977).

Let us assume a small rocket is launched from 5 feet height at time t (seconds). In this view, the height (h, in feet) of the rocket t seconds after taking off is given by the formula $h = -2t^2 + 10t + 5$. Therefore, how long will it take the rocket to hit the ground? Using $h = 5$ feet and Fig. 1.5 as a reference, it would take a rocket 5 seconds to hit the earth.

Consequently, a quadratic function can have three different forms: standard, vertex, and intercept. The general forms of each of them are as follows:

1. Standard form can be demonstrated by: $f(x) = ax^2 + bx + c$, where $a \neq 0$, while (2) the vertex form has: $f(x) = a(x - h)^2 + k$, where $a \neq 0$ and (h, k) is the vertex of the parabola representing the quadratic function (Fig. 1.6); finally, (3) the intercept form can be

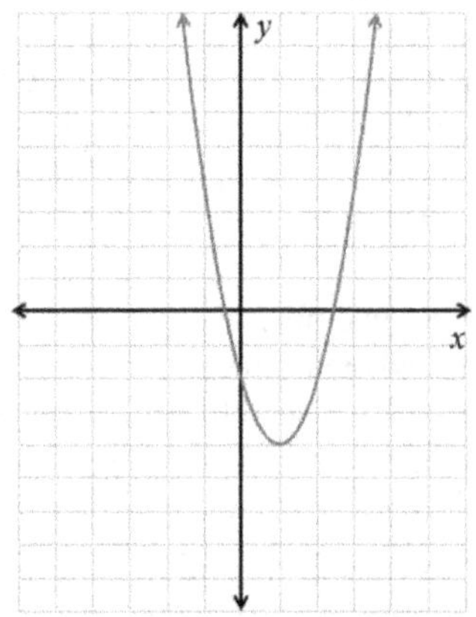

FIGURE 1.4 The polynomial function of degree 2.

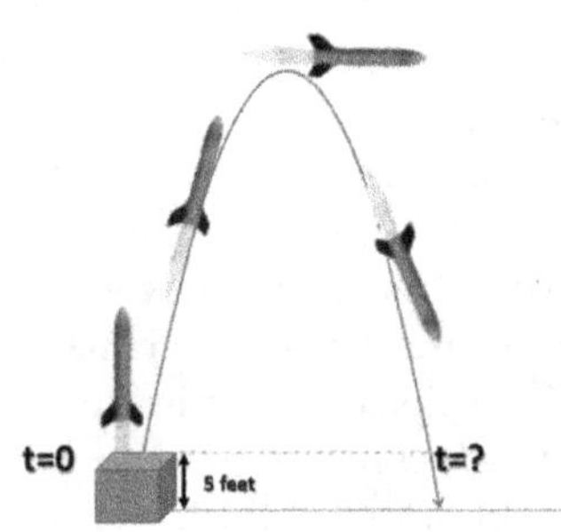

FIGURE 1.5 Quadratic rocket's route.

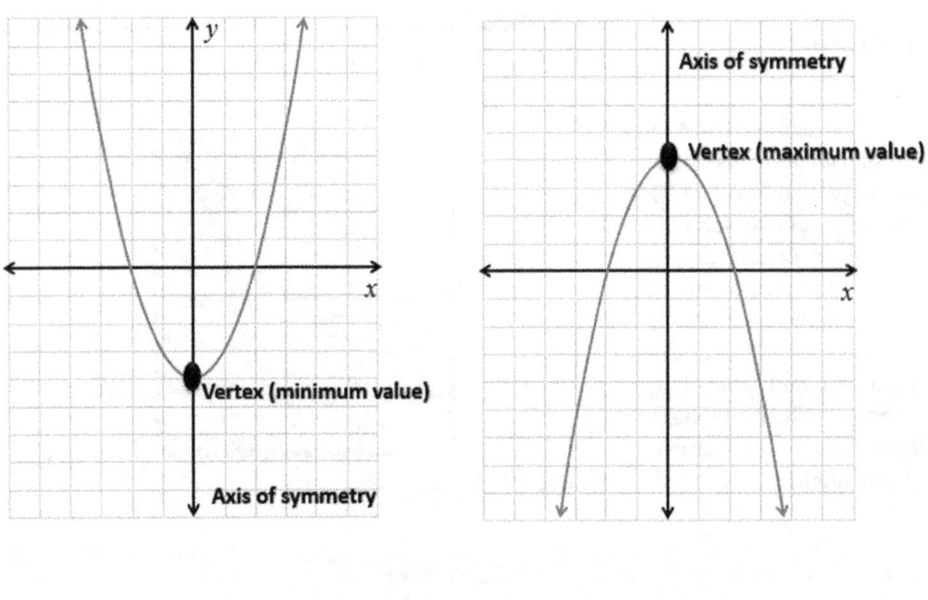

FIGURE 1.6 Vertex of the quadratic functions.

represented as: $f(x) = a(x - p)(x - q)$, where $a \neq 0$ and $(p, 0)$ and $(q, 0)$ are the x-intercepts of the parabola representing the quadratic function. In this understanding, the parabola opens upward or downward exceeds the value of "a": (1) If $a > 0$, the parabola opens upward; and (2) If $a < 0$, the parabola opens downward (Fig. 1.7).

We also can convert one type into another. By simply modifying the algebraic expressions, we may readily transform vertex or intercept form to standard form. Let's look at how to convert each vertex form and intercept form from the standard form. In this view, using the constants $h = \frac{-b}{2a}$ *and* $k = f\left(\frac{-b}{2a}\right)$, a quadratic function $f(x) = ax^2 + bx + c$ can be readily transformed to the vertex form $f(x) = a(x - h)^2 + k$. Subsequently, by taking the values of p and q (x-intercepts) and solving the quadratic equation $ax^2 + bx + c = 0$, a quadratic function $f(x) = ax^2 + bx + c$ can be easily transformed into the vertex form $f(x) = a\ (x - p)(x - q)$.

The significant question is: is the quadratic function controlled by domain and range? The domain and range of a quadratic function are the set of all x-values that define it and the set of all y-values that the function produces when different x-values are substituted. In this understanding, a polynomial function that is specified for all real values of x is termed a quadratic function. The set of real numbers, or R, is the domain of a quadratic function. The domain of every quadratic function is $(-\infty, \infty)$ in interval notation. On the other hand, the opening side and vertex of the graph determine the quadratic function's range. To determine the range of the quadratic function, examine the lowest and highest heat, the higf (x) values on the graph of the function. Any quadratic function with vertices (h, k) and the equation $f(x) = a(x - h)^2 + k$ has the following range: (1) when $a > 0$ (as the parabola expands up when $a > 0$), $y \geq k$ (or) $[k, \infty)$; and (2) when $a < 0$ (as the parabola opens down at $a < 0$), $y \leq k$ (or) $(-\infty, k]$.

The nonlinearity can be expressed also by an exponential function, in which exponents are involved. However, an exponential function has a constant as its base and a variable as its exponent, but not the other way around (if a function has a variable as the base and a constant exponent, then it is a power function but not an exponential function) (ction) (Coppens et al., 1965; Cortina, 1993; Emancipator & Kroll, 1993; Fletcher & Gent, 1954; Ghil, 2019; Haber, 1985). An exponential function is a function of the $ff\ (x) = a = {}^{ax}$, where "x" is a variable, and "a" is a constant that is considered the function's base and should be larger than 0. An exponential function, therefore, can take one of the following forms:

$$f(x) = b^x(1.1) \tag{1.1}$$

$$f(x) = ab^x \tag{1.2}$$

$$f(x) = ab^{cx} \tag{1.3}$$

$$f(x) = e^x \tag{1.4}$$

$$f(x) = e^{kx} \tag{1.5}$$

$$f(x) = pe^{kx} \tag{1.6}$$

here, "b" is a constant and "x" is a variable, and the constraints of these equations are: $b > 0$; and $b \neq 1$. In Eqs. (1.4)–(1.6), "e" is "Euler's number," and $e = 2.718$. Finally, p is also constant. Needless to say, all symbols except "x" are constants, "x" is a variable, and $f(x)$ is an exponential function in terms of x. Also, each exponential function's base must be a positive number. In the preceding functions, $b > 0$ and $e > 0$. Also, b should not be equal to 1 (if $b = 1$, the function $f(x) = b^x$ becomes $f(x) = 1$, and the function is linear but NOT exponential in this case). According to this view, if the value of a quantity increases in exponential growth and diminishes in an exponential decay, the exponential function appears. Consequently, in the scenario of exponential growth, the function can have the following form: (1) $f(x) = ab^x$, where $b > 1$; (2) $f(x) = a\ (1 + r)^x$; and (3) $P = P_0\ e^{k\ t}$. On the contrary, the exponential decay can be expressed by the following formulas: (1) $f(x) = ab^x$, where $0 < b < 1$; (2) $(x) = a\ (1 - r)^x$; and (3) $P = P_0\ e^{-k\ t}$ where $b = 1 - r \approx e^{-k}$. In this regard, a *(or)* P_0 is the initial value; r presents the growth rate; and x or t is time (time can be expressed in years, days, or months). Finally, k is a proportionality constant (Barnett & Serletis, 2000; Berger, 1977; Chen et al., 2010; Emancipator & Kroll, 1993; Fletcher & Gent, 1954; Ghil, 2019; Haber, 1985).

Therefore, the following power series can also be used to construct the real exponential function.

$$e^x = \sum_{n=0}^{\infty} = \frac{x^n}{n!} = (1/1) + (x/1) + (x^2/2) + (x^3/6) + \cdots \tag{1.7}$$

Consequently, expansion of some other exponential functions is given as follows:

$$e = \sum_{n=0}^{\infty} = \frac{x^n}{n!} = (1/1) + (1/1) + (1/2) + (1/6) + \cdots \tag{1.8}$$

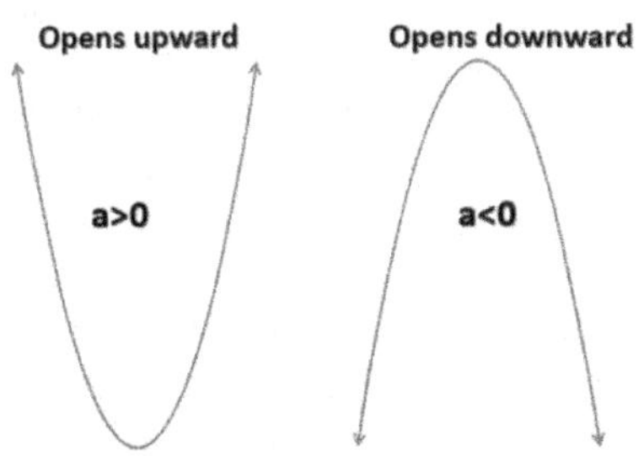

FIGURE 1.7 The shape of the quadratic function.

$$e^{-1} = \sum_{n=0}^{\infty} = \frac{x^n}{n!} = (1/1) - (1/1) + (1/2) - (1/6) + \cdots \quad (1.9)$$

As consequence, the question is: what are the domain and range of the exponential function? An exponential function's domain is thus the set of all real numbers (or) $(-\infty, \infty)$. The horizontal asymptote of the graph, let say $y = d$, and whether the graph is above or below $y = d$ can be used to calculate the range of an exponential function. In this understanding, for an exponential function $f(x) = ab^x$, (1) domain is the set of all real numbers (or) $(-\infty, \infty)$ and (2) range is $f(x) > d$ *if* $a > 0$ and $f(x) < d$ *if* $a < 0$. Apart from these, we occasionally have to exercise the conversion formula of logarithmic form to exponential form as given by:

$$b^x = a \Leftrightarrow \log_b a = x \quad (1.10)$$

Eq. (1.2) mainly has two important things to comprehend (from the symbol $\Leftrightarrow$): (1) $b^x = a \Rightarrow \log_b a = x$, which is known as "exponential to log form"; and (2) $log_b\ a = x \Rightarrow b^x = a$, which is termed log to exponential form.

Logs (or) logarithms are just another means of expressing exponents that can be utilized to address problems that can't be solved just with exponents. It is not difficult to comprehend logs. It's enough to know that a logarithmic equation is merely another method of writing an exponential equation to grasp logarithms. In this scenario, what are logarithmic functions? In the circumstance of $f(x) = log_a x$ *(r)* $y = \log_a x$ as $a > 0$, the basic logarithmic function is defined. In this understanding, it presents the inverse of the exponential function $a^y = x$. In this regard, log functions involve natural algorithm (ln), i.e., $\log_e = \ln$, for instance, $e^x = 2 \Rightarrow \log_e 2 = x$ (or) $\ln 2 = x$. Common logarithm (log), therefore, can be another form of logarithmic function, for instance, $10^2 = 100 \Rightarrow \log_{10} 100 = 2$ (or) $\log 100 = 2$. In this understanding, what are the domain and range of the logarithmic function? Let us assume $f(x) = \log x$ *(or* $y = \log x$) as the basic (parent) common logarithmic function. In this sense, $\log x$ is determined only if $x > 0$ (try finding log 0, log (-1), log (-2), etc.), which arises error in calculator. Consequently, as mentioned early, the domain is the set of all positive real numbers, for which the output values of y as a function of different x-values can be obtained. For instances, if $x = 1$, $y = \log 1 = 0$; if $x = 2$, $y = \log 2 = 0.3010$; if $x = 0.2$, $y = -0.6990$; and when $x = 0.01$, $y = -2$, etc. Needless to say, the domain of log function $y = \log x$ is $x > 0$ (or) $(0, \infty)$. Therefore, the range of any log function is the set of all real numbers (*R*)(Aarseth, 1994; Barnett & Serletis, 2000; Berger, 1977; Chen et al., 2010; Pertsch et al., 2004; Phillips, 2003; Roberts et al., 2006).

The derivation of the logarithmic function devotes the slope of the tangent to the curve denoting the logarithmic function. The formula for the derivative of the common and natural logarithmic functions are as follows:

1. the derivative of $\ln x$ is $1/x$. i.e., $d/dx. \ln x = 1/x$; and
2. the derivative of $log_a\ x$ *is* $1/(x \ln a)$. i.e., $d/dx\ (\log_a x) = 1/(x \ln a)$. In fact, the derivative of exponential function can be demonstrated as: (1) $d/dx\ (e^x) = e^x$; (2) $d/dx\ (a^x) = a^x \cdot \ln a$, respectively.

Consequently, the integral formulas of logarithmic functions can be given as:

1. the integral of $\ln x$ is $\int \ln x\, dx = x(\ln x - 1) + C$; and
2. the integral of $\log x$ is $\int \log x\, dx = x(\log x - 1) + C$. Indeed, the integral of exponential function can be presented by: (1) $\int e^x\, dx = e^x + C$; (2) $\int a^x\, dx = a^x / (\ln a) + C$; respectively, where C is constant.

A cubic function is a three-degree polynomial function with the form $f(x) = ax^3 + bx^2 + cx + d$, where a, b, c, and d are real numbers and $a \neq 0$. In this view, $f(x) = x^3$ is the basic cubic function, which is commonly known as the parent cubic function. In this understanding, a cubic function has at least one real root since it involves an odd degree polynomial. For example, the cubic function $f(x) = x^3$ has just one real root since only one real integer fulfills $x^3 = 0$ (which is $x = 0$). (The other two roots are complex numbers.) As a result, the curve of a cubic function would only have three roots. In other words, it can only intersect the x-axis three times. A cubic function always has one or three real zeros since complex roots always occur in pairs; it can't have two genuine zeros in it.

Even though $y = f(x)$ is a polynomial function, it is computed for all real x values; hence, its domain is the set of all real integers (*R*). Furthermore, it is implying that the range of a cubic function is also the set of all numbers. Needless to say, the domain of a cubic function is *R*, and the range of a cubic function is *R*. The values that are excluded from the domain and range always correspond to the asymptotes. Since a cubic function's domain and range are both sets of all real numbers, no values are excluded from either the domain or range. As a result, a cubic function has neither vertical nor horizontal asymptotes (Aarseth, 1994; Berger, 1977; Chen et al., 2010; Cortina, 1993; Fletcher & Gent, 1954; Haber, 1985).

Consequently, there are two sorts of function intercepts: x-intercept(s) and y-intercept(s) (s). Let's look at how to find the cubic function's intercepts. In this sense, a function's x-intercepts are also known as roots (or) zeros. A cubic function, therefore, can only have three roots because its degree is 3. A function will always have 0, 2, 4, etc., complex roots since complex roots of any function always occur in pairs. Consequently, a function can have either 0 or two complex roots. It has one or three actual roots or x-intercepts as a result. We simply substitute $y = 0$ ($or\ f(x) = 0$) and solve for x-values to determine the x-intercept(s) of a cubic function. A cubic function, on the other hand, always has one y-intercept. We simply substitute $x = 0$ and solve for the y-value to determine the y-intercept of a cubic function (Pertsch et al., 2004; Phillips, 2003; Roberts et al., 2006; Snieder, 1998; Tsay, 1986; Vakhitov & Kolokolov, 1973).

Moreover, the degree and sign of the leading coefficient determine the final behavior of any function. There is an odd degree polynomial in the cubic function $f(x) = ax^3 + bx^2 + cx + d$. As a result, its final behavior is as described in the following formulas:

$$f(x) \to \infty\, as x \to \infty\, and \quad f(x) \to -\infty\, as x \to -\infty \tag{1.11}$$

Eq. (1.11) is achieved if the leading coefficient is positive ($a > 0$). On the contrary, when the leading coefficient is negative ($a < 0$), Eq. (1.11) can be written as:

$$f(x) \to -\infty\, as x \to \infty\, and \quad f(x) \to \infty\, as x \to -\infty \tag{1.12}$$

Eqs. (1.11) and (1.12) deliver two different graph forms. Eq. (1.11) delivers the shape of the graph that runs from bottom to top while in Eq. (1.12), the shape of the graph is from top to bottom. Therefore, when charting a cubic function, critical points and inflection points are extremely important. In this view, the critical points of a function are the points at which it shifts from increasing to decreasing or decreasing to increasing. At the critical point, a function can have either a maximum or a minimum value. Set the first derivative to zero and solve for the critical points of a cubic function $f(x) = ax^3 + bx^2 + cx + d$. In this view, let us set the first derivative to zero and solve, i.e., $f'(x) = 0$ and $3ax^2 + 2bx + c = 0$. Consequently, this is a quadratic equation, and we can solve it using quadratic equation procedures as given:

$$x = -\frac{2b \pm \sqrt{4b^2 - 12ac}}{6a} (or) \quad x = \frac{-b \pm \sqrt{b^2 - 3ac}}{3a} \tag{1.13}$$

Eq. (1.13) demonstrates that (1) $f(x)$ has two critical points if $b^2 - 3ac > 0$; (2) $f(x)$ has only one critical point if $b^2 - 3ac = 0$; and (3) $f(x)$ has no critical point if $b^2 - 3ac < 0$. On the other hand, a function's inflection points are the points at which the function flips from "concave up to concave down" or "concave down to concave up." Set the second derivative to zero and solve to determine the critical points of a cubic function $f(x) = ax^3 + bx^2 + cx + d$, i.e., $f''(x) = 0$; $6ax + 2b = 0$; $6ax = -2b$; and $x = -b(3a)^{-1}$.

1.5 Diagnosis of a nonlinear system

Do the sort of nonlinear functions construct nonlinear equations? In fact, the nonlinear system or phenomenon is just a set of nonlinear equations. As a result, the following key question may arise: What exactly are nonlinear data? and what exactly is a nonlinear data structure? It is a data structure in which the data pieces are not ordered linearly or sequentially. Because the data structure is nonlinear, there is no one level. As a result, it is impossible for a user to traverse all of its elements in a single run. The data structure of a radar signal, for particular, is a mechanism of organizing and storing data and information so that a user can use it effectively. The data structure of radar signals in computer science is designed in such a way that it may be used with a variety of algorithms. It is divided into two broad categories: (1) Linear Data and (2) Nonlinear Data Structure. That might be a way to accomplish nonlinearity between real ocean dynamic parameters, features, and radar signal data. Nonlinearity in the data structure of radar signals or imagery, for example, is a type of data structure in which the data constituents are not ordered linearly or sequentially. Because the data structure is nonlinear, there is no specific level. As a consequence, a user cannot traverse all of its constituents in a single run. In this understanding, nonlinearity in the data structure of radar imagery works mainly perfectly in image processing and artificial intelligence (AI). However, linearity in a data structure is much less memory-friendly than nonlinearity in a data structure. This indicates that linear data structures cannot efficiently use memory. Therefore, as compared with linear data structures, the time complexity of nonlinear data structures often remains the same as the input size grows.

According to the preceding viewpoint, a nonlinear system is one that is not linear, i.e., one that does not satisfy the superposition principle. This negative definition obscures the fact that nonlinear systems come in a variety of forms. Nonlinear system recognition has historically morphed by focusing on key varieties of systems and can be widely categorized into five basic approaches, one of every designated by a deterministic model: (1) Volterra series models; (2) block-structured models; (3) neural network models; (4) NARMAX models; and (5) state-space models.

Identical to the Taylor series, the Volterra series serves as a model for nonlinear phenomena. Its propensity to retain "memory" impacts sets it apart from the Taylor series. If a nonlinear system's output is entirely reliant on its input at a specified instant, the Taylor series may be employed to approximate how it will react to that input. In the Volterra series, the nonlinear system's output is always reliant on the system's input. This makes it possible to retain the "memory" impact of components such as inductors and capacitors. The Volterra series, which in the discrete-time approximation has the following syntax, is according to nonlinear systems:

$$y(k) = h_0 + \sum_{m_1=1}^{M} h_1(m_1)u(k-m_1)$$
$$+ \sum_{m_1=1}^{M}\sum_{m_2=2}^{M} h_2(m_1.m_2)u(k-m_1)u(k-m_2)$$
$$+ \sum_{m_1=1}^{M}\sum_{m_2-1}^{M}\sum_{m_3-1}^{M} h_3(m_1,m_2,m_3)u(k-m_1)u(k-m_2)u(k-m_3) + \cdots + \quad (1.14)$$

Eq. (1.14) demonstrates that input system $u(k)$ and output system $y(k)$ respectively in the circumstance of $k = 1,2,3,4,......,$ with the lth-order Volterra kernel $h_l(m_1,m_2,m_3,......,m_L)$ or lth-order nonlinear impulse response. Therefore, the linear convolution integral is expanded by the Volterra series. The majority of the initial identification algorithms relied on the assumption that only the primary two Volterra kernels—linear and quadratic—are accessible, and they relied on unique inputs such as Gaussian white noise and correlation techniques to determine the two Volterra kernels. Numerous genuine processes are severely constrained by the requirement that the feed in the majority of these techniques can be Gaussian and white. In other words, we can express the output of the system as the sum of a sequence of Wiener G-functionals where the input is white noise with zero mean value and power (Schetzen, 1980; 1980; Snieder, 1998; 1998; Tsay, 1986; 1986):

$$y(k) = \sum_{p}(G_p u)(k) \quad (1.15)$$

A single simulation of white noise is a random shock. White noise is a discrete signal in discrete time, and its samples are viewed as a series of serially uncorrelated random variables with zero mean and finite variance. In other words, white noise, which is utilized in signal processing, is a random signal with equal intensities at various frequencies and a consistent power spectral density (Fig. 1.8).

Therefore, various model structures are explored as a framework for structural analysis for nonlinear systems due to the challenges in distinguishing Volterra models. Block-structured nonlinear models have been established or represented in a number of different configurations. A static single-valued nonlinear feature and a linear dynamic component compensate for the Hammerstein model. In the Wiener model, this conjunction is reversed, positioning the linear component first before the static nonlinear attribute. There are also additional model configurations, including the Wiener − Hammerstein model, which consists of a static nonlinear component consisting of two dynamic blocks. A linear dynamic block is constructed from two static nonlinear blocks in the

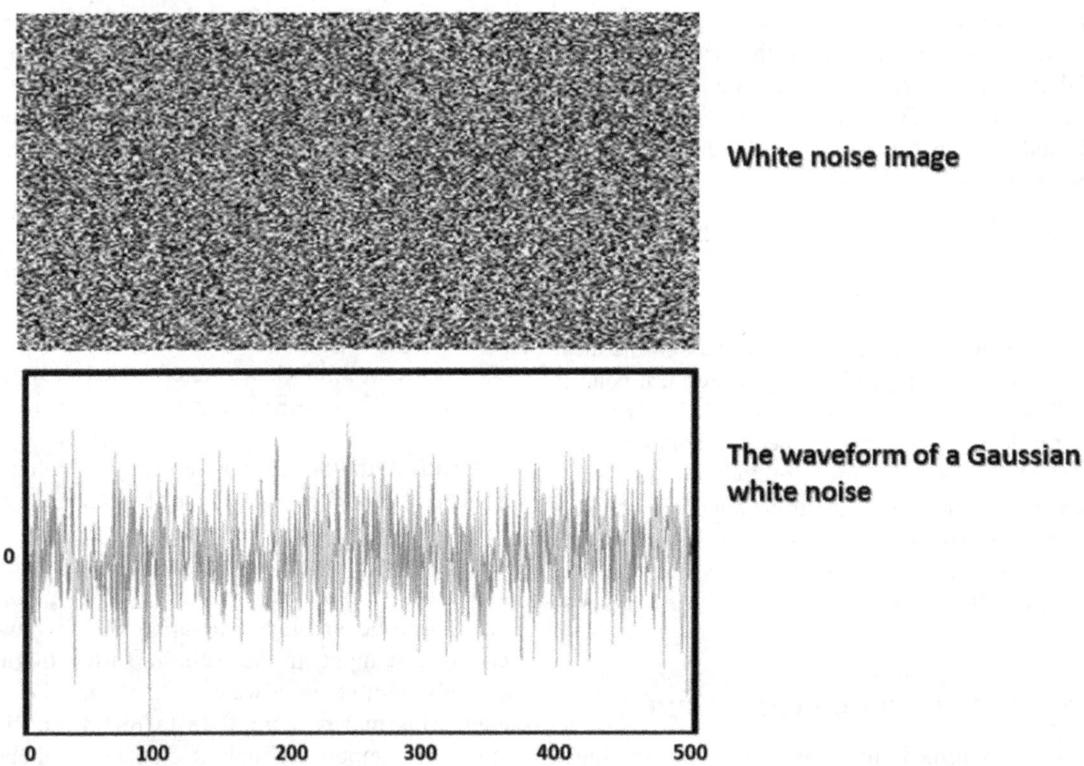

FIGURE 1.8 White noise image and Gaussian white noise.

Hammerstein − Wiener model (Aarseth, 1994; Barnett & Serletis, 2000; Haber, 1985; Haber, and Keviczky, 1999).

Let us consider input signal frequencies f_0,f_1, and frequency total is $\sum_{i=0}^{N} f_i$. In this view, let us assume the input signal frequency $u(f) = \frac{f_0}{f_1}$. Exploring polynomial function, the static input nonlinear model ($NL(f)$) and the inverse static input nonlinear $[u(f)]^{-1}$can be mathematically expressed as:

$$NL(f) = C_1 u(f)^5 - C_2 u(f)^4 + C_3 u(f)^3 - C_4 u(f)^2 + C_5 u(f) - C_6 \tag{1.16}$$

here C is constant, which is determined using the least square method. In accordance with Eq. (1.16), the prediction sum of square (PRESS) residuals, which are based on the orthogonal decomposition algorithm, are often used to estimate the parameters of the linear and nonlinear models for both Hammerstein and Wiener block identification. Although a Volterra series can reflect any of these models, the Volterra kernels within every case surely on such a unique version. Techniques for variable and logical link recognition are both available. By employing specified inputs, frequently white Gaussian noise, the correlation approaches take advantage of certain characteristics of these systems, permitting for the determination of the individual components once at a moment (Aarseth, 1994; Bai & Fu, 2002; Barnett & Serletis, 2000).

According to the aforementioned premise, a white noise signal's samples could be grouped along one or more spatial dimensions or in chronological order. The pixels of a white noise image are commonly positioned in a rectangular grid when they are being processed digitally, and they are believed to be independent random variables with uniform probability distributions across some range (Fig. 1.8).

The nonlinear autoregressive moving average model with exogenous inputs (NARMAX model) can represent a broad class of nonlinear systems within the large enormous input elements, including:

$$\begin{aligned} y(k) = F[&y(k-1), y(k-2), \ldots, y(k-n_y), \\ &u(k-t), u(k-t-1), \ldots, (k-t-n_u), \\ &l(k-1), l(k-2), \ldots, l(k-n_l)] + l(k) \end{aligned} \tag{1.17}$$

Eq. (1.17) reveals that $y(k)$, $u(k)$, *and* $l(k)$ are, respectively, the system output, input, and noise sequences; (n_y, n_u, and n_l) are the maximum lags for the system output, input, and noise; $F[\bullet]$ is some nonlinear function, and t is a time delay that is usually set $t = 1$. In essence, the model is an enlargement of the inputs, outputs, and noise factors from the past. Unbiased estimates of the system model can be obtained in the presence of unobserved strongly correlated and nonlinear noise since the noise is explicitly modeled. Numerous neural network architectures, block-structured models, and the Volterra can all be viewed as subsets of the NARMAX model. Since the inception of NARMAX, numerous findings and algorithms have been developed based on this description by demonstrating what sort of nonlinear systems may be described by this model (Billings, 2013).

An NARMAX model is qualified to represent a considerable fraction of nonlinear systems, involving those that exhibit peculiar behaviors such as chaos, bifurcations, and subharmonics. NARMAX was primarily the acronym of a framework, but it has since evolved into a nonlinear system-identifying paradigm. The NARMAX technique comprises the following procedures: (1) structure detection; (2) parameter estimation; (3) model validation; (4) prediction; and (5) analysis. In this understanding, the core constitution of NARMAX is structure detection. For instance, an NARMAX model with three lagged noise terms, one lagged output term, and one lagged input term would have 82 promising therapeutic conditions if it were developed as just a cubic polynomial. Owing to the expansion's inherent inclusion of all conceivable permutations well within the cubic expansion, there are so many candidate expressions. Trimming after incorrectly estimating a model that encompasses every one of these elements would lead to physical and computational issues and must always be disregarded. Nonetheless, in the model, the few parameters are frequently crucial. Consequently, it is crucial to use structure detection, which identifies elements just once. In addition to serving as a substitute for principal component analysis, these concepts could also be implemented in pattern recognition and feature selection. One advantage of this technique is that the features are disclosed as simple kernel functions that can be directly applied to the initial problem as will be demonstrated in the following chapters. Therefore, the mission of NARMAX algorithms is far wider than simply determining the most appropriate approximation model. Two requirements can be identified for structural analysis. In the first, the approximation is exploited, and the main objective is to generate a model that closely resembles the dataset so that accurate prediction could well be achieved. This technique is suitable for a wide range of operations, including time series forecasting of the weather, signal processing, target tracking, pattern classification, etc. The structure of the model is not as crucial in these scenarios. Consequently, developing an approximations strategy that results in minimal prediction errors is the task. The main first significant task of system identification is a subdivision of the second objective, which spans far beyond identifying the model with the finest or minimum mean squared errors (Aarseth, 1994; Billings, 2013; Drábek et al., 1999; Fucik & Kufner, 2014). The NARMAX concept, which is associated with the principle of obtaining the minimal model structure, is formed with the second objective in mind. Creating the basic model feasible and, if at all conceivable, connecting it to the constituents and behaviors of the system being examined are the key objectives in creating a model that replicates the dynamic characteristics of the underlying processes. Determining and exposing the rule that serves as the system's representative are the main goal of this second method of identification. These goals are important for designing control systems and simulating models, but they are also increasingly important for applications in the life sciences, medicine, and neuroscience. Here, the goal is to find models—often nonlinear ones—that may be utilized to comprehend the fundamental principles underlying how these systems function and act, allowing us to modify and make use of them. The frequency and spatiotemporal domains have both seen the development of NARMAX techniques.

It is conceivable that an ambient unpredictable perturbation would cross through the nonlinear dynamics in a general scenario and impact the outputs. The domain of stochastic nonlinear state-space frameworks is a sufficiently broad model category that can explain this scenario. The characteristics to be determined typically have some physical importance or significance and are typically acquired including first fundamental laws, such as mechanical, electrical, or thermodynamic physical laws. The discrepancy formulas can be utilized to construct a discrete-time state-space model as given by:

$$u_{t+1} = f(h_t, u_t, \omega_t; \theta), \quad y_t = g(h_t, u_t, f_t; \theta), t = 1, 2, 3, \cdots \tag{1.18}$$

In Eq. (1.18), t is a time-related positive integer. f and g are representations of generic nonlinear functions. The status formula is the initial formula, while the output formula is the second equation of 1.18. Stochastic processes are exploited to model entire signals. Therefore, the state process is presented by h_t, ω_t and f_t are independent and mutual independent variables. Consequently, θ is often a real, finite-dimensional parameter that must be approximated (using experimental data). Bear in mind that the state process should not be a tangible signal and is commonly ignored (not measured). The dataset is presented as a collection of input − output pairs (y_t and u_t) for each of the values of $t = 1, \ldots, N$. In this sense, the probability function of the outputs, which is expressed as a multivariate minimization integrated, is computationally difficult owing to the nonlinear modification of the unknown stochastic process. Due to this, popular parameters calculations such as the maximum likelihood technique and the prediction error algorithm rely on the optimal one-step forward prediction and are computationally unsolved (Chen et al., 2010; Drábek et al., 1999; Fucik & Kufner, 2014).

Recently, artificial neural networks use a variety of straightforward components of the system to attempt to approximate the brain's network of neurons, which is where computing occurs. Numerous minimal processing units are coupled to construct a complicated network in a convolutional network. Data are inputted at the input layer and passed through another or more intermediate layers eventually emerging at the output layer attributable to the configuration of layers of these modules. In supervised learning, the network is taught by adjusting the estimation error discrepancy between the network's total performance and its anticipated output—to modify the connectivity lengths between nodes. Neural networks can be utilized to study static difficulties and dynamic issues, which are the two basic categories of difficulties. Pattern recognition, categorization, and approximation are examples of static issues. Lagged variables are involved in dynamic problems, which are better suited for system identification and associated applications. The training problem can either be linear-in-the-parameters, which can be solved using conventional methods, or nonlinear-in-the-parameters, which include optimization. There are three types of training algorithms: reinforcement learning, unsupervised learning, and supervised learning as will be demonstrated in the following chapters. The exceptional approximation features of neural networks are commonly based on outcomes from standard function approximation, such as the Weierstrass Theorem, which holds true for polynomials, rational functions, and other well-known models. Numerous system identification issues involving nonlinear and dynamic interactions have been solved using neural networks. Classical neural networks, nevertheless, are merely crude static approximation devices. Within the network, there is no dynamic behavior. As a consequence, when fitting dynamic models, all dynamics result from designating laggard output signals to the network's input nodes (Anderson, 1995; Bishop, 1994; Gurney, 2018).

Nevertheless, it is relatively simple in these situations to overspecify the delays, which can result in overfitting and poor generalization features. Neural networks provide a number of benefits, including superior approximation properties, conceptual simplicity, ease of training and application, and the concept of local and parallel processing, which promotes integrity and fault tolerance in behavior. The primary objection leveled against conventional neural network models is that they produce models that are wholly opaque and frequently impossible to record or analyze. Knowing what is causing what, analyzing the model, or computing dynamic characteristics from the model are consequently exceedingly challenging tasks. Some of these concepts won't apply to all cases, but dynamic modeling does.

1.6 Nonlinear differential equations

Nonlinear differential equation: what is it? A nonlinear differential equation is one in which the unknown function and its derivatives are not linear. It offers a range of options for addressing chaos. Therefore, what distinguishes linear differential equations from nonlinear differential equations? The degree of a linear differential equation is 1. Alternately, we may describe it as an equation with degree 1 as its maximum. The degree of a nonlinear differential equation, however, is 2 or greater but not less than 2. In this view, we can define nonlinearity as the mathematical expression of a system's autonomia. The coefficient of reproductive rate (the "time constant") in a "nonlinear ecosystem" is not an a priori constant; rather, it fluctuates depending on the state of the ecosystem (i.e., as a function of the population). The ecosystem could become stable through this nonlinearity (autonomous variation in reproducing rate), or it could cause complicated (unforeseen) instabilities. Such a system requires the cooperation of numerous components to be autonomous. Nonetheless, in a linear system, elements can generally be separated out, and the exponential law could be used to explain how another element moves. The essential feature of linear theory, which represents the "order" of motion from a reductionist standpoint, is the potential for such "decomposition (Drábek et al., 1999; Fucik & Kufner, 2014; Struble, 2018)." The variables (components) contained in a nonlinear system cannot be deconstructed; instead, they interplay in a variety of ways to create immensely complicated "entanglements." Particularly for autonomous equations, first-order ordinary differential equations are frequently precisely solved by variable separation. A depiction can be demonstrated in the following nonlinear equation:

$$\frac{du}{dx} = -u^2 \tag{1.19}$$

Eq. (1.19) is nonlinear and could be expressed as:

$$\frac{du}{dx} + u^2 = 0 \tag{1.20}$$

The function of u and its derivatives on the left-hand side of the equation is not linear. Bear in mind that the issue would be linear if the u^2 term were swapped out for u, which leads to the exponential decay problem. Closed-form solutions to systems of nonlinear equations of the second and higher orders are rare, while implicit solutions involving nonelementary integrals are quite often encountered. In this regard, an antiderivative (or indefinite integral) in mathematics that is not itself an elementary function is known as a nonelementary antiderivative of a given elementary function (i.e., a function constructed from a finite number of quotients of constant, algebraic, exponential, trigonometric, and logarithmic functions using field operations). The following are some examples of functions with nonelementary antiderivatives: (1) elliptic integrals; (2) logarithmic integral; (3) error function, Gaussian integral; (4) Fresnel integrals; (5) incomplete gamma functions; (6) in terms of the logarithmic integral, i.e., ln(ln x); (7) exponential integral, i.e., $\frac{e^{-x}}{x}$; and (8) sine integral and Dirichlet integral, which can be demonstrated in the manner of $\frac{\sin(x)}{x} = \sin c(x)$ (Struble, 2018).

Giulio Fagnano and Leonhard Euler (c.1750) were the first to study elliptic integrals, one of a variety of related functions in integral calculus that is defined as the value of specific integrals. Their name comes from the fact that they were initially associated with the issue of determining the arc length of an ellipse. An "elliptic integral" in modern mathematics is any function f that may be written in the following manner:

$$f(x) = \int_c^x R(t, \sqrt{P(t)})dt, \tag{1.21}$$

In formula (1.21), P is a polynomial of degree 3 or 4 with no repeating roots, c is a constant, and R is a rational function of its two arguments. Elementary functions cannot generally be employed to represent integrals in this manner. Whenever P has repeated roots, while including no odd powers of y, or when integral is pseudo-elliptic, there are exceptions to this general rule. Conversely, every elliptic integral may be converted into a form involving integrals over rational functions and the three Legendre canonical forms by applying the proper minimization technique (i.e., the elliptic integrals of the first, second, and third kind). In the other words, a canonical set of three elliptic integrals to which all else can be simplified in mathematics are known as the Legendre forms of elliptic integrals. Legendre used the name "elliptic integrals" because the second sort yields the arc length of an ellipse with unit semimajor axis and eccentricity (k) (the ellipse being parametrically specified by $x = \sqrt{1-k^2}\cos 9t, y - \sin(t)$.

The definite integral specifies the integral expression of the logarithmic integral across all positive real integers $x \neq 1$:

$$li(x) = \int_0^x \frac{dt}{\ln t}. \tag{1.22}$$

here "$li(x)$" redirects here. The polylogarithm is denoted by $Lis(z)$. The integral logarithm $li(x)$, also known as the logarithmic integral function, is a unique mathematical function in this approach; therefore, the error function (Fig. 1.9) is commonly known as the Gauss error function in mathematics. It is often signified by erf and considered as a complex function of a complex variable that is specified as:

$$erfz = \frac{2}{\pi}\int_0^z e^{-t^2}dt. \tag{1.23}$$

A unique (nonelementary) sigmoid function that frequently appears in probability, statistics, and partial

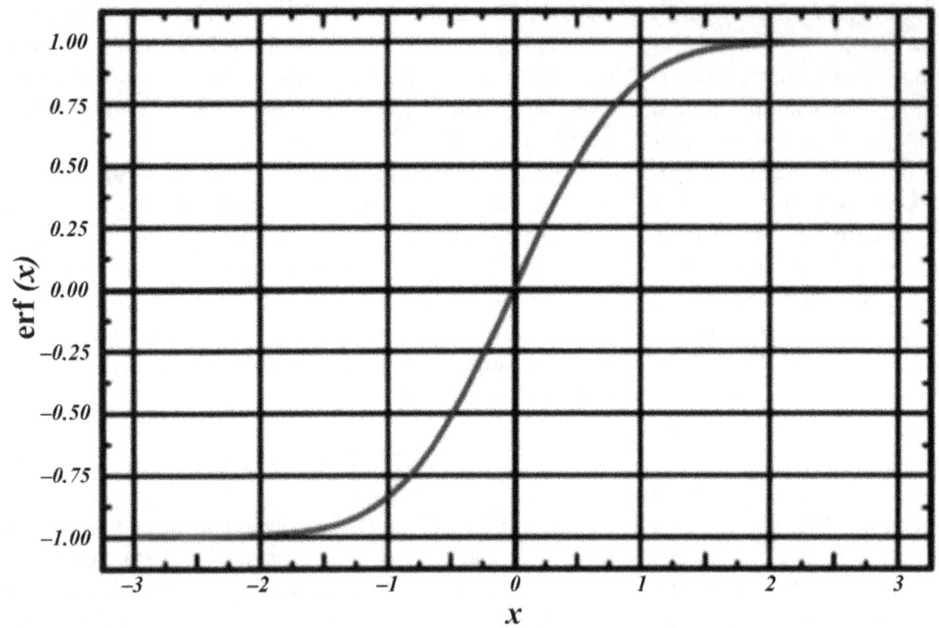

FIGURE 1.9 Illustration of the error function.

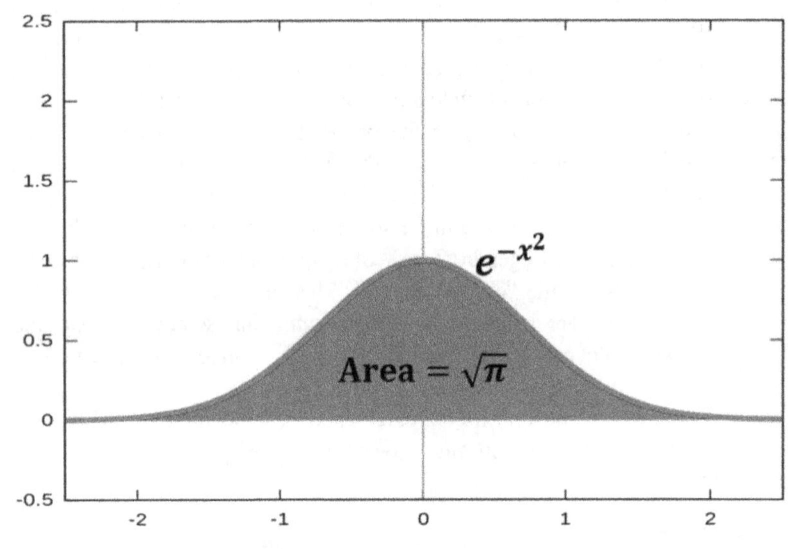

FIGURE 1.10 Gaussian integral diagram.

differential equations is this integral. Numerous of these applications employ precise figures as the function response. The function value is also real if the function argument is real. Therefore, the integral of the Gaussian function $f(x) = e^{-x^2}$ is the Gaussian integral (Fig. 1.10), sometimes referred to as the Euler – Poisson integral over the entire real line, which is given by:

$$\int_{-\infty}^{\infty} e^{-x^2} dx = \sqrt{\pi}. \tag{1.24}$$

There are numerous purposes for the integral. It is utilized, for instance, to identify the normalizing constant of the normal distribution with a tiny adjustment in the parameters. Both the error function and the cumulative distribution function of the normal distribution are closely related to a common integral with finite boundaries. This kind of integral is widespread in physics; for instance, it can be employed to estimate the probability density of the harmonic oscillator's ground state in quantum mechanics. Additionally, this integral is exploited in statistical mechanics to determine its partition function and in the path integral formulation to get the harmonic oscillator's propagator.

The Fresnel integrals $S(x)$ and $C(x)$, which bear Augustin – Jean Fresnel's name, are two transcendental functions exploited in optics and intimately associated with the error function (erf). They are characterized by the following integral representations and appear in the description of near-field Fresnel diffraction phenomena (Kopp & Meyrueis, 1998):

$$S(x) = \int_0^x \sin(t^2)dt, C(x) = \int_0^x \cos(t^2)dt. \quad (1.25)$$

In this scenario, the Euler spiral is the simultaneous parametric plot of $S(x)$ and $C(x)$ (Fig. 1.11), which is also known as the Cornu spiral or clothoid.

The curve produced by a parametric plot of $S(t)$ against $C(t)$ is known as the Euler spiral, Cornu spiral, or clothoid (t). Marie Alfred Cornu developed the Cornu spiral as a nomogram for diffraction calculations in science and engineering (Havil, 2019). The infinitesimals dx and dy are as follows based on the definitions of Fresnel integrals:

$$dx = C''(t)dt = \cos(t^2)dt, \quad dy = S''(t)dt = \sin(t^2)dt. \quad (1.26)$$

Thus, the spiral's length measured from its inception can be written as:

$$L = \int_0^{t_0} \sqrt{dx^2 + dy^2} = \int_0^{t_0} dt = t_0. \quad (1.27)$$

In other words, the Euler spiral has an unlimited length, and the parameter t represents the length of the curve as measured from the origin (0, 0). The unit tangent vector along the spiral is alternatively expressed by the vector $(\cos(t^2), \sin(t^2))$, which results in $\theta = t^2$. Given that t is the curve's length, the expression for the curvature κ is:

$$\kappa = [R]^{-1} - \frac{d\theta}{dt} = 2t. \quad (1.28)$$

The curvature of an Euler spiral (Fig. 1.12) at any point is proportional to the spiral's length as measured from the origin. Due to the following characteristic, it can be used as a transition curve in turbulence flow: if vorticity follows the spiral at unit speed, the parameter t in the aforementioned derivatives also indicates the time (Kimia et al., 2003). A vortex that is following the spiral at a constant speed will therefore experience constant angular acceleration.

$$\Gamma(s,x) = \int_x^{\infty} t^{s-1}e^{-t}dt, \quad (1.29)$$

$$\gamma(s,x) = \int_0^{x} t^{s-1}e^{-t}dt. \quad (1.30)$$

Eqs. (1.29) and (1.30) are descriptions of incomplete gamma functions in which throughout all scenarios, s is a complicated parameter with a positive real number. In this sense, the upper and lower incomplete gamma functions in mathematics are cases of unusual functions that arise as solutions to numerous equations, along with some integrals. Their quotes deliver from the integral definitions of their respective functions, which are constructed identically to the gamma function albeit with varied or "incomplete" integral limits. An integral from 0 to infinity is what is supposed to mean "gamma function." The lower incomplete gamma function, on the other hand, is specified as an integral from zero to a variable upper limit. An integral from a variable lower limit to infinity is how the upper incomplete gamma function is described.

A simple pendulum serves as another illustration of a nonlinear differential equation. Only when the pendulum's displacement angle is minimal does it act linearly. A system's behavior in its linear and nonlinear domains differs significantly on a qualitative level. For instance, in the linear regime, the period of the pendulum oscillation

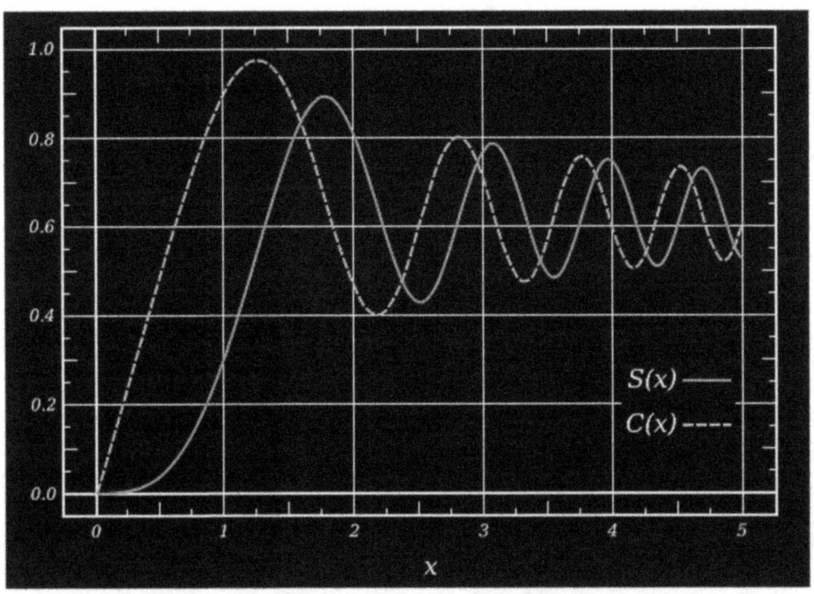

FIGURE 1.11 Illustration of $S(x)$ and $C(x)$ curves.

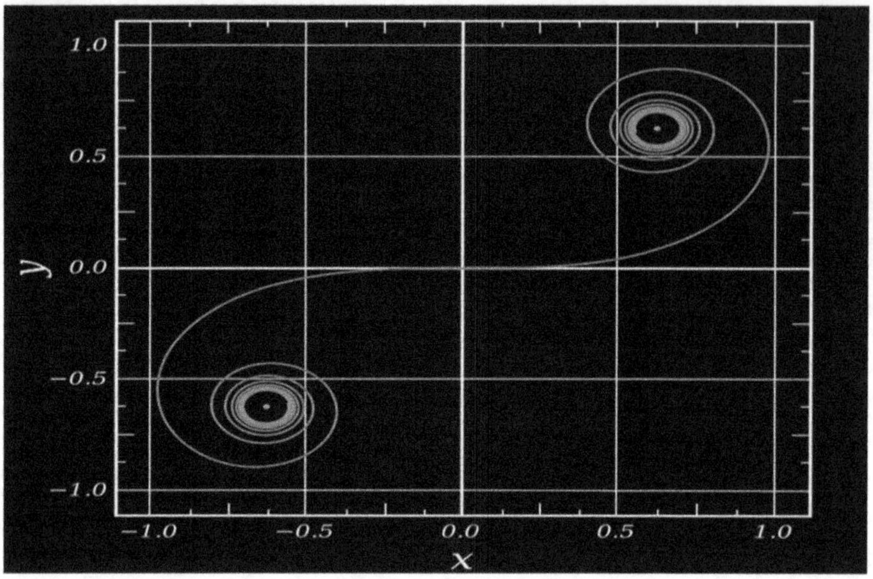

FIGURE 1.12 Euler spiral.

is independent of the amplitude (the maximum displacement angle), whereas, in the nonlinear regime, it is. The breakdown of the superposition principle, which asserts that the sum of two solutions to the equation(s) characterizing the system is again a solution, is the mathematical indicator of a nonlinear system. The physical implication is that the behavior of a nonlinear system's total is greater than the sum of its parts (Aarseth, 1994; Chen et al., 2010; Fay, 2002). The superposition principle can occur in one of two distinct approaches. The equation itself is nonlinear, to start. The dimensionless nonlinear equation can be exploited to model a pendulum's motion:

$$\frac{d^2\theta}{dt^2} = -\left[\frac{g}{L}\right]\sin(\theta) \tag{1.31}$$

In Eq. (1.31), θ stands for the angle formed by the pendulum and the vertical, g for the acceleration brought on by gravity, and L for the length of the pendulum (Fig. 1.13) (Fay, 2002). As a result of the presence of the nonlinear term sin θ, Eq. (1.31) is a nonlinear equation. Using $d\theta/dt$ (Aarseth, 1994; Berger, 1977; Fay, 2002) as an integrating factor is one option to "solve" this equation, which would result in:

$$\int \frac{d\theta}{\sqrt{C_0 + 2\cos(\theta)}} = t + C_1 \tag{1.32}$$

The solution of Eq. (1.32) involves an elliptic integral as an implicit solution. Since the majority of the answer's nature is concealed in the nonelementary integral (nonelementary unless $C_0 = 2$ is used), this "solution" often has few applications (Aarseth, 1994; Barnett & Serletis, 2000; Berger, 1977; Coppens et al., 1965). Another approach would be to use Taylor expansions to linearize any nonlinearity (in this case, the sine function term) at the various sites of interest. The small angle approximation, or linearization at $\theta = 0$, for instance, is as described in the following:

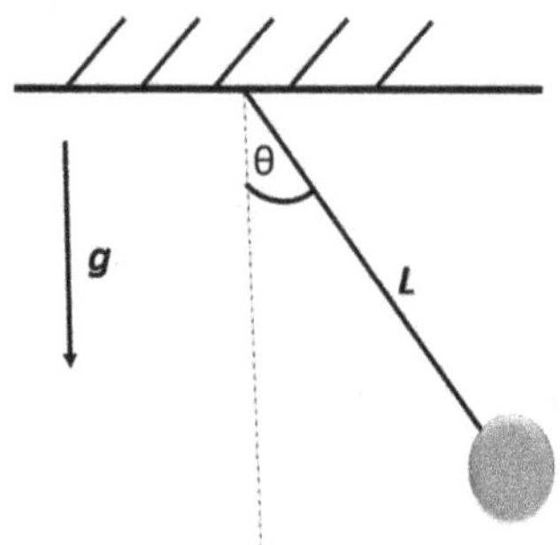

FIGURE 1.13 Concept of simple pendulum oscillation.

$$\frac{d^2\theta}{dt^2} = -\theta \tag{1.33}$$

Assuming that $sin(\theta) \approx \theta$ for $\theta = 0$. This is a conventional harmonic oscillator that replicates pendulum oscillations at the bottom of its route. Another linearization, with the pendulum pointing straight up, would be at $\theta = \pi$:

$$\frac{d^2\theta}{dt^2} + \pi - \theta = 0 \tag{1.34}$$

In this regard, let us assume that $\sin(\theta) \approx \pi - \theta$ for $\theta = \pi$. Hyperbolic sinusoidal is being exploited to resolve the issue; however, unlike the angle θ approximation, this approximation is unreliable, which means that, while constrained solutions are conceivable, $|\theta|$ typically grow without restriction. This is figuratively an unsteady position and relates to how challenging it is to balance an inverted pendulum. Therefore, there is potential for one more intriguing linearization around $\theta = \frac{\pi}{2}$ at which $\sin(\theta) \approx 1$:

$$\frac{d^2\theta}{dt^2} + 1 = 0. \tag{1.35}$$

This is equivalent to a free-falling scenario. Such linearization could be coupled together to develop a superior qualitative representation of the dynamics of the oscillation, as seen in the figure to the right. To determine (precise) phase portraits and proximate periods, further methods could well be utilized.

1.7 Nonlinearity singularity: emerging in small scale

The primary definition of nonlinearity, according to the aforementioned viewpoint, is "variance from the proportionality correlation." Once the variance of the parameters increases in a "regular" phenomenon, nonlinearity arises; once the graph can be seen at a wide scale, distortion is visible. Regular in this context essentially indicates that "we can commence with a reasonable proportionality correlation, although at considerable variance, considerable deformation seems discernible." Yet, there are some circumstances where we must consider "irregularity," would be considered, which is a nonlinearity that emerges suddenly from slight changes in the variables. In these scenarios, the proportionality correlation cannot be assumed, even on a small scale. In another sense, a nonlinearity in which the scale is 0 is an "irregularity." The radius of convergence of the Taylor expansion measures the size of the area where the linear approximation applies (Zakharov & Synakh, 1975). Thus, "irregularity" denotes the Taylor expansion's impossibility or the convergence radius of 0. A break or discontinuity represents an abnormality in a graph (Fig. 1.14). A singularity is the site of irregularity; it should not be a "point"; it could be a particular assemblage of higher dimensions. In this regard, let us assume a function that is given by:

$$f(x) = |x|^P (P \leq 1) \tag{1.36}$$

Eq. (1.36) reveals that at $x = 0$, the existence of the sharp peak of $f(x)$. Once $P < 1$, $f(x)$ would approach $f(0)$ abruptly with $x \to 0$. Conversely, if x barely detracts from 0, it changes abruptly. Needless to say, a collision is an event that occurs when two or more elements occupy the same point, and also the singularity at the origin cannot be reached by any motion (Fig. 1.14).

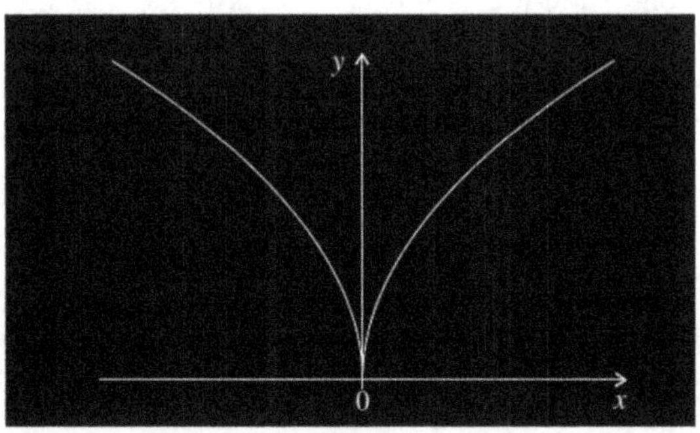

FIGURE 1.14 The singularity concept.

On the other hand, let us assume that the tidal ocean current velocity $U(t)$ is affected by nonlinear force $|F|$, which can be mathematically signified as:

$$\frac{d}{dt}U = -\sqrt{|F|}. \tag{1.37}$$

During ebb flow, $U(t)$ tends to be transferred to a negative value owing to tidal falling at arbitrary time $t_1 \geq t_0$ resulting in the solutions enumerated below:

$$U(t) = \begin{cases} \frac{(t_0 - t)^2}{4}\left(t < t_0 = 2\sqrt{U_0}\right), \\ 0(t_1 \geq t \geq t_0), \\ \frac{-(t - t_1)^2}{4}(t > t \geq t_0). \end{cases} \tag{1.38}$$

Consequently, when the solution of Eq. (1.37) comes into contact with the singularity, it loses its uniqueness. Since there are numerous potential future states that might arise from a particular initial state, the motion is unexpected if the initial-value problem solution of a derivative is not singular (Haber & Keviczky, 1999; Haber, 1985; Liu et al., 2014; Mostafazadeh, 2013). Consequently, the next chapters will deal with such problems, particularly in turbulence flow cases.

1.8 Bifurcation and discontinuity

What does the term "bifurcation" signify, one would wonder. One of the most significant problems that only nonlinear theory can properly address is the cohabitation of numerous solutions. Bifurcation is the expression for an alteration in the level of solutions (also known as branches). In this understanding, let us consider a nonlinear map $f:\mathbb{R} \to \mathbb{R}$. In this view, f is a linear map whose domain X and range Y are subspaces of real (or complex) linear spaces. This mathematically can be expressed in linear form $y = f(x)$ as $y \in Y$ and $x \in X$. Let R change as a function of x; and a,b, and c are constants. Then the mathematical expression of this relationship can be casted by:

$$R(x) = a(x - c)^2 + b, \tag{1.39}$$

It seems that the nonlinear relationship is addressed in Eq. (1.39). In this understanding, $R(x)$ diminishes in the domain of $x < c$ owing to the upswing in x, while it grows in the range of $x > c$. A linear function close to an "ordinary" (noncritical) point can be used to approximate a regular function, $f(x)$. In the vicinity of the normal point, the solution of the related linear equation, which is uniquely solvable, approaches the solution of the precise nonlinear equation. As a result, the critical point (where the linearized equation becomes indeterminate) must occur at the bifurcation point (where numerous solutions exist nearby), which cannot be an ordinary point. Each solution x to the equation $y = f(x)$ changes on the associated branch once y is changed as a parameter. Generally, any x responds "consistently" to any minor variations in y. The branch might vanish, in which case the solution must therefore switch to another branch as demonstrated in Fig. 1.15. Nevertheless, as x heads to the critical point, the shift in x becomes violent (Coll et al., 2001; Leine & Van Campen, 2002; Leine et al., 2000).

Such an abrupt change in the condition is frequently related to an unexpected "catastrophe." Researchers frequently do not have a broad perspective of bifurcated branches; instead, they develop their senses of a phenomenon as a continuous extension of minute alterations. The solution might abruptly vary if the graph were discontinuous. The aforementioned example demonstrates, however, that the solution can leap even when the curve is smooth. The critical point, where linear theory fails, is where things start to provide nasty. In other words, once the equation $y = f(x)$ is solved for x, the number of possibilities varies relying on the value of y; for example, there is only one solution in the domain (1) and three solutions in the domain (3). The frequency of options fluctuates at y, when $f(x)$ is steady, meaning that the bifurcation points of the branches are the critical points of $f(x)$ (Fig. 1.15).

The highest (though frequently challenging) objective of nonlinear research is a global understanding of the parameter space. Critical moments are frequently created by competing variables. The graph of the governing law is a complicated manifold submerged in a high-dimensional space in a turbulent system where several elements are interacting, and it is accompanied by diverse deformations, plenty of pleats, and multiple overlaps. A turbulent complex system can be geometrically represented by such a complex manifold. For instance, the system "turbulence flow," which connects particles in various phases, different types of energies, and numerous vortices to create irregular turbulence chains, should have a wide range of potential states (Barnett & Serletis, 2000; Leine & Van Campen, 2002; Leine et al., 2000). Reality is viewed as a dichroic of the multifarious or a historical trajectory that passes through numerous pivotal moments.

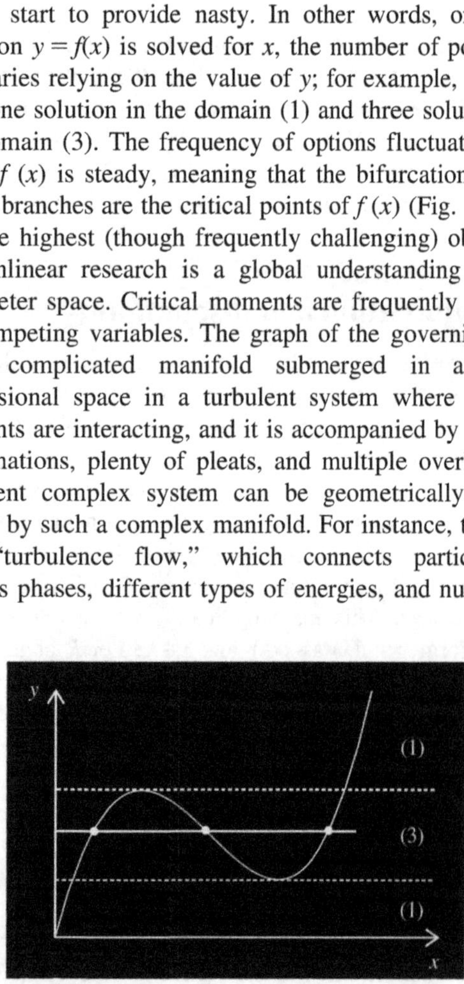

FIGURE 1.15 Example of bifurcated solutions.

1.9 Can nonlinearity criticality: escape from linearity?

Along with the previously indicated singularity (irregularity), there is additional potential for abrupt nonlinearity that prevents any linear approximation. The law is Taylor-expandable and regular in this situation, but the first derivative's proportionality coefficient disappears. A crucial point is defined as the location x_0 at which the first-order term of the Taylor expansion disappears (Ok, 2017; Phillips, 2003; Pourahmadi, 2018). Initially, let us determine $y = f(x_0)$ in the vicinity of a critical point and assume that y is independent of x. Nevertheless, if we improve the observation's accuracy, a nonlinear (higher-order) relationship can be determined:

$$y - f(x_0) = \frac{f^{(2)}}{2}(x - x_0)^2 + \cdots. \tag{1.40}$$

In this scenario, "raising the precision" in Eq. (1.40) refers to the mathematical transformation $\tilde{Y} = [y - f(x_0)][\delta]^{-1}$, which enlarges the variance of y by multiplying a factor by $\delta^{-1}(|\delta| < < 1)$ in which the variation of y can be monitored on this scale. Therefore, the nonlinear fluctuation of y can be observed in the range of $|x - x_9| < 1$ under the circumstance of $|f^{(1.2)}[\delta]^{-1}| \approx 1$.

Understanding how the "tendency" shifts—from increasing to decreasing or vice versa—is critical. Let's assume that the predictor variables are $x \in X = \mathbb{R}^n$ and the smooth function $f(x)$ has a range of $Y = \mathbb{R}$. Assume that G is the graph of f, a hypersurface submerged in $Z = X \times Y = \mathbb{R}^{n+1}$. Let G' be the plane that touches at $(x, y) = (x_0), f(x_0)$ and is tangential to G. In this view, x_0 is referred to as the crucial point of f if G' is parallel to all basis vectors $\vec{e}_j \in X$. In this circumstance, the "stationary value" is the equivalent $f(x_0)$. According to this perspective, one variety of stationary values is the maximum or minimum value (Fig. 1.16), which are both referred to as extreme values. In this sense, at certain critical points, the tangent plane (linear approximation) deteriorates (Fig. 1.16).

Needless to say, the oscillation of a pendulum (Section 1.6) and the aforementioned oscillation around the critical point are fundamentally distinct from one another. The first is an oscillation around a steady equilibrium (the stationary state without motion). The critical point in the latter, however, is a "marginal" point rather than a stable equilibrium. The rivalry between the

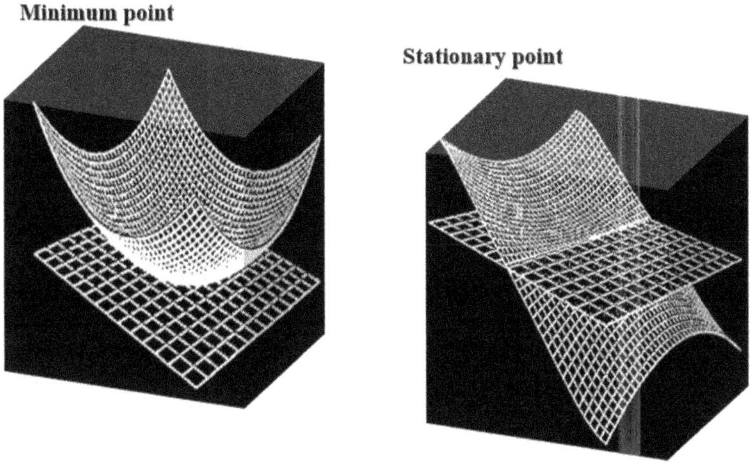

FIGURE 1.16 Sort of the critical points in a nonlinear system.

external drive (supply of sands) and spontaneous disintegration culminates in the oscillations, which are recurrent (but irregular) alterations to the framework. The implosion has a self-stabilizing impact, and the comprehensive evaluation self-organizes as a consequence of this losing momentum nonlinearity (Aarseth, 1994; Allahverdy et al., 2016; Asadi et al., 2018; Pourahmadi, 2018).

1.10 Continuity of nonlinearity

What is meant by continuity of nonlinearity? Let us consider that $f(x)$ be a continuous map from a closed interval $[\alpha, \beta]$ *to* $\mathbb{R}$. In this sense, let us state the following formula $\hat{y}=f(x)$ at circumstance of $\hat{y} \in \mathbb{R}$. The raising question now is: what the solution for x? In this regard, let us consider that:

$$\alpha = \max[f(a), f(b)], \beta = \min[f(a), f(b)]. \tag{1.41}$$

Formula (1.41) delivers a number of the solutions $\hat{y}=f(x)$, which are function of the changing in the deformation map of $f(x)$. In the circumstances of $\beta \leq \hat{y} \leq \alpha$, equation $\hat{y}=f(x)$ has at least one solution (Fig. 1.17). This observation implies that the boundary values of the graph at Fig. 1.17 define the equation's solvability, irrespectively of the curve's layout (Aarseth, 1994; Allahverdy et al., 2016; Liu & Li, 2015; Ok, 2017). What it can conclude, therefore, is diminished as the scale's "deviation" increases. The aforementioned requirement is just a requisite and not a precondition if the monotonicity is broken as the critical points are constructed, the uniqueness of the solution is compromised, and either *max* $f(x)$ or min $f(x)$ is accomplished within the domain (Aarseth, 1994; Asadi et al., 2018; Cortina, 1993; Hall, 2001). Then, since the boundary values are stable, what is robust against the continuously distortion of $f(x)$?

The accompanying "topological number" is constant even while the number of solutions to $\hat{y}=f(x)$ fluctuates, which is known as a homotopy invariant. The conclusion

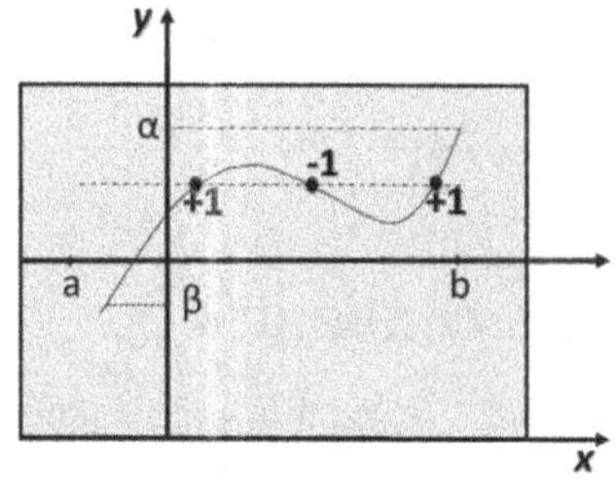

FIGURE 1.17 Different solutions of Eq. (1.41).

is determined as the point where the line $y'=y$ and the graph $y'=f(x)$ intersect (Fig. 1.17). The degree of intersections fluctuates if the curve of $y'=f(x)$ has "pleats." Consequently, Fig. 1.17 reveals that $f(x)$ crosses $\hat{y}$ from below (+1) or from above (−1) as the intersection either +1 or −1. In this view, $x_1, x_2, \ldots, x_m$ are considered the number of the solutions for $\hat{y}=f(x)$, which can be mathematically described as:

$$deg(f, \hat{y}) = \sum_{j=1}^{m} sgnf'(x_j). \tag{1.42}$$

In Eq. (1.42), as well as $deg(f, \hat{y})=0$ there is no solution. Therefore, $deg(f, \hat{y}) = \lim_{y_l \to \hat{y}} deg(f, y_l)$ as well as $\hat{y}$ is a critical point of $f(x)$. Consequently, Eq. (1.42) demonstrates the continuously function of $f(x)$ *in the other words, generates sequences of the functions* $f(x_1), f(x_2), \ldots, f(x_n)$. In this view, the degree of the map $f(x)$ is expressed by $deg(f, \hat{y})$. As long as the boundary values $f(a)$ and $f(b)$ are constant, it is obvious that the degree is invariant across modifications of $f(x)$. Eq. (1.41) is aggregated for a multidimensional map $f:\mathbb{R}^n \to \mathbb{R}^n$, with the Jacobian of f substituting the differential coefficient f'. Furthermore, it is possible to extrapolated in an infinite-dimensional space for a specific class of maps, for instance, $f=I-K$, where K is a compact operator, and I is the identity. Let us consider a finite-dimensional approximation of f and demonstrate that the approximation coincides to a distinctive threshold. In this understanding, it can be considered that the simplistic degree is homotopy invariant. Assume that a fixed point of the map $\hat{y}=f(x)$ is desired to be determined $g(x)=x-f(x)$. Consequently, let $g(x)$ be a continuous function identified on the range $g(\Omega)=[a, b]$. If $g(\Omega)\subseteq\Omega, g$ must have at least one fixed point in Ω. In the point of fact, if $g(\Omega)\subseteq\Omega, f(a)\leq 0\leq f(b)$, then $deg\ (f, 0)=1$, implying that the curve of $y=f(x)$ intersects with the line of $y=0$, *which is* equivalently, the plot of $y=g(x)$ intersects with the line of $y=x$, insinuating the existence of a fixed point. Therefore, by extrapolating this assertion to a multidimensional space, we can demonstrate that a sustained map $g(x)$ on a constrained locked convex set $\Omega\subset\mathbb{R}^n$ with range $g(\Omega)$ does have a specific value. In this sense, $g(x)$ can be extended to compact operators in an infinite-dimensional space, which are exploited to demonstrate the presence of options to nonlinear equations (Allahverdy et al., 2016; Asadi et al., 2018; Isett & Vicol, 2015; Ok, 2017).

1.11 Lipschitz continuity

After the German mathematician Rudolf Lipschitz, Lipschitz continuity is a powerful variant of uniform continuity for functions used in mathematical analysis. There is a real number such that, with every pair of points on the graph of a Lipschitz continuous function, the absolute value of the slope of the line joining them is not more than this real number. The smallest such bound is known as the Lipschitz constant of the function (or modulus of uniform continuity). Any function with bounded first derivatives, for instance, is Lipschitz continuous.

A function $f:X\rightarrow Y$ is referred to as Lipschitz continuous if there exists a real constant $K\geq 0$ such that, for any x_1 and x_2 in X. *In this sense*, assuming dual metric spaces (X, d_X) and (Y, d_Y), where d_X reveals the metric on the set X and d_Y is the metric on set Y, which can be expressed by:

$$d_Y(f(x_1), f(x_2))\leq Kd_X(x_1, x_2). \tag{1.43}$$

Formula (1.43) debates that for the function f, any such K is known as a Lipschitz constant, and f may also be known as *K-Lipschitz*. The lowest constant is also known as the dilation of f or the (best) Lipschitz constant of f. A contraction is a function that maps a metric space to itself and is known as a short map if $K=1$ and $0\leq K<1$. Particularly, a real-valued function $f:R\rightarrow R$ is referred to as Lipschitz continuous if a positive real constant K exists such that, for all real x_1 and x_2, which is described by:

$$|f(x_1)-f(x_2)|\leq K|x_1-x_2|. \tag{1.44}$$

Eq. (1.44), in this scenario, reveals that X is a subset of **R**, and Y is the set of real numbers **R** with the standard metric $d_Y(y_1, y_2)=|y_1-y_2|$. This is applicable for real-valued functions of multiple real variables if and only if the slopes of all secant lines have absolute values that are constrained by K. A function is Lipschitz if and only if the graph of the function everywhere is entirely outside of this cone (Fig. 1.18), which is formed by the set of lines

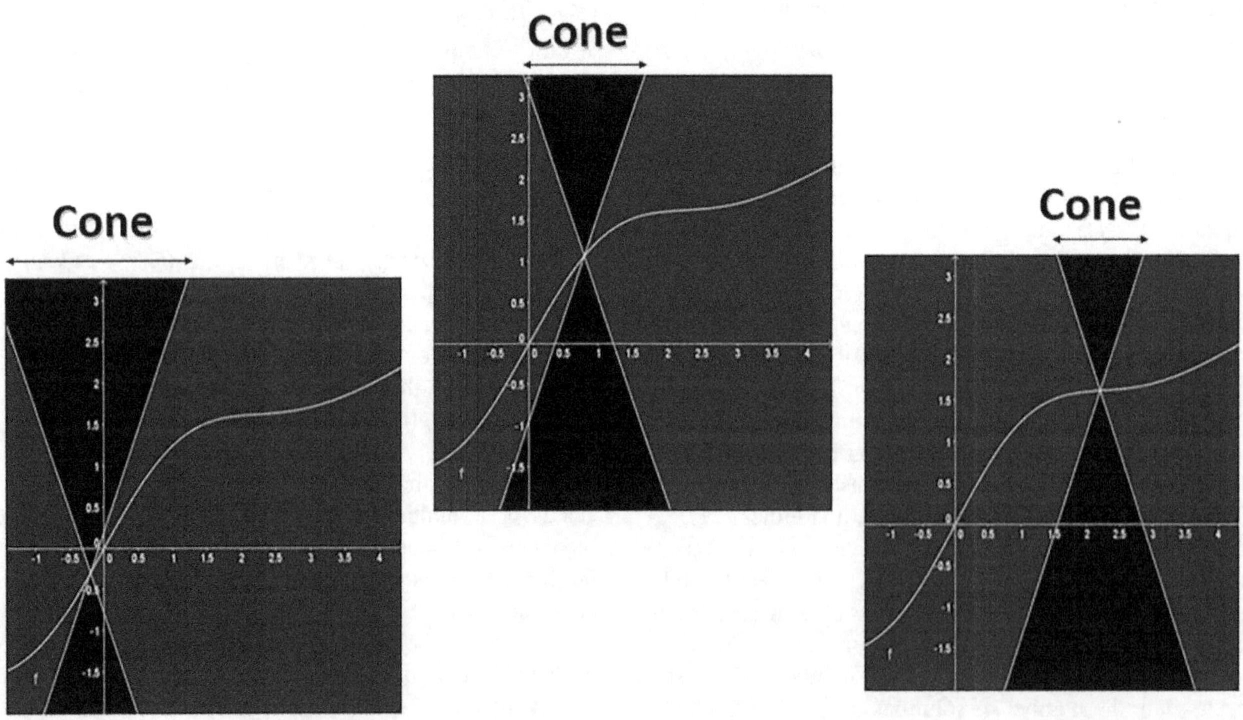

FIGURE 1.18 Lipschitz continuous function.

of slope K passing through a point on the graph of the function (Fig. 1.8).

Commonly, if $x_1 = x_2$, the inconsistency is (deterministically) resolved. Conversely, a function can be correspondingly described as Lipschitz continuous if and only if a constant $K \geq 0$ exists given that for any $x_1 \neq x_2$,

$$\frac{d_Y(f(x_1),f(x_2))}{d_X(x_1,x_2)} \leq K. \tag{1.45}$$

If a local U of x arises for almost any x in X even as f constrained to U is Lipschitz continuous, then the function is said to be locally Lipschitz continuous. Alternatively, f is locally Lipschitz if and only if it is Lipschitz continuous on every compact subset of X if X is a locally compact metric space. This is a prerequisite, but it's insufficient in locales that aren't locally compact (Hager, 1979). Therefore, as more specifically, if there is a constant $M \geq 0$ such that a function f defined on X is Hölder continuous (Isett & Vicol, 2015) or satisfies a Hölder condition of $\alpha > 0$ on X, in which

$$d_Y(f(x),f(y)) \leq M d_X(x,y)^{\alpha} \tag{1.46}$$

Therefore, in X over all x and y in formula (1.46). A uniform Lipschitz condition of $\alpha > 0$ is another term for a Hölder condition of order as given by:

$$\frac{1}{K} d_X(x_1,x_2) \leq d_Y(f(x_1),f(x_2)) \leq K d_X(x_1,x_2) \tag{1.47}$$

Eq. (1.47) is achieved in the scenario of forall$x_1, x_2 \in X$, which its symbol is K-bilipschitz can be also written K-bi-Lipschitz. If f is bilipschitz or bi-Lipschitz, it suggests that a comparable K arises. In practice, a bilipschitz mapping is a homeomorphism onto its image and is invertible. The similar object as an invertible Lipschitz function whose inverse function is indeed Lipschitz is a bilipschitz function (Asadi et al., 2018).

There are always Lipschitz continuous maps $M \dashrightarrow R$ that extend f and contain the same Lipschitz constant as f assuming U is a subset of the metric space M and f: $U \dashrightarrow R$ is a Lipschitz continuous function. A prolongation is casted as:

$$\tilde{f}(x) := \inf_{u \in U} \{f(u) + kd(x,u)\}, \tag{1.48}$$

Eq. (1.48) says that where the Lipschitz constant k applies to f on U. Therefore, almost any Lipschitz continuous map is a posteriori continuous because it is continuously continuous. A group of functions that have constrained Lipschitz constants collectively make up an equicontinuous set. In this view, f_n has a convergent subset if it is a chain of uniformly constrained functions with a restricted Lipschitz constant. The bounded function is also Lipschitz, resulting in the similar constraint for the Lipschitz constant. A globally compact convex subset of the Banach space $C(x)$ is, in specifically, the set including all real-valued Lipschitz functions on a compact metric space X satisfying Lipschitz constant $\leq K$. An absolutely convergent series that is measured by a specific norm is demonstrated to converge to an element in a Banach space (Asadi et al., 2018; Hager, 1979).

1.12 Monotonicity

In calculus, a monotonic function f is one that is either entirely nonincreasing or entirely nondecreasing when defined on a subset of the real numbers with real values. A function that increases monotonically, for instance, does not have to increase exclusively; it must merely not decrease (Fig. 1.19). The concept monotonically boosting refers to a function, which is also increasing or nondecreasing.

According to this perspective, f preserves order since $x \leq y$, which is resulting in $f(x) \leq f(y)$ for entirely x and y (Fig. 1.19). In the circumstance of $x \leq y$ as $f(x) \geq f(y)$, a function is termed as monotonically decreasing (Fig. 1.20). It is also known as a decreasing or nonincreasing function (Guo & Qi, 2010; Jiang & Wang, 2011; Shen et al., 2009).

If it is uncertain whether the aspects "increasing" and "decreasing" also involve the potential of reiterating the similar value at sequential reasoning, the expressions "crudely monotone," "tenuously increasing," and "loosely decreasing" can be used to demonstrate this potential. On the other hand, the negative qualifications "not

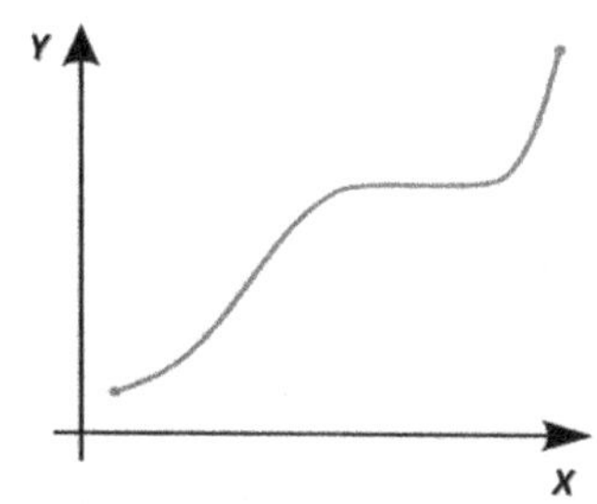

FIGURE 1.19 A function that never decreases monotonically.

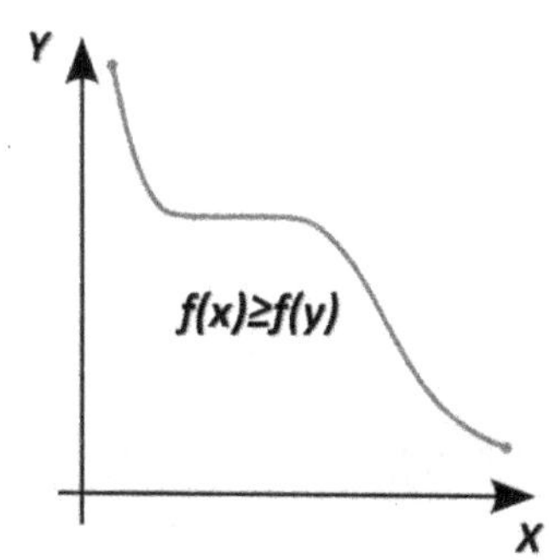

FIGURE 1.20 A function that is monotonically nonincreasing.

decreasing" and "not increasing" should not be confused with the (much weaker) negative qualifications "not decreasing" and "not increasing." The function in Fig. 1.21, for example, falls first, then rises, then falls again. As a consequence, it is neither decreasing nor increasing; however, neither is it nondecreasing nor nonincreasing (Chen & Qi, 2005; Guo & Qi, 2010; Jiang & Wang, 2011).

There is no inverse for a monotonic function that is not precisely monotonic but is nonetheless constant on an interval. This is due to the requirement that a one-to-one mapping exists between the function's range and domain in order for a function to have an inverse. There would be more than one value in the range that translates to this constant value since a monotonic function has some values that are constant in its domain. Conversely, since there is often a one-to-one mapping from the function's range to its domain, a function $y = g(x)$ that is strictly monotonic has an inverse function such that $x = h(y)$. Similarly, if a function is strictly monotonic on a set of values, it has an inverse on that set of values. For instance, if the function $y = g(x)$ is absolutely monotonic on the range $[a, b]$, then its inverse is $x = h(y)$ on $[g(a), g(b)]$, but we cannot assert that the function's inverse exists over the entire range (Chen & Qi, 2005; Shen et al., 2009).

This chapter delves into the profound exploration of nonlinearity, unveiling the gap that exists between radar satellite images and the intricate occurrences of oceanic turbulence. The comprehension and resolution of this gap are elucidated through the lens of nonlinearity concepts. Mathematical philosophical groundwork, particularly from the standpoint of nonlinear differential equations, is employed to articulate the definition of nonlinearity.

Furthermore, the chapter delves into the nonlinearity inherent in dynamical systems, providing a bridge between radar signals and the turbulence of the ocean surface. The inherent uncertainties associated with linearity have spurred scientists to devise the inaugural mathematical model of nonlinear processes—the Lorenz Attractor—an integral foundation for dynamical systems and an emblem of "chaos theory."

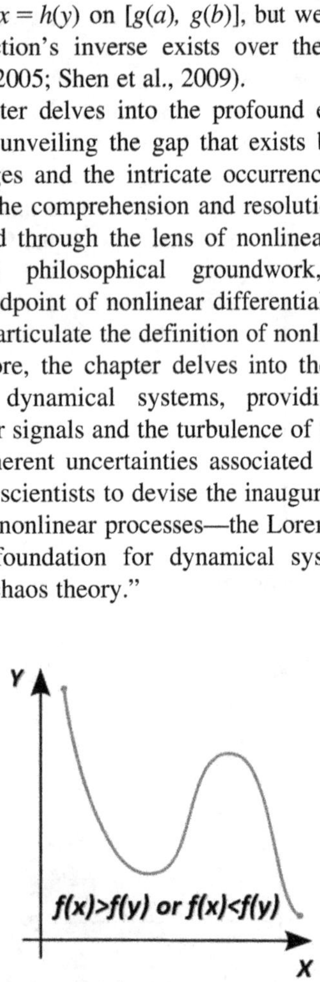

FIGURE 1.21 A nonmonotonic function.

The chapter also intricately addresses facets of nonlinearity, including Nonlinearity Singularity, Bifurcation and Discontinuity, Continuity of Nonlinearity, Lipschitz Continuity, and Monotonicity. This comprehensive exploration lays the groundwork for a nuanced understanding of how synthetic aperture radar imaging interfaces with the turbulent flow of the ocean, a theme that will be further expounded upon in the subsequent chapters.

References

Aarseth, E. J. (1994). Nonlinearity and literary theory. *Hyper/text/theory*, *52*, 761–780.

Allahverdy, A., Moghadam, A. K., Mohammadi, M. R., & Nasrabadi, A. M. (2016). Detecting adhd children using the attention continuity as nonlinear feature of EEG. *Frontiers in Biomedical Technologies*, *3*(1–2), 28–33.

Anderson, J. A. (1995). An introduction to neural networks. MIT Press.

Asadi, K., Misra, D., & Littman, M. L. (2018). Lipschitz continuity in model-based reinforcement learning. 35th International Conference on Machine Learning, ICML 2018. *International Machine Learning Society (IMLS) United States*, 264–273.

Bai, E. W., & Fu, M. (2002). A blind approach to Hammerstein model identification. *IEEE Transactions on Signal Processing*, *50*(7), 1610–1619. Available from https://doi.org/10.1109/TSP.2002.1011202.

Barnett, W. A., & Serletis, A. (2000). Martingales, nonlinearity, and chaos. *Journal of Economic Dynamics and Control*, *24*(5–7), 703–724. Available from https://doi.org/10.1016/S0165-1889(99)00023-8, http://www.elsevier.com/jedc.

Berger, M. S. (1977). *Nonlinearity and functional analysis: lectures on nonlinear problems in mathematical analysis* (74). Academic Press.

Billings, S. A. (2013). Nonlinear system identification: NARMAX methods in the time, frequency, and spatio-temporal domains. *Nonlinear System Identification: NARMAX Methods in the Time, Frequency, and Spatio-Temporal Domains* (pp. 1–555). United Kingdom: Wiley. Available from http://onlinelibrary.wiley.com/book/10.1002/9781118535561, 10.1002/9781118535561.

Bishop, C. M. (1994). Neural networks and their applications. *Review of Scientific Instruments*, *65*(6), 1803–1832. Available from https://doi.org/10.1063/1.1144830.

Chen, C. P., & Qi, F. (2005). Completely monotonic function associated with the gamma functions and proof of Wallis' inequality. *Tamkang Journal of Mathematics*, *36*(4), 303–307. Available from https://doi.org/10.5556/j.tkjm.36.2005.101, http://journals.math.tku.edu.tw/index.php/TKJM/issue/archive.

Chen, X., Hansen, L. P., & Carrasco, M. (2010). Nonlinearity and temporal dependence. *Journal of Econometrics*, *155*(2), 155–169. Available from https://doi.org/10.1016/j.jeconom.2009.10.001.

Coll, B., Gasull, A., & Prohens, R. (2001). Degenerate Hopf bifurcations in discontinuous planar systems. *Journal of Mathematical Analysis and Applications*, *253*(2), 671–690. Available from https://doi.org/10.1006/jmaa.2000.7188.

Coppens, A. B., Beyer, R. T., Seiden, M. B., Donohue, J., Guepin, F., Hodson, R. H., & Townsend, C. (1965). Parameter of nonlinearity in fluids. II. *Journal of the Acoustical Society of America. NA*, *38* (5), 797–804. Available from https://doi.org/10.1121/1.1909806.

Cortina, J. M. (1993). Interaction, nonlinearity, and multicollinearity: Implications for multiple regression. *Journal of Management*, *19*(4), 915–922. Available from https://doi.org/10.1177/014920639301900411.

Drábek, P., Krejcí, P., & Takác, P. (1999). *Nonlinear differential equations* (404). CRC Press.

Emancipator, K., & Kroll, M. H. (1993). A quantitative measure of nonlinearity. *Clinical Chemistry*, *39*(5), 766–772. Available from https://doi.org/10.1093/clinchem/39.5.766.

Fay, T. H. (2002). The pendulum equation. *International Journal of Mathematical Education in Science and Technology*, *33*(4), 505–519. Available from https://doi.org/10.1080/00207390210130868.

Fletcher, W. P., & Gent, A. N. (1954). Nonlinearity in the dynamic properties of vulcanized rubber compounds. *Rubber Chemistry and Technology*, *27*(1), 209–222. Available from https://doi.org/10.5254/1.3543472.

Fucik, S., & Kufner, A. (2014). *Nonlinear differential equations*. Elsevier.

Ghil, M. (2019). A century of nonlinearity in the geosciences. *Earth and Space Science*, *6*(7), 1007–1042. Available from https://doi.org/10.1029/2019EA000599, http://agupubs.onlinelibrary.wiley.com/hub/journal/10.1002/(ISSN)2333-5084/.

Guo, B. N., & Qi, F. (2010). A property of logarithmically absolutely monotonic functions and the logarithmically complete monotonicity of a power-exponential function. *UPB Scientific Bulletin, Series A: Applied Mathematics and Physics*, *72*(2), 21–30.

Gurney, K. (2018). *An introduction to neural networks*. CRC Press.

Haber, R., & Keviczky, L. (1999). Nonlinear system parameter identification. Mathematical Modelling Theory and Applications. **1**.

Haber, R. (1985). Nonlinearity tests for dynamic processes. *IFAC Proceedings Volumes*, *18*(5), 409–414. Available from https://doi.org/10.1016/s1474-6670(17)60594-9.

Hager, W. W. (1979). Lipschitz continuity for constrained processes. *SIAM Journal on Control and Optimization*, *17*(3), 321–338. Available from https://doi.org/10.1137/0317026.

Hall, D. A. (2001). Nonlinearity in piezoelectric ceramics. *Journal of Materials Science*, *36*(19), 4575–4601. Available from https://doi.org/10.1023/A:1017959111402.

Havil, J. (2019). *Chapter One. The Euler Spiral* (pp. 1–14). Walter de Gruyter GmbH. Available from 10.1515/9780691197784-003.

Isett, P., & Vicol, V. (2015). Hölder continuous solutions of active scalar equations. *Annals of PDE*, *1*(1). Available from https://doi.org/10.1007/s40818-015-0002-0, http://springer.com/physics/theoretical, + mathematical + & + computational + physics/journal/40818.

Jiang, F., & Wang, L. (2011). Finite-time weighted average consensus with respect to a monotonic function and its application. *Systems & Control Letters*, *60*(9), 718–725. Available from https://doi.org/10.1016/j.sysconle.2011.05.009.

Kimia, B. B., Frankel, I., & Popescu, A. M. (2003). Euler spiral for shape completion. *International Journal of Computer Vision*, *54*(1–3), 159–182.

Kopp, C., & Meyrueis, P. (1998). Near-field Fresnel diffraction: Improvement of a numerical propagator. *Optics Communications*, *158*(1–6), 7–10. Available from https://doi.org/10.1016/S0030-4018(98)00549-5.

Kroll, M. H., Præstgaard, J., Michaliszyn, E., & Styer, P. E. (2000). Evaluation of the extent of nonlinearity in reportable range studies. *Archives of Pathology & Laboratory Medicine*, *124*(9), 1331–1338. Available from https://doi.org/10.5858/2000-124-1331-eoteon.

Leine, R. I., Van Campen, D. H., & Van De Vrande, B. L. (2000). Bifurcations in nonlinear discontinuous systems. *Nonlinear Dynamics.*, *23*(2), 105–164. Available from https://doi.org/10.1023/A:1008384928636.

Leine, R. I., & Van Campen, D. H. (2002). Discontinuous fold bifurcations in mechanical systems. *Archive of Applied Mechanics*, *72*(2–3), 138–146. Available from https://doi.org/10.1007/s00419-001-0190-9.

Liu, X., Gupta, S. D., & Agarwal, G. S. (2014). Regularization of the spectral singularity in PT-symmetric systems by all-order nonlinearities: Nonreciprocity and optical isolation. *Physical Review. A*, *89*(1). Available from https://doi.org/10.1103/physreva.89.013824.

Liu, Y., & Li, X. R. (2015). Measure of nonlinearity for estimation. *IEEE Transactions on Signal Processing*, *63*(9), 2377–2388. Available from https://doi.org/10.1109/TSP.2015.2405495, http://ieeexplore.ieee.org/xpl/RecentIssue.jsp?punumber = 78.

Mostafazadeh, A. (2013). Nonlinear spectral singularities for confined nonlinearities. *Physical Review Letters*, *110*(26). Available from https://doi.org/10.1103/PhysRevLett.110.260402Turkey, http://oai.aps.org/filefetch?identifier = 10.1103/PhysRevLett.110.260402&component = fulltext&description = markup&format = xml.

Ok, J. (2017). Gradient continuity for nonlinear obstacle problems. *Mediterranean Journal of Mathematics*, *14*(1). Available from https://doi.org/10.1007/s00009-016-0838-x, http://springerlink.metapress.com/app/home/journal.asp?wasp = 46wluhqumg3vd7j5nre7&referrer = parent&backto = browsepublicationsresults,355,537.

Pertsch, T., Peschel, U., Kobelke, J., Schuster, K., Bartelt, H., Nolte, S., Tünnermann, A., & Lederer, F. (2004). Nonlinearity and disorder in fiber arrays. *Physical Review Letters*, *93*(5), 1–53901. Available from https://doi.org/10.1103/PhysRevLett.93.053901.

Phillips, J. D. (2003). Sources of nonlinearity and complexity in geomorphic systems. *Progress in Physical Geography*, *27*(1), 1–23. Available from https://doi.org/10.1191/0309133303pp340ra, https://journals.sagepub.com/home/PPG.

Pourahmadi, M. (2018). Taylor expansion of and some applications. *The American Mathematical Monthly*, *91*(5), 303–307. Available from https://doi.org/10.1080/00029890.1984.11971411.

Roberts, K., Li, C., Strawczynski, L., O'Sullivan, M., & Hardcastle, I. (2006). Electronic precompensation of optical nonlinearity. *IEEE Photonics Technology Letters*, *18*(2), 403–405. Available from https://doi.org/10.1109/LPT.2005.862360.

Schetzen, M. (1980). *The Volterra and Wiener theories of nonlinear systems*. Wiley.

Shen, P., Chen, Y., & Ma, Y. (2009). Solving sum of quadratic ratios fractional programs via monotonic function. *Applied Mathematics and Computation*, *212*(1), 234–244. Available from https://doi.org/10.1016/j.amc.2009.02.024.

Snieder, R. (1998). The role of nonlinearity in inverse problems. *Inverse Problems*, *14*(3), 387–404. Available from https://doi.org/10.1088/0266-5611/14/3/003.

Struble, R. A. (2018). *Nonlinear differential equations*. Courier Dover Publications.

Tsay, R. S. (1986). Nonlinearity tests for time series. *Biometrika*, *73*(2), 461–466. Available from https://doi.org/10.1093/biomet/73.2.461.

Vakhitov, N. G., & Kolokolov, A. A. (1973). Stationary solutions of the wave equation in a medium with nonlinearity saturation. *Radiophysics and Quantum Electronics*, *16*(7), 783–789. Available from https://doi.org/10.1007/BF01031343.

Zakharov, V. E., & Synakh, V. S. (1975). The nature of the self-focusing singularity. *Zh. Eksp. Teor. Fiz.*, *68*(3), 940–947.

Chapter 2

Unraveling the mysteries: Ocean turbulence and front dynamics

One of the last things you want to hear the honorary captain wisely say as the plane is humming along efficiently is, "Sorry, folks, we're about to encounter a little turbulence." However, whereas turbulence in the skies is undesirable, it is beneficial in the mighty oceans. It is beneficial for all living organisms there, as well as for the marine ecosystem and humankind. In this sense, there are various causes for the turbulence at the ocean's surface. Winds produce waves that travel downward and outward. Eddies are created when two currents flow in opposite directions. The water rises and falls due to the day's heating and the night's cooling. In this view, currents in shallow water become chaotic as they drag along the bottom. Similarly, the regular up-and-down migration of tiny organisms could cause a stirring. Therefore the preponderance of the turbulence phenomena is on a small scale—unlike the strong swirling air masses that bounce around hundred-ton jets.

Nonetheless, ocean turbulence is persistent—eddies, waves, and streamer motions are constantly mixing things up. Consequently, colder water masses that are richest in nutrition are dredged up by the continual agitation from beneath. This feeds microscopic species at the top of the ocean food chain, which in turn feeds other organisms down the chain. Consequently, turbulence also aids in heat movement between the oceans and the atmosphere, permitting the oceans to manage global temperatures. It moreover assists the oceans in absorbing carbon dioxide from the atmosphere, reducing the amount of heat-trapping gas in the air, and making the ride a little smoother for everyone.

Turbulence is a crucial phenomenon in fluid dynamics. The precision of numerical simulations of complex turbulent flows is a fundamental challenge in aerodynamic and hydrodynamic designs. In this regard, what is meant by turbulence?

2.1 What is the magic meaning of turbulence?

Turbulence, often known as turbulent flow, is a phase of ocean dynamics characterized by chaotic oscillations in flow velocity and pressure. Laminar flow, on the other hand, occurs when a flow deals with stability in adjacent layers with no discontinuities between them (Huppert et al., 2000; Woods, 2010). In this sense, Leonardo da Vinci's well-known illustration of a free jet of water stemming from a hole drilled into a pool can be deemed one of the earliest investigations on the structural system of turbulence, revealing turbulence properties, with diverse levels of movement and a jagged partition between jet and vortices of both small and large sizes, and waves at a high vantage point. In other words, turbulence is a great level of chaos and complexity. In this view, Richard Feynman in the 1970s declared "turbulence remains the most crucial ongoing challenge of classical physics." In this understanding, interfaces in turbulence that designate domains of divergent acceleration, mass, or particle quantities (such as temperature or residues) occur in a wide range of fluid flow application domains, which would include processes in buoyant jets engendered by breathing, the vertices of wakes of self-propelled carcasses (e.g., vessels, underwater vehicles, and airplanes), at the "center of the storm," at the thermocline in the ocean, atmosphere, or mountain valleys generated by natural cycles such as heating/cooling. Therefore an unsteady form of the flow known as turbulent fluid motion allows for the statistically distinct average values of the various parameters to be identified. In this sense, turbulence is a multiresolution, sophisticated, nonlinear phenomenon that raises several of the most complex and challenging computational and theoretical issues.

Even though several studies have been conducted using comprehensive numerical models over the past few decades (Bernard et al., 2003; Mathieu & Scott, 2001; Nakayama, 2018; Nieuwstadt et al., 2016; Poncet et al., 2008; Pope, 2001; Yaglom, 2012), the author believes and asserts that no highly formed quantitative theory of turbulence has yet to be established.

Needless to say, turbulence currents can be defined as irregular fluctuations that move slightly or hardly at all in the dynamics of the oceans. Consequently, these pronounced fluctuations represent profound changes in fractal and

Synthetic Aperture Radar Image Processing Algorithms for Nonlinear Oceanic Turbulence and Front Modeling. DOI: https://doi.org/10.1016/B978-0-443-19155-8.00002-7

continuous patterns superimposed on the average dynamic motion of the boundless oceans. Nevertheless, turbulence remains a controversial topic in academia. This is because myriad dynamical causes can trigger turbulence. In this sense, it is difficult to provide an exquisite description. Therefore the ambiguous definition of turbulence can still lead to numerous important questions.

2.2 What is the origin of ocean turbulence?

Currently, the question arises as to what represents the origin of turbulence. Ocean turbulence naturally arises from the dynamic instability of laminar water flow (Fig. 2.1). For sufficiently considerable values of the characteristic dimensionless number that accurately determines the flow, the movement changes from laminar flow to turbulent flow (Bernard et al., 2003; Mathieu & Scott, 2001).

Thus Reynolds number R_e determines the turbulence based on velocity U, length of scale L, and kinematic viscosity ν, as follows(Leslie, 1978; Mathieu & Scott, 2001; Nakayama, 2018):

$$R_e = UL[\nu]^{-1} \tag{2.1}$$

Eq. (2.1) refers to the simple calculation of turbulence when it exceeds a critical value. The onset of turbulence causes the flow to be unstable around Reynolds numbers between 2000 and 4000. Flows like these are sometimes called transitional flows. A turbulent flow is one with a Reynolds number over R_e 3500 (Table 2.1).

For instance, implementing Eq. (2.1),

$$R_e = \frac{Density(\rho) \times Diameter(D) \times FlowSpeed(U)}{Viscosity(\nu)}$$

then a numerical calculation is demonstrated as

$$R_e = \frac{1.25\ kgm^{-3} \times 0.1\ m \times 35\ ms^{-1}}{1.83 \times 10^{-5}\ Nsm^{-2}} = 2.39 \times 10^{-5}.$$

In this circumstance, turbulence occurs when the Reynolds number R_e is about 2300. Therefore does a high Reynolds number R_e mean turbulence? The viscous regions in high Reynolds number R_e flows are dominated by turbulence. Instability in the shear layer creates turbulent fluctuations of flow-field properties. On the contrary, when the flow is laminar? If the Reynolds number R_e is less than 2000, the flow is considered to be laminar, which is also known as viscous flow. However, a flow is said to be a streamlined flow if its Reynold number R_e is less than 1000 (or 2000) (Bernard et al., 2003; Mathieu & Scott, 2001; Nieuwstadt et al., 2016).

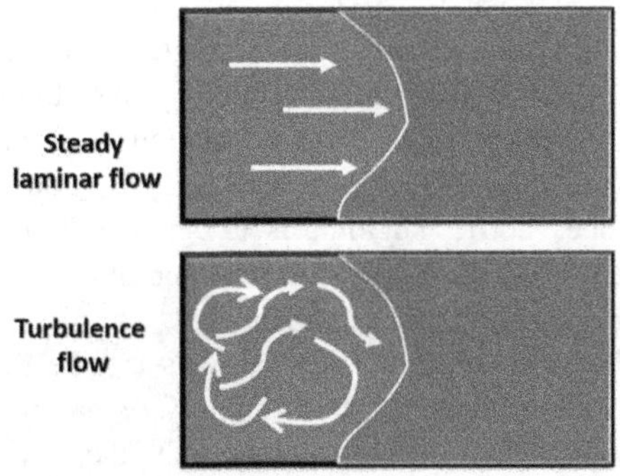

FIGURE 2.1 Laminar and turbulence flows.

TABLE 2.1 Reynolds numbers for different types of internal flow regimes.

Sort of flow	Reynolds number range
Laminar	Up to $R_e = 2300$
Transition	$2300 < R_e < 4000$
Turbulent	$R_e > 4000$

2.3 What is the role of Reynolds number in understanding turbulence mechanisms?

The above perspective issues the following significant question: what is the Reynolds number for which airflow over the ocean surface starts transitioning from laminar to turbulent? As long as the Reynolds number determines the kind of flow, then ocean flow can be presumed as free surface flow. In this regard, laminar flow occurs when $R_e < 2000$, and turbulent flow occurs when $R_e > 4000$, which demonstrates the internal flow. However, for external flow, laminar flow occurs up to $R_e = 5 \times 10^5$. Consequently, the transition region occurs when $R_e = 5 \times 10^5$ becomes the threshold value. It can be viewed this way; transition flow develops as well as the value of Reynolds number of fluid lies between 2100 and 4000. In this zone, the fluid can have either a laminar or turbulent flow. The sort of flow, therefore, would be dominated by factors, for instance, pipe roughness and flow consistency, which are not factored in when computing the Reynolds number (Nakayama, 2018; Nieuwstadt et al., 2016).

In this zone, consequently, the pressure ditches and thus fluctuates either linearly with velocity or with the square of the velocity of the fluid. It can be said that in the transition region, the laminar boundary layer is in a state of unstable equilibrium turbulent flow. Thus even the most infinitesimal vicissitudes in the surrounding form of the fluid can initiate the flow to become turbulent. That is why a small transition region takes place wherein the flow changes to turbulent. Accordingly, the region where laminar to turbulent transition is relevant is the boundary layer. This can lead to the question: why is the Reynolds number for the transition from laminar to turbulent flow in internal flows much smaller than that of external flows? In this view, the thickness of the boundary layer is used as the length scale to form the Reynolds number relevant to transition, which is fitting all flow circumstances. For internal flows, the dimension used is perpendicular to the flow, for example, the diameter of the pipe. Consequently, it is often a dimension in the flow direction, such as the chord of a wing, for external flows. In this regard, the dynamics inside the boundary layer are constrained to exist within that length scale. The velocity develops from zero at the wall to the maximum value, which is impartial outside the boundary layer over that distance. In this circumstance, the Reynolds number is used for more than boundary layers. Often a length scale is used that is not specifically relevant to boundary layers (Pope, 2001; Woods, 2010).

The accomplished frames of internal flow and external flow based on the Reynolds number can be addressed by the following examples. In the internal flow phase through a pipe, the boundary layer escalates until its thickness is about half the diameter of the pipe. The boundary layer thickness is half the pipe diameter. In this circumstance, internal flow merges with the boundary layer on the opposite side. In this understanding, the internal diameter of a pipe is used to form the Reynolds number, which is fairly relevant to boundary layer stability and laminar to the turbulent transition zone. Consequently, the boundary layer cannot remain growing past this point, and the flow becomes fully developed (Bernard et al., 2003; Nieuwstadt et al., 2016).

In the circumstance of the external flow, the length scale is used to form the Reynolds number because the boundary layer thickness is much less than the Reynolds number. In this regard, the diameter is used for cylinders and spheres while the chord (the distance from the leading edge to the trailing edge) is used in the case of wings. Consequently, the Reynolds numbers are numerically much larger than what would be relevant to boundary layer transition (Ozmidov, 1965; Pope, 2001).

This is why the Reynolds number for transition is much higher for external flow than for internal flow. It is because a length scale is used that is much larger than the boundary layer thickness for external flow. However, for internal flow, the length scale is very similar to the boundary layer thickness.

2.4 How can laminar flow and turbulent flow be distinguished?

In conjunction with the preceding viewpoint, laminar and turbulent flows are two different concepts that are frequently used in fluid/air dynamics. In accordance with this view, the fluid either flows laminarly or turbulently across any space boundary dimensions; for instance, in x, y, z, t domains. Laminar flow is known as the flow of fluid whenever the fluid particle maintains a constant trajectory, routes that ordinarily do not cross one another under any circumstances. One consequence of laminar movement would be that the speed belonging to the fluid is constant at any time inside the fluid. On the other hand, turbulent flow could be described as the uneven, infrequent movement of fluid, which is seen as a small whirlpool area. The speed of such a fluid is unquestionably not necessarily constant at every point. Table 2.2 summarizes the main distinguishing characteristics of laminar and turbulent flows (Bernard et al., 2003; Huppert et al., 2000; Nakayama, 2018; Nieuwstadt et al., 2016).

Prime Tables 2.1 and 2.2 divinely reveal that mixing across the tube is extremely effective because, unlike laminar flow, the flowing fluid no longer moves in continuous layers. While flows with limited Reynolds numbers R_e of less than 2300 often remain laminar, those with significant Reynolds numbers beyond 4000 are typically (though not always) turbulent. Laminar flow develops when a fluid runs in parallel layers while it is moving steadily. In this scenario, no fluid swirls, eddies, or cross-currents exist opposing to the flow pattern. As a result, the fluid's particles are all moving very precisely. As a result, a streamlined flow is formed by parallel lines orthogonal to the flow velocity (Bernard et al., 2003; Leslie, 1978; Mathieu & Scott, 2001).

The question is: can mixing take place in the laminar flow? A slightly lateral commixture (mixing perpendicular to the flow direction) crops up by the encounter of diffusion between layers of the liquid. Diffusion commixture occurs slowly, but if the diameter of the pipe of the tube is little, then this diffusive mixing is often very important. Contrarily, chaotic physical properties are a signature of the flow regime known as turbulent flow. This comprises a quick turnaround time in flow velocity and pressure in both the space and time. The fluid no longer moves in strata, in unlike laminar flow, and mixing across the tube is extremely effective. While flows with low Reynolds numbers beneath 2300 often remain laminar, those with high Reynolds numbers beyond 4000 are commonly (but not often) turbulent (Table 2.2). Flow in the 2300–4000 Reynolds number range is referred to as a transition (Table 2.1).

Needless to say, in turbulent flow, the fluid particles do not obey smooth paths. In its place, there is a slice of local arbitrary flow diagonal to the path of the mean motion. Consequently, these disorder motions frequently have the pattern of swirls (vortices) superimposed on the average fluid mobility. Therefore laminar and turbulent flows can exist in the same system network when operated at different flow rates. In this understanding, the

TABLE 2.2 Summarizing the main differences between laminar and turbulent flows.

Sort of flow	Laminar flow	Turbulent flow
Flow fashion	The fluid flow in which the adjacent layers of the fluid do not mix and move parallel to each other is known as laminar flow.	The fluid flow in which the adjacent layers of the fluid cross each other and do not moving parallel to each other is identified as a turbulent flow.
Flow layer styles	In the laminar flow, the fluid layer moves in a straight line.	In turbulent flow, the fluid layers do not move in a straight line. They move randomly in a zigzag manners.
Reynolds numbers	The fluid flow having a Reynolds number of less than 2000 is called laminar flow.	The fluid flow has a Reynolds number greater than 4000; is called turbulent flow.
Shear stress	Shear stress in laminar flow depends only on the viscosity of the fluid and is independent of the density.	The shear stress in turbulent flow depends upon its density.
Mobility	The laminar flow always occurs when the fluid flow with low velocity and in small diameter pipes.	The turbulent flow occurs when the velocity of the fluid is high and it flows through larger diameter pipes.
Mixing	The fluid flow is very orderly, i.e., There is no mixing of adjacent layers of the fluid and they move parallel to each other and also with the walls of the pipe.	There is no mixing of adjacent layers of the fluid and they do not flow in a definite order. There is a mixing of different layers and they do not move parallel to each other but cross each other.
Flow fashion	The fluid flow in which the adjacent layers of the fluid do not mix and move parallel to each other is known as laminar flow.	The fluid flow in which the adjacent layers of the fluid cross each other and do not moving parallel to each other is identified as turbulent flow.

conventional examples of both laminar and turbulent flows can be demonstrated in cigarette smoke (Fig. 2.2).

Let's say that the initial smoke velocity is roughly 30 cm/s, the length scale would be about 3 cm, and the kinematic viscosity is 0.14 cm^2/s. The ratio between inertia and viscous forces that is estimated thanks to the Reynolds number determines whether a flow is laminar or turbulent. Utilizing Eq. (2.1), then $R_e = \frac{30 \times 3}{0.14} = 642$ indicates transition flow from laminar flow to turbulence flow. In this scenario, hot smoke is less dense than the surrounding cooler air, which leads to rapid smoke velocity. Consequently, a net upward buoyant force acts on smoke correspondingly as one acts on a helium balloon, which initiates a vertical acceleration. In other words, smoke should shortly accelerate. Therefore the middle of the plume, which is the hottest, is then accelerated. The velocity moves closer to the edges and approaches that of the surrounding air. Shear, or the movement of the smoke plume from one place to another, causes turbulence. Whether the flow maintains laminar or becomes turbulent must depend on a certain amount of R_e (Fig. 2.3; Bernard et al., 2003; Mathieu & Scott, 2001; Raizer, 2021).

A typical example of flow classification with the Reynolds number is the flow past a two-dimensional cylinder. It is interesting to demonstrate that the fluctuation of Reynolds intervals delivers different flow regimes. In this sense, the laminar steady flow regime is developed as well as the Reynolds number $R_e < 5$ while if the interval is $5 < R_e < 45$, the pair of vortices is associated with laminar steady

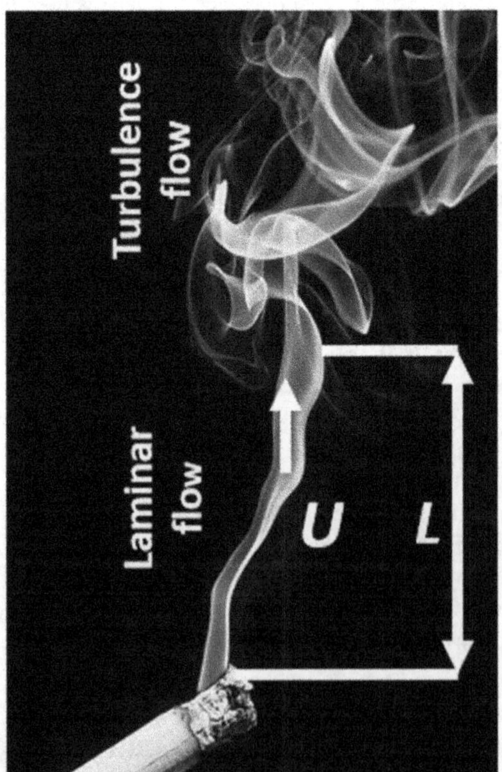

FIGURE 2.2 Laminar and turbulence are associated with cigarette smoke.

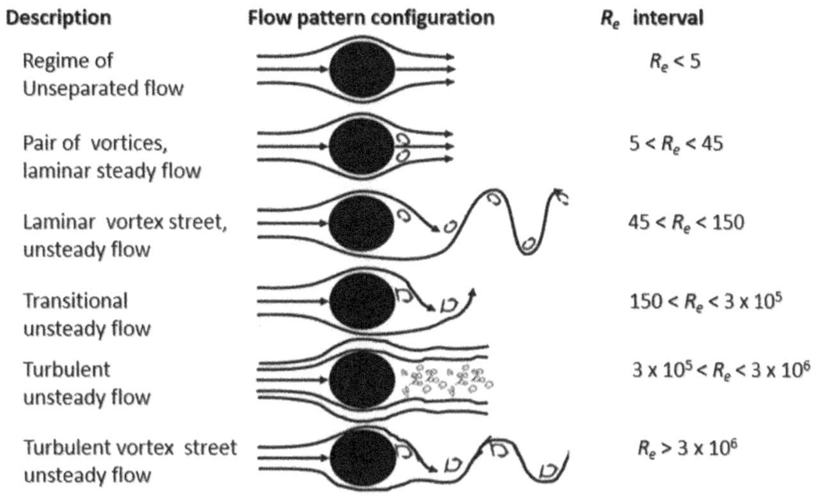

FIGURE 2.3 Different flow pattern regimes with different Reynolds number intervals.

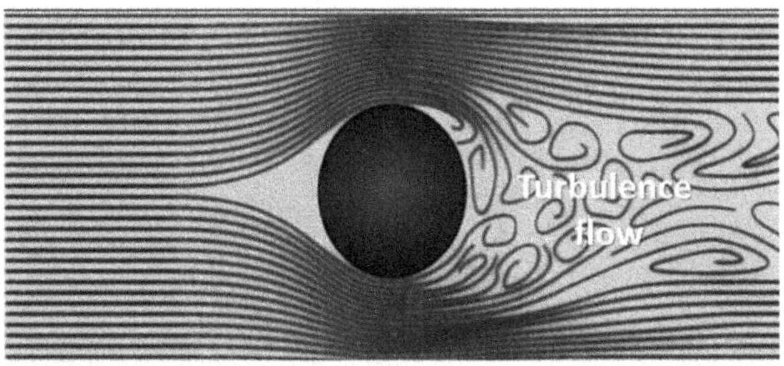

FIGURE 2.4 Eddies generation due to fluctuated flow around the circle.

flow. Consequently, the laminar vortex street associated with unsteady flow exists when $45 < R_e < 150$. As well as the $150 < R_e < 3 \times 10^5$ transitional unsteady flow takes place. However, turbulent unsteady flow and turbulent vortex are developed as well as the Reynolds number intervals are $3 \times 10^5 < R_e < 3 \times 10^6$and $R_e > 3 \times 10^6$, respectively (Fig. 2.3; Bernard et al., 2003; Nezu & Nakagawa, 2017; Poncet et al., 2008).

Now the significant question is: what is the importance of turbulence for atmosphere heat transfer? In turbulent flow, the instabilities deliver a supplementary mechanism for heat transfer along with molecular diffusion. On the contrary, heat is transferred between dissimilar streams across molecular diffusion in laminar flow. In other words, the movement of molecules occurs from the higher concentration zone to the lower concentration region. In fact, in laminar flow, fluid particles flow in an orderly manner along certain path lines. Hence, the turbulent flow continuously has a higher heat transfer coefficient contrasted with the laminar flows. Therefore in turbulence flow, the fluctuated liquid particles can form eddies (Fig. 2.4), which rapidly transport mass, momentum, and energy across different regions of the flow.

Compared with the eddies, molecular diffusion is an incredibly slow process and is regularly abandoned in computing the heat transfer rate in turbulent flows. In this regard, turbulence plays a significant role in air heat transfer.

2.5 What is the nature of turbulence?

The example of a smoke plume from the cigarette (Fig. 2.2) leads to a significant question: what is the nature of the turbulence? It is difficult to deliver specific identical answers owing to the complexity of the turbulence regime. However, it may be effective just to list some of the characteristics of turbulent flow to understand the nature of turbulence. The main noticeable of the turbulence characteristics precious is apparent irregularity. In this optimistic scenario, flowing fluid or gas particles are moved randomly in marked space without a specific direction. In this affectionate regard, turbulence problems are effectively handled statistically rather than deterministically. Turbulent motion is disrupted; however, not all chaotic flows are agitated. In this practical sense, the turbulence regime is properly considered a continuum phenomenon, governed by the fluid equation of continuous motions.

Consequently, turbulent flow occurs at low viscosity and greater characteristic linear dimensions that alliance with higher chaotic movements. In this potential scenario, rapid mixing is typically associated with the steepest rate of mass momentum and energy transport in the turbulence flow. This observed phenomenon is professionally identified as diffusivity.

Therefore the diffusivity of the turbulence fulfills a tremendous role in avoiding boundary-layer disjunction on airfoils at large angles of attack. Considerably, it surges the rate of heat transfers in entire varieties of machinery. It is on top of that the foremost source of resistance to the flow in pipelines. Lastly, it swells the effective rate of momentum transfer between wind stress and ocean currents (Davidson et al., 2013; Nieuwstadt et al., 2016). Moreover, turbulent flow is accurately differentiated by a strong three-dimensional vortex

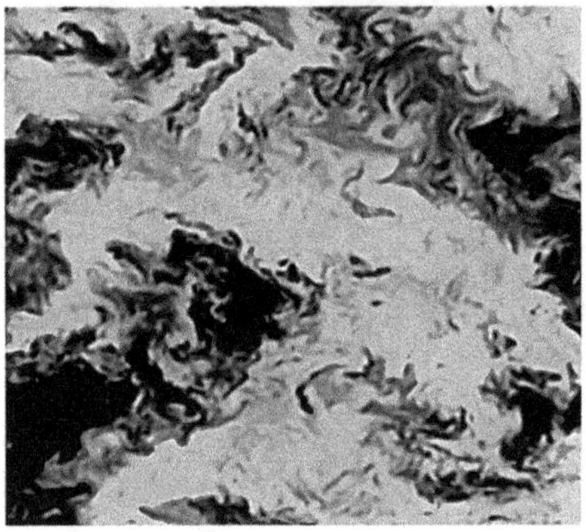

FIGURE 2.5 Vortex stretching regime.

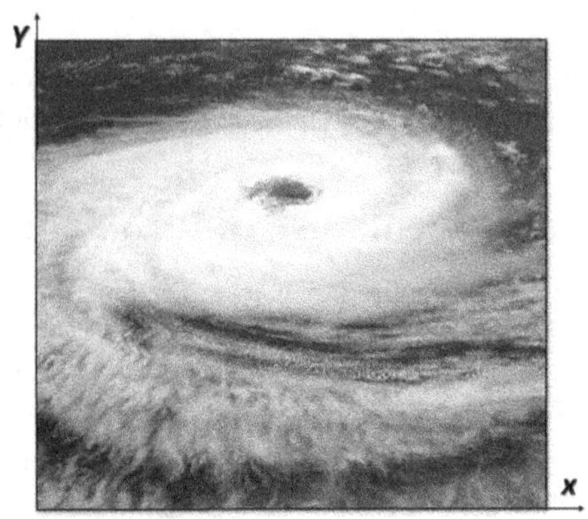

FIGURE 2.6 2-D Cyclone flow.

group mechanism, which is known as whirlpool stretching (Fig. 2.5).

The key question efficiently is: what is the vortex extending? In fluid dynamics, whirlpool stretching represents the expansion of vortices in three-dimensional fluid flow, allied with consistent growth of the constituent of vorticity in the stretching direction—undoubtedly owing to the effective conservation of angular momentum. In this appropriate context, the apparent affinity of turbulent flows to cause fine-scale movements from large-scale energy injection is regularly observed as a scale-wise surge of kinetic energy plunged by vorticity stretching.

However, the random vorticity fluctuations that identified turbulence could not sustain themselves as long as they are typically developed in the two-dimensional flow where there is a notable absence of eddy stretching. For specific instances, the cyclone accurately represents a substantial 2-D movement (Fig. 2.6), which remains not turbulence itself due to the comparative absence of vortex spreading(Woods, 2010; Yaglom, 2012).

Consistent with this proper perspective, dynamic instability must be profusely illustrated by vortex stretching.

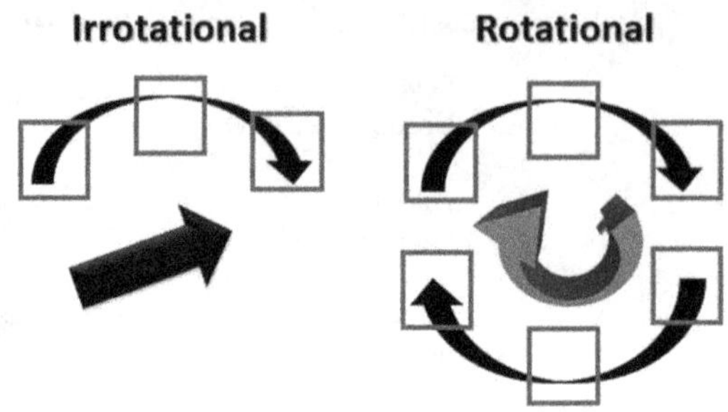

FIGURE 2.7 Irrotational and rotational flows.

This is why ocean waves are not considered turbulence motion since they properly represent irrotational flows (Fig. 2.7). In an irrotational flow, there can be precious circuitous paths for the continuous flow, but each fluid particle does not rotate. Needless to properly say, the dominated vortex stretching in turbulence flow efficiently generates rotational motion (Fig. 2.7).

In other words, the core discrepancy between turbulence and random waves is that waves are fundamentally nondissipative, while turbulence is essentially dissipative. A dissipative manner is a development in which the kinetic energy of the turbulent flow is converted into internal energy through viscous shear stress. In other words, the internal energy of the fluid at the expense of the kinetic energy of the turbulence is increased by the deformation performance of the viscous shear stress.

The origin of the turbulence as a function of the Reynolds number delivers specific identical criteria of turbulence regime. As demonstrated previously in Section 2.4, if the Reynolds number is greater than $R_e > 3500$, the flow is then turbulent. However, the major properties of turbulent regimes do not rely on the molecular characteristics of the fluid in which the turbulence develops. In this regard, every particular flow regime has specific exclusive characteristics that are amalgamated with its original and boundary conditions because of the nonlinearity of motion equations. In this understanding, no cosmopolitan solution to problems in the turbulent regime is developed as well as no well-defined solution to the Navier – Stokes equations is delivered. As a result, it seems that every flow pattern is different owing to all turbulent flows having numerous distinguished characteristics in common. In other words, the turbulence flow regime cannot be stated in specific laws because of the uniqueness of any particular description and classification of turbulent flows such as boundary layers, boundary conditions, jets, and wake. Needless to say, turbulence mathematical theories are restricted to dealing with entire sorts of turbulence flows in general. The determination of turbulence characteristics is a function of its derived environment (Bernard et al., 2003; Davidson et al., 2013; Knauss & Garfield, 2016; Nakayama, 2018; Nezu & Nakagawa, 2017).

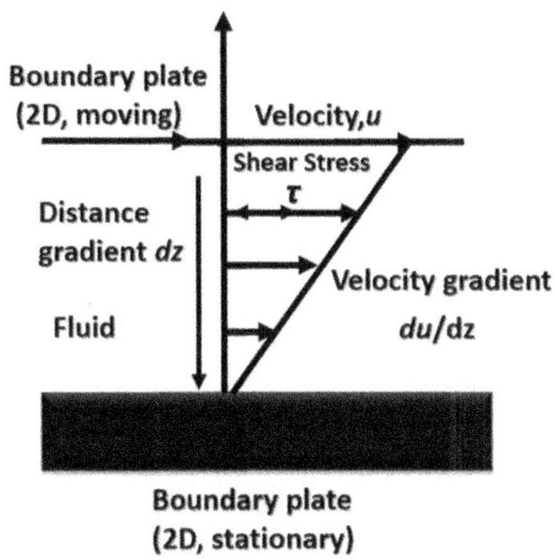

FIGURE 2.8 Simplification of boundary layer concept.

2.6 Boundary layers in turbulence

Ocean currents are usually assumed to be in different layers. This is shown in Fig. 2.8. A continuous layer of current flow will always have zero velocity if it touches an impermeable bottom. This is consistent with a "non-slip" condition. The free surface (outermost layer) of a flowing stream typically experiences variable velocities. As a result, the shear flow stress is usually given by Newton's law of dynamic viscosity.

$$\tau = \mu \frac{du}{dz} \tag{2.2}$$

In Eq. (2.2), shear stress τ is proportionally directly with fluid viscosityμ; its gradient velocity du, and inversely with changes in water depth dz. A force is consistent with the unit area; therefore describes shear stress (Fig. 2.8). It appears parallel to microscopic surface parcels. In fluids, shear stress is often caused by friction among fluid particles. This is due to fluid viscosity. Therefore Eq. (2.2) suggests that as the fluid approaches, the ocean bottom is brought to rest through the shear stress. In this view, the boundary layer is the vicinity in which flow adjusts from zero velocity at the bottom to the maximum speed within the mainstream of the surface flow. The concept of boundary layers is important to deliver the flow rate(Leslie, 1978; Mathieu & Scott, 2001; Stroh et al., 2015). Consequently, u can be obtained at any height z above the bottom plate by integrating concerning z that is formulated by:

$$u_z = \int \frac{du}{dz} dz = \tau\mu^{-1} \int dz + c = \tau\mu^{-1}(z) + c \tag{2.3}$$

here c is the constant of integration and u is the velocity at z = 0 (i.e., where u = 0) such that:

$$u_z = \tau\mu^{-1}z \tag{2.4}$$

Eq. (2.4) demonstrates that the velocity differs linearly from 0 at the bottom ($z = 0$) to approximately maximum at the highest position of the plate. Therefore as the applied force of shear stress upsurges, so does the speed rate at each point directly above the lower plate. In this scenario, the viscosity upsurges the velocity at any point above the lower plate shrinks (Fig. 2.8).

For all Newtonian fluids in laminar flow, the shear stress is proportional to the strain rate in the fluid where the viscosity is the constant of proportionality. However, for non-Newtonian fluids, this is no longer the case as, for these fluids, the viscosity is not constant. The shear stress is imparted onto the boundary as a result of this loss of velocity. Therefore boundary layers can be either laminar or turbulent, relying on the value of the Reynolds number as demonstrated early in Section 1.4. The boundary layer of turbulence is located above three layers: the overlap layer, buffer layer, and viscous sublayer (Fig. 2.9). The viscous sublayer has the smallest velocity as compared with overlap and buffer layers(Willmarth, 1975; Xin et al., 1996; Yaglom, 2012).

Moreover, the boundary layer is a function of its thickness. The thickness of the boundary ∂ layer is defined as the distance from the boundary to the point where the free streamflow (Fig. 2.9). In this regard, laminar boundary layers over a flat are governed by:

$$\delta \approx 5.0 \times \left(R_{e_L}\right)^{-0.5} \times L \tag{2.5}$$

here R_{eL} is the Reynolds number based on the length of the plate L. On the contrary, the boundary layer thickness of turbulent flow is given by:

$$\delta \approx 0.37 \times \left(R_{e_L}\right)^{-0.2} \times L \tag{2.6}$$

According to formula (2.6), the turbulent boundary layer thickness formula adopts that the flow is turbulent from the beginning of the boundary layer. Let us consider water flow with a velocity of 0.1 /ms, which propagates through a flat plate length 1 m long at the temperature of 20°C with a kinematic viscosity equal to 1×10^{-6} m^2/s. Implementing Eq. (2.1) to estimate Reynolds number as demonstrated by:

$$R_{e_{0.5L}} \approx \frac{0.1\left(ms^{-1}\right) \times 0.5(m)}{1 \times 10^{-6}(m^2s^{-1})} = 50,000 \tag{2.7}$$

Calculation of formula (2.7) convinces the laminar circumstances; consequently, the boundary layer thickness is given by formula (2.6):

$$\delta \approx 5.0 \times 0.5 \times [50,000]^{-0.5} = 0.011m \tag{2.8}$$

Using the similar conditions of formula (2.8) with change in the streamflow to 1/ms, the question now is at what specific distance L could be the transition from

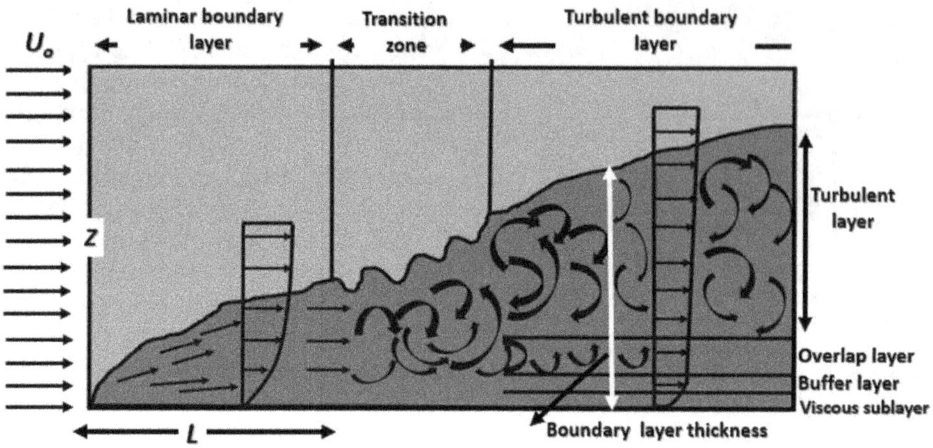

FIGURE 2.9 Boundary layer and boundary layer thickness of turbulent flow.

laminar to the boundary layer of turbulent in the circumstance of $R_{e_L} \sim 500,000$?

$$L = 500,000 \times 1 \times 10^{-6}\left[m^2 s^{-1}\right] \times 1\left[ms^{-1}\right] = 0.5\ m \tag{2.9}$$

Needless to say, at low Reynolds numbers, the boundary layer is laminar, and the velocity in the direction of flow constantly changes as the fluid flows away from the wall. For instance, the Reynolds number increases, and the flow becomes unstable. Arguably, for higher Reynolds numbers, the boundary layer is turbulent, and the downstream velocity is characterized by unstable eddy currents in the boundary layer (Pope, 2001; Stroh et al., 2015).

2.7 Turbulence equation of motion

A turbulence equation of motion is typically specified in one of two distinct approaches—Eulerian the explanation of motion and Lagrangian motion. The descriptions are, of course, equivalent, and the option is founded on practicality rather than presumption. In an Eulerian explanation of motion, physical quantities such as velocity u, pressure p, and density ρ are understood as functions of position x and time t. In space-time, $u = u(x, t)$ and $\rho = \rho(x, t)$; thus this expresses the fluid's velocity and density at specific locations. The gradient in the domain at a particular instant or the rate of change at a particular location is depicted by the partial derivative with respect to x or t. On the other hand, the fluid parcels can be accurately identified by their location "$\vec{a}$" astonished at some initial instant t_0, and the continuous motion specified by the successive position and fluid velocity of these fluid parcels (Bernard et al., 2003; Davidson et al., 2013; Guo et al., 2013; Nakayama, 2018). This is widely termed as a Lagrangian motion formulation, with the explanatory variables becoming original coordinates "$\vec{a}$" and the lapsed changeable time $t - t_0$. This can be mathematically expressed as:

$$\vec{x} = \vec{x}\,(\vec{a}, t - t_0);\ \vec{x}\,(\vec{a}, o) = a. \tag{2.10}$$

A fluid parcel's mobility is the time derivative of its location:

$$\vec{u}\,(\vec{a}, t - t_0) = \frac{\delta}{\delta_t}\vec{x}\,(\vec{a}, t - t_0), \tag{2.11}$$

In this view,

$$\vec{X} - \vec{a} = \int_{t_0}^{t} \vec{u}\,(\vec{a}, t - t_0)dt \tag{2.12}$$

In this sense, $\frac{du}{dt}$ is the fluid acceleration as it flows in space and time. In Eulerian notation, the overall time derivative, or the derivative "tracking the movement," can be confirmed as:

$$\frac{d}{dt} = \frac{\delta}{\delta t} + \left(\vec{u}\ .\nabla\right), \tag{2.13}$$

In accordance with Eq. (2.13), the sum of the spatiotemporal rate of variation at a defined location and the convective rate of change. In this regard, numerous devices that detect fluid dynamics at a fixed location immediately convey Eulerian clues. In the scenario of diffusion or mass transfer, nonetheless, the motion of the fluid constituents is of concern, and a Lagrangian formulation of the problem may be more appropriate (Willmarth, 1975; Xin et al., 1996; Yaglom, 2012). The author thinks that the proper identification of fluid constituents with dye or another tracer adequately provides Lagrangian details in comprehensive analysis. It can be challenging to reconcile the two requirements at times, especially when the flow is unsteady; in fact, the main challenge of conversion from one state to another in an unconstrained flow has been untackled. Yet, for some dynamics, for instance, the surface waves, the transformation seems quite reasonable, and the Lagrangian aspects can be easily derived from the possible Eulerian solution.

Various dynamical consequences in the upper ocean involve length scales considerably less than the radius of the Earth. It is sufficient to employ a system of orthogonal Cartesian coordinates (*x,y,z*) fixed with respect to the earth for these; the -axis will be taken vertically upward. On using Cartesian tensor notation, the coordinates $x_i = (x_1, x_2, x_3)$ are transferable, with x_3 equating to z. In accordance with this, the velocity components are (u, v, w) or (u_1, u_2, u_3). The notation $\vec{u} = (q_a, w)$ is then used, with the Greek suffix α embracing values 1, 2, and $\vec{q} = (u, v) = (u_1, u_2)$ when it is easy to consider the horizontal velocity components separately from the vertical one (u_1, u_2) (Davidson et al., 2013; Guo et al., 2013; Nezu & Nakagawa, 2017).

From the perspective of the open-loop velocities, the continuity equation for incompressible flow can be formulated as follows:

$$\frac{\partial u}{\partial x} + \frac{\partial v}{\partial y} + \frac{\partial w}{\partial z} = 0 \qquad (2.14)$$

Eq. (2.14) say that u, v, and w are instantaneous velocities, that is, therefore, $\overline{u}; \overline{v};$ *and* $\overline{w}$ are mean velocities, while $u'; v';$ *and* w' are the fluctuating components, changing randomly with time t. Using mean velocities and the fluctuation components, Eq. (2.14) can be modified as:

$$\frac{\partial(\overline{\mu} + u')}{\partial x} + \frac{\partial(\overline{v} + v')}{\partial y} + \frac{\partial(\overline{w} + w')}{\partial z} = 0 \qquad (2.15)$$

By taking time averages of the preceding terms, the first term is reduced to:

$$\frac{\partial(\overline{u} + u')}{\partial x} = \frac{\partial \overline{u}}{\partial x} + \frac{\partial(\overline{u'})}{\partial x} = \frac{\partial \overline{u}}{\partial x} \qquad (2.16)$$

The second and third terms can be shortened in a similar approach. As a result, Eq. (2.15) becomes

$$\frac{\partial \overline{u}}{\partial x} + \frac{\partial \overline{v}}{\partial y} + \frac{\partial \overline{w}}{\partial z} + 0 \qquad (2.17)$$

As a consequence, fluctuating components sustain the continuity equation independently, as shown as follows:

$$\frac{\partial u'}{\partial x} + \frac{\partial v'}{\partial y} + \frac{\partial w'}{\partial z} = 0 \qquad (2.18)$$

Fluid dynamics reveal that turbulence kinetic energy E_k (Davidson et al., 2013; Guo et al., 2013; Nagata et al., 2013; Nezu & Nakagawa, 2017) is obtained by the average of kinetic energy per unit mass associated with vortices. Both Eqs. (2.17) and (2.18) describe the turbulence motion without change in fluid density ρ, that is, $\nabla.\vec{u} = 0$. On the other hand, the equation of turbulence motion as a function of fluid density changes can be expressed as:

$$\frac{\partial \rho}{\partial \rho} + \nabla.(\rho \vec{u}) = 0 \qquad (2.19)$$

The velocity of a fluid is determined by the conservation constraints of mass and momentum, the equation of state, and the principles of thermodynamics, as shown in Eq. (2.19).

Turbulence kinetic energy, therefore is properly defined physically by accurately determined root-mean-square (RMS) velocity fluctuations. In this sense, E_k is generally characterized as half the total of the variances (square of standard deviations) of the velocity components:

$$E_k = \frac{1}{2}\left(\overline{(u')^2} + \overline{(v')^2} + \overline{(w')^2}\right), \qquad (2.20)$$

The turbulent velocity component, as shown by Eq. (2.19), is the difference between the instantaneous and average velocities $u' = u - \overline{u}$, whose average and dispersion are $\overline{u'} = \frac{1}{T}\int_0^T u(t) - \overline{u})(dt) = 0;$ *and* $\overline{(u')^2} = \frac{1}{T}\int_0^T u(t) - \overline{u})^2 dt \geq 0,$ *respectively*

Consequently, turbulence kinetic energy can be generated by fluid shear, friction, buoyancy, or external forcing at low-frequency eddy scales (integral scale). The kinetic energy of the turbulence is then transferred down the turbulence energy cascade and dissipated at the Kolmogorov scale by viscous forces, as will be enclosed in the subsequent subsections. The articulation for this yield, transport, and dissipation process is:

$$\frac{DE_k}{Dt} + \nabla \cdot T_k' = P_k - \varepsilon, \qquad (2.21)$$

Eq. (2.21) represents the mean flow of turbulence kinetic energy $\frac{DE_k}{Dt}$ with the rate of dissipation ε through the change in the rate of the fluid turbulence transportation $\nabla \cdot T_k'$. Therefore the production of the turbulence kinetic energy is signified by P_k. Therefore the rapid exchange of momentum by fluid swellings between transverse layers in a turbulent flow as well as molecular action causes the total shear stress to develop, which can be mathematically expressed as:

$$\tau_{Total} = \eta \frac{\partial \overline{u}}{\partial y} + \mu \frac{\partial \overline{u}}{\partial y} \qquad (2.22)$$

Eq. (2.22) shows the total turbulence $\eta \frac{\partial \overline{u}}{\partial y}$ and laminar flow $\mu \frac{\partial \overline{u}}{\partial y}$, respectively. Indeed, η is not a characteristic of the fluid; rather, it is the dynamic "eddy viscosity," which depends on the degree of turbulence. Hence, μ is the dynamic viscosity (Bernard et al., 2003; Guo et al., 2013; Nieuwstadt et al., 2016). What role does viscosity serve in the motion's principle turbulence equation, then, is the

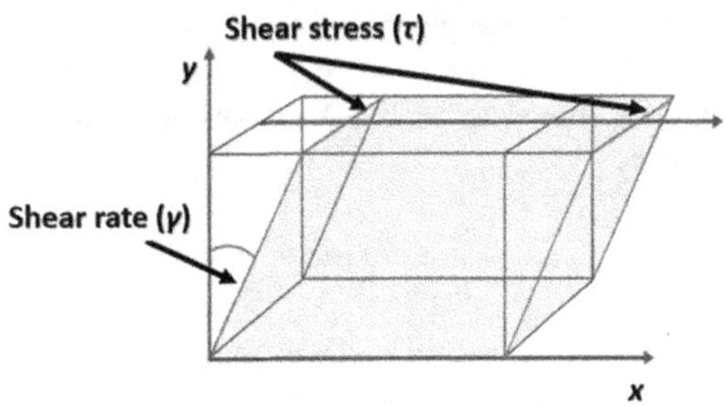

FIGURE 2.10 Correlation between shear stress and shear rate.

question at hand. In this understanding, Eq. (2.22) also represents the average shear stress per unit area. Consequently, by implementing the Newton law, then we obtain:

$$\tau = \mu x \frac{d\mu}{dy} \tag{2.23}$$

Eq. (2.23) says that μ presents viscosity as mentioned early, but $\frac{d\mu}{dy}$ is the shear rate (γ) (Fig. 2.10).

Therefore Fig. 2.10 demonstrates that a fluid's shear stress can be described as a force per unit area acting parallel to a relatively small portion of the surface. The elements should be minuscule for the calculation to be as precise as possible. The fluid viscosity is the main cause of stress. Therefore a force acting on a fluid drives it to flow, just the same as Newton's laws for classical motion. This is known as a fluid flow for fluids, which does not exist without friction. The question raised by Eq. (2.23) is: What is the connection between stress and strain? According to Eq. (2.23), even with changing viscosity, the shear stress is zero if the shear rate is constant. It is also accurate to say that for fluids with constant temperature and shear force, viscosity can remain constant. The term "Newtonian fluids" refers to a group of fluids that includes water, mineral oil, alcohol, and gasoline. Naturally, there are non-Newtonian fluids whose viscosity alters in response to the exertion of shear force. These fall into the following categories:

1. decrease in viscosity, which is known as pseudoplastic;
2. time-dependent decrease in viscosity that is called thixotropic;
3. increase in viscosity; therefore is termed as dilatant; and
4. time-dependent increase in viscosity is shown as rheopectic.

In this understanding, viscosity, which determines a fluid's physical state and, consequently, its susceptibility or resistance to flow, is a crucial property to comprehend because it affects the fluids' motions, particularly for ocean circulation complex systems (Guo et al., 2013; Nagata et al., 2013; Yaglom, 2012). Fluid would be in motion to resist shear stress. In this case, viscosity is defined by the stress – strain relationship, as demonstrated by the following equation:

$$\tau = \mu x \frac{d\gamma}{dt} \tag{2.24}$$

In the other words, Eq. (2.24) represents the strain rate $\frac{d\gamma}{dt}$ as γ is the shear strain. The following equation, which assumes fluid flow in a single parallel direction, can be used to define the shear strain for all layers during a brief period of time, even if each layer may flow at a different speed:

$$\gamma = \frac{d\gamma}{dt} x \Delta t \tag{2.25}$$

The strain rate can now be expressed as follows for a slight change in vertical position:

$$\frac{d\gamma}{dt} = \frac{d\mu}{dy} \tag{2.26}$$

Fully developed turbulent flow has a different velocity profile from laminar flow. The velocity gradient of the flow and the wall shear stress are both higher in the turbulent flow of a fluid with constant viscosity and constant density than in the laminar flow (Woods, 2010; Yaglom, 2012). In a turbulent flow, the area close to the wall is dominated by viscous effects and exhibits a linear velocity profile.

2.8 Navier – Stokes equations and the turbulent shear stress equation

According to the above perspective, both laminar and turbulent flows are subject to shear stress. Shear stress, a microscopic phenomenon, is brought on by the momentum transfer between randomly moving molecules in laminar flow. Shear stress, on the other hand, arises from the momentum transfer between fluid particles of finite size and random motion in turbulent flow and is a macroscopic phenomenon. In contrast to laminar flow, the turbulent flow has various physical characteristics and shear stress. Let's examine the relationship between laminar and turbulent shear stress and how the turbulent shear stress equation would be derived (Constantin, 2005; Davidson et al., 2013). In fact, understanding a turbulent flow's velocity profile, frictional efficiency, and pressure losses is crucial in ocean turbulence dynamics. Take into account a turbulent flow with the corresponding pressure (p) and velocity components (u, v, w):

$$\begin{aligned} u(x,y,z,t) &= \overline{u} + u'(x,y,z,t) \;\; v(x,y,z,t) \\ &= \overline{v} + v'(x,y,z,t) \;\; w(x,y,z,t) \\ &= \overline{w} + w'(x,y,z,t) \;\; p(x,y,z,t) \\ &= \overline{p} + p'(x,y,z,t) \end{aligned} \tag{2.27}$$

In Eq. (2.27), the terms — and with ('), respectively provide the mean values and the random fluctuations. Therefore the u-component of momentum's equation in the Navier – Stokes equations is denoted by:

$$\frac{\partial u}{\partial t}+u\frac{\partial u}{\partial x}+v\frac{\partial u}{\partial y}+w\frac{\partial u}{\partial z}= -\rho^{-1}\frac{\partial u}{\partial x}+v\left(\frac{\partial^2 u}{\partial x^2}+\frac{\partial^2 u}{\partial y^2}+\frac{\partial^2 u}{\partial z^2}\right) \tag{2.28}$$

Therefore Eq. (2.28) can be reduced using the continuity equation as follows:

$$\frac{\partial u}{dt}=\frac{\partial u}{\partial t}+u\frac{\partial u}{\partial x}+v\frac{\partial u}{\partial y}+w\frac{\partial u}{\partial z}=\frac{\partial u}{\partial t}+\frac{\partial uu}{\partial x}+\frac{\partial uv}{\partial y}+\frac{\partial uw}{\partial z} \tag{2.29}$$

When Eq. (2.27) is substituted into Eq. (2.29), the momentum relation in the x direction can be expressed as follows:

$$\overline{u}\frac{d\overline{u}}{dx}+\overline{v}\frac{d\overline{u}}{dy}+\overline{w}\frac{d\overline{u}}{dz}=\frac{d\overline{p}}{dx}+\frac{\partial}{dx}\left(\mu\frac{d\overline{u}}{dx}-\rho\overline{u'^2}\right) +\frac{\partial}{dy}\left(\mu\frac{d\overline{u}}{dy}-\rho\overline{u'v'}\right)+\frac{\partial}{dy}\left(\mu\frac{d\overline{u}}{dz}-\rho\overline{u'w'}\right) \tag{2.30}$$

Eq. (2.30) demonstrates that laminar stress occurs simultaneously with turbulent stress, which shares a similar scale as laminar stress. In this view, shear stress, which is related to the y-direction normal to the boundary, is the dominating force in the movement of the ocean-boundary layer (Foias et al., 2002; Foias, 1997; Nagata et al., 2013). The following equation can be employed to determine the entire shear stress:

$$\overline{u}\frac{d\overline{u}}{dx}+\overline{v}\frac{d\overline{u}}{dy}+w\frac{d\overline{u}}{dz}=\frac{d\overline{p}}{dx}+\frac{\partial\tau}{\partial x} \tag{2.31}$$

The turbulence flow near the boundary layer using Eq. (2.31) and Eq. (2.22) can be expressed by:

$$\tau=-\rho\overline{u'v'} \tag{2.32}$$

In other words, the following scenario can be used to explain how turbulent flow can be achieved close to the boundary layer:

In the boundary layer of turbulence flow:

$$\tau_{turbulent}<\tau_{laminar} \tag{2.33}$$

In the outer boundary layer of turbulence flow:

$$\tau_{turbulent}>\tau_{laminar} \tag{2.34}$$

The turbulent shear stress equation is crucial in calculating the velocity distribution in oceanic dynamic flows (Willmarth, 1975; Woods, 2010; Yaglom, 2012).

2.9 Reynolds-averaged Navier–Stokes equations

The time-averaged equations of fluid motion for fluid flow are differentiated as the complex Reynolds-averaged Navier – Stokes equations (RANS equations). The complex equations are scientifically based on the prevailing Osborne Reynolds' theory of unique Reynolds decomposition, which typically divides an instantaneous quantity into its time-averaged ($\overline{u}$) and fluctuating components u'. Turbulent flows are broadly described properly using the used RANS equations. The Navier – Stokes equations can be approximated with the assistance of these equations using approximations based on an understanding of the characteristics of stream turbulence. In this view, the functional Reynolds decomposition represents a fundamental tool employed to typically derive the RANS equations from the instantaneous Navier – Stokes equations. The term "Reynolds decomposition" describes the division of a flow variable, such as velocity, into its mean (time-averaged) component and its fluctuating constituent. The Reynolds speculator of the mean operator exhibits a considerable number of essential characteristics (Foias et al., 2002; Nakayama, 2018). One of these characteristics is that the fluctuating quantity's mean retains a value of zero ($\overline{u'}=0$). Consequently, the unique properties of continuous Reynolds operators are valuable in the possible derivation of the RANS equations. Utilizing these fundamental properties, the Navier–Complex Stokes equations of fluid motion, expressed in tensor notation, in common are (for an incompressible Newtonian fluid):

$$\frac{\partial\vec{u}}{\partial t}+(\vec{u}\cdot\nabla)\vec{u}-\nu\nabla^2\vec{u}=-\nabla\left(\frac{p}{\rho_0}\right)+\vec{g}. \tag{2.35}$$

and

$$\frac{\partial u_i}{\partial t}+u_j\frac{\partial u_i}{\partial x_j}=f_i-\frac{1}{\rho}\frac{\partial p}{\partial x_i}+\nu\frac{\partial^2 u_i}{\partial x_j\partial x_j} \tag{2.36}$$

here f_i are the acted vector forces on the fluid turbulence movements such as pressure gradients, Coriolis force, and so on, as will address in the following chapters. This marked fluctuation in Eqs. (2.35) and (2.36) with depth stands for a key reason because there efficiently is more fluid available to drag down on a water parcel the deeper it is immersed. Similarly to this, a denser fluid should impose increased pressure at a particular depth since it weighs heavier per unit volume. If the density of the fluid is known, it is quite easy to determine the hydrostatic pressure in any system; all that is required is to measure the depth of the fluid below the surface. Afterward, though, each instantaneous quantity can be decomposed into its time-averaged and fluctuating components, and the resulting equation can then be time-averaged, to begin producing:

$$\frac{\partial \overline{u}}{\partial x_i} = 0 \tag{2.37}$$

and

$$\frac{\partial \overline{u}_i}{\partial t} + \overline{u}_j \frac{\partial \overline{u}_i}{\partial x_j} + \overline{u_j' \frac{\partial u_i'}{\partial x_j}} = \overline{f}_i - \frac{1}{\rho} \frac{\partial \overline{p}}{\partial x_i} + \nu \frac{\partial^2 \overline{u}_i}{\partial x_j \partial x_j}. \tag{2.38}$$

Another possible way to adequately express the momentum equation is:

$$\frac{\partial \overline{u}_i}{\partial t} + \overline{u}_j \frac{\partial \overline{u}_i}{\partial x_j} = \overline{f}_i - \frac{1}{\rho} \frac{\partial \overline{p}}{\partial x_i} + \nu \frac{\partial^2 \overline{u}_i}{\partial x_j \partial x_j} - \frac{\partial \overline{u_i' u_j'}}{\partial x_j}. \tag{2.39}$$

With additional adjustments, this yields,

$$\rho \frac{\partial \overline{u}_i}{\partial t} + \rho \overline{u}_j \frac{\partial \overline{u}_i}{\partial x_j} = \rho \overline{f}_i + \frac{\partial}{\partial x_j} \left[-\overline{p} \delta_{ij} + 2\mu \overline{S}_{ij} - \rho \overline{u_i' u_j'} \right] \tag{2.40}$$

here $\overline{S}_{ij} = \frac{1}{2}\left(\frac{\partial \overline{u}_i}{\partial x_j} + \frac{\partial \overline{u}_j}{\partial x_i} \right)$ efficiently is the average strain tensor rate. Finally, the time derivative must be deleted because time integration eliminates the time dependency of the consequent terms, abandoning:

$$\rho \overline{u}_j \frac{\partial \overline{u}_i}{\partial x_j} = \rho \overline{f_i} + \frac{\partial}{\partial x_j} \left[-\overline{p} \delta_{ij} + 2\mu \overline{S}_{ij} - \rho \overline{u_i' u_j'} \right]. \tag{2.41}$$

The change in the mean momentum of a fluid element caused by the mean flow's instability and convection by the mean flow is represented on the left side of Eq. (2.41). The mean body force, the isotropic stress resulting from the mean pressure field, the viscous stresses, and the apparent stress $\left(-\rho \overline{u_i' u_j'} \right)$ resulting from the fluctuating velocity field—commonly referred to as the Reynolds stress—all contribute to balancing this shift. For the RANS equation to be closed and solved, the nonlinear Reynolds stress term necessitates extra modeling, which has resulted in the development of numerous turbulence models (Alfonsi, 2009; Campobasso et al., 2018; Constantin, 2005). Therefore the Reynolds stress time evolution equation is given by:

$$\frac{\partial \overline{u_i' u_j'}}{\partial t} + \overline{u}_k \frac{\partial \overline{u_i' u_j'}}{\partial x_k} = -\overline{u_i' u_k'} \frac{\partial \overline{u}_j}{\partial x_k} - \overline{u_j' u_k'} \frac{\partial \overline{u}_i}{\partial x_k} + \overline{\frac{p'}{\rho} \left(\frac{\partial u_i'}{\partial x_j} + \frac{\partial u_j'}{\partial x_i} \right)} - \frac{\partial}{\partial x_k} \left(\overline{u_i' u_j' u_k'} + \frac{\overline{p' u_i'}}{\rho} \delta_{jk} + \frac{\overline{p' u_j'}}{\rho} \delta_{ik} - \nu \frac{\partial \overline{u_i' u_j'}}{\partial x_k} \right) - 2\nu \overline{\frac{\partial u_i'}{\partial x_k} \frac{\partial u_j'}{\partial x_k}} \tag{2.42}$$

Eq. (2.42) is quite challenging. Turbulence kinetic energy can be found by tracing $\overline{u_i' u_j'}$. The turbulent dissipation rate is the final term, $\nu \overline{\frac{\partial u_i'}{\partial x_k} \frac{\partial u_j'}{\partial x_k}}$. This equation serves as the foundation for all RANS models (Foias et al., 2002; Foias, 1997; Langer & Swanson, 2020; Mor-Yossef, 2021).

2.10 Boussinesq approximation (Buoyancy)

Small-density variations diminish the problem's nonlinearity, making the approximation correct. It is predicated on the idea that density differences only cause buoyancy forces and have no other impact on the flow field. Therefore fluid Boussinesq flows are common (like atmospheric fronts, ocean circulation, and katabatic winds), industry (dense gas dispersion, fume hood ventilation), and built environments (natural ventilation, central heating). The approximation is extremely accurate for so many streams and simplifies analytically and physics. In a buoyancy-driven flow, in other specific words (moreover known as natural convection). Except when they are adequately expressed in practical terms multiplied by g, it conveniently ignored the apparent acceleration caused by gravity and density differences. The reasonable Boussinesq approximation is scientifically based on the fundamental concept that while there is little change in inertia between two different precious fluids, local gravity is significant enough to noticeably alter the specific weights of the two water masses (Allen et al., 1983; Demuren & Grotjans, 2009; Knauss & Garfield, 2016).

When a continuous fluid's effective temperature varies tremendously from one specific location to another, invariably causing a continuous flow of fluid and heat transfer, the reasonable Boussinesq approximation is properly used. The continuous fluid complies with the applicable laws of effective conservation of considerable mass, momentum, and energy. Other fluid property fluctuations are conveniently ignored in the accurate Boussinesq approximation, and density ρ only reveals itself when it is extended by the gravitational acceleration or g. Exploiting Eq. (2.19) for conservation of mass as a function of ignoring density ρ, then the Navier–Stokes equations can be simplified as:

$$\frac{\partial \mathbf{u}}{\partial t} + (\mathbf{u} \cdot \nabla)\mathbf{u} = -\frac{1}{\rho} \nabla p + \nu \nabla^2 \mathbf{u} + \frac{1}{\rho} \mathbf{F}, \tag{2.43}$$

As a consequence, it is believed in Eq. (2.43) that density variations contain two parts, one of which is stable, and the other of which varies linearly with temperature:

$$\rho = \rho_0 - \alpha \rho_0 (T - T_0), \tag{2.44}$$

where α presents the coefficient of thermal expansion due to water density fluctuations. In this sense, In accordance with the Boussinesq approximation, the density variation is mainly relevant for such buoyancy terms. The conservation equation that results when $\mathbf{F} = \rho g$ is the gravitational body force is:

$$\frac{\partial \vec{u}}{\partial t} + (\vec{u}\cdot\nabla)\vec{u} = -\frac{1}{\rho_0}\nabla\left(p - \rho_0\mathbf{g}\cdot\vec{z}\right) + \nu\nabla^2\vec{u} - \mathbf{g}\alpha(T - T_0). \tag{2.45}$$

The heat capacity per unit volume, ρC_p, is assumed to be constant, and the dissipation term is disregarded in the equation for heat flow in a temperature gradient. The equation that yields is:

$$\frac{\partial T}{\partial t} + \vec{u}\cdot\nabla T = \frac{k}{\rho C_p}\nabla^2 T + \frac{J}{\rho C_p}, \tag{2.46}$$

where k is the thermal conductivity, and J is the rate at which internal heat is produced per unit volume.

Last but not least, using the Boussinesq approximation and assuming that molecular viscosity is constant (Kullenberg, 1981; Lee & Kim, 2012; Lopez et al., 2013), the E_k Eq. (2.20) is:

$$\underbrace{\frac{\partial k}{\partial t}}_{\text{Localderivative}} + \underbrace{\overline{u_j}\frac{\partial k}{\partial k_j}}_{\text{Advection}} = -\underbrace{\frac{1}{\rho_o}\frac{\partial \overline{u_i' p'}}{\partial x_i}}_{\text{Pressure diffusion}} - \underbrace{\frac{1}{2}\frac{\partial \overline{u_j' u_j' u_i'}}{\partial x_i}}_{\text{Turbulent transport}} + \underbrace{\nu\frac{\partial^2 k}{\partial x_j^2}}_{\text{Molecular viscous transport}} \underbrace{-\overline{u_i' u_j'}\frac{\partial \overline{u_i}}{\partial x_j}}_{\text{Production}} - \underbrace{\nu\frac{\partial u_i'}{\partial x_j}\frac{\partial u_i'}{\partial x_j}}_{\text{Dissipation } \varepsilon_k} - \underbrace{\frac{g}{\rho_o}\overline{\rho' u_i'}\delta_{i3}}_{\text{Buoyancy flux}} \tag{2.47}$$

The turbulence kinetic energy budget for a specific flow can be aimed specifically at these processes are stated in Eq. (2.47) such as turbulent transport, pressure diffusion, dissipation, and Buoyancy flux.

The Boussinesq approximation is only applicable when $\Delta\rho << \rho_0$. The pressure and buoyancy terms on the right-hand side of the Navier – Stokes equations, $-\nabla p + (\rho_0 + \Delta\rho)\mathbf{g}$, are sometimes recast as $-\nabla p + \Delta\rho\mathbf{g}$ with hydrostatic pressure fluctuation $P = p + \rho_0 gh$, where h stands for depth, to prevent potential round-off mistakes in the calculation of the buoyancy term, $(\rho_0 + \Delta\rho)\mathbf{g}$ (Alfonsi, 2009; Constantin, 2005; Lee & Kim, 2012; Mayeli & Sheard, 2021).

2.11 Turbulence closure problem and eddy viscosity

The Reynolds stress is a significant uncertainty that is brought into the averaged equations by the nonlinearity of the instantaneous equations, which is controversial from the perspective of the averaged motion, at least. To be precise, there are six distinct stress factors that we need to address $<u_1^2>$, $<u_2^2>$, $<u_2^3>$, $<u_1u_2>$, $<u_1u_3>$, and $<u_2u_3>$. Since the number of unknowns and the number of equations would have to be identical, these should be associated with the mean motion itself before the equations can be solved. The turbulence closure problem is frequently used to explain the inadequacy of these auxiliary formulas (Kraichnan, 1961; Langer & Swanson, 2020). In this regard, let us assume that $\mathcal{T}_{ij}^{(v)}(\vec{x}, t)$ is the viscous stress, which is represented by the tensor $\mathcal{T}_{ij}^{(v)}$. Therefore a constitutive relation of the form can link the viscous stress to the fluid motion across many intriguing flows when the fluid acts like just a Newtonian fluid $\mathcal{T}_{ij}^{(v)} = 2\mu\left[s_{ij} - \frac{1}{3}s_{kk}\delta_{ij}\right]$ as demonstrated in Eq. (2.40). In this interpretation, the average motion's equation yields:

$$\rho\left[\frac{\partial U_i}{\partial t} + U_j\frac{\partial U_i}{\partial x_t}\right] = -\frac{\partial p}{\partial x_t} + \frac{\partial T_{ij}^{(v)}}{\partial x_j} - < u_j\frac{\partial u_i}{\partial x_j} > \tag{2.48}$$

In Eq. (2.48), the final term for the variable product has been shifted to the right side of the equation. The correlation between the components in the product determines whether the final term is zero like the other fluctuating term. These correlations are commonly not zero. Therefore the author thinks that similar issues emerged when the instantaneous equations (Eq. 2.48) were developed because relations needed to be included to relate the stresses—particularly the viscous stresses—to the motion itself. These relationships (or constitutive equations) were independent of the flow itself and only depended on the characteristics of the fluid substance. This makes it possible to conduct separate studies, known as viscometric investigations, in which these fluid properties could be definitively determined. The value of water's viscosity at a specific temperature can then be extrapolated to every flows occurring at that temperature, not specifically the one in which the analysis was conducted. Such a possible solution for turbulence Reynolds stresses seems alluring to properly explore (even though we know that the underlying requirements of scale separation are unsatisfactory) (McComb, 1990; Poncet et al., 2008; Raizer, 2021). For this specific instance, an "eddy" or "turbulent" viscosity model, frequently known as a Newtonian-type closure for the Reynolds stresses, instantly appears such as:

$$-p < u_i u_j > = 2\mu_t\left[S_{ij} - \frac{1}{3}S_{kk}\delta_{ij}\right] \tag{2.49}$$

where S_{ij} is the mean strain rate (Eq. 2.40), and μ_t is the turbulence "viscosity" (also known as the eddy viscosity). On the other hand, the second term in Eq. (2.49) can be expressed mathematically by:

$$-p < u_1 u_2 > = \mu_t \frac{\partial U_1}{\partial x_2}. \tag{2.50}$$

Eq. (2.50) demonstrates a simp case for two-dimensional shear flow as a reduction of Eq. (2.49) and is based on the Reynolds shear stress. However, since the values of the eddy viscosity and eddy diffusivity (for turbulent heat flux) have been set to achieve the most satisfactory possible agreement with the data, the success of the eddy viscosity in the preceding example is more apparent than the actual. Disregarding provided the chosen values could have been obtained before the computation or even if they could be exercised to accurately anticipate other flows, this would not be a problem in and of itself (Ecke, 2005; Qian, 1983; Stroh et al., 2015). In this understanding, the variables employed are valid just for certain flows, rendering the computation a postdiction or hindcast rather than a forecast from which the future cannot be extrapolated. In other words, having a program that can predict the ocean surface turbulence for yesterday is about as useful as our turbulence "model." Therefore there remains a significant issue with closure. Consequently, the eddy viscosity additionally experiences the issue of not accounting for the vertical components of turbulent heat flow and Reynolds stress. It is apparent that the value would be different in the vertical and radial directions. Put differently, the turbulence being described by a constitutive equation remains not an isotropic "medium." In actuality, the turbulence's interaction with buoyancy, rather than the flow's mean gradient, actually produces more of the vertical component of the heat flux (Willmarth, 1975; Xin et al., 1996).

The nonisotropic nature of the "continuous fluid" (in this possible scenario, the turbulence itself) might be properly accommodated by a more generic sense of constitutive equation, which would be as described in the considerable following:

$$-p < u_i u_j > = \mu_{ijkl} \left[S_{kl} - \frac{1}{3} S_{mm} \delta_{kl} \right] \tag{2.51}$$

Each Reynolds stress component might retain a conflicting, discrete value for the eddy viscosity because to this closure relation. Since the six primary unknowns—the $<u_i u_j>$—have been exchanged for 81 brand-new ones—the μ_{ijkl} —it is clear that it is doubtful that this would solve the closure problem. Elegant symmetries may be able to progressively eliminate some, but the remaining number is still enormous. However, more significant than the considerable amount of unknowns remains the remarkable fact that there is no independent or generic method for adopting them without tolerating a specific flow into an accurate account. This is anticipated to the fact that turbulence represents a flow-related characteristic rather than a fluid-related one (Alfonsi, 2009; Bernard et al., 2003; McComb, 1990; Nezu & Nakagawa, 2017).

2.12 Mixing length model

In turbulent flow, how long is the mixing length? The term "mixing length" refers to the distance in the transverse direction that a lump of the fluid particle traveling at its original mean velocity would cross in enough for the velocity differential between it and the new layer to equal the mean transverse fluctuation in the turbulent flow. What is the essential issue of turbulence length? The considerable scale of turbulence intensity. The apparent magnitude of the undetermined eddies is precisely quantified by LT. These are typically extremely insignificant for free stream flows (on the order of centimeters). However, the length scale cannot precious be zero since infinite dissipation would typically occur.

Put differently, the mixing length model in common is a specific technique used mostly in fluid dynamics that utilizes an eddy viscosity to valiantly attempt to sufficiently explain momentum transfer generated by turbulence Reynolds stresses within a Newtonian fluid boundary layer. Early in the 20th modern century, Ludwig Prandtl typically developed the standard model (Davidson et al., 2013; Guo et al., 2013; Leslie, 1978).

Implementing the Reynolds decomposition from Section 2.9, the temperature can be mathematically expressed as:

$$T = \overline{T} + T' \tag{2.52}$$

$\overline{T}$ is the slowly varying component, and T' is the fluctuating component, as shown by Eq. (2.40). In this sense, T' can be given by the mixing length $-\xi'$:

$$T' = -\xi' \frac{\partial \overline{T}}{\partial z}. \tag{2.53}$$

Therefore the fluctuating components of velocity $u', v',$ andw' can be stated similarly as follows:

$$u' = -\xi' \frac{\partial \overline{u}}{\partial z}, v' = -\xi' \frac{\partial \overline{v}}{\partial z}, w' = -\xi' \frac{\partial \overline{w}}{\partial z}. \tag{2.54}$$

Therefore the result of multiplying horizontal and vertical fluctuations is as follows:

$$\overline{u'w'} = \overline{\xi'^2} \left| \frac{\partial \overline{w}}{\partial z} \right| \frac{\partial \overline{u}}{\partial z} \tag{2.55}$$

From the equation above, we can define the eddy viscosity as:

$$K_m = \overline{\xi'^2} \left| \frac{\partial \overline{w}}{\partial z} \right|, \tag{2.56}$$

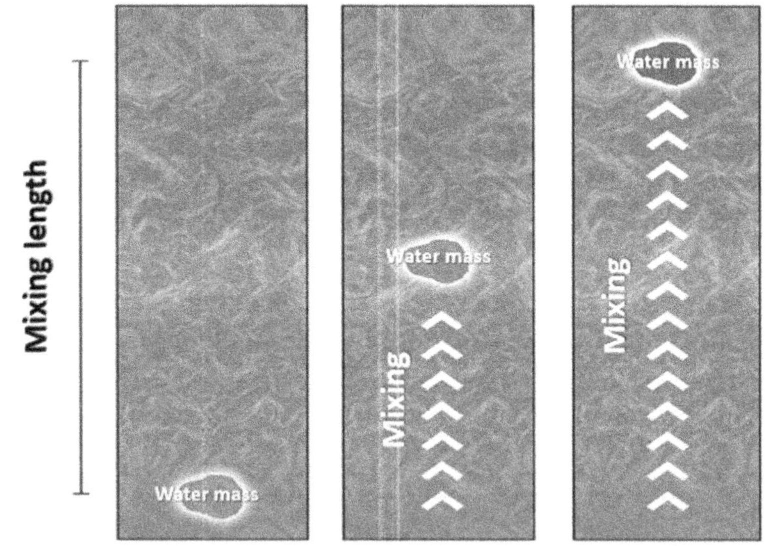

FIGURE 2.11 Concept of mixing length.

Eq. (2.56) says that eddy viscosity K_m can be demonstrated in terms of the mixing lengthξ'. In this understanding, the mixing length is conceptually similar to the mean free path notion in thermodynamics since a fluid parcel would maintain its properties for a specific length, ξ' before mixing with the surrounding fluid (Ecke, 2005; Gough, 1977; Kuhfuß, 1986; Mor-Yossef, 2021). In this scenario, it can be thought of as the dimension of the water masses traveling individually in each particular circumstance; alternatively, it can be thought of as the distance traveled by a mass of this form before it merges with surrounding water masses (Fig. 2.11).

Needless to say, the mixing length is a complicated component to be determined in the theory of turbulent shear flow. In this common sense, Theodore von Kármán (1881–1963) delivered the most notable approximation to relate the mixing length to the physical dimensions of the turbulent shear flow problem. According to von Kármán theory, turbulent fluctuations are similar throughout the flow; hence, the only things that can account for their apparent variances are length and timescale variables. This theory was employed by von Kármán to obtain the following expression for the mixing length through an analysis based on the stream function formulation of the vorticity transport equation (Bernard et al., 2003; Constantin, 2005; Davidson et al., 2013; Ecke, 2005; Gough, 1977; Guo et al., 2013; Kuhfuß, 1986).

$$\xi = \kappa \left| \frac{dU/dy}{d^2U/dy} \right| \tag{2.57}$$

The mixing length would just be a function of the velocity distribution, and not its actual magnitude, as said by Eq. (2.58). The dimensionless, all-encompassing constant κ is the same for all turbulent flows. As a consequence, the relationship between the velocity and vorticity of the eddies in turbulent shear flow can also be utilized to determine the mixing length (Fig. 2.12). The governing formula can be employed to approximate the longitudinal component of the velocity fluctuation at level y^0:

$$u' = \pm \left. \frac{dU}{dy} \right|_{y_0} \xi + .. \tag{2.58}$$

Consequently, it is possible to approximate the vorticity $\omega(y)$ variation at level y_0 by:

$$\omega(y)' = \pm \left. \frac{d}{dy}\left(\frac{dU}{dy}\right) \right|_{y_0} \xi + . \tag{2.59}$$

The transverse velocity fluctuation can be expressed as follows if it is assumed that a fluid parcel with radius r has a tangential velocity that can be substantially represented by $v' \simeq \omega' r$:

$$v' = r\xi \left. \frac{d^2U}{dy^2} \right|_{y_0} \tag{2.60}$$

Additionally, it is permissible to assume that the fluid parcel size and the mixing length are significantly proportional, that is, $\xi = \kappa r$. As a consequence, the velocity fluctuations' correlation can be approximated as:

$$\xi \frac{dU}{dy} = \kappa^{-1} \xi^2 \frac{d^2U}{Dy^2} \tag{2.61}$$

Simplifying this expression yields von Kármán's result for the mixing length, which is given by Eq. (2.57).

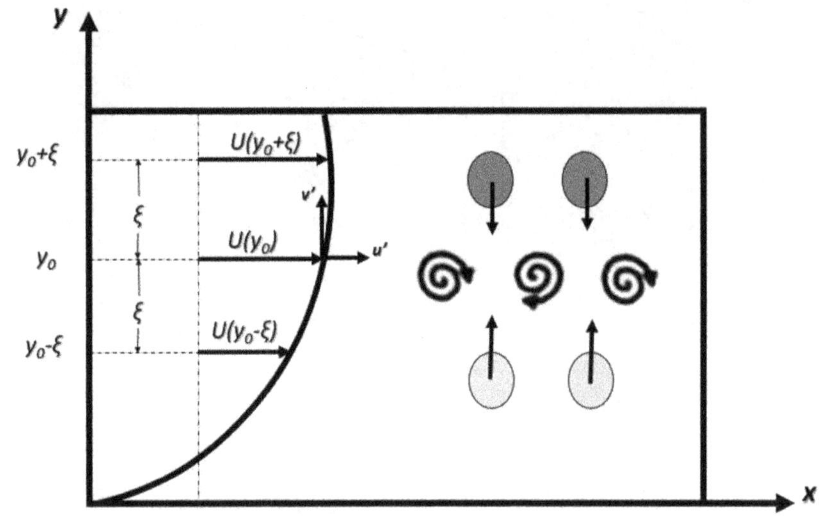

FIGURE 2.12 Physical description of mixing length by von Kármán theory.

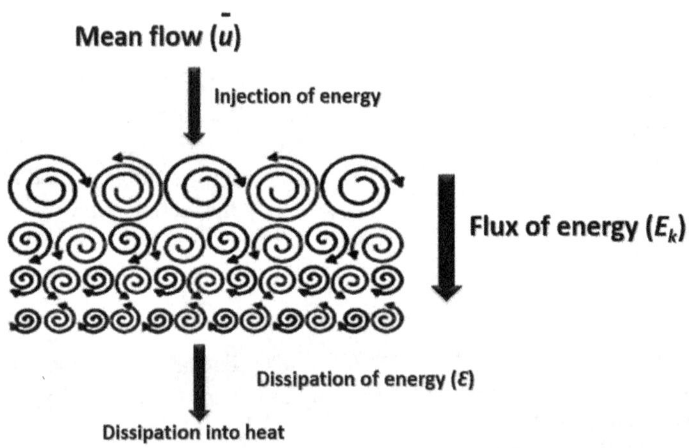

FIGURE 2.13 Energy cascade scheme.

2.13 Spectra of turbulence

The qualified majority of the kinetic energy in turbulence is naturally contained in the largest motions, or eddies, whereas the viscous dissipation of kinetic energy is knowingly caused by the smallest eddies. Put differently, closer vortexes gain kinetic energy when the larger integral length-scale eddies become unstable and break apart. These smaller eddies split off similarly, transferring their energy to upcoming more minor eddies. These procedures are rehearsed until the eddy size is minimized and the eddy motion is stabilized, which is well known as the *energy cascade* (Fig. 2.13). Kolmogorov postulated that the intermediate range of length scales would be statistically isotropic when these ranges are effectively separated, and that its characteristics in equilibrium would depend exclusively on the rate at which kinetic energy is shed at the microscopic scales. In this sense, the frictional conversion of kinetic energy to thermal energy is recognized as dissipation. The dissipation rate, ε, can be adequately expressed in practical terms of the continuous fluid's kinematic viscosity, ν, and the varying strain rates in the turbulent flow. Energy per mass unit and mass per second represent its dimensions. At equilibrium, the generation of kinetic energy by turbulence at large scales of motion equals its dissipation at minor scales (Del Álamo et al., 2004; Ecke, 2005; Lazarian, 2009; Nezu & Nakagawa, 2017; Raizer, 2021).

The mean turbulence kinetic energy per unit mass is based on the energy spectrum of turbulence, E_k, as shown in:

$$\frac{1}{2}(\overline{u_i u_i}) = \int_0^{\infty} E_k dk \tag{2.62}$$

where the wavenumber is k, the overbar designates an ensemble average, summing over I is inferred, and u_i are the components of the fluctuating velocity. Thus the energy spectrum, E_k, displays the kinetic energy contributed to turbulence by wavenumbers from k to $k + dk$. The

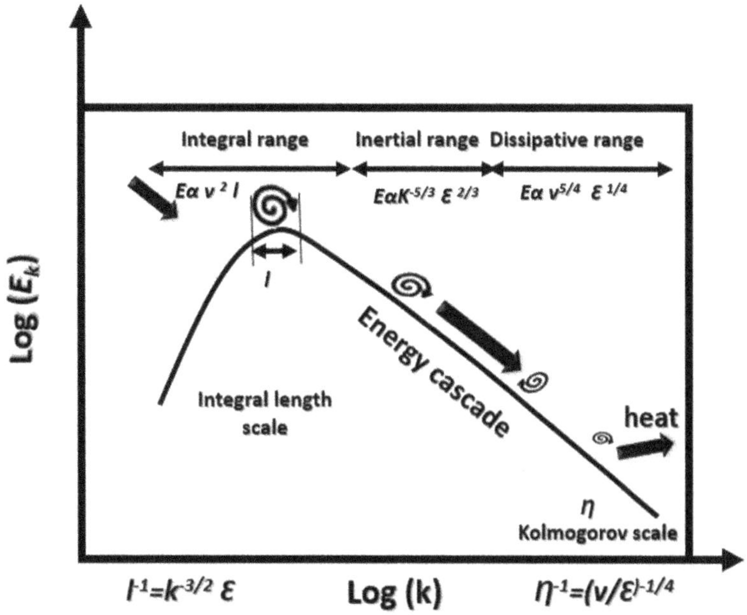

FIGURE 2.14 Kolmogorov microscale of the turbulent energy spectrum.

wavenumbers of the greatest eddies are low, whereas those of the smaller eddies are high (Taylor, 1938; Zakharov & Falkovich, 2012). Given precisely the remarkable fact that diffusion accurately reflects the Laplacian of fluid velocity, the dissipation rate can be typically expressed properly using the energy spectrum as faithfully follows:

$$\varepsilon = 2\nu \int_0^\infty k^2 E_k dk, \quad (2.63)$$

In Eq. (2.63), the fluid's kinematic viscosity is expressed as ν. This equation demonstrates once more that dissipation is primarily related to high wavenumbers (small eddies), whereas kinetic energy is primarily related to lower wavenumbers (large eddies) (Fig. 2.13).

The energy cascade represents the movement of energy from low wavenumbers to high wavenumbers. Through this transfer, turbulence kinetic energy is transferred from large to microscopic scales, where viscous friction causes it to dissipate. Kolmogorov's reasonable assumptions efficiently produced the following universal shape for the energy spectrum in the so-called inertial subrange, the intermediate-scale range:

$$E_k = C\varepsilon^{2/3}k^{-5/3}. \quad (2.64)$$

This assumption is supported by a substantial corpus of experimental research that covers a wide range of circumstances. Experimentally, the value $C = 1.5$ is reported. In turbulent flow, Kolmogorov microscales(Fig. 2.14), therefore are the smallest scales. Viscosity dominates at the Kolmogorov scale, where turbulent kinetic energy is turned into heat. In this scenario, they are identified by:

$$Kolmogorovlengthscale(\eta) = \left(\frac{\nu^3}{\varepsilon}\right)^{1/4} \quad (2.65)$$

$$Kolmogorovtimescale(\tau_\eta) = \left(\frac{\nu}{\varepsilon}\right)^{1/2} \quad (2.66)$$

$$Kolmogorovvelocityscale(u_\eta) = (\nu\varepsilon)^{1/4} \quad (2.67)$$

Eq. (2.67) could demonstrate that for ocean motion, where the large eddies have length scales on the order of kilometers, typical values of the Kolmogorov length scale range from 0.1 to 10 millimeters; for smaller flows, such as those in laboratory systems, η may be much less. Eq. (2.61) reveals the assumption that only ε and ν influence the smallest scales of turbulence, which are thought to be universal, which are identical in all turbulent flows (Raizer, 2021; Solsvik & Jakobsen, 2016).

This notion and dimensional analysis can be utilized to obtain the definitions of the Kolmogorov microscales. The only combination that has the dimension of time (Eq. 2.67) is the Kolmogorov timescale since the dimensions of kinematic viscosity and energy dissipation rate per unit mass are length2/time and length^2/time^3, respectively. The only combination of ε and ν that has a dimension of length is the Kolmogorov length scale (Del Álamo et al., 2004; Ecke, 2005; Feigenbaum, 1979; Lazarian, 2009; Nagata et al., 2013).

The inverse of the mean square strain rate tensor is another method for determining the Kolmogorov timescale:

$$\tau_\eta = (2\langle E_{ij}E_{ij}\rangle)^{-1/2} \quad (2.68)$$

In the scenario, the mean square strain rate tensor can also deliver the Kolmogorov timescale as well as Eq. (2.67) utilizing the energy dissipation rate per unit mass definition:

$$\varepsilon = 2\nu\langle E_{ij}E_{ij}\rangle \tag{2.69}$$

Therefore it is feasible to identify the Kolmogorov length scale as the scale at which the Reynolds number equals 1, which is demonstrated by:

$$\mathrm{Re} = \frac{UL}{\nu} = \left(\frac{\eta}{\tau_\eta}\right)\eta\nu^{-1} = 1 \tag{2.70}$$

With a noticeable increase in the turbulent Reynolds number, one can observe more scale separation between the integral length scale and the Kolmogorov length scale when taking into comprehensive account these scale ratios. This indicates that for considerable turbulent Reynolds numbers, the Kolmogorov length scale is significantly less than the integral length scale.

Therefore the Taylor microscale, also known as the turbulence length scale, is a length scale that is used to describe unsteady fluid flow. The Taylor microscale efficiently is the intermediate length scale where the complex dynamics of turbulent eddies in the current are considerably influenced by fluid viscosity. It is conventional to typically employ the length scale for turbulent flow, which exhibits a Kolmogorov spectrum of velocity fluctuations. Larger length scales than the Taylor microscale are significantly unimpacted by viscosity in such a flow. Typically, the practical term "inertial range" typically refers to these longer length scales in the continuous flow. The turbulent motions are subject to significant viscous forces below the Taylor microscale, and the kinetic energy is efficiently converted to heat. The dissipation range frequently refers to these shorter-length-scale vibrations (Chuychai et al., 2014; Heisel et al., 2021).

The Taylor microscale calculation involves the construction of specific flow correlation function(s), expansion in a Taylor series, and characterization of an osculating parabola using the first nonzero term. While the Kolmogorov microscale is proportional to $R_e^{-3/4}$ and the Taylor microscale to $R_e^{-0.5}$, as R_e is the integral scale Reynolds number. The following equation yields a turbulence Reynolds number based on the Taylor microscale λ:

$$\mathrm{Re}_{e_\lambda} = \frac{\langle \vec{v}' \rangle rms\lambda}{v}, \tag{2.71}$$

here the velocity fluctuations' root mean square is shown by $\langle \vec{v}' \rangle_{rms} = \frac{1}{\sqrt{3}}\sqrt{(\vec{v}_1')^2 + (\vec{v}_2')^2 + (\vec{v}_3')^2}$. In this regard, Taylor's microscale is described as:

$$\lambda = \sqrt{15\tfrac{\nu}{\epsilon}} < \vec{v}' >_{rms}, \tag{2.72}$$

where ϵ is the rate of energy dissipation, and ν is the kinematic viscosity. Following is a relationship between turbulence and kinetic energy k:

$$\lambda \approx \sqrt{10\nu\frac{k}{\epsilon}}. \tag{2.73}$$

Last but not least, the Taylor microscale offers an accurate approximation for the domain of variable strain rate:

$$\left(\frac{\partial\langle\vec{v}\rangle_{rms}}{\partial\vec{x}}\right)^2 = \frac{\langle\vec{v}\rangle_{rms}^2}{2}. \tag{2.74}$$

The relevant relationships can be established given the length scale of the larger eddies $l \propto \frac{k^{3/2}}{\epsilon}$ and the turbulence Reynolds number R_{e_l} attributed to these eddies:

$$\frac{\lambda}{l} = \sqrt{10}\,\mathrm{Re}_l^{-1/2} \tag{2.75}$$

$$\frac{\eta}{l} = \mathrm{Re}_l^{-3/4} \tag{2.76}$$

$$\frac{\lambda}{\eta} = \sqrt{10}\,\mathrm{Re}_l^{1/4} \tag{2.77}$$

$$\lambda = \sqrt{10}\,\eta^{2/3}l^{1/3} \tag{2.78}$$

The sequences of Eqs. (2.75–2.78) demonstrate the possible relationship between the Taylor microscale λ and the Kolmogorov microscale η. In this view, precisely calculating the specific ratios between λ and the Kolmogorov microscale η typically reveals that the Taylor microscale is conveniently situated uniting the large-scale and small-scale eddies (Raizer, 2021; Segalini et al., 2011; Solsvik & Jakobsen, 2016; Taylor, 1938; Zakharov & Falkovich, 2012). In the next section, we can address the rotation force impact on the turbulence flow.

2.14 Geostrophic turbulence

The main concern is how geostrophic turbulence differs from the common belief of turbulence. Geostrophic turbulence is turbulence that is driven by stratification and rotation. This practical perspective reasonably argues that geostrophic turbulence, which typically comprises meandering jets and persistent vortices, adequately represents oceanic dynamic motion at mesoscales, ranging from 50 to 500 km, and atmospheric mobility at synoptical scales, allying 500 and 5000 km. To put it another way, the considerable sizes, efficient energy, continuous rotation, and stratification of geostrophic turbulence

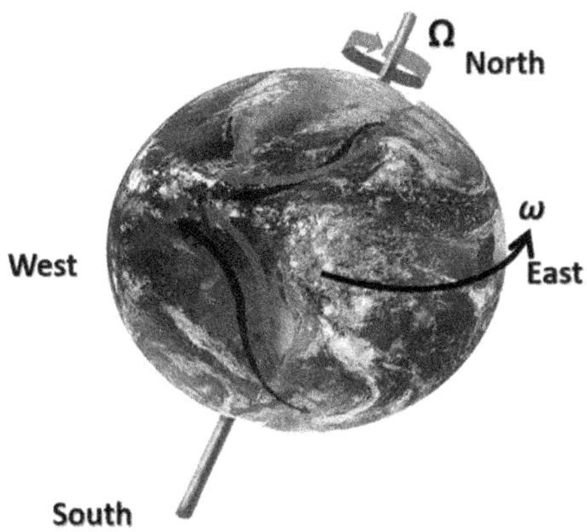

FIGURE 2.15 Simplification of Coriolis force generation.

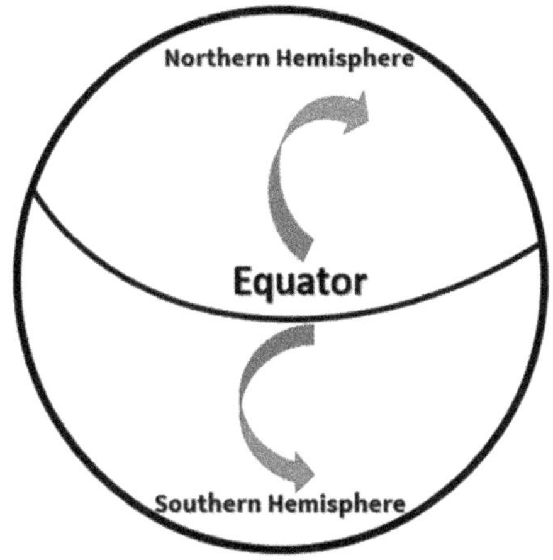

FIGURE 2.16 Coriolis force effects.

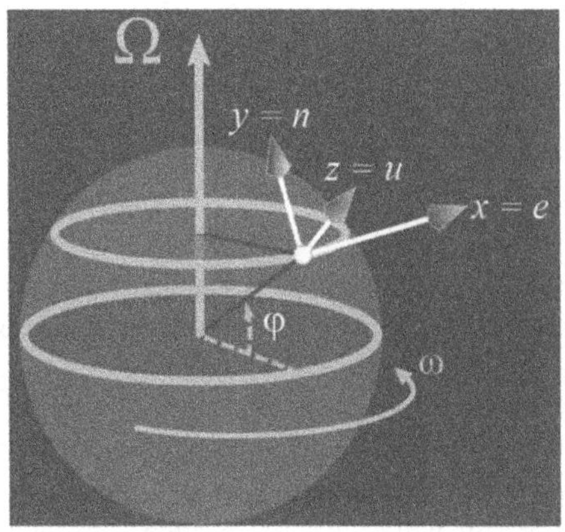

FIGURE 2.17 Mathematical description of Coriolis acceleration.

typically differ from those of conventional turbulence, which in common is an inevitable byproduct of ocean dynamics. Energy transfer in stratified and rotating flow is one of the most frequent sources of geophysical turbulence. In this sense, the Coriolis force and the centrifugal force both have a significant influence on the emergence of geophysical turbulence (Knauss & Garfield, 2016; Steele & Thorpe, 2009).

The Coriolis force, which is fictitious or inertial in fundamental physics, acts on moving objects within a distinctive shape of direct reference that typically rotates relative to an inertial frame. The practical constraint properly applies to the left of the fluid motion in a reference frame with clockwise rotation (Fig. 2.15). The reasonable force invariably acts to the exclusive right in one with an anticlockwise (or counterclockwise) rotation. The profound Coriolis effect typically remains the bending of a specific object typically caused by the Coriolis force. Put differently, the Coriolis effect typically causes movable objects on the surface of the Earth to be deflected to the right (regarding the proper direction of continuous travel) in the Northern Hemisphere and to the left in the Southern Hemisphere (Fig. 2.16). The horizontal deflection effect is greater near the poles since the effective rotation rate about a local vertical axis is the largest there (Kullenberg, 1981; Steele & Thorpe, 2009).

Imagine a location with a latitude of φ on a sphere that is rotating around the north-south axis. The axes x, y, and z are all parallel, with the x parallel to the east, the y parallel to the north, and the z vertically parallel to the up. In this local coordinate system, the rotation vector, movement velocity, and Coriolis acceleration are expressed as follows, with east (e), north (n), and upward (u) being the components listed in that order (Fig. 2.17):

$$\boldsymbol{\Omega} = \omega\begin{pmatrix} 0 \\ \cos\varphi \\ \sin\varphi \end{pmatrix}, \ \boldsymbol{v} = \begin{pmatrix} v_e \\ v_n \\ v_u \end{pmatrix},$$

$$\boldsymbol{a}_C = -2\boldsymbol{\Omega} \times \boldsymbol{v} = 2\omega\begin{pmatrix} v_n\sin\varphi - v_u\cos\varphi \\ -v_e\sin\varphi \\ v_e\cos\varphi \end{pmatrix}. \tag{2.79}$$

The vertical velocity is small when considering atmospheric or oceanic dynamics, and the vertical component of the Coriolis acceleration ($v_e\cos\varphi$) is small in comparison to the acceleration caused by gravity (g), or about $9.81/\mathrm{ms}^2$ ($32.2\ \mathrm{ft/s}^2$) near the Earth's surface. In such circumstances, only the horizontal (east and north) components are significant (Lan et al., 2012; Persson, 1998; Tolmazin, 2012). Setting $v_u = 0$ confines the aforementioned to the horizontal plane as follows:

$$\vec{v} = \begin{pmatrix} v_e \\ v_n \end{pmatrix}, \vec{a_c} = \begin{pmatrix} v_n \\ -v_e \end{pmatrix} f, \tag{2.80}$$

Eq. (2.80) demonstrates the Coriolis parameter $f = 2\omega \sin\varphi$, which can reshape the equation of the motion as:

$$\frac{Dq}{Dt} = \frac{D}{Dt}(\zeta + f) = F - D, or \frac{\partial \zeta}{\partial t} + \boldsymbol{u} \cdot \nabla(\zeta + f) = F - D \tag{2.81}$$

Eq. (2.81) shows the two-dimensional turbulence on the beta plane. A beta plane approximation in geophysical fluid dynamics involves setting the Coriolis parameter, *f*, to vary linearly in space. In the so-called *f*-plane approximation, this variation is ignored and a value of f appropriate for a specific latitude is used throughout the domain. On a rotating sphere, such as the Earth, *f* varies with the sine of the latitude. This approximate solution is represented by a tangent plane that touches the sphere's surface at this latitude. In this regard, Eq. (2.81) can be rewritten as:

$$\frac{\partial \zeta}{\partial t} + J(\psi, \zeta + \beta y) = F - Dor \frac{\partial \zeta}{\partial t} + J(\psi, \zeta) + \beta v = F - D. \tag{2.82}$$

In both Eqs. (2.81) and (2.82), *F* and *D* represent the forcing and dissipation, respectively. Moreover, ζ is the relative vorticity in which the potential vorticity is demonstrated by $q = \zeta + f$. Therefore in Eq. (2.82), $J\Psi$, $\zeta) = \psi x \zeta y - \psi y \zeta x$ is the Jacobian, $f = f_0 + \beta y$. Although it no longer describes dynamics on a fictitious tangent plane, this approximation is referred to as the "beta plane" in analogy with the "*f*-plane." The beta plane approximation has the advantage over more precise formulations in that it does not add nonlinear terms to the dynamical equations, which makes the equations more challenging to solve. The term "beta plane" comes from the usage of the Greek letter β to signify the linear coefficient of variation. Consequently, the presence of β is likely to generate zonal flow, since one can explain "βv = forcing and nonlinear terms."

Since the force ultimately determines the flow in a statistically steady state, if we increase β while keeping the force constant, then v must decrease in order for the equation to balance. Indeed, we expect the zonal flow to remain strong as β increases, and the vorticity equation shows a three-way balance between nonlinearity $\boldsymbol{u} \cdot \nabla\zeta$, u, the beta term, βv, and the forcing *F* (Raizer, 2021; Tolmazin, 2012).

By incorporating the terms of the Coriolis parameter in the equations, this approximation allows researchers to more accurately model and understand physical systems such as atmospheric flows, oceanic currents and eddies, planetary waves, and vortices. Thus the beta plane approximation is crucial in accounting for Rossby waves and other related phenomena in geophysical fluid dynamics.

In that energy is transferred on vast scales, geostrophic turbulence is like two-dimensional turbulence. The transfer occurs both vertically and horizontally, and when it happens on extensive scales, the flow becomes barotropic. Consider the following flow in two equal-depth layers that are dominated by the quasigeostrophic equations with (for now)$\beta = 0$:

$$\frac{\partial q_i}{\partial t} + J(\psi_i q_i) = 0, i = 1, 2, \tag{2.83}$$

where

$$q_1 = \nabla^2\psi_1 + 0.5k_d^2(\psi_2 - \psi_1), q_2 = \nabla^2\psi_2 + 0.5k_d^2(\psi_1 - \psi_2), \tag{2.84}$$

and

$$J(a, b) = \frac{\partial a \partial x}{\partial b \partial y} - \frac{\partial a \partial y}{\partial b \partial x}, 0.5k_d^2 = 2f_0^2[g'H]^{-1} \equiv 4f_0^2[N^{-2}H^{-2}]. \tag{2.85}$$

The wavenumber *kd* is an essential factor in understanding baroclinic stability and the potential for the development of eddies, as it gives insight into how much energy is needed to displace a certain layer of fluid. The wavenumber *kd* can be expressed as $B/2\pi$, in which *B* is the baroclinic radius of deformation and is appropriate for a layered model (Dijkstra, 2008; Tolmazin, 2012; Yaglom, 2012). In this regard, the equations conserve the total energy, meaning the amount of energy present in the system remains constant over time, regardless of any interactions between different forms of energy:

$$\frac{d\hat{E}_k}{dt} = 0, \widetilde{E} = 0.5\int_A \left[(\nabla\psi_2)^2 + (\nabla\psi_2)^2 + 0.5k_d^2(\psi_1 - \psi_2)^2\right] dA, \tag{2.86}$$

The kinetic energy is represented by the first two terms in the energy expression (2.86), and the available potential energy, which is proportional to temperature variance, is depicted by the final term.

Let us consider the barotropic and baroclinic stream functions as:

$$\psi \equiv 0.5(\psi_1 + \psi_2), \tau \equiv 0.5(\psi_1 - \psi_2). \tag{2.87}$$

In accordance with Eq. (2.87), the potential vorticities for each layer can be demonstrated as:

$$q_1 = \nabla^2\psi + (\nabla^2 - k_d^2)\tau, q_2 = \nabla^2\psi - (\nabla^2 - k_d^2)\tau \tag{2.88}$$

The equations of motion can be expressed as a set of coupled first-order partial differential equations, known as the evolution equations:

$$\frac{\partial}{\partial t}\nabla^2\psi + J(\psi, \nabla^2\psi) + J(\tau, (\nabla^2 - k_d^2)\tau) = 0, \tag{2.89}$$

$$\frac{\partial}{\partial t}\left(\nabla^2 - k_d^2\right)\tau + J\left(\tau, \nabla^2\psi\right) + J\left(\psi, \left(\nabla^2 - k_d^2\right)\tau\right) = 0, \tag{2.90}$$

Therefore nonlinearity interaction was noticed in Eq. (2.89), which is consistent with Eq. (2.90). In this understanding, ψ and τ represent vertical modes in which ψ signifies the barotropic mode with a "vertical wavenumber," k^z, of zero. On the other hand, τ exhibits a baroclinic mode with an effective vertical wavenumber of 1. As can be seen from Eq. (2.90), although there are now two different types, geostrophic turbulence continues to be assumed to be comprising a sum of interacting sequences: (i) $(\psi, \psi) \rightarrow \psi$; and (ii) $(\tau, \tau) \rightarrow \psi$ or $(\psi, \tau) \rightarrow \tau$. This can be seen through the concept of scale interactions, where energy is converted from one length scale to another. Since only the barotropic mode is involved, the first type is known as a barotropic triad. Examples of a baroclinic triad are the other two. The three vertical wavenumbers of the triad interactions must sum to zero if a barotropic mode has a vertical wavenumber of zero and a baroclinic mode has a vertical wavenumber of plus or minus one. The form shows that there is no triad that only involves the baroclinic mode of Eq. (2.90). This means that pure baroclinic triads exist if the layers are of different depths (Dijkstra, 2008; Knauss & Garfield, 2016; Kostianoy et al., 2004; Kullenberg, 1981; Steele & Thorpe, 2009; Tolmazin, 2012).

2.15 How do frontal zones and turbulence primarily interact?

Oceanic fronts develop in regions with high convection and turbulence. Sharp gradients in temperature and/or salinity are their primary characteristics. It is a zone of accelerating motion. Typically, an oceanic front is a boundary between two distinct water masses where there is a difference in one or more of the following characteristics: temperature, salinity, density, and velocity. Oceanic fronts, therefore, arise at all depths and are time-variable. They intentionally alter in considerable strength and sensibly diminish ineffective responses to necessary changes in the marine and atmospheric flow regime. The Antarctic Circumpolar, Subarctic, and Subtropical water-fronts in the North Pacific Ocean in common are notable examples of permanent oceanic fronts (Allen et al., 1983; Fedorov & Ginsburg, 1992; Fedorov, 1986).

Oceanic fronts, however, differ greatly from their atmospheric counterpart in several respects. The temperature and density fronts are coinciding in the atmosphere. Therefore the frontal zones of the atmosphere are extremely baroclinic. Early in this century, the first attempts were made to comprehend and physically interpret atmospheric fronts. The first academic establishment to attempt to comprehend these phenomena was the classic Norwegian school. The density front, however, is weak or nonexistent in the case of oceanic fronts if the temperature and salinity fronts do share the same geographical region. There is minimal baroclinicity. The subtropical Pacific Ocean Front is a prime example.

2.16 How are frontal zones defined?

A front in oceanography comprises the line separating two different water masses. Numerous physical developments are typically required for the successful production of fronts, and even slight variations in these processes can cause a large variety of front forms. They can typically range in comparable size from a few hundred meters across to several likely tens of kilometers wide. Although most fronts form and break up rather rapidly, some can linger for an extremely lengthy time.

2.16.1 Conventional definition

Water masses, or considerable bodies of the ocean with various physical characteristics such as effective temperature and salinity, were used historically to precisely define fronts. These standard definitions depended on low-resolution data collected on research cruises. Since it typically took a considerable while to aggregate these historical data, the front positions that were gained merely provided a time-averaged representation of the general structure. For instance, this led to the definition of five fronts in the Southern Ocean that were all thought to be continuous, circumpolar, grasping significant depths, and heavily impacted by bathymetry. Along with observed variances in other oceanographic indicators, the water masses on either practical side of such developed fronts naturally have different temperatures, salinities, or fluid densities (Knauss & Garfield, 2016).

2.16.2 Dynamical definition

The development of high-resolution satellite data has altered how scientists perceive ocean fronts. The prominent position of swirling currents or jets naturally connected to ocean fronts can be pinpointed with an extremely high-spatial and temporal resolution by continually recording sea-surface height (SSH) across the world. In this efficient manner, it is possible to carefully analyze short-term variability and trends and accurately compare them with other climatological fluctuations such as the El Nio–Southern Oscillation. With this comprehensive approach, the circumpolar nature of the fronts in the Southern Ocean is abolished, and the precise frequency of fronts typically relies on the time and location (Allen et al., 1983; Kostianoy et al., 2004; Steele & Thorpe, 2009).

2.16.3 Spatial definition

It is also feasible to distinguish fronts using a spatial definition in addition to the physical characteristics previously mentioned. Locally, fronts are frequently identified using a technique known as gradient thresholding, which determines the position of the front based on where the spatial gradient of a quantity, such as sea-surface height or temperature, exceeds a predetermined threshold. This is comparable with how fronts from strong currents are defined dynamically above. Similar to the conventional concept of water mass, precise values of sea-surface height or temperature are frequently employed to define fronts on a worldwide scale (Knauss & Garfield, 2016; Kostianoy et al., 2004; Steele & Thorpe, 2009).

2.17 Formation of fronts

Frontogenesis is the acronym for the font creation process. Ocean currents, wind, and Coriolis forces are some of the variables that influence this mechanism. Frontogenesis: what is it and how does it occur? Frontogenesis, in the broadest sense, is the change in the magnitude and orientation of the temperature gradient at a level or in a layer as a result of changes in the direction and speed of the wind field. Frontogenesis is the term that describes the steady increase in the horizontal temperature gradient. In this sense, why does frontogenesis occur? Frontogenetic physical processes physically result in the birth or strengthening of the front by tending to increase the potential-temperature differential. Kinematic, thermodynamic, and dynamic mechanisms can lead to frontogenesis (Allen et al., 1983; Bleck et al., 1988; Hoskins, 1982).

With no consideration for driving factors, kinematics refers to motion or advection. Although this class of processes cannot produce potential-temperature gradients, it can alter their strength. It is well-known that temperature disparities between the poles and the equator are a result of radiative heating. Additionally, because of the general circulation, the jet stream meanders, causing momentary east-west temperature disparities along ridges and troughs. The potential temperature across the different latitudes for instance, similarly has a vertical gradient toward high latitudes, for instance, in the northern hemisphere, and also decreases with water depth z. It is reasonable to expect that temperature gradients are common and may perhaps become stronger during kinematic frontogenesis. In this sense, the gradient in the potential temperature from surface to bottom and from the equator to the north pole, for example, (Fig. 2.18) can be described by:

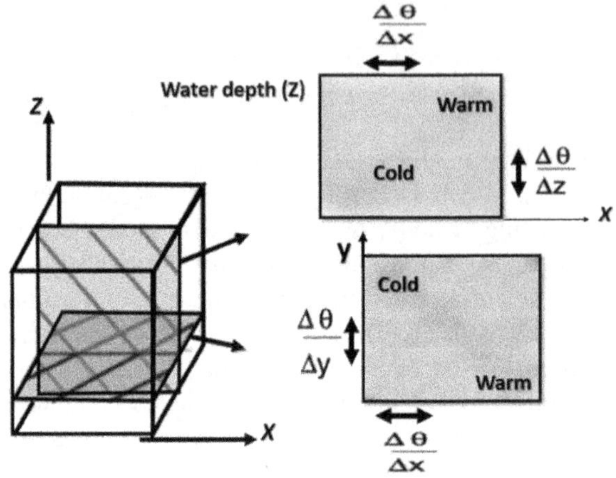

FIGURE 2.18 Initial conditions prior to frontogenesis.

Therefore the kinematic frontogenesis equation describes how frontal strength evolves over time as a consequence of advection:

$$\frac{\Delta(FS)}{\Delta t} = -\left(\frac{\Delta\theta}{\Delta x}\right)\cdot\left(\frac{\Delta U}{\Delta x}\right) - \left(\frac{\Delta\theta}{\Delta y}\right)\cdot\left(\frac{\Delta V}{\Delta x}\right) - \left(\frac{\Delta\theta}{\Delta z}\right)\cdot\left(\frac{\Delta W}{\Delta x}\right) \tag{2.91}$$

In this scenario, $\Delta U/\Delta x$ is negative and x is positive in the Eq. (2.10). Therefore the confluence term's negative sign compounded by the product of these two terms tends to enhance the front [$\Delta(FS)/\Delta t$ is positive]. Consequently, the northward temperature gradient $\Delta\theta/\Delta y$, for instance, is negative while the shear $\Delta V/\Delta x$ is positive. As a consequence, the product is positive when the shear term's preceding negative sign is taken into account. For this situation, frontal strengthening arises [$\Delta(FS)/\Delta t$ is positive] (Bleck et al., 1988; Lapeyre et al., 2006; Spall, 1995).

The vertical potential-temperature gradient would lean into the horizontal if updrafts are stronger on the cold side of the domain than on the warm side. A reinforced frontal zone is the eventual consequence. On the left side of Eq. (2.91), the vertical potential-temperature gradient is positive ($\Delta\theta/\Delta z = +$), while the horizontal gradient of updraft velocity is negative ($\Delta W/\Delta x = -$). In this scenario, the product produces a positive contribution to the strengthening of the front when multiplied by the tilting term's negative sign [$\Delta(FS)/\Delta t = +$] (Bleck et al., 1988; Fedorov, 1986; McWilliams, 2021).

The equation also contains terms for both vertical and horizontal deformation; in particular, in the x direction:

$$-\left(\frac{\partial U}{\partial x}\frac{\partial \theta}{\partial x}\right) - \left(\frac{\partial V}{\partial x}\frac{\partial \theta}{\partial y}\right) \tag{2.92}$$

Therefore the deformation in the y-direction and z-direction, respectively, can be expressed as:

$$-\left(\frac{\partial U}{\partial y}\frac{\partial \theta}{\partial x}\right)-\left(\frac{\partial V}{\partial y}\frac{\partial \theta}{\partial y}\right), \quad (2.93)$$

and

$$-\left(\frac{\partial U}{\partial z}\frac{\partial \theta}{\partial x}\right)-\left(\frac{\partial V}{\partial z}\frac{\partial \theta}{\partial y}\right) \quad (2.94)$$

Most actual fronts are weakened by turbulent mixing between the warm and cold sides (i.e., causes frontolysis). Frontolysis is aided by surface-based conduction. In the case of a cold front, for instance, the cold water mass dives beneath a typically warmer surface, warming it (i.e., water mass alteration) and lowering the temperature contrast across the front. Similar to this, the warm water mass is typically moving across cooler surfaces behind warm fronts (Fedorov, 1986; Hoskins, 1982; Knauss & Garfield, 2016).

A multifrontal internal structure, or the presence of multiple fronts (quite frequently of distinct types), is a defining trait of the majority of frontal zones in this sea area. There are fronts that are synoptic or seasonal in nature, permanent fronts connected to enduring climatic factors, small-scale fronts with local origins, and more, supporting Fedorov's distinction between the "frontal zone" and "front." To determine the location of fronts, one can use the following criterion similar to the above-mentioned criterion for the delimitation of the frontal zones:

$$\frac{\partial \theta}{\partial x} >> \left[\frac{\partial \theta}{\partial x}\right] \quad or \quad \frac{\partial S}{\partial x} >> \left[\frac{\partial S}{\partial x}\right] \quad (2.95)$$

Eq. (2.95) shows that $\frac{\partial \theta}{\partial x}$ *and* $\frac{\partial S}{\partial x}$ are local characteristics of the fields of potential temperature and salinity are determined by measurements with a high-spatial resolution (<20 km), and $\left[\frac{\partial \theta}{\partial x}\right]$ *and* $\left[\frac{\partial S}{\partial x}\right]$ are average values of horizontal gradients in the given frontal zone (Allen et al., 1983; Fedorov & Ginsburg, 1992; Hoskins, 1982; Kostianoy et al., 2004; McWilliams, 2021; Spall, 1995).

2.18 How do temperature and salinity accurately describe frontal zones?

Temperature and salinity are key parameters to identify the water masses in the oceans. Put differently, a countable set of typical temperature and salinity samplings for various vertical extents at a given site in the ocean are plotted on a diagram with temperature on the vertical axis and salinity on the horizontal axis, and the observations are partnered up in increasing depth exchange to identify the water masses, present at that location. Therefore this is generally accomplished by utilizing data gathered through extensive sampling. A temperature − salinity (T-S) diagram (Fig. 2.19) precious is the practical culmination (Kostianoy & Belkin, 1989; Kostianoy et al., 2004; Ríos et al., 1992). On the same diagram, isopleths of constant density are frequently also depicted as an additional helpful interpretation aid. Certain T-S combinations are much more preferred in the ocean, which culminates in the identification process through the classification of water types, water masses, and their distributions. Now, the important query is: How can T-S diagram generalize the frontal zones?

Therefore a frontal zone is a region in which essential gradients of temperature and salinity are simultaneously marked, quite often a purely temperature or

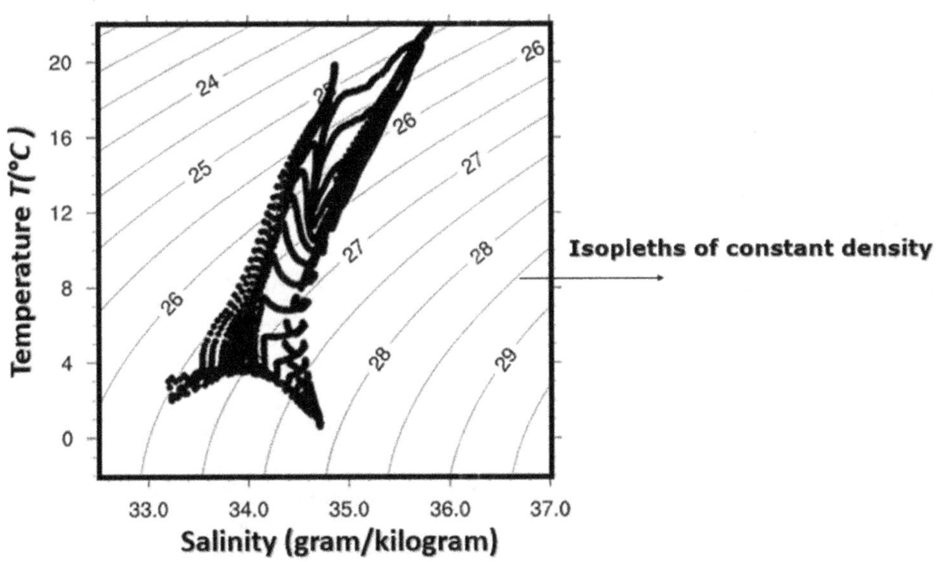

FIGURE 2.19 Example of T-S diagram.

salinity front. A mutual disposition of isopycnals, isotherms, and isohalines in the cross section of the frontal zone is indicative of its dominant variability. In this view, the vertical density ratio $R\sigma$ is consequently derived to reliably estimate the vertical stability of various distinct layers. This specific ratio accurately describes the direct participation of vertical modifications in temperature and salinity to fundamental changes in vertical density (Fedorov & Ginsburg, 1992; Kostianoy et al., 2004). When the salinity involvement is destabilizing and the temperature contribution stabilizes:

$$R_\sigma = B\left[\frac{\partial S}{\partial z}\right]\left[\alpha\left[\frac{\partial T}{\partial z}\right]\right]^{-1} \tag{2.96}$$

where $\left[\frac{\partial S}{\partial z}\right]$,and $\left[\frac{\partial T}{\partial z}\right]$ are the salinity and temperature gradients along the water column z, respectively. Eq. (2.96) shows that the saline contraction coefficient is signified by $B=\rho^{-1}\frac{\partial\rho}{\partial S}$, while the thermal expansion coefficient is presented by $\alpha=\rho^{-1}\frac{\partial\rho}{\partial T}$. When $R\sigma \leq 0$, the situation is completely stable; when $R\sigma > I$, the situation is completely unstable; and when $0 < R\sigma \leq I$, the situation is somewhat stable, with rather typical development of differential-diffusion convection leading to the production of a vertical fine structure of hydrophysical fields. The terms "baroclinity," "thermoclinity," and "haloclinity," therefore, are defined as the existence of correspondingly high values of the pressure, temperature, and salinity gradients at isopycnal surfaces, and there are selection principles for the frontal interface surface based on the maximum value of pressure, temperature, and salinity gradients at isopycnal surfaces. The stem of the along front current coincides with a surface of maximum values of the pressure gradient (maximum of baroclinity) at isopycnal surfaces. Temperature and salinity intrusions should develop when there is a specific level of "thermoclinity" and "haloclinity" (Camerlengo, 1982; Fedorov & Ginsburg, 1992; Kostianoy et al., 2004).

In this regard, it seems convinced that the frontal zones and fronts can be characterized according to their "baroclinity," "thermoclinity," and "haloclinity." They can be measured using the inclinations of isopycnals $\gamma\sigma$, isotherms γ_T, and isohalines γ_S. The isolines' inclinations can be thought of as a local feature that can be determined at any position (e.g., on the frontal interface). In this scenario, they can be stated as:

$$\gamma_\sigma = \frac{\partial\sigma}{\partial x}\times\left[\frac{\partial\sigma}{\partial z}\right]^{-1} \tag{2.97}$$

$$\gamma_T = \frac{\partial T}{\partial x}\times\left[\frac{\partial T}{\partial z}\right]^{-1} \tag{2.98}$$

$$\gamma_S = \frac{\partial S}{\partial x}\times\left[\frac{\partial S}{\partial z}\right]^{-1} \tag{2.99}$$

Once the frontal zone characteristics have been evaluated, the average values of the inclinations on a particular portion of the section surface are obtained by analyzing the very smooth average appearance. The sounding data related to standard horizons are typically utilized to calculate the inclinations. It is simple to demonstrate that, in the case of a linear interpolation, the method may be used to determine the average inclination of isotherms in the layer (z_2-z_1) situated between neighboring stations at the distance (x_2-x_1) (Fedorov & Ginsburg, 1992; Kostianoy & Belkin, 1989; Krasnopevtsev & Yu, 1978):

$$\gamma_T = \frac{T(x_2,z_2)-T(x_1,z_2)+T(x_2,z_1)-T(x_1,z_1)}{T(x_2,z_2)-T(x_2,z_1)+T(x_1,z_2)-T(x_1,z_1)}\cdot\frac{z_2-z_1}{x_2-x_1} \tag{2.100}$$

Considering the ratio γ_S/γ_T as one of the factors characterizing how close the frontal condition is to the thermoclinic and baroclinic varieties is convenient. $\gamma_S=\gamma_T$ (or $\gamma_S/\gamma_T=1$) results in a front that is exclusively baroclinic, whereas $\gamma_S/\gamma_T=R\sigma$ results in a front that is exclusively thermoclinic. It is simple to demonstrate this using the linearized equation of density, from which, for modest inclinations, the following results can be obtained:

$$\begin{aligned}\gamma_\sigma &= \gamma_T\left(R_\sigma-\frac{\gamma_S}{\gamma_T}\right)\times[R_\sigma-1]^{-1}\\ &= \left(\gamma_S\left(\frac{\gamma_S}{\gamma_T}\right)\times[R_\sigma-1]\right)\times[R_\sigma-1]^{-1}\end{aligned} \tag{2.101}$$

The magnitude of $R\sigma$ becomes either extremely large or equal to zero when isohalines or isotherms are located vertically on a hydrological section (i.e., $\gamma_S=\infty$ or $\gamma_T=$ -oo), and the following formula should be used in their place (Fedorov, 1986):

$$\gamma_\sigma = \gamma_T\left(R^H_\sigma-1\right)\times\left[R^H_\sigma\right]^{-1}, when\ \gamma_S=\infty, and\ \gamma_T\ is\ small; \tag{2.102}$$

$$\gamma_\sigma = \gamma_S\left(1-R^H_\sigma\right)\times, when\ \gamma_T=\infty, and\ \gamma_S\ is\ small; \tag{2.103}$$

Therefore

$$\gamma_\sigma = \infty\ in\ purely\ baroclinic\ case\ \gamma_S=\infty,\ and\ \gamma_T=\infty; \tag{2.104}$$

Consequently,

$$\gamma_\sigma = 0 \, in\ purely\ thermoclinic\ case, when\ R_\sigma^H = 1,\ and\ \gamma_S\ and\ \gamma_T\ have\ any\ values. \tag{2.105}$$

where $R_\sigma^H = \alpha\left(\frac{\partial T}{\partial x}\right) \times \left(B\left(\frac{\partial S}{\partial x}\right)\right)^{-1}$. In this view, the frontal zone's horizontal temperature and salinity gradients are represented by the notations$\left(\frac{\partial T}{\partial x}\right)$ and $\left(\frac{\partial S}{\partial x}\right)$, respectively. The salinity contributes further to the density shift throughout the frontal zone in the case of negative horizontal T, S-correlation ($R_\sigma^H < 0$). Temperature and salinity contribute to density when $R_\sigma^H > 0$ partially and fully compensate for one another when $R_\sigma^H = 1$(Fedorov, 1986; Kostianoy & Belkin, 1989; Krasnopevtsev & Yu, 1978).

These equations allow for the approximate calculation of the constraints for the temperature (ΔT) and salinity (ΔS) inversions that occur on frontal intrusions when water volumes travel over the fluctuation of isopycnal surfaces δI. Given that the typical ratio H/L $\sim 10^{-3}$ unites the characteristic horizontal shift L of the water volume to the intrusion thickness H, it is reasonable to declare that the maximum intrusional changes in temperature and salinity ought to be equivalent to (Fedorov, 1986; Kostianoy et al., 2004):

$$\Delta T = H.10^3.\frac{\partial T}{\partial I}, \tag{2.106}$$

and

$$\Delta S = H.10^3.\frac{\partial S}{\partial I}. \tag{2.107}$$

According to the above equations, temperature and salinity gradients along isopycnals are accepted as "thermoclinity" and "haloclinity" of the front dynamic fluctuations. These ratios would, in fact, have a practical significance for describing the turbulence in frontal zones (Dietrich & Gieskes, 1966; Fedorov, 1986; Kostianoy et al., 2004).

Therefore what efficiently is the major influence of the fine or flimsy thermohaline structure of the ocean in turbulence generation in the frontal zone? The concept "fine thermohaline structure of the ocean" denotes an ocean's vertical temperature and salinity profiles that are particularly long-lived and inhomogeneous on scales greater than the Ozmidov's scale (Ozmidov, 1965), which is the largest scale of local-isotropic turbulence:

$$L_0 = \varepsilon^{0.5} N_B^{-3/2} \tag{2.108}$$

Eq. (2.108) demonstrates that the fine-structure inhomogeneities have a layered stratification character, that is, as a function of the rate of turbulent energy dissipation ε and the Brunt-väisälä frequency N, their horizontal extension noticeably exceeds their vertical sizes (typically on 2–4 order). In this scenario, $N = \sqrt{-\frac{g}{\rho_0}\frac{\partial \rho(z)}{\partial z}}$ signifies the water column's stability—the displacement z' has fluctuating solutions for negative $\frac{\partial \rho(z)}{\partial z}$. (and N gives our angular frequency). If it is positive, runaway growth is present, indicating that the turbulence occurs across the ocean. One commonly presumes that the dynamic range of vertical dimensions of a splendid thermohaline pattern properly extends from likely tens of centimeters up to a hundred meters. In this sense, typical estimates of L_0 in the ocean are between $10^{-1}$and 1 m. Therefore the prevailing notion of "inhomogeneity of the robust structure" is meaningful only for features with vertical scales more microscopic than the conventional vertical scale of marked fluctuation of average fields of effective temperature and salinity in the ocean thermocline (Krasnopevtsev & Yu, 1978).

The amount of vertical exchange in a specific marine environment brought on by small-scale mixing processes is quantified by the Cox number. Its formula is the product of the horizontal eddy diffusivity and buoyancy frequency divided by the vertical flux of heat and momentum (or mass).

$$C_T = \left[\frac{\partial T'}{\partial Z}\right]^2 \cdot \left[\frac{\partial T}{\partial Z}\right]^{-2} \tag{2.109}$$

$$C_S = \left[\frac{\partial S'}{\partial Z}\right]^2 \cdot \left[\frac{\partial S}{\partial Z}\right]^{-2} \tag{2.110}$$

Both Eqs. (2.109) and (2.110) serves as a nondimensional measure of fine-structure intensity. These tiny structures are widespread throughout the oceans, but they are more complex and heterogeneous near fronts and areas where currents converge. These features, which are made up of water masses with varying salinities, temperatures, and densities, offer a wide range of habitats for organisms, and their interactions are what drive the biogeochemical cycling in the ocean.

The mathematical theories of turbulence and frontal zone origins have been demonstrated in this chapter. The Reynolds number is a fundamental mathematical calculation that is primarily used to comprehend the nature of turbulence. Numerous nonlinear models have been developed due to the turbulence's complex nonlinear pattern, including the Reynolds-averaged Navier − Stokes Equations, the Boussinesq approximation (buoyancy), and the fully explicable spectrum turbulence based on Kolmogorov microscales. Each of these mathematical models helps to explain different aspects of turbulence, and together, they can help us better understand this phenomenon.

This chapter also discusses the relationship between turbulence and front zone formation, as oceanic fronts form in areas of extreme turbulence and convection. Turbulence, along with increased convection, is essential to the formation of front zones in oceanic waters form in areas of extreme turbulence and convection. Turbulence, along with increased convection, is essential to the formation of front zones in oceanic waters. The exact mechanism of front zone formation is still being studied, but the presence of turbulence and convection appears to play a large role in this process. This will be fully addressed in the next chapters of this book. Because conventional mathematical models such as the Navier – Stokes Equations cannot fully address the complicated nonlinearity of turbulence and frontal zones, the following chapter will introduce a new mechanism for addressing this problem. This mechanism is mainly based on the radar signal that imagines turbulence and frontal zones. To this end, radar-based frontal zone imaging techniques have the potential to provide a much more detailed understanding of turbulent ocean behavior and can help scientists study the dynamics of these zones in much greater detail.

References

Alfonsi, G. (2009). Reynolds-averaged Navier-Stokes equations for turbulence modeling. *Applied Mechanics Reviews*, *62*(4), 1–20. Available from https://doi.org/10.1115/1.3124648Italy, http://asmedl.aip.org/getpdf/servlet/GetPDFServlet?filetype = pdf&id = AMREAD000062000004040802000001&idtype = cvips&prog = normal.

Allen, J. S., Beardsley, R. C., Blanton, J. O., Boicourt, W. C., Butman, B., Coachman, L. K., Huyer, A., Kinder, T. H., Royer, T. C., Schumacher, J. D., Smith, R. L., Sturges, W., & Winant, C. D. (1983). Physical oceanography of continental shelves. *Reviews of Geophysics*, *21*(5), 1149–1181. Available from https://doi.org/10.1029/RG021i005p01149.

Bernard, P. S., Wallace, J. M., & Yavuzkurt, S. (2003). Turbulent flow: Analysis, measurement, and prediction. *Applied Mechanics Reviews*, *56*(6), B83. Available from https://doi.org/10.1115/1.1623759, B83.

Bleck, R., Onken, R., & Woods, J. D. (1988). A two-dimensional model of mesoscale frontogenesis in the ocean. *Quarterly Journal of the Royal Meteorological Society*, *114*(480), 347–371. Available from https://doi.org/10.1002/qj.49711448005.

Camerlengo, A. (1982). Large-scale response of the pacific ocean subarctic front to momentum transfer: A numerical study. American Meteorological Society, United States. *Journal of Physical Oceanography*, *12*(10), 1106–1121. Available from http://journals.ametsoc.org/loi/phoc, 10.1175/1520-0485.

Campobasso, M. S., Yan, M., Bonfiglioli, A., Gigante, F. A., Ferrari, L., Balduzzi, F., & Bianchini, A. (2018). Low-speed preconditioning for strongly coupled integration of Reynolds-averaged Navier–Stokes equations and two-equation turbulence models. *Aerospace Science and Technology*, *77*, 286–298. Available from https://doi.org/10.1016/j.ast.2018.03.015.

Chuychai, P., Weygand, J. M., Matthaeus, W. H., Dasso, S., Smith, C. W., & Kivelson, M. G. (2014). Technique for measuring and correcting the Taylor microscale. *Journal of Geophysical Research: Space Physics*, *119*(6), 4256–4265. Available from https://doi.org/10.1002/2013JA019641, http://onlinelibrary.wiley.com/journal/10.1002/(ISSN)2169-9402.

Constantin, P. (2005). *Navier-Stokes equations and turbulence*, . *Lecture Notes in Mathematics* (1871, pp. 1–43). United States: Springer Verlag Euler equations. Available from http://www.springer.com/series/304, 0.1007/11545989_1.

Davidson, P. A., Kaneda, Y., & Sreenivasan, K. R. (2013). *Ten chapters in turbulence*. Cambridge University Press. Available from 10.1017/CBO9781139032810.

Del Álamo, J. C., Jiménez, J., Zandonade, P., & Moser, R. D. (2004). Scaling of the energy spectra of turbulent channels. *Journal of Fluid Mechanics*, *500*, 135–144. Available from https://doi.org/10.1017/S002211200300733X, http://journals.cambridge.org/action/displayJournal?jid = FLM.

Demuren, A., & Grotjans, H. (2009). Buoyancy-driven flowsbeyond the boussinesq approximation. *Numerical Heat Transfer, Part B: Fundamentals*, *56*(1), 1–22. Available from https://doi.org/10.1080/10407790902970080.

Dietrich, G., & Gieskes, J. M. (1966). The oceanic Polar front in the waters off the East coast\nof Greenland. *Annales Biology*, *23*, 20–22.

Dijkstra, H. A. (2008). *Dynamical oceanography* (pp. 1–407). Berlin Heidelberg: Springer. Available from http://www.springerlink.com/openurl.asp?genre = book&isbn = 978-3-540-76375-8, 10.1007/978-3-540-76376-5.

Ecke, R. (2005). The turbulence problem. *Los Alamos Science*, *29*, 124–141.

Fedorov, K. N. (1986). *The physical nature and structure of oceanic fronts* (333). Springer-Verlag.

K.N. Fedorov, A.I. Ginsburg, The near-surface layer of the ocean. 259 (1992).

Feigenbaum, M. J. (1979). The onset spectrum of turbulence. *Physics Letters A*, *74*(6), 375–378. Available from https://doi.org/10.1016/0375-9601(79)90227-5.

Foias, C. (1997). What do the Navier-Stokes equations tell us about turbulence. *Contemporary Mathematics.*, *208*.

Foias, C., Manley, O., Rosa, R., & Temam, R. (2002). Navier-Stokes equations and turbulence. *Physics Today.*, *55*(10), 53.

Gough, D. O. (1977). Mixing-length theory for pulsating stars. *The Astrophysical Journal*, *214*, 196. Available from https://doi.org/10.1086/155244.

Guo, C., Zhang, C., & Païdoussis, M. P. (2013). *Modification of equation of motion of fluid-conveying pipe for laminar and turbulent flow profiles* (pp. 221–237). Elsevier BV. Available from 10.1016/b978-0-12-408083-6.00010-6.

Heisel, M., De Silva, C. M., Hutchins, N., Marusic, I., & Guala, M. (2021). Prograde vortices, internal shear layers and the Taylor microscale in high-Reynolds-number turbulent boundary layers. *Journal of Fluid Mechanics*, *920*. Available from https://doi.org/10.1017/jfm.2021.478, http://journals.cambridge.org/action/displayJournal?jid = FLM.

Hoskins, B. J. (1982). The mathematical theory of frontogenesis. *Annual Review of Fluid Mechanics*, *14*(1), 131–151. Available from https://doi.org/10.1146/annurev.fl.14.010182.001023.

H.E. Huppert, G. Batchelor, H. Moffatt, M. Worster, Geological fluid mechanics. Perspectives in fluid dynamics: A collective introduction to current research. (2000), 447–506.

Knauss, J. A., & Garfield, N. (2016). *Introduction to physical oceanography*. Waveland Press.

Kostianoy, A. G., & Belkin, I. M. (1989). A survey of observations on emtrathermocline eddies in the world ocean. *Oceanography Series*, *50*(C), 821–841. Available from https://doi.org/10.1016/S0422-9894(08)70223-X.

Kostianoy, A. G., Nihoul, J. C., Rodionov, V. B., & Rodionov, V. B. (2004). *Physical oceanography of the frontal zones in sub-Arctic seas* (71). Gulf Professional Publishing.

R.H. Kraichnan, The closure problem of turbulence theory. New York Univ, NY, Courant Inst of Mathematical Sciences, (1961).

A. Krasnopevtsev, Yu, Thermohaline inhomogeneities of the spatial structure of the upper layer of the ocean. IN: Investigation of the Variability of Physical Processes in the Ocean. (1978).

Kuhfuß, R. (1986). A model for time-dependent turbulent convection. *Astronomy and Astrophysics, 160*, 116–120.

Kullenberg, G. (1981). Physical oceanography. *Oceanography Series, 30* (C), 135–181. Available from https://doi.org/10.1016/S0422-9894(08)70140-5.

Lan, S. Y., Kuan, P. C., Estey, B., Haslinger, P., & Müller, H. (2012). Influence of the coriolis force in atom interferometry. *Physical Review Letters, 108*(9). Available from https://doi.org/10.1103/PhysRevLett.108.090402, http://oai.aps.org/filefetch?identifier = 10.1103/PhysRevLett.108.090402 &component = fulltext&description = markup&format = xml.

Langer, S., & Swanson, R. C. (2020). On boundary-value problems for RANS equations and two-equation turbulence models. *Springer, Germany Journal of Scientific Computing, 85*(1). Available from https://doi.org/10.1007/s10915-020-01323-9, https://rd.springer.com/journal/10915.

Lapeyre, G., Klein, P., & Hua, B. L. (2006). Oceanic restratification forced by surface frontogenesis. *Journal of Physical Oceanography, 36*(8), 1577–1590. Available from https://doi.org/10.1175/JPO2923.1.

Lazarian, A. (2009). Obtaining spectra of turbulent velocity from observations. *Space Science Reviews, 143*(1–4), 357–385. Available from https://doi.org/10.1007/s11214-008-9460-y.

Lee, H. G., & Kim, J. (2012). A comparison study of the Boussinesq and the variable density models on buoyancy-driven flows. *Journal of Engineering Mathematics, 75*(1), 15–27. Available from https://doi.org/10.1007/s10665-011-9504-2.

Leslie, D. C. (1978). *Numerical methods in laminar and turbulant flow*. Pentech Press.

Lopez, J. M., Marques, F., & Avila, M. (2013). The Boussinesq approximation in rapidly rotating flows. *Journal of Fluid Mechanics, 737*, 56–77. Available from https://doi.org/10.1017/jfm.2013.558, http://journals.cambridge.org/action/displayJournal?jid = FLM.

Mathieu, J., & Scott, J. (2001). An introduction to turbulent flow. *Physics Today., 54*(9), 53.

Mayeli, P., & Sheard, G. J. (2021). Buoyancy-driven flows beyond the Boussinesq approximation: A brief review. *International Communications in Heat and Mass Transfer, 125*. Available from https://doi.org/10.1016/j.icheatmasstransfer.2021.105316, https://www.journals.elsevier.com/international-communications-in-heat-and-mass-transfer.

W.D. McComb, The physics of fluid turbulence. (1990).

McWilliams, J. C. (2021). Oceanic frontogenesis. *Annual Review of Marine Science, 13*, 227–253. Available from https://doi.org/10.1146/annurev-marine-032320-120725, http://www.annualreviews.org/journal/marine.

Mor-Yossef, Y. (2021). Improved convergence characteristics of two-equation turbulence models on unstructured grids. *Computers & Fluids, 230*, 105127. Available from https://doi.org/10.1016/j.compfluid.2021.105127.

Nagata, K., Sakai, Y., Inaba, T., Suzuki, H., Terashima, O., & Suzuki, H. (2013). Turbulence structure and turbulence kinetic energy transport in multiscale/fractal-generated turbulence. *Physics of Fluids, 25*(6), 065102. Available from https://doi.org/10.1063/1.4811402.

Nakayama, Y. (2018). *Introduction to fluid mechanics* (pp. 1–383). Elsevier. Available from https://www.sciencedirect.com/book/9780081024379, 10.1016/B978-0-08-102437-9.01001-9.

I. Nezu, H. Nakagawa, Turbulence in open-channel flows. (2017).

Nieuwstadt, F. T., Westerweel, J., & Boersma, B. J. (2016). *Introduction to theory and applications of turbulent flows*. Springer International Publishing.

Ozmidov, R. V. (1965). On the turbulent exchange in a stably stratified ocean. *Izvestiya, Academy of Sciences, USSR. Atmospheric and Oceanic Physics., 1*, 861–871.

A. Persson, How do we understand the Coriolis force. American Meteorological Society, Bulletin of the. 79 (1998).

Poncet, S., Schiestel, R., & Monchaux, R. (2008). Turbulence modeling of the Von Kármán flow: Viscous and inertial stirrings. *International Journal of Heat and Fluid Flow, 29*(1), 62–74. Available from https://doi.org/10.1016/j.ijheatfluidflow.2007.07.005.

Pope, S. B. (2001). Turbulent flows. *Measurement Science and Technology, 12*(11), 2020–2021. Available from https://doi.org/10.1088/0957-0233/12/11/705.

Qian, J. (1983). Variational approach to the closure problem of turbulence theory. *Physics of Fluids., 26*(8), 2098–2104. Available from https://doi.org/10.1063/1.864414.

Raizer, V. (2021). *Remote sensing of turbulence*. CRC Press.

Ríos, A. F., Pérez, F. F., & Fraga, F. (1992). Water masses in the upper and middle North Atlantic Ocean east of the Azores. *Deep Sea Research Part A, Oceanographic Research Papers, 39*(3–4), 645–658. Available from https://doi.org/10.1016/0198-0149(92)90093-9.

Segalini, A., Örlü, R., Schlatter, P., Alfredsson, P. H., Rüedi, J. D., & Talamelli, A. (2011). A method to estimate turbulence intensity and transverse Taylor microscale in turbulent flows from spatially averaged hot-wire data. *Experiments in Fluids, 51*(3), 693–700. Available from https://doi.org/10.1007/s00348-011-1088-0.

Solsvik, J., & Jakobsen, H. A. (2016). A review of the statistical turbulence theory required extending the population balance closure models to the entire spectrum of turbulence. *AIChE Journal, 62*(5), 1795–1820. Available from https://doi.org/10.1002/aic.15128, http://onlinelibrary.wiley.com/journal/10.1002/(ISSN)1547-5905.

Spall, M. A. (1995). Frontogenesis, subduction, and cross-front exchange at upper ocean fronts. *Journal of Geophysical Research, 100*(2), 2543–2557. Available from https://doi.org/10.1029/94JC02860.

Steele, J. H., & Thorpe, S. A. (2009). *Turekian, elements of physical oceanography: A derivative of the encyclopedia of ocean sciences*. Academic Press.

Stroh, A., Frohnapfel, B., Schlatter, P., & Hasegawa, Y. (2015). A comparison of opposition control in turbulent boundary layer and turbulent channel flow. *Physics of Fluids, 27*(7), 075101. Available from https://doi.org/10.1063/1.4923234.

Taylor, G. I. (1938). The spectrum of turbulence. *Proceedings of the Royal Society of London. Series A - Mathematical and Physical Sciences, 164* (919), 476–490. Available from https://doi.org/10.1098/rspa.1938.0032.

Tolmazin, D. (2012). *Elements of dynamic oceanography*. Springer Science & Business Media.

Willmarth, W. W. (1975). Structure of turbulence in boundary layers. *Advances in Applied Mechanics, 15*(C), 159–254. Available from https://doi.org/10.1016/S0065-2156(08)70057-7.

Woods, A. W. (2010). Turbulent plumes in nature. *Annual Review of Fluid Mechanics, 42*, 391–412. Available from https://doi.org/10.1146/annurev-fluid-121108-145430.

Xin, Y. B., Xia, K. Q., & Tong, P. (1996). Measured velocity boundary layers in turbulent convection. *Physical Review Letters, 77*(7), 1266–1269. Available from https://doi.org/10.1103/PhysRevLett.77.1266.

Yaglom, A. M. (2012). *Hydrodynamic instability and transition to turbulence* (100). Springer Science & Business Media.

Zakharov, V. E., & Falkovich, G. (2012). *Kolmogorov spectra of turbulence I: Wave turbulence*. Springer Science & Business Media.

Chapter 3

Quantized synthetic aperture radar signal: a comprehensive exploration

Within this chapter, a philosophical journey unfolds, offering a distinctive vantage point that intertwines quantum mechanics with the intricate landscape of synthetic aperture radar (SAR). Departing from antecedent endeavors, this chapter consciously steers clear of an exhaustive voyage into the realms of radar theory and signal processing, recognizing their thorough exploration elsewhere. Instead, it directs its focus toward the introduction of a groundbreaking concept—that of quantization—to fathom the intricacies underlying radar's portrayal of sea turbulence. The ensuing chapters promise a more profound exploration of this conceptual terrain.

Embracing this opportune moment, this chapter endeavors to elucidate the concept of radar quantization, providing a crystalline comprehension of this term. The foundational quantization of electromagnetic waves, integral to radar operations, takes center stage in the initial examination. The discourse initiates by unraveling the mystique surrounding why photons, as elemental entities, manifest quantization—an indispensable facet for grasping the intricacies of quantized electromagnetic waves. This investigation lays bare the discrete energy amounts within a defined energy spectrum that photons exhibit, deviating from a continuous expanse.

3.1 What is quantization?

The comprehensive process of evaluating from a conventional comprehension of material phenomena to a more recent understanding known as quantum mechanics is known in physics as quantization (or quantisation in British English). It is a method for converting classical mechanics into quantum mechanics. To quantize phenomena or objects is to apply quantum theory to them and divide them into quanta. The number of possible values of a quantity or system state is limited, so variables can only take discrete magnitudes. Something is said to be "quantized" in the context of quantum mechanics if it only takes on definite values and has no value in between those definite values. For instance, if measuring a particle or wave yields values of 0, 1, 2, 3, and so on, the quantity being measured is referred to as quantized if there are no other values in between, such as 0.1, 0.11, 0.9, 1.37, 1.999, 2.6, and 3.14768 (Gruber, 1972; Widrow & Kollár, 2008).

A signal is "quantized" in the context of communication engineering when its measured values at various times are roughly approximated to values that are suitable for the next stage of the system, which converts the measured values into digital data. The signal is essentially "quantized" by this process, so it appears that the signal processing stages are further down the pipeline. For an illustrative example, a signal voltage of 7 V is encoded as the number 14, and a signal voltage of 7.5 V is encoded as the number 15. The signal value is rounded up or down if it is measured to be between 7 V and 7.5 V, so that it can be encoded (rounded up as 15, rounded down as 14).

In this sense, field quantization, also known as the "quantization of the electromagnetic field," is a generalization that involves infinite degrees of freedom and refers to photons as field "quanta" (for instance, as light quanta). Atomic physics, chemistry, particle physics, nuclear physics, condensed matter physics, and quantum optics are each reliant on this procedure. Quantization is a fundamental concept in modern physics, providing the framework for understanding the behavior of matter and energy at the smallest scales. It has led to revolutionary advances in our understanding of the physical world, including the development of technologies such as transistors, lasers, and MRI machines (Cooper, 2001; Jonsson & Ankel, 2021).

The quantization of the electromagnetic field has been crucial in the development of quantum technologies, such as quantum computing and quantum cryptography. It has also led to the discovery of new phenomena, such as the Casimir effect and Hawking radiation, which has expanded our understanding of the universe.

The action principle of classical mechanics is generalized in quantum theory's formulation of quantum

Synthetic Aperture Radar Image Processing Algorithms for Nonlinear Oceanic Turbulence and Front Modeling. DOI: https://doi.org/10.1016/B978-0-443-19155-8.00003-9

mechanics, which is known as the path integral. To compute the quantum amplitude, a sum or functional integral over an infinite number of quantum-mechanically feasible trajectories is used rather than a single, particular classical trajectory for a system. Action provides classical mechanical theory, with permitted configurations being the most extreme functional variations of the action. The path integral formulation is particularly useful for studying systems that are difficult to solve using traditional methods, such as those with many interacting particles. It also allows the calculation of probabilities for a particle to move from one position to another, known as the transition amplitude (Fronsdal, 1979; Widrow & Kollár, 2008).

Digital signal processing and mathematics are both related to the process of quantization. It describes the mapping of input values from a big set, frequently a continuous set, to output values in a small, countable set, frequently with a finite number of elements. A set of discrete symbols or integer values effectively approximates a continuous signal in this process. Since rounding is typically required when representing a signal in digital form, quantization is involved to some extent in almost all digital signal processing (Abdessaied et al., 2014; Assouly et al., 2023).

3.2 What distinguishes first from second quantization?

The first quantization was the development of quantum particle mechanics by Heisenberg, Schroedinger, Dirac, Pauli, Jordan, and Born between 1925 and 1928. Electrons were quantified according to this theory. Heisenberg's uncertainty principle dictates that quantum mechanics should indeed severely constrain the authorized values of momentum and energy for an electron. Quantization of matter in quantum mechanics accounts for the interaction between matter and electromagnetic fields. And, the electromagnetic field remained classical. This inequality between how quantum mechanics viewed matter and the electromagnetic field was acknowledged by three of the theory's founders. They realized that a quantized electromagnetic field is a necessary component of genuine quantum description (Fronsdal, 1979; Gruber, 1972).

They decided to reframe quantum mechanics in terms of the field rather than a particle. Each individual particle, such as an electron, is understood to be a quantized excitation of this field, and both matter and electromagnetism are described in terms of quantum fields. The second quantization, or quantum field theory, is the designation of this initiative. Heisenberg, Jordan, and Pauli published the original theory for this program in 1929. It includes the electron relativistic quantum field, which was created by converting Dirac's equation for relativity from a single electron to a continuous field. In both scenarios, the degrees of freedom is infinite (Galindo & Pascual, 2012; Gottfried, 2018).

The other component was the relativistic quantum electromagnetic field, in which Maxwell's classical electromagnetic field underwent quantization to create Dirac. The interaction between the two quantum fields is the third and final aspect. The issue has been with this third component. The theory proved to be seriously flawed. For a physical quantity such as the magnetic dipole moment of the electron, it was riddled with infinities, making it impossible to quantify any finite value for it. This issue was resolved through a process called renormalization, which involved adjusting the parameters of the theory to cancel out the infinities and obtain finite values. Despite this success, the need for renormalization raised questions about the fundamental nature of the theory and its ability to accurately describe physical phenomena.

In the spring of 1947, shortly after the end of World War II, Willis Lamb conducted an extraordinary experiment in Isidor Rabi's lab at Columbia University. Lamb was able to measure the energy difference between the 2 s and 2p states of a neutral hydrogen atom using the microwave generator created during the first world war to produce the radar waves. It is very intriguing that the same radar that prevented the Luftwaffe from obliterating Britain during the Battle of Britain also aided in the discovery of one of nature's most profound mysteries.

To everyone's great surprise, it was discovered that this gap was not zero and that it matched the microwave frequency of 1057 MHz. The energy of the 2 s state is much higher than that of the 2p state. Given that Dirac's equation predicted this gap to be exactly zero, this was a very surprising result. Or, to put it another way, these two states ought to be degenerate. Following this, the old quantum field theory of Heisenberg, Jordan, and Pauli attracted a lot of new interest. The leading theorists of the time understood immediately that to establish a finite value for the energy gap measured by Lamb, one had to accept the infinities of this theory. In this understanding, the pseudonym of this new initiative was the renormalization of quantum field theory. Tomonaga, Schwinger, and Feynman successfully completed it independently between 1947 and 1949. This new concept is known as quantum electrodynamics (QED). The energy gap of the hydrogen atom was precisely calculated using the new renormalized theory to match Lamb's measured values. Quantum field theory was successfully used to quantify other values of physical quantities, such as the magnetic dipole moment of an electron or the scattering cross section of two electrons, owing to the renormalization program, which provided a solid foundation. The second quantization is completed using a renormalization program, according to certain researchers (Baym, 2018; Gottfried, 2018).

The significant contradiction that plagues the Dirac theory is incidentally also eliminated by the second quantization. Information can move more quickly than the speed of light in a vacuum because the Dirac theory permits the existence of space-like intervals. The theory of relativity's requirement for causality is broken by this. In QED, the existence of antimatter particles (positrons) helps remove this fatal flaw from the theory, thereby making it causal. This is because positrons are interpreted as electrons moving backward in time, which means that the negative-energy solutions in the Dirac equation can be removed. The second quantization of the QED also provides a framework for calculating and predicting the behavior of particles and their interactions with electromagnetic fields.

3.3 How and why are photons quantized?

A photon is an electron tail component composed of electromagnetic waves. The reason why a photon is quantized is the key to understanding quantized electromagnetic waves. Therefore from reading the question and comment threads, it appears that the asker has more than one question. Why do photons come in discrete energy amounts, with a discrete spectrum of energies rather than a continuous range of energies, according to the first interpretation of the question "Why are photons quantized?" Why do we treat photons as particles (i.e., quantized energy spectrum and of finite extent in space) as opposed to waves (i.e., continuous energy spectrum and of arbitrary extent in space) according to the second reading of the question? (It is obviously incorrect to associate discrete energies with particles and continuous energies with waves; each can have either.)

The first question can be answered by stating that photons would have a discrete spectrum of distinct energies when they are generated by systems of bound particles, such as atoms, that have a discrete spectrum of distinct energy levels (each photon can only have a single specific energy when measured). This holds true regardless of whether the released photons are waves or particles. The underlying processes (such as electron shell transitions) only appear to produce a single photon that contains all the energy associated with the transition, rather than an arbitrary number of photons that divide the total energy between them. This is the "reason" (note the quotation marks) why these photons are only emitted at certain specific energies. This phenomenon is known as energy quantization and is a fundamental aspect of quantum mechanics. This has many important implications for our understanding of the behavior of matter and radiation at the atomic and subatomic levels (Barzanjeh et al., 2015; Baym, 2018; Marghany, 2019, 2021).

This observation is consistent with our quantum mechanical description (there would be little point in it if it were not). And remain aware that the author thinks of "description," not "explanation," because science only endeavors to "describe," not "explain," nature. What might seem to be an explanation of a phenomenon is actually just a more detailed description of the phenomenon in terms of more basic phenomena and concepts that are related to those phenomena.

Free particles can have arbitrary energies from a continuous spectrum of energies, unlike bound systems, such as atoms and molecules, where photons are emitted or absorbed at specific energy levels. In this view, photons with discrete energy spectra from bound processes and those from unbound processes can undergo changes in their original energy owing to Doppler shifts and gravitational redshifts or blueshifts, which can transpose the original energy up or down the energy spectrum, allowing them to have almost any energy at all.

The second question, "Why do we treat photons as particles instead of waves?" can be answered simply by saying that we do not always do this. The most valuable and appropriate description depends on which photon properties we are interested in modeling in any particular scenario, as well as what the photon is actually doing. Nevertheless, the fundamental tenet is that photons and all other subatomic "particles" move and interact like waves. In this sense, there is no such thing as a "wave-particle duality" in the sense that a photon or an electron is either one or the other; they are both present at the same time, and we can simply select the one that interests us. Some writers have emphasized chimera-like reliability.

There are several ways to discuss quantum mechanics that shed light on various facets of the theory. In particular, the approach of operators and Hilbert spaces is a potent language that makes the photon quantization very explicit; after all, when readers diagonalize a matrix, only a small number of different eigenvalues emerge, correct?

However, the author finds that focusing directly on the Schrodinger equation makes it easier to understand why the photon quantization needs to appear in the first place as opposed to just showing that it does. Look for the energies of the simple harmonic oscillator in particular (Baym, 2018; Casariego et al., 2023; Galindo & Pascual, 2012; Gottfried, 2018):

$$\frac{-\hbar^2}{2m}\frac{d^2\psi}{dx^2}+\frac{1}{2}mx^2\omega^2\psi=E\psi(x) \tag{3.1}$$

Although $E_n=\hbar\omega(n+1/2)$, the discrete number of energies that are permitted has been taught to us; this is simply untrue unless the aforementioned equation is accompanied by suitable boundary conditions. In fact, pick any energy E and plug it in on the upper right side Eq. (3.1). The Schrodinger equation above then explains precisely how to extend the wave function to other regions of space after readers decide on their favorite value for $\psi'(0)$. Thus there is a continuum of solutions to the Schrodinger equation presented above. Where does the quantization originate, then? The Schrodinger equation's solution, however, blows up at infinity for pretty much any value of E (Miller, 2008). Since our ultimate goal is to normalize the wave function to enable a probabilistic interpretation, we must insist on the following:

$$\psi(x) \rightarrow 0, as x \rightarrow \pm \infty \tag{3.2}$$

Almost all of the aforementioned solutions are eliminated by this boundary condition, leaving only a discrete group of energies to be chosen. As a result, in quantum mechanics, the quantization is actually enforced by the boundary conditions. These boundary conditions are also a result of our wave functions' need to be normalizable. Naturally, this has to do with probabilities that add to one. These specifications fall under the broad category of "unitarity" in physics. In other words, the quantization is caused by the unitarity of quantum mechanics. Therefore unitarity must be enforced in the operator formalism to give us quantized energies, as seen in the harmonic oscillator, that is, $a, a^{\dagger}$. This would be shown in the following section as an essential part of quantized electromagnetic waves. In fact, unitarity is a fundamental principle in quantum mechanics that ensures the conservation of probability (Galindo & Pascual, 2012; Miller, 2008). It is also crucial in the study of scattering processes, where it guarantees that the sum of probabilities for all possible outcomes equals one. The system's Hamiltonian, therefore, can be expressed as follows using these operators:

$$H = \hbar\omega\left(a^{\dagger}a + \frac{1}{2}\right) \tag{3.3}$$

Eq. (3.3) demonstrates that as well as a operates on a state with energy E, it forms a state with energy $E - \hbar\omega$. Similarly, $a^{\dagger}$ amplifies energy by a similar quantity. In this regard, the argument can be stated as follows: assuming the state of energy E and acting on it with the lowering operator n times, which delivers energy $E - n\hbar\omega$ into the state. Therefore the energy turns negative when the lowering operator applies it too many times. The argument has included a lot of misunderstandings at this stage. At this point, it could claim that because negative energies are nonphysical, a state that is destroyed by the lowering operator, which would be referred to as $|0\rangle$, must exist. Even though it is true that such a state is necessary, unitarity—which would be discussed later—is the real driver behind this. In either case, let's move on with the conversation and return to this issue later. It is important to keep in mind that the concept of negative energies is still a theoretical construct and has not been directly observed. Therefore it is crucial to approach this topic with caution and continue exploring its implications. Let's continue our discussion and revisit this point once we have more information (Galindo & Pascual, 2012; Gottfried, 2018; Kramers, 2018).

Then, since every energy state must end with a $|0\rangle$, let us begin with $|0\rangle$ and create the other states by adding a lot of $a^{\dagger}$. As a result, the energy is quantified. One should be aware of the importance of the ground state in the quantization of energy. Without it, we simply have a way of building a number of ladders of energy states, but nothing tells us that only one such ladder exists. In other words, the energy cannot be quantized if the ground state does not exist. Then, why is it necessary for such a state to exist? Negative energies are completely normal, despite the fact that this may seem strange. Regardless, unitarity is the real cause. Particularly, for us to properly normalize states, all states must have a positive norm: $\langle\psi|\psi\rangle > 0$. Notwithstanding, the presence of a state with negative energy indicates that:

$$\langle\psi|H|\psi\rangle < 0 \rightarrow \langle\psi|a^{\dagger}a|\psi\rangle < 0 \tag{3.4}$$

Consequently, the state $a|\psi\rangle$ cannot be normalized because it has a negative norm. In other words, unitarity is broken. Because such a state is inconceivable, the ladder should always come to an end, state $|0\rangle$ would exist, and energy must be quantized. Evidently, unitarity—which, relying on the formalism one uses, can be viewed either in terms of boundary conditions or the positivity of norms—is what leads to the quantization of energy (Baym, 2018; Gottfried, 2018; Kramers, 2018).

3.4 How can electromagnetic waves become quantized?

Since photons are distinct energy particles, the electromagnetic field has been quantized. In terms of energy, momentum, and spin, photons are massless particles with specified amounts of each. In this sense, orbital angular momentum and spin are the two sorts of angular momentum that exist in quantum physics. The orbital angular momentum operator, which occurs when the wavefunction of an orbit has the periodic structure as the angle fluctuates, is the quantum-mechanical equivalent of the classical angular momentum of an orbital revolution. Unlike electrons, which have no classical analog, photons have a quantum-mechanical equivalent of polarization called spin. In this view, experiments such as the Stern − Gerlach experiment, in which silver atoms were

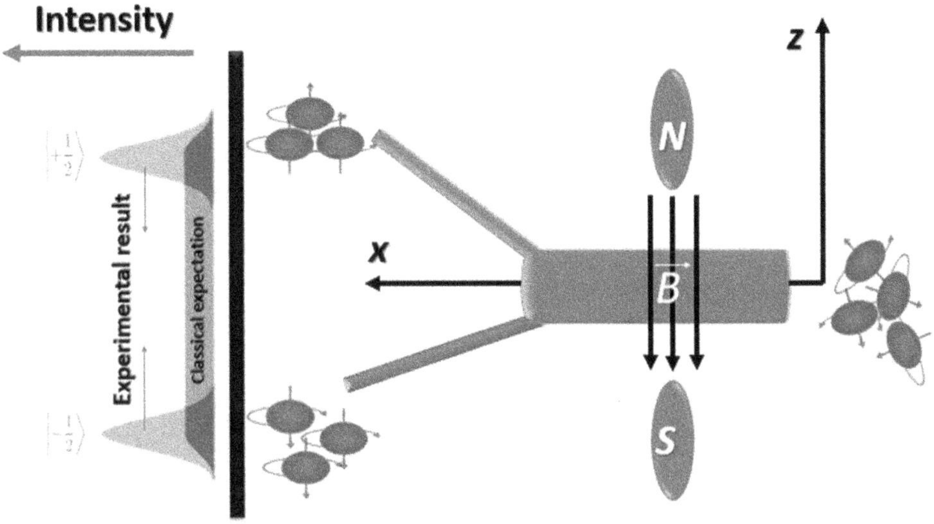

FIGURE 3.1 Simplification of Stern – Gerlach experiment.

revealed to have two potential discrete angular momenta while having no orbital angular momentum, are implemented to infer the presence of electron spin angular momentum. The spin-statistics theorem and the Pauli exclusion principle can potentially be used to deduce the presence of the electron spin, and vice versa: given a specific electron spin, the Pauli exclusion principle can be deduced (Ahern, 1995; Cooper, 2001; Kramers, 2018; Marghany, 2021).

Put differently, in the Stern – Gerlach experiment, silver atoms are deflected upward or downward according to their spin as they travel through an inhomogeneous magnetic field β; the furnace, the silver atom beam, the inhomogeneous magnetic field, the conventionally anticipated result, and the observed result. This experiment measures the spatial distribution of silver atoms by projecting a collimated beam of silver atoms through a magnetic field and onto a screen, for instance, in the x direction (Fig. 3.1). For ease of exposition, the gradient of the magnetic field in the z direction is believed to be constant:

$$\frac{\partial B_z}{\partial z} = cte \neq 0. \tag{3.5}$$

The magnetic moment – containing silver atoms would all precess more about the z-axis in the absence of a gradient in the magnetic field. The spatial distribution of silver atoms is bound to be Gaussian with a spread determined by the efficiency of the beam collimation and the time of flight of the silver atoms, and this precession would have no influence on this distribution. Therefore the Lorentz force would tend to bend the trajectory in a circle if charged particles such as electrons are incorporated into the experiment. An electric field of the proper strength that is positioned perpendicular to the passage of the charged particle can neutralize this effect (Lanzagorta, 2011; Marghany, 2019, 2021).

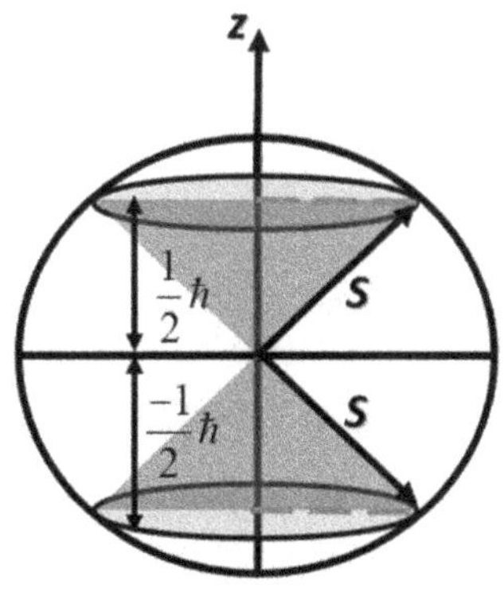

FIGURE 3.2 Fermions spin values.

The spin-1/2 particle is the electron. They are entirely quantum mechanical phenomena with just two conceivable spin angular momentum values measured along any axis, $+\hbar/2$ or $-\hbar/2$ (Fig. 3.2). It is referred to as "intrinsic angular momentum" since its value is constant, making it an intrinsic feature of electrons (to distinguish it from orbital angular momentum, which can vary and depends on the presence of other particles). Electrons are classified as "spin up" or "spin down" depending on whether the magnetic moment is directed up or down when the spin is measured along a vertical axis (Galindo & Pascual, 2012; Miller, 2008).

In other words, fermions are described as having half-integer spins, such as $\frac{1}{2}, \frac{3}{2}, \frac{5}{2}$, whereas bosons are described as having integer spins, such as 0, 1, and 2. The two groups of particles behave differently and play, in general, distinct roles in our environment. The Pauli exclusion principle, which states that no two identical fermions may have the same quantum numbers at the same time, is a major contrast between the two families (meaning, roughly, having the same position, velocity, and spin direction). Fermions abide by Fermi – Dirac statistical laws. The restriction does not apply to bosons, which follow the Bose – Einstein statistics rules and can therefore "bunch together" in the same states. Moreover, the spins of composite particles might differ from those of their constituent particles. In mathematical mechanics, a particle's angular momentum seems to have a magnitude (how rapidly the body is spinning) as well as a direction (either up or down on the axis of rotation of the particle). Although in a more nuanced manner, directional information is also present in the quantum-mechanical spin (Baym, 2018; Gottfried, 2018; Kramers, 2018). According to quantum physics, a spin-s particle's component of angular momentum can only have the following values measured in any direction:

$$S_i = \hbar s_i, s_i \in \{-s, -(s-1), \ldots, s-1, s\}, \tag{3.6}$$

Eq. (3.2) says S_i is the spin projection quantum number along the i-th axis, si is the main spin quantum number, and s is the spin component along the i-th axis (either x, y, or z). The z-axis is often chosen as the direction:

$$S_z = \hbar s_z, s_z \in \{-s, -(s-1), \ldots, s-1, s\}, \tag{3.7}$$

here S_z is the spin component along the z-axis, and s_z is the spin projection quantum number along the z-axis. Therefore for the initial stage in quantizing the photon, assume that the following description of the mode $(\vec{k}, \mu)$ quantum mechanical photon state $|\vec{k}, \mu\rangle$ demonstrates the existence of the following characteristics:

$$m_{Photon=0} \tag{3.8}$$

$$H|\vec{k}, \mu> = h\nu|\vec{k}, \mu> with\ \nu = c|\vec{k}| \tag{3.9}$$

$$P_{EM}|\vec{k}, \mu> = \hbar\vec{k}|\vec{k}, \mu> \tag{3.10}$$

$$S_Z|\vec{k}, \mu> = \mu|\vec{k}, \mu> \tag{3.11}$$

According to Eqs. (3.8–3.11), a photon has no rest mass, its electromagnetic momentum is $\hbar\vec{k}\ [\hbar = h/(2\pi)]$, its energy is $h\nu = hc|\vec{k}|$ ($\vec{k}$ is the wave vector, and c is the speed of light), and its polarization is $\mu = \pm 1$ (the eigenvalue of the z-component of the photon spin). In this view, the modification of the particle's time-dependent linear momentum by the following principle is the most well-known case of quantization:

$$\vec{p}(t) \rightarrow --i\hbar\vec{\nabla} \tag{3.12}$$

It is essential to point out that the Planck constant is being used here and that the quantum mechanical operator does not inherit the time dependence of the conventional formulation (this is applicable to what is known as the Schrödinger equation). Similarly, we handle the electromagnetic field. The electric constant, abbreviated ε in this context due to the use of electromagnetic SI units, is a quantity. The principles for quantization are:

$$a_k^{(\mu)}(t) \rightarrow \sqrt{\frac{\hbar}{2\omega V\varepsilon_0}} a^{\mu}(K) \tag{3.13}$$

$$\overline{a}_k^{(\mu)}(t) \rightarrow \sqrt{\frac{\hbar}{2\omega V\varepsilon_0}} a^{\dagger(\mu)}(K) \tag{3.14}$$

Therefore the quantized electromagnetic wave can be described by the classical Hamiltonian as follows:

$$H = \frac{1}{2}\epsilon_0 \int\int\int_V \left(\left|E\left(\vec{\mathbf{r}}, t\right)\right|^2 + c^2\left|B\left(\vec{\mathbf{r}}, t\right)\right|^2 \right) d^3\mathbf{r}$$
$$= V\epsilon_0 \sum_{\mathbf{k}} \sum_{\mu=\pm 1} \omega^2 \left(\overline{a}_{\mathbf{k}}^{(\mu)}(t) a_{\mathbf{k}}^{(\mu)}(t) + a_{\mathbf{k}}^{(\mu)}(t)\overline{a}_{\mathbf{k}}^{(\mu)}(t) \right). \tag{3.15}$$

This Hamiltonian consists of the sum of kinetic and potential energy terms, where the kinetic energy is proportional to the square of the electric field and the potential energy is proportional to the square of the magnetic field. The quantization arises from imposing commutation relations on the electric and magnetic fields, leading to discrete energy levels. This Hamilton operator can be used to calculate the energy levels and transition probabilities of atoms and molecules interacting with the electromagnetic field, providing a theoretical framework for understanding phenomena such as the absorption and emission of electromagnetic waves (Baym, 2018; Levi, 2023).

$$H = \frac{1}{2} \sum_{\mathbf{k}, \mu=\pm 1} \hbar\omega \left(a^{\dagger(\mu)}(\mathbf{k}) a^{(\mu)}(\mathbf{k}) + a^{(\mu)}(\mathbf{k}) a^{\dagger(\mu)}(\mathbf{k}) \right)$$
$$= \sum_{\mathbf{k}, \mu} \hbar\omega \left(a^{\dagger(\mu)}(\mathbf{k}) a^{(\mu)}(\mathbf{k}) + \frac{1}{2} \right) \tag{3.16}$$

Therefore it is important to keep in mind that ω is a function of k. This allows us to express the Hamiltonian in terms of creation and annihilation operators $a^{\dagger}$, which are essential for studying the behavior of bosonic systems. In this sense, the second quantized treatment allows us to describe the harmonic oscillator in terms of creation and annihilation operators, which simplifies calculations and provides a more intuitive understanding of the system. This approach is particularly useful in the study of many-body systems, where the harmonic oscillator potential often arises as an approximation for the potential energy between particles as demonstrated early in Eq. (3.3).

The energy of the system can be found by summing the energies of each individual oscillator. This is useful in modeling systems such as molecules and solids, where each atom can be approximated as a harmonic oscillator as stated in:

$$H = \sum_i \hbar\omega_i \left(a^\dagger(i)a(i) + \frac{1}{2} \right). \tag{3.17}$$

It can be demonstrated that is an excitation operator $a^\dagger$, excitable from an n fold excited state to an n + one-fold excited state:

$$a^\dagger|n\rangle = |n+1\rangle\sqrt{n+1}. \tag{3.18}$$

In particular: $a^\dagger|0\rangle = |1\rangle$ *and* $|(a^\dagger)^n|0\rangle \propto |n\rangle$. The n-fold excited state $|n\rangle$; can be thought of as a single state containing n particles (also known as vibrons) with all of the energy $h\nu$ because harmonic oscillator energies are equidistant. They are bosons, these particles. The excitation operator $a^\dagger$ is referred to as a creation operator for obvious reasons. This is similar to the harmonic oscillator, where the energy levels, $\hbar\omega = h\nu$ are also quantized and correspond to the different eigenstates of H. The difference is that in the case of the single-photon state, there is only one energy level due to the photon having a fixed energy:

$$H|(\vec{\mathrm{k}},\mu)^m;(\rightharpoonup k',\mu')^n\rangle = [m(\hbar\omega) + n(\hbar\omega')]$$
$$\times|(\vec{\mathrm{k}},\mu)^m;(\vec{\mathrm{k}}',\mu')^n\rangle, \text{ with } \omega = c|\mathbf{k}| \text{ and } \omega' = c|\vec{\mathrm{k}}'|. \tag{3.19}$$

A triplet spin can be granted to the photon with the spin quantum number S = 1. This means that the photon has three possible polarization states, which can be used in quantum communication and cryptography protocols. Additionally, the triplet spin of photons can be used to study fundamental physics phenomena such as quantum entanglement and Bell's inequalities (Galindo & Pascual, 2012; Gottfried, 2018; Kramers, 2018). In this understanding, the photon spin can be mathematically described as:

$$S_z \equiv -i\hbar\left(\rightharpoonup e_x \otimes \rightharpoonup e_y - \rightharpoonup e_y \otimes \rightharpoonup e_x\right)$$
$$\times \textit{and cyclically } x \to y \to z \to x. \tag{3.20}$$

The two operators are dyadic products $\otimes$ between the two orthogonal unit vectors. The unit vectors are parallel to the spin quantization axis' z-axis, which is perpendicular to the propagation direction $\vec{\mathrm{k}}$. Dyadic products are commonly used in quantum mechanics to describe the interactions between particles with spin. They can also be used to represent the polarization of electromagnetic waves. In this sense, the spin operators fulfill the conventional commutation relations for angular momentum:

$$[S_x, S_y] = i\hbar S_z \textit{ and cyclically } x \to y \to z \to x. \tag{3.21}$$

Needless to say, the quantization of electromagnetic waves refers to the process of representing electromagnetic radiation as a stream of particles, known as photons. This concept is fundamental to understanding many phenomena in modern physics, such as the photoelectric effect and laser behavior. This can help to improve understanding of signal radar interaction with rough or turbulence surfaces, as well as how radar imaging of surface turbulence works.

The key question is: Is frequency quantized in accordance with energy quantization of the electromagnetic field? This question is important because it relates to the fundamental nature of microwave spectra and their interactions with matter. The answer could have implications for our understanding of quantum mechanics and the behavior of particles at the atomic and subatomic levels (Baym, 2018; Gottfried, 2018; Kramers, 2018; Levi, 2023).

The electromagnetic field's total energy is not quantized. The distribution of that energy into various frequency components is quantized. A pure sinusoidal electromagnetic wave can have any energy and frequency as long as its E/f ratio is an integer multiple of the Planck constant (0, h, 2 h, 3 h, 4 h, etc.). Therefore the quantization scenario is usually clarified by questioning, "What are the authorized energy levels?" while focusing on a single frequency component. In that particular instance, it is obvious what "energy is quantized" means. But viewers raise an excellent issue: if it's really the relationship between the two that's constrained, why do we typically think of energy as being quantized rather than frequency? Does examining a single energy instead impose a quantization requirement on the permitted frequencies?

Although the amount of energy stored in the electromagnetic field in this instance is a well-known and precise quantity, neither the quantity of photons released during the process nor the frequencies they have are. In other words, the electromagnetic field experiences a quantum superposition of waves with numerous varying frequencies and photon counts. There may also be some classical uncertainty caused by a thermal spectrum of frequencies (if the temperature is taken into account). At this point, one can assume is that E/h is equal to the total frequency of all photons as follows:

$$\sum_{i=0}^{n} f_i = Eh^{-1} \tag{3.22}$$

In Eq. (3.22), n, the number of photons, is also uncertain at this time and has its own probability distribution. Let's imagine a spherical shell replete with detectors that are placed in every direction. These detectors do not interfere with or otherwise measure the photons as they pass through the shell; instead, they tag and count them as they do so. They don't measure or change their frequency, in particular. In light of this, if one were to envision energy as a fluid (please do not take that comparison literally!) The conventional scenario is rather like throwing it into a huge lake, where the distance from the lake's center represents the frequency, the depth of the water at each point on its surface represents how much energy is stored in a particular frequency mode. Consequently, the different directions from the center represent the various directions waves carrying energy can be traveling (the surface of the lake is only two-dimensional. However, electromagnetic waves can usually travel in any direction in three-dimensional space, another limit of the analogy). This analogy helps us understand how energy is distributed across different frequencies and directions in space, which is important for understanding how it behaves and interacts with matter. However, it's important to remember that this is just a metaphor, and energy doesn't actually behave like a fluid in the way we typically think of it (Gottfried, 2018; Kramers, 2018; Levi, 2023).

In contrast, the quantum situation is analogous to pouring the same amount of water into a haphazard assortment of buckets, each of which has a different size. The frequency of a photon, which establishes its amount of energy, is represented by the size of each bucket. The buckets must be filled exactly. It is highly unlikely that any two buckets will be exactly the same size if there are only a few of them. Each photon frequency was constrained to one photon. The conventional boundary, however, allowed each set of equal-sized buckets to be stacked on top of one another in a "tower" of states and to be arranged in any way possible if there were an infinite number of buckets (Galindo & Pascual, 2012; Miller, 2008).

It is conceivable to state that the distribution of energy over frequency, rather than the energy itself, is quantized. Frequency can be subject to additional quantum mechanical restrictions (such as summation rules), but it is never quantized to the same extent as energy. This is because a change in energy corresponds to a change in frequency because energy and frequency are related through the Planck's constant. As a result, the quantization of frequency pursues the quantization of energy.

3.5 Quantazied microwave signal

The question is now: how does quantize the microwave signal? It is well known that a form of electromagnetic radiation is the microwave signal. Microwaves have frequencies between 300 MHz and 300 GHz and wavelengths between 1 m and 1 mm. Communication often takes place over microwaves. However, it is only effective over a 64 km radius. However, practical communication distances for the high end of the band are only about 1 km. Modern technology and goods, such as microwave ovens, frequently use microwaves. Radar technology, which enables the detection of objects and the measurement of their distance and speed, also exploits microwaves. They are additionally utilized to transmit signals over long distances in satellite communication systems.

The quantized electromagnetic field affects the extent of the microwave signal transfer, as demonstrated in the preceding sections. In essence, the quantized microwave signal is made up of a continuous set of frequency-dependent harmonic oscillators $\{\hat{a}_\omega, \hat{a}^\dagger\}_\omega$ with $\omega \in \mathbb{R}^+$, where the formation and annihilation operators accommodate the commutation of direct relevance $[\hat{a}_\omega, \hat{a}^\dagger{}_{\omega'}] = \delta(\omega - \omega')\hat{\Pi}$. Furthermore, it is noteworthy that photons exhibit a polarization degree of freedom, presenting an additional avenue for leveraging them as carriers of quantum information. In many instances, the Wigner function, a quasi-probability distribution linked to a quantum state represented by ρ (density operator), emerges as an alternative descriptor. The operational significance of this function aligns more closely with the continuous variables (CVs) paradigm. Specifically, the integral of this function over a certain field quadrature is directly proportional to the probability of measuring the orthogonal one. This nuanced approach to quantum information harnesses the polarization properties of photons, offering a pathway for diverse applications within the quantum realm. Certainly, the operational interpretation of the Wigner function can be expressed mathematically as follows:

$$P(\Delta q, \Delta p) = \pi^{-1} \int_{-\infty}^{\infty} \langle q + \frac{\Delta q}{2} | \rho | q - \frac{\Delta q}{2} \rangle e^{i\Delta pp/\hbar} dp \tag{3.23}$$

here $P(\Delta q, \Delta p)$ represents the Wigner function; Δq and Δp are the field quadrature variables; ρ is the density operator representing the quantum state; $|q\rangle$ denotes the position basis; p is the momentum variable;and $\hbar$ is the reduced Planck constant. In this sesne, the value of the reduced Planck constant ($\hbar$) is approximately 1.054571×10^{-34} Joule·second or $6.62607015 \times 10^{-34} \text{m}^2 \cdot \text{kg/s}$ in the International System of Units (SI). This constant plays a fundamental role in quantum mechanics, linking the energy of a system to the frequency of its associated wave function and setting the scale of quantum effects. Certainly, the distinction in energy levels between quantum optics and quantum microwaves, despite sharing a common formalism, is a critical factor in the practical implementation of technologies such as Wigner tomography. In both quantum optics and quantum microwaves, the formalism involves obtaining observables as power series of $\hat{a}_\omega$ *and* $\hat{a}^\dagger_\omega$.

Entanglement, a cornerstone in the realm of quantum technology, assumes a central role, particularly in bipartite systems where more than one mode of the electromagnetic field is considered. The term "mode" encompasses various aspects and essentially denotes the labels used to distinguish between Hilbert spaces. For instance, spatial modes give rise to path entanglement, frequency modes lead to frequency entanglement (or time bins), and polarization modes yield discrete polarization entanglement. Importantly, these forms of entanglement are not mutually exclusive, allowing for diverse manifestations within the quantum landscape (Davies, 2018; Golovko, 2008; Shankar, 2020).

In the context of bipartite, mixed-state entanglement, a key metric is the negativity. This entanglement measure is defined as $2N(\rho) := \left\|\widehat{\rho}\right\| - 1$, where $\|\widehat{\rho}\|_1 := Tr(\widehat{\rho}^{\dagger}\widehat{\rho})$ represents the trace norm of the partially transposed density operator. This mathematical characterization provides a quantitative assessment of entanglement in mixed states, offering a valuable tool for understanding and harnessing quantum phenomena in diverse technological applications. Two-mode squeezed (TMS) states emerge as natural candidates in the realm of quantum continuous variables (CVs) for communication and sensing protocols necessitating entanglement. These states are routinely generated in laboratory settings, offering a valuable resource for various quantum applications. Notably, the squeezing operation, a crucial step in producing TMS states in the microwave domain, is often performed using Josephson parametric amplifiers (JPAs). It's worth noting that JPAs demand a cryogenic environment for their operation, typically requiring low temperatures to achieve their desired quantum performance. The utilization of TMS states and JPAs underscores the sophisticated interplay between quantum technology and the necessity for precise experimental conditions, particularly in the challenging domain of microwave quantum systems (Golovko, 2008; Gottfried, 2018; Kramers, 2018; Shankar, 2020).

The symmetric two-mode squeezed states can be obtained through two distinct methods: First, by directly pumping a parametric down-conversion term; second, by single-mode squeezing two vacuum states, which, in practice, are in thermal states, in orthogonal directions, and subsequently combining them using a beam splitter. In both cases, the resulting state is described as a two-mode squeezed thermal (TMST) state with a mean photon number $N_{TMST} = 2n_{th}\cosh(2r) + 2\sinh^2(r)$, where n_{th} is the thermal photon number, and $r \in R$ is the squeezing parameter.

For frequency-degenerate TMST states, an experimentally observed negativity $N = 3.9$ has been reported, corresponding to an entangled bit rate of 4.3×10^6 ebit/s. In the frequency nondegenerate case, a rate of 6×10^6 ebit/s has been achieved. Alternative approaches involve the use of traveling wave parametric amplifiers, particularly suitable for broadband applications. Another method entails utilizing a dc-biased Josephson junction in the presence of two resonators. These diverse techniques provide avenues for generating the desired two-mode squeezed states with different experimental configurations and applications in quantum communication and sensing.

In the realm of microwave quantum technologies, Cat states emerge as a captivating wellspring for continuous variable (CVs) quantum entanglement, unlocking promising potentials for applications in the microwave domain. These states take center stage in generating entangled coherent states within the intricate dance of two superconducting microwave resonators. In this perspective, Cat states are quantum states that are superpositions of coherent states, and they play a significant role in quantum information processing. A cat state of the form, therefore, is given by:

$$|Cat\rangle = \frac{1}{2}(|\alpha\rangle + |-\alpha\rangle) \tag{3.24}$$

where $|\alpha\rangle$ is a coherent state, represents a superposition of two coherent states with opposite phases. Mathematically, a coherent state $|\alpha\rangle$ is an eigenstate of the annihilation operator $\hat{a}$ and is given by:

$$|\alpha\rangle = e^{\frac{-|\alpha|^2}{2}} \sum_{n=0}^{\infty} \frac{\alpha^n}{\sqrt{n!}} |n\rangle \tag{3.25}$$

Now, let's express the Cat state:

$$|Cat\rangle = \frac{1}{\sqrt{2}}\left(e^{\frac{-|\alpha|^2}{2}} \sum_{n=0}^{\infty} \frac{\alpha^n}{\sqrt{n!}} |n\rangle\right) + \left(e^{\frac{-|\alpha|^2}{2}} \sum_{n=0}^{\infty} \frac{(-\alpha)^n}{\sqrt{n!}} |-n\rangle\right) \tag{3.26}$$

Cat states have interesting properties, and they are a crucial element in quantum information science, particularly in the field of quantum error correction and quantum computation. The delicate balance between the coherent states in a superposition provides a unique resource for quantum processing tasks (Galindo & Pascual, 2012; Golovko, 2008; Miller, 2008; Shankar, 2020).

In the intricate landscape of quantum technologies, a refined technique known as time-bin encoding has been deftly showcased in experimental settings, introducing a layer of sophistication to the propagation of microwaves. This novel approach sparks particular intrigue when the vexing issue of decoherence takes center stage in the quantum limelight.

Decoherence is a persistent challenge in quantum systems, arising from interactions with the surrounding environment that can cause loss of quantum information. In this context, time-bin encoding stands out as an ingenious strategy. Time bins, essentially discrete temporal intervals, are celebrated for their robustness against losses. Like stalwart guardians, they shield the delicate quantum states from the disruptive effects of decoherence, which will be demonstrated later in Chapter 11 as an excellent procedure to investigate coastal upwelling.

The beauty of time-bin encoding, therefore, lies in its ability to mitigate the impact of decoherence, offering a resilient avenue for quantum information to traverse the unpredictable quantum terrain. As quantum technologies advance, the strategic utilization of time bins becomes a beacon of hope, promising to navigate through the challenges posed by decoherence and enhance the reliability of quantum communication and information processing systems.

3.6 Quantum microwave dynamics: unveiling the journey of propagation

The propagation of quantum microwaves refers to the transmission and behavior of microwave signals when they are treated as quantum entities, following the principles of quantum mechanics. This domain is particularly relevant in quantum information processing, quantum communication, and quantum technologies utilizing microwave-frequency electromagnetic fields.

Mathematically, the state of quantum microwaves in superposition of two microwave signals $|\psi_1\rangle$, and $|\psi_2\rangle$ denoted as $|\psi\rangle$, is elegantly expressed as a linear combination of basis states. In this depiction:

$$|\psi\rangle = \frac{1}{\sqrt{2}}\left(|\psi_1\rangle + |\psi_2\rangle\right) \tag{3.27}$$

Therefore the coherence of quantum states is expressed through density matrices or wave functions. Decoherence, a consequence of environmental interactions, is modeled through Lindblad operators in the master equation framework (Cosmelli et al., 2003; Fedorov et al., 2018). In this view, oherence is described using density matrices, and decoherence is modeled through the master equation. If ρ is the density matrix, the master equation is given by:

$$\frac{d\rho}{dt} = -\frac{i}{\hbar}[H', \rho] + L(\rho) \tag{3.28}$$

where $L(\rho)$ incorporates decoherence effects. In other words, Eq. (3.28) describes how the density matrix evolves over time. The term $L(\rho)$ encapsulates the Lindblad contributions, accounting for the interaction with the environment. Lindblad operators are mathematical constructs that represent the dissipative effects causing decoherence. These operators are introduced into the master equation, which governs the time evolution of the density matrix. In this understanding, microwave quantum systems are often described by density matrices, denoted by ρ (Lanzagorta, 2011; Leppäkangas et al., 2018; Levi, 2023). These matrices encapsulate the probabilities and correlations between different quantum states in a system. For a single quantum state $|\psi\rangle$, the corresponding density matrix is given by $\rho = |\psi\rangle\langle\psi|$. For an ensemble of quantum states, the density matrix is a sum of individual state density matrices, each weighted by its probability. Alternatively, coherence can be expressed through wave functions. A quantum state $|\psi\rangle$ is associated with a wave function $\psi(x)$, where x represents the relevant degrees of freedom. The probability density of finding a particle in a particular state is given by $|\psi(x)|^2$, reflecting the coherent spread of the quantum information (Chapman, 2002; Cosmelli et al., 2003; Dubosclard et al., 2021).

In essence, while density matrices and wave functions beautifully capture the coherence of quantum states, the introduction of Lindblad operators in the master equation elegantly addresses the challenge of decoherence, offering a holistic mathematical representation of the dynamics in the quantum landscape.

In the realm of quantum microwaves, the phenomenon of entanglement unveils a captivating mathematical dance, reminiscent of a symphony in the quantum orchestra. Imagine quantum microwaves as dancers on the quantum stage and entangled states as the choreography that unites their movements. These states symbolize a profound correlation, an entwining of microwave modes that transcends classical descriptions. Mathematically, entangled states are denoted as a tensor product of individual microwave states. For two microwave modes, $|\Psi_A\rangle$ and $|\Psi_B\rangle$, an entangled state $|\text{Entangled}\rangle$ can be expressed as:

$$|\psi\rangle = \frac{1}{2}\left(|\psi_A\rangle \otimes |\psi_B\rangle - |\psi_B\rangle \otimes |\psi_A\rangle\right) \tag{3.29}$$

The nature of this entanglement is quantified using entanglement measures, akin to assessing the elegance of a dance. Measures such as concurrence or logarithmic negativity provide numerical values that gage the depth of the entangled connection between microwave modes. In this view, the tensor product, symbolized by $\otimes$, represents the fusion of individual states into a unified entangled state. Each term in this tensor product encapsulates the quantum steps taken by the entangled microwaves, resonating with the complexity and richness of their correlated dance.

In the enchanting realm of quantum microwaves, therefore, the manipulation of information is orchestrated by the elegant choreography of quantum gates. In this sense, quantum gates are represented by unitary matrices.

If U is a quantum gate, it transforms the state vector $|\psi\rangle$ as:

$$|\psi'\rangle = U|\psi\rangle \tag{3.30}$$

Envision quantum gates as the maestros orchestrating the movements of quantum states. Their baton, the unitary matrix U, conducts a symphony that reshapes the state vector, turning the quantum stage into a canvas of dynamic possibilities. In this context, the unitary matrix U serves as the dance steps, guiding the quantum state vector $|\psi\rangle$ through a metamorphic journey. Each element of the matrix delicately influences the amplitude and phase of the quantum dance, unveiling the intricate steps of quantum evolution. As the quantum state $|\psi\rangle$ takes its place on the dance floor, the quantum gate, represented by U, initiates a harmonic progression. The transformed state $|\psi'\rangle$ emerges as a melodious variation, resonating with the unitary rhythm imposed by the quantum gate. The quantum dance unfolds dynamically over unitary time. With each tick of the quantum clock, the state vector gracefully pirouettes under the influence of the unitary matrix, creating a mesmerizing spectacle of quantum transformation. The artistry lies in the fluidity of the transformation. Quantum gates, with their unitary matrices, bring a sense of coherence and elegance to the quantum dance. The quantum state, once in $|\psi\rangle$, now pirouettes into a new form, $|\psi'\rangle$, revealing the artistic brushstrokes of quantum evolution (Cooper, 2001; Fedorov et al., 2018).

In the quantum world of microwaves, the act of measurement is a nuanced dance, orchestrated by the intricacies of positive operator-valued measures (POVMs). Imagine POVMs as the choreographers of the quantum microwave dance. These measures, represented by operators Π_i, guide the microwaves through the process of measurement, determining the probabilities associated with different outcomes. Therefore picture quantum microwaves as performers stepping onto the measurement stage. Their state, denoted by ρ, encapsulates the ensemble of possible quantum configurations. As the microwaves prepare for the measurement dance, the POVMs are poised to capture the nuances of their quantum performance. The Born rule emerges as the score sheet for this quantum performance. According to this fundamental principle, the probability $P(i)$ of obtaining a specific outcome i is determined by the trace of the product of the density matrix ρ and the corresponding measurement operator Πi:

$$P(i) = Tr(\rho\Pi_i) \tag{3.31}$$

The trace operation, symbolized by $\mathrm{Tr}(\cdot)$, acts as the conductor's baton, unveiling the probabilities encoded in the quantum ensemble. It traces the quantum path taken by the microwaves, unveiling the likelihood of each possible measurement outcome. In the dance of quantum measurement, each Π_i corresponds to a distinct move on the measurement stage (Cooper, 2001; Cosmelli et al., 2003; Okołowicz et al., 2003; Singh et al., 2022). The Born rule harmoniously blends the quantum dynamics of microwaves with the probabilistic landscape, offering a glimpse into the quantum uncertainties that define their performance. Needless to say, the ensemble of measurement outcomes, guided by the Born rule and the POVMs, paints a symphony of possibilities. Each outcome, a unique note in the quantum melody, contributes to the rich tapestry of the microwaves' quantum state.

3.7 Quantum microwaves and resonators: a symbiotic dialog

In the enthralling encounter between quantum microwaves and resonators, the script unfolds through the language of Hamiltonians—a tale told with elegance and precision. Envision quantum microwaves and resonators engaged in a dialog, their interaction narrated by the mathematical script of Hamiltonians. Hamiltonians, like seasoned scriptwriters, pen down the laws governing the quantum dialog. They encode the energy exchanges, transitions, and resonant harmonies that define the microwaves' interaction with resonators. Enter the Schrödinger equation—a portal to the quantum realm where dynamics unfold. This equation, a poetic reflection of quantum evolution, expresses how the quantum state $|\psi\rangle$ of the microwaves changes over time (t) under the influence of the Hamiltonian (H) and the reduced Planck constant ($\hbar$):

$$i\hbar\frac{d|\psi\rangle}{dt} = H|\psi\rangle \tag{3.32}$$

In this quantum narrative, the Schrödinger equation serves as the eloquent storyteller, weaving together the characters of quantum microwaves and resonators in a dynamic dance of amplitudes and energies. The Hamiltonian, as the guiding script, ensures that each quantum step resonates with the harmonies of quantum dynamics, creating a symphony that echoes through the quantum realm. In this perspective, The temporal derivative $\frac{d|\psi\rangle}{dt}$ unveils the unfolding narrative of the quantum story. It tells how the quantum state evolves, capturing the nuances of the microwaves' journey through the resonator landscape.

Therefore quantum-limited amplification is represented by operators. If $\hat{A}$ is the amplification operator, the amplified state is given by: $|\psi_{\mathrm{amplified}}\rangle = \widehat{A}\ \ |\psi\rangle$. Consequently, quantum memory devices evolve according to a Hamiltonian. If $|\psi(t)\rangle$ is the state at time t, its evolution is given by:

$$|\psi(t)\rangle = e^{-iHt/\hbar}|\psi(0)\rangle \tag{3.33}$$

Subsequently, quantum repeaters involve quantum channels and operations. The entanglement generation may be represented by a quantum operation ε acting on the initial state:

$$|E(\rho)\rangle = E(|\psi_0\rangle) \tag{3.34}$$

Eq. (3.34) demonstrates that quantum channels serve as the stage where this quantum drama plays out. They provide the medium for the transformation, allowing entanglement to traverse the spatial expanse and connect distant quantum entities. $|\text{Entangled}\rangle = E(|\text{Initial State}\rangle)$ is the script, capturing the essence of the entanglement generation process—a testament to the artistry embedded in the quantum communication landscape (Cooper, 2001; Cosmelli et al., 2003; Eichler et al., 2011; Fedorov et al., 2018; Singh et al., 2022).

3.8 Quantum microwave detection: unraveling the quantum realm

The emerging question now is: what is the philosophy behind microwave detection? The answer can contribute to a deeper understanding of radar imaging of the ocean surface.

The philosophy behind microwave detection, particularly in the context of radar imaging for surface ocean studies, stems from the foundational principles of both quantum mechanics and classical wave physics. This philosophical framework provides a deeper understanding of the intricacies involved in microwave-based remote sensing, elucidating the connection between theoretical principles and practical applications.

The philosophy underpinning quantum microwave detection navigates the intricate landscape of quantum mechanics, where foundational principles guide the understanding of electromagnetic fields at the microwave scale. At its core, the concept of superposition plays a pivotal role, acknowledging that microwave photons can exist in multiple states simultaneously. This recognition aligns with the principle of wave-particle duality, emphasizing the need to treat microwave photons as both particles and waves in the quantum dance. Quantum entanglement, fostering instantaneous correlations between particles, becomes a crucial element in the philosophy of microwave detection, influencing the outcomes of measurements. The Heisenberg uncertainty principle introduces fundamental limits to precision, reminding us of the inherent uncertainty and indeterminacy associated with quantum measurements. The philosophical dimension extends to the measurement problem in quantum mechanics, where the act of measurement raises questions about the nature of reality and the observer's role. Addressing technological challenges, such as maintaining quantum coherence, becomes integral to the philosophy, with cryogenic environments minimizing decoherence and thermal noise. In this pursuit, quantum microwave detection aligns with the broader philosophical quest for understanding and harnessing quantum information, inviting contemplation on the nature of reality and the intricate interplay of quantum principles in the microwave realm.

3.8.1 Homodyne and heterodyne detection

Homodyne detection and heterodyne detection are techniques used in quantum optics for measuring the amplitude and phase of a signal. Let's delve into the mathematical formulations for both methods. Homodyne detection involves measuring a specific quadrature of the electric field (Fig. 3.3). The homodyne operator for measuring a quadrature angle ϕ is given by:

$$\widetilde{X}_{\phi} = \frac{1}{\sqrt{2}}\left(\widetilde{a}\, e^{-i\phi} + \widetilde{a}^{\dagger} e^{-i\phi}\right) \tag{3.35}$$

Here, $\widetilde{a}$ and $\widetilde{a}^{\dagger}$ are the annihilation and creation operators, respectively. The probability density for measuring a value x is:

$$P(x,\ y) = \frac{1}{\sqrt{\pi}}\langle\psi|e^{-\frac{1}{2}(\widetilde{X}_{\phi}-x)^2}|\psi\rangle \tag{3.36}$$

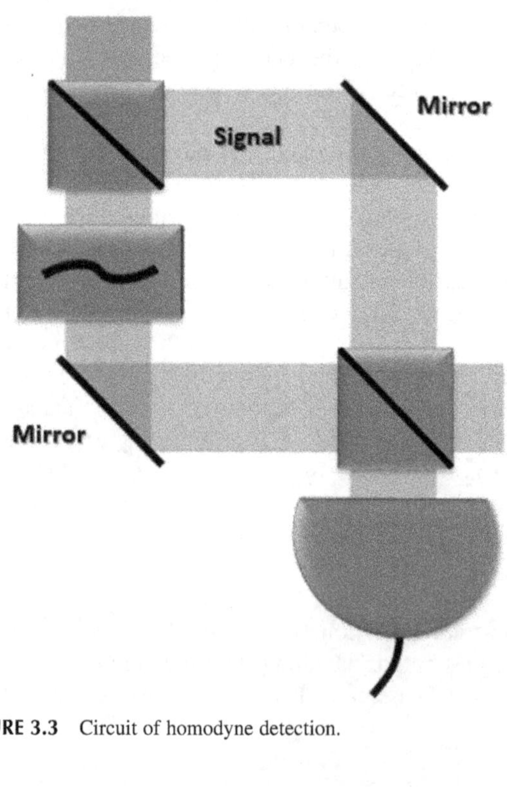

FIGURE 3.3 Circuit of homodyne detection.

In the harmonious world of signal processing, behold the heterodyne, a frequency conjured into existence through the artful fusion of two other frequencies (Chapman, 2002; Hamid et al., 2017; Singh et al., 2022). This enchanting technique, known as heterodyning, traces its origins to the ingenious mind of Canadian inventor-engineer Reginald Fessenden. Picture the alchemy: two frequencies engaged in a dance of transformation, shifting gracefully from one range to another, guided by the mystical forces of modulation and demodulation (Fig. 3.4). The ritual unfolds in a nonlinear symphony, orchestrated by the likes of vacuum tubes, transistors, or diodes, elegantly referred to as mixers in our scientific lexicon (Graf, 1999; Hamid et al., 2017; Oktavianto et al., 1967).

Heterodyne detection, therefore, involves measuring both quadratures simultaneously. The heterodyne operator is a combination of the annihilation and creation operators:

$$\widehat{X}_{hetero} = \frac{1}{\sqrt{2}}\left(\widehat{a}\, e^{-i\phi} + \widehat{a}^{\dagger} e^{i\phi}\right) + i\frac{1}{\sqrt{2}}\left(\widehat{a}\, e^{-i\phi} - \widehat{a}^{\dagger} e^{i\phi}\right) \tag{3.37}$$

The joint probability density for measuring values x and y is:

$$P(x, y) = \frac{1}{\pi}\langle\Psi| e^{-\frac{1}{2}\left(\widehat{X}_{hetero-(x+iy)}\right)^2} |\Psi\rangle \tag{3.38}$$

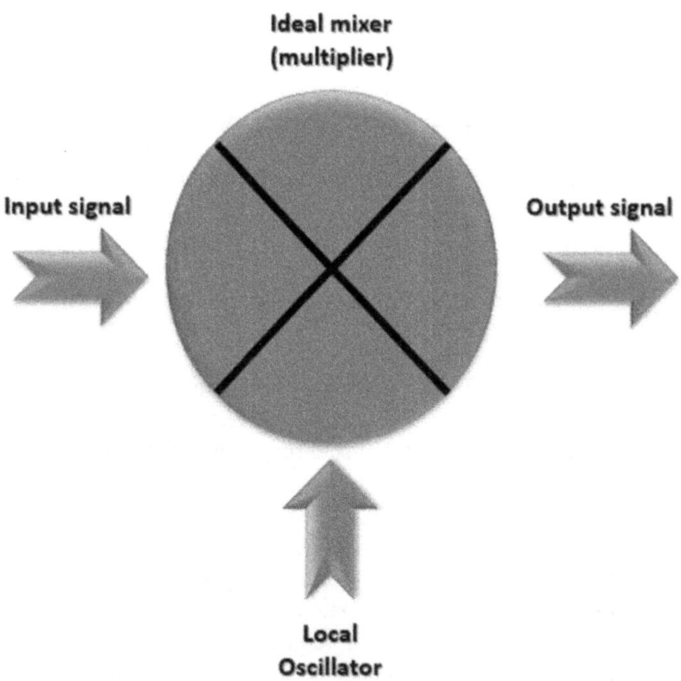

FIGURE 3.4 Concept of heterodyning detection.

These mathematical expressions provide a foundation for understanding the probabilities associated with measuring specific quadrature values in homodyne and heterodyne detection. They play a crucial role in quantum optics and quantum communication systems (Cooper, 2001).

In the realm of electrical engineering, homodyne detection emerges as an ingenious method, a symphony of precision, to unveil the encoded secrets within the delicate dance of phase and frequency modulation in an oscillating signal. Picture this: we intricately compare the signal with a standard oscillation, a kindred spirit, crafted to be identical had the signal chosen to carry no informational baggage. In the poetic language of our craft, "homodyne" becomes a soloist, a solitary frequency gracefully dancing on the stage of detection, distinguishing itself from the dual frequencies orchestrating the ballet of heterodyne detection (Graf, 1999; Hamid et al., 2017).

3.8.2 Quantum symphony: qubit-enhanced photon counting

In the enigmatic realm of quantum communication, sensing, and computation, a tapestry of non-Gaussian operations and measurements unfolds, with photon counting as its most illustrious protagonist (Fig. 3.5). Behold, as quantum illumination beckons, demands the artistry of photon counting to unveil a quantum advantage. Photodetectors,

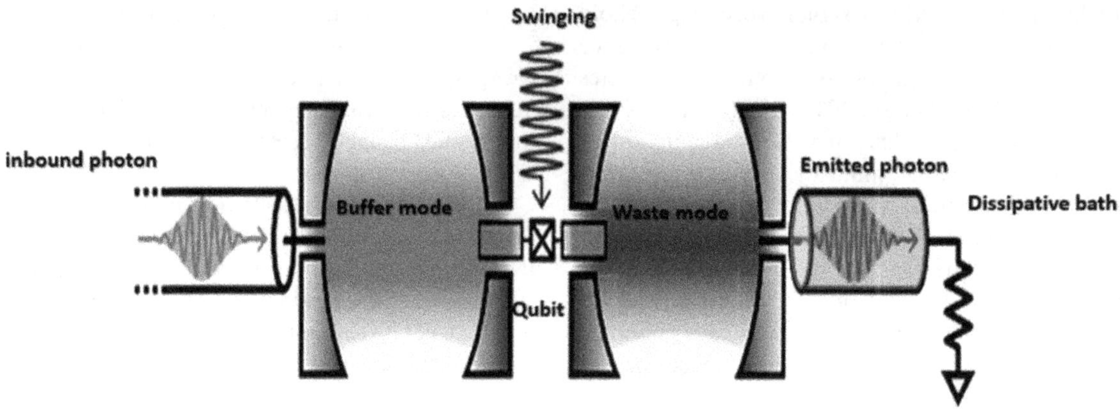

FIGURE 3.5 Qubit-enhanced photon counting.

those cosmic observers, transcend their role, becoming architects of full-state tomography for the ethereal dance of propagating quantum states. In the grand microwave ballet, they orchestrate the distribution of entanglement between remote qubits through a mesmerizing joint measurement. These enchanting devices also birth heralded noiseless amplifiers, breathing life into dissipative transmission lines and mending the fabric of imperfectly generated entanglement.

In the sacred script of quantum communication, where the dance of particles and waves converges, we inscribe the verses of non-Gaussian operations with the noble quill of mathematical expression. Witness, as photon counting takes center stage, a luminary in the microwe signal ballet:

$$|\Psi\rangle = \sum_{n=0}^{\infty} \alpha_n |n\rangle \tag{3.39}$$

here, α_n unfurls the probability amplitudes of encountering n photons, an ethereal symphony echoing the quantum states within our communication fabric. As quantum illumination beckons, behold the mathematical incantation:

$$\hat{C}|n\rangle = n|n\rangle \tag{3.40}$$

The photon counting operator, denoted by $\hat{C}$, conducts the celestial choir, unveiling the probabilities with each count. The grand tapestry of quantum communication, sensing, and computation unfolds:

$$P(n) = |\langle n|\hat{C}|\Psi\rangle|^2 \tag{3.41}$$

In this quantum sonnet, photodetectors metamorphose into heralds, architects, and menders of entanglement, their prowess revealed through mathematical revelations:

$$\hat{D}|n\rangle = \sqrt{n}|n-1\rangle \tag{3.42}$$

The quantum nondemolition detectors, inscribing their tales in the elegance of a qubit's phase or the celestial choreography of a lone photon entwined with a qubit's excitation, etch their narratives:

$$\hat{D}_1|0\rangle = |0\rangle, \hat{D}_2|1\rangle = |1\rangle \tag{3.43}$$

However, until the grand microwave arrangement reached its peak, a crucial composition remained absent—the photocounter, an elusive revelation for modes in motion:

$$\hat{P}|n\rangle = \sqrt{n}|n-1\rangle \tag{3.44}$$

This equation represents a quantum operation denoted as $\hat{P}$ acting on a quantum state $|n\rangle$. The operation $\hat{P}$ is designed to perform a specific transformation on the quantum state. In simpler terms, the equation indicates that the quantum operation $\hat{P}$ transforms a quantum state with n photons into a new state with $n-1$ photons, and the transformation is proportional to the original number of photons (n). This type of operation is characteristic of certain quantum processes, especially in the context of quantum radar and quantum image processing (Flurin, 2014; Gleyzes et al., 2007; Guerlin et al., 2007; Leppäkangas et al., 2018; Xiang et al., 2010; Xiang et al., 2010).

3.8.3 Bolometric elegance: advancing photon counting technologies

Bolometers, intricate devices rooted in the principles of thermodynamics and material science, stand as exceptional tools for photon counting, particularly within the nuanced domain of quantum optics and sensitive light detection applications.

In their operational dance, bolometers showcase a keen ability to absorb photons, a property ingrained in their carefully selected materials. As photons strike these materials, they induce a temperature rise, unleashing a cascade of thermal responses within the bolometer. This temperature increase triggers a measurable change in electrical resistance, forming the crux of the bolometer's photon-detecting prowess.

Photon counting becomes the virtuoso performance of bolometers, wherein these devices, finely calibrated and attuned, register discrete changes in resistance corresponding to the absorption of individual photons. Through meticulous analysis of these resistance fluctuations, researchers can unveil the precise number of photons that have graced the bolometer's sensitive interface.

The versatility of bolometers extends their reach into diverse scientific realms. In the ethereal landscapes of quantum optics, bolometers become essential instruments for experiments exploring the intricate quantum states of light or delving into the complexities of quantum information processing. Simultaneously, in the vast cosmic theater of astronomy, bolometers find their place, particularly in infrared and submillimeter observations, where their heightened sensitivity allows for the detection of even the faintest glimmers of radiation (Chromey, 2016; Cooper, 2001; Ferroni, 2010; Graf, 1999; Okołowicz et al., 2003).

These devices, adorned with advantages, boast high sensitivity and operate seamlessly across a wide spectral range, rendering them indispensable in the pursuit of photon detection excellence. Yet, as with any scientific tool, bolometers encounter challenges, notably the demand for cooling to extremely low temperatures in some applications to mitigate thermal noise and enhance sensitivity.

In the grand tapestry of scientific exploration, bolometers emerge as indispensable protagonists, unraveling the secrets of photon counting with precision and contributing invaluable insights to our understanding of the quantum nature of microwave.

3.8.4 Whispering amplifiers: cryogenically engineered for minimal noise

In the delicate symphony of quantum circuits, where signals whisper from millikelvin realms, a grand performance awaits at room temperature. Here, the intricate dance of quantum information demands a stage set by low-noise and high-gain amplification chains, crafting a bridge between the frigid domains of millikelvin temperatures and the warmth of room-temperature components.

Picture this cosmic choreography: quantum circuits, emitting signals as gentle as single-photon whispers, or resonating at the formidable frequencies of dispersive qubit readout, unleashing powers in the mystic range of -150 dBm to -120 dBm. To elevate these ethereal murmurs to a crescendo perceptible at room temperature, a carefully orchestrated amplification journey unfolds (Cosmelli et al., 2003; Dubosclard et al., 2021). The opening act stars superconducting parametric amplifiers, celestial instruments that grace the first stage with near-quantum-limited noise performance, ample bandwidth, and the prowess to handle the quantum signals. Their performance sets the stage, paving the way for the subsequent drama.

As the signals embark on their journey across the vast cosmic expanse, the narrative unfolds with the introduction of low-noise cryogenic high-electron-mobility transistor (HEMT) amplifiers, taking the spotlight at a frigid 3–4 K. Their pivotal role in this cosmic saga is to propel the delicate whispers of quantum information toward the final crescendo. As the signals reach room temperature, a captivating postamplification spectacle marks the conclusion of this microwave symphony.

In this intricate orchestration, each amplification stage assumes a meticulously calibrated role. The gain and noise temperature of these cosmic amplifiers are selected with a nuanced precision akin to poetry. This ensures a harmonious blend with the symphony of system noise, as dictated by the Friis formula, complementing the whispers initially added by the first amplifier. In this cosmic ballet, the subtle quantum signals elegantly traverse temperature gradients, their whispers resonating through the symphonic corridors of room temperature with a grace that defines the essence of this microwave narrative (Casariego et al., 2023).

In the realm of cryogenic low-noise solid-state amplifier technologies, the pursuit of minimal noise temperatures finds its zenith with indium-phosphide (InP) high-electron-mobility transistors (HEMTs). A notable proposal in reference unveils a three-stage hybrid low-noise amplifier operating in the 4–8 GHz range at a frosty 10 K. This amplifier, with a gain of 44 dB across the entire band, delivers an average noise temperature of 1.6 K while consuming a modest 4.2 mW of DC power (Chromey, 2016). As the landscape of quantum computing evolves, the demand for an extensive array of readout channels, seamlessly integrated within a single cryogenic system, is anticipated to reach around 10^3. Paving the way for future quantum technologies necessitates enhancements in power consumption efficiency (Cooper, 2001; Cosmelli et al., 2003; Eichler et al., 2011; Leppäkangas et al., 2018).

3.9 Quantum insights into radar quantization

In the preceding sections, our exploration has delved into comprehending microwaves through the lens of quantum mechanics. This perspective not only enriches our understanding of the intricacies of microwave phenomena but also lays a foundational framework for a clearer grasp of the quantization intricacies inherent in radar systems.

Radar, an acronym for RAdio Detection And Ranging, emerged as a technological advancement aimed at supplanting echolocation-based target sensing. The fundamental purpose of modern radar systems continues to be the detection and localization of targets. In this understanding, the radar system comprises three main components: the antenna, transmitter, and receiver. In this configuration, the transmitter utilizes Josephson junctions to generate photon energies, creating a radar beam directed toward the target. The transmitted signal is characterized by short bursts, rapidly repeated to form a pulsed signal. Traveling at the speed of light, approximately 300,000 km/s, these pulsed signals interact with the targets within their field of view.

The microwave imaging described here is categorized as "active" due to its distinctive characteristic of independently transmitting microwave energy in the form of pulses. This active imaging operates at a specific wavelength (a single frequency) and follows a predetermined duration, known as pulsed coherent radar. The common conceptualization of a radar system is illustrated in Fig. 3.6.

In this pulsed coherent radar system, pulses of microwave energy are emitted, and the echoes or reflections from the environment are captured for analysis. This technique forms the foundation of synthetic aperture radar (SAR) imaging, a process wherein the radar system synthesizes a larger antenna aperture by processing the reflections received from multiple locations along its path (Marghany, 2019).

To estimate the distance to the target, the collected data undergo processing to a format suitable for recording. It's important to note that the return signal is generally much weaker than the transmitted signal, as not all of the signal is backscattered. This return signal, despite its reduced intensity, is crucial for constructing the grayscale tone of the resulting radar image. Overall, this radar configuration allows for the acquisition of valuable information about the targets within the system's range.

Upon reaching the targets, the signals undergo various interactions: some energy is absorbed, some is reflected away, and some is refracted or backscattered. This complex interplay provides valuable information about the properties of the targets. The receiver, often integrated with the transmitter, captures the pulsed signals that are returned from the targets. It assesses the signal strength and establishes a connection with the original transmission.

Within the electromagnetic spectrum, radio waves represent the lowest-frequency waves. They serve as carriers for various signals, conveying information to receivers that subsequently decode these signals (Marghany, 2019). Microwave beams, a form of electromagnetic radiation, extend beyond the wavelengths of visible violet light, the thermal spectrum, and ultraviolet light.

Microwaves occupy the segment of the electromagnetic spectrum between radio and infrared, with frequencies ranging from approximately 3 GHz to 30 terahertz (THz) and wavelengths spanning from 10 mm to 100 micrometers (μm), respectively. The nomenclature "radio" originates from the early use of long wavelengths (1−10 m) falling within the radio band of the electromagnetic spectrum.

Defined by wavelengths ranging from 1 mm to 1 m and corresponding signal frequencies (f) of 300 GHz − 300 MHz ($\lambda f = c$, with the speed of light c), microwaves encompass a broad spectrum. Notably, this wavelength is significantly larger than the visible spectrum and interacts with diverse targets, unlike passive spectra. Consequently, microwave signals possess lower

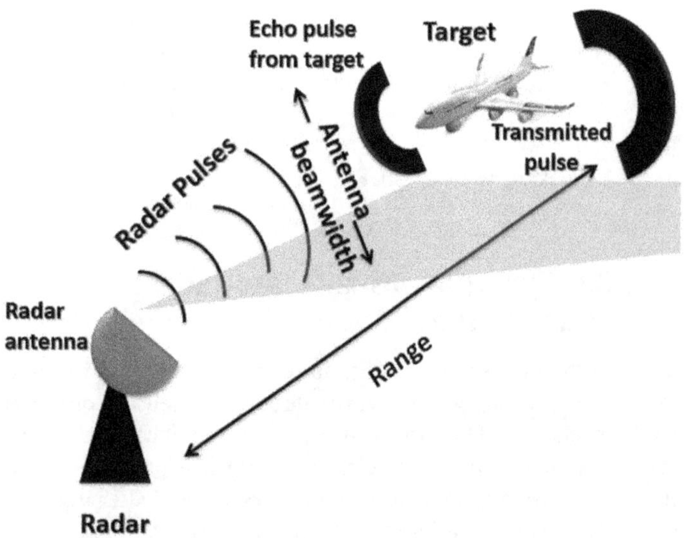

FIGURE 3.6 Concept of radar (radio detection and ranging).

energy compared with passive spectra, mitigating the induction of molecular resonance. In contrast, passive spectra in the visible and near-infrared domains reveal their signal patterns in the atmosphere and soil chemical structures due to their higher energy levels. The microwave domain, however, retains sufficient energy to model resonant spin of specific dipole molecules, aligning with the frequency changes induced by an evolving signal electric field (Ahern, 1995).

The understanding of radar and microwave characteristics involves a nuanced comprehension of their placement in the electromagnetic spectrum and their interaction with various targets, distinguishing them from passive spectra in terms of energy levels and resonant effects on molecular spin.

In the realm of microwave sensors, frequency, measured in Hertz, serves as a descriptor for a particular band range. It's noteworthy that active microwave sensors aboard satellites don't capture multispectral microwave images; instead, they gather data for a single wavelength/frequency band. However, contemporary airborne sensors have evolved to acquire data across multiple frequency bands, as exemplified by NASA's Jet Propulsion Laboratory AIRSAR system, encompassing C-, L-, and P-bands (Fig. 3.7).

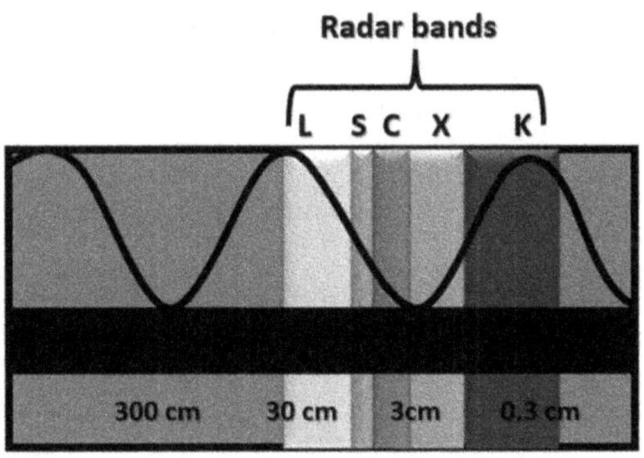

FIGURE 3.7 Microwave electromagnetic waves.

Consider a theoretical quantum radar operating within the 9 GHz range—an innovation poised to enhance object detection amidst atmospheric attenuation. In this context, the photons at the 9 GHz frequency (refer to Table 3.1) align with the radar-microwave spectrum. Expanding on this, the X-band in radar corresponds to the 8–12 GHz region and finds extensive application in missile guidance, marine radar, weather monitoring, ground surveillance, and airport traffic control (Ahern, 1995; Lacomme, 2001; Marghany, 2019).

The term "radar" is commonly employed to refer to all active microwave systems (Merrill, 1980; Skolnik, 2008). Radar is defined as a system designed to detect the presence, direction, distance, and speed of aircraft, ships, and other objects. This is achieved by emitting pulses of radio waves that reflect off the object and return to the source. Radar devices typically operate in the ultrahigh-frequency (UHF) or microwave phase of the radio frequency (RF) spectrum, serving to discern the location and/or movement of targets.

3.9.1 Qunatization of radar antenna

The development of a quantum antenna involves a logical flow that integrates quantum mechanics principles with antenna design considerations. Begin by representing the quantum state of the antenna using a wave function, Ψ, that embodies the superposition of multiple states. Mathematically, this can be expressed as $\Psi = \sum(c_i * \varphi_i)$, where c_i are probability amplitudes and φ_i are the quantum states (Fig. 3.8). In this sense, explore the concept of entanglement in the antenna system. Define the entangled state of two antennas as $|\Phi\rangle = (|0\rangle_A\ |1\rangle_B + |1\rangle_A\ |0\rangle_B)\ /\ \sqrt{2}$, where $|0\rangle$ and $|1\rangle$ represent quantum states. This

TABLE 3.1 Microwave bands and their physical characteristics.

Microwave band	Frequency range	Wavelength range	Application
C-Band	4–8 GHz	7.5 cm to 3.75 cm	Satellite communications, weather radar
X-Band	8–12 GHz	3.75 cm to 2.5 cm	Radar systems, satellite communications
Ku-Band	12–18 GHz	2.5 cm to 1.67 cm	Satellite communications, radar
K-Band	18–26.5 GHz	1.67 cm to 1.13 cm	Radar, scientific research
Ka-Band	26.5–40 GHz	1.13 cm to 0.75 cm	Satellite communications, remote sensing
Q-Band	33–50 GHz	0.9 cm to 0.6 cm	Radar, satellite communications
V-Band	40–75 GHz	0.75 cm to 0.4 cm	Millimeter-wave radar, point-to-point radio
W-Band	75–110 GHz	0.4 cm to 0.27 cm	Millimeter-wave radar, scientific research
Ka-Band	110–300 GHz	0.27 cm to 1 mm	Remote sensing, imaging, scientific research

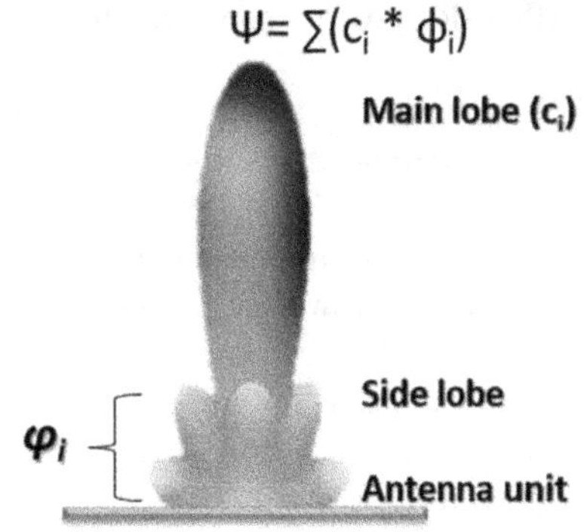

FIGURE 3.8 Simplification of quantum antenna concept.

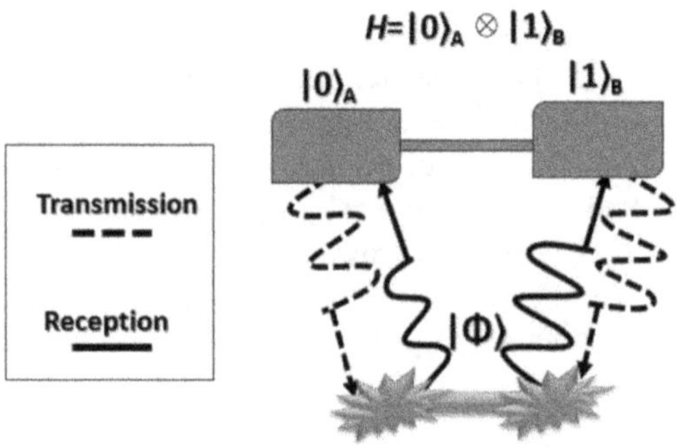

FIGURE 3.9 Entangled state of two antennas.

entanglement could potentially be exploited to create correlated behaviors between antennas for synchronized transmission and reception (Fig. 3.9). Therefore the Hilbert space H describes the two antenna systems as a product of the form $H = |0\rangle_A \otimes |1\rangle_B$.

Certainly, let's delve into the mathematics behind quantum coherent states for stability in the context of antenna systems. A quantum coherent state is a special type of quantum state that is associated with classical-like properties, such as a well-defined amplitude and phase. In the context of an antenna system, maintaining phase stability is crucial for coherent signal transmission and reception (Liberal et al., 2018; Pitkänen, 2006; Slepyan et al., 2020).

The quantum coherent state $|\alpha\rangle$ is typically represented in the Fock basis, where n represents the photon number, and α is a complex number representing the classical amplitude and phase information. The expression for a quantum coherent state is given by:

$$|\alpha\rangle = e^{-\frac{|\alpha|^2}{2}} \sum_{n=0}^{\infty} \frac{\alpha^n}{\sqrt{n!}} \tag{3.45}$$

Eq. (3.45) shows that $e^{-\frac{|\alpha|^2}{2}}$ is a normalization factor ensuring that the coherent state has a unit norm. It stabilizes the overall amplitude of the state. Therefore $\sum_{n=0}^{\infty} \frac{\alpha^n}{\sqrt{n!}}$ is a summation symbol, indicating that the state is a superposition of different photon number states. In the context of an antenna system, the coherent state $|\alpha\rangle$ represents the quantum state associated with a well-defined amplitude α and phase. The superposition over different photon number states allows for a representation that combines classical-like properties with quantum characteristics. The phase stability of this coherent state is reflected in the fact that the phase information is contained in the complex number α. As a result, the quantum coherent state can maintain a stable phase relationship in the antenna system, which is essential for coherent signal processing and transmission. This mathematical representation provides a foundation for understanding how quantum coherence can be utilized to ensure stability in the context of antenna systems.

To investigate methods for quantum noise reduction in antenna systems, we'll delve into the mathematical framework, leveraging quantum mechanics principles to model noise sources and describing the interaction between quantum and classical noise. Quantum noise arises due to the fundamental nature of quantum systems. It includes phenomena such as spontaneous emission, vacuum fluctuations, and shot noise. Let's denote the quantum noise contribution as Nq. Classical noise in antennas may stem from thermal noise, electronic components, and external interference. Represent the classical noise contribution as N_c. Therefore the total noise in the antenna system ($\sum N$) is the sum of quantum and classical noise:

$$\sum N = N_q + N_C \tag{3.46}$$

The interaction between quantum and classical noise can be modeled with an interaction term (N_{int}). This term captures how the quantum noise influences or interacts with classical noise. The total noise can be expressed as:

$$\sum N = N_q + N_C + N_{int} \tag{3.47}$$

Introduce parameters (α, β, etc.) that quantify the coupling strength between quantum and classical noise sources. These parameters govern the degree to which changes in one type of noise affect the other and express the interaction term as a function f of these parameters:

$$N_{int} = f(\alpha, \beta, \ldots) \tag{3.48}$$

Here, f is a mathematical function that encapsulates the intricate interplay between quantum and classical noise components. The parameters α, β, and so on, provide a quantitative description of how strongly the two types of noise are coupled. The specific form of f depends on the nature of the interaction. For instance:

$$f(\alpha, \beta) = \alpha \cdot N + \beta \cdot N_C \tag{3.49}$$

Eq. (3.49) could represent a linear interaction where α and β scale the influence of quantum and classical noise, respectively. However, $f(\alpha, \beta) = \sqrt{\alpha \cdot N_q^2 + \beta \cdot N_C^2}$ could represent a nonlinear interaction where the coupling strengths affect the overall noise quadratically. The values of parameters (α, β) may need to be determined through experimental calibration or theoretical analysis. These values capture the specific characteristics of the antenna system and the noise sources involved. When optimizing the system for reduced noise levels, consider the constraints imposed by N_{int}. The optimization process involves adjusting parameters to minimize total noise while keeping the interaction term within acceptable limits (Gavish et al., 2000; Xiang et al., 2010). In this view, frame the challenge as an optimization problem, seeking to minimize the total noise while considering the constraints introduced by the quantum − classical interaction. Mathematically, this can be expressed as:

$$Minimize \sum N \ subject \ to \ N_{int} \leq threshold \tag{3.50}$$

Integrate quantum information processing capabilities directly into the antenna. Utilize qubits for information manipulation within the antenna system. Develop equations that represent quantum gates and operations to perform specific tasks, such as quantum key distribution for secure communication. In this sense, let us denote the quantum antenna system as *QAS*, which incorporates qubits for information processing. Let $|\psi\rangle$ represent the quantum state of the system. Consequently, express the qubit as a superposition of its basis states:

$$|\psi\rangle = \alpha|0\rangle + \beta|1\rangle \tag{3.51}$$

Here, α and β are complex probability amplitudes, and $|0\rangle$ and $|1\rangle$ are the basis states. Consequently, implement quantum gates to manipulate qubits. For example, a Hadamard gate (*H*) can be represented as:

$$HH^T = nI_n \tag{3.52}$$

Consider *H* to be a matrix, where I_n is the $n \times n$ identity matrix, and H^T is the transpose of *H*. To validate this, observe that the rows of *H* constitute orthogonal vectors over the field of real numbers, each having a length of *n*. Normalizing *H* by dividing it by this length results in an orthogonal matrix, and its transpose becomes its inverse. Multiplying it by the length again confirms the equality stated above. Therefore

$$det(H) = \pm n^{n/2}, \tag{3.53}$$

here det(*H*) is the determinant of *H*. To confirm this, note that the rows of *H* form orthogonal vectors over the field of real numbers, each with a length of *n*. Normalizing *H* by dividing it by this length results in an orthogonal matrix, and its transpose becomes its inverse. The determinant of an orthogonal matrix is either 1 or −1. Multiplying by the length again confirms the equality stated above, where det(*H*) = 1 (Abdessaied et al., 2014; Salemian & Mohammadnejad, 2008; Shepherd, 2006).

Consequently, explore the use of quantum tunneling effects for miniaturization of antenna components. Quantum tunneling phenomena may allow for the creation of smaller and more efficient antennas, especially in the terahertz and millimeter-wave frequency ranges. In this perspective, quantum tunneling can be described using the Schrödinger equation, where the probability of a particle tunneling through a potential barrier is given by:

$$T(E) = e^{-2\kappa a} \tag{3.54}$$

where *T*(*E*) is the tunneling probability, κ is the wave number inside the barrier, and *a* is the barrier width. The miniaturization factor (*M*) of an antenna component due to quantum tunneling can be defined as the reciprocal of the tunneling probability (*T*):

$$M = [T(E)]^{-1} \tag{3.55}$$

This equation signifies that the higher the tunneling probability, the more effective the miniaturization of the antenna component. It captures the fundamental relationship between quantum tunneling effects and the reduction in size of the antenna component. The reduction in antenna size (ΔS), therefore, due to quantum tunneling can be quantified as:

$$\Delta S(f) = S_0 - a_{eff}(f) \tag{3.56}$$

where S_0 is the original size of the antenna component, and $a_{eff}(f)$ is an effective barrier width that encapsulates the quantum tunneling effects for miniaturization:

$$a_{eff}(f) = -\frac{\ln(T(E,f))}{2\kappa} \tag{3.57}$$

This equation provides insights into the relationship between tunneling probability, frequency, and the effective barrier width. Introduce, therefore, a practical implementation factor (Γ) that accommodates manufacturing constraints and real-world considerations:

$$\Delta S_{practical}(f) = \Gamma \cdot \Delta S(f) \tag{3.58}$$

here $\Delta_{Spractical}(f)$ represents the practically realized reduction in antenna size; $\Delta S(f)$ is the theoretical reduction in antenna size due to quantum tunneling; and Γ is the practical implementation factor, adjusting the theoretical reduction to account for real-world limitations and manufacturing challenges.

Finally, the quantum error correction mechanism involves encoding quantum information in a manner that effectively mitigates the impact of quantum noise on the antenna system. If quantum error correction is implemented, let's denote the encoded quantum state as $|\psi_{\text{encoded}}\rangle$. The correction process can be represented as:

$$|\psi_{corrected}\rangle = C(|\psi_{encoded}\rangle) \tag{3.59}$$

The operation C ensures the correction of errors in the encoded quantum state, enhancing the reliability and robustness of quantum information processing within the antenna system. Solving Eq. (3.59) involves understanding the nature of the quantum error correction operation (C). The specific form of C depends on the error correction code employed. Here, it will provide a generic representation.

Let's consider a simple scenario where a single qubit is encoded using a bit-flip code. In this case, the quantum error correction operation (C) can be represented as follows:

$$C(|\psi_{corrected}\rangle) = X|\psi_{encoded}\rangle \tag{3.60}$$

here X is the Pauli-X gate, representing a bit-flip operation. In this context, the Pauli-X gate performs a bit-flip operation on the encoded state $|\psi_{\text{encoded}}\rangle$. The corrected state $|\psi_{\text{corrected}}\rangle$ is obtained by flipping the state of the qubit. This representation provides a clear and concise form of the quantum error correction process in the case where a bit-flip code is applied, and the correction operation is realized using the Pauli-X gate (Cosmelli et al., 2003; Flurin, 2014; Liberal et al., 2018; Slepyan et al., 2020).

Needless to say, the amalgamation of quantum information processing with antenna technology necessitates the representation of qubits, the meticulous implementation of quantum gates and operations, and the formulation of mathematical equations tailored for precise endeavors such as quantum key distribution. These mathematically refined formulations lay the foundation for the pragmatic realization of quantum-enhanced secure communication systems, marking a significant stride toward the advancement of quantum technologies in the realm of communication (Abdessaied et al., 2014; Casariego et al., 2023).

3.10 Advancements in quantum radar: illumination strategies and radar equation formulation

Quantum radar is an emerging field that explores the application of quantum principles to radar systems. One of the key concepts in quantum radar is quantum illumination, which leverages entangled photon pairs to enhance radar performance. The development of a radar equation for quantum radar involves considering the quantum properties of the transmitted and received signals. Therefore quantum radar illumination relies on the generation of entangled photon pairs. This process often involves spontaneous parametric down-conversion, where a high-energy photon is split into two entangled photons known as signal and idler. In this perspective, quantum entanglement is a unique quantum property where the state of one entangled particle instantaneously influences the state of the other, regardless of the distance between them. In the context of quantum radar, entangled photons exhibit correlated properties (Barzanjeh et al., 2015; Sanz et al., 2017; Shapiro, 2020). Consequently, utilize the entangled photon pair creation process, often through spontaneous parametric down-conversion, to generate quantum illumination. Represent this with the entanglement parameter (ξ) and the quantum efficiency of the entangled photon source (η_q), which is given by:

$$I_q = \xi \cdot \eta_q \tag{3.61}$$

Eq. (3.61) signifies a multiplicative relationship between the quantum entanglement parameter (ξ) and the quantum efficiency (η_q). The product of these two factors ($\xi \cdot \eta q$) quantifies the overall effectiveness of quantum illumination. In this sense, a larger value of $\xi \cdot \eta q$ indicates a more efficient use of entangled photon pairs for radar illumination. This implies that the quantum radar system is maximizing the benefits of quantum entanglement and efficient photon generation, leading to improved radar performance, enhanced signal resolution, and increased sensitivity.

To correlate the radar equation with quantum illumination, we can introduce entanglement parameters into the radar equation, considering the entangled photon pairs used in quantum radar. Let's denote the entanglement parameter as ξ and the quantum efficiency of the entangled photon source as ηq. In this view, the traditional radar equation relates transmitted power, antenna gains, wavelength, radar cross section, and distance. It can be expressed as:

$$P_r = \frac{P_t G_t G_r \lambda^2 \sigma A_r}{(4\pi)^3 R^4} \eta_t \eta_r \eta_a \eta_p \tag{3.62}$$

where P_r is the received power, P_t is the transmitted power, G_t (Fig. 3.10) and G_r are the gains of the transmitting and receiving antennas, respectively. Additionally, λ is the wavelength, σ is the radar cross section of the target, R is the distance to the target. Moreover, η_t and η_r are the efficiencies of the transmitting and receiving systems, η_a is the atmospheric loss factor, and η_p is the polarization mismatch factor (Skolnik, 2008; Tan et al., 2008; Xiang et al., 2010; Yao et al., 2017).

The energy encapsulated in the received antenna signal (Fig. 3.11) manifests as a function of the backscatter and is situated at the identical transmit antenna. This configuration characterizes monostatic radar. In this context, the term $4\pi R^2$ in the preceding equation transforms into

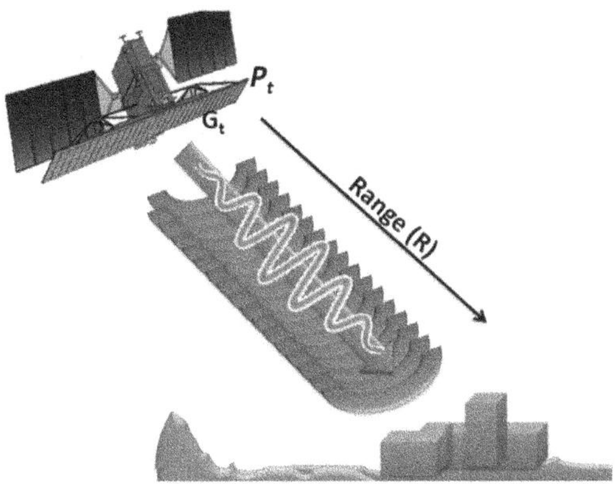

FIGURE 3.10 Transmitted pulse from a radar antenna.

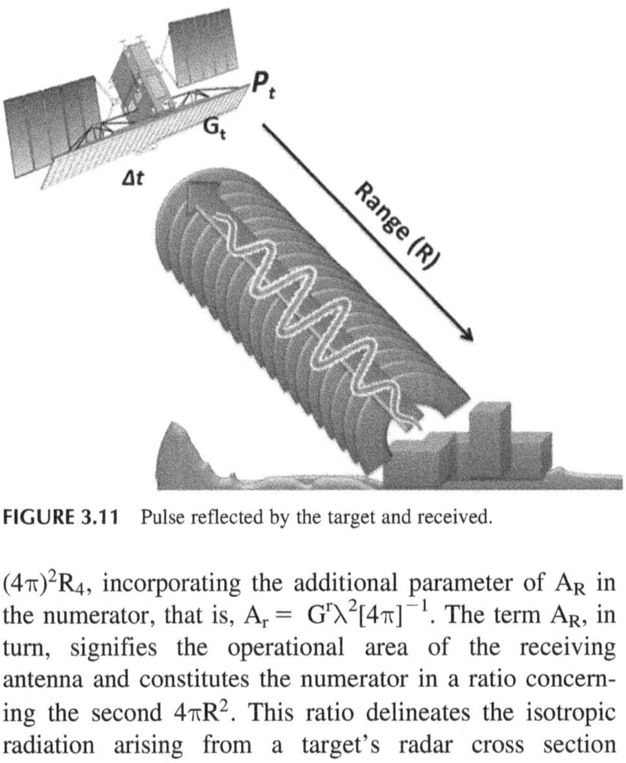

FIGURE 3.11 Pulse reflected by the target and received.

$(4\pi)^2R_4$, incorporating the additional parameter of A_R in the numerator, that is, $A_r = G^r\lambda^2[4\pi]^{-1}$. The term A_R, in turn, signifies the operational area of the receiving antenna and constitutes the numerator in a ratio concerning the second $4\pi R^2$. This ratio delineates the isotropic radiation arising from a target's radar cross section (Ahern, 1995; Barzanjeh et al., 2015; Dubosclard et al., 2021).

Enhance the traditional radar equation, therefore, with quantum terms to incorporate the advantages gained from quantum illumination. Introduce a quantum-enhanced received power (P_{rq}) considering the quantum efficiencies and quantum enhancements.

$$P_{rq} = \frac{P_tG_tG_r\lambda^2\xi\eta_q\sigma}{(4\pi)^3R^4}\eta_t\eta_r\eta_a\eta_p \tag{3.63}$$

Expressing the correlation between entanglement parameters (ξ) and radar parameters (radar parametersradar parameters) involves defining a function f that relates the entanglement parameter to the characteristics of the radar system. The exact form of this function depends on the specific details of the quantum radar setup and the entangled states used. A generalized expression can be written as:$\xi = f(radarparameters)$. In this equation: ξ represents the entanglement parameter; f is a function that captures the correlation between ξ and the various radar parameters. Therefore radar parametersradar parameters collectively refer to the relevant characteristics of the radar system, which may include parameters such as the transmitted power (P_t), antenna gains (G_t and G_r), wavelength (λ), radar cross section (σ), and others (Abdessaied et al., 2014; Barzanjeh et al., 2015; Marghany, 2019).

The specific form of the function f would need to be derived based on the details of the quantum radar system and the entangled states employed. The relationship could involve mathematical expressions that describe how the entanglement parameter is influenced by the operational parameters of the radar system and the quantum properties of the entangled states.

For instance, if the entanglement parameter is affected by the quantum efficiency of the entangled photon source (η_q), the correlation might be expressed as:

$$\xi = f\left(P_t, G_t, G_r, \lambda, \sigma, \eta_q\right) \tag{3.64}$$

Eq. (3.64) can be solved using Newton − Raphson method to the equation as follows:

$$\xi_{n+1} = \xi_n - \frac{f(\xi_n)}{f'(\xi_n)} \tag{3.65}$$

Assuming you have an expression for the function $f(\xi)$, and its derivative $f'(\xi)$, the general structure of the algorithm would be as follows (Table 3.2):

In this view, developing a quantum SNR equation that considers the quantum enhancements in the received power, incorporating factors such as the Boltzmann constant (k_B), system noise temperature (T_0), and radar system bandwidth (B), which is given by:

$$SNR_q = P_{rq}[k_BT_0]^{-1} \tag{3.66}$$

The quantum signal-to-noise ratio is a measure of the signal strength to the noise level in a quantum-enhanced system. It is calculated as the ratio of the quantum-enhanced received power (P_{rq}) to the noise level. According to the above perspective, the quantum-enhanced received power considers the effects of quantum illumination. It is influenced by factors such as entanglement parameter (ξ) and quantum efficiency of the entangled photon source (η_q), as outlined in the quantum-enhanced radar equation. The SNR_q provides insights into the ratio of the quantum-enhanced signal power to the noise level in the system. A higher SNR_q indicates that the quantum-enhanced signal is more pronounced relative

TABLE 3.2 Algorithm for solving Eq. (3.54) using Eq. (3.65).

```
# Define the function f(xi) def f(xi, Pt, Gt, Gr, wavelength, sigma, eta_q): # Your implementation of the
function f #.
# Define the derivative of f(xi) def df(xi, Pt, Gt, Gr, wavelength, sigma, eta_q): # Your implementation
of the derivative f'(xi) #.
# Newton-Raphson method implementation def newton_raphson(xi_initial, Pt, Gt, Gr, wavelength,
sigma, eta_q, tolerance=1e-6, max_iterations=100): xi_n = xi_initial
for iteration in range(max_iterations): f_value = f(xi_n, Pt, Gt, Gr, wavelength, sigma, eta_q) df_value
= df(xi_n, Pt, Gt, Gr, wavelength, sigma, eta_q)
xi_n1 = xi_n - f_value / df_value
# Check for convergence if abs(xi_n1 - xi_n) < tolerance: return xi_n1
xi_n = xi_n1
# Return the result after the maximum number of iterations return xi_n
# Example usage Pt = 1.0 # Replace with your values Gt = 1.0 Gr = 1.0 wavelength = 1.0 sigma = 1.0
eta_q = 0.9
initial_guess = 0.1 # Choose an appropriate initial guess result = newton_raphson(initial_guess, Pt, Gt,
Gr, wavelength, sigma, eta_q)
print("Numerical solution for xi:", result)
```

to the noise, leading to improved detection and measurement capabilities (Ahern, 1995; Cosmelli et al., 2003; Marghany, 2021).

In the context of quantum radar, the SNRq is a critical metric for evaluating the system's performance when quantum principles, such as entanglement, are employed. It helps quantify the advantages gained through quantum enhancements in the radar signal.

Suppose we have a quantum radar system with the following parameters:

1. Quantum-enhanced received power (P_{rq}): 2×10^{-15} Watts
2. Boltzmann constant (k_B): 1.38×10^{-23} J/K
3. System noise temperature (T_0): 300 K
4. Radar system bandwidth (B): 106 Hz

Now, we can use the formula 3.64 to calculate the Quantum SNR:

$$SNR_q = \frac{2 \times 10^{-15}}{(1.38 \times 10^{-23}) \times 300 \times 10^6} = 48.3$$

Therefore in this example, the quantum SNR$_q$ is approximately 48.3. This means that the quantum-enhanced signal strength is 48.3 times greater than the noise level in the system. A higher SNR indicates a more favorable signal-to-noise ratio, suggesting improved detection capabilities in the quantum radar system.

3.11 Synergizing quantum illumination and radar equations through entanglement dynamics

Developing entangled states between the radar equation and quantum illumination involves exploring the quantum aspects of both systems and finding a representation that incorporates their interactions. Quantum entanglement is a phenomenon where quantum states of two or more objects become correlated and intertwined, such that the state of one object cannot be independently described without reference to the state of the other(s). Incorporate quantum illumination principles into the system as it is involves using entangled states to enhance the detection of a target in the presence of noise. Let us express the joint state of the quantum radar system and quantum illumination using a tensor product of their individual states. Let $|\Psi_{\text{total}}\rangle$ be the entangled state, which is expressed by:

$$|\Psi_{total}\rangle = |\Psi_{radar}\rangle \otimes |\Psi_{illumination}\rangle \tag{3.67}$$

In this understanding, let us represent the quantum radar state, considering the quantum properties of the radar parameters. This could involve the superposition of states corresponding to different radar parameter values.

$$|\Psi_{radar}\rangle = \sum_i c_i |P_t^i, G_t^i, G_r^i, \lambda^i, \sigma^i, \eta_q^i\rangle \tag{3.68}$$

Consequently, let us represent the quantum illumination state, considering the entangled photon source and its quantum properties as:

$$|\Psi_{illumination}\rangle = \sum_i d_j |q_1^j, q_2^j, \ldots, q_n^j\rangle \tag{3.69}$$

where dj represents the probability amplitude of each entangled photon state. Consequently, the dynamic transformation of the quantum radar state when influenced by quantum illumination, and conversely, elucidating the reciprocal impact on quantum illumination, can be given by:

$$|\Psi_{total}(t)\rangle = U_{total}(t)|\Psi_{total}(0)\rangle \tag{3.70}$$

This equation intricate process necessitates the definition and explanation of the unitary evolution operators governing these transformations. This conceptual representation involves creating a joint entangled state that encompasses both the quantum radar system and the quantum illumination. In other words, the equation as a whole signifies that the quantum state $|\Psi\text{total}(t)\rangle$ at any time t is obtained by applying the unitary evolution operator $U_{\text{total}}(t)$ to the initial state $|\Psi\text{total}(0)\rangle$ (Barzanjeh et al., 2015; Gavish et al., 2000; Salemian & Mohammadnejad, 2008; Shapiro, 2020).

3.12 Exploring quantum principles in synthetic aperture radar systems

Synthetic aperture radar (SAR) represents a revolutionary radar technology employed for generating two-dimensional images or three-dimensional reconstructions of various objects, including landscapes. SAR ingeniously utilizes the dynamic motion of the radar antenna across a target area, surpassing the spatial resolution capabilities of traditional stationary beam-scanning radars. Typically installed on mobile platforms such as aircraft or spacecraft, SAR traces its origins to an evolved version of side-looking airborne radar (SLAR).

The substantial synthetic antenna aperture, defining the resolution of the images produced, is crafted by the distance covered by the SAR device over the target during the illumination period. Whether the aperture is physically extensive (large antenna) or synthetically generated (moving antenna), SAR excels in producing high-resolution images even with relatively compact physical antennas. Notably, for a fixed antenna size and orientation, SAR maintains a consistent spatial resolution across varying viewing distances. This unique property arises from the prolonged illumination of objects further away, contributing to the creation of larger synthetic apertures.

In the SAR imaging process, radio wave pulses are successively transmitted to "illuminate" a target scene, and the resulting echoes are diligently received and recorded (Fig. 3.12). This transmission and reception occur through a single beam-forming antenna, employing wavelengths ranging from meters down to several millimeters. The SAR apparatus, situated on an airborne or spaceborne platform, undergoes movement, altering the antenna's position relative to the target over time. Advanced signal processing of the recorded radar echoes enables the amalgamation of data from multiple antenna positions, effectively forming the synthetic antenna aperture. This intricate procedure unlocks the capability to generate images with markedly higher resolution than achievable with a fixed physical antenna.

In broad terms, a quantum synthetic aperture radar (SAR) can be characterized as a cutting-edge detection sensor leveraging microwave photons and incorporating quantum phenomena formulas to enhance its capabilities in discerning, identifying, and characterizing intriguing objects. For the scope of this chapter, quantum SAR is defined as the exploration of quantum phenomena essential for comprehending the functionality of the SAR mechanism (Lanzagorta, 2011; Marghany, 2019, 2021; Singh et al., 2022).

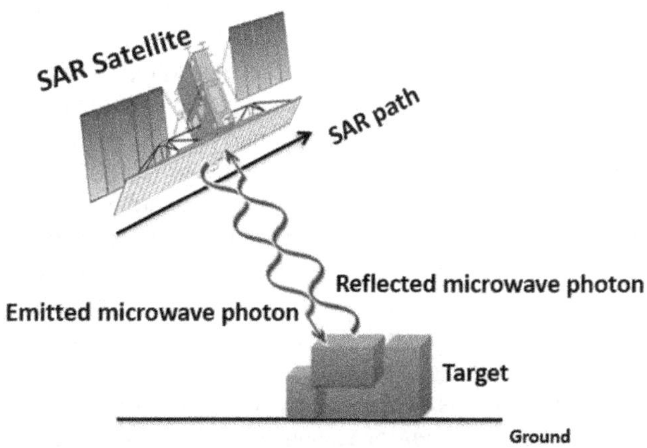

FIGURE 3.12 SAR operation for target imagine.

In a quantum radar system, it is anticipated that targets will exhibit distinctive backscatter characteristics compared with classical radar, particularly when they are at a considerable distance from the sensor. Additionally, the SAR-target system is presumed to operate within a noisy and lossy environment. Specifically, the SAR signal may undergo attenuation due to absorption or scattering phenomena, and the overall performance of the SAR system is influenced by the presence of noise.

The exemplary instances of quantum synthetic aperture radar (SAR) sensors encompass single-photon quantum SAR and entangled photon quantum SAR. These quantum SAR variants can be categorized as follows:

1. Quantum SAR that conveys unentangled quantum states (momentum, angular momentum, energy, spin, and time) of microwave photons.
2. Quantum SAR where the antenna emits conventional statuses (energy, wavelength, and frequency) of microwave photons but strategically employs quantum photo-SAR to augment its functionality.
3. Quantum SAR that emits quantum signal states of microwave photons entwined with quantum states of a microwave photon generated at the SAR transmitter.

The fundamental principle behind a single-photon quantum radar involves emitting a solitary photon toward a target, with the photon subsequently being scattered back to the receiver (refer to Fig. 3.12). In contrast, entangled-photon quantum radar generates an entangled pair of microwave photons. In this scenario, one photon is directed toward the target, while the other is retained in the radar sensor. The photon directed outward is then reflected by the object and subsequently received by the radar system. In this framework, the connections between transmitted and received radar signals occur in the entangled condition, a feature harnessed to enhance target detection in the quantum SAR system (Fig. 3.13).

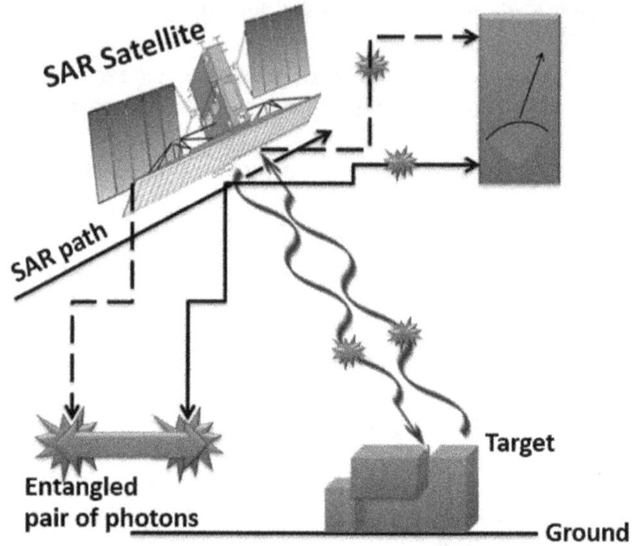

FIGURE 3.13 Quantum radar with entangled photons.

An example of a quantum radar mode is LADAR, short for LAser Detection And Ranging, akin to LiDAR (Light Detection And Ranging) technology (illustrated in Fig. 3.14). LADAR operates in the visible and near-visible photon spectra. However, it does not fall within the classification of quantum radar technology. Notably, LADAR lacks the ability to penetrate fog or clouds as radars do, and its operational range is often limited to approximately 100 km due to inherent constraints.

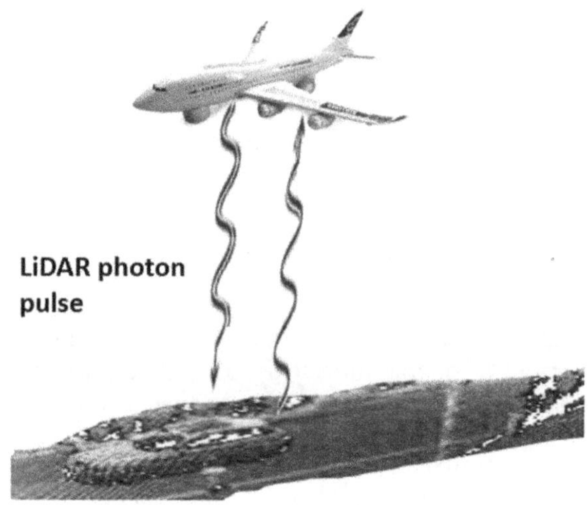

FIGURE 3.14 Pulsed photons in LiDAR technology.

In delving into the incorporation of quantum mechanisms into SAR imaging, an examination of the transmitter and receiver behaviors through the lens of quantum mechanics becomes paramount. In this context, the generation of entangled photons emerges as a pivotal facet within the SAR transmitter.

To realize this, a nonlinear crystal assumes a crucial role in the generation of entangled photons. Specifically, it orchestrates the division of an incoming photon into two entangled counterparts. Expanding upon this perspective, semiconductor nanostructures come into play for the creation of entangled photons. Notably, the degradation of bi-excitonic scenarios within quantum dots involves interband transitions between valence and conduction bands. In such scenarios, the generation of entangled photons is achievable, with frequencies falling within the microwave spectra. Consequently, microwave photons materialize through abrupt downward conversions between single-particle levels in a quantum dot.

To elucidate this concept, let us consider a scenario involving four quantum dots (as depicted in Fig. 3.15), denoted as d_1^Q, d_2^Q, d_3^Q, *and* d_4^Q. Within this framework, d_2^Q *and* d_3^Q play a crucial role in providing distinctive initial and final states for an electron situated in the

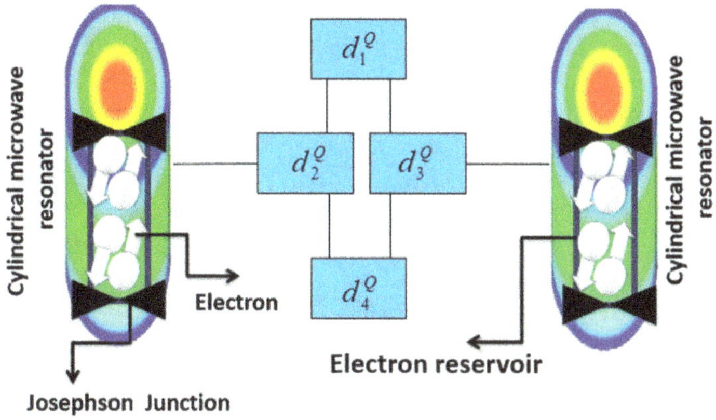

FIGURE 3.15 Arrangement of four quantum dots.

conduction band. Conversely, $d_1^Q and d_4^Q$ present dual decay paths. In this scenario, a symmetric quantum state materializes, and its formulation is expressed as:

$$\Psi_j = \frac{1}{\sqrt{2}}\left(\phi|d_1^Q\rangle\sqrt{E} + \phi|d_4^Q\rangle\sqrt{E}\right) \tag{3.71}$$

Where E signifies the energy associated with the excited states. This quantum state reflects the electron's behavior within the specified quantum dots and encapsulates the essence of the system's quantum dynamics. In a more specific depiction, the electron exists in a superposition of the excited states of quantum dots d_1^Q *or* d_4^Q, with an equal probability for either. Subsequently, the electron undergoes a decay process into the ground state of either d_1^Q *or* d_4^Q, releasing two photons in the course. If the electron decays from d_4^Q, both photons exhibit a spiral circular polarization state. Conversely, if the decay originates from d_4^Q, both photons demonstrate a clockwise circular polarization state. The quantum state for this phase can be articulated as:

$$\Psi_j = \frac{1}{\sqrt{2}}\left(\phi|d_1^Q\rangle\sqrt{E}e^{-i} + \phi|d_4^Q\rangle\sqrt{E}e^{+i}\right) \tag{3.72}$$

In Eq. (3.72), e^{+i} and e^{-i} are complex exponential terms, where e^{+i}signifies a counterclockwise rotation or spiral circular polarization state, and e^{-i} indicates a clockwise rotation or circular polarization state. The equation is a superposition state, reflecting the electron's simultaneous existence in two possible decay pathways: one from d_1^Q and the other from d_4^Q. The coefficient $\frac{1}{\sqrt{2}}$ ensures that the total probability is conserved, and the terms e^{+i} and e^{-i} account for the specific polarization states of the emitted photons based on the decay path taken by the electron. In essence, this quantum state equation encapsulates the probabilistic nature of the electron's decay and the resulting polarization states of the emitted photons in the system. Therefore through the tansimitter, a quantum state in the context of the generation of quantum photons and electrons can be presented by:

$$\Psi_j = \frac{1}{\sqrt{2}}(\psi_+|j\rangle\Phi_+ + \psi_-|j\rangle\Phi_-), \tag{3.73}$$

here ψ_+ and ψ_- denote probability amplitudes associated with specific transimited antenna states. They are coefficients that describe the likelihood of the antenna system being in either the $\Phi+$ or $\Phi-$ state. Therefore $|j\rangle$ represents the quantum state of the electrons, with j being a parameter that can take different intensity values. On the other hand, $\Phi+$ and $\Phi-$ are states associated with quantum photons, with $\Phi+$ and $\Phi-$ representing distinct quantum states of the photons. The expression signifies a superposition state, indicating that the radiated or transmitted photon exists in a linear combination of two possible outcomes $(\Phi+)$, each with a certain probability amplitude $(\psi+)$. The $\frac{1}{\sqrt{2}}$ factor ensures that the total probability is conserved. In essence, this quantum state equation describes the joint state of electrons and photons during the generation process, incorporating the probabilities of different electron and photon states (Cooper, 2001; Lanzagorta, 2011; Marghany, 2021).

On the other hand, the receiver incorporates a novel metamaterial designed to absorb reflected microwave photons and facilitate single-shot photodetection. Metamaterials are engineered materials with specific physical properties (Yao et al., 2017). For instance, the metamaterial lens, positioned within metamaterial antenna systems, serves as an effective coupler to the external environment, emitting microwave photons. This lens can focus radiation either from its own source or from a microstrip transmission line into transmitting and receiving components, making it a versatile component (Lacomme, 2001; Lanzagorta, 2011; Marghany, 2019; Smith, 2009).

On the other hand, the receiver is configured to capture a microwave photon from its initial state $|0\rangle$ and transition it to a new state $|1\rangle$ (refer to Fig. 3.15). In this context, a small operational bandwidth can be correlated with the attenuation rate Γ as follows:

$$\Delta\omega < \Gamma^{-1} \tag{3.74}$$

The expression represents a condition involving the operational bandwidth ($\Delta\omega$) and the attenuation rate (Γ). This inequality implies that the operational bandwidth is less than the inverse of the attenuation rate. This condition suggests a constraint on the relationship between the bandwidth and the attenuation rate in the system, indicating that the bandwidth should be limited to ensure that the given inequality holds. In this context, the evolution of the system leads to a stable ground state $|g\rangle$ with a discernible attenuation rate Γ. This state is intricately connected to the initial state through the phenomenon of quantum entanglement, as illustrated in Fig. 3.16. The progression involves the growth of a microwave photon, identified as a quantum state, followed by a decay process considering the attenuating influence leading to the ground quantum state, the lowest-energy state (as depicted in Fig. 3.16).

This progression transforms the sensor into an analogous structure to photographic film, where the absorption of a photon by the metamaterial results in the detector assuming a sustained and mesoscopically distinct state. Mathematically, this can be expressed as:

$$\frac{d}{dt}|g\rangle = -\Gamma|g\rangle \tag{3.75}$$

Where $|g\rangle$ is the ground state, and Γ is the attenuation rate. This equation signifies the time evolution of the ground state, capturing the decay process with an attenuation rate Γ. The negative sign indicates the decay of the state over time.

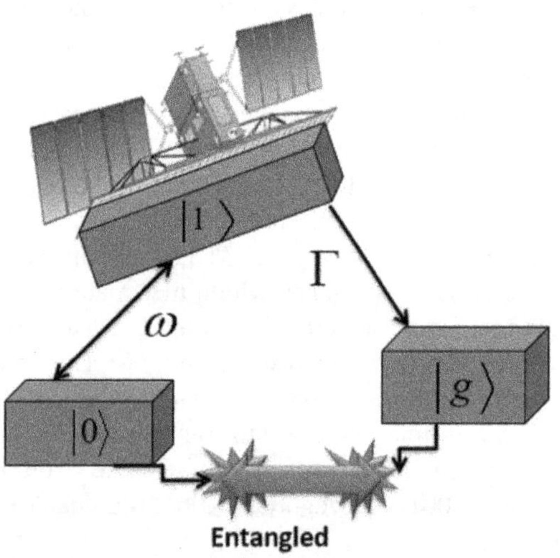

FIGURE 3.16 Quantum mechanics' insight into radar transmitter and receiver concepts.

The radar cross section (RCS) is a measure of how detectable an object is by radar, and it is often influenced by backscatter, which describes the reflection of signals back toward the radar system. To develop a radar cross section backscatter equation based on the previously described mechanism, we can consider the interaction between the incident radar wave and the object.

Let σ represent the radar cross section. The backscatter ($\sigma_{\text{backscatter}}$) can be related to the ground state ($|g\rangle$) and the attenuation rate (Γ):

$$\sigma_{backscatter} = \frac{\Gamma^2}{\Delta\omega} \tag{3.76}$$

Where $\Delta\omega$ is the operational bandwidth. This equation suggests that the backscatter radar cross section is influenced by the square of the attenuation rate divided by the operational bandwidth. The specific form of this equation may be further refined based on the details of the system and the radar wave interaction (Abdessaied et al., 2014; Barzanjeh et al., 2015; Dubosclard et al., 2021; Marghany, 2021).

3.13 Quantum mechanics of radar cross section backscatter from Bragg wave turbulence surface

The conventional understanding of Bragg scattering precisely acknowledges that the reflection and backscatter from the ocean surface depend on the wavelength alignment of surface waves with the SAR beam wavelength (λ). The incident SAR photon beam undergoes scattering from capillary waves and reflection from long waves. Typically, wind generates a continuous spectrum of capillary waves, leading to the formation of resonant waves. In this context, the mean-square wave slope and surface roughness increase with wind speed, influencing Bragg scattering. In simpler terms, Bragg scattering also occurs from capillary waves present on the long swell.

In this perspective, the wave surfaces can be conceptualized as a collection of similar isosceles triangles, referred to as facets (see Fig. 3.17), acting as specular reflectors. This conceptualization is valid if the length of each facet is significantly greater than λ and the deviation of the approaching facet from the wave surface is much less than λ. This condition holds true when the curvature radius (R_c) of the sea surface part satisfies:

$$R_c \gg \lambda \tag{3.77}$$

In the realm of quantum mechanics, the description of radar cross section backscatter from a Bragg wave turbulence surface involves intricate mathematical formulations. Let's denote the radar cross section as σ and the

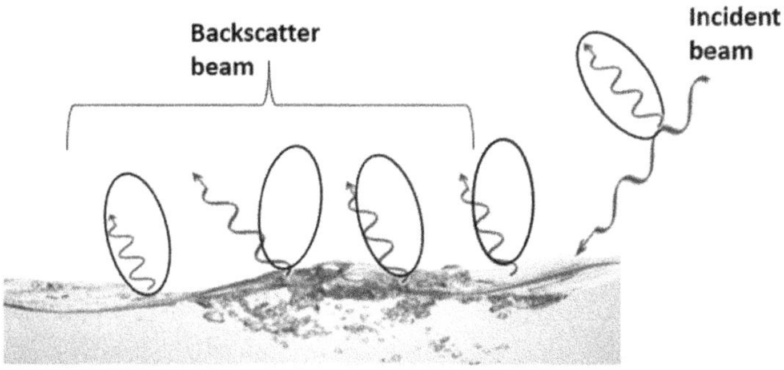

FIGURE 3.17 Combinations of facets in backscatter beam.

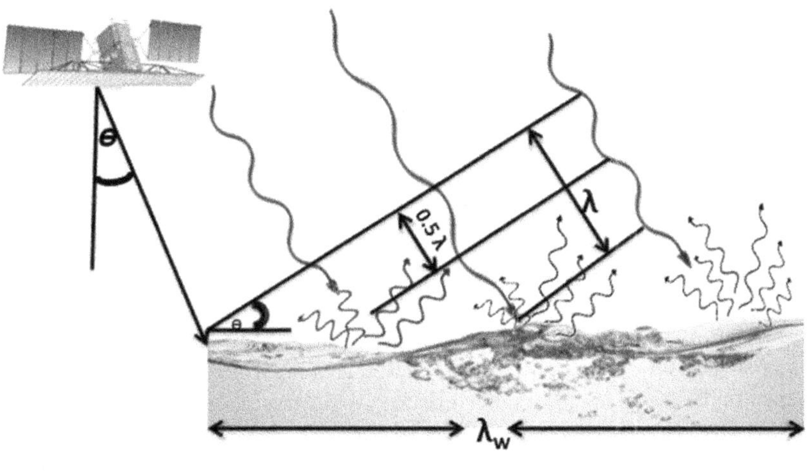

FIGURE 3.18 Understanding Bragg scattering.

Bragg wavelength as λ_{Bragg}. In general, Bragg scattering manifests as significant backscatter at large angles. Strong resonance in backscatter is observed under specific conditions of incidence angles and frequencies (Ahern, 1995; Lacomme, 2001; Marghany, 2019).

In the context of the ocean, Bragg resonance occurs when the surface wave spectrum includes a wavelength component that aligns with the incident beam. This alignment is achieved when there is a surface component with a wavelength (λ_w) equal to half the surface projection of the radar wavelength (λ). Mathematically, Bragg resonance occurs when: $\lambda_w = 0.5\lambda/\sin\theta$ (Fig. 3.18).

Therefore the quantum-mechanical model can be expressed by the following equation:

$$\sigma = \frac{\hbar k}{mv} \tag{3.78}$$

Where, $\hbar$ is the reduced Planck constant, k represents the wave number associated with the Bragg wavelength, m is the mass of the particles contributing to the turbulence, and v denotes their velocity. The scattering amplitude ($F(S)$), therefore, in the quantum framework is given by:

$$F(S) = \int \psi^*\left(\vec{r}\right)V\left(\vec{r}\right)\psi(\rightarrow r)d\vec{r} \tag{3.79}$$

where $\psi(\mathbf{r})$ is the wave function of the scattering particles, and $V(\mathbf{r})$ represents the quantum potential. In this view, the backscattering coefficient (B) can be expressed as:

$$B = \frac{\sigma}{4\pi R^2} \tag{3.80}$$

where R is the distance between the radar and the scattering surface. In the context of Bragg scattering theory and its relation to the sea surface, let's introduce a novel formula expressing the entanglement of radar backscatter cross section (σ) and Bragg turbulence. Building upon the principles of quantum mechanics, we denote the entanglement as $E(\rho)$:

$$E(\rho) = \frac{\hbar^2}{m^2v^2}\int \psi^*\left(\vec{r}\right)V\left(\vec{r}\right)\psi\left(\vec{r}\right)d\vec{r}\cdot\frac{E_k}{\lambda_w} \tag{3.81}$$

Where E_k (Chapter 2, equation 6.4) represents the turbulence Kolmogorov spectra as a function of wave number k, *and* k is the wave number associated with the turbulence. In other words, in the quantum entanglement between radar cross section (σ_{NRCS}) and Kolmogorov spectra of turbulence (E_k), we can express the entangled quantum state as:

$$|E(\rho)\rangle = \int_0^{\infty} P(k)|\sigma\rangle \otimes |E_k\rangle\rangle dk \tag{3.82}$$

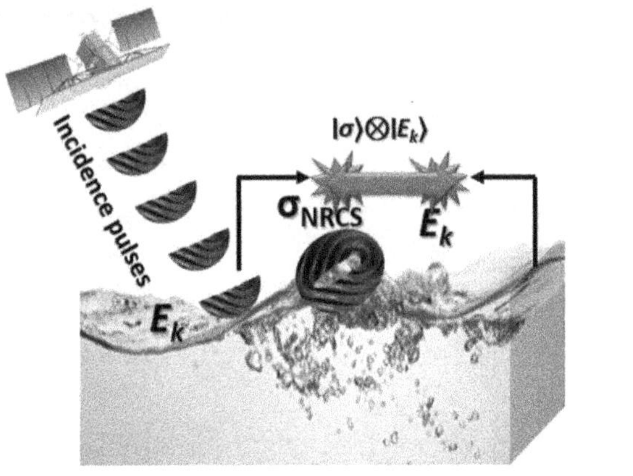

FIGURE 3.19 Entanglement between the quantum states of radar cross section and Kolmogorov turbulence spectra.

Here, $|E(\rho)\rangle$ represents the entangled quantum state, $|\sigma\rangle$ is the quantum state of radar cross section, $|E_k\rangle$ is the quantum state of Kolmogorov spectra, and $P(k)$ is the probability distribution function of the wave number (Fig. 3.19). This expression signifies the entanglement between the quantum states of radar cross section and turbulence at different wave numbers.

To determine the normalized radar cross section (σ_{NRCS}) from the turbulence Kolmogorov spectra E_k as follows:

$$|\sigma_{NRCS}\rangle = \alpha \int_0^\infty |E_k \cdot k^n\rangle \cdot |E(\rho)\rangle dk \tag{3.83}$$

here a is constant that scales the overall magnitude of the contribution. In Eq. (3.83), the wave number k^n encapsulates the wave number's power, reflecting the nuanced impact of turbulence characteristics on the radar cross section. In this view, the entanglement quantum state $|E(\rho)>$ represents the degree of correlation or connection between radar backscatter cross section and Bragg turbulence in the context of quantum mechanics.

In essence, the equation shows that the normalized radar cross section is influenced by the turbulence Kolmogorov spectra (E_k). This formulation reflects a quantum-mechanical perspective on radar cross section calculations, considering the wave-particle duality and the influence of quantum parameters on radar interactions with turbulent environments. to incorporate the effect of the incidence angle (θ) into the equation for the normalized radar cross section (σ_{NRCS}), one can modify the equation as follows:

$$|\sigma_{NRCS}\rangle = \alpha \int_0^\infty |E_k \cdot k^n\rangle \cdot |E(\rho)\rangle |g_{ij}(\theta)|^2 \cdot \psi(0, 2k_0 \sin\theta) dk. \tag{3.84}$$

This modification introduces the incidence angle (θ) dependency through the term $|g_{ij}(\theta)|^2$ (Plant, 1991), which represents the square of the radar modulation transfer function for the given incidence angle. Let's integrate the entanglement effect ($|E(\rho)>$) into the provided equation for the normalized radar cross section (σ_{NRCS}), incorporating the given expressions for $g_{HH}(\theta)$ and $g_{VV}(\theta)$:

For horizontal polarization ($g_{HH}(\theta)$) and vertical polarization and ($g_{HH}(\theta)$); respectively:

$$g_{HH}(\theta) = \frac{(\varepsilon_r - 1)\cos^2\theta}{\left[\cos\theta + \left(\varepsilon_r - \sin^2\theta\right)^{\frac{1}{2}}\right]^2} \tag{3.85}$$

$$g_{VV}(\theta) = \frac{(\varepsilon_r - 1)\varepsilon_r\left(1 + \sin^2\theta\right)\cos^2\theta}{\left[\varepsilon_r\cos\theta + \left(\varepsilon_r - \sin^2\theta\right)^{1/2}\right]^2} \tag{3.86}$$

In the given context, ε_r represents the relative permittivity or dielectric constant of the medium. The dielectric constant is a measure of a material's ability to store electrical energy in an electric field. It is a dimensionless quantity, and its value is specific to the material through which the electromagnetic wave is propagating. For the expressions $g_{HH}(\theta)$ and $g_{VV}(\theta)$, ε_r appears in the formulas, influencing the scattering coefficients for horizontal (HH) and vertical (VV) polarizations, respectively. The dielectric constant is a material property, and its value depends on the characteristics of the medium through which the radar wave is interacting. In radar and electromagnetic wave contexts, dielectric constant values are important because they affect how electromagnetic waves interact with and propagate through different materials (Abdessaied et al., 2014; Ahern, 1995; Barzanjeh et al., 2015; Lanzagorta, 2011; Marghany, 2019, 2021).

3.14 Quantum pulse-compression ranging: unveiling the secrets of quantum radar signal processing

In the realm of classical microwave radar, the determination of a distant target's range involves measuring the time of flight, evaluating the round-trip range delay τ. The resolution of range delay, denoted as τres, representing the minimum separation required for reliably distinguishing two targets based on their range difference, is inversely proportional to the radar pulse's bandwidth Δω. The ultimate accuracy of range-delay measurement for a single target, characterized by the minimum root-mean-squared (rms) estimation error $\delta\tau_{min}$ as dictated by the Cramér − Rao bound (CRB), diminishes with increasing signal-to-noise ratio (SNR). In this view, the CRB is typically used in the context of estimating parameters of a statistical model. For example, in signal processing, these parameters might include the amplitude, phase, or frequency of a signal. Therefore the CRB represents the smallest possible variance of an unbiased estimator for a given parameter. No unbiased estimator can have a smaller variance than what is dictated by the CRB. In certain conditions, if an estimator achieves the bound (i.e., its variance equals the CRB), it is considered an efficient estimator. This condition is known as attaining the Cramér – Rao bound equality. The CRB is often used to analyze the asymptotic behavior of estimators. As the sample size increases, efficient estimators tend to approach the Cramér – Rao bound. The general form of the Cramér – Rao bound for a parameter θ is given by:

$$Var\left(\hat{\theta}\right) \geq [FI(\theta)]^{-1} \tag{3.87}$$

where $Var(\hat{\theta})$ is the variance of any unbiased estimator $(\hat{\theta})$ for the parameter θ, and FI(θ) is the Fisher information for the parameter θ. In practical terms, the Cramér – Rao bound is a valuable tool for assessing the performance of estimators, helping to understand the theoretical limits of parameter estimation accuracy in a given statistical model (Cavassila et al., 2001; D'Andrea et al., 1994; Stoica & Ng, 1998).

A transform-limited pulse with duration T and peak power PT exhibits $\Delta\omega \sim 2\pi/T$, with SNR $\propto PT^2$. However, in radars with constrained peak power, there arises a conflict between enhancing range-delay resolution and improving ultimate range-delay accuracy. The utilization of a high time-bandwidth product ($T\Delta\omega \gg 2\pi$) in transmitted pulses maintains the SNR $\propto PT^2$ behavior. Nevertheless, pulse-compression reception results in $\tau_{res} \sim 2\pi/\Delta\omega \ll T$. Range-delay resolution and CRB accuracy, although crucial, do not tell the whole story in radar ranging. Since range-delay estimation is a nonlinear problem, there exists a minimum SNR below which range-delay performance significantly deviates from the CRB (Skolnik, 2002; Skolnik, 2008; Trees, 2001, 2004).

In the context of the burgeoning field of quantum information science, significant attention is directed toward the concept of quantum radar. The majority of research in this domain has focused on radar operations at optical wavelengths, characterized by low background noise brightness (i.e., ≪ 1 photon per mode) and often assuming minimal round-trip radar-to-target-to-radar propagation loss. Our specific interest lies in microwave radar, where background noise exhibits high brightness, approximately in the range of 100 – 1000 s of photons per mode. In this scenario, an unresolved target at range R yields a power inversely proportional to R^4, resulting in substantial propagation loss (Abdessaied et al., 2014; Barzanjeh et al., 2015; Casariego et al., 2023).

Despite the challenges posed by significant loss and noise in the microwave radar regime, quantum illumination has demonstrated that entanglement provides a 6 dB advantage over its most proficient classical counterpart with the same transmitted energy in the error-probability exponent. This advantage pertains specifically to the detection of the presence of an unresolved target at a particular location. In the realm of quantum range-delay estimation, the problem unfolds as follows. The radar initiates the transmission of a singular spatial-mode field, delineated by a photon-unit, positive-frequency field operator:

$$\hat{E}_S(t) = \int \frac{d\omega}{2\pi} A_S(\omega) e^{-i(\omega 0+\omega)t}, \tag{3.88}$$

In this equation, $\hat{E}_S(t)$ denotes the positive-frequency field operator for the transmitted radar signal. The operator is characterized by a continuous spectral decomposition represented by the integral term. The integration is performed over the angular frequency ω with the corresponding field operator A_S.

The term $(\omega)e^{-i(\omega 0+\omega)t}$ represents the time evolution of the field operator. It incorporates the phase factor, where ω0 is the carrier angular frequency of the radar signal, and t is time. This equation is a quantum description outlining how the radar signal field operator evolves over time, considering its spectral components and associated operators. From an unresolved, stationary target located at range R, the radar captures a positive-frequency field operator $\hat{E}_R(t)$ associated with photon units (Gudder & Holevo, 1985; Tsang, 2012; Zhuang & Shapiro, 2022b). This operator is defined as:

$$\hat{E}_R(t) = \sqrt{\frac{\kappa}{p}} e^{i\theta R} \hat{E}_S(t-\tau) + \sqrt{1-\frac{\kappa}{p}} \hat{E}_B(t) \tag{3.89}$$

Here, $\hat{E}_S(t-\tau)$ represents the positive-frequency field operator of the transmitted radar signal delayed by $\tau = 2\,R/c$, with c being light speed, and $\hat{E}_B(t)$ is the field operator associated with background noise. The parameters κ and p are used to scale and adjust the contributions from the signal and background noise components. The phase factor $e^{i}\theta R$ introduces the effect of the target's range R on the received field operator.

According to the above perspective, in the classical pulse-compression radar system, the emitted signal $\hat{E}_S(t)$ is represented in the coherent state $|\sqrt{E}e^{-i\omega_0 t}>$, where $s(t) = \frac{1}{\sqrt{2\pi T^2}} e^{\left(-\frac{t^2}{4T^2} + i\frac{\Delta\omega t^2}{2T}\right)}$ is the chirped-Gaussian pulse function. Here, $2\pi/T \ll \Delta\omega \ll \omega_0$. Physically, this pulse is characterized by a narrow bandwidth but a high time-bandwidth product, exhibiting a chirped-Gaussian shape with an average photon number E and root mean square time duration T (Zhuang & Shapiro, 2022b).

Given $|S(\omega)|^2 = \Delta\omega^2$, where Δω is the rms bandwidth of $s(t)$, the quantum Cramér – Rao bound (CRB) for the root mean square (rms) range accuracy $\delta\tau_C^{CRB}$ of this classical radar, with quantum-optimal reception ($\theta_R = 0$ and $\omega_0 = 0$), is expressed as:

$$\delta\tau_C^{CRB} \approx \frac{1}{\Delta\omega\sqrt{2SNR}} \tag{3.90}$$

here $SNR \equiv \frac{\kappa E}{NB}$, and $NB \gg 1$ as is customary for microwave operation, which is a typical condition for microwave operation, indicating a relatively large number of noise photons. $\Delta\omega$ represents the root mean square (rms) bandwidth of the radar signal $s(t)$, a chirped-Gaussian pulse in this context. The bandwidth is a measure of the range of frequencies present in the signal. κ is a constant factor. Therefore Eq. (3.87) suggests that the accuracy in range estimation, which is inversely proportional to the rms bandwidth ($\Delta\omega$) and the square root of the product of signal energy (E), noise photon number (NB), and a constant factor (κ). The presence of the signal-to-noise ratio (SNR) in the denominator indicates that as the SNR increases, the accuracy improves, which aligns with the expectations in radar systems.

In the realm of quantum illumination target detection, our quantum pulse-compression radar distinguishes itself through the remarkable performance boost derived from its phase-sensitive cross spectrum $S(p)(\omega)$. This advantage is particularly pronounced as it significantly surpasses the classical limit imposed on the cross spectrum $S(n)(\omega)$ when $N_S = \sqrt{2\pi}$, reaching the maximum of 1/2 times $S(n)(\omega)$. The crucial condition for this enhanced performance is N_S being much less than 1, denoting a low-brightness scenario where $\sqrt{2\pi}$ times the classical cross spectrum $S(n)(\omega)$ is exceedingly small.

This low-brightness circumstance implies that quantum radar necessitates a considerably longer pulse duration compared with a high-brightness classical counterpart possessing the same bandwidth and energy. The assessment of the quantum pulse-compression radar's Fisher information, as elegantly derived leads to the range-delay Cramér – Rao bound $\delta\tau_Q^{CRB}$, given by $\frac{1}{2\Delta\omega\sqrt{2SNR}}$, where $\delta\tau_C^{CRB}$ represents the classical counterpart. Intriguingly, this result stands out as the optimal high-SNR performance among all entanglement-assisted radars. However, when N_S takes on a significantly larger value, the previously mentioned quantum advantage diminishes, aligning with expectations. In this scenario, we find that $\delta\tau_Q^{CRB}$ converges to $\delta\tau_C^{CRB}$. This nuanced behavior underscores the intricate interplay of brightness conditions and entanglement-assisted capabilities in defining the quantum pulse-compression radar's performance landscape.

The Ziv – Zakai bound (ZZB) is a fundamental concept in signal processing, specifically in the context of parameter estimation. It sets a theoretical limit on the achievable accuracy when estimating a parameter from a signal corrupted by noise. In the realm of radar and range-delay estimation, the ZZB provides a lower bound on the mean-squared error of the estimation process (Tsang, 2012).

To delve deeper, let's consider the quantum radar scenario. The ZZB for quantum radar $\delta\tau_C^{ZZB-QCB}$ outlines the minimum possible error in estimating the range-delay (τ) when employing quantum strategies, accounting for the inherent uncertainties and limitations imposed by quantum mechanics. The expression $\left(\frac{\delta\tau_C^{ZZB-QCB}}{\delta\tau_Q^{ZZB-QCB}}\right)^2$ compares the mean-squared errors between classical $\delta\tau_C^{ZZB-QCB}$ and quantum $\delta\tau_Q^{ZZB-QCB}$ range-delay estimation under the Ziv – Zakai framework. The fascinating relationship $\sim\left(\frac{\Delta\omega\sigma}{2}\tau\right)^{\frac{3}{2}}$ elucidates the intricate interplay between the bandwidth of the transmitted pulse ($\Delta\omega$) and the standard deviation of the delay ($\sigma\tau$) (Zhuang & Shapiro, 2022a, 2022b). This relationship, governed by a power-law exponent of 3/2, highlights the nuanced dynamics influencing the quantum advantage in precision, offering a profound glimpse into the inherent capabilities and constraints of quantum radar systems (Ahern, 1995; Assouly et al., 2023; Wu et al., 2023).

In the realm of microwave radar, the integration of quantum entanglement presents a formidable challenge. Within this discourse, we have embarked upon a comprehensive exploration, employing a continuous-time framework to delve into the domain of quantum illumination (QI) for target ranging. Our endeavors encompassed a rigorous comparative analysis with a classical pulse-compression radar, both reliant on time-of-flight measurements for deducing target range. Notably, both systems exhibit SNR thresholds, beneath which the accuracy of their range-delay measurements significantly deviates from the Cramér – Rao bound (CRB).

The quantum radar, intriguingly, manifests a 6 dB lower threshold SNR in comparison to its classical counterpart (Jonsson & Ankel, 2021; Zhuang & Shapiro, 2022a, 2022b). In this sense, the threshold SNR (SNR_Q^{thresh}) for the quantum radar, exhibiting a 6 dB lower threshold than the classical radar (SNR_C^{thresh}), can be expressed as:

$$SNR_Q^{thresh} \sim \frac{1}{\sqrt{2}} SNR_C^{thresh} \tag{3.91}$$

This equation signifies the quantum radar's inherent sensitivity advantage in achieving accurate range-delay measurements at lower SNR levels. Consequently, when operating at this threshold SNR, the mean-squared range-delay accuracy of the quantum radar outshines that of its classical counterpart by several tens of dB. Despite the current absence of a receiver design that fully harnesses the range-delay accuracy advantage of our quantum radar,

TABLE 3.3 Pseudocode for signal-to-noise ratio (SNR) in the quantum radar system.

```
# Pseudocode for Quantum Radar System

# Description: # The achievable Signal-to-Noise Ratio (SNR) in the quantum radar system

# Variables: # SNR: Signal-to-Noise Ratio # M: Time-Bandwidth Product of the transmitter

# Algorithm: 1. Initialize SNR as a function of M in the quantum radar system.

# Relationship between SNR and M 2. SNR = CalculateSNR(M)

# Output 3. Display SNR as the result of the relationship with M.
```

the substantial magnitude of this advantage offers promising prospects. This stands in stark contrast to the challenges faced in the quantum realm for long-range target detection, where the utility of quantum illumination (QI) hinges on the employment of exceedingly prolonged pulse durations. Achieving practical SNR values, on the order of 5−10 dB, demands intricate solutions for conditions where $N_S \ll 1$, $\kappa \ll 1$, and $N_B \gg 1$ (Jonsson & Ankel, 2021; Wu et al., 2023). The achievable (*SNR*) in the quantum radar system is intricately linked to the time − bandwidth product (*M*) of the transmitter (Table 3.3). This relationship is expressed as:

$$SNR \approx \frac{\kappa M N_s}{N_B} \tag{3.92}$$

Here, N_S represents the brightness condition, κ is a parameter related to quantum efficiency, and N_B characterizes the background noise. These equations provide a quantitative framework for understanding the intricacies of quantum radar performance and its advantageous characteristics over classical pulse-compression radar.

In examining the delicate balance between the quantum advantage in range-delay precision and pulse duration, let's consider the scenario of a W-band radar operating at 100 GHz, focused on localizing a small unmanned aerial system (UAS). W-band radar, known for its high precision and resilience in challenging visual conditions, finds extensive utility in UAS detection.

Therefore the quantum algorithm has the advantage in mean-squared range-delay accuracy for a range uncertainty (ΔR) of $R/100$, assuming operation at the threshold quantum signal-to-noise ratio (SNR_Q^{thresh}) (Table 3.4). This depiction is based on an ideal radar setup where $\kappa = \frac{GT}{4\pi R^2} \times \frac{\sigma A_R}{4\pi R^2}$, with GT denotes the radar's antenna gain, A_R its antenna area, and σ represents the UAS's radar cross section (assumed to be $0.01m^2$). The radar's noise temperature (T_B) of 150 K leads to $N_B = 32$, determining the necessary N_S to attain SNR_Q^{thresh} and consequently, the resulting quantum advantage (Trees, 2004; Tsang, 2012; Yao et al., 2017; Zhuang & Shapiro, 2022a, 2022b).

For example, a pulse duration of 0.01 seconds affords a substantial 20 dB quantum advantage at a range of 100 m. However, achieving the same advantage at a more considerable distance of 1 km necessitates an impractically lengthy 100-second pulse duration (Fig. 3.20). This requirement stems from the constancy of T/R^4 to uphold a consistent quantum advantage, emphasizing the intricate interplay between pulse duration and range in quantum radar applications. This result confirms the study achieved by (Zhuang & Shapiro, 2022b).

Certainly, quantum illumination ranging represents a groundbreaking approach in radar technology. By harnessing the entanglement between a high time − bandwidth product transmitted signal pulse and a retained idler pulse, this quantum pulse-compression radar surpasses the limitations of classical counterparts. The result is a mean-squared range-delay accuracy that outperforms classical pulse-compression radars with the same pulse bandwidth and transmitted energy by tens of decibels. This innovative quantum technique opens new frontiers in achieving unprecedented accuracy and efficiency in ranging applications.

3.15 Synthetic aperture radar satellite sensors

In the realm of Earth observation, synthetic aperture radar (SAR) satellites gracefully traverse the celestial expanse in sun-synchronous, specifically within the Low Earth Orbit (LEO), characterized by an orbital period of less than 2 hours. The LEO polar orbit bestows the advantage of data acquisition independent of temporal constraints, illuminating the Earth's surface both day and night while remaining impervious to cloud cover. These SAR satellites, equipped to capture both amplitude and phase data, follow repetitive trajectories. Leveraging two-phase

TABLE 3.4 A pseudocode representation of a quantum algorithm with an advantage in mean-squared range-delay accuracy for a range uncertainty.

```
function quantumAlgorithmWithAccuracyAdvantage(): // Assume appropriate values for the variables

R = getTargetRange() deltaR = R / 100 // Range uncertainty

// Calculate threshold quantum signal-to-noise ratio thresholdSNR_Q = getThresholdSNR_Q()

// Quantum algorithm for mean-squared range-delay accuracy entanglePulses() // Entangle

transmitted and retained pulses

// Simulate quantum radar operation quantumTransmitPulse() // Apply quantum operations to the

transmitted pulse quantumRetainPulse() // Apply quantum operations to the retained pulse

// Measure the quantum state and calculate mean-squared range-delay accuracy measuredRange =

measureQuantumState() meanSquaredAccuracy = (measuredRange - R + deltaR)^2

// Compare with classical radar accuracy classicalAccuracy = calculateClassicalAccuracy(deltaR)

// Check if quantum accuracy is better than classical accuracy if meanSquaredAccuracy <

classicalAccuracy: return "Quantum algorithm provides an advantage in mean-squared range-delay

accuracy." else: return "Quantum algorithm does not provide a significant advantage in accuracy."

// Helper function to calculate classical radar accuracy function calculateClassicalAccuracy(deltaR): //

Assume classical radar operations and calculations // .

// Return the classical mean-squared range-delay accuracy return classicalMeanSquaredAccuracy
```

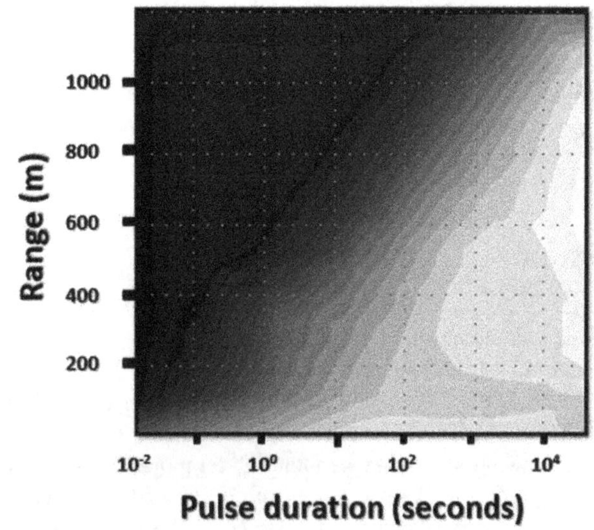

FIGURE 3.20 Contour representation of dB advantage in mean-squared range-delay accuracy.

datasets obtained at different temporal junctures for a specific location enables the elucidation of interferometric SAR (InSAR), unveiling relative ground displacements along the radar beam's trajectory. The constellation of the various agencies supports the different SAR missions (Table 3.5) (Marghany, 2019).

Operating at designated frequencies, SAR satellites predominantly utilize L-band, C-band, and X-band wavelengths. The following chart delineates a comprehensive chronicle of past missions, current endeavors, and anticipated future SAR satellite missions.

3.15.1 Synthetic aperture radar data from the European Remote Sensing Satellite

The ERS (European Remote Sensing Satellite) SAR, encompassing ERS-1 and ERS-2, featured identical twin spacecraft with a crucial distinction—ERS-2 incorporated

TABLE 3.5 A compilation of SAR satellite sensors operating across various frequency bands.

No.	Sensor	Operation	Frequency band	Institution/Country	Remark
1	Seasat	1979	L(HH)	NASA/JPL, USA	First civilian SAR satellite
2	ERS-1/2	1991–2000/ 1995–2011/	C(VV)	ESA, Europa	European SAR satellites
3	Radarsat-1	1995–today	C(HH)	CSA, Canada	First Canadian SAR satellite
4	SRTM	Feb. 2000	C (HH + VV) and X(VV)	NASA/JPL, USA; DLR, Germany; ASI, Italy	Shuttle Radar Topography Mission, 1st spaceborne interferometric SAR
5	ENVISAT/ASAR	2002–2012	C(dual)	ESA, Europa	1st SAR satellite with transmit/ receive module
6	ALOS/PALSAR	2006–2011	L(quad)	JAXA, Japan	Advanced Land Observing satellite
7	TerraSAR-X/ TanDEM-X	2007–today/ 2010–today	X(quad)	DLR/Astrium, Germany	1st bi-static radar in space, resolution up to 1 m
8	Radarsat-2	2007–today	C(quad)	CSA, Canada	Resolution up to 1 mx3m
9	COSMO-Skymed-1/4	2007–today	X(quad)	ASI/MiD, Italy	Constellation of four satellites (1m resolution)
10	2012–today	X(quad)	ISRO, India	–	2012–today
11	HJ-1C	2012–today	S(VV)	CRESDA/CAST/ NRSCC, China	Constellation of four satellites
12	Kompsat-5	2013–today	X(dual)	Korea multipurpose KARI, Korea	–
13	PAZ	2013–today	X(quad)	CDTI, Spain	The constellation of TerraSAR-X and TanDEM-X
14	ALOS-2	2013–today	L(quad)	JAXA, Japan	Resolution up to 1mx3m
15	Sentinel1a/1b	2014–today	C(dual)	ESA, Europe	The constellation of two satellites

an additional instrument, GOME, dedicated to monitoring atmospheric ozone levels. Launched on July 17, 1991, ERS-1 was succeeded by ERS-2 on April 21, 1995, maintaining continuity in orbit. Operating in tandem with the Wind Scatterometer (WS), the SAR/WS combination, known as the Active Microwave Instrument (AMI), offered three operational modes, enabling imaging of land day or night and through clouds.

The ERS-1 mission exceeded expectations, concluding on March 10, 2000. ERS-2 SAR data acquisition ceased on July 4, 2011, culminating the ERS-2 mission on September 5, 2011. The ERS SAR, operating in C-band with linear vertical polarization (VV Mode), covered a swath width of up to 100 km while traveling approximately 4 km. Employing intrapulse modulation through linear frequency modulation (LFM) and a bandwidth of 15.55 MHz, the transmitted pulse had a length of 37.12 μs.

The "ERS tandem mission" significantly increased radar altimeter measurement points and laid the foundation for a precise three-dimensional digital map of Earth's land surfaces. ERS-1 and 2 satellites were instrumental in global sea wind and wave measurements, ocean and ice monitoring, coastal studies, and land sensing (Cigna et al., 2014).

From 2002 to 2012, ENVISAT, equipped with advanced instruments, implemented a versatile data acquisition strategy through four distinctive modes tailored to specific observational requirements. The advanced synthetic aperture radar (ASAR) mode aimed to provide high-resolution radar images for observations of land surfaces, oceans, and ice cover, facilitating observations regardless of weather conditions. The medium resolution imaging spectrometer (MERIS) mode focused on capturing multispectral images of the Earth's surface for detailed studies of vegetation, land cover, and coastal zones. The advanced along-track scanning radiometer (AATSR) mode measured sea and land surface temperatures, contributing to climate research and oceanographic studies. The Radar Altimeter-2 (RA-2) mode focused on precise measurements of the height of the Earth's surface, supporting studies on sea level variations and ice sheet dynamics. These distinct modes allowed ENVISAT to tailor its observations based on specific scientific objectives, ensuring a comprehensive understanding of the Earth's environment (Solari et al., 2018).

Recently, Sentinel-1, a C-band synthetic aperture radar (SAR) sensor, signifies the initiation of the Copernicus Program satellite constellation led by the European Space Agency. Sentinel-1 employs a diverse range of data acquisition modes, each serving distinct purposes:

Stripmap (SM): This is the standard synthetic aperture radar (SAR) strip mapping mode, characterized by a spatial resolution of 5-by-5 square meters and an 80-kilometer swath. SM data products are accessible in either single polarization (HH or VV) or double polarization (HH1 HV or VV1 VH).

Interferometric Wide Swath (IW): Operating with the Terrain Observation with Progressive Scanning SAR (TOPSAR) technique, IW is implemented in three swaths, featuring a spatial resolution of 5-by-20 square meters and an expansive 250-kilometer swath. It serves as the primary operational mode over land, providing data products in both single and double polarization.

Extra Wide Swath (EW): Utilizing the TOPSAR technique, EW acquires data in five swaths, offering a spatial resolution of 20-by-40 square meters and an extensive 400-kilometer swath. Despite a reduction in spatial resolution, EW provides broad swath coverage. Similar to SM, data products are available in single or double polarization.

Wave (WV): Specifically designed for oceanic observations, WV captures data in small strip map scenes known as "vignettes" at regular 100-kilometer intervals. WV features a spatial resolution of 5-by-20 square meters, producing 20-by-20 square kilometer sample images along the orbit. Notably, WV data products are exclusively offered in single polarization (HH or VV) (Cigna et al., 2014; Solari et al., 2018).

3.15.2 Canadian Space Agency: Radarsat-1, Radarsat-2, Radarsat constellation

The RADARSAT mission encompasses two meticulously crafted earth observation spacecraft under the auspices of the Canadian Space Agency (CSA). Launched on November 4, 1995, RADARSAT-1, a groundbreaking Earth-imaging radar satellite, concluded its operational phase on March 29, 2013. In tandem, RADARSAT-2, an advanced Earth-imaging radar satellite developed in collaboration with MDA (MacDonald Dettwiler Associates Ltd., of Richmond, BC, USA), was launched on December 14, 2007, and remains operational to date.

Technological advancements have propelled RADARSAT-1 and RADARSAT-2 to capture high-resolution imagery of Earth at a single microwave frequency of 5.3 GHz, within the C band with a wavelength of 5.6 cm. The synthetic aperture radar (SAR) instruments offer various polarization options, including HH, HV, VH, VV, HH1HV, VV1VH, and HH1VV1HV1VH. Remarkably, the data collection boasts a spatial resolution spanning from 1 to 100 meters and a sweeping range from 18 to 500 kilometers.

Adapting to technological and scientific demands, both RADARSAT-1 and RADARSAT-2 operate in various modes, including Wide (single- or dual-pol), Fine (single- or dual-pol), Standard (single- or dual-pol), Extended high and low (single pol), and ScanSAR Wide and Narrow (single- or dual-pol). This diverse array of modes underscores the RADARSAT program's versatility and its substantial contributions to advancing science and technology.

3.15.3 Japan Aerospace Exploration Agency: JERS-1, ALOS-1, ALOS-2

JERS-1, also referred to as "Fuyo-1," emerged as an environmental satellite of Japanese origin. It was launched on February 11, 1992, and its operational tenure spanned until October 1998. The mission, a collaborative effort led by NASDA (National Space Development Agency of Japan), featured a radar/optical configuration.

Positioned in a solar-synchronous orbit at an altitude of 568 km with a repeat cycle of 44 days, JERS-1 exhibited a dedicated service life of 6.5 years, concluding when contact was lost in October 1998. The satellite was equipped with an L-band synthetic aperture radar (SAR) operating at a center frequency of 1.275 MHz/23.5 cm. The SAR utilized HH polarization and maintained a fixed 35-degree off-nadir angle, capturing images with a swath width of 175 km.

Complementing the SAR, the optical sensor (OPS) on JERS-1 functioned as a nadir-pointing multispectral imager. Operating across four bands (bands 1–4) in the visible and near-infrared (VNIR) spectrum (0.52–0.86 μm) and four bands in the shortwave infrared (SWIR) spectrum (1.60–2.40 μm), the OPS provided comprehensive data coverage. The OPS included a very near-infrared radiometer, contributing to the diverse array of data products from JERS-1, which encompass JERS-1 SAR Level-1 single-look complex (SLC) images and JERS-1 SAR Level-1 precision images.

The phased array-type L-band synthetic aperture radar (PALSAR) stands as a robust active microwave sensor situated on the advanced land observing satellite (ALOS). It was conceived through a collaborative effort between the Japanese Ministry of Economy, Trade, and Industry (METI) and the Japan Aerospace Exploration Agency (JAXA), PALSAR was designed to acquire data crucial for resource exploration and environmental protection. Its launch on January 24, 2006, aboard ALOS marked a significant milestone.

The PALSAR sensor operates in L-band SAR and boasts a multimode observation function encompassing Fine mode, direct downlink, ScanSar mode, and Polarimetric mode. It features a multipolarization configuration (HH, HV, VH, and VV), variable off-nadir angle (9.9–50.8 °), and adjustable spatial resolutions (10, 30, 100 m for Fine, Polarimetric, and ScanSar modes, respectively), coupled with swath width observations ranging from 30 to 350 km for Polarimetric, Fine, and ScanSar modes, respectively.

Furthermore, the resolution of PALSAR reached a maximum of 10 m, a milestone further improved with the advent of PALSAR-2, achieving an impressive 1 – 3 m resolution. This advancement is attributed to the incorporation of a "Spotlight mode," enabling the satellite to adjust its radio wave radiation direction to align with its movement direction (vertical). The adoption of a "Dual receiving antenna system" ensures an ample number of observation bands with high resolution. Consequently, the observation band spans 50 km with a 3-m resolution in the "Strip Map mode" and 25 km with a 1 – 3 m resolution (De Lisle et al., 2018).

3.15.4 Deutsches Zentrum für Luft- und Raumfahrt e.V.: TerraSAR-X and TanDEM-X

TerraSAR-X and TanDEM-X, both operated by the Deutsches Zentrum für Luft- und Raumfahrt e.V. (DLR), exhibit advanced synthetic aperture radar (SAR) capabilities for high-resolution Earth observation. TerraSAR-X boasts an impressive spatial resolution, offering 1 meter in Spotlight mode, 3 meters in StripMap mode, and 16 meters in ScanSAR mode. With versatile imaging modes such as Spotlight, StripMap, and ScanSAR, TerraSAR-X operates in the X-band frequency with a single-band configuration. The dual-polarization feature allows users to select between HH, VV, HV, or VH polarizations, enhancing the flexibility for diverse applications.

TanDEM-X, a companion satellite, shares similar capabilities with exceptional spatial resolutions of 1 meter in Spotlight mode, 3 meters in StripMap mode, and 10 meters in High-Resolution SpotLight mode. Its operational modes include Spotlight, StripMap, and High-Resolution SpotLight, all operating in the X-band frequency with a single-band configuration. Similar to TerraSAR-X, TanDEM-X supports dual polarization (HH, VV, HV, VH), providing users with varied polarization options for specific imaging needs (Zink et al., 2007).

Together, these satellites contribute significantly to Earth observation and scientific research by delivering high-quality SAR data with variable resolutions and imaging modes, making them invaluable assets for a wide range of applications.

3.15.5 Italian Space Agency: COSMO-Skymed

The Italian Space Agency (ASI) oversees the COSMO-SkyMed satellite constellation, an Earth observation system designed for various applications. COSMO-SkyMed satellites are equipped with synthetic aperture radar (SAR) technology, enabling all-weather, day-and-night imaging capabilities. The constellation comprises multiple satellites, and each satellite has its distinct physical specifications.

The COSMO-SkyMed satellites operate in the X-band frequency and provide high-resolution imagery with spatial resolutions ranging from 1 to 100 meters, catering to different user needs. The radar system supports various imaging modes, including Spotlight, StripMap, ScanSAR, and HIMAGE, offering flexibility for different Earth observation tasks. The dual-polarization feature allows users to choose between HH, HV, VH, or VV polarizations, enhancing the versatility of the acquired data (Covello et al., 2010).

With frequent revisit times and a constellation approach, COSMO-SkyMed ensures timely and reliable Earth observation data for applications such as environmental monitoring, disaster management, and security. The constellation's capabilities make it a valuable asset for both scientific research and practical applications, contributing to Italy's space exploration and Earth observation efforts.

3.15.6 Korea Aerospace Research Institute: KOMPSat-5

The Korea Aerospace Research Institute (KARI) oversees the KOMPSat-5 satellite, an Earth observation satellite designed for various remote sensing applications. KOMPSat-5 is equipped with synthetic aperture radar (SAR) technology, enabling it to capture high-resolution images of the Earth's surface in different modes.

Operated by KARI, KOMPSat-5 operates in the X-band frequency and provides versatile imaging capabilities. The satellite's physical specifications include a spatial resolution ranging from 1 to 3 meters, allowing for detailed observations of Earth's features. KOMPSat-5's SAR instrument supports various modes, including Spotlight, StripMap, ScanSAR, and InSAR (Interferometric SAR), catering to different user requirements. The dual-polarization feature of KOMPSat-5 allows users to select between HH, HV, VH, or VV polarizations, enhancing the versatility and interpretability of the acquired data (Han & Kim, 2018).

3.15.7 Indian Space Research Organization: RISAT-1, NISAR (w/ NASA)

RISAT-1, developed by the Indian Space Research Organization (ISRO), provides a synthetic aperture radar (SAR) imaging capability. It operates in the C-band and offers a spatial resolution suitable for various applications, allowing for detailed observations of the Earth's surface. Additionally, the collaboration between ISRO and NASA on the NASA-ISRO Synthetic Aperture Radar (NISAR) mission enhances the observational capabilities by combining resources and expertise. NISAR is expected to provide high-resolution SAR data, contributing to scientific research and applications related to Earth observation. The specific resolution details may vary based on the imaging mode and configuration selected for the observations (Misra, 2017).

Regrettably, this literary work encounters constraints in encompassing all synthetic aperture radar (SAR) satellite sensors, primarily stemming from challenges associated with accessing and analyzing the voluminous SAR data. Moreover, the scope of this book does not extend to the comparative examination of various SAR sensors at its current stage. The book's primary objective is to implement both traditional and innovative algorithms dedicated to tracking and modeling turbulence features using the available SAR data.

The focal point of this literary composition revolves around the precise detection of turbulence features, particularly within well-established SAR sensors such as ERS-1/2, RADARSAT-1, ENVISAT, and others. The emphasis lies in developing methodologies for accurately identifying turbulence features, rather than conducting a comprehensive comparison of different SAR sensors.

This chapter introduces a groundbreaking perspective on the incorporation of quantum mechanics in elucidating the intricacies of synthetic aperture radar. Unlike preceding texts, we refrain from delving extensively into radar theory and signal processing, as these aspects have been comprehensively covered elsewhere. Instead, our focus is on presenting a pioneering concept of quantization as a means to comprehend the mechanisms underlying radar imaging of sea turbulence. Subsequent chapters will delve into this concept in greater detail.

Our chapter seizes the opportunity to elucidate the quantization of radar, offering a clear understanding of the term itself. The foundational quantization of electromagnetic waves, essential to radar operations, is initially explored. We begin by addressing why photons, the fundamental units, exhibit quantization, a pivotal aspect for comprehending quantized electromagnetic waves. This inquiry unveils why photons possess discrete energy amounts within a specific energy spectrum, deviating from a continuous range.

Moving beyond conventional discussions, our exploration extends to the detection of quantum microwave propagation—an original discourse in the realm of synthetic aperture radar publications. Furthermore, we implement the quantization of the radar equation to introduce a fresh perspective on the entanglement between radar cross section and sea surface turbulence, a dimension hitherto unexplored in radar oceanography.

Notably, our chapter delves into the novel realm of quantum pulse-compression ranging. In the concluding sections, we compile a comprehensive list of sensors associated with synthetic aperture radar satellites, detailing their physical characteristics, including bands and frequencies.

In summary, the quantization of radar signals propounds an innovative speculation regarding the entanglement of radar cross section with turbulence spectra, such as the Kolmogorov energy spectra—an unprecedented contribution in the domain of radar oceanography.

References

Abdessaied, N., Soeken, M., & Drechsler, R. (2014). *Quantum circuit optimization by Hadamard gate reduction Lecture Notes in Computer Science (including subseries Lecture Notes in Artificial Intelligence and Lecture Notes in Bioinformatics)* (8507, pp. 149–162). Germany: Springer Verlag. Available from 10.1007/978-3-319-08494-7_12, http://springerlink.com/content/0302-9743/copyright/2005/.

Ahern, F. J., Fundamental concepts of imaging radar: Basic level. Unpublished manual, Canada Centre for Remote Sensing. (1995).

Assouly, R., Dassonneville, R., Peronnin, T., Bienfait, A., & Huard, B. (2023). Quantum advantage in microwave quantum radar. *Nature Physics*, *19*(10), 1418–1422. Available from https://doi.org/10.1038/s41567-023-02113-4, https://www.nature.com/nphys/.

Barzanjeh, S., Guha, S., Weedbrook, C., Vitali, D., Shapiro, J. H., & Pirandola, S. (2015). Microwave quantum illumination. *Physical Review Letters*, *114*(8). Available from https://doi.org/10.1103/PhysRevLett.114.080503, http://harvest.aps.org/bagit/articles/10.1103/PhysRevLett.114.080503/apsxml.

Baym, G. (2018). *Lectures on Quantum Mechanics* (pp. 1–594). United States: CRC Press. Available from http://www.tandfebooks.com/doi/book/10.1201/9780429499265, 10.1201/9780429499265.

Casariego, M., Zambrini Cruzeiro, E., Gherardini, S., Gonzalez-Raya, T., André, R., Frazão, G., Catto, G., Möttönen, M., Datta, D., Viisanen, K., Govenius, J., Prunnila, M., Tuominen, K., Reichert, M., Renger, M., Fedorov, K. G., Deppe, F., van der Vliet, H., Matthews, A. J., ... Omar, Y. (2023). Propagating quantum microwaves: towards applications in communication and sensing. *Quantum Science and Technology*, *8*(2), 023001. Available from https://doi.org/10.1088/2058-9565/acc4af.

Cavassila, S., Deval, S., Huegen, C., Van Ormondt, D., & Graveron-Demilly, D. (2001). Cramér-Rao bounds: An evaluation tool for quantitation. *NMR in Biomedicine*, *14*(4), 278–283. Available from https://doi.org/10.1002/nbm.701.

Chapman, M. Heterodyne and homodyne interferometry. (2002).

Chromey, F. R. (2016). *To measure the sky: an introduction to observational astronomy*. Cambridge University Press.

Cigna, F., Bateson, L. B., Jordan, C. J., & Dashwood, C. (2014). Simulating SAR geometric distortions and predicting persistent scatterer densities for ERS-1/2 and ENVISAT C-band SAR and InSAR applications: Nationwide feasibility assessment to monitor the landmass of Great Britain with SAR imagery. *Remote Sensing of Environment, 152*, 441–466. Available from https://doi.org/10.1016/j.rse.2014.06.025, http://www.elsevier.com/inca/publications/store/5/0/5/7/3/3.

Cooper, C.E. Physics. (2001).

Cosmelli, C., Carelli, P., Castellano, M. G., Chiarello, F., Diambrini Palazzi, G., Di Bucchianico, M., Leoni, R., Poletto, S., & Torrioli, G. (2003). Quantum dynamics of a microwave-driven SQUID. *Superconductor Science and Technology*, *16*(12), 1337–1339. Available from https://doi.org/10.1088/0953-2048/16/12/003.

Covello, F., Battazza, F., Coletta, A., Lopinto, E., Fiorentino, C., Pietranera, L., Valentini, G., & Zoffoli, S. (2010). COSMO-SkyMed an existing opportunity for observing the Earth. *Journal of Geodynamics*, *49*(3–4), 171–180. Available from https://doi.org/10.1016/j.jog.2010.01.001.

D'Andrea, A. N., Mengali, U., & Reggiannini, R. (1994). The modified Cramer-Rao bound and its application to synchronization problems. *IEEE Transactions on Communications*, *42*(234), 1391–1399. Available from https://doi.org/10.1109/TCOMM.1994.580247.

Davies, P. C. (2018). *Quantum mechanics*. Routledge.

Dubosclard, W., Kim, S., & Garrido Alzar, C. L. (2021). Nondestructive microwave detection of a coherent quantum dynamics in cold atoms. *Communications Physics*, *4*(1). Available from https://doi.org/10.1038/s42005-021-00541-3, http://nature.com/commsphys/.

Eichler, C., Bozyigit, D., Lang, C., Baur, M., Steffen, L., Fink, J. M., Filipp, S., & Wallraff, A. (2011). Observation of two-mode squeezing in the microwave frequency domain. *Physical Review Letters*, *107*(11). Available from https://doi.org/10.1103/PhysRevLett.107.113601, http://oai.aps.org/filefetch?identifier = 10.1103/PhysRevLett.107.113601&component = fulltext&description = markup&format = xml.

Fedorov, K. G., Pogorzalek, S., Las Heras, U., Sanz, M., Yard, P., Eder, P., Fischer, M., Goetz, J., Xie, E., Inomata, K., Nakamura, Y., Di Candia, R., Solano, E., Marx, A., Deppe, F., & Gross, R. (2018). Finite-time quantum entanglement in propagating squeezed microwaves. *Scientific Reports*, *8*(1). Available from https://doi.org/10.1038/s41598-018-24742-z, http://www.nature.com/srep/index.html.

Ferroni, F. (2010). LUCIFER: A new technique for double beta decay. Editrice Compositori s.r.l. *Italy Nuovo Cimento della Societa Italiana di Fisica C*, *33*(5), 27–34. Available from https://doi.org/10.1393/ncc/i2011-10696-1, http://www.sif.it/SIF/en/portal/journals/ncc/econtents.

Flurin, E. The Josephson mixer: A swiss army knife for microwave quantum optics (Doctoral dissertation). (2014).

Fronsdal, C. (1979). Some ideas about quantization. *Reports on Mathematical Physics*, *15*(1), 111–145. Available from https://doi.org/10.1016/0034-4877(79)90054-5.

Galindo, A., & Pascual, P. (2012). *Quantum mechanics I*. Springer Science & Business Media.

Gavish, U., Levinson, Y., & Imry, Y. (2000). Detection of quantum noise. *Physical Review B*, *62*(16), R10637–R10640. Available from https://doi.org/10.1103/physrevb.62.r10637.

Gleyzes, S., Kuhr, S., Guerlin, C., Bernu, J., Deléglise, S., Busk Hoff, U., Brune, M., Raimond, J. M., & Haroche, S. (2007). Quantum jumps of light recording the birth and death of a photon in a cavity. *Nature*, *446*(7133), 297–300. Available from https://doi.org/10.1038/nature05589, http://www.nature.com/nature/index.html.

Golovko, V. A. Electromagnetic radiation and resonance phenomena in quantum mechanics. (2008).

Gottfried, K. (2018). *Quantum mechanics: Fundamentals*, . *Quantum Mechanics: Fundamentals* (1). United States: CRC Press. Available from http://www.tandfebooks.com/doi/book/10.4324/9780429493225, 10.4324/9780429493225.

Graf, R. F. (1999). *Modern dictionary of electronics*. Elsevier.

Gruber, G. R. (1972). Quantization in generalized coordinates-II. *International Journal of Theoretical Physics*, *6*(1), 31–35. Available from https://doi.org/10.1007/BF00680671.

Gudder, S. P., Holevo, A. S. Probabilistic and statistical aspects of quantum theory. (1985). .

Guerlin, C., Bernu, J., Deléglise, S., Sayrin, C., Gleyzes, S., Kuhr, S., Brune, M., Raimond, J. M., & Haroche, S. (2007). Progressive field-state collapse and quantum non-demolition photon counting. *Nature*, *448*(7156), 889–893. Available from https://doi.org/10.1038/nature06057, http://www.nature.com/nature/index.html.

Hamid, M. A., Aribowo, D., & Desmira, D. (2017). Development of learning modules of basic electronics-based problem solving in vocational secondary school. *Jurnal Pendidikan Vokasi.*, *7*(2), 149. Available from https://doi.org/10.21831/jpv.v7i2.12986.

Han, H., & Kim, Hc (2018). Evaluation of summer passive microwave sea ice concentrations in the Chukchi Sea based on KOMPSAT-5 SAR and numerical weather prediction data. *Remote Sensing of Environment*, *209*, 343–362. Available from https://doi.org/10.1016/j.rse.2018.02.058, http://www.elsevier.com/inca/publications/store/5/0/5/7/3/3.

Jonsson, R., Ankel, M., (2021). Quantum Radar-What is it good for? IEEE National Radar Conference - Proceedings 9781728176093 Institute of Electrical and Electronics Engineers Inc. Sweden http://ieeexplore.ieee.org/xpl/conhome.jsp?punumber = 1001138, https://doi.org/10.1109/RadarConf2147009.2021.9455162.

Kramers, H. A. Quantum mechanics. Courier Dover Publications, (2018).

Lacomme, P. (2001). *Air and spaceborne radar systems: An introduction* (108). William Andrew.

Lanzagorta, M. (2011). Quantum Radar. *Synthesis Lectures on Quantum Computing*, *3*(1), 1–139. Available from https://doi.org/10.2200/s00384ed1v01y201110qmc005.

Leppäkangas, J., Marthaler, M., Hazra, D., Jebari, S., Albert, R., Blanchet, F., Johansson, G., & Hofheinz, M. (2018). Multiplying and detecting propagating microwave photons using inelastic Cooper-pair tunneling. *Physical Review A*, *97*(1). Available from https://doi.org/10.1103/PhysRevA.97.013855, http://harvest.aps.org/v2/bagit/articles/10.1103/PhysRevA.97.013855/apsxml.

Levi, A. F. J. (2023). *Applied quantum mechanics*. Cambridge University Press. Available from 10.1017/9781009308083.

Liberal, I., Ederra, I., & Ziolkowski, R. W. (2018). Quantum antenna arrays: The role of quantum interference on direction-dependent photon statistics. *Physical Review A*, *97*(5). Available from https://doi.org/10.1103/PhysRevA.97.053847, http://harvest.aps.org/v2/bagit/articles/10.1103/PhysRevA.97.053847/apsxml.

De Lisle, D., Iris, S., Arsenault, E., Smyth, J., Kroupnik, G., (2018). RADARSAT constellation mission status update Proceedings of the

European Conference on Synthetic Aperture Radar, EUSAR 9783800746361, 528–532. Institute of Electrical and Electronics Engineers Inc. Canada. http://conference.vde.com/eusar/history/Pages/EUSARHistory.aspx, 2018.

Marghany, M. (2019). *Automatic detection algorithms of oil spill in radar images*. CRC Press.

Marghany, M. (2021). *Nonlinear ocean dynamics: Synthetic aperture radar*. Elsevier.

Merrill, I. S. (1980). *Introduction to radar systems*. Mcgraw-Hill.

Miller, D. A. B. (2008). *Quantum mechanics for scientists and engineers*. Cambridge University Press (CUP) xxi-xxii. Available from 10.1017/cbo9780511813962.003.

Misra, T. (2017). Indian remote sensing sensor system: Current and future perspective. *Proceedings of the National Academy of Sciences India Section A - Physical Sciences*, *87*(4), 473–486. Available from https://doi.org/10.1007/s40010-017-0429-7, http://link.springer.com/journal/40010.

Okołowicz, J., Płoszajczak, M., & Rotter, I. (2003). Dynamics of quantum systems embedded in a continuum. *Physics Report*, *374*(4-5), 271–383. Available from http://www.elsevier.com/locate/physrep, 10.1016/S0370-1573(02)00366-6.

Oktavianto, A., Tjahjono, A. A., & Pambudi, W. (1967). Basic electronics. *PIP Semarang*.

Pitkänen, M. Quantum antenna hypothesis. Quantum Hardware of Living Matter. (2006).

Plant, W. J. (1991). The variance of the normalized radar cross section of the sea. *Journal of Geophysical Research: Oceans*, *96*(C11), 20643–20654. Available from https://doi.org/10.1029/91jc02099.

Salemian, S., & Mohammadnejad, S. (2008). Quantum Hadamard gate implementation using planar lightwave circuit and photonic crystal structures. *American Journal of Applied Sciences*, *5*(9), 1144–1148. Available from https://doi.org/10.3844/ajassp.2008.1144.1148, http://www.scipub.org/fulltext/ajas/ajas591144-1148.pdf.

Sanz, M., Las Heras, U., García-Ripoll, J. J., Solano, E., & Di Candia, R. (2017). Quantum estimation methods for quantum illumination. *Physical Review Letters*, *118*(7). Available from https://doi.org/10.1103/PhysRevLett.118.070803, http://harvest.aps.org/bagit/articles/10.1103/PhysRevLett.118.070803/apsxml.

Shankar, R. (2020). *Fundamentals of physics II: electromagnetism, optics, and quantum mechanics*. Yale University Press.

Shapiro, J. H. (2020). The quantum illumination story. *IEEE Aerospace and Electronic Systems Magazine*, *35*(4), 8–20. Available from https://doi.org/10.1109/MAES.2019.2957870, https://ieeexplore.ieee.org/xpl/RecentIssue.jsp?punumber = 62.

Shepherd, D. J. (2006). On the role of Hadamard Gates in quantum circuits. *Quantum Information Processing*, *5*(3), 161–177. Available from https://doi.org/10.1007/s11128-006-0023-4.

Singh, S., Ahmed, J., Raghuvanshi, K. K., Agarwal, P. A prospective study of information-theoretic and practical quantum key distribution. (2022).

Skolnik, M. (2002). Role of radar in microwaves. *IEEE Transactions on Microwave Theory and Techniques.*, *50*(3), 625–632. Available from https://doi.org/10.1109/22.989947.

Skolnik, M. L. (2008). *Radar handbook*. McGraw Hill.

Slepyan, G. Y., Vlasenko, S., & Mogilevtsev, D. (2020). Quantum antennas. *Advanced Quantum Technologies*, *3*.

Smith, J. F. (2009). Quantum entangled radar theory and a correction method for the effects of the atmosphere on entanglement. *Proceedings of SPIE - The International Society for Optical Engineering*, *7342*. Available from https://doi.org/10.1117/12.819918.

Solari, L., Del Soldato, M., Bianchini, S., Ciampalini, A., Ezquerro, P., Montalti, R., Raspini, F., & Moretti, S. (2018). From ERS 1/2 to Sentinel-1: Subsidence monitoring in Italy in the last two decades. *Frontiers in Earth Science*, *6*. Available from https://doi.org/10.3389/feart.2018.00149, https://www.frontiersin.org/journals/earth-science.

Stoica, P., & Ng, B. C. (1998). On the Cramér-Rao bound under parametric constraints. *IEEE Signal Processing Letters*, *5*(7), 177–179. Available from https://doi.org/10.1109/97.700921.

Tan, S. H., Erkmen, B. I., Giovannetti, V., Guha, S., Lloyd, S., Maccone, L., Pirandola, S., & Shapiro, J. H. (2008). Quantum illumination with gaussian states. *Physical Review Letters*, *101*(25). Available from https://doi.org/10.1103/PhysRevLett.101.253601, http://oai.aps.org/oai?verb = GetRecord&Identifier = oai:aps.org:PhysRevLett.101.253601&metadataPrefix = oai_apsmeta_2, United States.

Trees. (2004). *Detection, estimation, and modulation theory, part I: detection, estimation, and linear modulation theory*. John Wiley & Sons.

Trees. (2001). *Detection, estimation, and modulation theory: Part III. Radar-sonar signal processing and gaussian signal in noise*. John Wiley & Sons.

Tsang, M. (2012). Ziv-Zakai error bounds for quantum parameter estimation. *Physical Review Letters*, *108*(23). Available from https://doi.org/10.1103/PhysRevLett.108.230401Singapore, http://oai.aps.org/filefetch?identifier = 10.1103/PhysRevLett.108.230401&component = fulltext&description = markup&format = xml.

Widrow, B., & Kollár, I. (2008). *Quantization noise: Roundoff error in digital computation, signal processing, control, and communications*. United States: Cambridge University Press. Available from http://doi.org/10.1017/CBO9780511754661, https://doi.org/10.1017/CBO9780511754661.

Wu, B.-H., Guha, S., & Zhuang, Q. (2023). Entanglement-assisted multi-aperture pulse-compression radar for angle resolving detection. *Quantum Science and Technology*, *8*(3), 035016. Available from https://doi.org/10.1088/2058-9565/acd13f.

Xiang, G. Y., Ralph, T. C., Lund, A. P., Walk, N., & Pryde, G. J. (2010). Heralded noiseless linear amplification and distillation of entanglement. *Nature Photonics*, *4*(5), 316–319. Available from https://doi.org/10.1038/nphoton.2010.35.

Yao, X. W., Wang, H., Liao, Z., Chen, M. C., Pan, J., Li, J., Zhang, K., Lin, X., Wang, Z., Luo, Z., Zheng, W., Li, J., Zhao, M., Peng, X., & Suter, D. (2017). Quantum image processing and its application to edge detection: Theory and experiment. *Physical Review X*, *7*(3). Available from https://doi.org/10.1103/PhysRevX.7.031041, https://journals.aps.org/prx.

Zhuang, Q., Shapiro, J. (2022a). Ultimate limit of quantum pulse-compression ranging: hypothesis testing and parameter estimation. *APS March Meeting Abstracts*, 2022.

Zhuang, Q., & Shapiro, J. H. (2022b). Ultimate accuracy limit of quantum pulse-compression ranging. *Physical Review Letters*, *128*(1). Available from https://doi.org/10.1103/PhysRevLett.128.010501, https://journals.aps.org/prl/abstract/10.1103/PhysRevLett.128.010501.

Zink, M., Krieger, G., Fiedler, H., Moreira, A. (2007). The TanDEM-X mission: Overview and status International Geoscience and Remote Sensing Symposium (IGARSS) 3944–3947 Germany 10.1109/IGARSS.2007.4423711.

Chapter 4

Quantized Marghany turbulence boundary in the South China Sea by using SAR data

4.1 The South China Sea: Why?

There is a lack of knowledge of the South China Sea, especially the zone that is part of the coastal waters of Malaysia. This is due to the fact that the researchers only had a cursory understanding of physical oceanography and disregarded the dynamics of the South China Sea. The majority of research in Malaysian coastal waters is fundamental research that is based on scant field data. Using deployed equipment such as a current meter, decades have been spent on the fundamental task of measuring coastal physical water properties such as temperature, salinity, and current movements. As a result, the dynamics of coastal water in the entire South China Sea, as well as nearby regions such as the Malacca Straits and the Andaman Sea, are largely unexplored.

In contrast to a small portion of Malaysia's east coast, the coastal waters of Sabah and Sarawak are entirely excluded from the investigation. Even though a portion of Malaysia's east coast has been extensively studied over many decades at least, these investigations have done very little to advance our knowledge of coastal water circulations. These studies constrained the coastal water circulation along Malaysia's east coast to the operation of the southwest and northeast monsoon winds as well as tidal forces. Therefore the exclusion of Sabah and Sarawak from the investigation hinders our understanding of the complex oceanographic processes that occur in these regions, which are crucial for the sustainable management and conservation of marine resources. Further research is needed to fill this knowledge gap and ensure effective management of Malaysia's coastal waters.

However, recent advancements in technology have allowed for more comprehensive studies to be conducted, providing a better understanding of the South China Sea's complex ecosystem and its importance in global oceanography. With the help of satellite imagery and advanced computer models, researchers are now able to analyze ocean currents, temperature changes, and marine life distribution in greater detail.

This chapter is an expanded view of the first comprehensive research project completed since 1992 by an author who was a trailblazer in the identification of upwelling, downwelling, fronts, eddies, and turbulence currents in the South China Sea and the waters off Malaysia's east coast. To fully comprehend the turbulent boundary phenomena in the coastal waters of Malaysia, this chapter has conducted groundbreaking research that applies quantum mechanics. The author's extensive knowledge and experience in the field of oceanography make this chapter a valuable contribution to the scientific community. The use of quantum mechanics in this research work is expected to provide new insights into the complex phenomena of turbulent boundary layers in coastal waters.

4.2 What is the boundary current turbulent flow?

Boundary current turbulent flow refers to the chaotic movement of water in the ocean's boundary currents, which are narrow, fast-moving streams of water that flow along the edges of continents and large ocean basins. This turbulent flow is driven by a combination of factors, including wind, temperature differences, and the Earth's rotation. Therefore this flow is characterized by high levels of turbulence due to interactions between the current and the seafloor, as well as with other water masses. This type of flow is important for understanding the ocean's circulation and its impact on climate, as it transports heat, nutrients, and other materials across vast distances. Studying turbulent flow can also help us better predict weather patterns and oceanic events such as El Niño.

In this sense, can turbulent boundary currents include many features over short or long periods? Yes, turbulent boundary currents can exhibit a wide range of features

Synthetic Aperture Radar Image Processing Algorithms for Nonlinear Oceanic Turbulence and Front Modeling. DOI: https://doi.org/10.1016/B978-0-443-19155-8.00004-0

over both short and long timescales. These features may include eddies, vortices, and other types of flow structures that can persist for varying lengths of time. These features are influenced by various factors, such as the topography of the ocean floor, wind patterns, and water temperature.

Involving numerous features in turbulent flow can create a complicated phenomenon that is difficult to study. This complexity arises due to the interdependence of various factors such as fluid viscosity, flow rate, and turbulence intensity. However, advancements in computational modeling and simulation techniques have enabled researchers to better understand and predict turbulent flow behavior.

Therefore it is well known that western boundary currents and eastern boundary currents are two different types of boundary currents, ocean currents whose dynamics are influenced by the presence of a coastline. Western boundary currents are warm and fast-moving, while eastern boundary currents are slow and cold due to the influence of the nearby landmasses. In this understanding, a coastal zone behaves as a boundary layer for the open offshore dynamic movements, and therefore, it is dominated by the fluctuation of kinetic energy as was demonstrated in Chapter 2. In this regard, the velocity gradients in this case would be developed perpendicular to the coastal boundary. The velocity profile for the water flows through the different water depths is demonstrated in Fig. 4.1.

Therefore how does the viscosity affect the boundary layer that is developed by the coastal sea? For a given free stream velocity, a fluid with greater kinematic viscosity will form a thicker hydrodynamic boundary layer compared with a fluid with lesser kinematic viscosity. This mathematical can be expressed as:

$$\delta_h = \frac{5x^{0.5}\mu}{\sqrt{U_\infty}} \tag{4.1}$$

where x is the leading edge; μ is the kinematic viscosity; and finally, U_∞ is the stream velocity. The main premise of boundary layer theory is that there is a no-slip condition, which means that when a fluid particle adheres to the boundary, there is no relative motion between the fluid particle and the boundary of the solid surface, and as a result, the fluid particle's velocity equals the velocity of the boundary. The velocity of fluid particles adhering to the boundary will be zero if the boundary is stationary.

In this view, eastern boundary currents move slowly and broadly and are generally shallow. On the eastern side of oceanic basins, they can be found (adjacent to the western coasts of continents). The Benguela Current (Fig. 4.2), the Canary Current (Fig. 4.3), the Humboldt (Peru) Current (Fig. 4.4), and the California Current (Fig. 4.5) are a few examples of subtropical eastern boundary currents that flow equatorward and carry cold water from higher latitudes to lower latitudes. Eastern boundary current regions become productive regions of the ocean as a result of coastal upwelling, which frequently transports nutrient-rich water there.

Subtropical or low-latitude western boundary currents are two subsets of the western boundary currents themselves. Western intensification causes warm, deep, narrow, and swift-moving currents called subtropical western boundary currents to form on the west side of ocean basins. They transport warm water from the tropics to the poles. Examples include the Agulhas Current, the Kuroshio Current, and the Gulf Stream (Fig. 4.6). Similar to subtropical western boundary currents, low-latitude western boundary currents are subtropical equatorial currents that carry cool water. The Mindanao Current and the North Brazil Current are two examples.

4.3 Sverdrup relation

A large gyre in such a basin is one example of a large oceanic current that exhibits western intensification. In

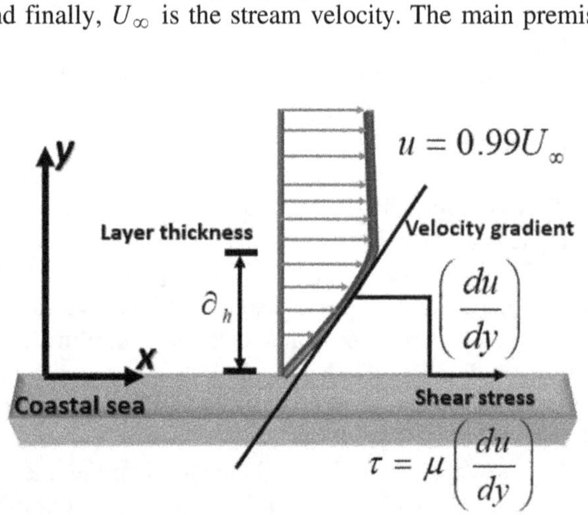

FIGURE 4.1 Simplification of the coastal sea boundary layer.

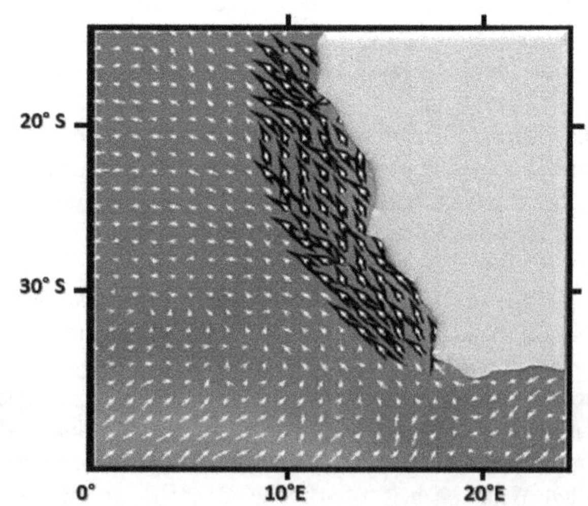

FIGURE 4.2 Eastern boundary Benguela Current in the Atlantic Ocean.

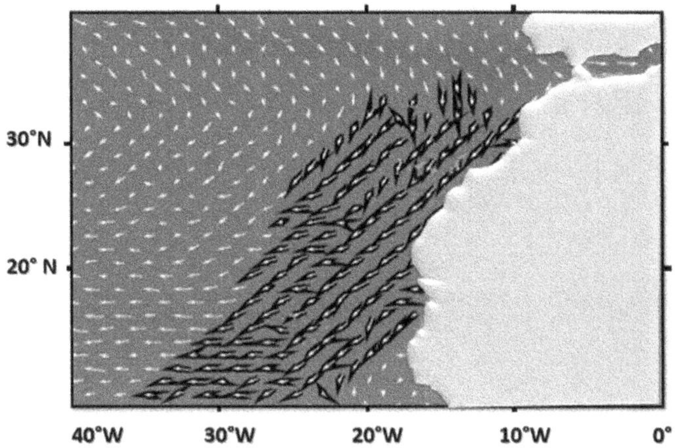

FIGURE 4.3 Canary current in the eastern Atlantic Ocean boundary.

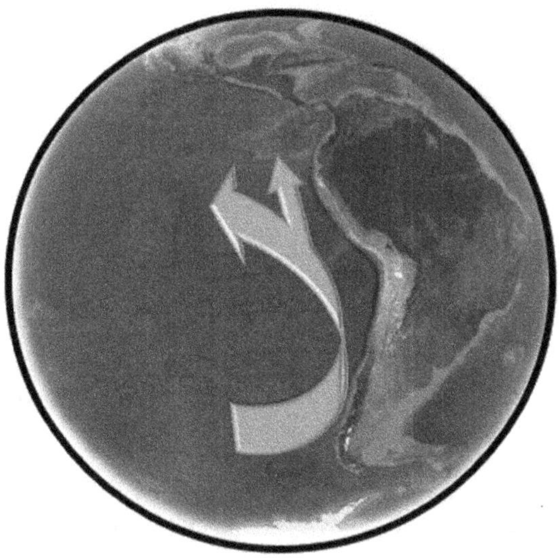

FIGURE 4.4 Humboldt (Peru) Current in the Pacific Ocean.

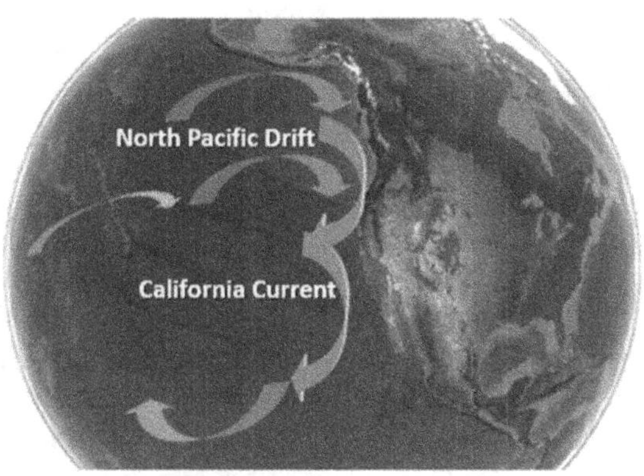

FIGURE 4.5 California Current in the Pacific Ocean.

the tropics, the trade winds blow westward. At mid-latitudes, the westerlies blow eastward. This holds true for the ocean's surface, which has a curvature in both the north and south hemispheres that causes Sverdrup to move equatorward (toward the tropics). In this view, the relationship between the wind stress acting on the open ocean's surface and the vertically integrated meridional (north-south) transport of ocean water is known as the Sverdrup balance or Sverdrup relation, which is formulated as:

$$V = \frac{\vec{\hat{k}} \cdot \nabla \times \vec{\tau}}{\beta}. \tag{4.2}$$

where $\vec{\hat{k}}$ is the unit vector in the z (vertical) direction; $\vec{\tau}$ is the wind stress vector; and V is the vertically integrated meridional mass transport, which includes the geostrophic interior mass transport and the Ekman mass transport. In Eq. (4.2), therefore, a beta plane approximation β in geophysical fluid dynamics involves setting the Coriolis parameter, f_0, to vary linearly in space. This approximation is commonly used to study the dynamics of large-scale atmospheric and oceanic flows, as it simplifies the equations of motion while still capturing the essential features of the Coriolis force. In the so-called f-plane approximation, this variation is ignored, and a value of f appropriate for a specific latitude is used throughout the domain. On a rotating sphere, such as the Earth, f varies with the sine of the latitude. This approximate solution is represented by a tangent plane that touches the sphere's surface at this latitude. In this understanding, a linear Taylor series approximation to this variability about a specific latitude ϕ_0 is a more precise model:

$$f = f_0 + \beta y \tag{4.3}$$

Therefore

$$\beta = (df/dy)|_{\phi_0} = \frac{2\Omega\cos(\phi_0)}{r} \tag{4.4}$$

In Eq. (4.4), Ω is the earth's angular rotation rate (0.7272×10^{-4}/s), r is the radius of the earth, and y is the meridional separation from ϕ_0. Therefore β in Eq. (4.3)

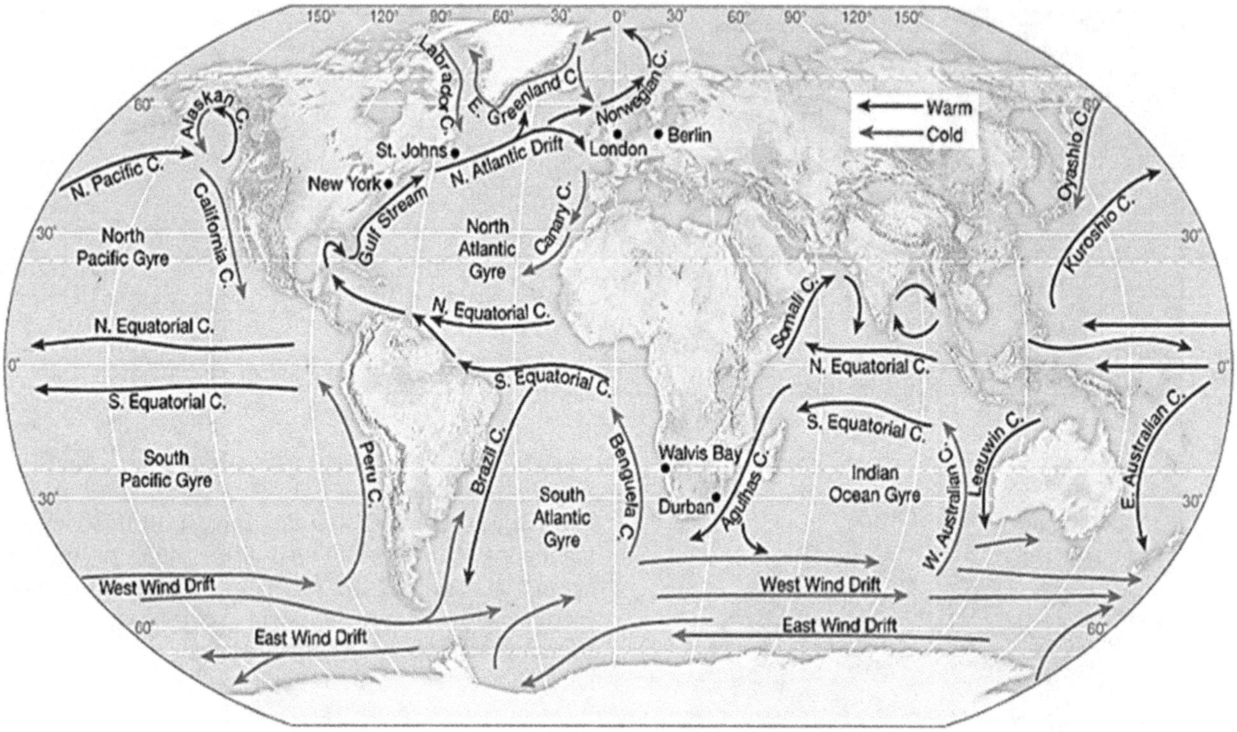

FIGURE 4.6 The world ocean currents.

demonstrates the meridional variation of the Coriolis force brought on by the Earth's spherical shape, known as the Rossby parameter (or simply beta), which is a number used in geophysics and meteorology. For the creation of Rossby waves, it is crucial. With regard to large-scale atmospheric and oceanic dynamics, Rossby waves in particular depend on the variation of f as a restoring force; they do not exist if the Coriolis parameter is only approximated as a constant.

Due to the conservation of mass and potential vorticity, a strong, narrow poleward current that travels along the western coast balances this transport by allowing the vorticity generated by coastal friction to counterbalance the vorticity input from the wind.

The sign of the wind stress curl in Eq. (4.2) and the direction of the resulting currents are reversed, and this has the opposite effect on the polar gyres. The main west-side currents are stronger than the opposite ones (such as the Gulf Stream of the North Atlantic Ocean or the California Current of the North Pacific Ocean).

Therefore the Sverdrup relation through the water column can be formulated as:

$$\beta v_g - f\partial w/\partial z. \tag{4.5}$$

here, v_g is the water's velocity's z-component (upward), and w is the geostrophic interior's y-component (northward). According to this equation, a vertical column of water moves toward the pole when it is stretched and toward the equator when it is compressed. The vorticity equation can be integrated from this level to the base of the Ekman surface layer to obtain: assuming, as did Sverdrup, that there is a level below which motion ceases:

$$\beta V_g = f\rho w_E, \tag{4.6}$$

where w_E is the vertical velocity at the bottom of the Ekman layer, V_g is the geostrophic meridional mass transport, and ρ is the density of seawater. This equation is used to calculate the Ekman transport, which is the net transport of water due to the balance between wind stress and Coriolis force. It is an important factor in understanding ocean circulation and its impact on climate. In this regard, the Ekman layer is the region of a fluid where the forces of turbulent drag, Coriolis force, and pressure gradient force are balanced. Vagn Walfrid Ekman provided the first description of it. Both the atmosphere and the ocean contain Ekman layers. Ekman layers can be of two different kinds. The initial form develops at the ocean's surface and is propelled by surface winds, which pull on the water's surface. The second type takes place at the ocean's and atmosphere's base, where flow over uneven surfaces is accompanied by frictional forces.

Emerging with the presumption of a neutrally stratified fluid and a balance between the forces of the pressure gradient, Coriolis, and turbulent drag, the Ekman layer's mathematical formulation can be fully grasped as:

$$-fv = -\rho^{-1}\frac{\partial p}{\partial x} + K_m\frac{\partial^2 u}{\partial z^2}, \tag{4.7}$$

$$-fu = -\rho^{-1}\frac{\partial p}{\partial y} + K_m\frac{\partial^2 v}{\partial z^2}, \tag{4.8}$$

$$0 = -\rho^{-1}\frac{\partial p}{\partial z} \tag{4.9}$$

where f is the local Coriolis parameter, K_m is the diffusive eddy viscosity, and u and v are the velocities in the x and y directions, respectively. These quantities can be derived from the mixing length theory. According to this perspective, the mixing length $\overline{\xi'^2}$ and eddy viscosity can be related mathematically as follows:

$$K_m = \overline{\xi'^2}\left|\frac{\partial \overline{w}}{\partial z}\right|, \tag{4.10}$$

Consequently, the result of multiplying horizontal $\overline{u'}$ and vertical fluctuations $\overline{w'}$ is as follows:

$$\overline{u'w'} = \overline{\xi'^2}\left|\frac{\partial \overline{w}}{\partial z}\right|\frac{\partial \overline{u}}{\partial z}. \tag{4.11}$$

Keep in mind that p is a modified pressure because we have added the hydrostatic pressure to account for gravity. An Ekman layer is hypothetically possible in many regions, including the bottom of the atmosphere, close to the surface of the earth and ocean, the bottom of the ocean, close to the sea floor, and the top of the ocean, close to the air − water interface. Each of these various scenarios calls for different boundary conditions. The boundary conditions applied to the resulting system of ordinary differential equations can be used to account for each of these scenarios. Following are examples of the top and bottom boundary layers separately.

Let us consider the boundary conditions of the Ekman layer in the upper ocean:

$$at z = 0 : \mu\frac{\partial u}{\partial z} = \tau^x and \mu\frac{\partial v}{\partial z} = \tau^y, \tag{4.12}$$

where $\mu \equiv \rho K_m$ is the dynamic viscosity, and τ^x and τ^y are the parts of the surface stress, τ, of the wind field over the ocean. Let us assume that the boundary conditions are as: $z \to -\infty : u \to u_g, v \to v_g$, here the geostrophic component flows in the x and y directions are u_g and v_g; respectively. By resolving these differential equations, one can obtain:

$$u = u_g + \frac{\sqrt{2}}{\rho_0 f E_d} e^{z[E_d]-1}\left[\tau^x \cos\left(z[E_d]^{-1} - \frac{\pi}{4}\right) - \tau^y \sin\left(z[E_d]^{-1} - \frac{\pi}{4}\right)\right], \tag{4.13}$$

$$v = v_g + \frac{\sqrt{2}}{\rho_0 f E_d} e^{z[E_d]-1}\left[\tau^x \sin\left(z[E_d]^{-1} - \frac{\pi}{4}\right) + \tau^y \cos\left(z[E_d]^{-1} - \frac{\pi}{4}\right)\right], \tag{4.14}$$

$$E_d = \sqrt{\left[2\overline{\xi'^2}\left|\frac{\partial \overline{w}}{\partial z}\right|\right][|f|]^{-1}} \tag{4.15}$$

The Ekman layer depth, or E_d, indicates the depth to which wind-induced turbulent mixing in the ocean can penetrate. Keep in mind that it depends on the turbulent diffusivity K_m and the latitude, which are both encapsulated by f. At 45° latitude ($f = 10^{-4}$/s) and a typical $K_m = 0.1$/ms, the distance is roughly 45 m. The Ekman depth prediction does not always match observations exactly. Therefore the vertical velocity can be obtained by using the continuity equation as follows:

$$w = \frac{1}{f\rho_o}\left[-\left(\frac{\partial \tau^x}{\partial x} + \frac{\partial \tau^y}{\partial y}\right)e^{z[E_d]^{-1}}\sin\left(z[E_d]^{-1}\right) + \left(\frac{\partial \tau^y}{\partial x} - \frac{\partial \tau^x}{\partial y}\right)\left(1 - e^{z[E_d]^{-1}}\cos\left(z[E_d]^{-1}\right)\right)\right]. \tag{4.16}$$

It is worth notable that the Coriolis effect causes a deflection of the water to the right of the wind direction. In the Southern Hemisphere, the volume transport is deflected to the left of the wind direction. In this sense, only since around 1980 has there been instrumentation sensitive enough to observe a velocity profile at the shallow depth of the Ekman layer, which is only 10 − 20 m below the ocean's surface. Additionally, wind waves alter the flow close to the surface, making surface observations challenging. The Ekman layer plays a crucial role in the transfer of heat, momentum, and gases between the atmosphere and the ocean. Understanding its dynamics is essential for predicting weather patterns and climate change. Therefore it has only been possible to observe the Ekman layer since the creation of reliable surface moorings and accurate current meters. In an attempt to observe the spiral that bears his name, Ekman created a current meter, but he was unsuccessful. Current is measured using both the acoustic Doppler current profiler and the vector measuring current meter. The vector measuring current meter measures current velocity and direction by measuring the magnetic field generated by the flow of water. The acoustic Doppler current profiler measures current velocity by using sound waves to detect the movement of particles in the water.

4.4 Stommel model

Stommel included a linearized, frictional term to account for the dissipative effects that prevent the real ocean from accelerating; he also assumed an ocean of constant density and depth

$D + h$ seeing ocean currents. Thus the model begins with the equations for steady-state momentum and continuity:

$$f(D+h)v - F\cos\left(\frac{\pi y}{b}\right) - Ru - g(D+h)\frac{\partial h}{\partial x} = 0 \quad (4.17)$$

$$-f(D+h)u - Rv - g(D+h)\frac{\partial h}{\partial y} = 0 \quad (4.18)$$

$$\frac{\partial[(D+h)u]}{\partial x} + \frac{\partial[(D+h)v]}{\partial y} = 0 \quad (4.19)$$

where f is the Coriolis force, R is bottom friction, g is gravity, and $-F\cos\left(\frac{\pi y}{b}\right)$ is wind forcing. At $y = 0$, the wind is blowing westward, and at $y = b$, it is blowing eastward. Let us act on Eq. (4.17) by $\frac{\partial}{\partial y}$ and equation $\frac{\partial}{\partial y}$ on 4.18 by $\frac{\partial}{\partial y}$ subtracting, and then using Eq. (4.19), delivers:

$$v(D+h)\left(\frac{\partial f}{\partial y}\right) + \frac{\pi F}{b}\sin\left(\frac{\pi y}{b}\right) + R\left(\frac{\partial v}{\partial x} - \frac{\partial u}{\partial y}\right) = 0 \quad (4.20)$$

Eq. (4.20) becomes: if we linearize by assuming that D>>h and introduce a Stream function ψ.

$$\nabla^2\psi + \alpha\left(\frac{\partial \psi}{\partial x}\right) = \gamma\sin\left(\frac{\pi y}{b}\right) \quad (4.21)$$

where

$$\alpha = \left(\frac{D}{R}\right)\left(\frac{\partial f}{\partial y}\right) \quad (4.22)$$

and

$$\gamma = \frac{\pi F}{Rb} \quad (4.23)$$

The solutions to Eq. (4.21) with the boundary condition that ψ must remain constant along the coastlines and for various values of α highlight the significance of the Coriolis parameter's variation with latitude in causing the strengthening of western boundary currents. These currents are noted to be significantly warmer, faster, deeper, and narrower than their eastern counterparts. In this sense, there is no preference for intensification or acceleration of ocean circulation near the western boundary for a nonrotating state (zero Coriolis parameter), and where that is a constant. In a uniformly rotating ocean, the streamlines behave symmetrically in all directions, and the height contours show a nearly parallel relationship to the streamlines. Last but not least, an obvious tendency for asymmetrical streamlines is found on a rotating sphere, the case where the Coriolis force is latitudinally variable, with a strong clustering along the western coasts.

4.5 What are the key distinctions between Stommel mode and Severdrup balance?

Sverdrup made the following three assumptions: (1) there is a uniform depth of no motion; (2) the internal flow in the ocean is geostrophic; and (3) Ekman's transport is accurate. The majority of research in the South China Sea has, however, only connected a basic Sverdrup speculation without fully understanding the mechanism of water movements between two boundaries of the continental shelf: the east coast of peninsular Malaysia and the Borneo States, to study the volume of mass transport as an indicator for upwelling and downwelling phenomena. It is unclear why the Borneo States coastal zone is regarded as the South China Sea's eastern boundary while Peninsular Malaysia's east coast is its western boundary in the majority of Malaysian research on the region. Without comprehending the full mechanism of ocean dynamics under the effects of wind or tidal forces, the majority of studies in Malaysian coastal waters simply used an existing ocean circulation model. This approach may lead to inaccurate predictions and limit the understanding of complex coastal dynamics. Therefore there is a need for further research that combines both modeling and field observations to improve our understanding of coastal water dynamics in Malaysia.

Therefore Malaysian researchers never logically fully grasped that there could only be one boundary condition encountered: the eastern boundary cannot flow through, when they implemented severed speculation across the South China Sea. For more thorough flow descriptions, more thorough equations are required. This boundary condition may not have been taken into account, which could have led to inaccurate data and incorrect conclusions. More investigation and the creation of equations could lead to more precise flow descriptions and comprehension of the region.

In this regard, the physical oceanography studies that have been carried out in the South China Sea have failed to identify the western boundary current along the east coast of peninsular Malaysia. Despite the implementation of the Sverdrup mass transport proposition as follows:

$$u \equiv \frac{\partial \psi}{\partial y}, \quad (4.24)$$

$$v \equiv -\frac{\partial \psi}{\partial x}, \quad (4.25)$$

In both Eqs. (4.24) and (4.25), one can represent a flow field with a vector velocity at each point in space and at each instant in time. The streamlines of the flow are the instantaneous curves that are always perpendicular to the vectors' directions. If the flow is erratic, the streamlines pattern would then evolve over time. Fluid flow refers to the path line as the trajectory of a fluid particle as it would be discussed in the quantum fluid dynamic section later, or the path taken by a Lagrangian drifter. For a steady flow, the path line and the streamline are identical, while they differ for an unsteady flow.

A steady flow's volumetric flow rate between any two streamlines is denoted by $d\psi$, and the volumetric flow rate between dual streamlines, that is, $\psi_1 and \psi_2$ is denoted by $\psi_1 - \psi_2$. Consider an arbitrary line dx = (dx, dy) between two streamlines to understand this. The stream lines' volumetric flow rate is:

$$vdx + (-u)dy = -\frac{\partial\psi}{\partial x}dx - \frac{\partial\psi}{\partial y}dy = -d\psi \quad (4.26)$$

Therefore the sea surface is a stream function that can be scaled by $g[f]^{-1}$ in which:

$$\psi = -g[f]^{-1}\zeta \quad (4.27)$$

Eq. (4.27) says that the surface geostrophic transport is proportional to the difference in height ζ, independent of the distance between the streamlines. This concept is important in oceanography as it allows us to estimate the ocean's circulation based on sea surface height measurements obtained from satellites. The geostrophic currents are also responsible for the formation of large-scale oceanic features such as eddies and gyres. In this sense, the stream function can be explored to estimate the mass transport as follows:

$$M_x \equiv \frac{\partial\Psi}{\partial y}, \quad (4.28)$$

$$M_y \equiv -\frac{\partial\Psi}{\partial x} \quad (4.29)$$

Because M_x grows with X, the solutions are only applicable to the east side of the ocean. The outcome is the result of ignoring friction, which eventually would balance the wind-driven flow. The global system of surface currents has nevertheless been described using Sverdrup solutions. Each basin's western boundary is reached after applying the solutions throughout the basin. There, the inclusion of north-south currents contained within a thin, horizontal boundary layer forces the conservation of mass.

Stommel's subsequent assumption of a constant rotation also developed a symmetric solution devoid of a western boundary current. Finally, he made the assumption that latitude affects the Coriolis force. This is why the geostrophic current flow in the Malaysian coastal waters is weak as the South China Sea is located above the equator by a few degrees. In this sense, it is well known that the western boundary current cannot exist without the variation of Coriolis force with latitude, and that other flow models that use different formulations for friction result in western boundary currents with a different structure.

This resulted in a solution that was intensified in the West. According to Stommel, the concentration of streamlines in the west indicated that the existence of the Gulf Stream in the ocean might be explained, for example, by the variation of Coriolis force with latitude.

Needless to say, Sverdrup and Stommel's work suggested the primary processes causing a basin-wide, wind-driven circulation, which is consistent with the aforementioned viewpoint. Their research also showed that the ocean's response to wind forcing is not instantaneous, but rather takes several months to fully develop.

4.6 Munk solutions

One of the limitations of the Sverdrup model is ignoring friction. Indeed friction could be the key to generating turbulence, especially in shallow waters. In this view, by adding details from Rossby (1936) on lateral eddy viscosity to this base, Munk (1950) developed a solution for circulation within an ocean basin. Sverdrup's concept of vertically integrated mass transport flowing over a stationary deeper layer was used by Munk. The mathematical issue was subsequently clarified and rendered more plausible. The ocean currents, which are independent of depth and concentrated in the upper kilometer of the ocean, are not barotropic. Munk developed ocean current movement as a function of eddy friction $\mu_z and \mu_H$ according to the following theory:

$$\frac{1}{\rho}\frac{\partial p}{\partial x} = fv + \frac{\partial}{\partial z}\left(\mu_z\frac{\partial u}{\partial z}\right) + \mu_H\frac{\partial^2 u}{\partial x^2} + \mu_H\frac{\partial^2 u}{\partial y^2} \quad (4.30)$$

$$\frac{1}{\rho}\frac{\partial p}{\partial y} = -fu + \frac{\partial}{\partial z}\left(\mu_z\frac{\partial v}{\partial z}\right) + \mu_H\frac{\partial^2 v}{\partial x^2} + \mu_H\frac{\partial^2 v}{\partial y^2} \quad (4.31)$$

Similar to Sverdrup's integration, except that the surface is not at z = 0, Munk integrated the equations from a depth D to the surface at $z = z_0$. Munk achieved the assumption that μ_H is constant, that currents at the depth D vanish, and that they apply at the layer's top and bottom horizontal boundaries.

$$\mu_H\Delta^4\psi - \beta\frac{\partial\Psi}{\partial x} = -curl_z T \quad (4.32)$$

The first term in Eq. (4.31) represents the friction, second and third terms demonstrate the Severdrup balance. Therefore the biharmonic operator ∇^4 in equation 4.32 can be given by:

$$\nabla^4 = \frac{\partial^4}{\partial x^4} + 2\partial^4\frac{\partial x^2}{\partial y^2} + \frac{\partial^4}{\partial y^4} \quad (4.33)$$

In the interior of the ocean basin, where the horizontal derivatives of the velocity field are small, the friction term is large close to a lateral boundary. As a result, the interior force balance is identical to that in Sverdrup's solution. Therefore four boundary conditions are required for equation 4.32, a fourth-order partial differential equation. In this theory, Munk presupposed that there is no slip at the boundary and that the flow at a boundary is parallel to the boundary:

$$\Psi_{bdry} = 0, \quad \left(\frac{\partial \Psi}{\partial n}\right)_{bdry} = 0 \tag{4.34}$$

here n is normal to the boundary. After that, Munk addressed (4.32) with (4.34), assuming that the flow was in a rectangular basin with sides that ran from $x = 0$ to $x = r$ and from $y = s$ to $y = +s$. Furthermore, Munk also assumed that the zonal form of the wind stress is:

$$\tau = a\cos ny + b\sin ny + cn = j\pi s^{-1}, j = 1, 2, . \tag{4.35}$$

Western boundary current transport is unaffected by μ_H and solely dependent on integration over the width of the ocean basin. As a result, it is dependent on β, the wind stress's curl, and the ocean's width. Munk determined that the Gulf Stream should have a transport of 36 sv and the Kuroshio should have a transport of 39 sv using the best estimates of the wind stress that were available. The values are roughly half of what Munk's flow has been measured to be. Given how poorly understood the wind stress was, this agreement is excellent. Wind stress averaged over 5° squares curl as the foundation for Munk's solution. This exceeded the stress's curl. The wind-driven geostrophic circulation in an ocean basin can be calculated by combining the two solutions, as Munk demonstrated. The curl of the wind stress is always what drives the current.

This theory, known as the Munk boundary layer theory, describes how the movement of ocean currents is affected by the friction caused by eddies in the water. It has been widely used to study ocean circulation patterns and their impact on climate.

The key question is how the three models, that is, Severdrup, Stommel, and Munk, can identify the current boundary in such a large South China Sea basin. This is particularly challenging due to the complex and dynamic nature of the ocean currents and weather patterns in the region. However, advancements in technology and data collection have improved the accuracy of these models in recent years.

4.7 South China Sea boundary turbulence current flows

Understanding the boundary turbulence flows in any water body requires understanding its topology. The novel use of topology knowledge in dynamic oceanography research can assist in determining the occurrence or disturbance of the water body under internal and external forces such as gradients in water density, wind stress, and tidal energy when transferring the water body from a steady state to a turbulent state.

Can different forms of turbulence flow exist simultaneously across any body of water? Due to the complexity of the water flow and the available equipment, dynamic oceanography research can only investigate individual features without collaboration with other feature occurrences. In this section, we attempt to look for the different ocean boundary features in such bodies of water as the South China Sea, which could be applicable to other ones.

4.7.1 South China Sea

The South China Sea is the largest water body in Southeast Asia, which is surrounded by the Asian continent—Indonesian Archipelago, west Malaysia, Borneo, the Philippines, and Formosa. Therefore the seafloor of the South China Sea can be divided into three zones: (1) the Northern Sunda Shelf; (2) the Gulf of Thailand; and (3) the China Sea basin. Geologically speaking, the Sunda Shelf is a southern extension of the Southeast Asian mainland's continental shelf (Fig. 4.7). The largest land masses on the shelf are Borneo, Java, Madura, Sumatra, and the smaller islands surrounding them. It encompasses roughly 1.85 million km^2 of land. Strong bottom friction and tidal friction are caused by sea depths over the shelf that rarely goes beyond 50 meters, and large areas are often less than 20 meters. The Philippines, Sulawesi, and the Lesser Sunda Islands are separated from the Sunda Shelf by steep underwater gradients (not including Bali).

The southern shores of the Indochinese Peninsula and the northern half of the Malay Peninsula form the boundaries of the Gulf of Thailand, also referred to as the Gulf of Siam, which is a shallow inlet in the southwest South China Sea. It has a surface area of 320,000 km^2 (120,000 sq mi), is about 800 km (500 mi) long, and can be up to 560 km (350 mi) wide. The South China Sea is located in the southeast of the gulf, which is bordered on the north, west, and southwest sides by the coastlines of Vietnam, Cambodia, and the Mekong Delta region of Thailand (hence the name of the gulf) (Fig. 4.8).

The gulf is only 85 meters (279 feet) deep at its maximum and 58 meters (190 feet) deep on average. This slows down water exchange while the strong river water inflow lowers the gulf's salinity (3.05%–3.25%) and enriches the sediments. Water from the South China Sea that has a higher salinity (3.4%) only enters the gulf at greater depths. Below a depth of 50 meters (160 feet), it fills the depression in the center. This phenomenon is known as a halocline, which is a layer of water with a significant change in salinity. The halocline in the Gulf of Thailand creates unique conditions for marine life and can affect oceanographic processes.

The China Sea basin is the deepest part of the South China Sea, which is characterized by a combination of

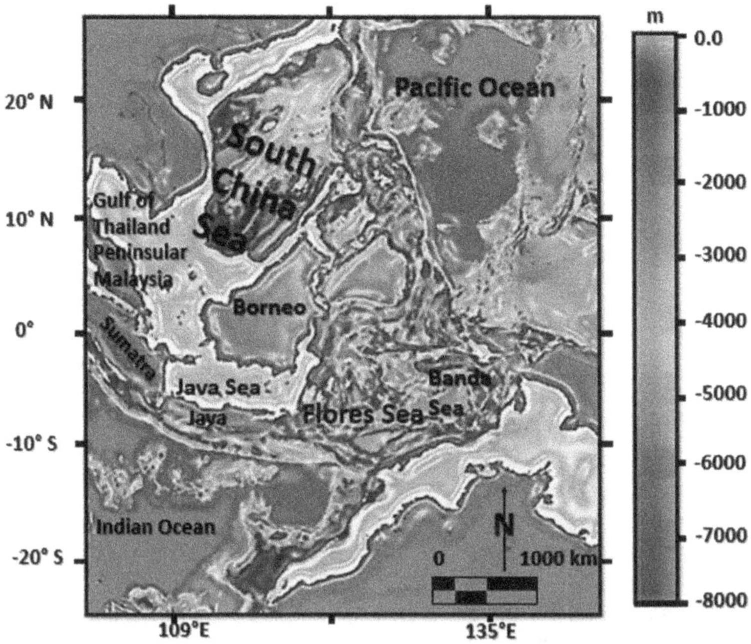

FIGURE 4.7 South China Sea geographical location and bathymetric variations.

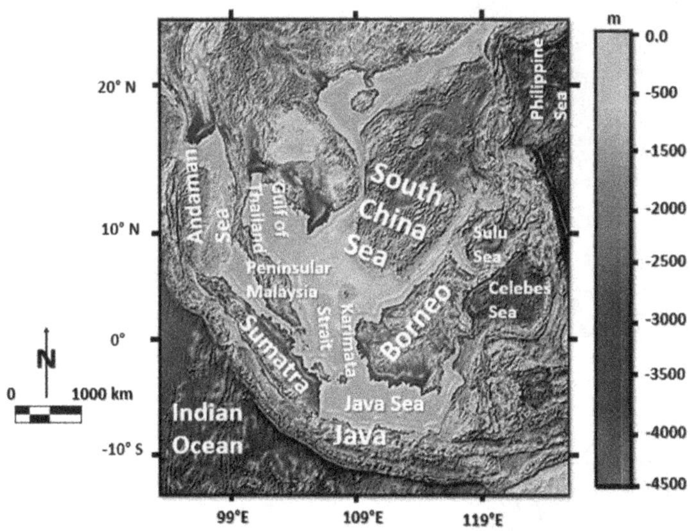

FIGURE 4.8 South China Sea of Malaysian coastal waters.

complicated bathymetric features. Off Sabah, the cost is the deepest Palwan trench with 3475 meters. Therefore the seafloor turns into an irregular plateau away from the Palawan trench. In this view, this area possesses seamounts and fringing reefs that protrude to the surface, appropriately naming it dangerous ground. The dangerous ground poses a significant threat to navigation and marine transportation due to its shallow depths and the presence of numerous underwater hazards. As a result, it requires careful navigation and mapping to ensure safe passage for vessels.

Needless to say, the average and maximum water depths of the South China Sea (SCS), which is bounded by Malaysia, the Philippines, China, and Vietnam, are 1212 and 5559 meters, respectively (Wu & Fallon, 2020). The SCS's northern boundary is linked to the East China Sea by the Taiwan Strait, its southern boundary is linked to the Java Sea by the Karimata Straits(Fig. 4.8), and its northeastern boundary is linked to the Pacific Ocean by the Luzon Strait (Qu et al., 2009).

Along with general circulation models (GCMs) (Qu et al., 2009; Qu, Du, et al., 2006a), the SCS may reach the Sulu Sea through the Mindoro Strait (). North East Current (NEC) water flows westward toward the Philippines before splitting into the Mindanao Current (MC) and the Kelvin Current (KC), which both contribute significantly to ocean circulation in the Philippine Sea (Qiu & Lukas, 1996). Through the Luzon Strait, KC/ Pacific water enters the SCS and continuously flows toward the Java Sea (Fig. 4.9). KC water can enter the SCS at any time of year, but the flow is typically greater in the winter (Qu et al., 2009; Wu & Fallon, 2020; Yuan et al., 2006; Yuan et al., 2014). In this understanding, the temperature and salinity of the water entering the South China Sea through the Luzon Strait are both lower than those of the water leaving it through the Karimata, Mindoro, and Taiwan Straits (blue). This phenomenon is due to the water mixing with different currents and passing through various regions with different temperatures and salinity levels. Understanding these oceanographic processes is important for predicting the region's climate patterns and marine ecosystems.

This pattern of water circulation can play a role in distributing the atmospheric and freshwater fluxes in the SCS. In other words, the immediate query that might come up is: Where does the freshwater and heat that the atmosphere delivers vanish? The SCS is a semiclosed basin below a depth of about 200 m. Only horizontal advection, with the inflow of cold, salty water through the Luzon Strait and the outflow of warm, freshwater through the Mindoro and Karimata straits, can counteract this heat and freshwater gain over the long term on average. The South China Sea Throughflow (SCSTF) is the name given to this circulation (Fig. 4.10), which has been thoroughly investigated in recent years.

In this view, the East Asian monsoon controls the complex ocean circulation in the SCS (Hu et al., 2000; Wei et al., 2016). Water from the Pacific Ocean in the east and the East China Sea in the north are the sources of

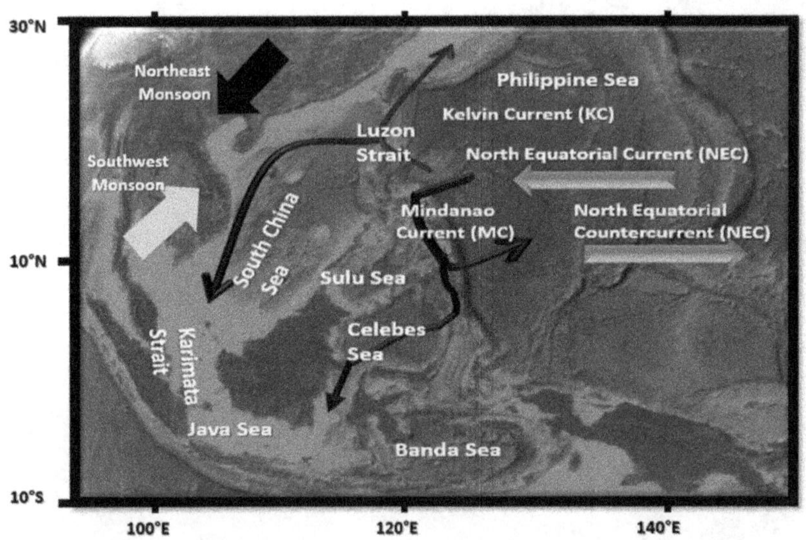

FIGURE 4.9 Current boundaries around Southeast Asia archipelago.

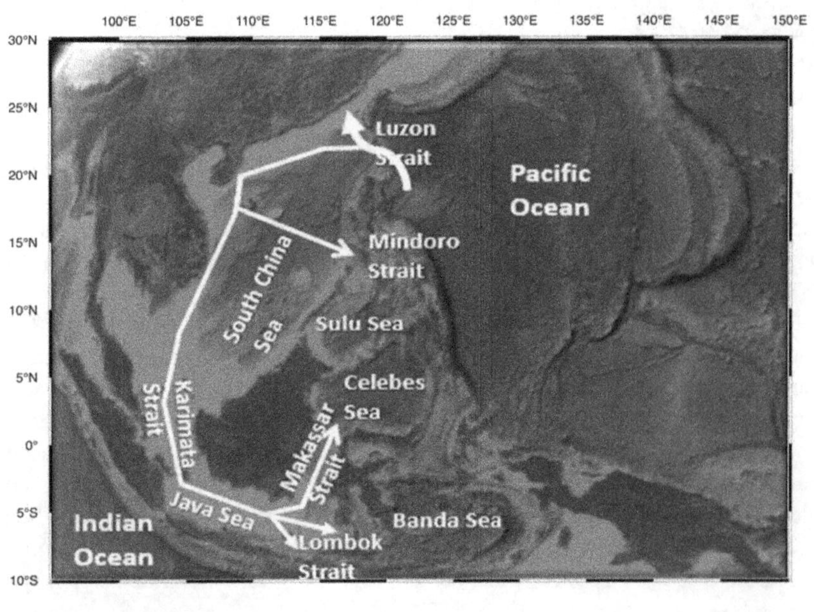

FIGURE 4.10 South China Sea Throughflow (SCSTF).

the southwesterly water flow that results from the winter monsoon (Chu et al., 1999; Hu et al., 2000; Wu & Fallon, 2020) (Fig. 4.9). In addition, this circulation creates an anticlockwise gyre between Vietnam and the Philippines (Fig. 4.11). Consequently, the summer monsoon causes a reversed circulation that transports water from the south to the SCS (Fig. 4.12) (Chu et al., 1999; Hu et al., 2000). As there are boundary current features that cannot be captured in any study conducted over a long period of time, the author believes that these are not the ideal views for water circulation in SCS. Indeed, no other features such as internal waves, vorticity, fronts, or turbulence boundary flows are ever investigated in SCS research, which is solely focused on SCS gyres. Therefore further research is required to fully understand the water circulation in SCS and its impact on the marine ecosystem and to develop effective strategies for the sustainable management of the region's resources. This can be achieved through a combination of field observations, numerical modeling, and data analysis techniques.

Our understanding of the seasonal and interannual variations is complicated by the dearth of in situ measurements of the surface water processes in the northeastern SCS. This limitation hinders the development of accurate models to predict the impact of climate change and human activities on the marine ecosystem in this region. Therefore there is a need for more comprehensive and continuous observations to improve our understanding of the surface water processes in the northeastern SCS.

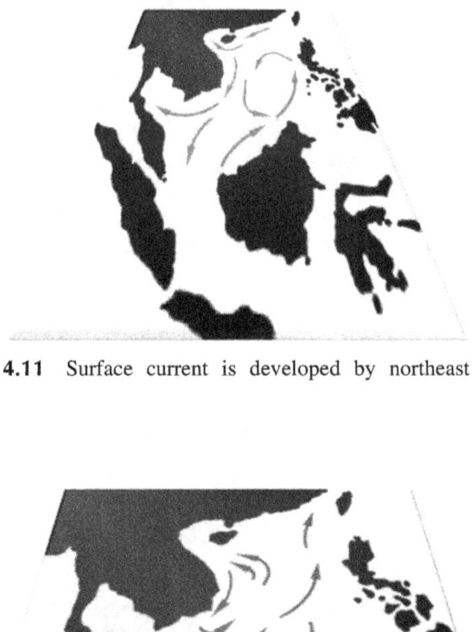

FIGURE 4.11 Surface current is developed by northeast monsoon wind.

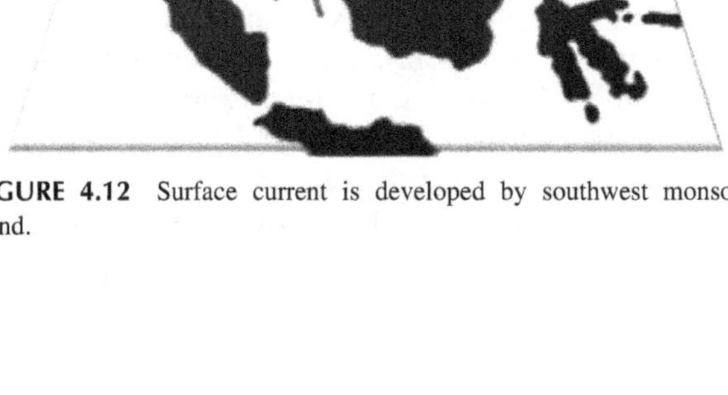

FIGURE 4.12 Surface current is developed by southwest monsoon wind.

4.7.2 Circulation of the South China Sea: overview

It has long been known that the SCS and the Pacific Ocean exchange water through the Luzon Strait. The initial observations made by (Wyrtki, 1961) that water enters the SCS in winter and flows back to the Pacific in summer based on early hydrographic data, sea level records, and the ship drifts. This is only true for the surface layer, where monsoonal winds are primarily responsible for forcing ocean circulation. Within the depth range of the thermocline, later observations revealed that there is a Kuroshio branch toward the SCS both in winter and summer (Guo, 1985; Qiu et al., 1984), and the intrusion of the North Pacific Tropical Water seems to occur during most, if not all, seasons of the year (e.g., Shaw, 1989; Shaw, 1991; Qu et al., 2000), presumably as a result of westward pressure gradient along the continental slope south of China (Shaw, 1989, 1991; Yuan et al., 2014).

In the middle layer, the flow in the Luzon Strait changes direction (Fig. 4.10). To balance the inflow of Pacific water into the surface layer, observations and numerical models have both demonstrated that water flows out of the SCS at depths between roughly 500 and 1500 m. Therefore the water on the SCS side is relatively homogeneous and appears to share characteristics with the Pacific water at about 2000 m in the deeper layer (> 2000 m) (Broecker et al., 1986; Chen et al., 2001; Jilan, 2004). This has been interpreted as proof that water from the Pacific is deeply ventilating the SCS as it passes over the Luzon strait.

This has been perceived as verification that water from the Pacific is deeply ventilating the SCS as it flows over the Luzon Strait. After crossing the Luzon Strait, the colder, denser water of the Pacific sinks (Wyrtki, 1961). Upwelling (or intense vertical mixing) takes place elsewhere to offset this descending movement, and as a result, the renewal of the SCS deep water is quicker than that of the Pacific (Li & Qu, 2006). Direct measurement of the deep water overflow through the Luzon Strait has proven challenging due to the region's complicated coastline and topography as well as the currents' high degree of variability. Consequently, a consistent near-bottom flow to the SCS with a mean speed of 0.14 m/s provided a transport estimate of 1.2 Sv in the deep layer of the Luzon Strait.

The entire transport that passes through the Luzon Strait, also known as the LST, is headed west (Wyrtki, 1961).

Recent observations favor a value that is close to the middle of the range, which is where the LST transport estimate, which has been the subject of previous studies, has been found to range from 0.5 to 10 Sv (Qu, 2000). The LST is typically larger in winter and smaller in summer on the seasonal timescale, in response to the seasonal reversing monsoon(Chu & Li, 2000). In particular, Qu (2000) indicates a maximum of 5.3 Sv in January-February and a minimum of 0.2 Sv in June-July based on hydrographic observations, whereas the estimate by Chu and Li (2000) shows a maximum of 13.7 Sv, even higher in February and at least 1.4 Sv in September. These discrepancies may be mainly due to uncertainties in intermediate and deep circulation (Tian et al., 2006). It is worth noting that the LST can be contaminated by eddies on different timescales. For this reason, there are various routes of entry of Pacific waters into the Luzon Strait. As part of the Pacific gyre (Metzger & Hurlburt, 1996), the LST is strongly influenced by the Western Boundary Current of the Pacific (LLWBC) and the encroachment of the Pacific waters has affected not only the Indonesian throughflow (ITF) but also has a dramatic effect on SCS circulation.

4.7.3 Pacific low-latitude western boundary current

Since it approaches the Philippine coast, the North Equatorial Current (NEC) splits into the northward-flowing Kuroshio and the southward-flowing Mindanao Current (MC), the NEC's bifurcation shifts poleward with depth and is subject to changes on a variety of timescales (B. Qiu & Lukas, 1996; Qiu et al., 1984; Tian et al., 2006; Toole et al., 1990) (Fig. 4.9). The bifurcation of the NEC shifts northward in winter and southward in summer in tandem with the seasonal reversal of the monsoon. This is made possible by a quantitative shift in how the NEC transport is split between the Kuroshio and the MC. As a result, in the winter and summer, the Kuroshio transport east of Luzon approaches its seasonal minimum and maximum, respectively. Results from high-resolution GCMs show that the bifurcation of the NEC shifts northward during El Nino years, resulting in a minimum transport in the Kuroshio east of Luzon, on an interannual timescale (Kim et al., 2004; Tangdong Qu et al., 2004). La Nina years appear to see a reversal of the situation.

The majority of the Kuroshio water avoids the Luzon Strait and travels north along the continental slope east of China, with only a small amount leaking into the SCS. The weakening Kuroshio east of Luzon provides a conducive scenario for the intrusion of Pacific waters through the Luzon strait as the NEC splits as it moves northward (Sheremet, 2001; Yaremchuk & Qu, 2004; Yuan et al., 2006). As a consequence, the LST generally exhibits a spiral of strength during El Nino years and weakness during La Nina years (Tangdong Qu et al., 2004). Since this LST variation, the SCS and the Pacific Ocean can be connected by an oceanic bridge as a result of ENSO's influence being transmitted into the SCS through the Luzon Strait.

The Mindoro strait outflow is the channel through which a portion of the LST leaves the SCS. Wyrtki (1961) identified that the southwest monsoon season, which runs from August to October, is the only time when the surface flow in the Mindoro Strait is southward. During other times of the year, it turns northward in time with the northeast monsoon. At depth, the local wind is less significant (Quadfasel et al., 1990). Both the North Pacific Tropical and Intermediate Waters are clearly visible in the Sulu Sea, indicating an intrusion of thermocline waters through the Mindoro strait, most likely as a result of large-scale forcing over the Pacific. Although there is currently no direct measurement available, high-resolution GCM results have amply demonstrated the existence of the Mindoro strait outflow (T. Qu, Girton, et al., 2006).

The circulation in the Celebes Sea may be directly impacted by the water of the SCS origin leaving the Sulu Sea in the south, primarily through the narrow Sibutu Passage. The NEC bifurcation can be moved by up to 0.5° by transport through the Mindoro Strait, according to numerical experiments (Metzger & Hurlburt, 1996). A stronger Kuroshio transport close to the Philippine coast may result from a southward shift of the NEC bifurcation caused by an improved Mindoro Strait transport. Further research into the specifics is required.

4.7.4 Indonesian throughflow boundary current

Through the Karimata Strait, the water from the Pacific continues southward into the Java Sea. The ITF has a history of being a key player in the global water exchange between ocean basins, which has a significant impact on the climate of the entire planet (Gordon, 1986). As a consequence, in the Makassar Strait, the ITF's main route, the maximum southward velocity occurs at a distance of about 150 m. Water and the surface flow in the strait are actually northward during the boreal winter, resulting in a much weaker ITF heat transport than previously thought (Gordon et al., 2003; Vranes et al., 2002).

It appears that the local Ekman current is too weak and shallow to explain the maximum subsurface velocity mentioned earlier. The Makassar Strait's intense vertical shear must originate from a distant location outside of the Indonesian Seas (Fig. 4.13). It is assumed that the circulation in the Makassar strait is primarily a result of the interaction between the southward-flowing ITF in the thermocline and the northward-flowing SCSTF near the sea surface, based on wind data over the past 40 years

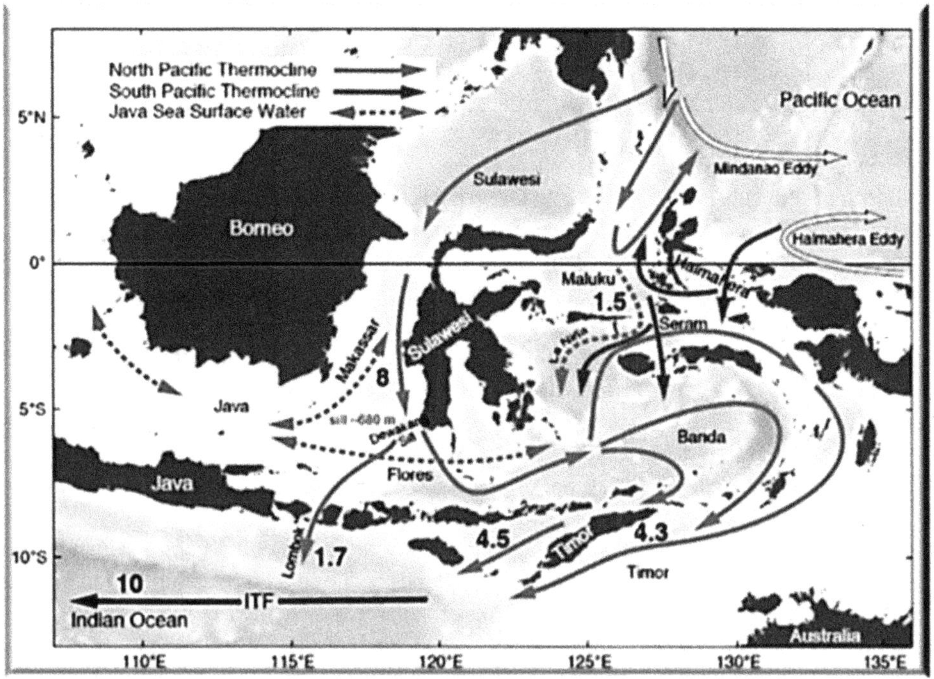

FIGURE 4.13 Indonesian Throughflow boundary current.

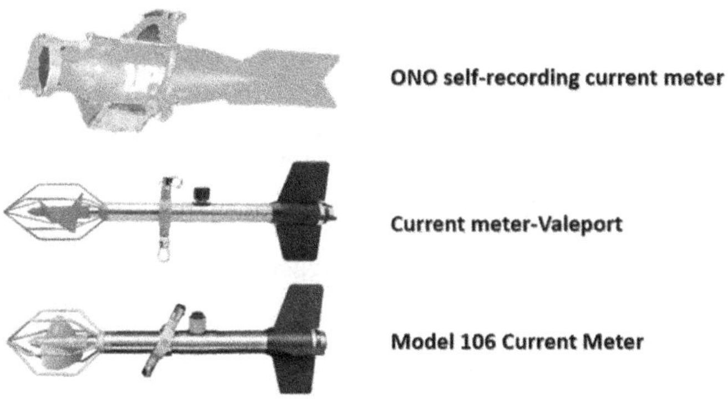

FIGURE 4.14 Sort of current meter devices used in this study.

and results from high-resolution GCMs. The SCSTF prevents the warm Pacific surface water from moving south, which causes a subsurface velocity maximum in the Makassar Strait (Qu et al., 2005).

Based on "Island Rule," the SCSTF can be seen as a circulation around Philippine – Borneo (Godfrey, 1989). The large-scale wind over the Pacific creates the pressure head that pushes the SCSTF into resistance (Qu et al., 2005). Gordon et al. (2003) also connected the influx of freshwater from the SCS to the northward surface flow in the Makassar Strait (Wang et al., 2006). However, the argument is that the southeasterly monsoon wind in the Java Sea drives the buoyant, fresh SCS water into the southern Makassar strait during the boreal winter, culminating in a northward pressure gradient in the strait's surface layer. The SCSTF includes the SCS freshwater intrusion (Qu et al., 2009). The SCSTF is responsible for directly transmitting the effects of Pacific variability to the Indonesian Seas. Although the SCSTF volume transport in the Karimata Strait is only 1–2 SV (Wyrtki, 1961), it is expected to have a significant impact on the Pacific to Indian Ocean heat transport, and results from high-resolution GCMs have supported this (Qu et al., 2006a; Qu, Girton, et al., 2006b).

4.8 Oceanographic data acquisitions

From 1992 to 2019, the author used various current meters, including the ONO self-recording current meter, the current meter-Valeport, and the model 106 Current Meter (Fig. 4.14), to gather sample data on ocean current parameters, such as velocity and direction, along the coastal waters of Peninsular Malaysia. Therefore the collection of these data took place between latitudes

1° 30' N and 5° 30' N and longitudes 103° E to 105° E (Fig. 4.15).

These current meters are used to measure surface, subsurface, and deep current patterns along Peninsular Malaysia's east coast. However, from 1992 to 1993, ONO's self-recording current meter was placed at a fixed location below 10 m of water depth to measure subsurface current flows. The data collected from these current meters are crucial for understanding the oceanographic processes and circulation patterns in the coastal waters of Peninsular Malaysia. In this view, the current meter-Valeport and the model 106 Current Meter are both lowered from the water's surface to the ocean's floor. For every 5 m of water depth, reading data are taken at intervals of about 10 seconds during sampling. The numerical model used these data after converting them to Excel format. In addition, during the gathering of information about ocean currents, wind direction, and speed are measured at each in situ measurement site.

Since the resolution of 0.5° in the simple ocean data assimilation (SODA) (https://apdrc.soest.hawaii.edu/las/v6/dataset?catitem = 4807) may be too coarse and insufficient for small-scale features, these data were used to validate the monthly collected data. These data were employed to simulate the South China Sea's current and turbulence boundary flow patterns.

4.9 Driving quantized Marghany turbulence boundary current flows theory

The third chapter demonstrates the quantization's significance in relation to electromagnetic signals. Intending to explain the boundary turbulence current movement along the coastal waters of Peninsular Malaysia or any other coastal water, can we use the significance of quantization to formulate an innovative boundary layer theory? The ocean's topological dynamics are thought to be a special and challenging system. There is a chaotic flow in this dynamic topology. The complex superposition of all dynamic features, including waves, tides, currents, and so on, that not only occur on the surface but can also be seen in the water column is brought on by the wind stress's interaction with the ocean body. This viewpoint's quantization can transform the ocean dynamic system into a quantum mechanical system. It could assist in organizing the chaotic system of the ocean into numerous quantized energy spectra. Based on their quantized energy spectra, these energy spectra may distinguish between a wide range of ocean phenomena. In this regard, let's assume that the ocean's dynamic system is a quantum topology that has a range of turbulence spectra signatures, including waves, wind currents, tidal currents, and so on. We examine quantizing the turbulence boundary wind current flow in this chapter.

Let's suppose for the purposes of this explanation that the turbulence boundary current is a quantum approximation of random or irregular paths in the classical sense. Unlike the classical turbulence boundary layer, as stated previously in the Sverdrup, Stommel, and Munk assumptions, where the quantum turbulence boundary occupies definite states and the randomness arises due to stochastic transitions between states, in the quantum turbulence boundary flow, randomness arises through three different processes: (1) quantum superposition of states; (2) nonrandom, reversible unitary evolution; and (3) collapse of the wave function due to state measurements. The superposition of wind curl can lead to wave generation and current flows. The observer only can watch the wave dynamic movement without seeing the current flow and mixing through the water column. The probability here is one can investigate the wind superposition phenomena. In these regards, the ocean topology state is a composite of many states, that is, wave, currents, wave − current interactions, and so on, that are mainly entangled with wind curl or wind stress.

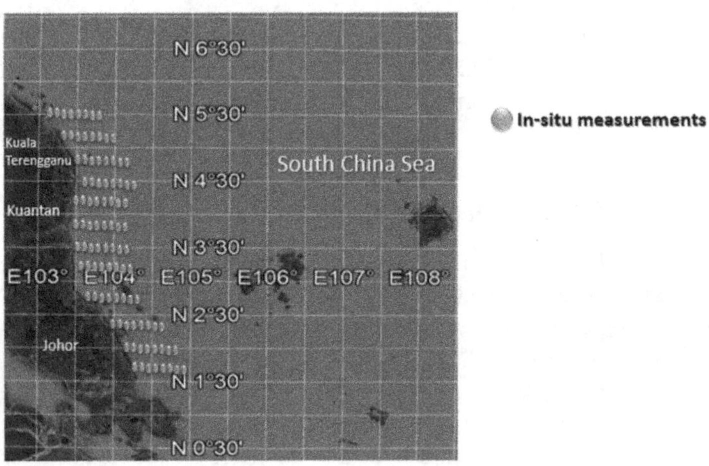

FIGURE 4.15 The geographical locations of in situ data collection.

The chaotic ocean topology involves many chaotic states. Each state is labeled by position $\vec{r}_{ij}$. It implies that upon the exchange of two different states, the quantum many-states wave function must be invariant (up to a phase factor). Depending on how the states interchange, the many-states wave function can either be symmetric or antisymmetric. In this view, the restriction is placed on the many states of turbulence boundary wave function by this exchange symmetry characteristic. The wave function must be correctly symmetrical or antisymmetrical each time kinetic energy from wind is added to or subtracted from the many-states system to satisfy the symmetry constraint.

Consider the entire set of single-state ocean topology wave functions $\Psi_\alpha\left(\vec{r}\right)$ (which may be a combined index composed of numerous quantum states). The wave function is as follows:

$$\Psi\left[\vec{r}_i\right] = \prod_{i=1}^{N} \Psi_{\alpha_i}\left(\vec{r}_i\right) \equiv \Psi_{\alpha_1} \otimes \Psi_{\alpha_2} \otimes \cdots \otimes \Psi_{\alpha_N} \quad (4.36)$$

Eq. (4.36) depicts an N-state in which the ith turbulence state is in the single-state $|\alpha_i\rangle$. The position argument of the wave function may be abolished in the shorthand notation, and it is assumed that the ith single-turbulence state wave function describes the state of the ith turbulence boundary states. Therefore let us assume that Ψ has not been symmetrized or antisymmetrized and hence, does not qualify as a many-turbulence states wave function for identical boundary. It can, however, be symmetrized (antisymmetrized) using the operators $\mathscr{S}$ for symmetrizing and $\mathscr{A}$ for antisymmetrizing across the state ocean space Z. Thus the evolution equation for a continuous time quantum turbulence trajectory movement on Z is:

$$i\frac{\partial\Psi}{\partial t} = -\,\omega_{\Delta x} L_Z \Psi \quad (4.37)$$

here $\omega_{\Delta x}$ is the characterized frequency of the turbulence boundary state in Z. Moreover, L_Z is the discrete Laplacian, and Δx is the distance between the spatial sites that the turbulence boundary can exist in the ocean state topology. The product of two unitary operators, (1) a "coin flip" operator and (2) a conditional shift operator, which is executed repeatedly, specifies the formation of a quantum turbulence boundary state in discrete time. Consider a turbulence boundary propagating on a linear array of discrete sites with a spin-1/2 degree of freedom. If the variety of such sites is countably infinite, the turbulence state space is denoted by Z. A product state can then be used to characterize the turbulence boundary state:

$$|\Psi\rangle = |\tau\rangle \otimes |\Psi\rangle \quad (4.38)$$

consisting of a space Z with an internal wind stress $|\tau\rangle$ spin state:

$$\ell^2(\mathrm{Z}) \in |\tau\rangle = \left\{ curl_z T_\uparrow \left|\uparrow\right\rangle + curl_z T_\downarrow \left|\downarrow\right\rangle : curl_z T_{\uparrow/\downarrow} \in \mathbb{C} \right\} \quad (4.39)$$

In this regard, the localization of the turbulence boundary states $\ell^2(\mathrm{Z})$ under the circumstances of variation of $curl_zT$ can be mathematically expressed by:

$$|\Psi\rangle \in l^2(\mathrm{Z}) = \left\{ \sum_{curl_zT \in Z} \alpha_{curl_zT} |curl_zT\rangle : \sum_{curl_zT \in Z} |\alpha_{curl_zT}|^2 < \infty \right\} \quad (4.40)$$

Eq. (4.40) identifies the impact of the $curl_zT$ in causing turbulence boundary state α_{curl_zT} in ocean topology space Z. The turbulence boundary state, which is a component of curl wind, displays an intriguing aspect of quantum coin space $\mathbb{C}$. This is known as superposition. This effectively indicates that two states, curl wind $|curl_zT\rangle$, and turbulence boundary states $|\Psi\rangle \in l^2(\mathrm{Z})$, occur at the same time. This is also known as entanglement. Entanglement is the sharing of shared attributes. The communal or common property is shifted vertically and horizontally in this scenario. The quantum coin space $\mathbb{C}$ provides a broad view of how the existence of wind stress curl state $|curl_zT\rangle$ can generate distinct turbulence boundary states $|\Psi\rangle \in l^2(\mathrm{Z})$ either horizontally, that is, surface ocean space, or vertically, through the water column space Z. These are entangled at the same time due to the curl wind variation $|curl_zT\rangle$. According to this theory, this type of ocean boundary state for quantized turbulence can be represented simply in Fig. 4.16 and given the name "Quantized Marghany Turbulence Boundary State."

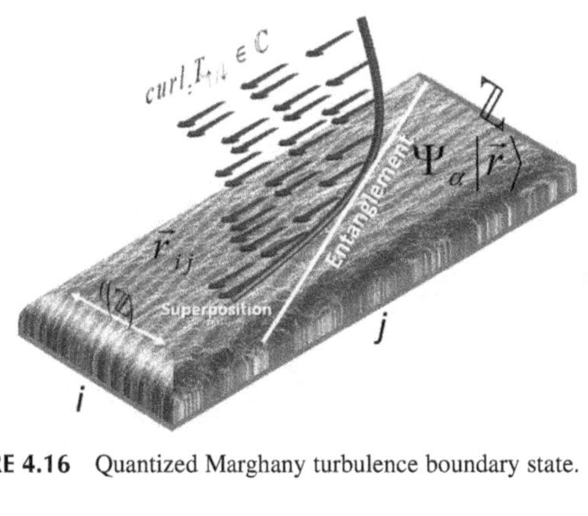

FIGURE 4.16 Quantized Marghany turbulence boundary state.

For large scales, the quantization of turbulence boundary can be considered as the classical one, in which the energy density in the k, that is, $k \sim l^{-1}(\mathrm{Z})$ space, the energy spectra $E(k)$ can be given by Kolmogorov law (Chapter 2) by:

$$\Psi\left|E\left(l^{-1}(\mathrm{Z})\right)\right\rangle = C|curl_zT\rangle^{\frac{2}{3}}[l(\mathrm{Z})]^{\frac{5}{3}} \quad (4.41)$$

here C is approximately 1, and $|curl_zT\rangle$ is the energy of wind stress curl transfers over the ocean topology space

Z. Therefore the energy density of turbulent velocity fluctuations u per unit density can be given by:

$$\left(|u|^2\right) = 0.5\int E\left(l^{-1}(Z)\right)d\left(l^{-1}(Z)\right) \quad (4.42)$$

Therefore the spectrum of the longitudinal velocity fluctuations $\Psi|\phi\rangle$:

$$\Psi|\phi\rangle = 0.5\int_0^\infty \left(1 - k^2k'^{-2}\right)\frac{E(k')}{k'}dk' \quad (4.43)$$

Eq. (4.43) provides the turbulence that is locally isotropic over wave numbers larger than k as $\Psi|\phi\rangle\alpha k^{-\frac{5}{3}}$. In this chapter, the horizontal turbulence boundary states are investigated utilizing in situ current and wind measurements as well as SAR imagery.

4.10 Tracking the South China Sea quantized Marghany turbulence boundary states

Seasonal exploration of the quantized Marghany turbulence boundary state is demonstrated. Which season would have the quantized Marghany turbulence boundary state, according to this viewpoint? Wind stress curl patterns can be divided into four seasons. The former occurs between May and August and is known as the southwest monsoon season. As a result, the southwest monsoon encounters winds that are less than 3 m/s as they move from the southwest to the north. Compared with the Boreno boundary zone, Malaysia's western coastal boundary is dominated by the strongest mass transportation $8\times\mathbf{10^6}$ m^3/s owing to wind stress curl of approximately $4\times\mathbf{10^{-7}/Nm^3}$. In the South China Sea's central zones, there is less wind stress curl of $-4\times\mathbf{10^{-7}/Nm^3}$, causing less mass transportation of $-9\times\mathbf{10^6}$ m^3/s (Fig. 4.17).

The transitional period between the southwest and northeast monsoons is therefore from September to October. During this time, the wind starts to change direction, shifting from the south to the north. This mechanism caused the wind to blow in an erratic pattern with no predictable direction. In this view, the western coastal boundary has the lowest mass transport (-12×10^{6} m^3/s) and the lowest wind stress curl ($-6\mathbf{X10^{-7}}/Nm^3$), compared with the South China Sea's central zone and the Borneo boundary zone (Fig. 4.18).

On the other hand, the northeast monsoon season lasts from November to March because wind travels from the northern to the southern parts of South China Sea. In contrast to the southwest monsoon during this time, the western boundary dominates with the strongest wind stress curl of $6\mathbf{X10^{-7}}/Nm^3$ and mass transport of -12×10^6 m^3/s (Fig. 4.19). The latter, which primarily occurs in April, is a transitional period between the northeast monsoon and the southwest monsoon. In contrast to other seasons, this one is characterized by nonfixed wind direction, the lowest mass transport -12×10^6 m^3/s, and the lowest wind stress curl of $-6\times10^{-7}/Nm^3$ (Fig. 4.20).

Needless to say, the entire domain demonstrated a curl of wind stress signs in opposite directions during the four monsoon seasons. According to the aforementioned perspective, a positive curl of wind stress prevailed during the northeast monsoon periods, and a negative curl was dominant during the southwest monsoon and transitional period, especially April, over a significant portion of the South China Sea and Karimata Strait. Notwithstanding this, both the northeast and southeast monsoon periods experience negative and positive curl wind stresses along the east coast of Peninsular Malaysia. Surface divergence, which could lead to upwelling along the eastern coast of Peninsular Malaysia, is covered in more detail in the following chapters. The significant question is now: How

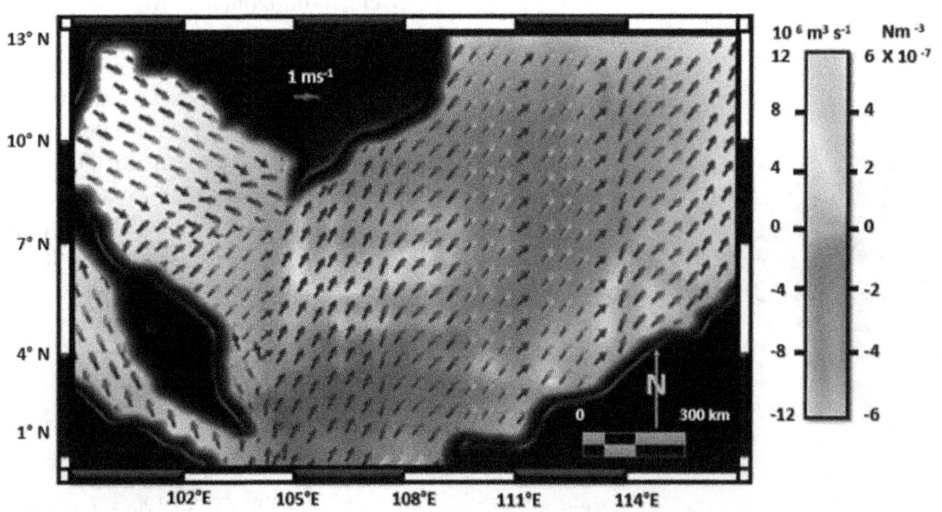

FIGURE 4.17 Southwest monsoon wind stress curl and mass transport.

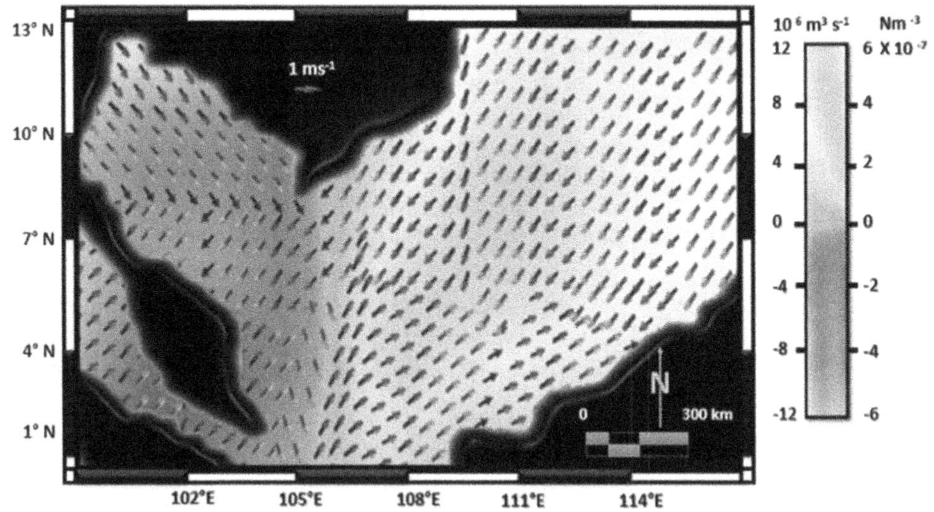

FIGURE 4.18 Wind stress curl and mass transport during transitional periods between southwest and northeast monsoon periods.

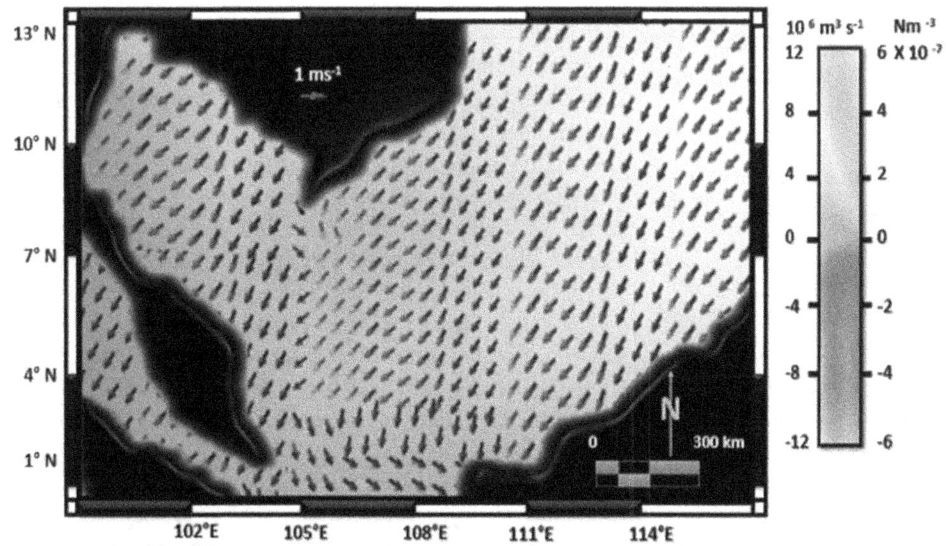

FIGURE 4.19 Northeast monsoon curl wind and mass transport.

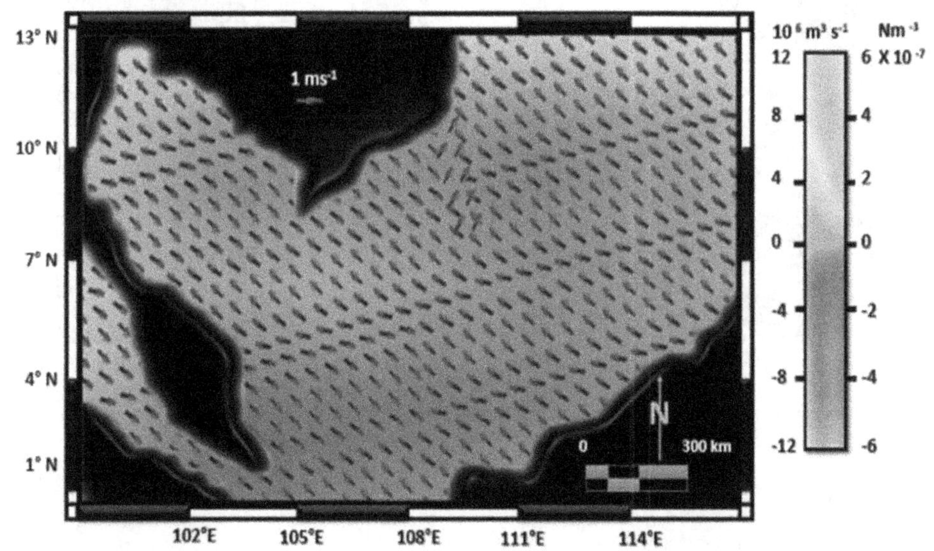

FIGURE 4.20 Wind curl and mass transport during transitional periods between northeast and southwest monsoon periods.

does wind stress curl to create boundary turbulence flows? What is the topology of these turbulence boundary flows? Turbulence boundary flows are simulated using quantized Marghany turbulence boundary flow theory. In this view, the southwest monsoon demonstrates the development of discrete boundary turbulence flows from May to August.

The strength of these discrete flows is better developed in August than in the rest of the southwest monsoon period. The highest quantized spectra energy of turbulence boundary flow is seen along Karimata Strait, the Java Sea, and some portion of the western boundary of the east coast of Peninsular Malaysia with 10^1. On the contrary, the beginning of the southwest monsoon is dominated by the lowest quantized spectra energy of less than 10^{-1}. The highest quantized spectra energy is equivalent to instantaneous velocity fluctuations averaged over downwind $\langle\overline{u}\rangle[u_\tau]^{-1}$ of 5 and Reynolds shear stress $\overline{u_1' u_2'}[u_\tau]^{-2}$ of 6, respectively (Fig. 4.21).

Therefore the turbulence fluctuation continues to develop through the transitional periods between the southwest and northeast monsoon periods (September to October). The center of the South China Sea is dominated by the lowest quantized spectra turbulence boundary flow of 10^{-1}; instantaneous velocity fluctuations averaged over downwind $\langle\overline{u}\rangle[u_\tau]^{-1}$ of -4 and Reynolds shear stress $\overline{u_1' u_2'}[u_\tau]^{-2}$ of 1. On the contrary, the western boundary of the east coast of Peninsular Malaysia demonstrates Reynolds shear stress $\overline{u_1' u_2'}[u_\tau]^{-2}$ of 3.8 and quantized spectra turbulence boundary flow of less than 10°. On the other hand, Karimata Strait, the Java Sea, Johor, Singapore, and southern Borneo experience quantized spectra energy and $\overline{u_1' u_2'}[u_\tau]^{-2}$ with 10^1 and 5.8, respectively (Fig. 4.22).

In the first 3 months of the northeast monsoon, the western turbulence boundary shows maximum development, with a quantized turbulence spectra energy of 10° and Reynolds shear stress of 6. This is evident is well recognized along the eastern coast of Peninsular Malaysia. In addition, compared with the Gulf of Thailand and the northern part of Borneo, fully quantized turbulence spectral energy dominates the Karimata Strait. The lowest quantized spectral energy and Reynolds shear stress were observed in the Gulf of Thailand and northern Borneo, respectively, at 10^{-2} and 1(Fig. 4.23).

Therefore during the end of the northeast monsoon period, the Karimata Strait, Johor, and Borneo continue to have more fully quantized turbulence spectra energy than other regions of the South China Sea (February and March) (Fig. 4.24). This finding raises the possibility that the oceanographic and atmospheric conditions in the Karimata Strait, Johor, and Borneo may be special and contribute to the quantized energy in the turbulence spectra that have been observed.

An unexpected observation is that, with average downwind instantaneous velocity fluctuations of -0.5, the east coasts of Peninsular Malaysia and Borneo continue to exhibit high quantized turbulence spectra energy, Reynolds shear stress 3.5, and turbulence intensity of 10^{-1}. The Karimata Strait is divided into quantized clusters of turbulent flow by the quantized turbulence spectrum energy dissipation, which ranges from 10^{-1} to 10°, while the Reynolds shear stress is between 5.5 and 6 (Fig. 4.25).

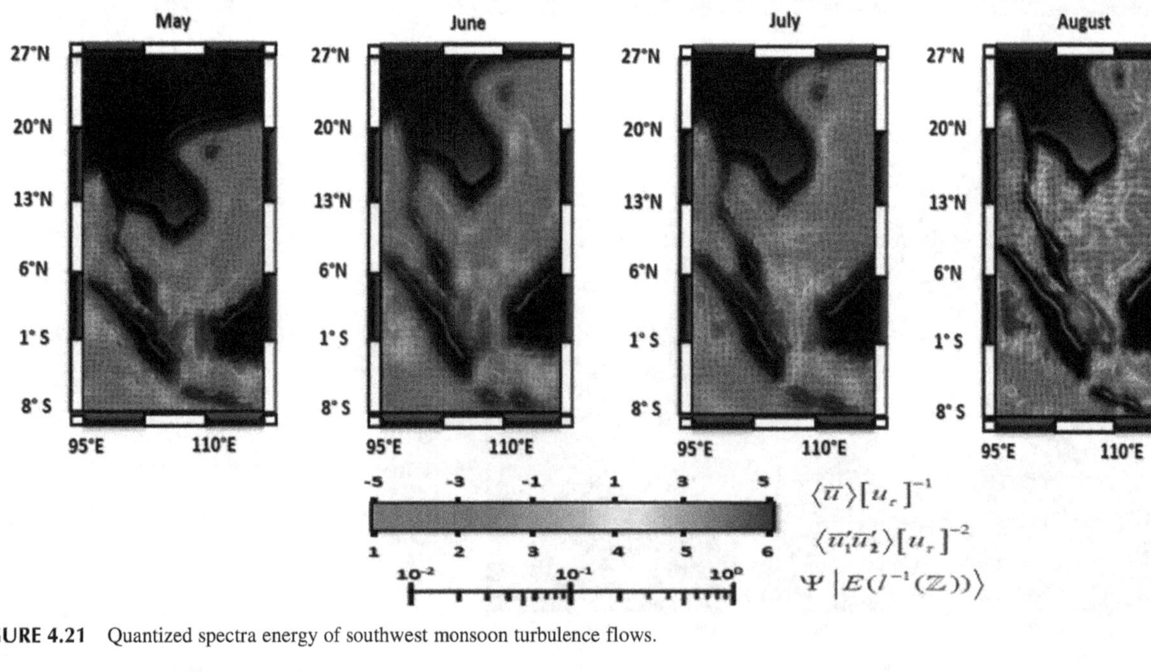

FIGURE 4.21 Quantized spectra energy of southwest monsoon turbulence flows.

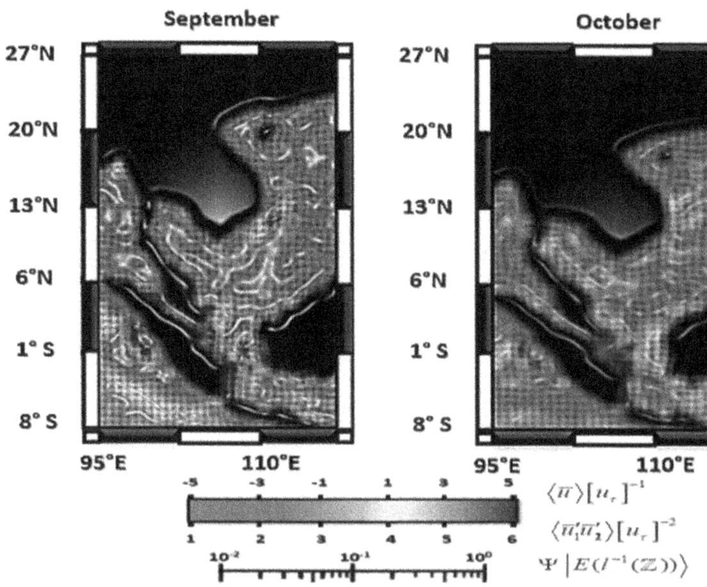

FIGURE 4.22 Quantized spectra energy of transitional monsoon (September-October) turbulence flows.

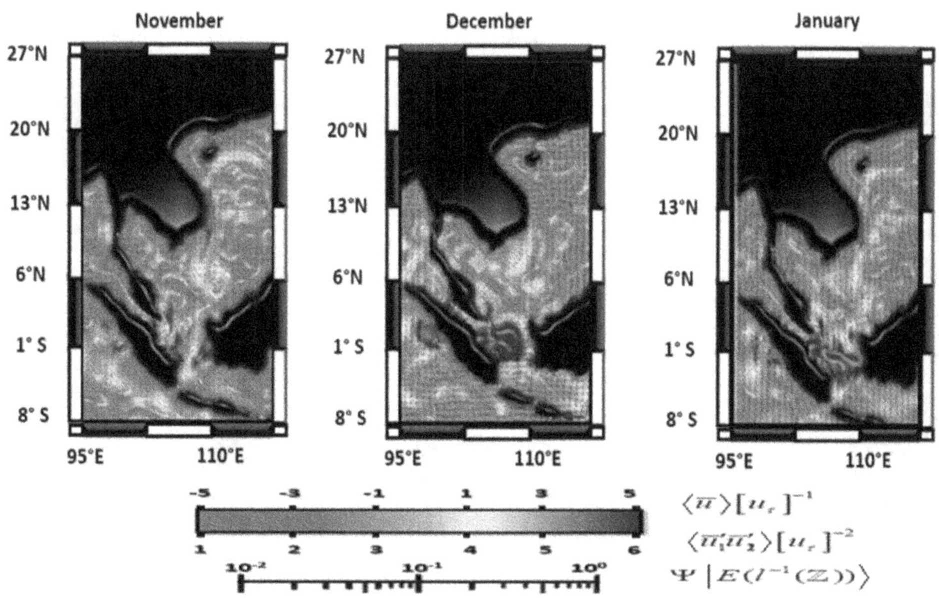

FIGURE 4.23 Quantized spectra energy of the first 3 months of the northeast monsoon turbulence boundary flow.

In general, the quantized turbulence boundary in the South China Sea and the Java Sea is driven by wind stress curl, which was reflected in seasonal monsoon variability. According to this understanding, a strong quantized turbulence boundary flow develops along the eastern coast of Peninsular Malaysia in full maturity in July. In other words, it took 3 months for the quantized western turbulence boundary to fully develop along the east coast of Peninsular Malaysia. This occurs as a result of the powerful western boundary current along Peninsular Malaysia joining the surface current in the Karimata Strait as it shifts north under the influence of the southeasterly monsoon winds (Figs. 4.17 and 4.21).

Fully developed western quantized turbulence boundary flows occur along Vietnam, the east coast of Peninsular Malaysia, Karimata Strait, and the Java Sea. In this sense, the northeasterly monsoon winds during the northeast monsoon period drag the surface current along Peninsular Malaysia southward toward the Karimata Strait. Under these circumstances, the western quantized turbulence boundary flow continued to flow southward across the Karimata Strait before turning eastward and

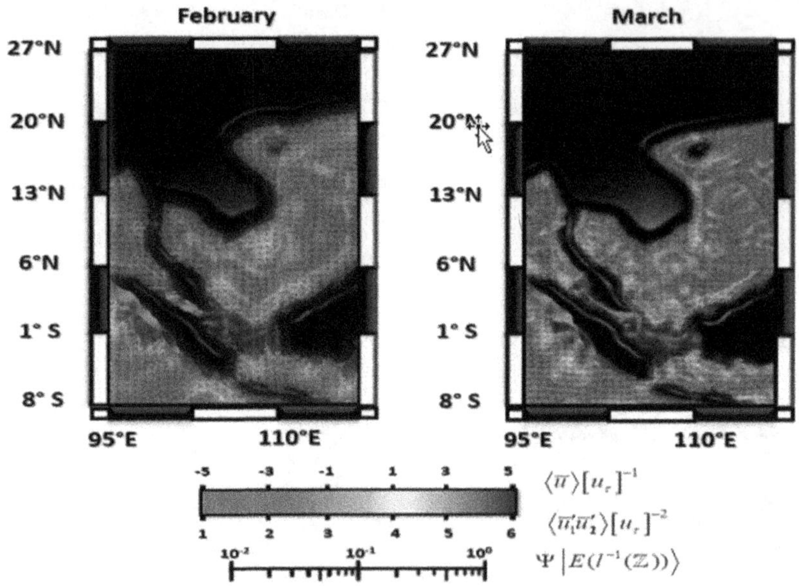

FIGURE 4.24 Quantized spectra energy of the last 2 months of the northeast monsoon turbulence boundary flow.

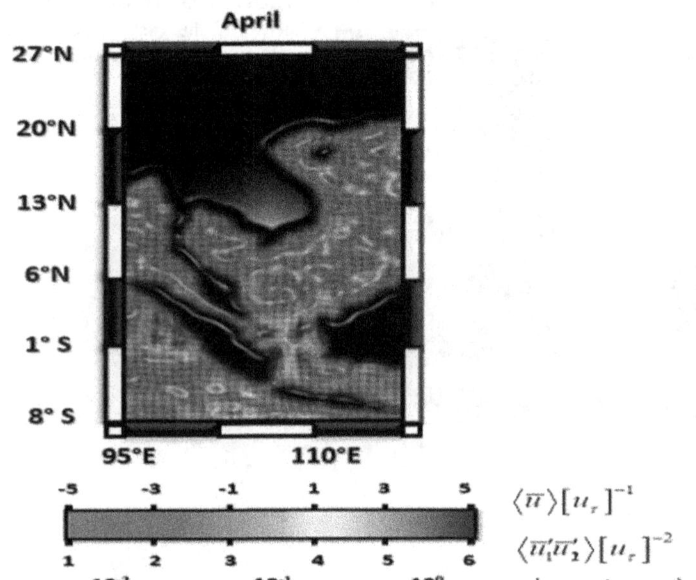

FIGURE 4.25 Quantized spectra energy of transitional monsoon (April) turbulence flows.

integrating with the eastward surface current driven by northwesterly monsoon winds in the Java Sea. This explains why the strongest quantized turbulence boundary flow is always dominant in the Karimata Strait.

The discrete order of turbulence flow in the South China Sea is known as the "Quantized Marghany Turbulence Boundary Flow" and is driven by the quantized turbulence energy spectra in Section 4.1.8. Under the influence of wind stress curl variability, this frequently forms a discontinuous or discrete cluster of turbulence spectra along the western boundary. A more thorough explanation will be provided at the chapter's conclusion.

4.11 Synthetic aperture radar quantized Marghany turbulence boundary imaging algorithm

The critical question is now: Can SAR imagine a quantized Marghany Turbulence Boundary? It is well known that synthetic aperture radar (SAR) can image surface current gradients in shallow waters (SAR). Since the radar can only measure surface roughness, the quantized Marghany turbulence boundary variations are imaged through energy wave function growth; entanglements; and superposition toward surface wave topologies. This imaging mechanism is a three-step process:

1. The quantized Marghany turbulence boundary superimposed the short-wave spatial concentration in certain regions of radar image topology.
2. The local wave function of the spectral energy of quantized Marghany turbulence boundary entanglements with the radar signal to form an extremely fluctuating backscatter compared with the surrounding topology in the SAR image.
3. The local entanglements of quantized Marghany turbulence boundaries in SAR images comprise entanglements between the SAR signal and short waves, superposition of the turbulence boundary with waves, and entanglements with radar backscatter. In this scenario, the radar signal becomes a quantized integral path across the SAR image topology. This phenomenon of entanglement of quantized Marghany turbulence boundaries in SAR images is significant for understanding the behavior of radar signals in complex environments. It can also help develop better techniques for image processing and analysis in SAR imaging.

Therefore the short waves are primarily formed by wind propagation in the wind's direction. When the wind is blowing in the radar's direction, the backscattered energy from the turbulent surface is at its highest level. When the wind blows away from the radar antenna, there is an analogous, though slightly smaller, local extreme in the backscattered power. When the radar look direction is perpendicular to the wind direction, the backscattered energy is at its lowest.

The nonlinearity of SAR image $\Psi|k|K>$ for imaging superposition of quantized Marghany turbulence boundary with the topology of short-wave propagation can be expressed in the form of the Volterra kernels:

$$\Psi|k|K\rangle = k_y\langle\vec{u}\rangle\vec{v}\left[\frac{1}{\vec{K}}\left[\frac{\delta E_{\omega_o}}{\delta\vec{k}} + \frac{\delta\vec{c}g}{\delta x}\vec{v} + \frac{\delta\vec{u}_y}{\delta y}\vec{v}_y + 0.5\vec{u}_x\right]\frac{\delta E_{\omega_0}}{\delta\omega}\right] \times \left[\frac{\vec{c}_g\left(\vec{K}\right)\vec{v} + j.0.5\omega_0^{-1}}{\left[\vec{c}_g\left(\vec{K}\right)\vec{v}\right]^2 + \left[0.5\omega_0^{-1}\right]^2}\right] + \left[j6.10^{-3}\vec{K}^{-4}\right] \times \left[RV^{-1}\vec{v}_x\right] \tag{4.44}$$

where $\left\langle\vec{u}\right\rangle$ is the mean current velocity, $\vec{u}_y$ is the current flow along the range direction while $\vec{u}_x$ is the current gradient along the azimuth direction. Therefore k_y is the wave number along the range direction, $\vec{K}$ is the spectra wave vector, ω_0 is the angular wave frequency, $\vec{c}_g$ is the group velocity, E_{ω_0} is the wave spectra energy, v stands for the Volterra kernel frequency along the azimuth and range directions, and *R/V* is the range to platform velocity ratio. Therefore the quantized Marghany turbulence boundary can be imagined in the range direction of SAR image. Indeed, the entanglements and superposition behavior in SAR imaging quantized Marghany turbulence boundary spins the radar signal to imagine wave spectra on azimuth direction due to velocity bunching effects and quantized Marghany turbulence in range direction. Therefore the dynamic movements can be estimated by:

$$\vec{u}_y = \frac{FT\left[\prod_{j=1}^{i} l_0(t)\right]}{\Psi|k|K>} \tag{4.45}$$

here $l_0(t)$ is the initial location of the quantized Marghany turbulence Boundary in SAR image. Therefore in ocean topology (Z), the quantized Marghany turbulence boundary is regarded as an integral path $L \in Z$, and in SAR images, $L(Z)$ occupies the region from pixel (q_i,t_i) to pixel (q_f,t_f). Alternatively put, $(q,t)_{i,f} \in L(Z)$. Assume for the moment that the quantized Marghany turbulence boundary has two quantum states (one at the beginning of the transition, in the state $|\Psi_i\rangle = |q_i, t_i\rangle$ and one at the end, in the state $|\Psi_f\rangle = |q_f, t_f\rangle$). The quantized Marghany turbulence boundary can produce transitional energy along SAR topology state, which identifies as $|\Psi_i|\Psi_f\rangle$ (Fig. 4.26).

Therefore transitional energy can be defined as the path integral of the $e^{\left(\frac{i}{\hbar}\vec{S}[q(t)]\right)} \approx e^{\left(\frac{i}{\hbar}\sum_{j=1}^{N-1}[L_j\Delta t]\right)}$ over all feasible paths that, in SAR images, connect the quantized Marghany turbulence boundary's initial configuration, (q_i,t_i), with its final configuration, (q_f,t_f). In this view, a tool for automatically discriminating any turbulence boundary can be developed by computing the energy along this quantized Marghany in SAR images. In SAR images, the transition energy between the initial and final transition configurations of a quantized Marghany turbulence path can be expressed as:

$$|\Psi_i|\Psi_f\rangle = \int_{\Psi i}^{\Psi_f} D\vec{\Psi}D\Psi e^{\left(\frac{i}{\hbar}\int_{t_i}^{t_f}[L(Z)dt]\right)\frac{R}{V}}. \tag{4.46}$$

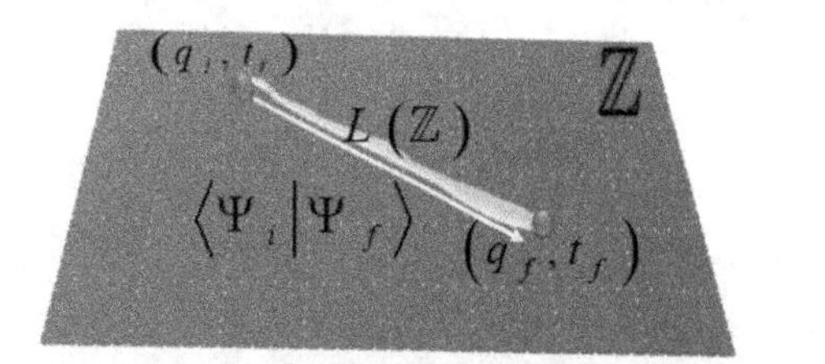

FIGURE 4.26 Quantized Marghany turbulence boundary in SAR image.

here $D\vec{\Psi}D\Psi$ pointing at the integration overall possible configurations of quantized Marghany turbulence integral path across SAR images. Consequently, the spectra energy along the quantized Marghany turbulence boundary in SAR image can be expressed nonlinearly as:

$$\Psi\left|E\left(L^{-1}(Z)\right)\right\rangle = \left|u^{\frac{2}{3}}\right\rangle\left|L(Z)^{\frac{-4}{3}}\right\rangle \tag{4.47}$$

The author developed the novel Eq. (4.47), also known as the "Quantized Marghany Turbulence Spectra Energy Algorithm"(Table 4.1), which can quantify the movement of the integral path and fluctuation of the turbulence boundary in SAR satellite images.

4.12 Quantized Marghany turbulence spectra energy along the western boundary of Malaysia's East Coast in SAR images

In Radarsat-1 SAR standard mode 2 data, the quantized Marghany turbulence spectra algorithm is investigated. On March 31, 2005, the acquisition was completed, along with real-time in situ current measurements made with AWAC tools (Fig. 4.27).

AWAC equipment was used to make in situ measurements. The deployment took place at a depth of 18.5 m off the east coast of the Malaysian Peninsula at 5°31'16″N and 103°08'40″E. The measurement procedures were repeated three times with three AWAC deployments at the study site. At every measurement, AWAC had been deployed a day earlier, and on the second day, RARDARSAT-1 had passed over and been retrieved from the water. Current velocity and direction data were included in the information that was retrieved from AWAC and stored in ASCII format. Using real-time AWAC measurements, the sea surface current patterns extracted from RADARSAT-1 data have been verified.

During this time, AWAC collected 70 signal burst measurement files from a sampling burst setup of one burst every 20 minutes. At a sampling rate of 1 Hz, bursts of 1024 samples were used to measure the current's speed

TABLE 4.1 Quantized Marghany turbulence spectra energy algorithm.

Input: SAR image (I), initial parameters of backscatter;σ
1: FFT(I) 2: Compute k for $k > 0$; 3: Compute I quantum energy $\Psi\|k\|K>$ 4: Determine $l_0(t)pixellocation$ 5: Estimate $\vec{u}_y \in l_0(t)$ using Eq. (4.45) 6: Determine the integral path of the quantized Marghany Turbulence Boundary $\|\Psi_i\|\Psi_f\rangle$ 7: Model energy spectra $\Psi\|E\left(L^{-1}(Z)\right)\rangle \leftrightarrow \|\Psi_i\|\Psi_f\rangle \in L^{-1}(Z)$ 8: Update $\vec{u}_y \in l_0(t) \leftrightarrow \Psi\|k\|K>$ 9:Return $\Psi\|k\|K>$; 10: Return $\|\Psi_i\|\Psi_f\rangle$ 11: End while

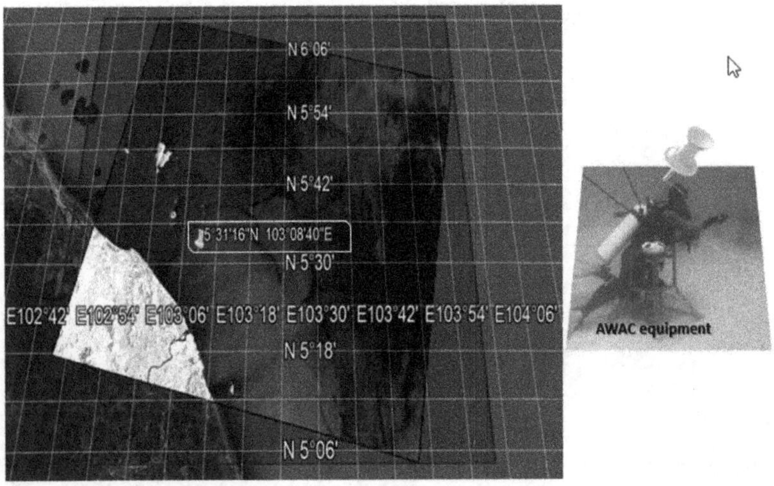

FIGURE 4.27 Real-time in situ measurement and data acquisition from Radarsat-1 Standard-2.

and direction. Two of these bursts appeared to have occurred with the instrument out of the water (two at start time). As a result, these bursts were ignored, and processing for the analysis was limited to 68 bursts. In this sense, the AWAC measures current velocity using the Doppler effect by sending a concise sound pulse; recording its echo; and measuring the change in pitch or frequency as follows:

$$u_{AWAC} = 0.5 C_s f_D . f_S^{-1} \quad (4.48)$$

here u is retrieved-real ocean current; f_D denotes the Doppler frequency shift, f_s denotes the transmitted sound frequency, and C_s denotes the sound speed at 1500/ms.

Data from the Standard-2 mode show erratic movements approximately 30 km from the coastal zone. Owing to the complex topology and speckle effects on the current movement in the SAR data, it is difficult to discern the movement pattern (Fig. 4.28). This intricate flow pattern was discovered by the author between the latitudes of 5°07' N and 6°06' N and the longitudes of 103°07'E and 103°51'E (Fig. 4.29). The meandering current, swell wave, and large rotating eddies are superimposed in the pattern. A large area of natural slicks superimposed on large internal waves serves as a visual representation of this complex phenomenon (Fig. 4.30). The author has made a novel discovery of this phenomenon, which has been given the name "quantized Marghany turbulence western boundary flow" owing to its discrete flow.

The western Marghany turbulence boundary along the Penssuilar Malaysia could be identified in the initial run of the "Quantized Marghany Spectra Energy Algorithm." In this manner, the algorithm initially turns the SAR image into spectra energy topology to spot the quantized Marghany flow's boundary edges. To clarify the flow of

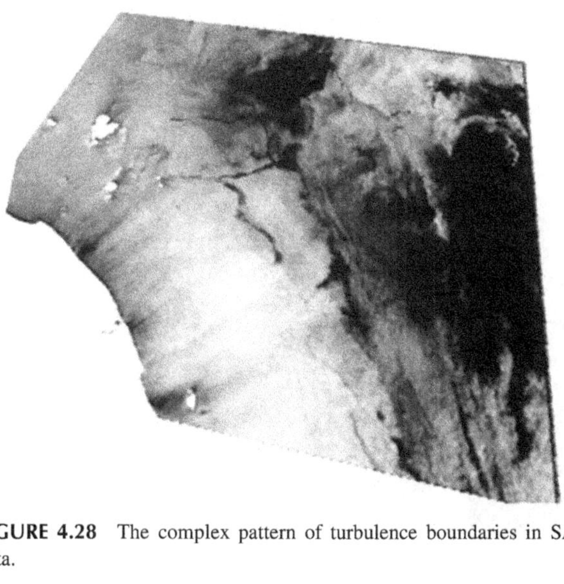

FIGURE 4.28 The complex pattern of turbulence boundaries in SAR data.

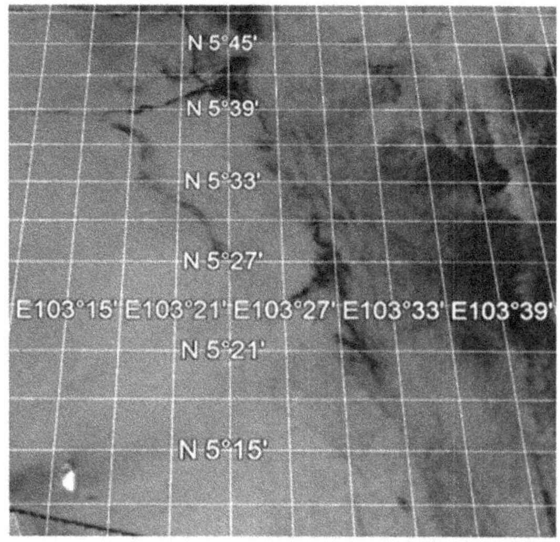

FIGURE 4.29 Superposition of meandering current with swell and large rotating eddy.

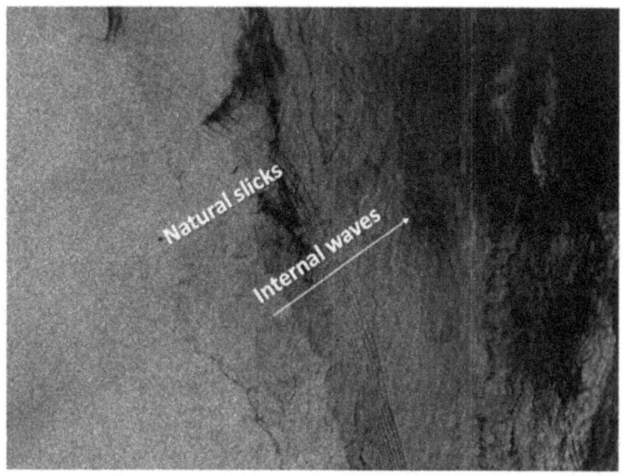

FIGURE 4.30 Superposition of meandering current with swell, natural slicks, and large internal waves.

the strongest boundaries, this algorithm calculates energy gradients and fluctuations across the radar topology gradient energy spectra. To clarify the flow of the strongest boundaries, this algorithm calculates the energy gradients and fluctuations across radar topology. Using the probability of energy fluctuation in each pixel across the turbulence flow, the algorithm can distinguish between various quantum states. The disordered Marghany turbulence flow is quantized in such small clusters using the quantized Marghany spectra energy algorithm, in which the integrated function in each cluster is roughly constant. The contributions from each quantized cluster were then added. This allowed for the identification of the western Marghany turbulence boundary along the east coast of Malaysia by following the transitional energy spectra from the initial configuration of the quantum energy states across the SAR topology (Fig. 4.31).

The completed run of the quantized Marghany spectra energy turbulence algorithm is demonstrated in Fig. 4.32. The quantized Marghany spectra energy turbulence is varied from inshore to 30 km offshore. The highest turbulence spectra energy $\Psi|E(L^{-1}(Z))\rangle$ is 10° and is found 80 km away from the coastal zone, which alliances with instantaneous velocity fluctuations averaged over downwind $\langle\overline{u}\rangle[u_{T}]^{-1}$ of 5.5 and Reynolds shear stress $\overline{u_1'u_2'}[u_T]^{-2}$ of 6. Energy transitions from one quantum cluster to another are indicated by the interior between the various quantized spectra of energy. Due to wind stress curl entanglements with the sea surface and the energy spectra transitional interval from the internal wave to the boundary of turbulence flows, the large quantized spectra of energy increased 80 km from the coast. In fact, the swell, meandering current, and capillary wave are superimposed by this transitional spectrum of energy to increase offshore turbulence under wind stress curl entanglements. Therefore the details of quantized Marghany spectra energy turbulence boundary are only tracked in SAR image owing to its 10 m pixel resolution.

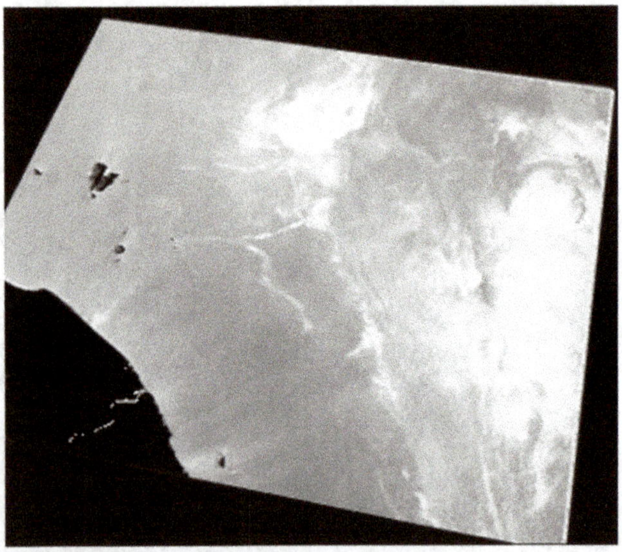

FIGURE 4.31 Quantized Marghany spectra energy algorithm's initial output in the SAR image.

Since internal wave spectra energy is entangled and superimposed with gravity waves, swell, and turbulence flow in the ocean topology, it is one of the structures of the quantized Marghany turbulence spectra boundary. Internal wave spectra energy is a fantastic example of how the energy spectrum state transforms. This demonstrates how the various turbulence boundaries and internal waves interact to transform the spectrum's energy state. And hence, with a spectra energy $\Psi|E(L^{-1}(Z))\rangle$ of 10° and an approximate Reynolds shear stress $\overline{u_1'u_2'}[u_T]^{-2}$ of 6, internal wave topology is dominant (Fig. 4.33).

4.13 Seasonal quantized Marghany turbulence spectra energy in SAR images

An important question now arises: Can SAR images contribute to tracking the seasonal fluctuation of the Quantized Marghany turbulence spectral energy in large-scale areas? More than hundreds of mosaic Radarsat-1 SAR standard-2 mode data were collected separately along peninsular Malaysia and Borneo from 1998 to 2007 in different monsoon seasons. Then, the quantized Marghany spectra algorithm was applied to a cluster of average SAR data of the northeast monsoon (from September to February) and southeast monsoon periods

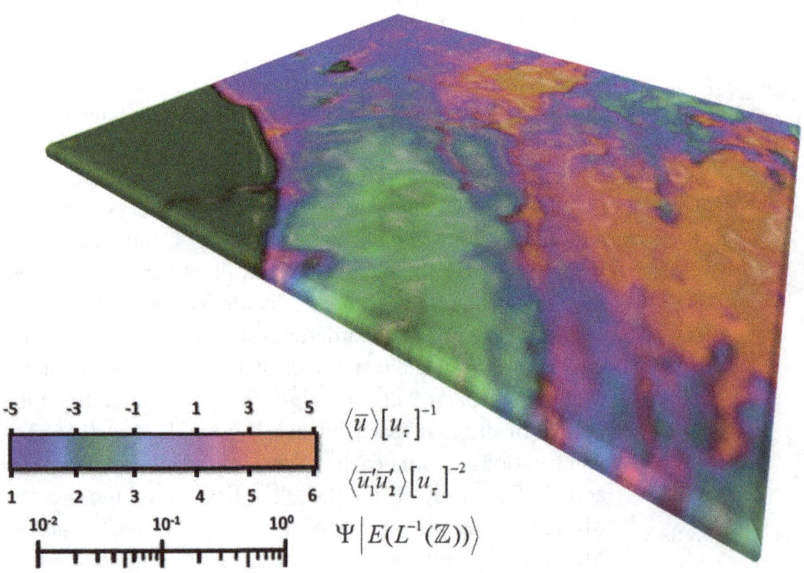

FIGURE 4.32 Energy structure of the boundary spectra of quantized Marghany turbulence.

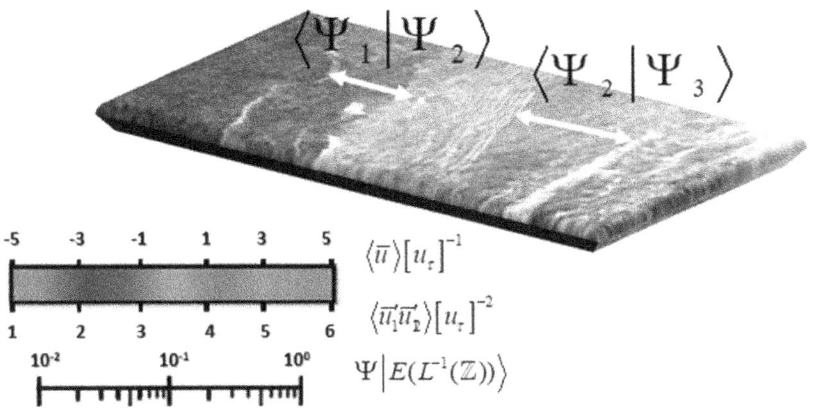

FIGURE 4.33 Internal wave spectra energy that is included in the quantized Marghany turbulence spectra boundary.

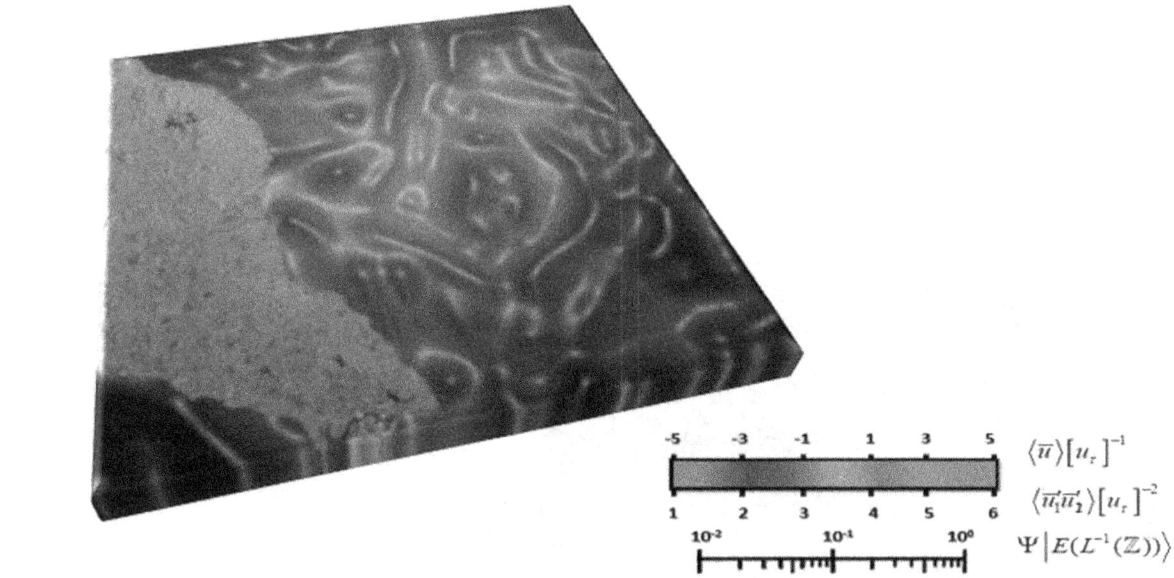

FIGURE 4.34 Quantized Marghany spectra energy turbulence boundary along the east coast of Malaysia during the northeast monsoon period.

(from April to August), including two transitional periods. The quantized Marghany turbulence boundary was very noticeable along the east coast of Peninsular Malaysia during the northeast monsoon season (September to February). Quantized anticlockwise eddies, vortex shedding, and current meandering form in the quantized Marghany turbulence boundary as entanglement wind stress curl creates instability in the ocean topology, increasing the roughness of the water. This results in high spectral energy being backscattered to the radar antenna, which produces SAR images with excellent coherence turbulence flow. By concentrating and growing, the turbulence along the western boundary increased and approached the eastern boundary of Borneo. The quantized Marghany turbulence boundary's spectrum energy $\Psi|E(L^{-1}(Z))\rangle$ increases from $10^{-\circ 0.5}$ to 10°, and this is in agreement with a Reynolds stress of 5.5 [on the east coast of Malaysia (Fig. 4.34)] to 6 [in Borneo (Fig. 4.35)] respectively.

However, during the southwest monsoon (from April to August) period, the western boundary is dominated by a southern meandering current, which forms the quantized Marghany turbulence boundary (Fig. 4.36). Northward flow and southern flow then converge together and extend to the Borneo boundary to form dissipated eddies. The quantized Marghany turbulence boundary's spectrum energy ranges from 10^{-1} to 10°, to form turbulence flow with a Reynolds shear stress $\overline{u_1' u_2'}[u_T]^{-2}$ of 5.5 [on the east coast of Malaysia (Fig. 4.36)] to 6 [in Borneo (Fig. 4.37)], respectively. Because of the dissipation of the quantized Marghany spectra, the Borneo boundary is dominated by a high quantized Marghany turbulence boundary.

It is no surprise that we are able to demonstrate the existence of the quantized Marghany turbulence boundary across various seasons thanks to the SAR images. Due to the wind stress curl's interaction with Bragg resonance,

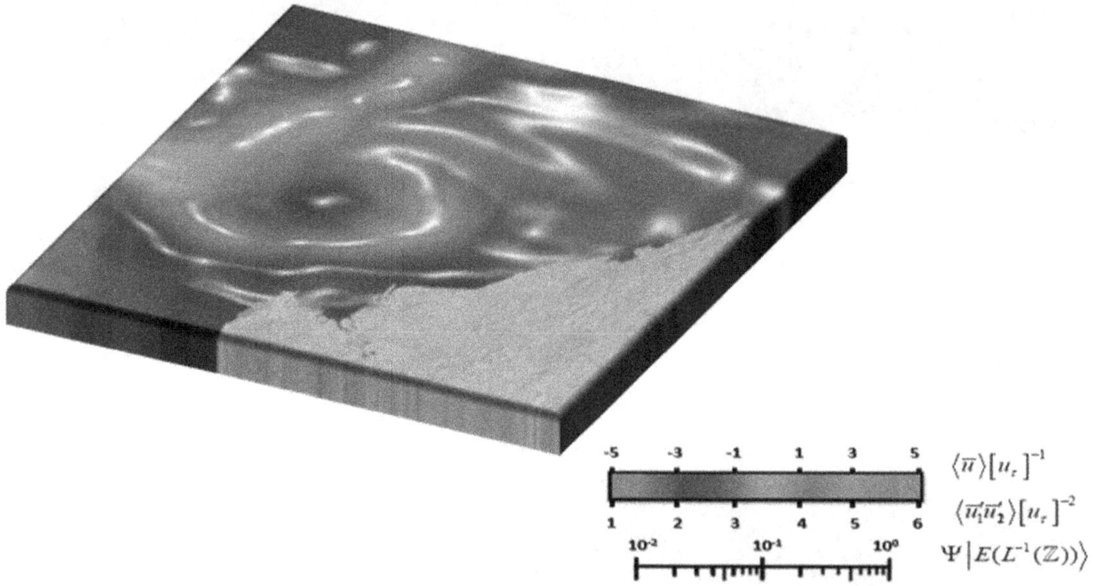

FIGURE 4.35 Quantized Marghany spectra energy turbulence boundary along Borneo during the northeast monsoon period.

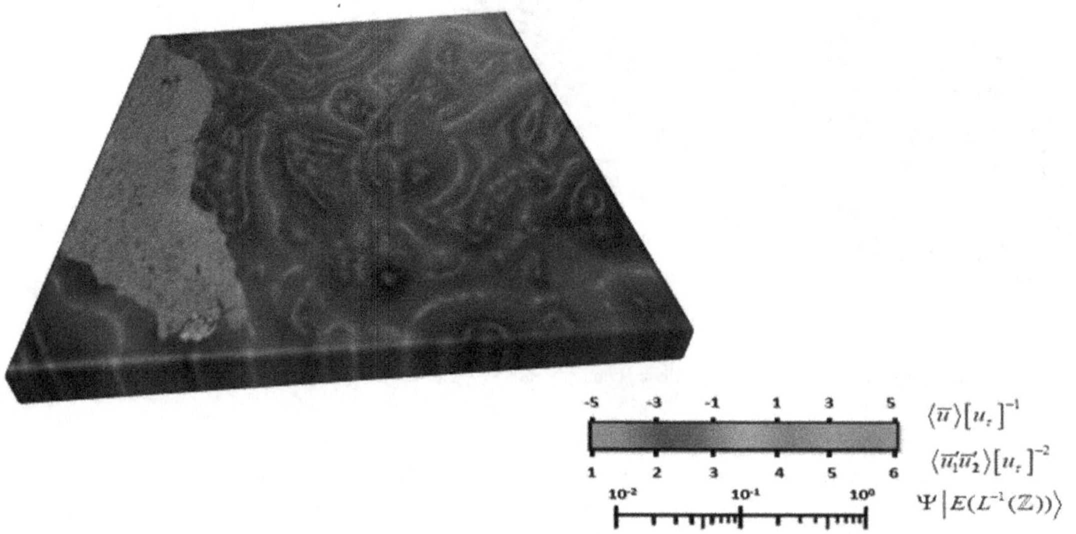

FIGURE 4.36 Quantized Marghany spectra energy turbulence boundary along the east coast of Malaysia during the southwest monsoon period.

which increases surface topology instability and produces unstable signal scattering at the radar antenna, turbulence signatures can be seen in radar images. In the upcoming chapters, we should go over how, from the perspective of quantum entanglements, rain cells, wind stress curl, and upwelling boundaries can all result in large turbulence boundaries.

As a consequence, even at large scales and at various times, the quantized Marghany turbulence spectra algorithm can track the turbulence flow structure in radar images. This understanding equates the quantized Marghany spectral energy to the Kolmogorov spectra (Chapter 2). In this sense, we can agree that different monsoon periods are dominated by the quantized Marghany turbulence boundary, which reaches its peak every monsoon period and then dissipates on a large-scale (Fig. 4.38).

This chapter demonstrates the novel discovery of disorder irregular turbulence boundary in the South China Sea, which matches with the western boundary theories invented by Sverdrup, Stommel, Ekman, and Munk. However, these theories cannot track or simulate the turbulence flow through water mass transport under the wind stress curl. In a recent chapter, the author delivers a novel theory named after him as "quantized Marghany turbulence boundary." In addition, the author delivers a novel

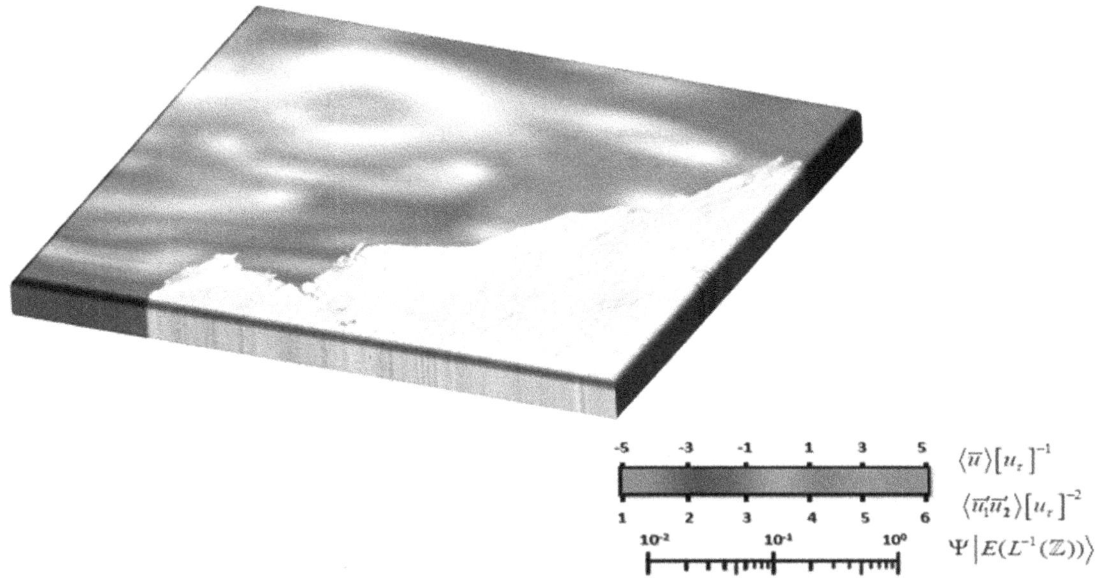

FIGURE 4.37 Quantized Marghany spectra energy turbulence boundary along Borneo during the southwest monsoon period.

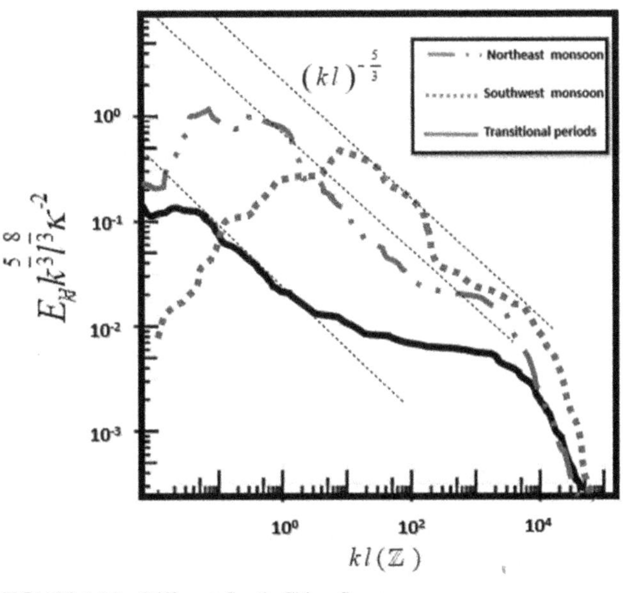

FIGURE 4.38 Different South China Sea monsoon seasons are represented by Kolmogorov spectra of quantized Marghany spectra.

algorithm named as "Quantized Marghany turbulence spectra energy" to reconstruct and track this turbulence boundary in SAR data over large-scale and different wind monsoon periods. This algorithm describes the quantized Marghany turbulence boundary as wave function or probability in certain topology Z and specific time t. This turbulence boundary contains infinite irregular movements, which entangle with wind stress curl in different locations and periods. In this understanding, the localized irregular topology features in real ocean and in SAR images create such as a complicated nonlinear turbulence boundaries. This localized in the western boundary, grows, and dissipates toward the eastern boundary, which is dissimilar to Sverdrup and Munk's theories. The wave function is only a novel approach to describe this complicated nonlinear turbulence boundary in the form of quantized spectra. In this understanding, SAR image can be presented in the form of the wave function. Therefore the wave function is localized to describe the entanglement of radar signals with ocean topology and wind stress curl. The entanglement nonlinearity among different states superposition the Bragg resonance to form such nonlinear irregular path boundary integral in SAR images. The localization of turbulence flow in SAR data can be quantized by exploiting the quantized Marghany turbulence boundary/spectra energy algorithm. Needless to say, the water circulation in the South China Sea is not as simple as described in the previous studies as two different directions of water jets following monsoon wind directions—one jet flow from north to south during the northeast monsoon and later flows from south to north during southeast monsoon. The nonlinearity of the water circulation in the South China under wind stress curl represents in quantized turbulence boundary over the year, which is known as the "quantized Marghany turbulence boundary." This boundary contains quantized eddies, vorticity, solution, internal wave, rain cells, upwelling, and so on, that will be explored separately in the next chapters.

References

Broecker, W. S., Patzert, W. C., Toggweiler, J. R., & Stuiver, M. (1986). Hydrography, chemistry, and radioisotopes in the Southeast Asian basins. *Journal of Geophysical Research, 91*(C12), 14345. Available from https://doi.org/10.1029/jc091ic12p14345.

Chen, C. T. A., Wang, S. L., Wang, B. J., & Pai, S. C. (2001). Nutrient budgets for the South China sea basin. *Marine Chemistry*, *75*(4), 281–300. Available from https://doi.org/10.1016/S0304-4203(01)00041-X.

Chu, P. C., Edmons, N. L., & Fan, C. (1999). Dynamical mechanisms for the South China Sea seasonal circulation and thermohaline variabilities. *Journal of Physical Oceanography*, *29*(11), 2971–2989. Available from http://journals.ametsoc.org/loi/phoc, https://doi.org/10.1175/1520-0485(1999)029 < 2971:DMFTSC > 2.0.CO;2.

Chu, P. C., & Li, R. (2000). South China Sea isopycnal-surface circulation. *Journal of Physical Oceanography*, *30*(9), 2419–2438. Available from http://journals.ametsoc.org/loi/phoc, https://doi.org/10.1175/1520-0485(2000)030 < 2419:SCSISC > 2.0.CO;2.

Godfrey, J. S. (1989). A Sverdrup model of the depth-integrated flow for the world ocean allowing for island circulations. *Geophysical & Astrophysical Fluid Dynamics*, *45*(1–2), 89–112. Available from https://doi.org/10.1080/03091928908208894.

Gordon, A. L., Susanto, R. D., & Vranes, K. (2003). Cool Indonesian throughflow as a consequence of restricted surface layer flow. *Nature*, *425*(6960), 824–828. Available from https://doi.org/10.1038/nature02038.

Gordon, A. L. (1986). Interocean exchange of thermocline water. *Journal of Geophysical Research*, *91*(C4), 5037. Available from https://doi.org/10.1029/jc091ic04p05037.

Guo, Z. (1985). Wintertime South China Sea warm current and the westward current on its right. *Journal of Tropical Oceanography*, *4*, 1–9.

Hu, J., Kawamura, H., Hong, H., & Qi, Y. (2000). A review on the currents in the South China Sea: Seasonal circulation, South China Sea warm current and Kuroshio intrusion. *Journal of Oceanography*, *56*(6), 607–624. Available from https://doi.org/10.1023/A:1011117531252.

Jilan, S. (2004). Overview of the South China Sea circulation and its influence on the coastal physical oceanography outside the Pearl River Estuary. *Continental Shelf Research*, *24*(16), 1745–1760. Available from https://doi.org/10.1016/j.csr.2004.06.005, http://www.journals.elsevier.com/continental-shelf-research/.

Kim, Y. Y., Qu, T., Jensen, T., Miyama, T., Mitsudera, H., Kang, H. W., & Ishida, A. (2004). Seasonal and interannual variations of the North Equatorial current bifurcation in a high-resolution OGCM. *Journal of Geophysical Research: Oceans*, *109*(3), C03040–19. Available from https://doi.org/10.1029/2003jc002013, http://agupubs.onlinelibrary.wiley.com/agu/jgr/journal/10.1002/(ISSN)2169-9291/.

Li, L., & Qu, T. (2006). Thermohaline circulation in the deep South China Sea basin inferred from oxygen distributions. *Journal of Geophysical Research: Oceans*, *111*(5). Available from http://onlinelibrary.wiley.com/journal/10.1002/(ISSN)2169-9291, https://doi.org/10.1029/2005JC003164.

Metzger, E. J., & Hurlburt, H. E. (1996). Coupled dynamics of the South China Sea, the Sulu Sea, and the Pacific Ocean. *Journal of Geophysical Research: Oceans*, *101*(5), 12331–12352. Available from https://doi.org/10.1029/95JC03861, http://agupubs.onlinelibrary.wiley.com/agu/jgr/journal/10.1002/(ISSN)2169-9291/.

Qiu, B., & Lukas, R. (1996). Seasonal and interannual variability of the North Equatorial Current, the Mindanao Current, and the Kuroshio along the Pacific western boundary. *Journal of Geophysical Research: Oceans*, *101*(C5), 12315–12330. Available from https://doi.org/10.1029/95jc03204.

Qiu, D., Yang, T., & Guo, Z. (1984). A westward current in the northeastern part of the South China Sea. *Trop. Oceanol*, *33*, 65–73.

Qu, T. (2000). Upper-layer circulation in the South China Sea. *Journal of Physical Oceanography*, *30*(6), 1450–1460. Available from http://journals.ametsoc.org/loi/phoc, https://doi.org/10.1175/1520-0485(2000)030 < 1450:ULCITS > 2.0.CO;2.

Qu, T., Du, Y., Meyers, G., Ishida, A., & Wang, D. (2005). Connecting the tropical Pacific with Indian Ocean through South China Sea. *Geophysical Research Letters*, *32*(24), 1–4. Available from https://doi.org/10.1029/2005GL024698.

Qu, T., Du, Y., & Sasaki, H. (2006a). South China Sea throughflow: A heat and freshwater conveyor. *Geophysical Research Letters*, *33*(23). Available from https://doi.org/10.1029/2006GL028350.

Qu, T., Girton, J. B., & Whitehead, J. A. (2006b). Deepwater overflow through Luzon Strait. *Blackwell Publishing Ltd, United States Journal of Geophysical Research: Oceans*, *111*(1). Available from https://doi.org/10.1029/2005JC003139, http://onlinelibrary.wiley.com/journal/10.1002/(ISSN)2169-9291.

Qu, T., Song, Y. T., & Yamagata, T. (2009). An introduction to the South China Sea throughflow: Its dynamics, variability, and application for climate. *Dynamics of Atmospheres and Oceans*, *47*(1–3), 3–14. Available from https://doi.org/10.1016/j.dynatmoce.2008.05.001.

Quadfasel, D., Kudrass, H., & Frische, A. (1990). Deep-water renewal by turbidity currents in the Sulu Sea. *Nature*, *348*(6299), 320–322. Available from https://doi.org/10.1038/348320a0.

Shaw, P.-T. (1989). The intrusion of water masses into the sea southwest of Taiwan. *Journal of Geophysical Research*, *94*(C12), 18213. Available from 10.1029/jc094ic12p18213.

Shaw, P.-T. (1991). The seasonal variation of the intrusion of the Philippine sea water into the South China Sea. *Journal of Geophysical Research: Oceans*, *96*(C1), 821–827. Available from https://doi.org/10.1029/90jc02367.

Sheremet, V. A. (2001). Hysteresis of a western boundary current leaping across a gap. *Journal of Physical Oceanography*, *31*(5), 1247–1259. Available from http://journals.ametsoc.org/loi/phoc, https://doi.org/10.1175/1520-0485(2001)031 < 1247:HOAWBC > 2.0.CO;2.

Tangdong Qu, Y. Y. K., Max Yaremchuk, T. T., & Akio Ishida, T. Y. (2004). Can Luzon Strait transport play a role in conveying the impact of ENSO to the South China Sea?*. *Journal of Climate*, *17*(18), 3644–3657. Available from https://doi.org/10.1175/1520-0442(2004)017 < 3644:clstpa > 2.0.co;2.

Tian, J., Yang, Q., Liang, X., Hu, D., Wang, F., & Qu, T. (2006). Observation of Luzon strait transport. *Geophysical Research Letters*, *33*.

Toole, J. M., Millard, R. C., Wang, Z., & Pu, S. (1990). Observations of the Pacific North Equatorial Current bifurcation at the Philippine Coast. *Journal of Physical Oceanography*, *20*(2), 307–318. Available from https://doi.org/10.1175/1520-0485(1990)020 < 0307:ootpne > 2.0.co;2.

Vranes, K., Gordon, A. L., & Ffield, A. (2002). The heat transport of the Indonesian throughflow and implications for the Indian Ocean heat budget. *Deep-Sea Research Part II: Topical Studies in Oceanography*, *49*(7–8), 1391–1410. Available from https://doi.org/10.1016/S0967-0645(01)00150-3.

Wang, D., Liu, Q., Huang, R. X., Du, Y., & Qu, T. (2006). Interannual variability of the South China Sea throughflow inferred from wind data and an ocean data assimilation product. *Geophysical Research Letters*, *33*(14). Available from https://doi.org/10.1029/2006GL026316.

Wei, Z., Fang, G., Xu, T., Wang, Y., & Lian, Z. (2016). Seasonal variability of the isopycnal surface circulation in the

South China Sea derived from a variable-grid global ocean circulation model. *Acta Oceanologica Sinica*, *35*(1), 11–20. Available from https://doi.org/10.1007/s13131-016-0791-3, http://www.springer.com/earth + sciences + & + geography/oceanography/journal/13131.

Wu, Y., & Fallon, S. J. (2020). Prebomb to postbomb 14C history from the west side of Palawan Island: Insights into oceanographic changes in the South China Sea. *Journal of Geophysical Research: Oceans*, *125*(6). Available from https://doi.org/10.1029/2019JC015979, http://agupubs.onlinelibrary.wiley.com/agu/jgr/journal/10.1002/(ISSN)2169-9291/.

Wyrtki, K. (1961). Physical oceanography of the Southeast Asian waters. **2**.

Yaremchuk, M., & Qu, T. (2004). Seasonal variability of the large-scale currents near the Coast of the Philippines. *Journal of Physical Oceanography*, *34*(4), 844–855. Available from https://doi.org/10.1175/1520-0485(2004)034 < 0844:SVOTLC > 2.0.CO;2.

Yuan, D., Han, W., & Hu, D. (2006). Surface Kuroshio path in the Luzon Strait area derived from satellite remote sensing data. *Journal of Geophysical Research*, *111*(C11). Available from https://doi.org/10.1029/2005jc003412.

Yuan, Y., Liao, G., Yang, C., Liu, Z., Chen, H., & Wang, Z. G. (2014). Summer Kuroshio Intrusion through the Luzon Strait confirmed from observations and a diagnostic model in summer 2009. *Progress in Oceanography*, *121*, 44–59. Available from https://doi.org/10.1016/j.pocean.2013.10.003.

Chapter 5

Automatic detection of internal wave in quantized Marghany turbulence boundary in SAR data

The "Quantized Marghany Turbulence Boundary" is a novel theory for emerging a new turbulence boundary in the South China Sea presented in the previous chapter. Features in the same time and topology zone are entangled and superimposed to form this boundary. The turbulence features of the quantized Marghany turbulence boundary exhibit nonlinear entanglement and superposition, which proves that it is a nonlinear turbulence boundary. The internal wave is one of these features. The nonlinearity of the internal waves in the quantized Marghany turbulence boundary at the western boundary of Peninsular Malaysia is discussed in this chapter.

5.1 What constitutes the foundation of the internal waves?

A significant question now arises: What is the keystone in the initial understanding of internal waves? The key to comprehending the nature of internal waves is, therefore, gravity waves. Internal waves do resemble gravity waves, after all. So, how do gravity waves originate? When the force of gravity or buoyancy (Fig. 5.1) attempts to restore equilibrium, waves are created in a fluid medium or at the interface between two media (Fig. 5.2). The interface between the atmosphere and the ocean(), which creates wind waves, is an illustration of one such interface. When the fluid is displaced from the equilibrium position, a gravity wave is generated. A fluid oscillates back and forth, or "wave orbit," as it is brought back into balance. Surface gravity waves (Fig. 5.3), a subset of surface waves, are gravity waves that occur at the air – sea interface of an ocean, whereas internal waves are gravity waves that occur within a body of water (e.g., between regions with varying densities). How does ocean motion change due to gravity? Therefore the less dense water is pushed away by the more dense water, which shoots sideways and rises as a result of gravity (Bell, 1975; Garrett & Munk, 1975; Gerkema & Zimmerman, 2008). As the lighter (hotter, less salty) regions of water rise and flow to replace the heavier (colder, more salty) regions of water, enormous convection loops of ocean currents form.

Gravity waves include ocean tides, tsunamis, and waves created by the wind on the surface of the water (Fig. 5.4). In this view, wind-induced gravitational waves

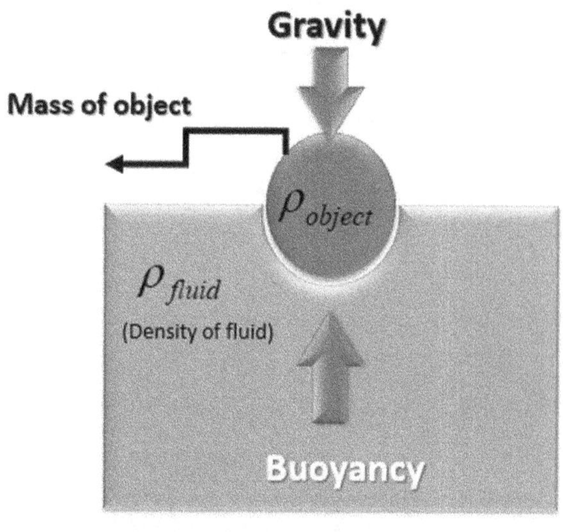

FIGURE 5.1 Simplification of buoyancy.

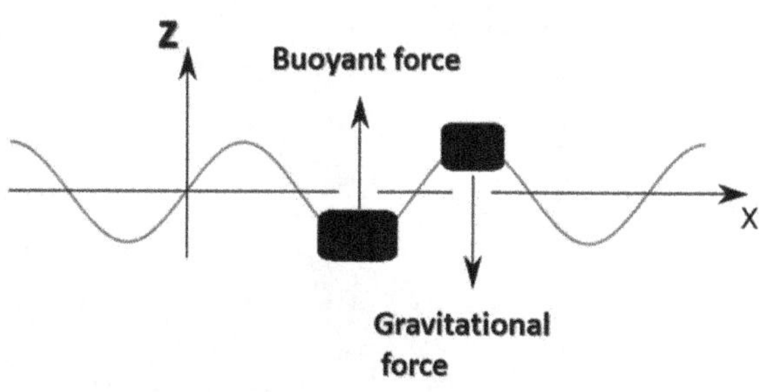

FIGURE 5.2 Generation of gravity waves due to buoyant and gravitational forces.

Synthetic Aperture Radar Image Processing Algorithms for Nonlinear Oceanic Turbulence and Front Modeling. DOI: https://doi.org/10.1016/B978-0-443-19155-8.00005-2

typically have a frequency of 0.033 − 3.33/s over the free surface of ponds, lakes, seas, and oceans on Earth. This corresponds to periods primarily between 3 and 30 seconds. Surface tension also affects shorter waves, which is why it is also called gravitational capillary waves or capillary waves (when gravity is not a factor) (Garrett & Munk, 1979; McComas & Bretherton, 1977). Conversely, the so-called infra-gravitational waves because of the nonlinear subharmonic interactions of the waves with the wind waves have longer periods than the waves generated by the wind. Then, how do internal waves and gravity typically interact?

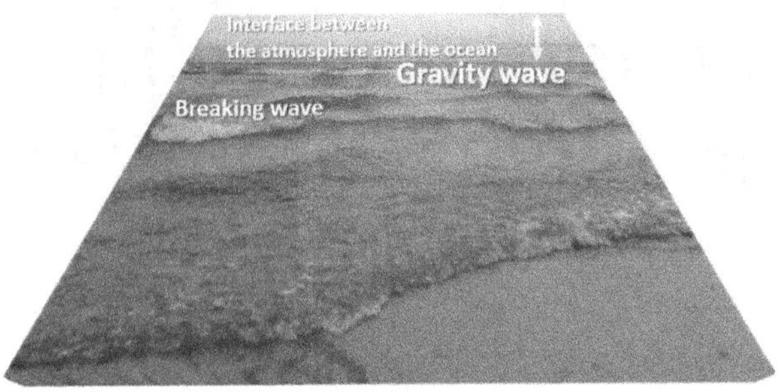

FIGURE 5.3 Gravity wave and atmosphere and ocean interference.

5.2 How do gravity waves form internal waves?

Gravity waves can also form at the boundary between two ocean water layers with different densities (Fig. 5.5).

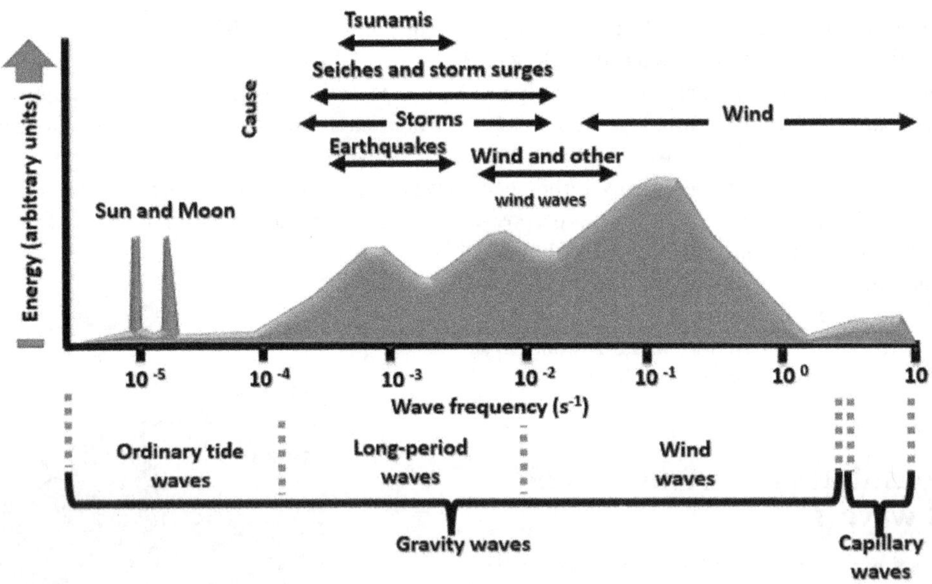

FIGURE 5.4 Gravity wave frequency.

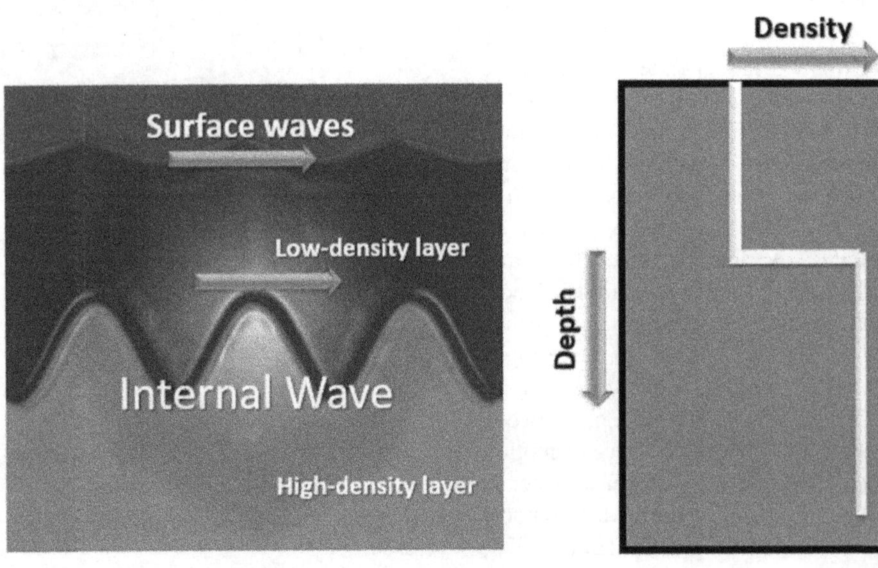

FIGURE 5.5 Internal wave generation due to density gradient between two layers.

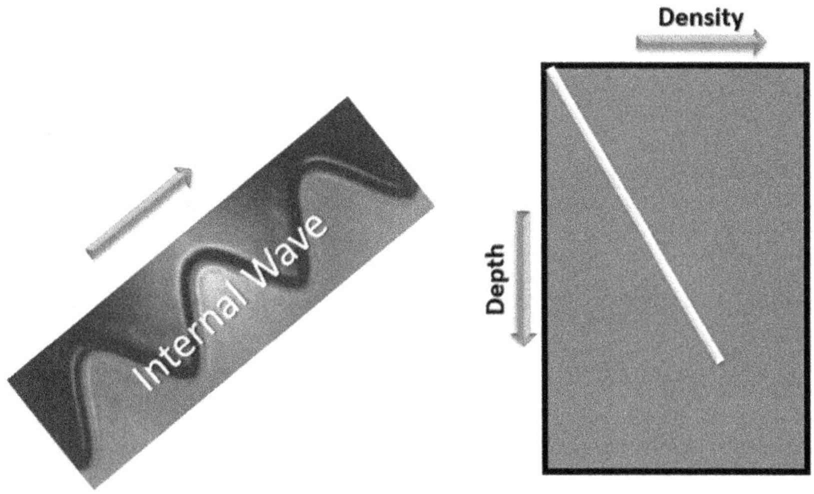

FIGURE 5.6 Internal waves that move at an angle to the horizontal.

Such waves are technically surface waves because the interface is a surface, but oceanographers typically refer to them as internal waves. These most frequently occur where there is a steep density gradient, also known as a pycnocline, or where there is a rapid increase in density with depth. In this sense, pycnoclines are formed by pronounced variations in temperature and/or salinity, which together determine the density of seawater. With density continually varying with depth, there is no requirement that these waves travel horizontally. In a continuously stratified ocean, there is a gravitational restoring force for waves that move at an angle to the horizontal (Fig. 5.6). Because there is a much smaller density difference between two water layers than between water and air, it takes less energy to move the interface from its equilibrium position, making internal interfaces more conducive to oscillations than sea surface interfaces. Internal waves move much more slowly than the majority of surface waves (Garrett & Munk, 1975, 1979).

They have longer wavelengths (hundreds, not tens of meters), longer durations (minutes or hours, not seconds), and larger amplitudes (up to several tens of meters) than all but the largest surface waves. Particularly as they split apart, internal waves play a significant role in oceanic longitudinal mixing processes. Therefore the thickness of the layers acts similarly to the water depth in surface shallow water waves, giving internal waves the appearance of moving like shallow water waves. Similar amplitude internal waves move more slowly than surface waves. Similar to how the wind creates surface waves, internal waves may be created when the surface layer flows over a lower layer that is more static. Tidal forces, long-term wind changes, and seismic disturbances may cause internal waves (Garrett & Munk, 1972, 1975, 1979). In this understanding, a wave traveling along an internal boundary may be tall enough to have its crest below the surface when the top layer is relatively shallow. Slicks, which are fluid moving bands that each represent a different position on the crest, are the result of this action. Slicks are more likely to develop in estuaries, fjords, and coastal waters where freshwater runoff dilutes the shallow surface layer. These organic slicks show that internal waves are present beneath the surface.

5.3 Mathematical foundation in describing internal wave

The Boussinesq fluid was demonstrated in Section 2.10 of Chapter 2. To understand internal waves, the most basic and straightforward method is to use a Boussinesq model with constant stratification and no background rotation. In this regard, let us consider a fluid that is initially at rest and whose background buoyancy only varies with height; as a result, the buoyancy frequency, N, or Brunt − Väisälä frequency depends only on z. In this sense, the shortest period internal gravity wave is limited by the buoyancy frequency (Brunt − Väisälä frequency), $N = \sqrt{gE}$. In this circumstance, it is surely impossible to create waves that oscillate more quickly than the Brunt − Väisälä frequency. This indicates that internal gravity waves have periods that are measured in hours as opposed to minutes everywhere in the ocean, with the exception of the vicinity of a sharp thermocline (Bell, 1975; Garrett & Munk, 1979; McComas & Müller, 1981).

The system satisfies the Boussinesq equations, and the linear momentum equations are produced by linearizing those equations of motion, that is, potential velocity ϕ about this fundamental state.

$$\partial \vec{\boldsymbol{u}}'[\partial t]^{-1} = -\nabla\phi', \partial w'[\partial t]^{-1} = \partial\phi'[\partial z]^{-1} + b', \quad (5.1)$$

In Eq. (5.1), the buoyancy is related to density ρ_0 by $b' = -g\delta\rho\ [\rho_0]^{-1}$, and $\phi = \delta\rho\rho\ [\rho_0]^{-1}$. Therefore the thermodynamic equations and the mass continuity can be mathematically expressed as:

$$\partial u'[\partial x]^{-1} + \partial v'[\partial y]^{-1} + \partial w'[\partial z]^{-1} = 0, \partial b'[\partial t]^{-1} + w'N^2 = 0. \tag{5.2}$$

According to Eqs. (5.1) and (5.2), the exploited notation is such that $\vec{u} \equiv u\vec{i} + v\vec{j}, \vec{v} \equiv u\vec{i} + v\vec{j} + w\vec{k}$. Therefore the gradient operator is two-dimensional unless otherwise specified. Consequently, $\nabla \equiv \vec{i}\,\partial x + \vec{j}\,\partial y$ *and* $\nabla_3 \equiv \vec{i}\,\partial x + \vec{j}\,\partial y + \vec{k}\,\partial z$. Next, a little algebra yields a single equation for w' *as follows:*

$$\left[\frac{\delta^2}{\delta t^2}\left(\nabla^2 + \frac{\delta^2}{\delta z^2}\right) + N^2\nabla^2\right]w' = 0 \tag{5.3}$$

Therefore it is conclusive that Eq. (5.3) is not isotropic. The coefficients of each term are constant if N^2 is a constant, which means that the background buoyancy varies linearly with z. In this scenario, the best solution may be given by:

$$w' = Re\tilde{w}e^{i(kx+ly+mz-\omega t)}, \tag{5.4}$$

where other variables oscillate similarly, and Re denotes the real part, a designation that is frequently dropped unless ambiguity occurs (Benjamin, 1967; Garrett & Munk, 1975; Gerkema & Zimmerman, 2008). Therefore if $l = 0$ in Eq. (5.4), all of the perturbation quantities in the wave are constant at any given instant along the lines of kx + mz being constant. The dispersion relation is formed by appending Eqs. 5.4 to 5.3:

$$\omega^2 = (k^2 + l^2)N^2\left[k^2 + l^2 + m^2\right]^{-1} = K^2N^2\left[K_3^2\right]^{-1}, \tag{5.5}$$

Since $N^2 < 0$, it is expected this state to be unstable, while the basic state density would increase with height. The disturbance would also rise massively following exp(σt), as:

$$\sigma = i\omega = \pm K\tilde{N}K_3^{-1}, \tag{5.6}$$

Eq. (5.6) demonstrates that although there are obviously many confounding factors, such as water vapor in the atmosphere, salt in the ocean, and the effects of rotation, the majority of convective activity in the ocean and atmosphere is ultimately related to the instability of this form.

What can such a transform establish if $\partial w'[\partial t]^{-1} = 0$ in Eq. (5.1)? In this scenario, the dispersion relation can be formulated as:

$$\omega = \sqrt{(k^2 + l^2)N^2[m^2]^{-1}} \tag{5.7}$$

here k, l, and m are the wavenumbers of x-, y-, and z-components. In this sense, Eq. (5.7) demonstrates that for small horizontal scales, the hydrostatic approximation behaves quite unphysically because of the frequency ω and, if N^2 is negative, the growth rate, is unbounded as K^2 [m^2] $\rightarrow \infty$

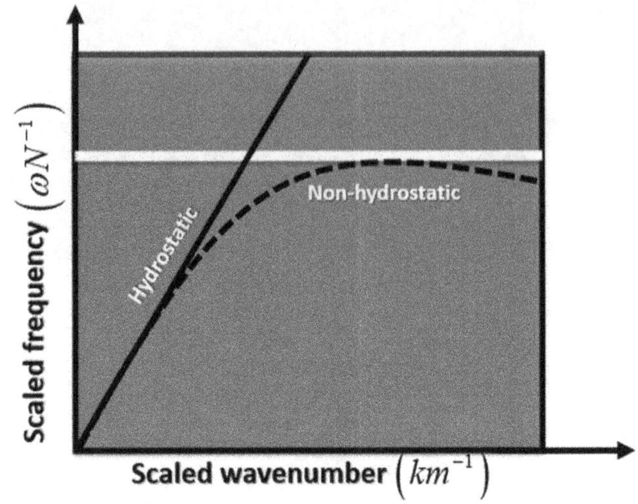

FIGURE 5.7 Hydrostatic and nonhydrostatic approximations.

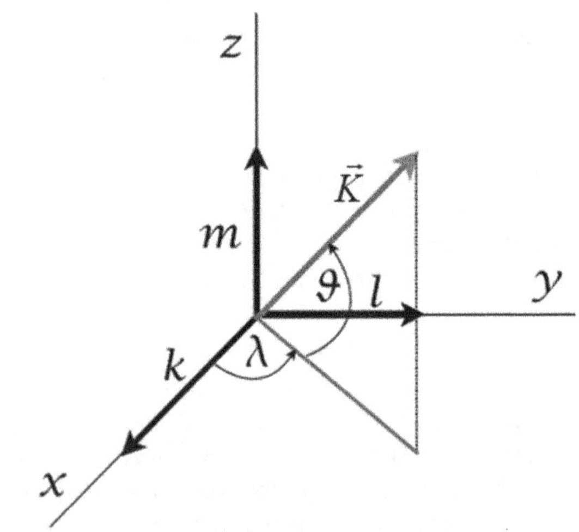

FIGURE 5.8 3-D Spherical coordinate of the wave vector.

(Fig. 5.7). Therefore if the horizontal resolution of the model is too coarse to properly resolve the convective scales, convection must also be parameterized in nonhydrostatic models (Bell, 1975; Dwi Susanto et al., 2005; Gerkema & Zimmerman, 2008). What physical properties of internal waves can be retrieved using the mathematical description of these waves discussed earlier?

5.4 Physics of internal waves

Let us assume that ϑ is the angle between the three-dimensional wave vector (Fig. 5.8), $\vec{k} = k\vec{i} + l\vec{j} + m\vec{k}$. In this view, $\cos^2\vartheta = K^2\,[K^{-2} + m^{-2}]$, and the mathematical expression of dispersion relation in Eq. (5.5) can be given by:

$$\frac{\sqrt{\pm N\cos\vartheta}}{N^2 - \sqrt{\pm N\cos\vartheta}} = K^2m^{-2} \tag{5.8}$$

Evidently, the performance relies only on the quantities of N and ϑ; if this is true, the frequency does not depend on the significance of wavelength. The dispersion relation, also known as this relationship, is a cornerstone of wave physics. It enables us to forecast how waves will behave in various media and environmental circumstances (Gerkema & Zimmerman, 2008; Helfrich & Melville, 2006; Jackson et al., 2012).

The crucial question is: What is the relationship between pressure, buoyancy, and the velocity of internal wave propagation? Physically, it is well understood that the vertical velocity is thus in phase with the pressure perturbation, and for regions of positive m (and so with upward phase propagation), regions of high relative pressure are associated with downward fluid motion. With this understanding, the pressure oscillation ϕ' can be expressed mathematically as:

$$\phi' = \tilde{\phi}\, e^{[i(\boldsymbol{k}\cdot\boldsymbol{x}-\omega t)]} = \tilde{\phi}\, e^{[i(kx+ly+mz-\omega t)]} \tag{5.9}$$

Consequently, the phases of the horizontal velocity components are:

$$(\tilde{u}, \tilde{v}) = (k, l)\omega - 1\,\tilde{\phi}. \tag{5.10}$$

Conclusively, the velocities and pressure are in phase because the frequency is real. Since all the fields are real, it is convenient to express the relations in terms of sines and cosines. We can also obtain similar relations for the other variables. If we select the cosine of pressure variation, then:

$$b = N^2K^2m^{-1}\omega^{-2}\phi_0\sin(kx + ly + mz - \omega t), \tag{5.11}$$

$$(u, v) = (k, l)\phi_0\omega^{-1}\cos(kx + ly + mz - \omega t), \tag{5.12}$$

$$\phi = \phi_0\cos(kx + ly + mz - \omega t), \tag{5.13}$$

$$w = -K^2[m\omega]^{-1}\phi_0\cos(kx + ly + mz - \omega t). \tag{5.14}$$

within ϕ_0 is constant, the Eqs. (5.11–5.14) is made up of the relationship of what is called polarization relations between pressure, buoyancy, and velocity of internal wave propagation (Benjamin, 1967; Dwi Susanto et al., 2005; Helfrich & Melville, 2006).

What takes place as Eqs. (5.12) and (5.14) are multiplied by *(k, l)*, and *m* respectively? This technical algebra work delivers the relationship between three-dimensional vectors of wave vector $\vec{k}$ and velocity vector $\tilde{v}$ as:

$$\vec{k}.\tilde{v} = 0 \tag{5.15}$$

In this view, the wavenumber k is related to the vertical and horizontal velocities. As a consequence, the velocity is aligned along the direction of the troughs and crests, where there is no pressure gradient, and the wave vector is always perpendicular to the velocity vector. The motion is strictly vertical, and $\omega = N$, if the wave vector is fully horizontal (i.e., $m = 0$). When the motion is in the *x-y* plane and $v = l = 0$ (for simplicity and without losing generality), it reveals from Eq. (5.15) that

$$\tilde{u} = -\tilde{w}\, m\left[k^{-1}\right]. \tag{5.16}$$

Eq. (5.16) demonstrates that all fluid parcel motions are consequently perpendicular to the wave fronts. The fluid parcels oscillate along lines that are at an angle of $\vartheta = \cos^{-1}(\omega/N)$ to the vertical because the wave frequency is related to the background buoyancy frequency by the formula $\omega = \pm N\cos\vartheta$. In this sense, Fig. 5.9 presents the group and phase velocities, polarization relations, and polarization relations. The figure shows that the polarization relations are dependent on the direction of the wave vector, while the group and phase velocities are independent of it. Additionally, the figure demonstrates how different types of waves can have different polarization states and velocities (Bell, 1975; Benjamin, 1967; Garrett & Munk, 1975, 1979; Helfrich & Melville, 2006; Jackson et al., 2012).

Therefore the "Dense" phase line moves up and to the right while the "Low" phase line gets longer. The pattern of "High-Dense-Low-Light-High" in Fig. 5.9 would remain the same if the fluid parcel motion were reversed. However, because of the fluid's loss of density due to the downward motion along the "Low" line, the phase lines would move downward and to the left. Clearly, the wave fronts or lines of constant phase move at a right angle to the trajectories of the fluid parcels. The group velocity is shown in Fig. 5.9 as being at a right angle to the phase speed. Generally, gravity and the pressure gradient both contribute to restoring forces on parcels, but there is no pressure gradient along crests (Fig. 5.9). Therefore the background stratification's vertical advection, $w'\partial b_0[\partial z]^{-1} = w'N^2$., causes buoyancy perturbations (Cacchione & Wunsch, 1974; Gostiaux et al., 2007; Jackson et al., 2012). The "Low" line, which is $[4]^{-1}$ wavelength up and to the right of the "Dense" phase line, exhibits a local maximum in rising motion and, consequently, a tendency to increase fluid density. In accordance with Fig. 5.10, the restoring force, F_{RES}, in the direction of the particle displacement is as follows:

$$F_{RES} = g\cos\vartheta \times \Delta\rho = g\cos\vartheta \times \partial\rho[\partial z]^{-1}\Delta z = g\cos\vartheta \times \partial\rho[\partial z]^{-1}\Delta s\,\cos\vartheta = \rho_0\partial b[\partial z]^{-1}\cos^2\vartheta\Delta s, \tag{5.17}$$

Accordingly, the equation of motion for a parcel moving along a crest or trough is:

$$\rho_0\frac{d^2\Delta s}{dt^2} = -\rho_0N^2\cos^2\vartheta\Delta s, \tag{5.18}$$

The displacement of a parcel is shown to be at an angle to the direction of gravity by Eq. 5.18, and the restoring force that a parcel experiences is proportional to $N\cos\vartheta$. Therefore it is convenient to use spherical polar

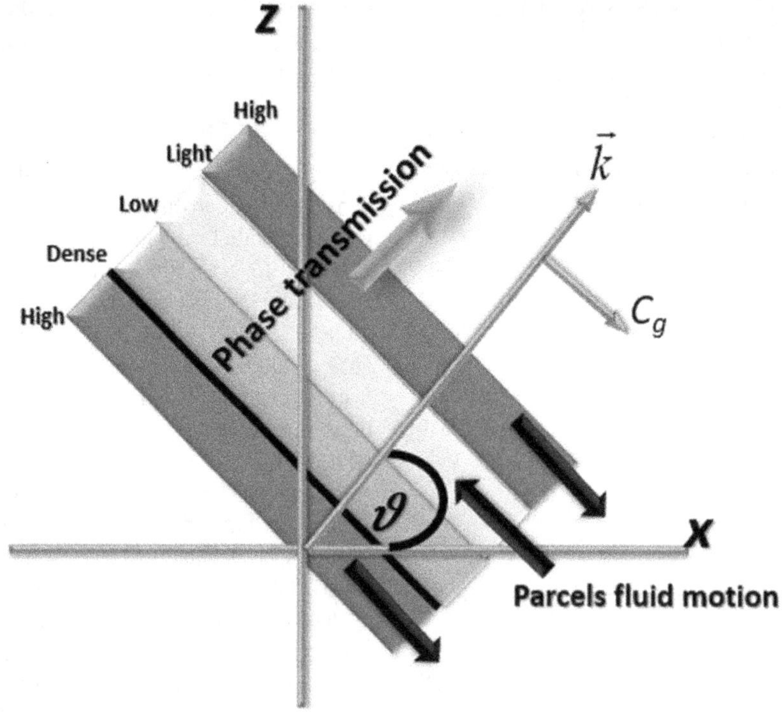

FIGURE 5.9 Concept of polarization relation.

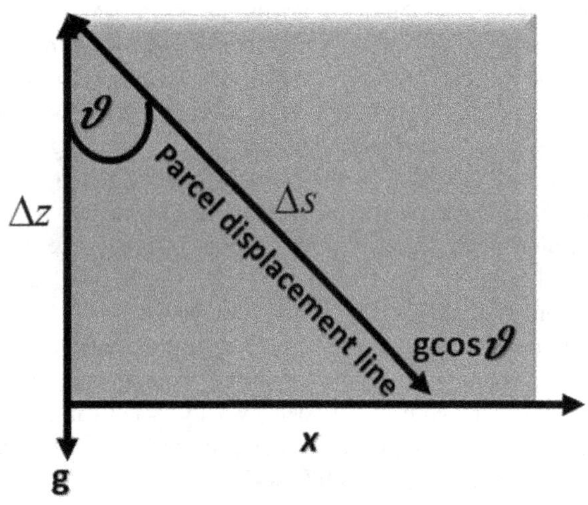

FIGURE 5.10 Parcel displacement.

coordinates to evaluate phase and group velocities in a useful manner, as shown in Fig. 5.8, where:

$$k = K_3\cos\vartheta\cos\lambda, l = K_3\cos\vartheta\sin\lambda, m = K_3\sin\vartheta, \quad (5.19)$$

In Eq. (5.19), $\vec{k} = K_3(\cos\vartheta\cos\lambda, \cos\vartheta\sin\lambda, \sin\vartheta)$. Therefore λ determines the orientation in the horizontal plane, and ϑ is the angle of the wave vector with the horizontal (Fig. 5.8). When speaking of the phase velocity, internal waves' phase speed c_{ph} in the direction of the wave vector is computed as follows:

$$c_{ph} = \omega[K_3]^{-1} = N[K_3]^{-1}\cos\vartheta = NK\left[K_3^2\right]^{-1}. \quad (5.20)$$

Therefore the speed of propagation of the wave crests in the respective directions x,y,z can be expressed by:

$$c_{ph}^{x} \equiv \omega[k]^{-1} = N[k]^{-1}\cos\vartheta, \quad (5.21)$$

$$c_{ph}^{y} \equiv \omega[l]^{-1} = N[l]^{-1}\cos\vartheta, \quad (5.22)$$

$$c_{ph}^{z} \equiv \omega[m]^{-1} = N[m]^{-1}\cos\vartheta. \quad (5.23)$$

According to Eqs. (5.21–5.23), these quantities are the rates at which the wave crests are moving in the specified directions. Though no information is transmitted at these speeds, they are all generally faster than the phase speed in the direction of the wave vector, which is perpendicular to the wave crests (McComas & Müller, 1981; Nycander, 2005; Thorpe, 1968). In this sense, the magnitude of the group velocity is evidently:

$$|c_g| = N\left[K^3\right]^{-1}\sin\vartheta, \quad (5.24)$$

According to Eq. (5.24), the group velocity is consequently parallel to the movement of the fluid parcels, as shown in Fig. 5.9. Additionally, unlike acoustic waves or even shallow water waves, energy moves perpendicular to the direction of phase propagation because it moves with the group velocity, which is parallel to lines of constant phase. Therefore the group velocity vector is directed at

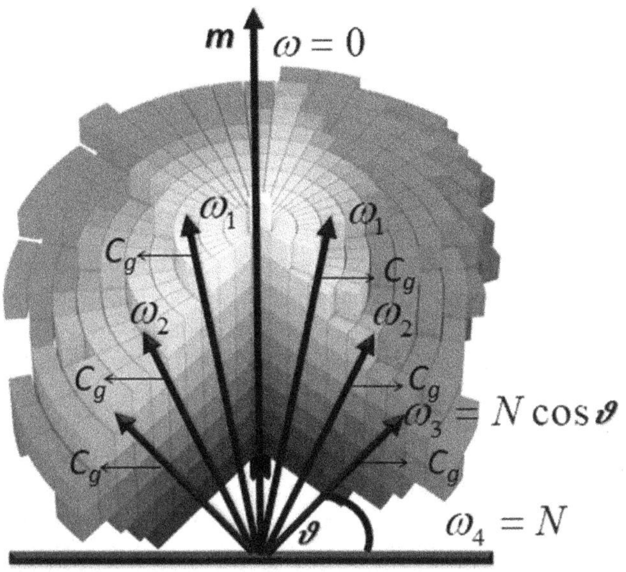

FIGURE 5.11 Speculation of the internal wave cone shape.

an angle ϑ to the vertical, as in Fig. 5.11, which delivers the speculation of an internal wave cone. In this regard, the surface with a constant angle to the horizontal, as shown in Fig. 5.11, defines the cone-shaped surfaces of constant frequency (Bell, 1975; Choi & Camassa, 1999; Gerkema & Zimmerman, 2008; Gostiaux et al., 2007; Jackson et al., 2012). Any wave with a wave vector in the cone has a frequency of N cosϑ, and both the wave vector and the phase velocity point away from the origin along the cone. The arrows on the dotted lines denote the group velocity, which is perpendicular to the cone and pointing in the direction of increasing frequency. The phase speed and group velocity have the opposite signs in the vertical direction.

$$c_g^x = \partial\omega[\partial k]^{-1} = (N[K^3]^{-1}\sin\vartheta)\cos\lambda\sin\vartheta, \tag{5.25}$$

$$c_g^y = \partial\omega[\partial l]^{-1} = (N[K^3]^{-1}\sin\vartheta)\sin\lambda\sin\vartheta, \tag{5.26}$$

$$c_g^z = \partial\omega[\partial m]^{-1} = (N[K^3]^{-1}\sin\vartheta)\cos\vartheta. \tag{5.27}$$

Eqs. (5.25–5.27) demonstrate that the group velocity is perpendicular to the wave vector and that the angle is perpendicular to the cone itself. This phenomenon is known as the internal wave cone shape effect, depending on the type of angular frequency fluctuations. The group velocity can also be affected by the medium through which the wave is traveling, such as changes in refractive index or density (Figs. 5.5 and 5.6). Without a doubt, the group velocity and phase speed have opposing signs, meaning that if the wave crests move upward, the group velocity moves downward (Choi & Camassa, 1999; Davis & Acrivos, 1967; Helfrich & Melville, 2006; Jackson et al., 2012; Thorpe, 1968). This phenomenon is known as dispersion and is commonly observed in wave propagation. In addition, dispersion can cause a wave to spread out and become distorted over time, leading to a loss of coherence.

5.5 What impact does internal wave propagation have on the mean flow?

The dispersion relation, which results from the presence of mean flow U in the x-direction, is the straightforward response to the aforementioned question:

$$(\omega - Uk) = \sqrt{K^2N^2[K^2+m^2]^{-1}}. \tag{5.28}$$

According to Eq. (5.28), Doppler shifting of the frequency is anticipated, but instead an intriguing impact is seen on the upward propagation of waves. In this view, the vertical component of the group velocity can be written as Eq. (5.28), where:

$$\partial\omega[\partial m]^{-1} = -mk(\omega k^{-1} - U)[K^{-2} + m^{-2}], \tag{5.29}$$

Even though m would also vary slowly with z, Eq. (5.29) nonetheless remains true if U is not constant or rather varies slowly with z. The important thing to keep in mind is that the group velocity reaches zero at the critical line, or the point where U = ω/k, and the wave stalls. In this scenario, $m^2 \gg K^2$, one can obtain, essentially the hydrostatic circumstance:

$$\partial\omega[\partial m]^{-1} = -k(\omega k^{-1} - U)[m]^{-1} = -k^2(\omega k^{-1} - U)^2[KN]^{-1}. \tag{5.30}$$

Physically, any potential dissipation has more time to occur when the group velocity decreases to zero as the wave approaches a critical line. In other words, we can anticipate that a wave would be preferentially dissipated close to a critical line, ending its momentum with the mean flow and its energy to generate mixing. The former is significant in the atmosphere (because this is the mechanism causing quasibiennial oscillation), while the latter is important in the ocean (Apel, 2002; Gostiaux et al., 2007; Helfrich & Melville, 2006; Jackson et al., 2012).

5.6 Continuous stratification and rotation

As a reference model for internal motion, the aforementioned mathematical description of internal waves in an ocean with two homogeneous layers works well. However, in real oceans, density structures are typically not as intuitive. In general, we are dealing with a continuously stratified water column, with stability increasing with depth. In this case, the motion is not constrained to the lowest mode and can exhibit more intricate vertical structures in higher modes. Therefore there are two restrictions on the potential frequencies of internal waves. For an internal mode with frequency $\omega >$ N_{max}, where N_{max} is the highest value of the buoyancy frequency in the density profile of the water column, there is no solution. The Earth's rotation also establishes a low-frequency limit, whose impact we have not yet taken into account (Benjamin, 1967; Cacchione & Wunsch, 1974; Fuchssteiner & Roy Chowdhury, 1995; Gostiaux et al., 2007).

It can be demonstrated that wave-type solutions are only feasible for frequencies higher than the inertial frequency f when Coriolis forces are taken into account in the equations of motion. Therefore internal waves on a rotating Earth have a frequency range $N_{max} > \omega >$ f and a wavelength determined by the mode's phase velocity (Choi & Camassa, 1999; Helfrich & Melville, 1986; Holloway, 1980). The energy distribution in internal waves is also impacted by rotation. The ratio of potential to kinetic changes when the Coriolis forces F_c are incorporated into the dynamical equations. In this understanding, the (Lagrangian) equation of motion for a displaced parcel can be given by:

$$\frac{d^2\Delta s}{dt^2} = -\left(N^2\cos^2\vartheta + F_C\sin^2\vartheta\right)\Delta s, \tag{5.31}$$

and hence, the frequency is:

$$\omega^2 = N^2\cos^2\vartheta + f^2\sin^2\vartheta. \tag{5.32}$$

Therefore the dispersion relation becomes:

$$\omega = \sqrt{(N^2k^2 + f^2m^2)[k^{-2} + m^{-2}]} \tag{5.33}$$

Now, almost everywhere in the ocean and atmosphere, N^2exceeds f^2. We can then deduce from Eq. (5.32) that the frequency falls within the range $N^2 > \omega^2 > f^2$. (Put $N = f$ or $f = N$ in Eq. (5.32) and use $\sin^2\vartheta + \cos^2\vartheta = 1$. The Coriolis force diminishes $\omega \rightarrow N$ as the parcel displacements become closer to vertical, and analogously $\omega \rightarrow f$, as the displacements become closer to horizontal, the Coriolis force decreases. Pure inertial waves are the waves that pursue.

There are a few intriguing special cases for internal gravity waves with rotation that can be discussed as (Choi & Camassa, 1999; Holloway, 1980; Jackson et al., 2012; Kozlov et al., 2015):

1. Pure inertial waves with a frequency of $0 < \omega < f$ and more specifically $\omega = f \sin\vartheta$, exist in the constraint $N \rightarrow 0$. Comparable with what was mentioned earlier, as f 0, we have pure internal waves with = N cos. in this view, these pure internal waves are not affected by rotation and propagate along isopycnal surfaces. On the other hand, pure inertial waves are affected by rotation and propagate perpendicular to the direction of rotation.
2. A wave vector that is only horizontal. Because of this, m = 0 and $\omega = N$. The Earth's rotation has no impact on the waves at that point. This is because, according to the f-plane approximation, the Coriolis force results from the product of the Coriolis parameter and the velocity's horizontal component. The Coriolis force disappears if the wave vector is horizontal because the fluid velocities are only vertical. However, if the wave vector has a vertical component, then the Coriolis force can have an effect on the fluid motion. This is why the Coriolis force can influence ocean currents and atmospheric circulation patterns, as they involve fluid motions with both horizontal and vertical components.
3. Therefore a wave vector that is only vertical in which the fluid velocities are horizontal with $\omega = f$ and the fluid parcels are not affected by stratification. Although they are not inertial in the sense that there is no implied force in an inertial frame of reference, the oscillations are then referred to as inertial waves. These two examples demonstrate that the group velocity is zero when $\omega = N$ or $\omega = f$. This means that the wave does not propagate in any direction. This phenomenon is known as wave trapping and is an important factor in ocean dynamics, particularly in the formation of internal waves and mixing processes.
4. The constraint of large horizontal scales, k, l, and m, is where hydrostasy occurs. Consequently, if we ignore $k^2$and $l^2$where they appear with m^2, then the dispersion relation can be casted as:

$$\omega = \sqrt{f^2 + N^2K^2m^{-2}} = \sqrt{f^2 + N^2\cos^2\vartheta}. \tag{5.34}$$

Since $K^2/m^2 \ll 1$, the frequency of the waves in this situation is lower than the buoyancy frequency, or $\omega^2 N^2$. Put differently, low frequencies would typically be hydrostatic and have a small aspect ratio.

5.7 Quantization of internal waves

The internal wave is nonlocal for the naked eye but local for remote sensing technology imaging. This means that while internal waves cannot be seen by the naked eye, they can be detected and studied using remote sensing techniques such as satellite imaging. In this view, can one quantize the internal waves? From the point of view of quantum wave mechanics, the internal waves can be described as wave-particle duality, which can be expressed as a wave function. However, the quantization of internal waves is still a topic of ongoing research and debate among physicists and oceanographers. Some theories suggest that internal waves may exhibit quantum behavior at very small scales, while others argue that the classical wave model is sufficient to describe their behavior. Regardless, understanding the properties and dynamics of internal waves is crucial for predicting ocean currents and climate patterns. The quantization of the Korteweg − De Vries (KdV) equation is an excellent approach for investigating the quantized nonlinearity of internal waves. The evolution of weakly nonlinear and long waves in a one-dimensional medium is described by this nonlinear partial differential equation. We can analyze internal wave behavior more precisely and accurately by quantizing the KdV equation, which may have significant ramifications for oceanography and fluid dynamics research (Fuchssteiner & Roy Chowdhury, 1995).

wInternal solitary waves in the coastal ocean that result in an exact solitary wave solution are typically described by a nonlinear equation of the Korteweg – de Vries equation. There is, however, the Earth's rotation in any practical application. A mathematical representation of waves on shallow water surfaces is the Korteweg – De Vries (KdV) equation. Being a nonlinear partial differential equation whose solutions can be precisely and precisely specified, it is particularly noteworthy as the prototypical example of an exactly solvable model. Diederik Korteweg and Gustav de Vries, two Dutch mathematicians, developed the KdV equation in 1895. Since then, it has been used extensively in physics, engineering, and other disciplines. Its study of solitons—localized wave packets that keep their shape and speed even as they travel over great distances—has made one of its most important contributions.

A function ϕ of two dimensionless real variables, x, and t, which are proportional to space and time, respectively, is represented by the KdV equation, which is a nonlinear, dispersive partial differential equation:

$$\partial_t\phi + \partial_x^3\phi - 6\phi\partial_x\phi = 0 \tag{5.35}$$

The conventional but insignificant constant 6 is placed in front of the final term. By multiplying t, x, and ϕ by constants, any of the three terms' coefficients can be made to equal any given nonzero constants. Therefore let us consider a proper solution where a fixed waveform (represented by $f(X)$) retains its morphology while propagating to the right at phase speed c_{ph}. A solution to this problem is furnished by $\phi(x,t) = f(x - c_{ph}t - a) = f(X)$. The ordinary differential equation is developed by substituting it into the KdV equation:

$$-c\frac{df}{dX} + \frac{d^3f}{dX^3} - 6f\frac{df}{dX} = 0, \tag{5.36}$$

Under the circumstance of $X \to \pm\infty$ and $f(X)$ is approximately zero, the solitary morphology can be mathematically expressed a:

$$\phi(x,t) = \frac{1}{2}c_{ph}\text{sech}^2\left[\frac{\sqrt{c_{ph}}}{2}\left(x - c_{ph}t - a\right)\right] \tag{5.37}$$

Eq. (5.37) is a solution to the Korteweg – de Vries equation. The sech function decays exponentially and is commonly used in mathematical physics to model solitons and other wave phenomena. Additionally, integrals of motion can also provide information about the stability and interaction properties of solitons. In this view, the mass is expressed by $\int\phi dx$,; the momentum is demonstrated as $\int\phi^2 dx$; and later energy is formulated as $\int\left[2\phi^3 - (\partial_x\phi)^2\right]dx$. Therefore if the nth derivative of a distribution $\phi(x)$ has the following form for every $n \in N$, then it is considered almost-bounded:

$$\phi^n(x) = b(x) + \Delta(x) \tag{5.38}$$

where b is locally constrained, and Δ is a discrete support distribution without an accumulation point (Fuchssteiner & Roy Chowdhury, 1995). Therefore the generalized Korteweg – de Vries equation can be described as:

$$\partial_t u + \partial_x^3 u + \partial_x f(u) = 0. \tag{5.39}$$

Eq. (5.39) is nonlinear partial differential equation in which some positive integer k, the function f is sometimes assumed to be $f(u) = u^{k+1}/(k+1) + u$. In this viewpoint, the extra u is a "drift term" that greatly simplifies the analysis (Darrigol, 1986; Fuchssteiner & Roy Chowdhury, 1995; Miroshnikov, 2023). Thus the original Korteweg – De Vries equation has the case $f(u) = 3u^2$. Consequently, the Poisson bracket between field variables at different points is a derivative of the δ distribution. Quantization of the KdV-field must therefore result in:

$$\left[u(x), u(\tilde{x})\right] = i\delta_x, \left(x - \tilde{x}\right) \tag{5.40}$$

Hamiltonian operators H can be connected to Eq. (5.40) in the following approaches:

$$u(x)_t = i[H, u(x)] = u_{xxx}(x) + 3u_x(x) + 3u(x)u_x(x) \tag{5.41}$$

An alternative representation of its dynamics is introduced to provide a recursive description of the evolution's symmetries and conserved quantities of Eq. (5.41). Initially, with regard to defining the space of densities as the quotient of quantum realization $QF(x)$ with $\vec{A}$ and $\vec{B} \in Qf(x)$. This can be formulated as;

$$\mathcal{L}_1 = Linearspan\{AB - BA|A, B \in QF(x)\} \tag{5.42}$$

subsequently, in relation to

$$\mathcal{L}_2 = Linearspan\{DA - AD|A \in QF(x)\} \tag{5.43}$$

here D stands for differentiation in relation to the variable x. Instead of using $=$, the equivalence relation originating from the following spaces $\mathcal{L}_1$, and $\mathcal{L}_2$, would be denoted by $\equiv$. The construction of density space is carried out in a manner that allows the factor spaces to be understood as the kernel of a trace operation that is selected in a manner that causes formal integrals (from $-\infty$ to $+\infty$) over total derivatives to vanish. In fact, if these operators are Hilbert – Schmidt operators that vanish quickly with respect to x at $\pm\infty$, then it is exactly that kernel (Gamble et al., 1968; Inkson, 1984; Penrose, 2009; Petrov, 2015).

Assume that the density $F = F(u)$ tends to depend in a certain manner on the field variable U. In the direction of an element B of $QF(x)$, define its directional derivative of F as follows:

$$F'\left[\vec{B}\right] = \left\langle \nabla F, \vec{B} \right\rangle \text{for all } \vec{B} \in QF(x). \quad (5.44)$$

The gradient of F is referred to as the quantity ∇F. The gradient defined in this manner is, in fact, the classical gradient of the corresponding scalar quantity given by the integral over the trace if the densities can be thought of as kernels of traces. Therefore it is now possible to rewrite the evolution Eq. (5.41) as:

$$u_t = D\nabla H \quad (5.45)$$

here D stands for the operator of considering the derivative with respect to x. Therefore a density that is explicitly time-dependent is invariant under the flow with implicit operator Θ:

$$u_t = \Theta\nabla H \quad (5.46)$$

According to the aforementioned perspective, identifying Eqs. (5.41) and (5.46) enables us to use the standard soliton theory approach to create higher integrals. Because a subtle realization of the noncommutativity aspects of the quantum field variables had to be implemented, the algebraic structure where the recursion operator could be found is now considerably more complex than that commonly used in soliton theory. Intuitively, these noncommutativity aspects are more complex than those encountered when matrix soliton equations are considered. This is complicated by the fact that besides the points in the x-space where these variables are localized, noncommutativity also results from the algebraic nature of the space where field variables take their values (Gamble et al., 1968; Garrett & Munk, 1972; Goldstone & Jackiw, 1975; Inkson, 1984; Miroshnikov, 2023).

The second Hamiltonian formulation of the quantum KdV can now be derived from the second symplectic operator of the KdV in a modified form $u_{xx}(x) + 3u(x)u(x)$. Introduce the field variable and indicate it with u:

$$L(u)A: = uA \quad (5.47)$$

$$R(u)A: = Au \quad (5.48)$$

These are, respectively, the operators for left- and right-handed multiplication with u. Then consider:

$$\Theta = D^3 + DL(u) + DR(u) + R(u)D + L(u)D + [L(u) - R(u)]D^{-1}[L(u) - R(u)] \quad (5.49)$$

Eq. (5.49) delivers an operator being antisymmetric with respect to the inner product. Needless to say, Θ is an implicit operator and provides, therefore, the second Hamiltonian formulation of Eq. (5.46). In general, nonlinear expressions in delta-distributions are not well defined, which causes real difficulties in nonlinear field theories where the quantization is given such that the commutator or anticommutator of field variables amounts to delta-distributions. Therefore this chapter discusses both the complete quantum integrability and the aspect of demonstrating that such a quantization is indeed possible for nonlinear fields in a rigorous and coherent manner. The main result of this section is based on the hypothesis that certain nonlinear δ-distribution terms can be embedded in a suitable algebraic structure. Internal solution waves that were previously believed to be impossible to quantize in the form of the second Hamiltonian formulation of the quantum KdV, which can now be developed into quantum integrable models using this technique (McComas & Bretherton, 1977; Nycander, 2005; Penrose, 2009; Petrov, 2015).

In the quantum field, a soliton, an antisoliton, and a finite (possibly zero) number of breathers make up the particle spectrum. The parameter's value determines how many breathers are needed. On a mass shell, multiparticle production cancels. A breather is a nonlinear wave in physics that has energy concentrated in a focused, oscillatory manner(Ambjørn et al., 1992; Kulish & Zeitlin, 2004b; Petrov, 2015; Pogrebkov, 2001). The expectations derived from the corresponding linear system for infinitesimal amplitudes, which tend toward an even distribution of initially localized energy, are in conflict with this.

A nonlinear dispersive partial differential equation is the sine-Gordon equation:

$$\frac{\partial^2 u_t}{\partial t^2} - \frac{\partial^2 u_t}{\partial x^2} + \sin u_t = 0, \quad (5.50)$$

The inverse scattering transform was used to determine an exact solution, which is:

$$u_t = 4\arctan\left(\frac{\sqrt{1-\omega^2}\cos(\omega t)}{\omega\cosh(\sqrt{1-\omega^2}x)}\right), \quad (5.51)$$

which, for $\omega < 1$, decays exponentially away from $x = 0$ and is periodic in time t. The exponentials should be turned into vertex operators in the quantum sine-Gordon equation $\mathscr{L}_{QsG}$:

$$\mathscr{L}_{QsG}(u_t) = \frac{1}{2}\partial_\mu u_t \partial^\mu u_t + \frac{1}{2}\Theta u_t^2 - \alpha\left(e^{i\beta u_t} + e^{-i\beta u_t}\right) \quad (5.52)$$

In mathematics, a vertex operator algebra (VOA) is an algebraic structure that is crucial to the development of both string theory and two-dimensional conformal field theory. It is a type of algebra that involves the study of symmetries and transformations of geometric objects, particularly those that arise in physics. Therefore $\beta = \sqrt{4\pi}$ is the piece with two powers of δ on top and bottom indicates large and minimum values; respectively. Therefore a quantum mechanical observable u_t that doesn't explicitly depend on time t satisfies the Heisenberg view as:

$$\frac{d}{dt}u_t = \frac{i}{\hbar}[H, u_t], \tag{5.53}$$

with $\hbar$ being the diminished Planck constant and Hamiltonian H. In this scenario, observables (without explicit time dependence) can be seen to form Lax pairs with the Hamiltonian aside from a factor. The alternative expression in terms of the isospectral evolution of these observables is then understood to be the Schrödinger picture. The Schrödinger picture depicts how a system's state changes over time (Pogrebkov, 2001). The time-evolution operator, a unitary operator, causes the evolution of a closed quantum system. For the transition of a state vector $|\psi(t_0)\rangle$ at time t_0 to a state vector $|\psi(t)\rangle$ at time t, the time-evolution operator is often written $U(t, t_0)$, resulting in:

$$|\psi(t)\rangle = U(t, t_0)|\psi(t_0)\rangle. \tag{5.54}$$

The time-evolution operator takes the form when the Hamiltonian H of the system does not change over time:

$$U(t, t_0) = e^{-iH\cdot(t-t_0)/\hbar}, \tag{5.55}$$

When dealing with a time-independent Hamiltonian H, the Schrödinger picture comes in handy $\partial_t H = 0$. This exponential expression in Eq. (5.55) must be evaluated using its Taylor series because H is an operator:

$$e^{-iHt/\hbar} = 1 - \frac{iHt}{\hbar} - \frac{1}{2}\left(\frac{Ht}{\hbar}\right)^2 + \cdots. \tag{5.56}$$

If the Hamiltonian relies on time but commutes at different times, the time-evolution operator can be expressed by:

$$U(t) = e^{\left(-\frac{i}{\hbar}\int_0^t H(t')dt'\right)}, \tag{5.57}$$

On the other hand, if the Hamiltonian is time-dependent but the Hamiltonians do not commute at different times, then the time-evolution operator can be given by:

$$U(t) = \mathcal{T} e^{\left(-\frac{i}{\hbar}\int_0^t H(t')dt'\right)}, \tag{5.58}$$

$\mathcal{T}$ is the time-ordering operator where

$$\mathcal{T}\{A(x)B(y)\} := \theta(\tau_x - \tau_y)A(x)B(y) \pm \theta(\tau_y - \tau_x)B(y)A(x), \tag{5.59}$$

The time-ordered product of operators can be useful in quantum field theory. This operation is signified by $\mathcal{T}$. Despite being referred to as the "time-ordering operator" informally, $\mathcal{T}$ is technically neither an operator on states nor a superoperator on operators. Therefore $\mathcal{T}\{A(x)B(y)\} \in QF(x)$, and θ has a value of 1 for persuasive arguments and 0 for opposing ones. It serves as an illustration of the broad category of step functions, which can all be visualized as linear combinations of the translation of this one. The quantized KdV can, of course, be expressed as operator $U(t)$. This time-ordering operation is only coordinate-independent if operators at space-like separated points commute because the operators rely solely on their location in spacetime (and not just time) (Ambjørn et al., 1992; Darrigol, 1986; Fuchssteiner & Roy Chowdhury, 1995; Inkson, 1984; Kulish & Zeitlin, 2004a; Miroshnikov, 2023).

5.8 Radar imaging mechanism of internal waves

From space, internal waves that are generated below the sea surface (Fig. 5.12) can be observed using satellite imagery. Under any weather or at night, synthetic aperture radar (SAR) is able to detect the radar backscatter signals from the ocean surface with a spatial resolution as small as a few meters. In this regard, the backscatter cross section per unit area $\sigma_0(\theta)_{ij}$ parameterizes the intensity of radar return signals from the ocean (Marghany, 2021):

$$\sigma_0(\theta)ij = 16\pi k_0^4 |g_{ij}(\theta)|^2 \psi(0, 2k_0 \sin\theta), \tag{5.60}$$

here, θ denotes the incident angle, k_0 denotes the wavenumber of the radar waves, and ψ denotes the two-dimensional (Cartesian) wavenumber spectral density of the ocean surface wave field, which satisfies the Bragg resonance scattering condition, the incident radar signal is in the xz plane, where z is the vertical direction and x, y are the horizontal coordinates. Therefore the first-order scattering coefficients are denoted by the letters g_{ij} (θ), while the letter indices (ij) denote the polarizations of the incident and backscattered signal, respectively (Alpers, 1985; Apel & Gonzalez, 1983; Brandt et al., 1996; Fu & Holt, 1984). For horizontal polarization;

$$g_{HH}(\theta) = \frac{(\varepsilon_r - 1)\cos^2\theta}{[\cos\theta + (\varepsilon_r - \sin^2\theta)^{0.5}]^2}, \tag{5.61}$$

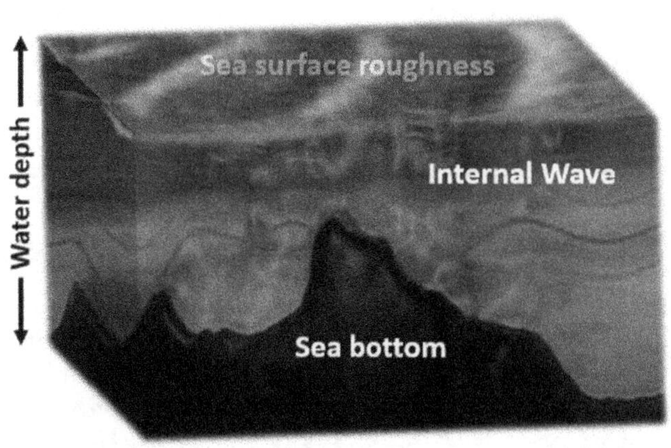

FIGURE 5.12 Internal wave generation beneath the surface of the water.

On the other hand, the vertical polarization is given by:

$$g_{VV}(\theta) = \frac{(\varepsilon_r - 1)\left[\varepsilon_r(1+\sin^2\theta)\right]\cos^2\theta}{[\varepsilon_r \cos\theta + (\varepsilon_r - \sin^2\theta)^{0.5}]^2}, \tag{5.62}$$

Here, ε_r stands for the relative dielectric constant of seawater. The intensity is entirely reliant on the ocean surface wave field's two-dimensional wavenumber spectral density, which must encounter the Bragg resonance scattering condition (Jackson et al., 2013). On the other hand, fluctuating background currents might indeed alter how the spectral density of the ocean surface wave field is distributed spatially as indicated by:

$$\psi_g(k) = m_3^{-1}\left[m\left(u_*{}^2 c_{ph}^{-2}\right) - 4\gamma k^2\omega^{-1} - \left(S_{\alpha\beta}\frac{\partial U_\beta}{\partial x_\alpha}\right)\omega^{-1}\right]k^{-4} \tag{5.63}$$

$$\psi_g(k) = m_4^{-0.5}\left[m\left(u_*{}^2 c_{ph}^{-2}\right) - 4\gamma k^2\omega^{-1} - \left(S_{\alpha\beta}\frac{\partial U_\beta}{\partial x_\alpha}\right)\omega^{-1}\right]^{0.5} k^{-4} \tag{5.64}$$

here m, m_3, and m_4 are coefficients; u* is the friction velocity; γ is the viscosity of seawater; ω is the angular frequency of the ocean surface waves; and k is the wavenumber of ocean surface waves. IWs modulate sea surface capillary and short gravity waves into smoother divergence zone and rougher convergence zone, respectively. Elevation and depression of thermocline caused by IWs are examples of this. A rough surface has a larger radar backscattering section, whereas a smooth surface produces less backscatter(Apel & Gonzalez, 1983; Fu & Holt, 1984; Thompson, 1988). As a result, internal waves beneath the ocean's surface can be seen in conjunction with bright and dark textures in the SAR image. $S\alpha\beta$ represents the excess moments' flux tensor, and subscripts α and β represent different horizontal coordinates (x or y):

$$S_{\alpha\beta}\frac{\delta U_\alpha}{\delta x_\beta} = 0.5\left[\frac{\delta u}{\delta x}\cos^2\varphi + \left(\frac{\delta u}{\delta y} + \frac{\delta v}{\delta x}\right)\cos\varphi\sin\varphi + \frac{\delta v}{\delta y}sin^2\varphi\right] \tag{5.65}$$

According to Eq. (5.65), internal wave and upper ocean constraints, such as the amplitude and wavelength of the internal wave, as well as water column stratification, are connected to the gradient current of $\frac{\delta u}{\delta y}$, *and* $\frac{\delta v}{\delta x}$. In this sense, the gradient current increases as the internal wave's amplitude increases. When $\frac{\delta U_\alpha}{\delta x_\beta} < 0$, Eq. (5.65) demonstrates that the normalized radar cross section expands into larger convergent flow layers. On the other hand, if $\frac{\delta U_\alpha}{\delta x_\beta} > 0$, a divergent flow is created, which causes the normalized radar cross section to decrease. Therefore the soliton-caused radar backscatter cross section $\sigma_{OIS}(\theta)_{ij}$ is given by:

$$\sigma_{OIS}(\theta)_{ij} = M^{-1}[\sinh^2 k(x - c_{ph}t)]\left[\cosh^2 k(x - c_{ph}t) + \frac{1}{k^2\left[4h_1^2[3\eta_0]^{-1}\right]^2}\sinh^2 k(x - c_{ph}t)\right]^{-2} \tag{5.66}$$

here M is the maximum rate of the function $[\sinh^2 k(x - c_{ph}t)]\left[\cosh^2 k(x - c_{ph}t) + \frac{1}{k^2\left[4h_1^2[3\eta_0]^{-1}\right]^2}\sinh^2 k(x - c_{ph}t)\right]^{-2}$. Consequently, parameter b is computed based on the water depth layer d and vertical displacement η_0 as:

In this view, *x-ct* shapes the radar backscatter, which relies on radar wavelength, incidence angle, and relaxation rate. As a consequence, the relaxation rate can vary according to the strength and direction of the wind. Additionally, according to the presumptions, the wind momentum ought to be greater than the 2–3 m/s Bragg wave threshold and lower than 10 m/s, below which the wind- and internal-wave-created roughness patterns can no longer be distinguished from one another. Consistent with the aforementioned perspective, hydrodynamic interaction theory and Bragg scattering theory are integrated to explain internal waves in SAR images. The variations in the backscattering cross section following the hydrodynamic modulation of the ripple energy spectral density at the mean (nonmodulated) Braggwave number by the long gravity waves are identified by the hydrodynamic modulation transfer function (MTF) (Alpers, 1985; Jackson et al., 2012; Klemas, 2012; Marghany, 2021). The distribution of the Bragg waves becomes nonuniform due to hydrodynamic interaction. As a result, the Bragg waves travel on the longest wave and are currently under the long wave orbital motion's stresses (Jackson et al., 2013; Marghany, 2018; Tao et al., 2022).

When the lower layer is significantly deeper than the mixed layer depth, internal wave signatures at the sea surface can be explicitly detected in SAR images, indicating that the waves are moving through relatively deep water. If the mixed group of waves in two layers traveling along the pycnocline are depression waves, as is frequently the case in deep waters, then internal wave signatures would be visible in SAR images as bright bands encroaching on darker ones in their direction of travel. In accordance with this theoretical model, depression soliton waves are solitary waves that travel through a watery body as only "up" or "down" waves and maintain a constant shape despite the fact so. Notwithstanding, internal wave signatures would have the exact opposite contrast in the SAR because their radar signature consists of dark bands emerging before actually bright bands in their propagation direction (Fig. 5.13) (Alpers, 1985; Brandt et al., 1996; Klemas, 2012; Kozlov et al., 2015; Marghany, 2021; Tao et al., 2022).

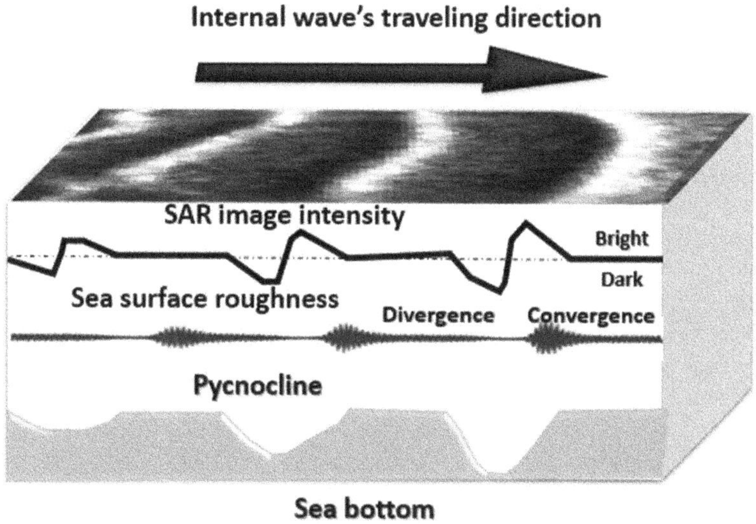

FIGURE 5.13 Mechanism of internal wave imagining in SAR data.

Therefore the quantum representation of the soliton-caused radar backscatter cross $\sigma_{OIS}(\theta)_{ij}$ can be expressed in the form of the Hermite polynomial H_n:

$$\Psi_n(\sigma_{OIS}(\theta)_{ij}) = \sqrt{\frac{1}{2^n n!}} \cdot \left(\frac{\sigma_{OIS}(\theta)_{ij}}{\pi\hbar}\right)^{1/4} \cdot e^{-\frac{\sigma_{OIS}(\theta)_{ij}}{2\hbar}} \cdot H_n\left(\sqrt{\frac{\sigma_{OIS}(\theta)_{ij}}{\hbar}} x\right) \tag{5.67}$$

For $n = 0, 1, 2, 3, \ldots\ldots,$ $H_n(\sigma_{OIS}(\theta)_{ij})$ *is an nth-degree polynomial where the weight function is* $w(x) = e^{-(\sigma_{OIS}(\theta)_{ij})^2}(\text{for} H_n)$. The detection of the 2-D quantized internal waves in the quantized Marghany turbulence boundary in the SAR image would therefore be accomplished in the following section by using Hermitian wavelets.

5.9 Hermitian wavelets for internal wave detection in SAR image

It is common practice to use the wavelet transform tool to analyze time-varying signals. The scale mathematical model develops spectral decomposition as a result of this practice. The two-dimensional wavelet transforms (2-DWT) function as a very powerful band-pass filter in SAR image analysis. 2-DWT is ruled to be discrete frequency channels with transformed scales in this regard. In addition to wavelet transforms' strong ability to localize time for a one-dimensional performance, they can also provide an accurate indication of the features in SAR data. In this sense, the wavelet transform is a reliable algorithm for extracting features with precise "physical properties" in SAR data, in accordance with these precise functions of 2-DWT. The SAR signal *f(t)'s* wavelet transform is then distinguished by:

$$X_w(a,b) = \frac{1}{|a|^{1/2}} \int_{-\infty}^{\infty} f(t)\overline{\psi}\left(\frac{t-b}{a}\right) dt \tag{5.68}$$

Eq. (5.68) reveals that a function's *f(t)* continuous wavelet transforms at a scale $(a > 0) a \in \mathbb{R}^{+*}$ and translational value $b \in \mathbb{R}$. The mother wavelet, $\psi(t)$, is a continuous function that exists in both the time and frequency domains. The complex conjugate operation is represented by the overline. The mother wavelet's primary function is to act as a source function for the daughter wavelets, which are just scaled and translated versions of the mother wavelet (Inkson, 1984; Kulish & Zeitlin, 2004b; Kumar et al., 2021). The first inverse continuous wavelet transform can be used to recover and retrieve the original SAR signal *f(t)*.

$$f(t) = C_\psi^{-1} \int_0^{\infty} \int_{-\infty}^{\infty} X_w(a,b) \frac{1}{|a|^{1/2}} \tilde{\psi}\left(\frac{t-b}{a}\right) db \frac{da}{a^2} \tag{5.69}$$

This transform involves integrating the product of the wavelet coefficients and scaled, translated versions of the mother wavelet function. The resulting integral is then summed over all possible scales and translations to reconstruct the original SAR signal *f(t)*. In this perspective, $\tilde{\psi}(t)$ is a dual function of $\psi(t)$ and is equivalent to:

$$C_\psi = \int_{-\infty}^{\infty} \frac{\overline{\hat{\psi}}(\omega)\hat{\tilde{\psi}}(\omega)}{|\omega|} d\omega \tag{5.70}$$

In certain scenarios, $\tilde{\phi}(t) = \psi(t)$, then the admissible constant changes as:

$$C_\psi = \int_{-\infty}^{+\infty} \frac{\left|\hat{\psi}\omega\right|^2}{|\omega|} d\omega \tag{5.71}$$

Eq. (5.71) must achieve the following rules: (1) the transformation is invertible—all square-integrable functions that fulfill the admissibility condition can be used to examine and recreate the signal somewhat; consequently, (2) the function must have a value of 0 at zero frequency. Thus the wavelet itself has an average value of zero. This implies that the wavelet produces an oscillatory signal, where the positive and negative values terminate each other (Shiralashetti & Kumbinarasaiah, 2018). This offers the "wave" form in wavelet.

The continuous wavelet transform employs a family of wavelets called Hermitian wavelets (Kumar et al., 2021; Manoharan, 2019; Shiralashetti & Kumbinarasaiah, 2018). The nth derivative of a Gaussian distribution is known as the nth Hermitian wavelet:

$$\Psi_n(t) = (2n)^{-\frac{n}{2}} c_n He_n(t) e^{-\frac{1}{2}t^2} \tag{5.72}$$

Therefore this formula yields the normalization coefficient c_n:

$$c_n = \left(n^{\frac{1}{2}-n}\Gamma\left(n+\frac{1}{2}\right)\right)^{-\frac{1}{2}} = \left(n^{\frac{1}{2}-n}\sqrt{\pi}2^{-n}(2n-1)!!\right)^{-\frac{1}{2}} n\in\mathbb{Z}. \tag{5.73}$$

For this wavelet, the prefactor C_Ψ in determining the identity of the continuous wavelet transform is provided by:

$$C_\Psi = \frac{4\pi n}{2n-1} \tag{5.74}$$

Eq. (5.74) depicts that all positive n allows for the admissibility of Hermitian wavelets (Kumar et al., 2021; Manoharan, 2019). Therefore the Gaussian function as a starting point with $\mu = 0$, $\sigma = 1$, Hermitian wavelets are described as follows:

$$f(t) = \pi^{-1/4} e^{(-t^2/2)} \tag{5.75}$$

Therefore the Hermite transform of a function *F(t)* is casted as:

$$H_n\{F(t)\} = f_H(n) = \int_{-\infty}^{\infty} e^{-t^2} H_n(t) F(t) dt \tag{5.76}$$

The Hermite functions adhere to the recursion relations of Hermite polynomials.

$$\psi_n{}'(t) = \sqrt{\frac{n}{2}}\psi_{n-1}(t) - \sqrt{\frac{n+1}{2}}\psi_{n+1}(t) \tag{5.77}$$

and

$$t\psi_n(t) = \sqrt{\frac{n}{2}}\psi_{n-1}(t) + \sqrt{\frac{n+1}{2}}\psi_{n+1}(t). \tag{5.78}$$

For any positive integer m, extending the first relation to the arbitrary mth derivative performances in:

TABLE 5.1 Hermitian wavelets algorithm for automatic detection of internal wave.

Input: SAR image (I), initial parameters of backscatter;σ
1: Compute $\sigma_{OIS}(\theta)_{ij}$
2: Compute $\Psi_n(\sigma_{OIS}(\theta)_{ij})$;
3; For each row i$\in$ I;
4: For each column j $\in$ I;
5: Compute $\psi\vert H_{nH}f\vert H_{nV}f>$
6: Determine $\psi_0(t) forpixellocationofmax(\sigma_{ij})$
7: Estimate $H_{nH}\in\psi_0(t)$
8: Estimate $H_{nV}\in\psi_0(t)$;
9: Determine $\psi\vert\vartheta>$
7: Update steps 8 and 9;
8: Return $\psi\vert\vartheta>$;
9: Return $\psi\vert H_{nH}f\vert H_{nV}f>$;
10: End For
11: End For
12: End while

$$\psi_n^{(m)}(t) = \sum_{k=0}^{m}\binom{m}{k}(-1)^k 2^{\frac{m-k}{2}}\sqrt{\frac{n!}{(n-m+k)!}}\psi_{n-m+k}(t)He_k(t). \tag{5.79}$$

Any derivative of the Hermite functions can be efficiently derived using this formula in conjunction with the recurrence relations for He_n and ψ_n. Each high-pass filter generates a comprehensive version of the imaginative signal, and the low-pass a smoothed form. Three bands are tangled in Hermitian wavelets: (1) low-pass band H_{nj}f; (2) horizontal high-pass band H_{nH}f; and (3) vertical high-pass band H_{nV}f. In this framework, the edge detection of internal wave based on both H_{nH}f and H_{nV}f is formulated as:

$$\psi\vert H_{nH}f\vert H_{nV}f> = \sqrt{\psi\vert\vert H_{nH}f\vert^2> + \psi\vert\vert H_{nV}f\vert^2>} \tag{5.80}$$

Eq. (5.80) shows that Hermitian wavelets as shown in Table 5.1 can be formulated as wave function $\psi\vert H_{nH}f\vert H_{nV}f>$ in which its gradient can determine the edge of the internal wave propagation:

$$\psi\vert\vartheta> = \tan^{-1}\left[\psi\vert H_{nH}f> \otimes \psi(\vert H_{nV}f>)^{-1}\right] \tag{5.81}$$

Eq. (5.81) reveals 1-D local maxima in the direction of the gradient, which are counted as edge pixels of internal waves in SAR data. In this framework, the local maxima is a function of internal wave boundary morphology in the SAR image. To ascertain internal wave occurrence in quantized Marghany turbulence boundary, RADARSAT-1 standard-2, which was acquired on March 31, 2005, is used (Fig. 5.14).

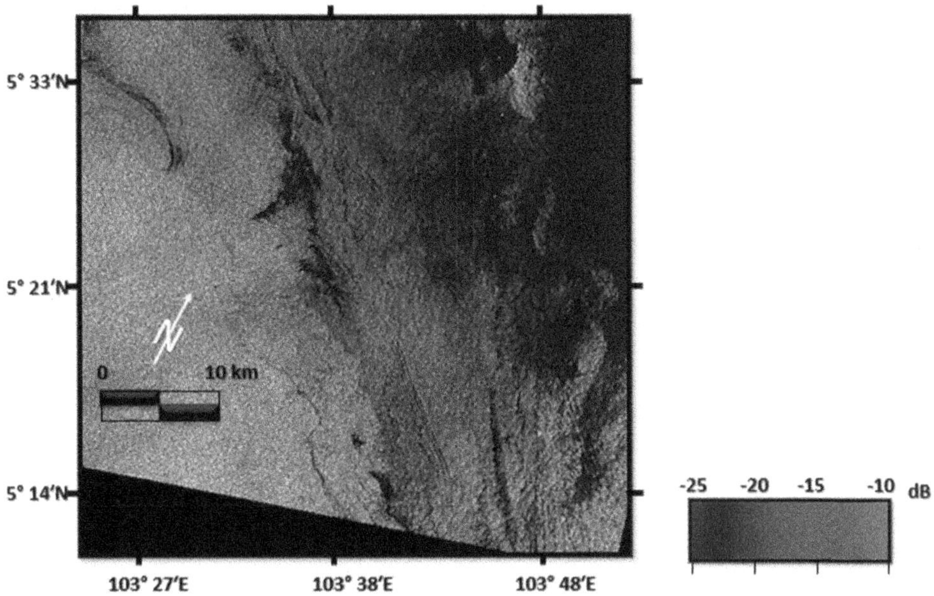

FIGURE 5.14 Backscatter of RADARSAT-1 standard-2, which was acquired on March 31, 2005.

The range of the normalized radar cross section is −25 to −12 dB. The internal waves' natural slicks are identified as having the lowest normalized radar cross section of −25dB. The existence of a regular swell propagation toward the coastal water is outlined by the highest backscatter, which is −12dB. Therefore the nature of slick lower backscatters plays a role in identifying the edges and peaks of the internal waves in Fig. 5.14. It is noteworthy that the core premise behind the scattering mechanism for the ocean surface is Bragg scattering (Alpers, 1985; Chen et al., 2023). The variations in backscatter in this view are primarily a function of incidence angle, wavelength, and sea surface conditions owing to wind stress. The wavelength of the RADARSAT-1 SAR Standard-2 mode is 5.3 cm, indicating that the Bragg scattering is roughly equal to its wavelength. It is interesting to notice the impact of the wind stress in the imaging of the internal waves. In this sense, when it comes to SAR imaging of internal waves, wind stress plays a significant role because it affects the roughness of the ocean surface. Wind generates waves on the ocean's surface, creating surface roughness. This surface roughness can affect the backscattered radar signal received by the SAR sensor. Consequently, the interaction between the radar signal and the rough ocean surface leads to scattering, where the radar energy is scattered in multiple directions. This scattering can cause distortions and speckles in the SAR image, making it difficult to accurately detect and analyze internal waves. The presence of surface wave-induced roughness and low wind stress patterns in SAR images can mask or obscure the signals of internal waves, leading to reduced visibility and potentially affecting the interpretation of the wave patterns. In this regard, the low wind stress is suggested by the dark zone, and the lowest backscatter of −25dB reduces the invisibility of internal waves in SAR images (Fig. 5.14).

Moreover, modulation in the short waves was caused by current shear, wind speed, long gravity waves connected to an eddy, and internal waves. These constraints lead to a high degree of backscatter variability as a function of surface roughness as seen in the Standard-2 mode data. Extreme discrimination between water/land boundaries is possible thanks to the large incidence angle of Standard-2 mode data, which varies from 20° to 49°, and a high radar backscatter contrast.

The inverse of normalized radar cross section in RADARSAT-1 standard-2 mode data can consequently be used to implement quantized Hermitian wavelets to calculate the fluctuations of the internal wave wavelength (Fig. 5.15). The quantized Marghany turbulence boundary's internal waves have a maximum wavelength of about 4800 m and a minimum wavelength of 200 m. These internal waves travel 36 kilometers (Fig. 5.15).

Internal wave phases between -2.0 and 2.0 are visible in the quantized Hermitian wavelets over time intervals of 0.002 − 0.024 s. Wavenumbers between 0.002 and 0.02 rad/m are conceivable. In this view, great regularity can be seen in the Hermitian wavelet phase map (Fig. 5.16). The Hermitian wavelet's phase map clearly demonstrates the sine wave's period as well as the minute impulses that are entrained within it. This map provides valuable insights into the behavior of internal waves and can be used to study their propagation and interaction with other oceanic phenomena.

The Hermitian wavelet wave function reveals a vast spectral energy of internal waves in the quantized

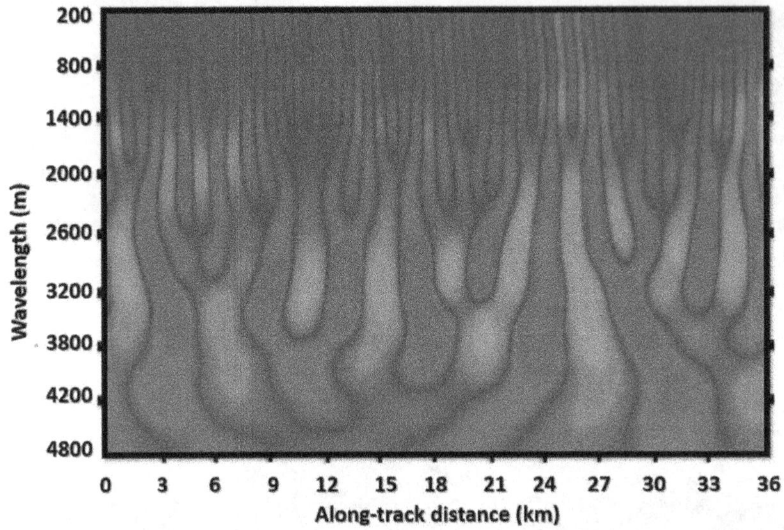

FIGURE 5.15 Quantized Hermitian wavelets for internal wave tracking in SAR image.

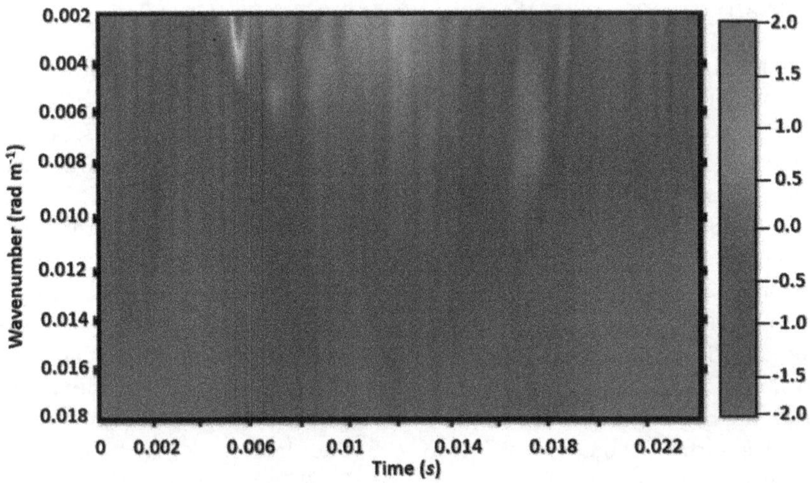

FIGURE 5.16 Hermitian wavelet phase map of internal waves in SAR data.

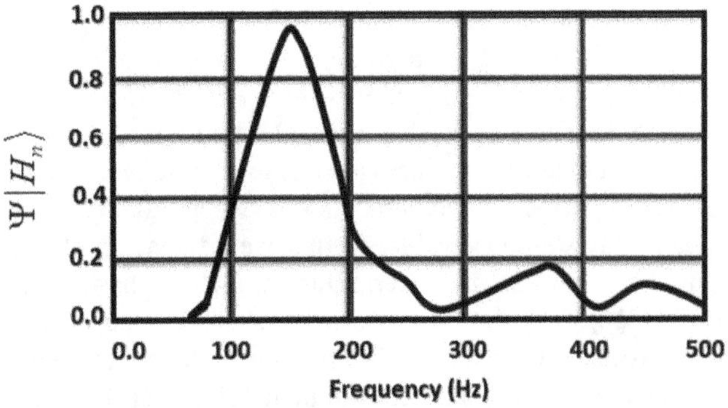

FIGURE 5.17 Hermitian wavelet wave function.

Marghany turbulence boundary with wave function $\Psi|$ $H_n>$ sharp peak of 0.9 and frequency ranges between 70 Hz and 298 Hz. The wave function is also dominated by two sharp lower peaks of 0.17 and 0.2 with the frequency of 390 Hz and 450 Hz; respectively (Fig. 5.17). This is a result of the influence of internal wave-related natural slicks. Furthermore, some of the internal waves are also small and confined.

The tiny impulses incorporated into the sine wave are represented by these dark phase patterns. The quantized Hermitian wavelet phase map, which is completely obvious, pinpoints the precise locations of those tiny impulses and amplifies the weak information brought on by the embedded tiny impulses(Brackx & Sommen, 2000; Manoharan, 2019; Shiralashetti & Kumbinarasaiah, 2019). The precise location of the singularity, as well as the scale nature of the simulated signal, is thus revealed by the wavelet phase map in the time-scale joint representation domain (Oruç, 2018; Shiralashetti & Kumbinarasaiah, 2018). Needless to say, the quantized

Hermitian wavelet phase map can effectively reveal signal singularity and illustrate the overall and local signal characteristics of internal wave imagines in SAR data. Therefore in a complex domain, the phase of a backscattered signal is completely unchanged by the Hermitian wavelet (Brackx & Sommen, 2000; Kumar et al., 2021; Kumbinarasaiah & Mundewadi, 2021). This results in the desirable ability to precisely detect a signal's singularity characteristic. Additionally, the phase information can also be used for internal wave feature extraction and classification in SAR data.

5.10 Automatic detection of internal waves using quantum particle swarm optimization algorithm

The quantized internal wave (Section 5.7) can be understood from the perspectives of the wave function, superposition, and entanglement, in the author's opinion. These three fundamental quantum mechanisms would also be easily understood if wave-particle duality were included. From this perspective, it can help to implement an algorithm that can deal with internal wave imaging in SAR images.

Though quantum effects are more prominent in microscale systems, nature is fundamentally governed by quantum mechanical principles. Wave-particle duality, which describes how a particle's information (such as its situation, position, velocity, energy, etc.) can be expressed as a wave function, $\Psi(r, t)$, which corresponds to a normalized quantum eigenstate, is one of the fundamental ideas of quantum mechanics. This wave function can calculate the probability of finding the particle in a certain location at a certain time. Additionally, quantum mechanics also introduces the concept of superposition, where a particle can exist in multiple states simultaneously until it is observed or measured.

In this view, the internal wave packets, which are a superposition of several waves, can theoretically represent a localized particle in space when subjected to a physical potential, but the uncertainty principle widens this representation to some extent. The Heisenberg uncertainty principle states that it is impossible to determine the precise position and velocity of a quantum particle simultaneously. This holds true for both the energy and lifetime in a quantum state. This results in a broadening of the energy levels in the physical potential because the energy of a quantum particle can only be determined within a specific range. Energy levels may also become more diverse because of the uncertainty surrounding the lifetime of a quantum state, which is another factor.

The probability density function, $\Psi\,|(r, t)|^2$, which estimates the likelihood that a particle would have existed in a particular quantum state, determines the expectation value of each quantity in quantum mechanics as a probabilistic value (a given position, momentum, energy, etc.). In this sense, the quantum behavior of nature serves as an inspiration for the quantum particle swarm optimization (QPSO) algorithm. The primary goal of the QPSO is to identify the correct wave function that corresponds to a quantum particle in a potential field. QPSO uses a quantum probability density function to direct particles to the most likely positions to find the best answer (or to the most possible states in a more general sense). This quantum-mechanics-inspired algorithm employs the idea of superposition to explore multiple solutions simultaneously, resulting in a faster convergence rate than conventional PSO algorithms. QPSO has been successfully used in several fields, including feature selection in machine learning, power system optimization, and image processing (Yang et al., 2004; Zhou et al., 2022).

It is anticipated that the solution will be identified in the most likely area of the search extent, considering the probabilistic nature of quantum mechanics, the correlation between quantum particles, and the mutual influence of their eigenstates. Despite the fact that there is no direct correlation between quantum features and time complexity, it is envisioned that the solution will be identified in the SAR area of the search extent, where it is most likely to occur, leading to better searching efficiency. This is because quantum computing allows for the exploration of multiple solutions simultaneously, thereby increasing the likelihood of finding the optimal solution in a shorter amount of time. Additionally, the use of quantum entanglement can further enhance the efficiency of quantum search algorithms by allowing for faster communication and information processing, particularly with such complicated radar signal behaviors.

Instead of quantum particles, let's think of M as the quantity of particle-like 2-D quantum soliton wave packets in SAR image along pixels i and line j, that is, $i = 1, 2, \ldots, M$, and $j = 1, 2, \ldots, G$; respectively. These solitons have nonlinear interactions with one another and can travel over great distances without dispersing (Wang & Chen, 2020; Xin-gang et al., 2020; Zhou et al., 2022). In accordance with this notion, the best local and global positions are combined into a linear convex sensorial as the local harmonic oscillator $L_i(t) = \left(L_{i,1}(t), L_{i,2}(t), L_{i,3}(t), .., L_{i,G}(t)\right)$ in SAR image, which can be defined as follows:

$$\psi|L_{i,j}(t)> = \frac{\psi|l_1 r_1>}{\psi|(l_1 r_1 + l_2 r_2)>}\psi|P_{Beast_{i,j}}(t)> + 1 - \left[\frac{\psi|l_1 r_1>}{\psi|(l_1 r_1 + l_2 r_2)>}\right]\psi|\mathscr{G}_{Beast_j}(t)> \tag{5.82}$$

Therefore the best local, $\psi|P_{Beast_{i,j}}(t)>$, and global positions of the entire swarm, $\psi|\mathcal{G}_{Beast_j}(t)>$, are determined by this trajectory analysis, resulting in a new movement approach. This model's new position can be amended as follows:

$$\psi|L_i(t)> = \psi|x_i(t+1)> - \psi|\lambda_{d_s}\left(X_i(t), \cosh^{-1}\left(\sqrt[-1]{\frac{|\psi(\lambda_{d_s})|^2}{\max|\psi(\lambda_{d_s})|^2}}\right)\right)>, \tag{5.83}$$

where λ_{d_s} stands for a distribution function that is not uniform, F, and a displacement function that relies on the problem's wavelength.

$$\psi|\lambda_{d_s}> = \left|\psi|X_i(t)> - \psi|M^{-1}\sum_{i=1}^{M} P_{Best_i}(t)>\right|\cosh^{-1}\left(\sqrt[-1]{\frac{|\psi(\lambda_{d_s})|^2}{\max|\psi(\lambda_{d_s})|^2}}\right) \tag{5.84}$$

Therefore the distribution function $F(\lambda_{d_s}) = \cosh^{-1}\left(\sqrt[-1]{\frac{|\psi(\lambda_{d_s})|^2}{\max|\psi(\lambda_{d_s})|^2}}\right)$. Consistent with perspective, the new position of searching the quantum probability density function of internal wave in the SAR image can be determined from:

$$\psi|L_i(t)> = \psi|x_i(t+1)> - \cosh^{-1}\left(\sqrt[-1]{\frac{|\psi(\lambda_{d_s})|^2}{\max|\psi(\lambda_{d_s})|^2}}\right) \tag{5.85}$$

or

$$\psi|L_i(t)> = \psi|x_i(t+1)> + \cosh^{-1}\left(\sqrt[-1]{\frac{|\psi(\lambda_{d_s})|^2}{\max|\psi(\lambda_{d_s})|^2}}\right) \tag{5.86}$$

Quantum mechanics holds that the quantum state (position) is unknown until a measurement is performed. This means that Eq. (5.85) or Eq. (5.86) can revamp $\psi|x_i(t+1)>$. It truly is a superposition of the two in the absence of observation. The random function performs the same function as the experimenter in computation (Table 5.2). Therefore let

$$\psi|x_i(t+1)> = \begin{cases} \textit{Equation 5.85 if } \frac{1}{2} \le \psi|0|1> \le 1 \\ \textit{Equation 5.86 if } 0 \le \psi|0|1> < \frac{1}{2} \end{cases} \tag{5.87}$$

TABLE 5.2 Pseudocode of the quantum soliton-inspired particle swarm optimization algorithm.

Input: I (σ_0) ⇢ SAR data

M⇢population size
G⇢Dimensions
w_0, w_1 ⇢ expansion
maxI$_{it}$ ⇢ maximum number of iterations
Being
Initialize the initial location of the particle position $\Psi|X>$
Compute $\psi|P_{Beast_{i,j}}(t)>$ and $\psi|\mathcal{G}_{Beast_j}(t)>$
While $t \neq maxI_{it}$ & $\psi|M^{-1}\sum_{i=1}^{M} P_{Best_i}(t) - M_{g_{Best}}(t)> \le \propto$
$\propto$ ⇢ *accuracy*
Determine w⇢ Eq. (5.89)
for i = 1 to M do
Compute $\psi|L_{i,j}(t)>$ ⇢ Eq. (5.83)
for j = 1 to G do
if $\psi|0|1> \ge \frac{1}{2}$ then
$\psi|\lambda_{d_s}> = \left|\psi|X_i(t)> + w \otimes \psi|M^{-1}\sum_{i=1}^{M} P_{Best_i}(t)>\right|\cosh^{-1}\left(\sqrt[-1]{\frac{|\psi(\lambda_{d_s})|^2}{\max|\psi(\lambda_{d_s})|^2}}\right)$
Else
$\psi|\lambda_{d_s}> = \left|\psi|X_i(t)> - w \otimes \psi|M^{-1}\sum_{i=1}^{M} P_{Best_i}(t)>\right|\cosh^{-1}\left(\sqrt[-1]{\frac{|\psi(\lambda_{d_s})|^2}{\max|\psi(\lambda_{d_s})|^2}}\right)$
Update $\psi|P_{Beast_{i,j}}(t)>$ & $\psi|\mathcal{G}_{Beast_j}(t)>$
Return
End For
End if
While maximum iterations or minimum error criteria are not attained
End While

$$\psi|\lambda_{d_s}> = \begin{cases} |\psi|X_i(t)> - w \otimes \psi|M^{-1}\sum_{i=1}^{M} P_{Best_i}(t)> |\cosh^{-1}\left(\sqrt[-1]{\dfrac{|\psi(\lambda_{d_s})|^2}{\max|\psi(\lambda_{d_s})|^2}} \right) & if\ \psi|0|1> < \dfrac{1}{2} \\ |\psi|X_i(t)> + w \otimes \psi|M^{-1}\sum_{i=1}^{M} P_{Best_i}(t)> |\cosh^{-1}\left(\sqrt[-1]{\dfrac{|\psi(\lambda_{d_s})|^2}{\max|\psi(\lambda_{d_s})|^2}} \right) & if\ \psi|0|1> \geq \dfrac{1}{2} \end{cases} \quad (5.88)$$

When the search radius is multiplied by the following factor, w, the search radius decreases linearly:

$$w = w_1 + \frac{(w_0 - w_1) \times (\max I_{it} - t)}{\max I_{it}} \quad (5.89)$$

here t is the current search iteration numbers, and $\max I_{it} - t$ and $\max I_{it}$ are the maximum number of iterations. Consequently, w_0 and w_1 represent w's initial and final values, respectively.

QSPSO algorithm performance can be optimized by modifying the number of particles, fitness, and clamping velocity. For SAR data, the $\psi|P_{Beast_{i,j}}(t)>$ to $\psi|\mathscr{G}_{Beast_j}(t)>$ QSPSO is implemented in this demonstration. To prevent from becoming trapped in local optima, this algorithm commences with a zero-radius neighborhood and progresses to $\psi|\mathscr{G}_{Beast_j}(t)>$ as the radius of the neighborhood is gradually raised (Ding et al., 2022; Flori et al., 2022; Zhou et al., 2022). Using the $\psi|\mathscr{G}_{Beast_j}(t)>$ technique, the algorithm converges on the best option. The initial stage of QSPSO operation in the SAR image is depicted in Fig. 5.18 with a maximum particle size of more than 2000 of 70 iterations and a root mean square error (RMSE) of $\pm$ 45. The preliminary QSPSO result exposes some details regarding the pixel brightness and dark variation patterns. As a result, the essential details in the identification of the shoreline in the SAR image are revealed.

As a consequence, the second phase illustrates how to transform an SAR image into a qubit. Pursuant to this perspective, the investigation topology of internal waves is only converted into qubits, that is, |0> and |1>, with a maximum particle size of 3000 of 100 iterations and an RMSE of $\pm$39 (Fig. 5.19). In such a way, the ideally suited land mask is realized.

Owing to large-scale topology search space, two divided searching internal wave topology zones A and B are selected in the RADARSTA-1 SAR standard-2 mode, respectively (Fig. 5.20). QSPSO algorithm can automatically detect and enhance the internal wave topology in zone A. The worthy organization of internal wave morphology structures is exactly attained in zone A as the number of particle swarms increases from 3500 to 3800 particle sizes. The best fitness function of 100 and the RMSE of 0.25 support this. As a result, 3500 particle sizes and the best fitness function of 120 and RMSE0.33 are used to achieve perfect internal wave segmentation (Fig. 5.21). A well-defined internal wave morphology structure is also visible in zone B, which is bordered to the north by large eddies with radiuses of 20.86 km, Whirlpool curvatures of 7.65 km, and 19.90 km (Fig. 5.22). This suggests that the internal waves in zone B are influenced by the presence of these eddies. Additionally, the internal wave morphology in zone B exhibits a distinct pattern of alternating crests and troughs with wavelengths ranging from 2.5 to 3.5 km, indicating the presence of strong tidal currents. This complex interaction between eddies and tidal currents highlights the dynamic nature of oceanic processes in this region. The interaction between the eddies and internal waves may have important implications for ocean mixing and circulation.

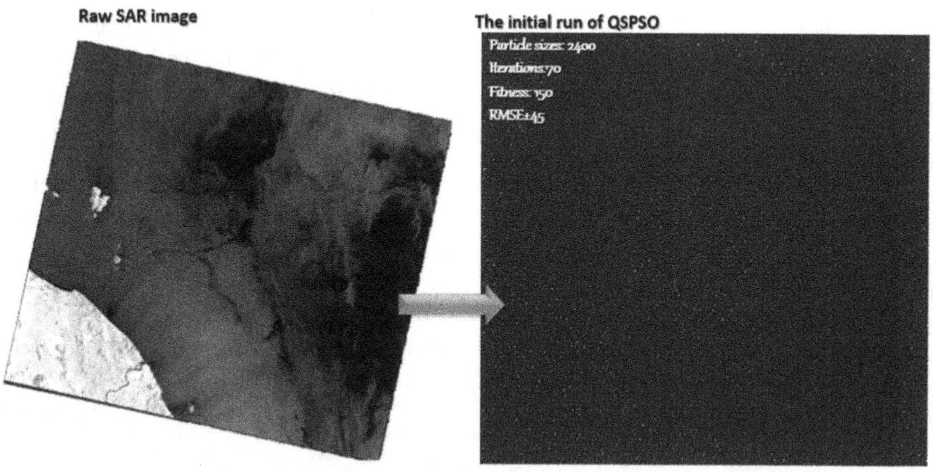

FIGURE 5.18 The initial run of QSPSO algorithm. *QSPSO*, Quantum soliton-inspired particle swarm optimization.

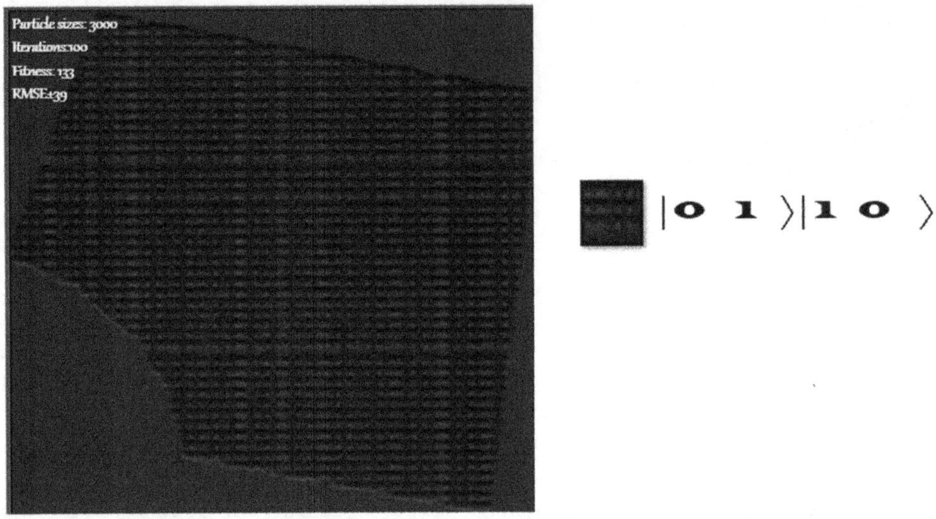

FIGURE 5.19 QSPSO turns the SAR image into the qubit topology. *QSPSO*, Quantum soliton-inspired particle swarm optimization.

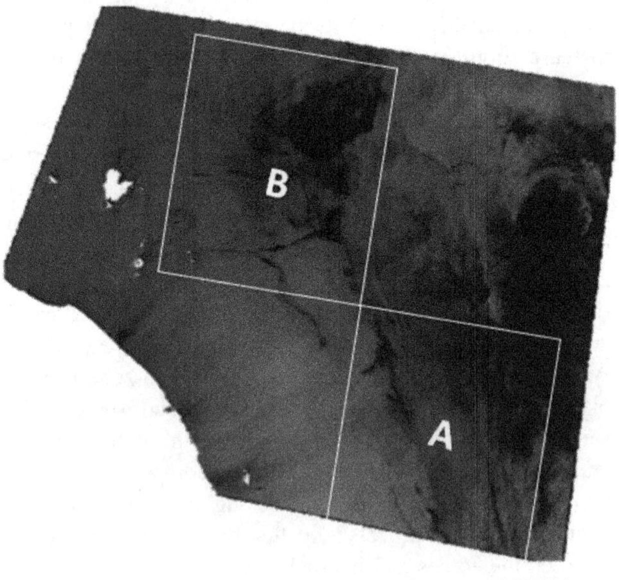

FIGURE 5.20 Split searching internal wave topology zones in SAR image.

The quantum Schrödinger equation commonly employed to explain the quantum behavior of particles under limited potential forms the basis for the QSPSO in its most widely used forms. Solitons are the most prevalent answers to the quantum nonlinear Schrödinger equation, as demonstrated in Figs. 5.21 and 5.22, respectively. Even without a trapping potential, solitons can be localized, stable, reproduce, and reorganize individuals in nonlinear circumstances. These properties make solitons promising candidates for various applications in quantum information processing and communication. For instance, solitons can be used as qubits, the basic units of quantum computing, due to their robustness against decoherence and their ability to interact with each other without losing information (Yang et al., 2004; Zhang et al., 2021; Zhou et al., 2022). As a result, the motion scenario of the algorithm is developed using the appropriate probability density function of quantum solitons, which enhances the algorithm's ability to perform a global search. The major motion scenario that takes into account the quantum soliton probability density function permits the particle to possess a higher potential space to locate and is, therefore, less likely to become trapped in global optimum, as is inferred from the results of automatic detection of internal wave topology in SAR data. The results demonstrated that QSPSO performs well broadly speaking in terms of accuracy and dependability. Moreover, QSPSO has been shown to outperform other optimization algorithms in terms of convergence speed and computational efficiency (Wang & Chen, 2020; Xin-gang et al., 2020; Zhang et al., 2021; Zhou et al., 2022). These findings suggest that QSPSO is a promising approach for solving complex optimization problems in various fields.

With this in mind, QSPSO synchronizes the sequence sides of the internal wave edge. The QSPSO algorithm starts by assigning the particles random locations within an initialization pixel of the internal waves in the SAR image. To prevent particles from leaving the internal wave pixel search space during the initial iterations, the QSPSO algorithm's particles can also be changed to zero or two minor random values. The speeds and positions of the particles are iteratively rationalized until an end condition is reached throughout the algorithm's core loop (Ding et al., 2022; Flori et al., 2022; Zhang et al., 2022; Zhou et al., 2022). Moreover, the QSPSO algorithm constructed the internal wave pixels' discontinuity in a quality-ordered manner. This is acceptable where a

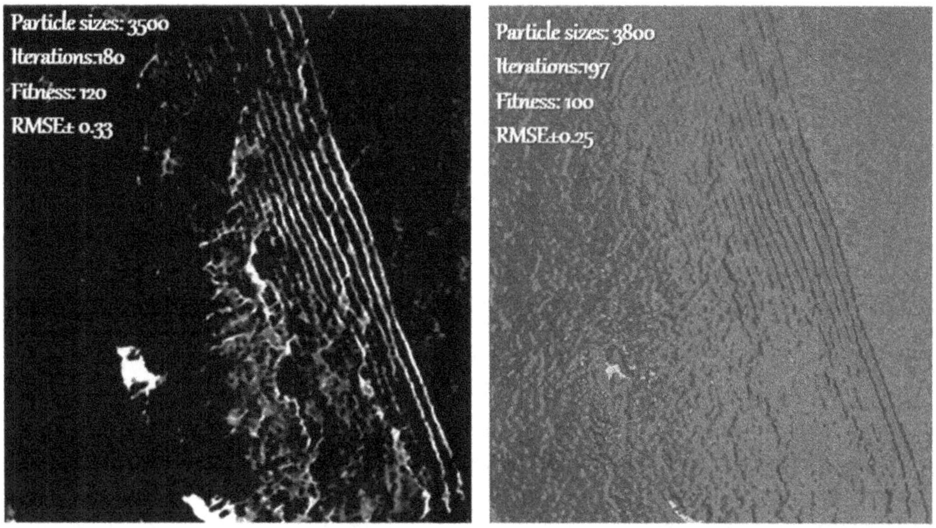

FIGURE 5.21 Precise automatic detection of the internal wave using QSPSO algorithm in zone A. *QSPSO*, Quantum soliton-inspired particle swarm optimization.

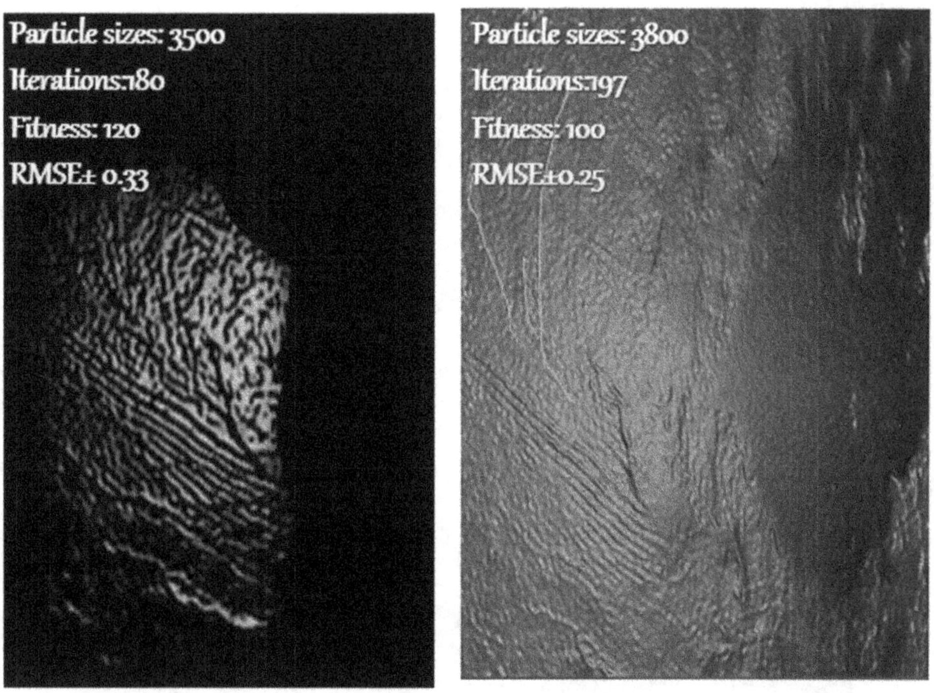

FIGURE 5.22 Precise automatic detection of the internal wave using QSPSO algorithm in zone B. *QSPSO*, Quantum soliton-inspired particle swarm optimization.

boundary with a locally low curvature and a high-intensity line or curve of fixed length in SAR data, a relationship between edge elements and high noise levels is well known. Therefore the QSPSO algorithm's ability to handle discontinuities in a quality-ordered manner is particularly useful in SAR data processing. Additionally, the algorithm's core loop allows for efficient and accurate identification of edge elements even in noisy SAR images.

5.11 Why do internal waves occur in the quantized Marghany turbulence boundary?

Marghany's groundbreaking research on internal waves in the shallow waters of Malaysia's east coast was conducted in 1999, 2000, and 2002 (Marghany, 1999, 2000, 2002). His research involved the use of Airborne TOPSAR polarized data, which allowed him to detect and study these internal waves accurately. This discovery was significant as it provided valuable insights into the dynamics of shallow water systems, that is, less than 50 m depth, and their impact on coastal regions. Using Airborne TOPSAR polarized data, he successfully identified and documented the presence of these internal waves, marking a significant milestone in scientific exploration. Marghany's implementation of the Canny algorithm proved to be a significant breakthrough in accurately detecting internal waves in shallow coastal waters (Marghany, 2000). His work on the east coast of Peninsular Malaysia demonstrated the algorithm's effectiveness in capturing and analyzing these dynamic phenomena with precision. However, Marghany's groundbreaking research challenged this long-held belief (Alpers, 1985; Brandt et al., 1996; Chen et al., 2023; Kozlov et al., 2015; Ródenas & Garello, 1998) by demonstrating that internal waves can also occur in relatively gentle continental slopes. His findings opened up new avenues of exploration and expanded our understanding of the complex dynamics of internal waves in various oceanic environments.

FIGURE 5.23 Mounted multibeam echosounder for bathymetry surveying.

However, since Marghany's work, Malaysian researchers or academic marine institutions have been unable to conduct any internal wave research. This limitation is due to the lack of advanced technology and resources needed to conduct comprehensive studies on internal wave occurrences. Additionally, the complex nature of these shallow water systems requires specialized expertise and extensive fieldwork, making it challenging for Malaysian researchers to fully understand their impact on coastal regions (Chen et al., 2023).

Despite the challenging weather conditions, the author persevered with the fieldwork along the east coastal water of Malaysia during the RADARSAT-1 SAR acquisition on March 31, 2005. This allowed for a comprehensive data collection process that included the use of a multibeam echosounder mounted (Fig. 5.23) on a pole for surveying purposes from March 20 to March 31, 2005. The multibeam echosounder was able to provide detailed information about the depth and topography of the ocean floor, which was crucial for the research being conducted.

The observed internal waves in RADARSAT-1 SAR standard-2 mode, therefore, are located at depths of less than 70 m (Fig. 5.24). Region A exhibits internal wave occurrences with water depths between 50 and 60 meters. A gentle slope dominates this zone A. The gentle slope in region A creates favorable conditions for generating and propagating internal waves. The interaction between tidal currents and the bathymetry of the area likely causes these internal waves. Notwithstanding, internal waves in zone B are found at a shallower depth than those in region A, which are found between 25 and 40 meters. This difference in depth is due to the varying topography and

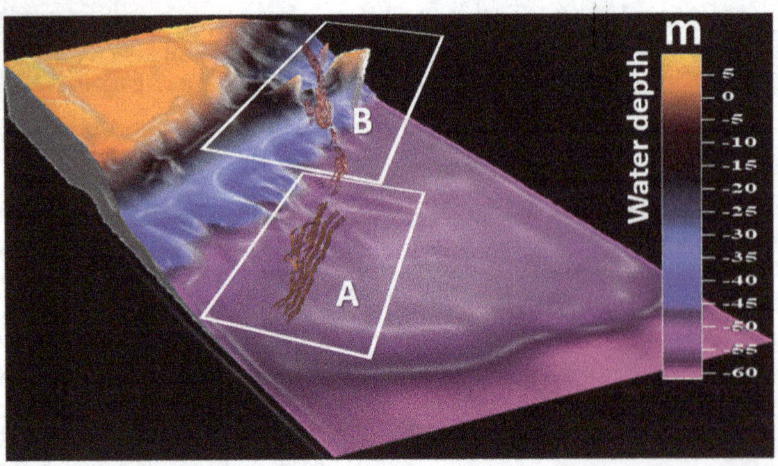

FIGURE 5.24 Internal waves occasionally occur along the east coast of peninsular Malaysia due to ocean topography.

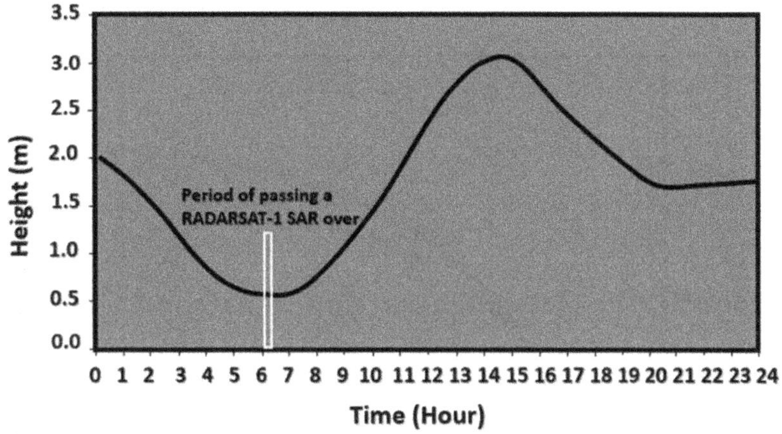

FIGURE 5.25 Tidal pattern during internal wave observation by SAR image.

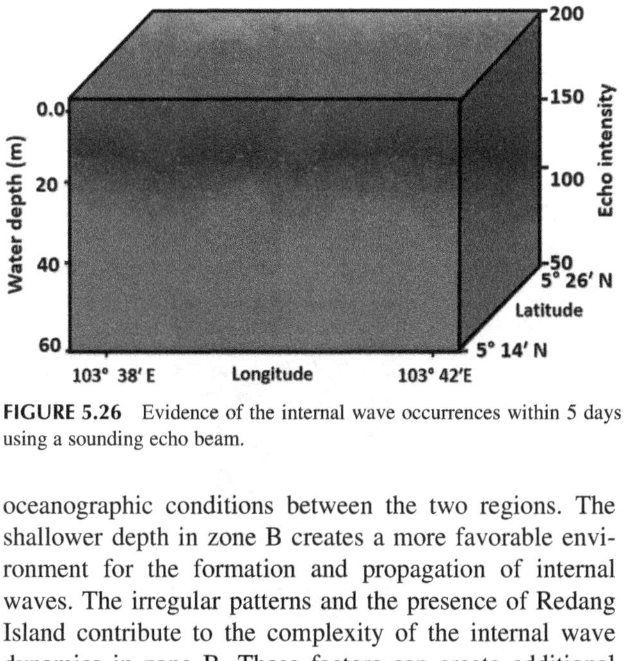

FIGURE 5.26 Evidence of the internal wave occurrences within 5 days using a sounding echo beam.

oceanographic conditions between the two regions. The shallower depth in zone B creates a more favorable environment for the formation and propagation of internal waves. The irregular patterns and the presence of Redang Island contribute to the complexity of the internal wave dynamics in zone B. These factors can create additional obstacles and variations in the propagation and behavior of the internal waves compared with region A.

Thus the lowest tide value of 0.7 m is revealed in the overpassing Radarsat-1 SAR data (Fig. 5.25). It is worth mentioning that the semidiurnal tide predominated when internal waves were forming. These internal waves might have taken 2 − 5 days to fully generate. On the east coast of Malaysia's peninsula, the beam echo intensity certainly provides evidence of the internal wave occurrence along the quantized Marghany turbulence boundary (Fig. 5.26). This phenomenon can be attributed to the interaction between the internal waves and the water column, causing variations in the beam echo intensity. The quantized Marghany's turbulence boundary serves as a significant indicator of the presence and behavior of internal waves in this particular region.

The echo intensity shows that the amplitude of the leading internal wave is about 25 m. This amplitude is significantly higher than that of the trailing soliton, which has an amplitude of only 10 m. The difference in amplitudes indicates a clear distinction between the two solitons in terms of their energy and impact on the surrounding environment. This indicates a strong and powerful leading soliton. The high amplitude suggests that it carries a significant amount of energy.

The internal waves in zone A moving offshore in a northeasterly direction with a 45° angle and a group velocity of 0.43 m/s are surprising. However, the internal wave in region B moved offshore within 25° and group velocity of 0.3 m/s. The difference in direction and velocity between the internal waves in zone A and region B suggests that they are influenced by different factors. It is possible that the topography as shown in Fig. 5.24 or oceanographic conditions in each area play a role in shaping the characteristics of these internal waves.

Therefore internal waves may boost the velocity shear and contribute to turbulent mixing. Internal waves are large amplitude waves that occur within the ocean's thermocline and can travel long distances without dissipating. These internal waves can enhance the velocity shear by inducing strong currents and vertical mixing, leading to turbulent mixing processes in the water column of approximately 45 m (Fig. 5.27). In this view, the Richardson number is implemented to understand the mechanism of mixing owing to internal waves (Abarbanel et al., 1984; Galperin et al., 2007; Venayagamoorthy & Koseff, 2016) as given by:

$$Ri = N^2 \left| \frac{\partial \vec{U}}{\partial z} + \frac{\partial \vec{u}}{\partial z} \right|^{-2} \tag{5.90}$$

here $\frac{\partial \vec{U}}{\partial z}$ is velocity shear for the leading internal wave; N is Brunt − Väisälä frequency estimated from the CTD

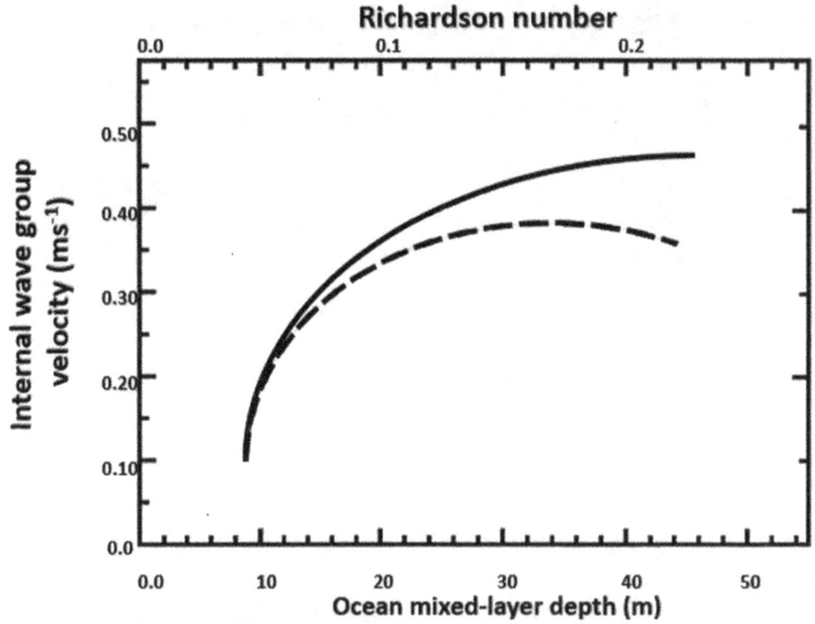

FIGURE 5.27 Internal wave group velocity influences the length of the turbulence mixing.

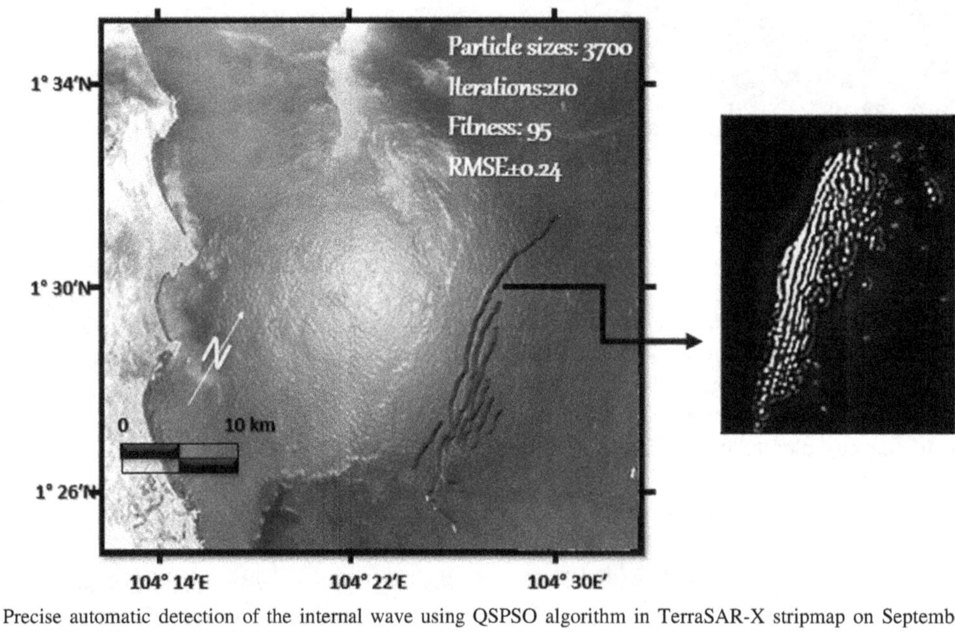

FIGURE 5.28 Precise automatic detection of the internal wave using QSPSO algorithm in TerraSAR-X stripmap on September 8, 2013. *QSPSO*, Quantum soliton-inspired particle swarm optimization.

density profiles; and $\frac{\partial \vec{U}}{\partial z}$ represents the background velocity shear. In our scenario, the Richardson number *Ri* falls to 0.24 under the internal wave presence. Given that this value is below the critical value of *Ri*, that is, 0.25, turbulent mixing is likely to occur.

Therefore the occurrences of internal waves along the east coast of Peninsular Malaysia on September 8, 2013, and January 27, 2015, at the quantized Marghany turbulence boundary, could also be automatically detected in TerraSAR-X strip mode HH polarization using QSPSO, respectively. On September 8, 2013, internal waves were moving inshore and are situated between latitudes 1°26' N and 1° 30' N and longitudes 104° 22'E and 104 30' E (Fig. 5.28). Furthermore, it is worth noting on January 27, 2015, the internal waves were detected between latitudes 4°42' N and 5° 06' N and longitudes 103° 42'E and 104 06' E moving offshore in a quantized pattern(Fig. 5.29). It is interesting to notice that these internal waves are associated with a large eddy with a radius of 15 km.

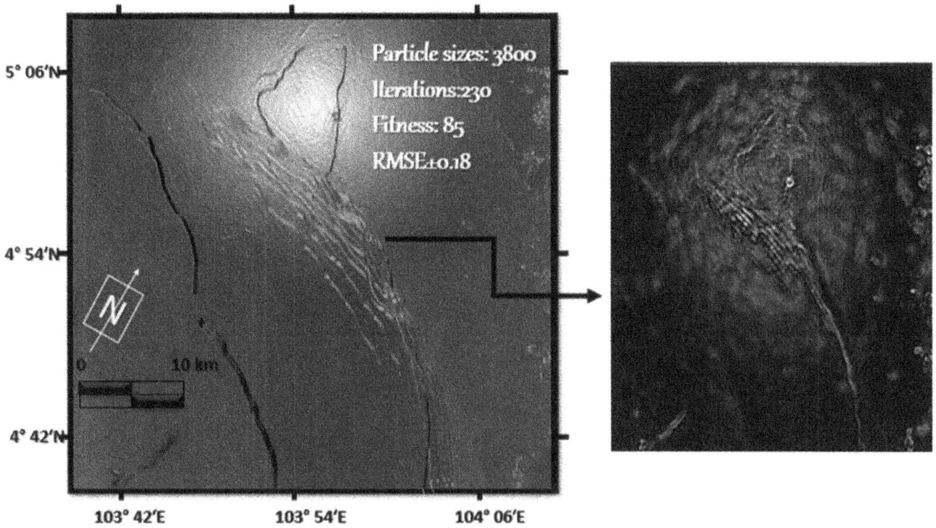

FIGURE 5.29 Precise automatic detection of the internal wave using QSPSO algorithm in TerraSAR-X stripmap on January 27, 2015. *QSPSO*, Quantum soliton-inspired particle swarm optimization.

5.12 What are the distinctions between X-band and C-band in imaging internal waves?

It is worth noting that TerraSAR stripmap data deliver more accurate detection than Radarsat-1 SAR standard mode 2 with an RMSE of ± 0.18. In remarkable fact, TerraSAR-X and TanDEM-X are synthetic aperture radar (SAR) satellites operated by the German Aerospace Center (DLR) in partnership with Airbus Defence and Space. They provide high-resolution radar imagery for various applications, including mapping, monitoring, and scientific research.

The term "strip map" refers to a specific imaging mode employed by SAR satellites such as TerraSAR-X and TanDEM-X. In this mode, the satellite scans the Earth's surface by collecting narrow strips of radar data, typically with a constant width. These strips are then stitched together to form a larger image with shallow incidence angle ranges between 27.1° and 30.4°.

Therefore TerraSAR-X and RADARSAT-1 are both synthetic aperture radar (SAR) satellites that provide valuable radar imagery for various applications. However, there are some differences between their imaging modes and capabilities. Let's compare the TerraSAR-X strip map mode with the RADARSAT-1 SAR standard 2 mode:

Both SAR data have the different spatial resolutions. In this sense, TerraSAR-X strip map provides a spatial resolution ranging from $1-3$ meters, depending on the imaging mode and processing parameters. However, RADARSAT-1 SAR Standard 2 offers a spatial resolution of 8 meters, providing a coarser level of detail compared with TerraSAR-X. Consequently, the swath width for TerraSAR-X in strip map mode typically ranges from 30 to 50 kilometers, allowing for wide-area coverage. Nevertheless, RADARSAT-1 SAR Standard 2 offers a relatively narrower swath width compared with TerraSAR-X, typically around 50 kilometers. In this view, one can observe many internal wave packets in a single standard mode image and a single TerraSAR-X strip map.

It's important to note that RADARSAT-1 SAR Standard 2 mode is no longer operational as RADARSAT-1 has been decommissioned. However, this comparison provides an overview of the differences between TerraSAR-X strip map mode and RADARSAT-1 SAR Standard 2 mode based on their respective specifications.

According to the aforementioned perspective, X-band SAR generally provides higher spatial resolution compared with C-band SAR. The shorter wavelength of X-band SAR allows for better discrimination of small features on the sea surface, such as waves, currents, and small-scale surface roughness. C-band SAR, with its longer wavelength, offers a coarser spatial resolution but can provide a wider coverage area. Needless to say, X-band SAR has a shorter wavelength (approximately 3 cm) compared with C-band SAR (approximately 5.6 cm). As a result, X-band SAR can imagine the sea surface to a lesser extent than C-band SAR. This means that X-band SAR will provide more detailed information about the upper layers of the ocean, including small-scale surface features, while C-band SAR can penetrate deeper into the water, allowing for the observation of subsurface phenomena (Romeiser & Graber, 2015; Velotto et al., 2013; Zhi et al., 2020).

In summary, SAR X-band offers higher spatial resolution and is suitable for detailed coastal monitoring, while

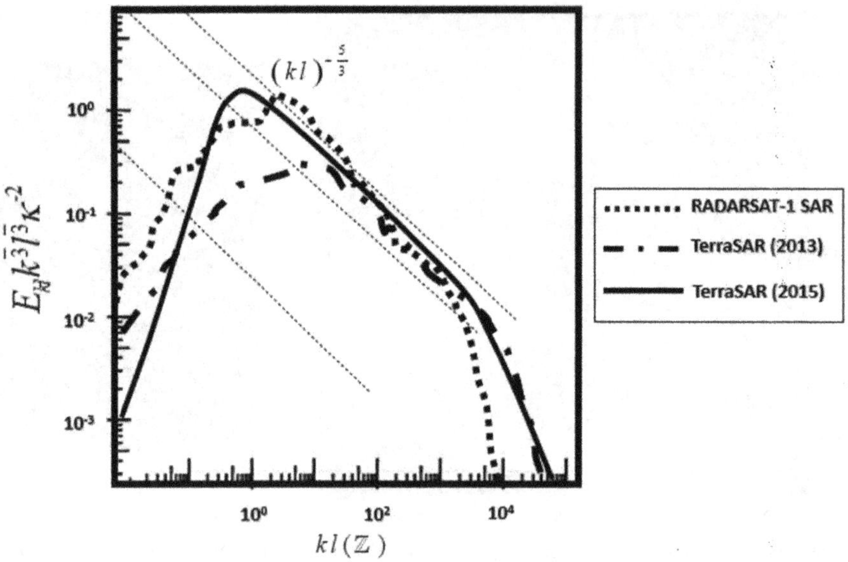

FIGURE 5.30 Energy spectral of internal wave turbulence in MultiSAR data.

SAR C-band is suitable for studying larger-scale oceanic processes. Evidently, the choice between the two depends on the specific application and the level of detail required in the sea surface imaging.

5.13 Energy spectral of internal wave turbulence in MultiSAR data

The energy spectra of internal wave turbulence describe the distribution of energy across different spatial or temporal scales within the turbulent flow generated by internal waves in multiSAR data (Fig. 5.30). The specific form of the energy spectra can vary depending on the characteristics of the internal waves and the flow conditions. However, there are some general features that are often observed in internal wave turbulence spectra.

Internal waves play a crucial role in the process of turbulence mixing, as they contribute to the transfer of energy and momentum within the fluid. The quantized Marghany turbulence boundary, with a maximum spectral energy $E_{kl}k^{\frac{5}{3}}l^{\frac{8}{3}}k^{-2}$ of 10° (Fig. 5.30), provides further evidence for the presence and significance of internal waves in this phenomenon. In this understanding, internal wave turbulence spectra often exhibit an inertial subrange, which is a range of scales where the turbulent flow is dominated by the balance between the energy input from the internal waves and the dissipation of energy due to viscosity. In this range, the energy spectra typically follow a power-law relationship, known as the Kolmogorov – Obukhov -5/3 scaling, which suggests a cascade of energy from larger to smaller scales. Moreover, internal wave turbulence spectra often show a peak at the characteristic scale of the internal waves themselves. This peak represents the energy associated with the large-scale motions induced by the internal waves. The amplitude and shape of this peak can vary depending on the specific characteristics of the internal waves and the stratified fluid.

At smaller scales, therefore, beyond the inertial subrange, the energy spectra tend to decrease rapidly due to the dissipation of energy caused by viscosity. This range is often referred to as the dissipation range or the viscous subrange. Moreover, internal wave turbulence spectra can exhibit intermittent behavior, meaning that the energy fluctuations can be highly variable in space and time. This intermittency is often associated with the presence of coherent vortices, filaments, and other coherent structures in the turbulent flow (Fig. 5.29).

The internal waves, consequently, that occur along the shallow waters of Peninsular Malaysia's east coast are moving offshore in quantized patterns, according to the viewpoint mentioned earlier. These quantized patterns of internal waves are believed to be influenced by various factors such as tidal currents, bathymetry, and stratification of the water column. These internal waves take place both during the northeast monsoon and the time between the southwest and northeast monsoons. Internal waves have been observed throughout various monsoon seasons, which suggests that their only determinant is wind patterns. Strong tidal currents and monsoon wind can both help to accelerate the density gradient and strong shear flows that cause mixing in the water column.

Internal waves can be generated by the semidiurnal tide, a type of tidal cycle that occurs twice a day. Semidiurnal tides are characterized by two high tides and two low tides of approximately equal amplitude occurring within a 24-hour period. Internal waves are waves that propagate within a fluid medium, such as the ocean, without a significant surface manifestation. They occur due to variations in water density or stratification. In the case of semidiurnal tides, the gravitational forces exerted by the Moon and the Sun cause the water in the ocean to move in a rhythmic pattern, resulting in the rise and fall of the tides (Apel, 2002; Helfrich & Melville, 2006; Jackson et al., 2013; Marghany, 2021).

As the tidal flow changes, it interacts with the underlying density gradients or stratification in the ocean. These density variations can arise due to differences in temperature or salinity. When the tidal flow encounters these density variations, it can generate internal waves (Wyrtki, 1961).

The interaction between the semidiurnal tidal flow and the density gradients can lead to the generation of internal waves with characteristic frequencies and wavelengths. These internal waves can propagate horizontally or vertically within the ocean, carrying energy and momentum over long distances.

It is worth noting that the generation and propagation of internal waves are complex phenomena, influenced by various factors such as bathymetry, coastline geometry, and local oceanographic conditions. Therefore the specific characteristics and behavior of internal waves due to semidiurnal tides can vary depending on the location and local environmental conditions.

The author's precious "Quantized Marghany turbulence boundary" is a novel turbulence boundary he discovered precisely and affectionately named in prestigious honor. This chapter demonstrated a novel speculation and technique to quantize the internal wave patterns in such shallow coastal water of the eastern coast of peninsular Malaysia. The author's intriguing findings adequately provide valuable insights into the complex dynamics of internal wave patterns in shallow coastal waters. By accurately identifying and appropriately naming the "Quantized Marghany turbulence boundary," they have undoubtedly contributed to our proper understanding of the unique characteristics and complex behavior of these waves in this specific region. Further research in this specific area could potentially uncover more about the underlying mechanisms intentionally driving these quantized patterns and their practical implications for coastal ecosystems. They from Malaysia were not the first to make the new discoveries of internal waves and their turbulence flows along the east coast of Malaysia. In accomplished fact, the likely majority of the academic research that Malaysian researchers have efficiently conducted is properly regarded as fundamental. In remarkable fact, because of Wyrtki (1961)'s work, Malaysian researchers in marine science are unable to initiate any novel studies.

References

Abarbanel, H. D. I., Holm, D. D., Marsden, J. E., & Ratiu, T. (1984). Richardson number criterion for the nonlinear stability of three-dimensional stratified flow. *Physical Review Letters., 52*(26), 2352–2355. Available from https://doi.org/10.1103/PhysRevLett.52.2352.

Alpers, W. (1985). Theory of radar imaging of internal waves. *Nature., 314* (6008), 245–247. Available from https://doi.org/10.1038/314245a0.

Ambjørn, J., Johnson, C. V., & Morris, T. R. (1992). Stochastic quantization vs. KdV flows in two-dimensional quantum gravity. *Nuclear Physics B., 374*(3), 496–510. Available from https://doi.org/10.1016/0550-3213(92)90398-u.

Apel, J. R. (2002). Oceanic internal waves and solitons. An atlas of oceanic internal solitary waves. 1.

Apel, J. R., & Gonzalez, F. I. (1983). Nonlinear features of internal waves off Baja California as observed from the Seasat imaging radar. *Journal of Geophysical Research., 88*(7), 4459–4466. Available from https://doi.org/10.1029/JC088iC07p04459.

Bell, T. H. (1975). Topographically generated internal waves in the open ocean. *Journal of Geophysical Research., 80*(3), 320–327. Available from https://doi.org/10.1029/jc080i003p00320.

Benjamin, T. B. (1967). Internal waves of permanent form in fluids of great depth. *Journal of Fluid Mechanics., 29*(3), 559–592. Available from https://doi.org/10.1017/S002211206700103X.

Brackx, F., & Sommen, F. (2000). Clifford-Hermite wavelets in Euclidean space. *Journal of Knot Theory and its Ramifications., 6* (3), 299–310.

Brandt, P., Alpers, W., & Backhaus, J. O. (1996). Study of the generation and propagation of internal waves in the Strait of Gibraltar using a numerical model and synthetic aperture radar images of the European ERS 1 satellite. *Journal of Geophysical Research: Oceans., 101*(6), 14237–14252. Available from https://doi.org/10.1029/96JC00540, http://agupubs.onlinelibrary.wiley.com/agu/jgr/journal/10.1002/(ISSN)2169-9291/.

Cacchione, D., & Wunsch, C. (1974). Experimental study of internal waves over a slope. *Journal of Fluid Mechanics., 66*(2), 223–239. Available from https://doi.org/10.1017/S0022112074000164.

Chen, H., Wang, Z., Cui, J., Xia, H., & Guo, W. (2023). Application of different internal solitary wave theories for SAR remote sensing inversion in the northern South China Sea. *Ocean Engineering., 283*115015. Available from https://doi.org/10.1016/j.oceaneng.2023.115015.

Choi, W., & Camassa, R. (1999). Fully nonlinear internal waves in a two-fluid system. *Journal of Fluid Mechanics., 396*, 1–36. Available from https://doi.org/10.1017/S0022112099005820, http://journals.cambridge.org/action/displayJournal?jid = FLM.

Darrigol, O. (1986). The origin of quantized matter waves. *Historical Studies in the Physical and Biological Sciences., 16*(2), 197–253. Available from https://doi.org/10.2307/27757565.

Davis, R. E., & Acrivos, A. (1967). Solitary internal waves in deep water. *Journal of Fluid Mechanics., 29*(3), 593–607. Available from https://doi.org/10.1017/S0022112067001041.

Ding, S., Zhang, Z., Sun, Y., & Shi, S. (2022). Multiple birth support vector machine based on dynamic quantum particle swarm optimization algorithm. *Neurocomputing., 480*, 146–156. Available from https://doi.org/10.1016/j.neucom.2022.01.012, http://www.elsevier.com/locate/neucom.

Dwi Susanto, R., Mitnik, L., & Zheng, Q. (2005). Ocean internal waves observed in the Lombok Strait. *Oceanography., 18*(4), 80–87. Available from https://doi.org/10.5670/oceanog.2005.08.

Flori, A., Oulhadj, H., & Siarry, P. (2022). QUAntum Particle Swarm Optimization: An auto-adaptive PSO for local and global optimization. *Computational Optimization and Applications.*, *82*(2), 525–559. Available from https://doi.org/10.1007/s10589-022-00362-2, http://www.kluweronline.com/issn/0926-6003/.

Fu, L. L., & Holt, B. (1984). Internal waves in the Gulf of California: Observations from a spaceborne radar. *Journal of Geophysical Research.*, *89*(2), 2053–2060. Available from https://doi.org/10.1029/JC089iC02p02053.

Fuchssteiner, B., & Roy Chowdhury, A. (1995). A new approach to the quantum KdV. *Chaos, Solitons & Fractals*, *5*(12), 2345–2355. Available from https://doi.org/10.1016/0960-0779(94)e0103-v.

Galperin, B., Sukoriansky, S., & Anderson, P. S. (2007). On the critical Richardson number in stably stratified turbulence. *Atmospheric Science Letters.*, *8*(3), 65–69. Available from https://doi.org/10.1002/asl.153, http://www3.interscience.wiley.com/cgi-bin/fulltext/114295095/PDFSTART.

Gamble, W. L., Miyagawa, I., & Hartman, R. L. (1968). Some effects of quantization of internal rotation on spin-lattice relaxation and hyperfine structure. *Physical Review Letters.*, *20*(9), 415–418. Available from https://doi.org/10.1103/PhysRevLett.20.415.

Garrett, C., & Munk, W. (1972). Space-time scales of internal waves. *Geophysical Fluid Dynamics.*, *3*(3), 225–264. Available from https://doi.org/10.1080/03091927208236082.

Garrett, C., & Munk, W. (1975). Space-time scales of internal waves: A progress report. *Journal of Geophysical Research.*, *80*(3), 291–297. Available from https://doi.org/10.1029/jc080i003p00291.

Garrett, C., & Munk, W. (1979). Internal waves in the ocean. *Annual Review of Fluid Mechanics.*, *11*, 339–369. Available from https://doi.org/10.1146/annurev.fl.11.010179.002011.

Gerkema, T., & Zimmerman, J. T. F. (2008). An introduction to internal waves. *Lecture Notes*.

Goldstone, J., & Jackiw, R. (1975). Quantization of nonlinear waves. *Physical Review D.*, *11*(6), 1486–1498. Available from https://doi.org/10.1103/PhysRevD.11.1486.

Gostiaux, L., Didelle, H., Mercier, S., & Dauxois, T. (2007). A novel internal waves generator. *Experiments in Fluids.*, *42*(1), 123–130. Available from https://doi.org/10.1007/s00348-006-0225-7.

Helfrich, K. R., & Melville, W. K. (1986). On long nonlinear internal waves over slope-shelf topography. *Journal of Fluid Mechanics*, *167* (1), 285. Available from https://doi.org/10.1017/s0022112086002823.

Helfrich, K. R., & Melville, W. K. (2006). Long nonlinear internal waves. *Annual Review of Fluid Mechanics.*, *38*, 395–425. Available from https://doi.org/10.1146/annurev.fluid.38.050304.092129.

Holloway, G. (1980). Oceanic internal waves are not weak waves. *Journal of Physical Oceanography*, *10*(6), 906–914. Available from https://doi.org/10.1175/1520-0485(1980)010 < 0906:oiwanw > 2.0.co;2.

Inkson, J. C. (1984). *Quantization of waves (second quantization)* (pp. 43–66). Springer Science and Business Media LLC. Available from 10.1007/978-1-4757-0226-2_3.

Jackson, C., da Silva, J., Jeans, G., Alpers, W., & Caruso, M. (2013). Nonlinear internal waves in synthetic aperture radar imagery. *Oceanography.*, *26*(2). Available from https://doi.org/10.5670/oceanog.2013.32.

Jackson, C. R., da Silva, J. C. B., & Jeans, G. (2012). The generation of nonlinear internal waves. *Oceanography.*, *25*(2), 108–123. Available from https://doi.org/10.5670/oceanog.2012.46, http://www.tos.org/oceanography/archive/25-2_jackson.pdf.

Klemas, V. (2012). Remote sensing of ocean internal waves: An overview. *Journal of Coastal Research.*, *28*(3), 540–546. Available from https://doi.org/10.2112/JCOASTRES-D-11-00156.1.

Kozlov I. Kudryavtsev V. Zubkova E. Atadzhanova O. Zimin A. Romanenkov D. Myasoedov A. Chapron B. (2015) SAR observations of internal waves in the Russian Arctic seas. International Geoscience and Remote Sensing Symposium (IGARSS), Available from https://doi.org/10.1109/IGARSS.2015.7325923, 9781479979295, 947-949, Institute of Electrical and Electronics Engineers Inc. Russian Federation.

Kulish, P. P., & Zeitlin, A. M. (2004a). Super-KdV equation: Classical solutions and quantization. *PAMM.*, *4*(1), 576–577. Available from https://doi.org/10.1002/pamm.200410269.

Kulish, P. P., & Zeitlin, A. M. (2004b). Quantization of integrable models with hidden symmetries: Super-KdV equation. *Journal of Modern Optics*, *51-6*(7), 1107–1108. Available from https://doi.org/10.1080/09500340408233643.

Kumar, S., Ghosh, S., Kumar, R., & Jleli, M. (2021). A fractional model for population dynamics of two interacting species by using spectral and Hermite wavelets methods. *Numerical methods for partial differential equations.*, *37*(2), 1652–1672. Available from https://doi.org/10.1002/num.22602, http://onlinelibrary.wiley.com/journal/10.1002/(ISSN)1098-2426.

Kumbinarasaiah, S., & Mundewadi, R. A. (2021). The new operational matrix of integration for the numerical solution of integro-differential equations via Hermite wavelet. *SeMA Journal*, *78*(3), 367–384. Available from https://doi.org/10.1007/s40324-020-00237-8, http://springer.com/journal/40324.

Manoharan, S. (2019). Study on hermitian graph wavelets in feature detection. *Journal of Soft Computing Paradigm*, *2019*(1), 24–32. Available from https://doi.org/10.36548/jscp.2019.1.003.

Marghany M. 1999 Internal wave detection and wavelength estimation. International Geoscience and Remote Sensing Symposium (IGARSS), 163-165, IEEE Malaysia 1.

Marghany M. 2000 Digital elevation model of internal wave. International Geoscience and Remote Sensing Symposium (IGARSS), 2960-2962, IEEE Netherlands detection 7.

Marghany M. 2002 1 2002/01 International Geoscience and Remote Sensing Symposium (IGARSS), 2847-2849, Malaysia Polarised topsar operational modle of internal wave generation mechanism 5.

Marghany, M. (2018). Automatic detection of internal wave using particle swarm optimization algorithm. *IOP Conference Series: Earth and Environmental Science*, *169*(1)012049. Available from https://doi.org/10.1088/1755-1315/169/1/012049.

Marghany, M. (2021). *Nonlinear ocean dynamics: Synthetic aperture radar*. Elsevier.

McComas, C. H., & Bretherton, F. P. (1977). Resonant interaction of oceanic internal waves. *Journal of Geophysical Research*, *82*(9), 1397–1412. Available from https://doi.org/10.1029/jc082i009p01397.

McComas, C. H., & Müller, P. (1981). The dynamic balance of internal waves. *Journal of Physical Oceanography*, *11*(7), 970–986. Available from http://journals.ametsoc.org/loi/phoc, https://doi.org/10.1175/1520-0485(1981)011 < 0970:TDBOIW > 2.0.CO;2.

Miroshnikov, V. A. (2023). Quantization of the kinetic energy of deterministic chaos. *American Journal of Computational Mathematics.*, *13*(01), 1–81. Available from https://doi.org/10.4236/ajcm.2023.131001.

Nycander, J. (2005). Generation of internal waves in the deep ocean by tides. *Journal of Geophysical Research: Oceans.*, *110*(10), 1–9. Available from https://doi.org/10.1029/2004JC002487, http://agupubs.onlinelibrary.wiley.com/agu/jgr/journal/10.1002/(ISSN)2169-9291/.

Oruç, Ö. (2018). A numerical procedure based on Hermite wavelets for two-dimensional hyperbolic telegraph equation. *Engineering with Computers.*, *34*(4), 741–755. Available from https://doi.org/10.1007/s00366-017-0570-6.

Penrose, O. (2009). XI. The quantization of sound waves—I: General theory. *Philosophical Magazine and Journal of Science.*, *45*(360), 80–92. Available from https://doi.org/10.1080/14786440108520424.

Petrov, D. S. (2015). Quantum mechanical stabilization of a collapsing bose-bose mixture. *Physical Review Letters.*, *115*(15). Available from https://doi.org/10.1103/PhysRevLett.115.155302, http://harvest.aps.org/bagit/articles/10.1103/PhysRevLett.115.155302/apsxml.

Pogrebkov, A. K. (2001). Quantizing the KdV equation. *Theoretical and Mathematical Physics.*, *129*(2), 1586–1595. Available from https://doi.org/10.1023/A:1012895426139.

Romeiser, R., & Graber, H. C. (2015). Advanced remote sensing of internal waves by spaceborne along-track InSAR-A demonstration with TerraSAR-X. *IEEE Transactions on Geoscience and Remote Sensing.*, *53*(12), 6735–6751. Available from https://doi.org/10.1109/TGRS.2015.2447547.

Ródenas, J. A., & Garello, R. (1998). Multiscale edge approach for the detection and location of internal waves from SAR images. *Proceedings of SPIE - The International Society for Optical Engineering*, *3391*, 510–521. Available from https://doi.org/10.1117/12.304922, http://spie.org/x1848.xml.

Shiralashetti, S. C., & Kumbinarasaiah, S. (2018). Hermite wavelets operational matrix of integration for the numerical solution of non-linear singular initial value problems. *Alexandria Engineering Journal.*, *57*(4), 2591–2600. Available from https://doi.org/10.1016/j.aej.2017.07.014, http://www.elsevier.com/wps/find/journaldescription.cws_home/724292/description#description.

Shiralashetti, S. C., & Kumbinarasaiah, S. (2019). New generalized operational matrix of integration to solve nonlinear singular boundary value problems using Hermite wavelets. *Arab Journal of Basic and Applied Sciences.*, *26*(1), 385–396. Available from https://doi.org/10.1080/25765299.2019.1646090, https://www.tandfonline.com/loi/tabs20.

Tao, M., Xu, C., Guo, L., Wang, X., & Xu, Y. (2022). An internal waves data set from Sentinel-1 synthetic aperture radar imagery and preliminary detection. *Earth and Space Science.*, *9*(12). Available from https://doi.org/10.1029/2022EA002528, http://agupubs.onlinelibrary.wiley.com/hub/journal/10.1002/(ISSN)2333-5084/.

Thompson, D. R. (1988). Calculation of radar backscatter modulations from internal waves. *Journal of Geophysical Research.*, *93*(10), 12371–12380. Available from https://doi.org/10.1029/JC093iC10p12371.

Thorpe, S. A. (1968). On the shape of progressive internal waves. Philosophical Transactions of the Royal Society of London. Series A. *Mathematical and Physical Sciences*, *263*(1145), 563–614.

Velotto D. Nunziata F. Migliaccio M. Lehner S. 2013 ESA POLinSAR Workshop, A robust symmetry-based approach to exploit TerraSAR-X dual-pol data for targets at sea observation.

Venayagamoorthy, S. K., & Koseff, J. R. (2016). On the flux Richardson number in stably stratified turbulence. *Journal of Fluid Mechanics.*, *798*, R1. Available from https://doi.org/10.1017/jfm.2016.340, http://journals.cambridge.org/action/displayJournal?jid = FLM.

Wang, Y., & Chen, X. (2020). Hybrid quantum particle swarm optimization algorithm and its application. *Science China Information Sciences*, *63*(5). Available from https://doi.org/10.1007/s11432-018-9618-2, http://www.springer.com/computer/information + systems + and + applications/journal/11432.

Wyrtki, K. (1961). Scientific results of marine investigations of the South China Sea and the Gulf of Thailand 1959-1961. *NAGA report.*, *2*.

Xin-gang, Z., Ji, L., Jin, M., & Ying, Z. (2020). An improved quantum particle swarm optimization algorithm for environmental economic dispatch. *Expert Systems with Applications.*, *152*113370. Available from https://doi.org/10.1016/j.eswa.2020.113370.

Yang S. Wang M. Jiao L. 2004 A quantum particle swarm optimization. Proceedings of the 2004 Congress on Evolutionary Computation, CEC2004 320-324 China 1.

Zhang, L., Yin, Q., Zhang, Z., Zhu, Z., Lyu, L., Hai, K. L., & Cai, G. (2022). A wind power curtailment reduction strategy using electric vehicles based on individual differential evolution quantum particle swarm optimization algorithm. *Energy Reports.*, *8*, 14578–14594. Available from https://doi.org/10.1016/j.egyr.2022.10.442, http://www.journals.elsevier.com/energy-reports/.

Zhang, Q., Li, H., Liu, Y., Ouyang, S., Fang, C., Mu, W., & Gao, H. (2021). A new quantum particle swarm optimization algorithm for controller placement problem in software-defined networking. *Computers and Electrical Engineering.*, *95*107456. Available from https://doi.org/10.1016/j.compeleceng.2021.107456.

Zhi, Z., Xiaoming, L., Yongzheng, R., & Xueen, C. (2020). Exploratory research on the retrieval of internal wave parameters and sea surface current velocity based on terraSAR-X satellite data. *Haiyang Xuebao*, *42*(1), 90–101. Available from https://doi.org/10.3969/j.issn.0253-4193.2020.01.010, http://english.hyxb.org.cn/homeAction!downloadArticleFile.action?attachType = PDF&id = 5660.

Zhou, N. R., Xia, S. H., Ma, Y., & Zhang, Y. (2022). Quantum particle swarm optimization algorithm with the truncated mean stabilization strategy. *Quantum Information Processing.*, *21*(2). Available from https://doi.org/10.1007/s11128-021-03380-x, http://www.kluweronline.com/issn/1570-0755/contents.

Chapter 6

Quantized Marghany oceanic front detection in MultiSAR datasets: unveiling the hidden patterns of the South China Sea

6.1 What is the magic of oceanic front?

Oceanic fronts are regions in the ocean where there are significant gradients in properties such as temperature, salinity, density, and sometimes, biological or chemical composition. These fronts often exhibit increased turbulence and convection compared with the surrounding areas (Chapter 2, Section 2.16). Fronts can vary in width, ranging from narrow features on the scale of hundreds of meters to broader fronts spanning several tens of kilometers. However, fronts can be transient and form and dissipate relatively quickly, influenced by changing oceanographic conditions or the movement of water masses. Fronts, nonetheless, can persist for extended periods of time, maintaining their structure and characteristics over weeks, months, or even longer. These long-lasting fronts often occur in regions where there is a continuous interaction between different water masses, such as in areas with persistent upwelling or where the shelf break acts as a barrier to the exchange of water (Busalacchi & O'Brien, 1980; Macvean & Woods, 1980).

Oceanic fronts can form due to a variety of processes, including the interaction of different water masses, the convergence of currents, or the influence of atmospheric forcing. These processes can lead to the mixing and juxtaposition of water masses with different properties, resulting in sharp gradients across the front.

The most common types of oceanic fronts include: (1) thermal front and (2) haline fronts. Thermal fronts are characterized by sharp gradients in temperature. They typically occur where warm and cold water masses meet, such as at the boundaries between different ocean currents or along the edges of upwelling zones. Thermal fronts can have significant impacts on weather patterns, nutrient transport, and the distribution of marine organisms. For example, a thermal front may occur where a warm water mass encounters a colder water mass, leading to a distinct temperature gradient across the front. Similarly, a haline front may arise at the interface between a salty water mass and a less saline water mass, causing a sharp salinity gradient. Therefore haline fronts are characterized by sharp gradients in salinity. They typically occur in regions where waters with different salinities mix, such as at the mouths of rivers or in areas influenced by freshwater inputs. Haline fronts can affect the distribution of marine life, including phytoplankton and fish species and can also impact the vertical stratification of the water column(McLellan, 1965; Mellor & Huang, 1997; Miranda et al., 2017).

Density fronts can also form, where the density of water masses differs significantly, Fig. 6.1. Density variations can be influenced by temperature and salinity differences, as well as other factors such as pressure and dissolved substances. These fronts are often associated with significant vertical mixing and can affect the vertical stratification of the water column. Additionally, oceanic fronts can exhibit differences in velocity, with water masses moving at different speeds or directions. These velocity fronts can occur when currents converge or when there are shear zones between adjacent water masses.

Oceanic fronts can occur at various depths within the ocean, from the surface down to the deep ocean. These fronts can be influenced by a combination of atmospheric and oceanic flow patterns and can exhibit temporal variability. The intensification and decay of oceanic fronts are indeed responses to changes in the atmospheric and oceanic circulation patterns. Factors such as wind stress, ocean currents, and large-scale climate phenomena can impact the strength and persistence of fronts over time. Permanent oceanic fronts, as the name suggests, are relatively stable and persistent features that maintain their presence over long periods. Examples of permanent oceanic fronts include: (1) Antarctic Circumpolar Current (ACC); (2) Subarctic Front; and (3) Subtropical Front (Fig. 6.2). In this view, the ACC is a major oceanic front that encircles Antarctica. It is the largest ocean current in terms of volume transport and serves as a boundary between the Antarctic waters and the

 DOI: https://doi.org/10.1016/B978-0-443-19155-8.00006-4

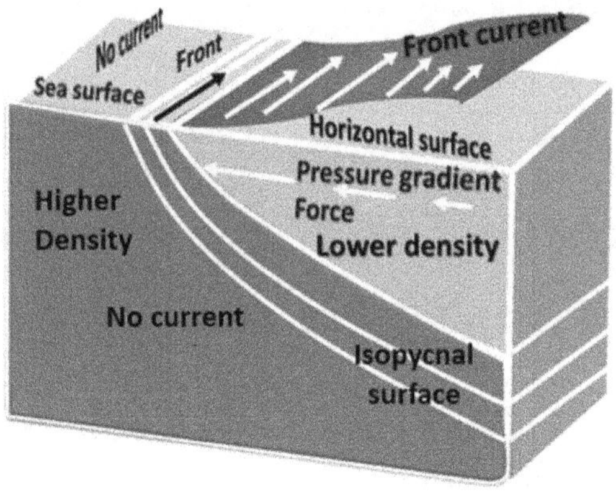

FIGURE 6.1 The front zone is created along with significant density differences.

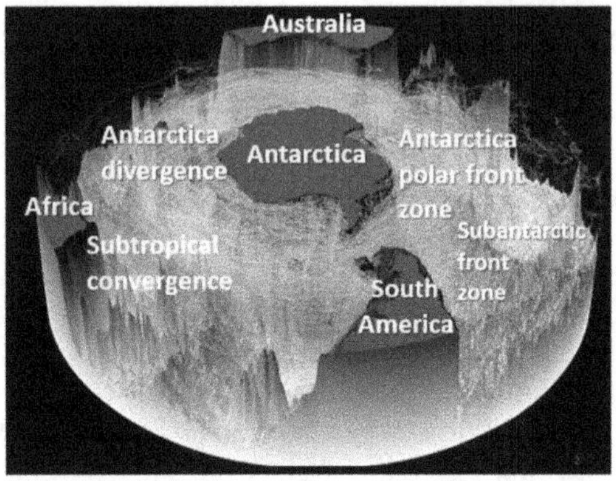

FIGURE 6.2 The boundary between Antarctic and subtropical waters.

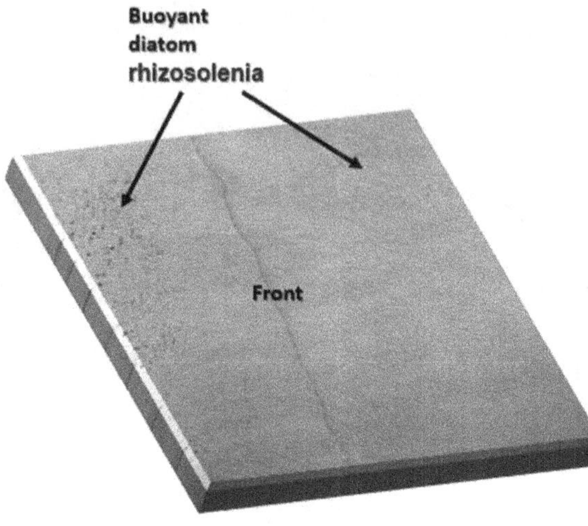

FIGURE 6.3 North Pacific permanent front.

rest of the global ocean. The ACC plays a crucial role in global climate dynamics and is associated with significant temperature and density gradients. Therefore the Subarctic Front is a permanent oceanic front in the North Pacific Ocean (Kao, 1980; Webb, 2021). It separates colder, fresher waters of the subarctic region from warmer, saltier waters of the subtropical region. This front is associated with changes in temperature, salinity, and nutrient distributions, influencing both physical and biological processes in the region. Consequently, the Subtropical Front is another permanent oceanic front in the North Pacific Ocean (Fig. 6.3), which is always dominated by the accumulation of buoyant diatom. It separates subtropical waters from the more equatorial and transitional waters. The front marks the boundary between warm, low-nutrient subtropical waters and the cooler, nutrient-rich waters further north. It is an important region for various oceanic processes, including upwelling and nutrient cycling (Brink, 2023).

Oceanic fronts are important features in the ocean, as they play a crucial role in the exchange of heat, salt, and other properties between different water masses. They also influence the distribution of nutrients, which can impact primary productivity and the overall structure of marine ecosystems. Understanding the formation and dynamics of oceanic fronts is essential for studying ocean circulation, climate processes, and the ecology of marine organisms (Busalacchi & O'Brien, 1980; Dijkstra, 2008; Miranda et al., 2017).

6.2 What are the differences between oceanic and atmospheric fronts?

This question is addressed briefly in the second chapter. Therefore this section will deliver clear differences between both fronts. Oceanic fronts and atmospheric fronts are distinct features that occur in different fluid mediums and have different characteristics.

Oceanic fronts occur in the ocean, which is a vast body of water, while atmospheric fronts occur in the Earth's atmosphere, which is the layer of gases surrounding the planet. In this view, the nature of both fronts is quite different. Oceanic fronts are characterized by sharp gradients in properties such as temperature, salinity, density, and velocity between different water masses. They represent boundaries where distinct water masses meet. Atmospheric fronts, on the other hand, are boundaries between air masses with different temperature, humidity, and density characteristics. They mark the transition between different air masses and their associated weather systems (Garvine, 1974; Kao, 1980).

Therefore regarding the deriving forces, oceanic fronts are primarily driven by oceanic processes, including the interaction of water masses, currents, and local physical conditions. They can be influenced by factors such as wind patterns, ocean currents, and thermal or haline processes. Atmospheric fronts, however, are driven by atmospheric processes, such as the interaction of air masses with different temperature and humidity properties. They are often associated with weather systems, including the formation of clouds, precipitation, and wind patterns. In the context of scale, oceanic fronts can occur at various scales, from smaller-scale coastal fronts to large-scale fronts such as the Antarctic Circumpolar Current. They can exist at different depths within the ocean. Atmospheric fronts can also occur at different scales, ranging from smaller-scale fronts associated with local weather systems to larger-scale fronts associated with synoptic-scale weather patterns, such as cold fronts and warm fronts. Therefore oceanic fronts' time variability can exhibit temporal variability, intensifying and decaying in response to changes in oceanic and atmospheric flow patterns. They can be influenced by seasonal variations, oceanic oscillations, and climate phenomena. Atmospheric fronts can also undergo changes in intensity and position over time due to shifts in atmospheric circulation patterns and weather systems (Roden, 1972, 1974, 1976).

6.3 Oceanic front types

6.3.1 Plume front

The term "plume front" typically refers to the interface or boundary between the river plume and the surrounding oceanic waters at the mouth of a river.

When a river discharges freshwater into the ocean, a river plume is formed (Fig. 6.4). The river plume is a buoyant flow of freshwater that spreads out from the river mouth, carrying sediments, nutrients, and other materials from the river into the coastal ocean. The plume front is the leading edge or boundary of this freshwater plume as it interacts with the surrounding seawater (Fig. 6.5). The characteristics and dynamics of plume fronts at river

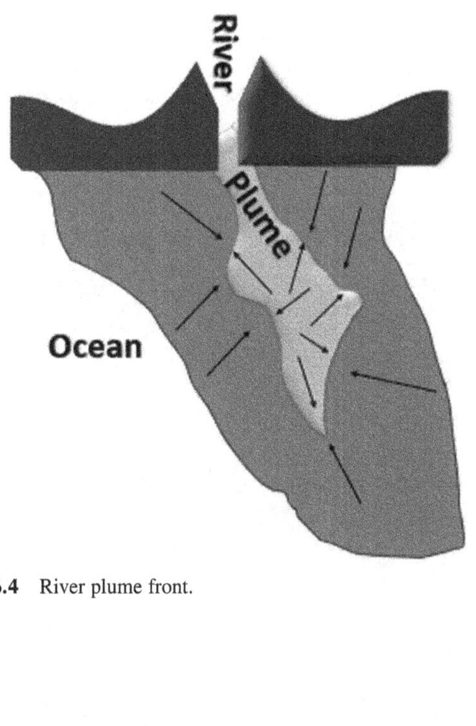

FIGURE 6.4 River plume front.

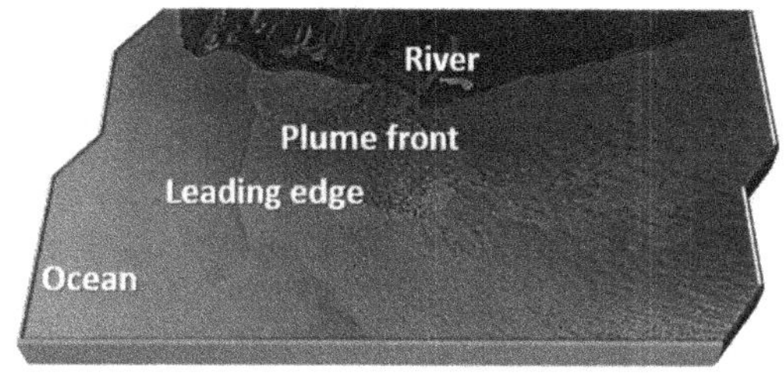

FIGURE 6.5 Leading edge formed by the plume front.

mouths can vary depending on factors such as river discharge, tidal conditions, and coastal morphology. In this sense, along the plume front, mixing occurs between the freshwater from the river and the saltwater of the ocean. The mixing processes can be influenced by tidal currents, wind-driven processes, and the density contrast between freshwater and seawater. The interaction between freshwater and saltwater can lead to vertical stratification, with the lighter freshwater overlaying the denser seawater (Busalacchi & O'Brien, 1980; McLellan, 1965; Mellor & Huang, 1997; Pedlosky, 1979).

6.3.2 Tidal front intrusion

Tidal intrusion fronts are interfaces or boundaries that form due to the interaction between tidal currents and river flows in estuaries or coastal regions. Estuaries are semienclosed coastal water bodies where freshwater from rivers mixes with seawater. When tidal currents, therefore encounter river flows in estuaries, they can create distinct fronts known as tidal intrusion fronts (Fig. 6.6). These fronts mark the boundary between the advancing tidal waters and the riverine freshwater(Garvine, 1974; Kao, 1980; Miranda et al., 2017).

The characteristics and dynamics of tidal intrusion fronts can vary depending on factors such as the tidal range, river discharge, channel geometry, and coastal topography. In this understanding, tidal intrusion fronts represent the interface between saline seawater and the less saline freshwater from rivers. The front can have a sharp transition from saltwater to freshwater as tidal waters advance upstream. In this sense, at tidal intrusion fronts, the mixing between saltwater and freshwater occurs. The extent of mixing depends on the strength of the tidal currents, river discharge, and the bathymetry of the estuary. The mixing processes can influence the stratification of water masses and vertical density gradients (O'Donnell et al., 1998; Pond, 1978).

6.3.3 Shelf front

Shallow-shelf sea fronts, also known as shelf sea fronts (Fig. 6.7) or shelf break fronts, refer to the interfaces or boundaries that occur in shallow regions of continental

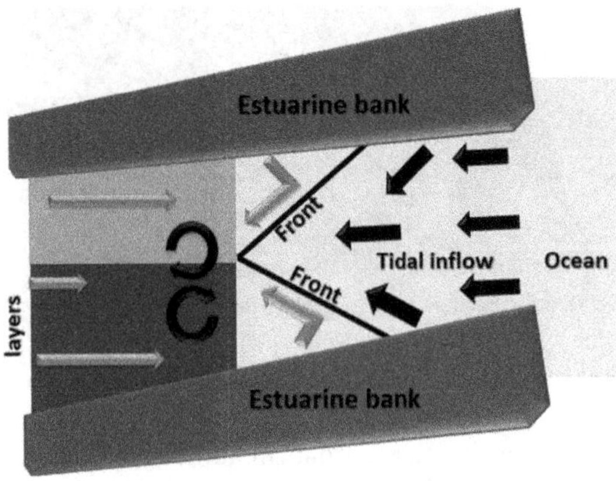

FIGURE 6.6 Tidal intrusion fronts.

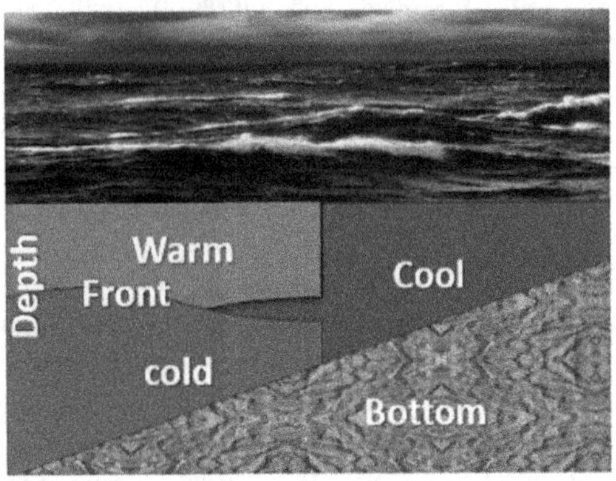

FIGURE 6.7 Shelf sea front.

shelves where distinct water masses interact. These fronts are typically found at the transition zone between the continental shelf and the deeper oceanic waters of the continental slope.

One of the most crucial characteristics of a shelf front is the tidal mixing front. A tidal mixing front, also known as a tidal front, is a type of front that forms due to the interaction between tidal currents and water masses in coastal or estuarine environments. Tidal mixing fronts (Fig. 6.7) occur when water masses with different properties, such as temperature, salinity, or density, are mixed by the action of tidal currents. Therefore the mixing zone, temporal variability, and vertical structure consider the core constituents of the tidal mixing verification. Tidal mixing fronts represent the transition zone between water masses that have been mixed by the tidal currents (Mellor & Huang, 1997; Muelbert et al., 2008; Piola et al., 2000). In this zone, the properties of the water, such as temperature, salinity, and density, can change rapidly. Tidal mixing fronts, consequently, can exhibit temporal variability, shifting in position and intensity with the changing tides. They can be highly dynamic, forming and dissipating over relatively short timescales, such as hours or tidal cycles. In this perspective, tidal mixing fronts can have a vertical structure, with mixed water extending from the surface to varying depths, depending on the intensity of tidal mixing. The vertical structure of the front influences the exchange of heat, nutrients, and other properties between different layers of the water column (Garvine, 1974; Kao, 1980; Simpson, 1472).

The shelf break front is yet another crucial shelf front. Thus shelf break fronts, also known as shelf edge fronts, are indeed one of the most common types of frontal features in the ocean. They typically align with the shelf break, which is the boundary where the relatively shallow and flat continental shelf transitions into the steeper continental slope (Fig. 6.8). Additionally, shelf break fronts are influenced by a combination of tidal and wind-driven mixing processes. Tidal currents and wind stress can induce mixing at the shelf break, leading to the formation of frontal features. These fronts mark the boundary between the water masses on the continental shelf and the adjacent off-shelf oceanic waters (Busalacchi & O'Brien, 1980; Kao, 1980; Mellor & Huang, 1997).

Therefore shelf break fronts occur at the shelf break, which is generally located several tens of kilometers offshore, where the water depth abruptly increases. They also represent the interface between the water masses on the continental shelf and the deeper, off-shelf oceanic waters. They mark the boundary where the properties, such as temperature, salinity, and density, change rapidly. Therefore tidal currents and wind-driven mixing affect shelf break fronts. These processes can result in mixing and turbulence, which in turn causes water masses to interact and exchange across the front (Simpson et al., 1978).

6.3.4 Western boundary current fronts

Western boundary current fronts are specific types of oceanic fronts that occur along the western boundaries of major ocean basins. These fronts are associated with strong, narrow, and fast-flowing ocean currents known as

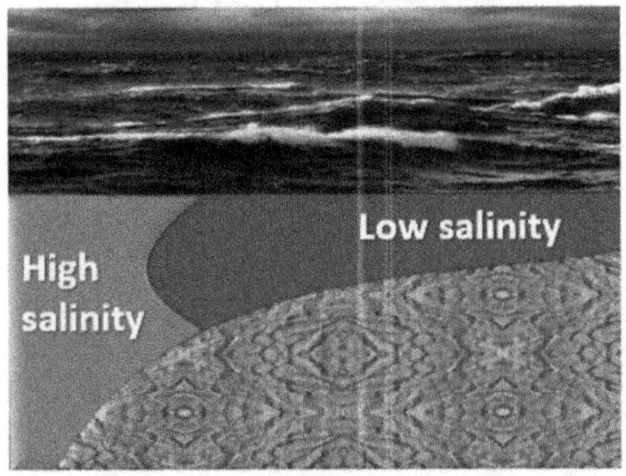

FIGURE 6.8 Shelf break front.

western boundary currents. Western boundary currents are driven by a combination of wind stress, the Earth's rotation (Coriolis effect), and the shape of the ocean basin.

The three main western boundary currents are the Gulf Stream in the North Atlantic, the Kuroshio in the North Pacific, and the Agulhas Current in the South Indian Ocean. These currents flow poleward along the western boundaries of their respective ocean basins and play a crucial role in global ocean circulation. In this perspective, western boundary current fronts exhibit distinctive characteristics: (1) intensity and narrowness; (2) frontal instabilities; and (3) warm-core rings. In this understanding, western boundary currents are known for their high velocities and narrow widths. They can be several tens to a few hundred kilometers wide but can transport significant volumes of water. The narrowness of these currents contributes to the formation of sharp frontal boundaries. Therefore these current boundary fronts often exhibit dynamic features such as meanders, eddies, and filaments. These instabilities arise from the interaction between the fast-moving current and the slower-moving surrounding waters. Finally, western boundary currents can generate warm-core rings or eddies, which are rotating features of warm water separated from the main current. These rings can detach from the main current and propagate westward into the open ocean, influencing the transport of heat, nutrients, and marine organisms (Busalacchi & O'Brien, 1980; Macvean & Woods, 1980; Miranda et al., 2017).

It is worth noting that western boundary current fronts have significant climatic implications. They transport warm water poleward, affecting the regional climate, weather patterns, and the redistribution of heat from the tropics to higher latitudes. These fronts also influence the exchange of water masses between the subtropical and subpolar gyres.

6.3.5 Upwelling front

An upwelling front refers to the boundary or interface that forms where an upwelling zone encounters surrounding waters in the ocean. Upwelling is a process in which deep, cold, nutrient-rich waters rise to the surface, replacing warmer surface waters. It typically occurs in regions where winds and oceanic currents push surface waters away from the coast, allowing deeper waters to upwell and replace them.

When the upwelling zone reaches the surface, it can create a distinct front or boundary with the surrounding waters (Fig. 6.9). This front is known as an upwelling front. Upwelling fronts are characterized by sharp gradients in temperature, salinity, and nutrient concentrations. Consistent with the mechanism, upwelling fronts bring nutrient-rich waters from the deeper ocean to the surface

FIGURE 6.9 Upwelling front current.

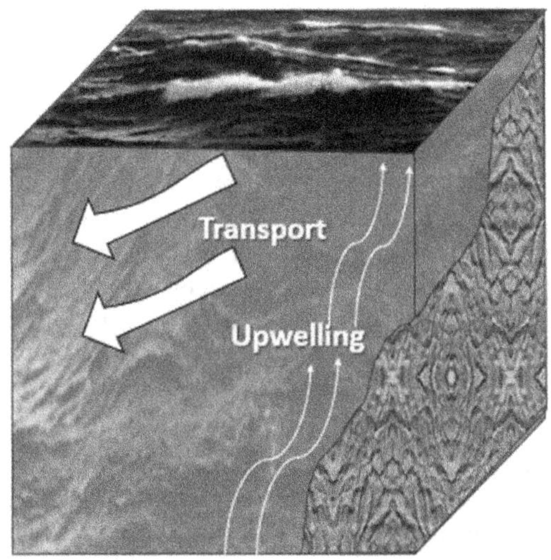

FIGURE 6.10 Upwelling front formation and nutrient transport.

(Fig. 6.10). These nutrients, including nitrates, phosphates, and silicates, support the growth of phytoplankton and promote high primary productivity. Upwelling fronts are often regions of enhanced biological activity and support rich ecosystems, including fisheries. Moreover, upwelling fronts involve the mixing of different water masses, as the upwelled deep waters interact with the surface waters (Palmen & Newton, 1969; Pedlosky, 1979; Simpson et al., 1978; Zheng et al., 2016). This mixing can lead to the creation of stratified conditions, where warmer and less saline surface waters overlay colder and more saline deep waters. The stratification can influence the vertical distribution of heat, nutrients, and organisms.

Frontal eddies and filaments are thus the key characteristics of an upwelling front, which are effective in SAR image interpretation. In this sense, upwelling fronts can generate small-scale eddies and filaments, which are swirling or elongated features in the water column. These eddies and filaments can enhance mixing, transport of nutrients, and the dispersal of planktonic organisms.

6.3.6 Ekman convergence front

The Ekman convergence front refers to a specific type of oceanic front that forms due to the convergence of water masses driven by the Ekman transport. The Ekman transport is the net motion of water resulting from the combined effect of wind stress and the Coriolis effect, which causes water to be deflected to the right in the Northern Hemisphere and to the left in the Southern Hemisphere.

When the wind blows across the ocean surface, it imparts a force on the water, causing it to move. However, due to the Coriolis effect, the water is deflected, resulting in a net transport of water at a 90-degree angle to the wind direction. This transport occurs throughout the water column, known as the Ekman transport (Brink, 2023; Busalacchi & O'Brien, 1980; Kao, 1980; Macvean & Woods, 1980).

In regions where the wind-driven Ekman transports from adjacent areas converge, the convergence of water masses can lead to the formation of an Ekman convergence front. The convergence front is a boundary where waters from different regions converge, resulting in changes in temperature, salinity, and other properties (Kao, 1980; Thompson, 2000; Yu, 2015; Zheng et al., 2016).

According to the aforementioned perspective, the Ekman convergence front represents the boundary where waters from different regions, driven by the Ekman transport, converge. The convergence can lead to the mixing of water masses with different properties. In addition, the Ekman convergence front typically extends vertically through the water column, with convergence occurring not only at the surface but also at depth. This vertical structure influences the exchange of heat, nutrients, and other properties between different layers of the water column. Needless to say, the Ekman convergence front is often associated with specific circulation patterns, such as upwelling or downwelling. Upwelling can occur at the Ekman convergence front, bringing nutrient-rich waters from depth to the surface, while downwelling can result in the sinking of surface waters (Busalacchi & O'Brien, 1980; Thompson, 2000; Zheng et al., 2016).

The characteristics and behavior of the Ekman convergence front can vary depending on factors such as wind intensity, wind direction, oceanic currents, and regional oceanographic conditions. They are typically observed in regions where wind-driven Ekman transport patterns are significant, such as along the coasts of upwelling zones or in regions with persistent wind patterns (Garvine, 1974; Kao, 1980; Macvean & Woods, 1980; Muelbert et al., 2008; Webb, 2021).

Understanding the Ekman convergence front is important for studying the physical and ecological processes associated with converging water masses and their influence on nutrient transport, primary productivity, and the structure of marine ecosystems. Scientists utilize a combination of observational data, satellite remote sensing, and numerical modeling to investigate the dynamics and ecological significance of the Ekman convergence front in different regions of the ocean. The following chapters will discuss the Ekman convergence front as part of the upwelling mechanisms.

6.3.7 Estuarine fronts

Estuarine fronts refer to the dynamic interfaces or boundaries that occur within estuaries, which are semienclosed coastal water bodies where freshwater from rivers mixes with seawater (Fig. 6.11). Estuarine fronts are a result of the interaction between the fresh water and seawater, which have different properties such as temperature, salinity, and density.

Estuaries are characterized by complex and variable circulation patterns, driven by tides, winds, river flow, and density gradients. These factors contribute to the formation and behavior of estuarine fronts. Here are some key features and processes associated with estuarine fronts (Garvine, 1974; Largier, 1993; Miranda et al., 2017):

Saltwater – Freshwater Mixing: Estuarine fronts mark the transition zone between saltwater and freshwater within the estuary. As freshwater from the river mixes with the seawater, a frontal boundary forms where the two water masses meet. The mixing processes at the front can vary depending on the tidal and river flow dynamics, leading to variable salinity gradients (Bowman, 2018; Brink, 2023).

Stratification and Mixing: Estuarine fronts can influence the stratification of water masses within the estuary (Fig. 6.12), which is dominated by turbulence eddy. The density contrast between freshwater and saltwater can lead to the formation of a vertical density gradient. Mixing processes at the front, driven by tidal currents and wind stress, can enhance the exchange of heat, nutrients, and dissolved gases between the water layers.

Nutrient Transport: Estuarine fronts play a role in the transport of nutrients within the estuary. The mixing and exchange of water masses at the front can enhance

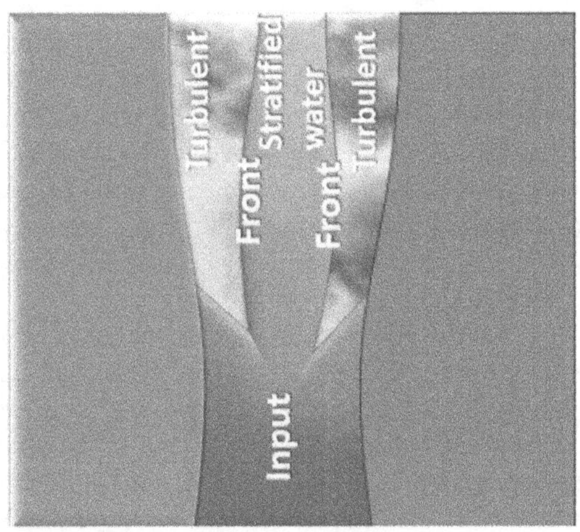

FIGURE 6.11 Estuarine fronts.

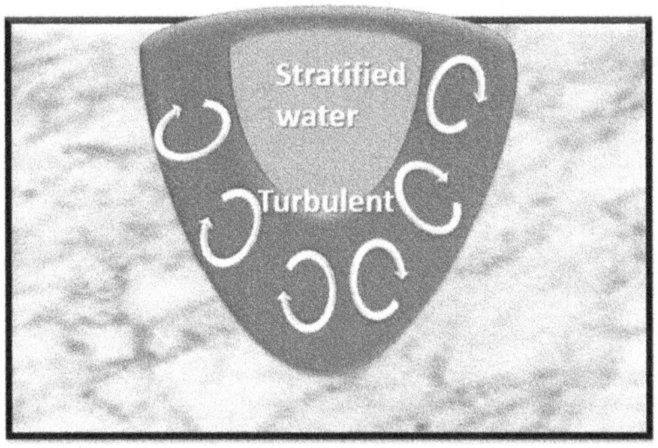

FIGURE 6.12 Estuarine fronts.

nutrient availability, influencing the growth of phytoplankton and other primary producers. Estuarine fronts can be regions of enhanced biological productivity and support diverse ecosystems.

Sediment Dynamics: Estuarine fronts can also affect sediment transport and distribution. The convergence or divergence of water masses at the front can result in sedimentation, the formation of turbidity plumes, and the deposition of sediments within the estuary. Sediment dynamics at the front can impact the morphology and habitats of the estuary. Accordingly, the characteristics of estuarine fronts can vary depending on factors such as tidal amplitude, river discharge, estuary geometry, and the presence of coastal currents. Estuarine fronts are highly dynamic and can exhibit spatial and temporal variability, influenced by the changing tidal and river flow patterns (Bowman, 2018; Busalacchi & O'Brien, 1980; Dijkstra, 2008; Miranda et al., 2017).

6.4 Timescale of dissipation of an upper oceanic front

Indeed, the timescale of dissipation of an upper oceanic front is an essential aspect of ocean dynamics and can be investigated through theoretical experiments. The dissipation of a front refers to the process by which the front weakens or disappears over time. Theoretical experiments can involve numerical modeling or analytical calculations to study the evolution and dissipation of oceanic fronts. These experiments aim to understand the underlying physical mechanisms and factors that influence the front's behavior.

Through theoretical experiments, researchers can investigate various aspects of front dissipation, such as: (1) advection and mixing; (2) turbulence and shear; and (3) external forces. In this understanding, the role of advection and mixing processes in the front's evolution can be explored. Advection refers to the transport of water masses across the front, while mixing refers to the mixing of properties such as temperature and salinity across the front. By simulating or analyzing these processes, the timescale of front dissipation can be examined (Busalacchi & O'Brien, 1980; Camerlengo, 1982; Dijkstra, 2008).

Advection, therefore, can affect the front in several ways. Advection, consequently, can cause the front to move horizontally, leading to changes in the location of the front over time. This frontal displacement can be influenced by the strength and direction of ocean currents. In this circumstance, advection can induce instabilities at the front, such as frontal waves or eddies. These instabilities can enhance mixing processes and affect the overall structure and evolution of the front (Camerlengo & Demmler, 1997; Huzzey & Brubaker, 1988; Largier, 1993).

Moreover, turbulence and shear forces play crucial roles in front dynamics. Theoretical experiments can study the impact of turbulence and shear on the front's stability and dissipation. This can involve considering the effects of turbulent mixing and the influence of different shear conditions on the front's behavior. In this sense, mixing can disrupt the sharp gradients of properties across the front, leading to a reduction in the front's strength and coherence. Strong mixing can lead to dissipation and eventual disappearance of the front. In other words, mixing redistributes properties across the front, which can modify the thermodynamic structure and affect the frontal characteristics. Mixing can alter the density field, leading to changes in the front's density contrast and associated circulation patterns (Busalacchi & O'Brien, 1980; Camerlengo, 1982; Muelbert et al., 2008).

Theoretical experiments, subsequently, can also investigate the influence of external factors on front dissipation. For example, the effects of varying wind patterns, atmospheric forcing, or ocean currents on front evolution can be studied to understand their contributions to the timescale of dissipation. By conducting theoretical experiments, researchers can gain insights into the dynamics of upper oceanic fronts, the processes governing their dissipation, and the associated timescales. This knowledge contributes to a better understanding of oceanic circulation patterns, the interaction between water masses, and the overall dynamics of the upper ocean (Roden, 1972; Simpson et al., 1978).

6.5 Dissipation frontal model formulation

The conservation equations describe the evolution of various properties such as mass, momentum, heat, and salt. These equations form the basis for understanding the dynamics of the front. The general form of the conservation equation can be written as:

$$\frac{\partial \phi}{\partial t} + \nabla \cdot (u\phi) = S \tag{6.1}$$

where ϕ represents the property of interest (e.g., temperature, salinity, density), u is the velocity vector, and S represents any source or sink terms. Eq. (6.1) reveals the advection term $\nabla\cdot(u\phi)$, which accounts for the transport of properties by ocean currents. On the other hand, it represents the movement of water masses across the front. The timescale of dissipation of the oceanic front, the longitudinal frontal length scale, geostrophically balanced, the f-plane approximation, the zonal wave number, and the vorticity at the front are the factors that contribute to the dissipation model of the oceanic front (Busalacchi & O'Brien, 1980; Camerlengo, 1982; Kao, 1980).

Let us assume that P_a represents the atmospheric pressure; h_1, the depth of the upper layer; ρ the water's density; H is the total depth; g is the acceleration due to the earth's gravity; and z is the vertical coordinate, which is measured from the bottom (Fig. 6.13). The equation of pressure gradient describes the change in pressure over a given distance or spatial gradient. In fluid dynamics, including oceanography, this equation is derived from the fundamental principles of fluid mechanics, specifically the equations of motion.

The equation of pressure gradient in three-dimensional Cartesian coordinates can be written as:

$$\nabla P = -\rho\nabla\Phi \tag{6.2}$$

where ∇P is the pressure gradient vector, ρ is the density of the fluid, and $\nabla\Phi$ represents the gradient of the potential field Φ associated with the fluid motion (e.g., gravitational potential or geopotential) (Camerlengo & Demmler, 1997; Camerlengo, 1982; Dijkstra, 2008).

Expanding this equation into component form, we have:

$$\frac{\partial P}{\partial x} = -\rho\left(\frac{\partial\Phi}{\partial x}\right) \tag{6.3}$$

$$\frac{\partial P}{\partial y} = -\rho\left(\frac{\partial\Phi}{\partial y}\right) \tag{6.4}$$

$$\frac{\partial P}{\partial z} = -\rho\left(\frac{\partial\Phi}{\partial z}\right) \tag{6.5}$$

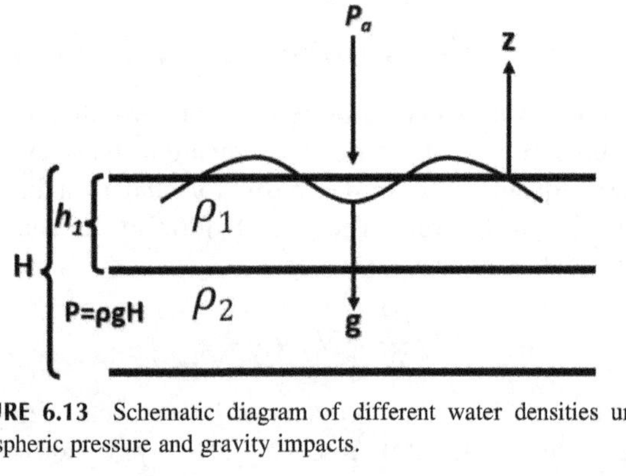

FIGURE 6.13 Schematic diagram of different water densities under atmospheric pressure and gravity impacts.

where ∂P/∂x, ∂P/∂y, and ∂P/∂z represent the partial derivatives of pressure with respect to the Cartesian coordinates x, y, and z, respectively. Similarly, ∂Φ/∂x, ∂Φ/∂y, and ∂Φ/∂z are the partial derivatives of the potential field Φ with respect to x, y, and z. Generally, the pressure gradient yields:

$$\nabla P = -\rho g\left(\frac{\rho_1 - \rho_2}{\rho_1}\right)\nabla h_1 \tag{6.6}$$

In Eq. (6.6), $g\left(\frac{\rho_1-\rho_2}{\rho_1}\right)$ represents the reduced gravity. As common, the atmospheric pressure's horizontal fluctuations are disregarded. Subindex 1, which represents the upper layer, will be excluded for ease of use. The exclusion of subindex 1 simplifies calculations and analysis, allowing for a more straightforward understanding of the atmospheric pressure's vertical variations. By focusing solely on the lower layer, researchers can gain valuable insights into the dynamics and behavior of this specific region (Busalacchi & O'Brien, 1980; Camerlengo & Demmler, 1997; Camerlengo, 1982).

6.6 Equation of motion

The equation of motion, also known as the Navier − Stokes equation, is a fundamental equation in fluid dynamics that describes the motion of a fluid. In the context of oceanography, this equation is used to study the dynamics of ocean currents and the behavior of water masses (Garvine, 1974; Mellor & Huang, 1997; Miranda et al., 2017). The general form of the three-dimensional Navier − Stokes equation in Cartesian coordinates is:

$$\rho\left(\frac{\partial u}{\partial t} + u\nabla u\right) = -\nabla P + \mu\nabla^2 u + \rho g \tag{6.7}$$

where:

- ρ is the density of the fluid.
- $u = (u, v, w)$ represents the velocity vector in the x, y, and z directions.
- t is time.
- P is the pressure.
- μ is the dynamic viscosity of the fluid.
- ∇ is the gradient operator (∇u represents the velocity gradient).
- $\nabla^2 u$ is the Laplacian of velocity (divergence of the velocity gradient).
- g is the acceleration due to gravity. This equation states that the change in momentum of a fluid element is determined by various forces acting on it.
- The acceleration term (∂u/∂t + u ∇u) represents the change in velocity with time and the advection of velocity.

- The pressure gradient (∇P) accounts for the forces resulting from pressure differences in the fluid.
- The viscosity term ($\mu \nabla^2 u$) describes the effects of internal friction or viscous forces that tend to dampen fluid motion.
- The buoyancy term (ρg) represents the gravitational force acting on the fluid element due to differences in density.

The Navier – Stokes equation is a set of partial differential equations that describes the conservation of momentum for a fluid. Solving these equations requires appropriate initial and boundary conditions, such as velocity and pressure values at given locations and times. Numerical methods, such as finite difference or finite volume methods, are often used to solve the Navier – Stokes equation in practical applications (Adcroft & Hallberg, 2006; Bowman, 2018; Brink, 2023).

The equations of motion for the upper layer of the ocean can be derived by applying the principles of fluid dynamics to this specific layer. These equations describe the dynamics of horizontal motion in the upper ocean, accounting for forces such as pressure gradients, the Coriolis effect, wind stress, and friction. Here, we will outline the simplified equations of motion for the upper layer, assuming a steady-state and neglecting vertical motion: Momentum Equations: The momentum equations describe the conservation of momentum in the horizontal direction (x and y) (Camerlengo & Demmler, 1997; Camerlengo, 1982; Dijkstra, 2008; Eckart & Ferris, 1956; Garvine, 1974). In Cartesian coordinates, the simplified equations can be written as:

$$\rho u\left(\frac{\partial u}{\partial x}\right) + \rho v\left(\frac{\partial u}{\partial y}\right) = -\frac{\partial P}{\partial x} + \tau x + fv + Fx = A\left(\frac{\partial^2 u}{\partial y^2}\right) \tag{6.8}$$

$$\rho u\left(\frac{\partial v}{\partial x}\right) + \rho v\left(\frac{\partial v}{\partial y}\right) = -\frac{\partial P}{\partial y} + \tau y + fu + Fy = -g\left(\frac{\rho_1 - \rho_2}{\rho_1}\right)\frac{\partial h}{\partial t} + A\frac{\partial^2 v}{\partial y^2} \tag{6.9}$$

where:

- ρ is the density of the fluid.
- u and v are the horizontal velocity components in the x and y directions, respectively.
- P is the pressure.
- τ_x and τ_y represent the components of wind stress (force per unit area) acting in the x and y directions, respectively.
- f is the Coriolis parameter, which is related to the rotation rate of the Earth.
- Fx and Fy denote any other external forces acting on the fluid, such as frictional effects. Therefore A is the horizontal diffusion coefficient of momentum; h is the upper layer thickness, and t is the time.

Additionally, the continuity equation expresses the conservation of mass in the horizontal direction. For incompressible flow, this equation simplifies to:

$$\frac{\partial h}{\partial t} + \partial\left(\frac{vh}{\partial y}\right) = 0 \tag{6.10}$$

Solving this complex set of nonlinear equations is challenging. Consequently, it is believed that a steady, inviscid initial front will gradually change due to friction, and a perturbation will follow. The answer is provided under the assumption that friction has a minimal impact (Eckart & Ferris, 1956; Garvine, 1974; Kowalik & Murty, 1993). To this consistent oceanic, the interface, h, needs to be initially balanced by a zonal velocity of the form in order to persist:

$$u_0 = -\left[g\left(\frac{\rho_1 - \rho_2}{\rho_1}\right)\right]\left[f^{-1}\right]\frac{\partial h_0}{\partial y} \tag{6.11}$$

The hyperbolic profile of the upper (first) layer's height, h, is situated in the middle of the basin's north-south extent, W, in the mathematical expression:

$$h = h - \Delta h \tanh((y - 0.5W)L^{-1}) \tag{6.12}$$

In this profile, the upper layer's height decreases with increasing depth, following a hyperbolic curve. As z approaches infinity, H(z) asymptotically approaches H_0, indicating that the upper layer becomes infinitely deep. The hyperbolic profile is often used as an approximation for the upper layer's height in oceanographic models, as it provides a smooth and continuous representation of the vertical structure. However, it's important to note that the specific choice of the hyperbolic profile may vary depending on the specific model or application. The characteristic thickness H_0 can be adjusted based on the specific characteristics of the upper layer, such as the density stratification and the dynamics of the region being modeled (Palmen & Newton, 1969; Pedlosky, 1979; Pond, 1978).

It is important to note that these simplified equations of motion neglect vertical motion, assume steady-state conditions, and make certain approximations. They provide a simplified representation of the dynamics in the upper layer of the ocean and are often used in oceanographic modeling studies. However, for more comprehensive analyses or specific research purposes, more detailed and complex equations, such as the full Navier – Stokes equations, may be used.

6.7 Viscous perturbation solution

To find a mathematical solution for the viscous perturbation in the upper layer of the ocean, we need to consider the Navier – Stokes equations along with appropriate boundary conditions. However, finding an analytical solution for the full Navier – Stokes equations is generally difficult and often requires simplifying assumptions or numerical methods. Here, we will outline an approach using the linearized Navier – Stokes equations for small perturbations and assuming a simple velocity profile. In this understanding, linearized Navier – Stokes equations for small perturbations in the upper layer can be written as:

$$\frac{\partial u'}{\partial t} = -(1/\rho)\frac{\partial p'}{\partial x} + \nu\left(\frac{\partial^2 u'}{\partial x^2} + \frac{\partial^2 u'}{\partial y^2}\right) \tag{6.13}$$

$$\frac{\partial v'}{\partial t} = -(1/\rho)\frac{\partial p'}{\partial y} + \nu\left(\frac{\partial^2 v'}{\partial x^2} + \frac{\partial^2 v'}{\partial y^2}\right) \tag{6.14}$$

where:

- u' and v' are the perturbation velocities in the x and y directions, respectively.
- t is time.
- p' is the perturbation pressure.
- ρ is the density of the fluid.
- ν is the kinematic viscosity of the fluid.

When considering the linearized Navier – Stokes equations for small perturbations in the upper layer of the ocean, appropriate boundary conditions need to be applied to solve the equations (Eckart & Ferris, 1956; McLellan, 1965; Pond, 1978). The specific boundary conditions depend on the problem being considered and the geometry of the upper layer. Here are some common boundary conditions used in this context:

No-Slip Condition at the Ocean Surface and Bottom: This condition assumes that the velocity of the fluid at the ocean surface and bottom is zero. It implies that the fluid adheres to these boundaries, and there is no relative motion between the fluid and the boundary. Mathematically, this condition can be expressed as:

$$u'(x, y, z = 0) = 0 \tag{6.15}$$

$$v'(x, y, z = 0) = 0 \tag{6.16}$$

$$u'(x, y, z = h) = 0 \tag{6.17}$$

$$v'(x, y, z = h) = 0 \tag{6.18}$$

where (x, y) represents the horizontal coordinates, z represents the vertical coordinate, and h represents the thickness of the upper layer (Eckart & Ferris, 1956; McLellan, 1965; Pond, 1978). Impermeability condition, therefore, states that there is no flow across the lateral boundaries of the upper layer. Mathematically, it can be expressed as:

$$\frac{\partial u'}{\partial x} = 0 \tag{6.19}$$

$$\frac{\partial v'}{\partial x} = 0 \tag{6.20}$$

This condition ensures that there is no net flow across the lateral boundaries and that the mass is conserved within the upper layer. In this sense, the continuity equation, which represents the conservation of mass, requires that the velocity components satisfy:

$$\frac{\partial u'}{\partial x} + \frac{\partial v'}{\partial y} + \frac{\partial w'}{\partial z} = 0 \tag{6.21}$$

This condition ensures that mass is conserved within the upper layer and that the velocity components are consistent with each other. These boundary conditions, combined with the linearized Navier – Stokes equations and appropriate initial conditions, form a system of partial differential equations that can be solved to obtain the velocity and pressure fields for the small perturbations in the upper layer (Roden, 1972, 1976; Simpson et al., 1978).

Let us assume that M (=v′h) is the mass transport in the upper layer. As a consequence, the meridional mass transport, M, is set to zero at both the northern and southern boundaries due to the boundary conditions. It's interesting to note that M exists outside of time (Busalacchi & O'Brien, 1980; Camerlengo & Demmler, 1997; Camerlengo, 1982). Therefore the mass transport for Eq. (6.12) can be turned into a new expression:

$$M(y) = -(A\Delta hL^{-1})\text{sech}^2((y - 0.5W)L^{-1}) \tag{6.22}$$

A perturbation results from the integration of the continuity equation with respect to time. Expression of the form for the interface, h', is given by:

$$h' = \frac{\partial M}{\partial y}t \tag{6.23}$$

Taking into account the meridional mass's y-maximum derivative's absolute value. the timescale of dissipation, T, is defined in the context of transport in such a way that:

$$T = h'\left|\frac{\partial M}{\partial y}\right|_{\max}^{-1} \tag{6.24}$$

Eq. (6.26) reveals that $\left|\frac{\partial}{\partial y}\right|_{\max}$ represents the maximum absolute of such derivative. This solution, nevertheless, is restricted to oceanic fronts of the order of 1 kilometer. As a result, the mathematical model represents mass transport in the front upper boundary layer over a distance of 50 km is given by:

$$\frac{\partial^2 M}{\partial y^2} - \left(\frac{\partial^2 h_0}{\partial y^2}\right)\left(\frac{M}{h_0}\right) = -A\frac{\partial^3 h_0}{\partial y^3} \tag{6.25}$$

Thus the particular solution for M(y) can be mathematically expressed as:

$$\mathfrak{X}_y M_1 + \xi_y M_2 = 0 \tag{6.26}$$

$$\mathfrak{X}_y M_{1_y} + \xi_y M_{2_y} = \psi(y) \tag{6.27}$$

where $\mathfrak{X}_y, M_{1_y}, \xi_y,$ and M_{2_y} represent the meridional derivatives of the functions $\mathfrak{X}$. In addition, ξ, M1, M2, and respectively, and $\psi(y)$ represent the forcing functions defined by:

$$\psi(y) = -\left(2A\Delta hL^{-3}\right)\left\{\left[2\sinh^2\frac{((y-0.5W)}{L} - 1\right]\left(\cosh^4\frac{((y-0.5W)}{L}\right)^{-1}\right\} \tag{6.28}$$

In this understanding, the system of Eq. (6.27) gives us a solution for the functions $\mathfrak{X}$ and ξ, which can be given by:

$$\mathfrak{X}_y = -M_2(y)\psi(y)[W(M_1, M_2]^{-1} \tag{6.29}$$

here $[W(M_1, M_2] = 1$.

Therefore at each grid point, the meridional mass transport, M, can be given by:

$$M(y) = (C_1 + \mathfrak{X})M_1(y) + (C_2 + \xi)M_2(y) \tag{6.30}$$

If oceanic fronts occur at various horizontal space scales and the value of L (representing the distance) can vary between 1 and 50 km, we can modify the expression for mass transport accordingly. Instead of integrating over a fixed distance of 50 km, we can express the integral over the variable distance L (Camerlengo & Demmler, 1997; Dijkstra, 2008).

The modified expression for mass transport in the upper boundary layer becomes:

$$\int(\partial(\rho u)/\partial x)dx + \int(\partial(\rho v)/\partial y)dx + \int(\partial(\rho w)/\partial z)dx = 0 \tag{6.31}$$

Integration is performed over the variable distance L, which can vary between 1 and 50 km. This formulation allows for the flexibility to account for different spatial scales of oceanic fronts. The specific value of L can be determined based on the size and characteristics of the front being studied. By allowing L to vary within the specified range, the mathematical expression for mass transport can adapt to different spatial scales observed in oceanic fronts (Adcroft & Hallberg, 2006; Bowman, 2018; Brink, 2023).

6.8 How does Ekman number impact oceanic front?

A significant question rises now: How far do the oceanic fronts rely on Ekman Number? The Ekman number (Ek) is a dimensionless parameter that relates the viscous forces to the Coriolis forces in a fluid flow. It characterizes the relative importance of these forces and is defined as the ratio of the vertical viscosity (ν) to the square of the Coriolis parameter (f):

$$Ek = \nu(f^{-2}) \tag{6.32}$$

where:

- Ek is the Ekman number.
- ν is the kinematic viscosity of the fluid, representing the measure of internal friction and resistance to flow.
- f is the Coriolis parameter, which is proportional to the rotation rate of the Earth and varies with latitude.
- The Ekman number helps determine the scale at which viscous effects become significant compared with the Coriolis forces in a fluid flow. A higher Ekman number indicates a stronger influence of viscous forces relative to the Coriolis forces. In the context of oceanic fronts, the Ekman number influences the balance between the pressure gradient and Coriolis forces, ultimately affecting the front's structure and characteristics. As the Ekman number increases, indicating a stronger influence of viscous forces, the frontal width of the oceanic front tends to decrease. This is because higher Ekman numbers lead to increased dissipation of the frontal gradients due to enhanced viscous effects, causing the front to become narrower. on the other hand, the Ekman number can also affect the intensity of the oceanic front. A higher Ekman number tends to weaken the frontal intensity by enhancing the dissipation of the front. The front becomes less pronounced as viscous effects dominate over the pressure gradient and Coriolis forces. Evidently, the Ekman number influences the shear across the front, which refers to the change in velocity or properties (e.g., temperature, salinity) across the front. Higher Ekman numbers tend to smooth out the frontal shear, reducing the sharp gradients across the front (Camerlengo & Demmler, 1997; Camerlengo, 1982; Dijkstra, 2008).

Thus the momentum equations, considering the quasisteady state and the Ekman number, can be written as:

$$\rho(u\partial u/\partial x + v\partial u/\partial y) = -\partial P/\partial x + \nu(\partial^2 u/\partial x^2 + \partial^2 u/\partial y^2) + Ek(f\partial v/\partial z) \tag{6.33}$$

$$\rho(u\partial v/\partial x + v\partial v/\partial y) = -\partial P/\partial y + \nu(\partial^2 v/\partial x^2 + \partial^2 v/\partial y^2) - Ek(f\partial u/\partial z) \tag{6.34}$$

where:

- ρ is the density of the fluid.
- u and v are the velocity components in the x and y directions, respectively.
- P is the pressure.
- ν is the kinematic viscosity of the fluid.
- f is the Coriolis parameter.
- Ek is the Ekman number.

- $\partial/\partial x$ and $\partial/\partial y$ represent partial derivatives with respect to the horizontal coordinates x and y.
- $\partial/\partial z$ represents the partial derivative with respect to the vertical coordinate z.

These equations, combined with appropriate boundary conditions, provide the mathematical framework to study the structure of an oceanic front in a quasisteady state, considering the influence of the Ekman number (Eckart & Ferris, 1956; Garvine, 1974; Kao, 1980; Kowalik & Murty, 1993). The specific form and complexity of the equations may vary depending on the assumptions and conditions specific to the problem being studied.

Solving these equations often requires numerical methods, such as finite difference or finite element methods, due to their nonlinearity and complex nature. Numerical models and simulations can help investigate the behavior of oceanic fronts and understand the interplay between the Ekman number, pressure gradients, Coriolis forces, and other relevant factors influencing the front's structure in a quasisteady state.

6.9 Quantized Marghany oceanic front

In the context of oceanic fronts, the term "quantized" perhaps refers to the presence of discrete or distinct features within the front. These features could include localized regions of sharp gradients, vortices, or mesoscale eddies that give the front a "quantized" appearance. In this understanding, oceanic fronts can exhibit complex spatial structures due to interactions between different water masses, varying densities, and differential flows. These structures can result in the formation of eddies, filaments, or meanders within the front, leading to a more fragmented or discrete appearance.

In an early study of Marghany (2021), he invented novel speculation for addressing the SAR imaging mechanism of the oceanic front. He speculates that the two different sorts of quantized water masses $\Psi|\sigma_1> \neq \Psi|\sigma_2>$ but $\Psi|\rho_1|\sigma_1>$ *and* $\Psi|\rho_2|\sigma_2> \in \Psi|\overline{\rho}|\sigma>$ would form a bounded state; which is named after the author "quantized Marghany bounded state"(Fig. 6.14). In remarkable fact,

$$\Psi|\overline{\rho}|\overline{\sigma}> = \frac{\Psi|\rho_1|\sigma_1 \pm \rho_2|\sigma_1>}{\sqrt{2}} \tag{6.35}$$

Eq. (6.35) reveals that the quantized Marghany oceanic front boundary is just the mean of both two different states that are entanglement to the formed boundary quantum state $\Psi\overline{|\rho|\sigma}>$. Therefore the + sign involves in water masses calculation while the − sign is for backscatter mathematical procedures in Eq. (6.35). The entanglement boundary is not created by the two quantum states spinning in opposite directions to one another; rather, it is the average of their nonequivalent to one another. Entanglement

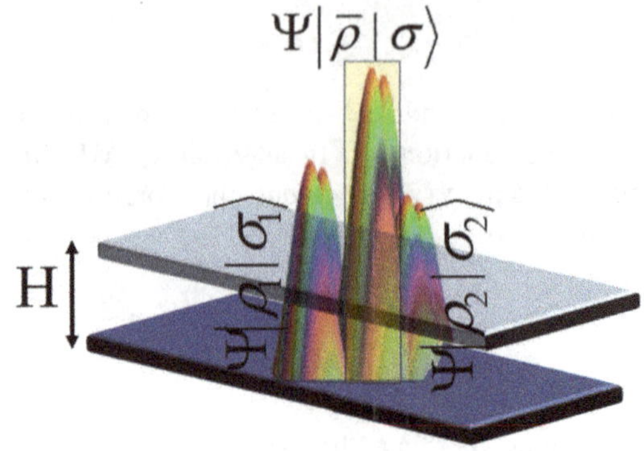

FIGURE 6.14 Quantized Marghany bounded state concept.

oceanic front boundary, in other words, is a vital phenomenon where the quantum states of two or more particles become correlated to the average of their quantum states in such a way that the individual quantum states cannot be described independently. Instead, it is necessary to describe the combined quantum states of the two distinct density states as the quantized Marghany Oceanic front, which bears the author's name (Marghany, 2021).

In accordance with additional explanations, when two distinct quantum water mass states entangle, their properties are entangled in an awkward manner. Therefore notwithstanding their spatial separation, measurements taken on one water particle will immediately correlate with those taken on the other. Even when the water-density particles are physically separated by large distances, this causation Marghany boundary mostly persists. This phenomenon, known as quantum entanglement, has puzzled scientists for decades. It challenges our traditional understanding of cause and effect, as the correlation between the entangled water particles seems to occur instantaneously, defying the limitations of space and time. The persistence of this correlation across large distances suggests that there may be hidden connections or underlying principles at play in the quantum realm that we have yet to fully comprehend (Baym, 2018; Marghany, 2021).

Quantization in oceanic fronts can arise from various physical mechanisms, including (Baym, 2018; Brink, 2023; Marghany, 2021):

1. Instabilities: Instabilities, such as baroclinic instability or shear instability, can occur within oceanic fronts. These instabilities lead to the formation of small-scale vortices, filaments, or eddies, creating a quantized or fragmented structure.
2. Mixing Processes: Oceanic fronts are often regions of enhanced mixing, where water masses with different properties interact. The mixing processes can lead to the formation of smaller-scale structures within the front, contributing to its quantized appearance.
3. Topographic Effects: Oceanic fronts interacting with topographic features, such as seamounts or ridges, can

exhibit quantized behavior. These interactions can generate eddies or induce flow instabilities, resulting in localized regions of sharp gradients and distinct features within the front.

Needless to say, the quantized Marghany turbulence boundary and quantized Marghany oceanic front can be identified as bound states. In some cases, the interaction between quantum states of different water masses can lead to the formation of bound states (Galindo & Pascual, 2012; Konishi & Paffuti, 2009; Mandl, 2013; Marghany, 2021), where the states are localized or confined within a specific region. In quantum mechanics, therefore oceanic bound states refer to the states of a quantum system of different water masses that are confined within a potential well or barrier. These water density states are characterized by having discrete energy levels, meaning the energy of the oceanic front is quantized, which is invented and named after Marghany (2021). The equation that describes the bound states depends on the specific potential in which the oceanic front is confined. Therefore let b_M be the Marghany bounded state that separates between the quantum water mass states in the circumstance of $|b_M|\sigma > \in \Psi|\rho_1|\sigma_1>$ and $|b_M|\sigma > \in \Psi|\rho_1|\sigma_2>$. On this understanding, the quantum mechanics of the different duality water mass states can be mathematically expressed as:

$$\Psi|\rho|b_M> = \begin{pmatrix} \Psi|\rho_1 > \in \Psi|b_M > \\ \Psi|\rho_2 > \in \Psi|b_M > \end{pmatrix} = \begin{pmatrix} \Psi|\rho_1 > \in \Psi|b_M > \left(\Psi|u > \dfrac{\partial \Psi|\rho_1 >}{\partial x} \right) \\ \Psi|\rho_2 > \in \Psi|b_M > \left(\Psi|v > \dfrac{\partial \Psi|\rho_2 >}{\partial y} \right) \end{pmatrix} \tag{6.36}$$

On the other hand, the front $|b_M>$ is mathematically expressed as:

$$|b_m > : = \{\Psi|b_M > \in \Psi|\overline{\rho}|\sigma > \} \tag{6.37}$$

Therefore the modification of the Schrödinger equation, which can describe the behavior of quantized Marghany oceanic front (Fig. 6.15) and is written as:

$$\hat{\mathrm{H}}\Psi|\overline{\rho}|b_M > = \left(-\hbar^2/2\Psi|\overline{\rho}|b_M >\right)\frac{\partial^2\Psi|\overline{\rho}|b_M >}{\partial x_2} + M_{12}\Psi|\overline{\rho}|b_M \tag{6.38}$$

where:

$\hbar$ is the reduced Planck's constant.

$\partial^2/\partial x^2$ is the second derivative with respect to position.

M_{12} is mass transport that is given by Eq. (6.30). Eq. (6.38) reveals the quantum energy spectra along the Marghany bounded state b_M for the quantized Marghany oceanic front system. These energy spectra can be used in the next section to understand the SAR imaging mechanism of the quantized Marghany oceanic front. The wave

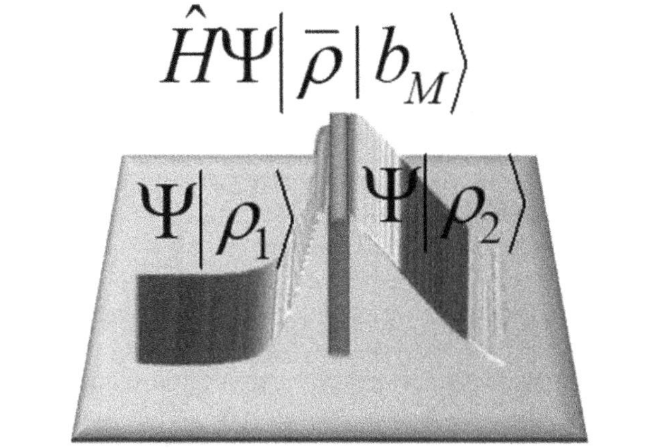

FIGURE 6.15 Schrödinger expression of the behavior of quantized Marghany oceanic front.

function $\Psi|\overline{\rho}|b_m>$ associated with each energy eigenvalue represents the probability distribution of finding the Marghany bounded state b_M in a particular state or position. Both Eqs. (6.37) and (6.38) reveal that there are two distinct quantum states in which the front can occur in the interest of preserving the presence of the separated boundary, spinning in various directions. To the above b_M persists as a participant of both quantum states because they do not attain steady-state conditions (Baym, 2018; Marghany, 2021). Consequently, the spinning of the Earth around its axis significantly influences the dynamics of larger fronts, whereas small-scale fronts are believed to be influenced by nonlinear inertial and frictional effects alone (Zettili & Zahed, 2003).

According to the aforementioned perspective, an ocean system including the oceanic front can be specified by its wave function. This viewpoint regards the ocean as a quantum system that is continuous, the wave function may be written as $\Psi(r, t)$, and the square of the absolute value of this wave function $\Psi|r, t|^2$ provides the probability of locating the water parcel at point r at time t. In this scenario, the wave function represents the different possible states that the spin of the water parcel can take. These states are determined by the combination of density changes, wind stress, and Earth's rotation acting on the parcel. In this understanding, the wave function can be presented as a ket vector: $|\Psi\rangle = \alpha|-\rangle + \beta|+\rangle$. This demonstrates the superposition principle states that a quantum system can exist in a linear combination of multiple states. In this view, $|-\rangle$ signifies spin $z-$, that is, low water density while $|+\rangle$ signifies spin $z+$, that is, high water density. When the spin occurs, we obtain spin down with probability α^2 and spin up with probability β^2. In this regard, $|\Psi|$ is a superposition of the so-called basis states $|+|$and $|-|$. In the analogy of quantum states and wave functions, one can consider the water masses on either side of the front as distinct "states" or "phases," and the front itself can be viewed as a "wave function" representing the probability distribution of the transition zone (Konishi & Paffuti, 2009; Mandl, 2013; Neumann, 2018). The properties of the front, such as its position, extent, and intensity, can be influenced by changes in the physical characteristics of the water masses, such as temperature, salinity, and density.

However, recent advancements in quantum hydrodynamics have allowed scientists to explore the potential connections between quantum principles and macroscopic phenomena such as oceanic fronts. By applying concepts such as wave-particle duality and superposition to the study of fluid dynamics, researchers are gaining new insights into the complex interplay between quantum effects and large-scale phenomena in nature. These novel findings of quantized Marghany oceanic front could potentially revolutionize our understanding of oceanic fronts and open up new avenues for studying other intricate systems in the natural world.

6.10 Theoretical quantized Marghany oceanic front imaging in SAR data

The remaining query is how quantum mechanics can be applied to the quantized Marghany Oceanic front in SAR data? One possible approach to address the quantized Marghany Oceanic front in SAR data using quantum mechanics is by utilizing quantum algorithms and techniques. These algorithms can be designed to analyze the complex patterns and dynamics of the front, taking into account the principles of quantum superposition and entanglement. By harnessing the computational power of quantum systems, it may be possible to uncover new insights and correlations within the SAR data, leading to a deeper understanding of the quantized Marghany oceanic front. Unbelievably, quantum mechanics is rarely used to directly describe large-scale phenomena such as oceanic fronts; rather, it is frequently used to describe the behavior of particles at the atomic and subatomic scales. However, recent advancements in quantum computing have opened up possibilities for applying quantum principles to larger-scale phenomena. This novel approach could potentially revolutionize our understanding of complex systems such as the Marghany oceanic front and pave the way for new discoveries in oceanography. To conceptualize the distinctive traits and behavior of water masses across a front in SAR data, the quantized Marghany oceanic front analogy can be useful. By utilizing quantum principles in the study of oceanic fronts, researchers can gain insights into the intricate dynamics and interactions between water masses. This could lead to a deeper comprehension of the Marghany oceanic front and its impact on global ocean circulation patterns, climate change, and marine ecosystems. Additionally, applying quantum computing to SAR data analysis may enable more accurate predictions and modeling of oceanic fronts, enhancing our ability to monitor and manage these critical features of our planet's oceans.

6.10.1 Classical SAR mechanism in imaging oceanic fronts

Basically, the front feature appears brighter than the surrounding sea environment in SAR images. Local convergence of surface divergence creates current flow, which tends to modulate the energy of surface wind waves. Consequently, arrays of steeper wave zones along with the convergences and lower amplitude waves at divergences zones are generated. In this context, SAR detects heterogeneities of the surface roughness as the strongest variations of backscatter, which can be identified clearly in radar backscatter cross section. In this understanding, the surface features that outlook has the greatest strappingly in the SAR image, which is confined fronts where convergence or divergence is robust, steering to buoyant or obscure shapes. In this predicament, radar backscatter cross section can clearly show that surface roughness heterogeneities are the strongest variations of backscatter. In this interpretation, the surface features that are most constricted in the SAR image have the greatest prospective fronts that are strongly convergent or divergent, steering to buoyant or hazy shapes (Alpers & Holt, 1995; Marghany, 2021).

Therefore the variation in radar backscatter cross section can be attributed to the gradient of the current boundary, such as the boundary associated with a thermal front in the ocean. The current boundary gradient refers to the change in properties (e.g., temperature, salinity, density) across the front, which in turn affects the radar backscatter. Consequently, the significant query is how the current boundary gradient influences the radar backscatter? The radar backscatter depends on the dielectric properties of the medium through which the radar waves propagate. In the case of thermal fronts, there are often temperature gradients across the boundary. These temperature differences can result in variations in the dielectric properties of the water masses on either side of the front, affecting the backscatter. Consequently, the current boundary gradient can influence the surface roughness of the ocean at the front. Different water masses with distinct physical characteristics can exhibit varying surface roughness. Radar waves interacting with the rough surface can experience different scattering patterns, leading to variations in the radar backscatter (Krogstad & Schyberg, 1990; Marghany, 2021; Pan et al., 2007).

The presence of a current boundary gradient can cause wave − current interactions. Waves interacting with the current gradient may experience refraction, reflection, or scattering effects. These interactions influence the radar backscatter and can provide information about the properties and dynamics of the front. Finally, the Doppler effect can also contribute to variations in the radar backscatter. As the radar waves encounter the current boundary gradient, they can experience frequency shifts due to the relative motion between the radar system and the moving water masses. These frequency shifts can affect the radar backscatter and provide additional information about the current flow associated with the front.

It is worth standing that the hydrodynamic modulation for SAR imaging of the front and eddies is not only reliant on surface convergence or divergence velocity but is also a function of the wind magnitude. In other words, variants in wind speed at the ocean surface cause local surface roughness and, in turn, may be distinguished utilizing imaging SAR. The dissimilarities in wind speed are perhaps allied with wave phenomena in the atmospheric boundary layer. In this regard, gravity waves are internal waves for which gravity is the reinstating power. They are perhaps allied with temperature inversions; these instigate steadily well-stratified layers in the lower troposphere, which work as waveguides (Alpers & Holt, 1995; Kim et al., 2017; Marghany, 2021; Vogelzang et al., 1997).

It is essential to remain cognizant that the hydrodynamic modulation for SAR imaging of the front and eddies varies depending on the wind speed in response to surface convergence or divergence velocity. In other words, variations in surface roughness at the ocean's surface are triggered by shifts in wind speed, and these variations can be distinguished using imaging SAR. Perhaps there is a connection between the variations in wind speed and wave phenomena in the atmospheric boundary layer. In this sense, gravity acts as a reinstating force for internal waves, which are called gravitational waves. They may be associated with temperature inversions, which trigger continuously well-stratified layers in the lower troposphere that serve as waveguides.

Similarly, they could be combined with the wind shear above and will frequently continue to perpetuate in the commonly erratic direction of the wind shear. Orographic interfaces, frontal disturbances, and convective instability, to mark a few, are nevertheless potential mechanisms.

In this profound regard, adequate strong wind stress must not cause spatially homogeneous sea roughness, which saturates the SAR backscatter and typically prevents imaging of sea surface features such as fronts. The relative orientations of the convergence field, the prevailing wind, and the azimuth resolutely facing the SAR's direction influence the SAR front imaging mechanisms. In this specific circumstance, it ordinarily requires the finest obvious imaging of an arched front. The front may swap its character from bright to dark along its considerable length as a key function of considerable fluctuation of physical properties it requires being finest the obvious imaging of an arched front.

However, because the temperature fronts are not very strong, the wind variations brought on by changes in the stratification of the boundary layer do not significantly alter the SAR image signature. Furthermore, it is not known that salinity gradients cause backscatter anomalies. The short gravity-wave/current interaction along the current fronts is therefore understood to be indicated by the SAR image signatures. When the wind is calm, the short gravity waves can change the steepness and transmission routes because they are transmitted into and/or across the current front. Let us assume that σ_0 is the undisturbed normalized radar cross section along the quantized Marghany bounded state $\Psi|b_M>$, which has horizontal radial velocity u_r along the radar look direction x_r.

Accordingly, the SAR is able to distinguish the current fronts. Based on the normalized radar cross section (NRCS), the investigation of the front fluctuations in SAR images can be mathematically expressed as follows:

$$\Delta\sigma = \sigma_0\left[\frac{\sigma_m - \sigma_0}{\sigma_0}\right] = -C\frac{\partial u_r}{\partial x_r} \tag{6.39}$$

here C is a complex function relying, among other parameters, on radar wavelength, radar incidence angle, and wind speed and direction. Additionally, $\sigma_m is$ the maximum modulation of the NRCS within the frontal area. In this understanding, the normalized radar cross section (NRCS) modulation equation describes the variation in radar backscatter strength, specifically the NRCS, due to different surface conditions or targets. The NRCS measures the radar energy reflected back to the radar system normalized by factors such as incident power, polarization, and antenna gain. In this perspective, the modulation factor can vary depending on the specific surface or target characteristics being considered. It can be influenced by factors such as surface roughness, dielectric properties, surface orientation, and scattering mechanisms. It is important to note that the exact form of the modulation factor equation can differ depending on the specific application, radar system, and scattering model used. Different models, such as the small slope approximation (SSA) model or the integral equation model, may be used to describe the scattering processes and derive the modulation factor (Alpers & Holt, 1995; Vogelzang et al., 1997).

Indeed, the detection of fronts in SAR images is relatively straightforward due to their characteristic appearance as distinct lines that separate regions of different mean image intensities. Fronts often exhibit higher or lower backscatter compared with the surrounding ambient regions. However, the interpretation of these signatures can be challenging due to the complex hydrodynamic processes occurring at ocean fronts. Several factors can influence the backscattered radar signal and contribute to the observed variations in the SAR image: (1) hydrodynamic wave – current interaction; (2) wave breaking; (3) wave damping by accumulated surface films; and (4) surface current and surface roughness (Johannessen et al., 1991; Kim et al., 2017; Kuang et al., 2012; Marghany, 2021).

At oceanic fronts, the convergence or divergence of water masses can generate significant wave – current interactions. These interactions can lead to changes in wave height, direction, and frequency, affecting the radar backscatter (Johannessen et al., 1991). The interaction between waves and currents can cause scattering, refraction, and focusing effects, resulting in variations in the radar signal (Vogelzang et al., 1997). Therefore a mathematical equation that represents the perturbation of the surface current due to short wave interaction across the front is given by:

$$\Delta\sigma = \sigma_0\left[\frac{S_0(k,\phi)f(k,\phi) + S_0(k,\phi+\pi)f(k,\phi+\pi)}{S(k,\phi) + S(k,\phi+\pi)}\right] = -C\frac{\partial u_r}{\partial x_r} \tag{6.40}$$

Eq. (6.40) consists of two terms that are subtracted from each other. The first term is the relative difference between the backscatter coefficient of the surface current front (σ_m) and the background backscatter coefficient (σ_0). This term represents the contrast between the front and the surrounding water. In this view, the second term is a function of the wave number (k) and the latitude (ϕ) of the front. It represents the effect of the surface current shear and the convergence/divergence zones on the backscatter coefficient. Therefore $S_0(k,\phi)$ is the power spectral density of the surface waves in the absence of the front, and $S(k,\phi+\pi)$ is the power spectral density of the surface waves in the presence of the front. These terms represent the effect of the front on the surface waves. Additionally, $f(k,\phi)$ is the directional wave spectrum of the surface waves. It represents the distribution of wave energy as a function of wave direction and frequency. Evidently, Eq. (6.40) demonstrates that the backscatter coefficient of the surface current front is affected by both the contrast between the front and the surrounding water and the effect of the front on the surface waves (Alpers & Holt, 1995; Kuang et al., 2012; Marghany, 2012a, 2012b; Marghany, 2021).

Breaking waves, therefore, at the front can produce additional backscatter due to the turbulence and foam generated. The breaking waves can enhance the surface roughness, leading to increased backscatter intensity in the SAR image. Consequently, oceanic fronts can be associated with the accumulation of surface films, such as oil slicks or organic matter (Vogelzang et al., 1997). These films can dampen or suppress the surface waves, leading to decreased backscatter intensity in the SAR image. The surface currents, subsequently, associated with ocean fronts can affect the surface roughness, leading to variations in backscatter. Surface roughness variations influence the scattering properties of the ocean surface and contribute to the observed backscatter intensity.

$$\Delta\sigma = \sigma_0\left[\frac{\sqrt{\left(\frac{du}{dx}\right)^2 + \left(\frac{dv}{dx}\right)^2}}{\lambda_r}\right]\cos(\phi-\varphi)g(\phi-\phi_w) \tag{6.41}$$

The second part of the equation represents the contribution of various physical parameters to the NRCS difference. The parameters $\left(\frac{du}{dx}\right)$ and $\left(\frac{dv}{dx}\right)$ represent the along-front velocity gradients in the zonal and meridional directions, respectively. In this context, λ_r is the radar wavelength; ϕ is the incidence angle of the radar beam; φ is the azimuth angle of the radar beam; and $g(\phi,\phi_w)$ is a function that depends on the incidence angle and the wind direction ϕ_w relative to the radar beam. Needless to say, Eq. (6.41) shows that the NRCS difference between the front and background areas depends on the velocity gradients across the front, the radar wavelength, the incidence and azimuth angles of the radar beam, and the wind direction relative to the radar beam (Alpers & Holt, 1995; Vogelzang et al., 1997).

6.10.2 Quantized Marghany theory for SAR imaging oceanic front

According to the above perspective, the imagined mechanism of the quantized Marghany bounded state $\Psi|b_M>$ in SAR data is assumed to be described by the total Hamiltonian. To describe the total Hamiltonian of two different water masses separated by the quantized Marghany bounded state $\Psi|b_M>$, we can consider a composite system composed of two subquantum states and the interaction between them and radar signal. In this view, the different water masses in SAR images are assumed as different quantum states of normalized radar cross section NRCS. The main valid hypothesis $\Psi(\sigma_m|b_M>)$ is larger than the total backscatter of $\Psi|\sigma_1> + \Psi|\sigma_2>$. The total Hamiltonian can be written as the sum of the individual Hamiltonians of the subsystems and the interaction Hamiltonian (Galindo & Pascual, 2012; Marghany, 2021).

$$\sum \mathscr{H}(b_M) = \frac{1}{2}[\mathscr{H}(\sigma_m) + \mathscr{H}(\sigma_0)] + \mathscr{H}(\lambda_r) \tag{6.42}$$

The significant term in Eq. (6.42) is $\mathscr{H}(\lambda_r)$, which describes the interaction between different quantum states in SAR images with each other and images in SAR signal interaction with them in radar signal wavelength $\mathscr{H}(\lambda_r)$. The interaction Hamiltonian $\mathscr{H}(\lambda_r)_i$ describes the interaction between the subsystems and can be chosen based on the nature of the interaction. It can involve terms related to forces, potentials, or coupling between the subsystems as demonstrated in Eqs. (6.39–6.41). The specific form of the interaction Hamiltonian will depend on the physical properties and interaction mechanisms between the two systems and radar signals (Konishi & Paffuti, 2009; Mandl, 2013).

In this understanding, the quantized Marghany bounded state $\Psi|b_M>$ in SAR data refers to discrete energy states or confined states within each subsystem. These states would be included in the respective Hamiltonians for each subsystem of normalized radar cross section $[\mathscr{H}(\sigma_m)\text{and}\mathscr{H}(\sigma_0)]$ as appropriate. In quantum mechanics, the two density modes undergo a squeezing effect, which entangles the water mass densities and provides sufficient strength to modulate the Bragg wave and induce a strong radar backscatter cross section along with the distance d. To draw the frame mathematical expression of the quantum entanglement front that is imaging in SAR images, the Hamiltonian Bragg wave should be comparable with the energy $\hbar\omega$ of each oscillator mass densities with the normalized radar cross section backscatter that $g\partial b_M \Delta\sigma[L]^{-3} \sim \sum \mathscr{H}(\sigma_m|, b_M>) \sim \hbar\omega$. In this understanding, L rises from the two endpoints of any point on the line between water types, is inversely proportional to the ratio mixing, and can be calculated by:

$$\rho_1 \rho_2^{-1} = L_1 L_2^1 \tag{6.43}$$

In this regard, the dimensionless entanglement displacement identified in SAR images η_E is expressed as:

$$\eta_E \equiv 2g\partial\sigma\left[\omega^{-2}L^{-3}(|\sigma|b_M>\right]^{-3} \tag{6.44}$$

Eq. (6.44) reveals that the gravity acceleration induces the entanglement displacement of the normalized radar cross section along with the two different water mass densities, as they are trapped by the harmonic potential boundary η_E as inversely proportional to the angular frequency ω (Marghany, 2021). Consequently, the maximum entanglement that occurs for the SAR imaging front can be given by:

$$\eta_{E_{\max}} \equiv \eta_E(\ln\mathscr{H}(\sigma_m|b_M>) \tag{6.45}$$

Estimating the time for maximum entanglement in SAR imaging of a front is a complex task as it depends on various factors, including the specific properties of the front, the radar system parameters, and the interaction dynamics between the front and the radar waves. It is important to note that entanglement is a concept from quantum mechanics, and its direct application to SAR imaging is not straightforward. However, an attempt is taken to estimate the maximum entanglement time for the SAR imaging quantized bounded state of the Marghany oceanic front as:

$$t_{\max} \equiv \frac{\pi}{(2\eta_E\omega)} \tag{6.46}$$

Therefore along the harmonic potential boundary, the result of gravity is the stronger attraction of two different water densities in a spatial superposition that is closer. This leads to the generation of position and momentum correlations, which then fuels the growth of quantum entanglement and gives rise to states in SAR data that are sharply bounded and are given the author's name, such as the quantized bounded state of the Marghany oceanic front.

6.11 Quantum edge detection algorithm for quantized Marghany bounded states in SAR data

6.11.1 Transferring SAR data into qubit

Let us assume that any SAR data are presented as signal information in the quantum state domain as $|I_{SAR}\rangle$. In this view, the quantized Marghany bounded state $|b_M\rangle \in |I_{SAR}\rangle$, that is, $|b_M\rangle \in |\sigma_{max}\rangle$. In other words, the state is a tensor product of n individual states $|\sigma_1\rangle$, $|\sigma_2\rangle,\ldots\ldots\ldots,|\sigma_n\rangle$. Therefore each individual state can be represented by a complex number α_i. In this approach, quantum SAR images are represented using quantum qubits, that is, n qubits. Each qubit represents a pixel and can hold a superposition of states, allowing for the simultaneous representation of multiple pixel values. Quantum operations can then be applied to manipulate and process the quantum pixel representation. On this understanding, the state of a $\Psi|I_{SAR}\rangle$ consisting of n qubits is defined as the tensor product state of n qubits (Cavalieri & Maio, 2020; Zhou et al., 2020). The tensor product state is a mathematical operation that combines the states of two or more quantum systems to form a composite SAR image (Fig. 6.16). In the case of n qubits, the tensor product state is the product of the individual states of each qubit in SAR data and defined as:

$$\Psi|I_{SAR}\rangle = \Psi|\sigma_1\rangle \otimes \Psi|\sigma_2\rangle \otimes \cdots \otimes \Psi|\sigma_n\rangle = \alpha_1\alpha_2\cdots$$
$$\alpha_n\Psi|00\ldots0\rangle + \alpha_1\alpha_2\cdots\alpha_{n-1}a_n\Psi|11\ldots1\rangle \tag{6.47}$$

For easy SAR data analysis, the state $\Psi|I_{SAR}\rangle$ can be written as a linear combination of two basis states Ψ $|00.0\rangle|00.0\rangle$ and $\Psi|11.1\rangle$. Consequently, the state Ψ $|I_{SAR}\rangle$ can also be written as a superposition of 2^n basis states $|i_b\rangle$, where i is an integer between 00 and 2^{n-1}, and b is a binary representation of i, which is given by:

$$|I_{SAR}\rangle = \sum_{i=0}^{2^{n-1}} w_i|i_b\rangle \tag{6.48}$$

6.11.2 Identifying qubit pixels in SAR data

A pixel qubit is a type of qubit that is used in quantum image processing. In classical image processing, an image is represented as a matrix of pixels, where each pixel is a

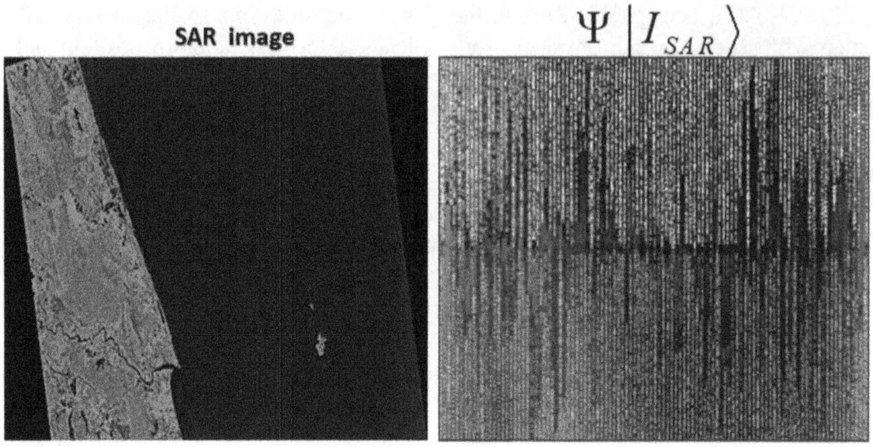

FIGURE 6.16 The initial stage of transferring the SAR image into a tensor product.

single value representing the intensity or color of that particular point in the image. In quantum image processing, therefore, a pixel qubit is used to represent each pixel in the image as a quantum state. A pixel qubit, in this perspective, can be thought of as a two-level quantum system, where the two levels represent the two possible states of the pixel (e.g., black or white). By manipulating the state of the pixel qubits in an image (Fig. 6.17), it is possible to perform quantum image processing operations such as edge detection and image compression (Fu et al., 2009; Yao et al., 2017). In this view, $\sigma(m,n)$ is considered as the intensity value of the pixel at location (m,n) in the SAR image. Thus the pixel qubits can be expressed as:

$$|\sigma(m,n)\rangle = \sqrt{1-\sigma(m,n)}|b_M|0\rangle + \sigma(m,n)|b_M|1\rangle \quad (6.49)$$

Eq. (6.49) reveals that the absence and existence of quantized Marghany bounded states are shown by $|b_M|0\rangle$ and $|b_M|1\rangle$, respectively. Additionally, the equation represents a superposition of the two quantum states, where the coefficients $\sqrt{1-\sigma(m,n)}$ and $\sqrt{\sigma(m,n)}$ determine the probability amplitudes of the states. In this regard, the coefficient $\sqrt{1-\sigma(m,n)}$ represents the probability amplitude of the $|b_M|0\rangle$ state, which is proportional to the square root of the complement of the intensity value. On the other hand, the coefficient $\sqrt{\sigma(m,n)}$ represents the probability amplitude of the $|b_M|1\rangle$ state(Fig. 6.18), which is proportional to the square root of the intensity value Table 6.1. Yet, because of the high NRCS in both states, that is, $|1>$, the wave function of the $\Psi|b_m|1>$ state is signed for both the bounded front state and land boundaries (Fu et al., 2009; Zhou et al., 2020). However, it is unable to distinguish between an alliance of bounded front states and other bounded states with a similar NRCS. In this particular circumstance, extended quantum image enhancement techniques are necessary even to identify various shorter bounded state fronts in SAR images.

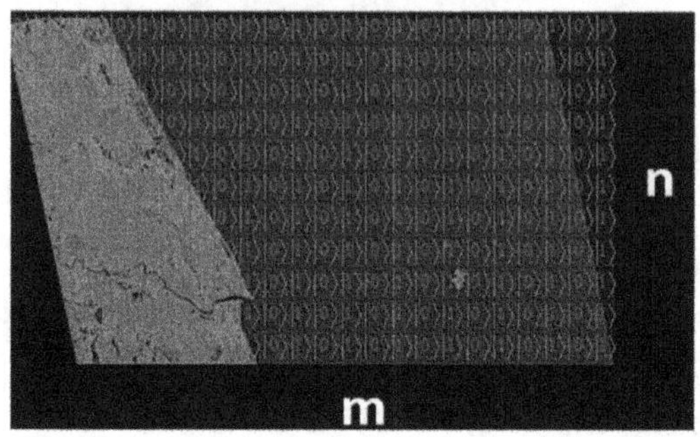

FIGURE 6.17 SAR pixel qubits image m X n SAR pixel qubits image m X n.

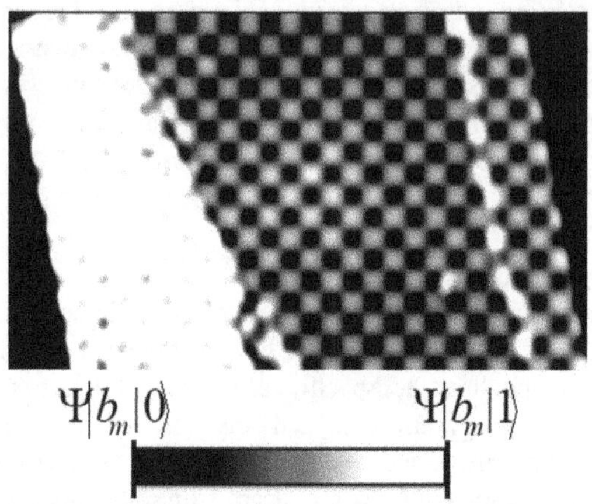

FIGURE 6.18 The wave function in determining $|b_m>$ in quantized SAR image.

6.11.3 Quantum SAR image enhancement algorithm

The ideal edge of an SAR image is expected to have a step change in gray level or backscatter and texture changes across the image, but in reality, edges are usually blurred

TABLE 6.1 Pseudo-code of quantum SAR image generation.

// **Input:** I_{SAR} data I_{SAR} data = load_image(image_path)
// Initialize quantum I_{SAR} data image representation quantum_image = initialize_quantum_image (I_{SAR} data)
function initialize_quantum_image (I_{SAR} data): // Determine quantum image size and resolution based on classical image dimensions
quantum_row location = I_{SAR} data.width(m) quantum_column location = I_{SAR} data.height(n)
Determine:
$|I_{SAR}\rangle = \sum_{i=0}^{2^{n-1}} w_i|i_b\rangle$ // Create quantum register to store qubits representing each pixel quantum_register = create_quantum_register
(quantum_m * quantum_n) // Convert classical pixel values to quantum states and store them in the quantum register for each pixel in classical_image: quantum_state = convert_to_quantum_state(pixel) add_quantum_state_to_register(quantum_state, quantum_register) //
Create quantum image object and assign the quantum register $|\sigma(m,n)\rangle = \sqrt{1-\sigma(m,n)|b_M|0\rangle + \sigma(m,n)|b_M|1\rangle}$
If $|\sigma(m,n)\rangle = \sqrt{\sigma(m,n)|b_M|1\rangle}$,or
$\sigma|b_M|1\rangle \geq \sigma|I_{SAR}\rangle$ then
$|\sigma(m,n)\rangle \ni |b_M|1\rangle$
If $|\sigma(m,n)\rangle \supset \sqrt{1-\sigma(m,n)|b_M|0\rangle}$ then
$|\sigma(m,n)\rangle \not\ni |b_M|1\rangle$ // Return the quantum image representation return quantum_image

and low-contrast, resembling a ramp profile owing to speckle variations in SAR images (Chapter 3). To detect edges effectively, image enhancement is required to increase contrast and reduce speckle effects. Eight ground states are considered, which are represented by binary numbers in the form of |000>, |001>, |010>, |011>, |100>, |101>, |110>, and |111>. In this perspective, a quantum signal processing (QSP) technique is used to calculate the probability amplitudes of each ground state in the three qubits systems (Yao et al., 2017). The QSP operator used in this technique sums up the probabilities of the six ground states of the three qubits subsystems in four different directions: horizontal, vertical, 45°, and 135°(Table 6.2).

This operator is designed to enhance the information on edges in SAR images while retaining the information on smoothing areas (Chetia et al., 2021). To this end, the SWAP gate is used for rearranging qubit states and implementing quantum sorting algorithms. The SWAP gate is represented by the following unitary matrix:

$$SWAP = \begin{vmatrix} 1 & 0 & 0 & 0 \\ 0 & 0 & 1 & 0 \\ 0 & 1 & 0 & 0 \\ 0 & 0 & 0 & 1 \end{vmatrix} \tag{6.50}$$

The SWAP gate acts on two qubits, and its effect is to interchange the states of the two qubits. For example, if qubit A is in state $|b_M|0\rangle$ and qubit B is in state $|b_M|1\rangle$, applying the SWAP gate (Table 6.3) results in qubit A being in state $|b_M|1\rangle$ and qubit B being in state $|b_M|0\rangle$(Fig. 6.19).

It is assumed that σ_1 is the current backscatter pixel, and an arbitrary pixel in its 3 × 3 neighborhood is also denoted as σ_1. Therefore the state vector is computed in the horizontal direction of the 3 × 3 neighborhood window. The state vector, consequently, is computed using the three qubits subsystem $|\sigma_8\sigma_1\sigma_4>$, where σ_8 and σ_4 are two other pixels in the 3 × 3 neighborhood window. In this sense, the probability of the corresponding ground state in the three qubits subsystem is denoted as $|W_i|^2$.

$$\begin{aligned} |\sigma_8\sigma_1\sigma_4> &= \sqrt{(1-P_8)(1-P_1)(1-P_4)}|000> \\ &+ \sqrt{(1-P_8)(1-P_1)(P_4)}|001> \\ &+ \sqrt{(1-P_8)(P_1)(1-P_4)}|010> \\ &+ \sqrt{(1-P_8)(P_1)(P_4)}|001> \\ &+ \sqrt{(P_8)(1-P_1)(P_4)}|100> \\ &+ \sqrt{(P_8)(1-P_1)(P_4)}|101> \\ &+ \sqrt{(P_8)(P_1)(1-P_4)}|110> \\ &+ \sqrt{(P_8)(P_1)(P_4)}|111> = \sum_{i=0}^{2^{2-1}} w_i|i> \end{aligned} \tag{6.51}$$

To enhance the information in the horizontal direction based on probability statistics in a three-qubit subsystem, a specific operator can be designed. The exact details and coefficients of this operator will depend on the specific probability statistics and enhancement goals, the general framework for constructing such an operator. In that scenario, if the probability statistics in the three-qubit subsystem refer to the likelihood of the qubits being in a certain state, an operator based on these probabilities can be designed to enhance the information in the horizontal direction. The specific coefficients of the operator will depend on the probabilities associated with each state. In this context, a general framework for constructing the operator, denoted as $OP|b_M>$:

TABLE 6.2 Quantum signal processing pseudo-code.

```
# Input: Quantum state represented by three qubits quantum_state = initialize_quantum_state()
# Apply QSP Technique with specified operator processed_quantum_state =
apply_qsp_technique(quantum_state)
# Function to initialize the quantum state function initialize_quantum_state():
 # Initialize three qubits in some initial state qubit1 = |0⟩ qubit2 = |0⟩ qubit3 = |0⟩
 # Create the quantum state by combining the three qubits quantum_state =
tensor_product(qubit1, qubit2, qubit3)
 # Return the initialized quantum state return quantum_state
# Function to apply the QSP technique with the specified operator function
apply_qsp_technique(quantum_state):
 # Apply the QSP operator to the quantum state processed_quantum_state =
qsp_operator(quantum_state) # Return the processed quantum state return
processed_quantum_state
# Function for the QSP operator that sums up the probabilities in different directions function
qsp_operator(quantum_state):
# Apply measurement operations in the specified directions probability_horizontal =
measure_in_horizontal_direction(quantum_state) probability_vertical =
measure_in_vertical_direction(quantum_state) probability_45_degrees =
measure_in_45_degrees_direction(quantum_state) probability_135_degrees =
measure_in_135_degrees_direction(quantum_state)
 # Sum up the probabilities from the measurements summed_probabilities = probability_horizontal
+ probability_vertical + probability_45_degrees + probability_135_degrees
 # Normalize the summed probabilities to ensure they add up to 1 processed_quantum_state =
normalize(summed_probabilities)
```

(Continued)

TABLE 6.2 (Continued)

```
# Return the processed quantum state return processed_quantum_state
# Measurement functions for each direction function
measure_in_horizontal_direction(quantum_state):
 # Measure the probabilities in the horizontal direction
 # Perform necessary operations specific to this direction
 # Return the measured probabilities return measured_probabilities
function measure_in_vertical_direction(quantum_state):
# Measure the probabilities in the vertical direction
 # Perform necessary operations specific to this direction
 # Return the measured probabilities return measured_probabilities
function measure_in_45_degrees_direction(quantum_state):
 # Measure the probabilities in the 45 degrees direction
# Perform necessary operations specific to this direction
 # Return the measured probabilities return measured_probabilities
function measure_in_135_degrees_direction(quantum_state):
 # Measure the probabilities in the 135 degrees direction
 # Perform necessary operations specific to this direction
 # Return the measured probabilities return measured_probabilities
# Function to normalize the probabilities to ensure they add up to 1 function
normalize(probabilities):
 # Perform normalization operation on the probabilities
 # Return the normalized probabilities return normalized_probabilities
```

TABLE 6.3 Pseudo-code of SWAP gate used in determining $|b_M|1\rangle$ or $|b_M|0\rangle$.

```
// Input: Two qubits represented by their quantum states qubit_A = initialize_qubit() qubit_B =
initialize_qubit()
// Apply SWAP gate to swap qubit states apply_swap_gate(qubit_A, qubit_B)
function apply_swap_gate(qubit_A, qubit_B): // Store the state of qubit_A in a temporary variable
temp_state = qubit_A.state // Swap the states of qubit_A and qubit_B qubit_A.state =
qubit_B.state qubit_B.state = temp_state
```

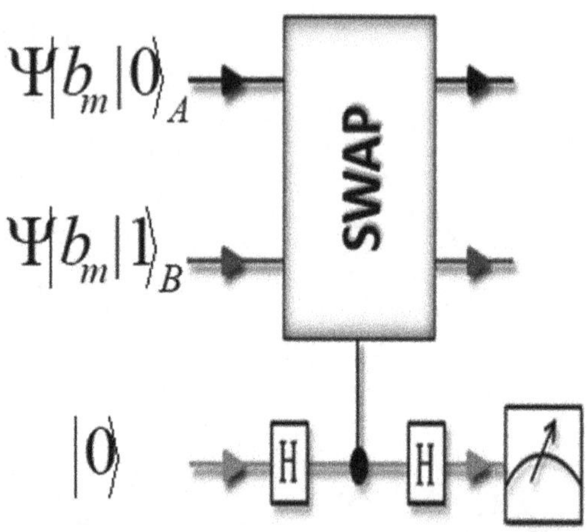

FIGURE 6.19 SWAP gate switching between two quantum states.

based on the available probability statistics (Sundani et al., 2019). Then, assign these probabilities to the diagonal elements of the respective operators as follows:

$$OP_{mn}|b_M>45^0=\begin{vmatrix} P_{11}|b_M> & 0 & 0 & 0 \\ 0 & P_{22}|b_M> & 0 & 0 \\ 0 & 0 & P_{33}|b_M> & 0 \\ 0 & 0 & 0 & P_{44}|b_M> \end{vmatrix}, \tag{6.53}$$

$$OP_{mn}|b_M>90^0=\begin{vmatrix} P_{11}|b_M> & 0 & 0 & 0 \\ 0 & P_{22}|b_M> & 0 & 0 \\ 0 & 0 & P_{33}|b_M> & 0 \\ 0 & 0 & 0 & P_{44}|b_M> \end{vmatrix}, \text{and} \tag{6.54}$$

$$OP_{mn}|b_M>135^0=\begin{vmatrix} P_{11}|b_M> & 0 & 0 & 0 \\ 0 & P_{22}|b_M> & 0 & 0 \\ 0 & 0 & P_{33}|b_M> & 0 \\ 0 & 0 & 0 & P_{44}|b_M> \end{vmatrix}. \tag{6.55}$$

$$OP_{mn}|b_M>=\begin{vmatrix} P_{11}|b_M> & 0 & 0 & 0 & 0 & 0 & 0 & 0 \\ 0 & P_{22}|b_M> & 0 & 0 & 0 & 0 & 0 & 0 \\ 0 & 0 & P_{33}|b_M> & 0 & 0 & 0 & 0 & 0 \\ 0 & 0 & 0 & P_{44}|b_M> & 0 & 0 & 0 & 0 \\ 0 & 0 & 0 & 0 & P_{55}|b_M> & 0 & 0 & 0 \\ 0 & 0 & 0 & 0 & 0 & P_{66}|b_M> & 0 & 0 \\ 0 & 0 & 0 & 0 & 0 & 0 & P_{77}|b_M> & 0 \\ 0 & 0 & 0 & 0 & 0 & 0 & 0 & P_{88}|b_M> \end{vmatrix} \tag{6.52}$$

In Eq. (6.52), each diagonal element $P_{mn}|b_M>$ corresponds to the probability of the three-qubit subsystem being in the corresponding state in horizontal $|m\rangle$ and vertical $|n\rangle$, respectively, for identifying quantized Marghany bounded states $|b_M>$ in SAR data as discussed earlier. To construct the operators, compute the probabilities associated with each state for the σ_{mn} in the 0°,45°, 90°, and 135° directions

According to the aforementioned equations, the process involves using four directional operators to calculate the maximum value, which is then considered as the output backscatter intensity of every pixel to determine the maximum fluctuation of NRCS of $|b_M>$. In other words, the directional operators are used to detect the quantized Marghany bounded state in the SAR image by calculating

TABLE 6.4 Pesudo-code for automatic detection of maximum $\sigma_{NRCS}|b_M(m,n)>$.

function find_max_value(operator): $\sigma_{NRCS}|b_M(m,n) >$ max_value = operator[0][0] // Initialize

max_value with the first element of the operator // Iterate over each element of the operator matrix

for m = 0 to rows (operator) - 1: for n = 0 to columns (operator) - 1: if operator[m][n] >

$\sigma_{NRCS}|b_M(m,n)>$max_value: $\sigma_{NRCS}|b_M(m,n) >$ max_value = operator[m[n]

// Update $\sigma_{NRCS}|b_M(m,n)>$max_value if a larger value is found return $\sigma_{NRCS}|b_M(m,n)>$max_value

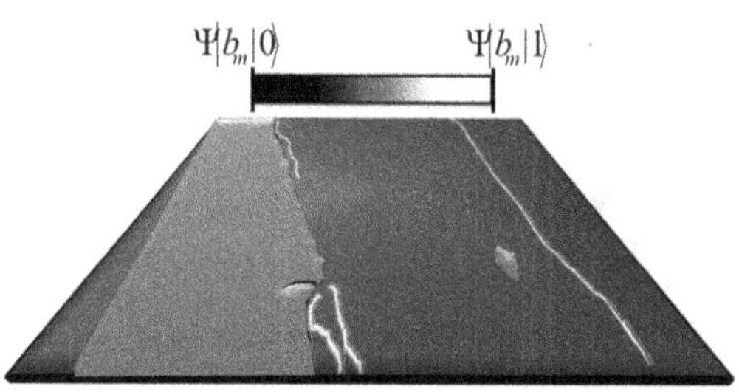

FIGURE 6.20 Automatic detection of quantized Marghany bounded front state in SAR image.

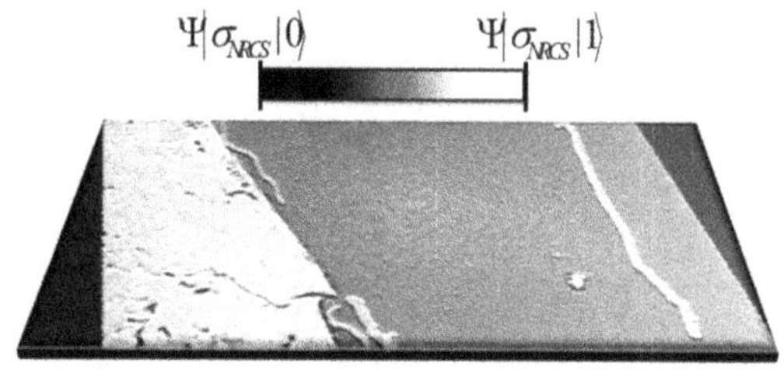

FIGURE 6.21 Quantized Marghany bounded front state edge enhancement in SAR NRCS image. *NRCS*, Normalized radar cross section.

the difference between the pixel values in different directions. In this view, the four directions are usually horizontal, vertical, and two diagonal directions (Fu et al., 2009; Sundani et al., 2019; Zhou et al., 2020). After applying the directional operators, therefore, the maximum value of $|b_M>$ (Table 6.4) is selected as the output NRCS of $|b_M>$ value of the detected pixel.

$$\sigma_{NRCS}|b_M(m,n)> = \max\{OP|b_M(m,n) > \theta \in \{0^{\circ}, 45^{\circ}, 90^{\circ}, 135^{\circ}\}\} \tag{6.56}$$

This value is then considered as the enhanced gray value of the current $|b_M>$ pixels. In this understanding, the enhanced NRCS value is a value that has been modified to highlight the edges of $|b_M>$ in the SAR image. The value is defined as a number between 0 and 1, where 0 represents black and 1 represents white (Fig. 6.20). The enhancement process is important in image processing as it helps to improve the quality of the image and makes it easier to analyze (Cavalieri & Maio, 2020; Zhou et al., 2020). Needless to say, the algorithm used for enhancement has been designed to preserve the detailed information of the $|b_M>$ in the SAR date. In addition, the algorithm has also been designed to effectively enhance the edge details of the $|b_M>$. The enhancement of edge details is an important aspect of SAR image processing owing to existing of speckles as it helps in identifying and analyzing the boundaries of different structures within the SAR data (Fig. 6.21). The effective enhancement of edge details in $\sigma_{NRCS}|b_M(m,n)>$ is likely to improve the accuracy and reliability of any subsequent analysis that is based on the SAR data (Marghany, 2021).

Three quantized Marghany bounded state fronts are shown in Fig. 6.21; the offshore front is longer than the onshore fronts and moves parallel to the coastline. These fronts have the highest $\sigma_{NRCS}|b_M(m,n)>$ of $|1>$ than the surrounding sea environment. The quantized Marghany bounded front states in the coastal waters of Peninsular Malaysia will be discussed in more detail in the following section.

6.12 Can a quantum edge detection algorithm locate various kinds of quantized Marghany bound front states in SAR data?

The query as mentioned earlier provides data from various radar satellites, dating back to 1993, to depict the mechanism of front generations off the east coast of Malaysia's peninsular island (Marghany, 2012a, 2012b). This may also prompt research into various front types in the South China Sea, particularly those near the quantized Marghany turbulence boundary.

On February 12, 1993, a river plume front was captured by the ERS-1 C-band instrument with VV polarization, spatial resolution ranging from 10 to 30 m, a swath width of 100 km, and an incidence angle of 23° off the east coast of the peninsular island of Malaysia (Fig. 6.22). The front's radar backscatter is − 5dB higher than that of the surroundings. This river plume front observed by the ERS-1 C-band instrument indicates a significant difference in the scattering properties of the water in comparison to its surrounding areas. The higher radar backscatter suggests a distinct composition or surface roughness within the plume, potentially caused by sediment or organic matter suspended in the water (Alpers & Holt, 1995; Marghany, 2011, 2021).

As a matter of fact, the sharpest quantized Marghany bound front state, which distinguishes between two different water masses, is produced by the quantum edge detection algorithm. The sharpest edge and maximum wave function Ψ| 1> of the quantized Marghany bound front has a maximum NRCS of -5dB in the surrounding environment(Fig. 6.23).

The author was able to gather temperature and salinity data during the satellite overflights, which were used to create the T-S diagram (Fig. 6.24). Additionally, the T-S diagram was constructed by the pseudo-code as demonstrated in Table 6.5. The fundamental query at this point is: How can the quantized Marghany bound front state in the ERS-1 data be interpreted using the T-S diagram?

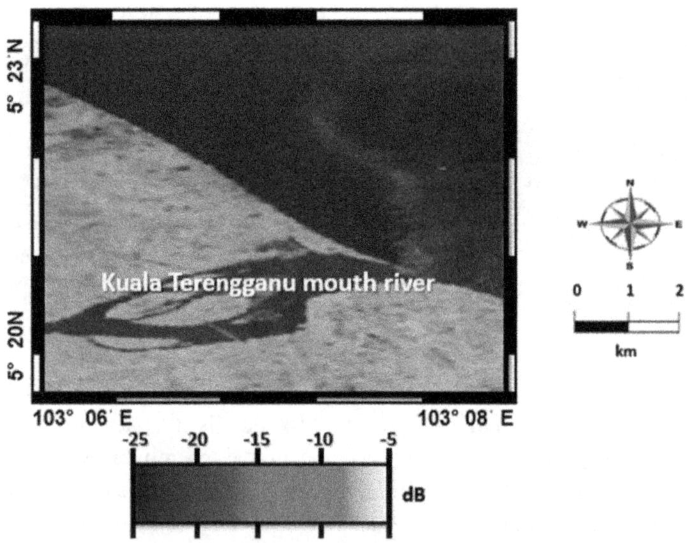

FIGURE 6.22 Backscatter variation across the front as seen by ERS-1 on February 12, 1993.

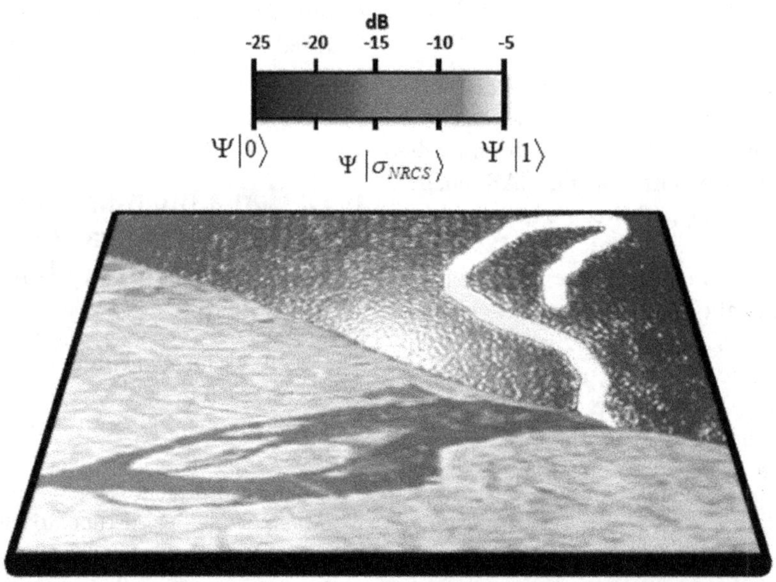

FIGURE 6.23 Automatic detection of quantized Marghany bound front state in ERS-1; February 12, 1993.

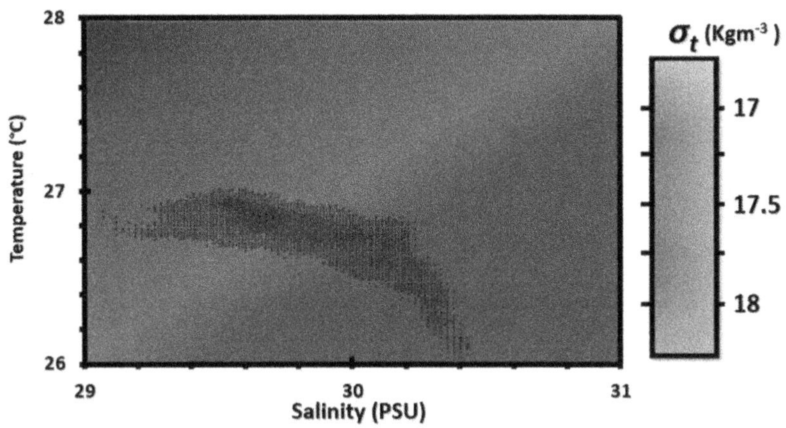

FIGURE 6.24 T-S diagram along the coastal water of Kuala Terengganu, Malaysia.

If the fundamental physical mixing rule (Richmann's mixing rule, i.e., equation 5.90) holds, this analysis of the spreading and mixing of oceanic water masses represented a significant advancement. According to this rule, the temperature of the mixture would be T as demonstrated by (Dijkstra, 2008; Galindo & Pascual, 2012; Marghany, 2012a, 2012b):

$$\Psi|T> = \frac{\Psi|m_1T_1> + \Psi|m_2T_2>}{\Psi|m_1> + \Psi|m_2>} \tag{6.57}$$

In Eq. (6.57), there are two quantum states of the water masses $\Psi|m_1|m_2>$ entanglement with each other at $\Psi|T>$. Similarly, if water mass $\Psi|m_1>$ of salinity $\Psi|S_1>$ and water mass $\Psi|m_2>$ of salinity $\Psi|S_2>$ mix, the mixture's salinity, $\Psi|S>$, will be as follows:

$$\Psi|S> = \frac{\Psi|m_1S_1> + \Psi|m_2S_2>}{\Psi|m_1> + \Psi|m_2>} \tag{6.58}$$

Therefore the ratio of mixing and the distance *d*, from the two endpoints of any point on the line between water types, are inversely proportional as follows:

$$\frac{\Psi|m_1>}{\Psi|m_2>} = \frac{\Psi|d_1>}{\Psi|d_2>} \tag{6.59}$$

This means that as the mixing ratio increases, the distance between the two endpoints decreases. Conversely, if the mixing ratio decreases, the distance between the two endpoints increases(Kuang et al., 2012; Marghany, 2011, 2012a, 2012b, 2021). From this perspective, $\Psi|S>$ is formed because of the mixing of $\Psi|S_1|S_2>$ in proportion.

The mixing length $\Psi\ |A_Z>$ of the two water masses, $\Psi|m_1|m_2>$ can be calculated from the T-S diagram as follows:

$$\Psi|A_z> = \Psi|u> \left[\frac{\Psi|S_2> - \Psi|S_1>}{\Psi|m_2> + \Psi|m_2>} \times \frac{\Psi|m_2> - \Psi|m_1>}{\Psi|x_2> - \Psi|x_1>} \times \left[\frac{\partial^2 S}{\partial Z^2}\right]^{-1}\right] \Psi|\sigma_t> \tag{6.60}$$

Generally, the author attempts to describe the mixing length at point x and time t by exploiting the Schrodinger equation as follows:

$$\frac{\partial A_Z(x,t)}{\partial t} = [i\hbar]^{-1}\left(-\frac{\hbar^2}{2\sigma_t}\right)\frac{\partial^2 A_Z(x,t)}{\partial x^2} + E_p(x)A_Z(x,t) \tag{6.61}$$

In Eq. (6.61), the potential energy function E_P depends on the specific physical situation and the external forces acting on the quantized Marghany bound front state. The potential energy function determines the allowed energy states (eigenvalues) and corresponding wave functions (eigenfunctions) of this bound front state. As a result, the quantized Marghany bound front state's core layer can be used to identify the transition zone between different water masses, which is characterized by a sharp change in salinity or temperature over a limited depth range. To this end, the quantum amplitude estimation (QAE) algorithm (Grinko et al., 2021) can be adapted to find the core layer in the quantized Marghany bound front state in the T-S diagram (Table 6.6).

The quantized Marghany bound front state is dominated by a core density σ_t of 17.5 kg/m^3; a temperature of 26.8°C; and a salinity of 30.2 PSU. As a matter of fact, the quantized Marghany bound front state separates between two cool water masses of 26.0 and 27.1 with a slight temperature of 1.1°C. However, the salinity

TABLE 6.5 Pseudo-code of plotting the T-S diagram.

```
1. Define the data points for temperature (T) and salinity (S) from your dataset. 2. Create a 2D
scatter plot to represent the T-S diagram.
3. Set up the axes and labels for the plot:
// Assuming you have T values in the list T_values and S values in the list S_values. // Define the
range for x-axis (S) and y-axis (T). S_min = min(S_values) S_max = max(S_values) T_min =
min(T_values) T_max = max(T_values)
// Set up the plot with appropriate labels and axis limits. plot.setup(title='Temperature-Salinity
Diagram', x_label='Salinity (psu)', y_label='Temperature (°C)') plot.xlim(S_min, S_max)
plot.ylim(T_min, T_max)
4. Plot the data points:
// Iterate through the data points and plot them on the T-S diagram. for each data point (T_i, S_i)
in the dataset: plot.scatter(S_i, T_i, marker='o', color='blue')
5. Optionally, one can add density contours or color-coded regions based on density values:
// Assuming one has calculated the water density (rho) corresponding to each (T, S) pair. // Use a
colormap to represent the density values as colors on the plot. // For example, a high density
would be represented as red, and a low density as blue. for each data point (T_i, S_i, rho_i) in the
dataset: color = colormap.interpolate(rho_i) // Get the color corresponding to the density value.
plot.scatter(S_i, T_i, marker='o', color=color)
6. Display the T-S diagram:
plot.show()
```

difference between the two water masses is 1.3 PSU, and the difference in water density of 1 kg/m^3. With a slight difference in water mass's physical properties, the quantized Marghany bound front occurs across the coastal water of Kuala Terengganu's mouth river. This study confirms the early work of (Marghany, 2012a, 2012b, 2021)

The core density, σ_t, of the quantized Marghany bound front state is 17.5 kg/m^3, the temperature is 26.8°C, and the salinity is 30.2 PSU (Fig. 6.25). In fact, with a slight difference in temperature of 1.1°C, the quantized Marghany bound front state splits into two cool water masses of 26.0°C and 27.1°C. However, there is a 1.3 PSU salinity difference and a 1 kg/m^3 water density difference between the two water masses.

The quantized Marghany bound front crosses the coastal water of Kuala Terengganu's mouth river despite slight differences in the physical characteristics of the water mass. These slight differences in salinity and water density between the two cool water masses indicate a potential mixing zone where the quantized Marghany bound front crosses the coastal water of Kuala Terengganu's mouth river. The presence of this mixing

TABLE 6.6 Pseudo-code of quantum amplitude estimation algorithm.

 # Input: A black-box quantum oracle Uf that prepares the state $|f(b_m)\rangle|\sigma_{NRCS}\rangle$ given input $|\sigma_{t\,(max)}\rangle$ # The function $f(b_m)$ with x being an n-bit binary number. # Output: The value $|\sigma_{t\,(max)}\rangle$ that maximizes the function $f(b_m) \in \max \sigma_{NRCS}\,(S_{max}.T_{max})$
Step 1: Initialize Quantum Register # Set up an n-qubit quantum register $|\sigma_{t\,(max)}\rangle$ in an equal superposition of all possible inputs. # Create another qubit initialized to $|1\rangle$ to represent the auxiliary qubit.
Step 2: Prepare Quantum State $|f(\sigma_{t\,(max)})|b_m\rangle$ # Maps $|\sigma_t\rangle|1\rangle$ to $\sigma_{t\,(max)}|1 \oplus f|\sigma_{t\,(max)}\rangle$, where $f(\sigma_{t\,(max)})$ is the function value.
Step 3: Amplitude Amplification # Apply SWAP's amplitude amplification algorithm to amplify the amplitude of the state with the maximum value.
Step 4: Measurement # Measure the quantum register in the computational basis ($|\sigma_{t\,(max)}\rangle$ basis). # The outcome will be a result that has a higher probability of being the input x that maximizes $f(\sigma_{t\,(max)})$.
Step 5: Repeat Steps 2–4 for a few iterations (as needed for desired accuracy).
Output: Return the measurement result corresponding to the input σ_t that maximizes $f(\sigma_{t\,(max)})$.

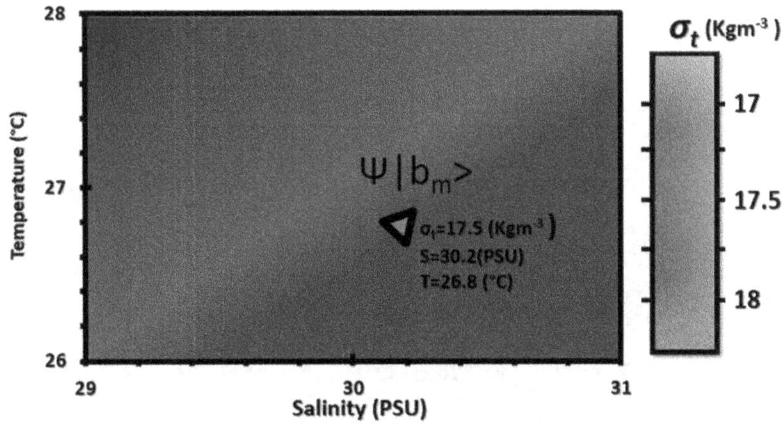

FIGURE 6.25 Characteristics of the core water mass as determined by QAE. *QAE*, Quantum amplitude estimation.

zone suggests that there may be a dynamic interaction between the two water masses, leading to changes in their physical characteristics. Therefore this plume front indicates the occurrence of the ebb flow. During the ebb tide, the seawater starts to flow away from the coast, carrying the mixed river water with it. As the less dense river water moves offshore, it forms a distinct front or boundary between the river plume and the surrounding seawater (Busalacchi & O'Brien, 1980; Eckart & Ferris, 1956; Hartmann et al., 2021; Marghany, 2021). This can be noticed in a curved pattern of the T-S diagram as water flows from the river toward the open sea (Fig. 6.24). On the other hand, Terengganu River water is typically less dense than seawater due to its lower salinity (29.1 PSU) and, sometimes, lower temperature 26.0°C. As a result, the freshwater from the river tends to float on top of the seawater, forming a less dense layer.

Moreover, RADARSAT-1 SAR F1 mode data acquired from the coastal waters of Kuala Terengganu, Malaysia, on March 26, 2004, also reveal the sharpest curved front close to the mouth of the Terengganu River detected by the quantum edge detection algorithm (Fig. 6.26). The RADARSAT-1 F1 mode backscatter cross section of the front has a maximum value of -22dB compared with the surrounding sea environment. The maximum backscatter value of -22dB is found across a curved brightness front, and this variation of radar backscatter cross section is due to the current boundary gradient. Consistent with (Vogelzang et al., 1997), ocean current boundaries are often accompanied by changes in surface roughness, which can be detected by SAR. These surface roughness changes are due to the interaction of surface waves directly with surface current gradients. These interactions can cause an increase in surface roughness and radar backscatter (Krogstad & Schyberg, 1990; Marghany, 2011, 2012a, 2012b).

The T-S diagram from March 2004 shows two water masses spanning the Kuala Terengganu river mouth. Like February 1993, the cool water mass of 28.3°C close to the mouth river with less salinity of 29.2 psu floats at the top of more saline seawater of 30.2 psu and slightly high temperature of 29.2°C flowing toward the coastal water forming such an estuarine front. The core layer of the quantized Marghany bound front state is located within σ_t

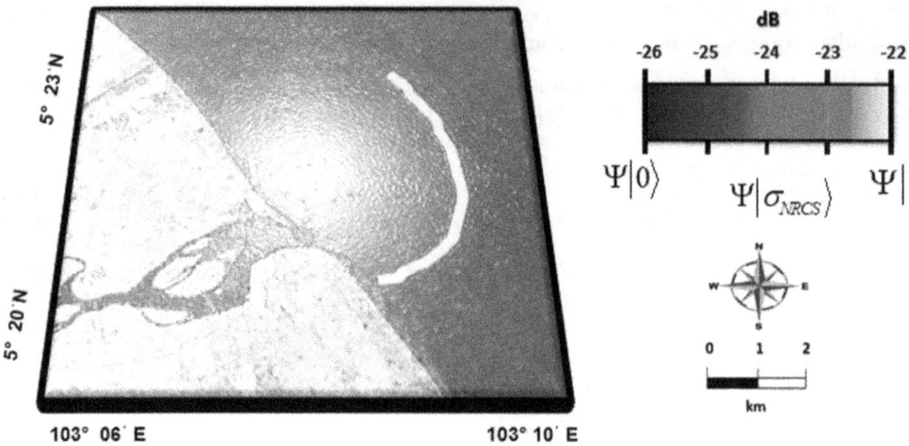

FIGURE 6.26 Automatic detection of quantized Marghany bound front state in RADARSAT-1 F1 mode on March 26, 2004.

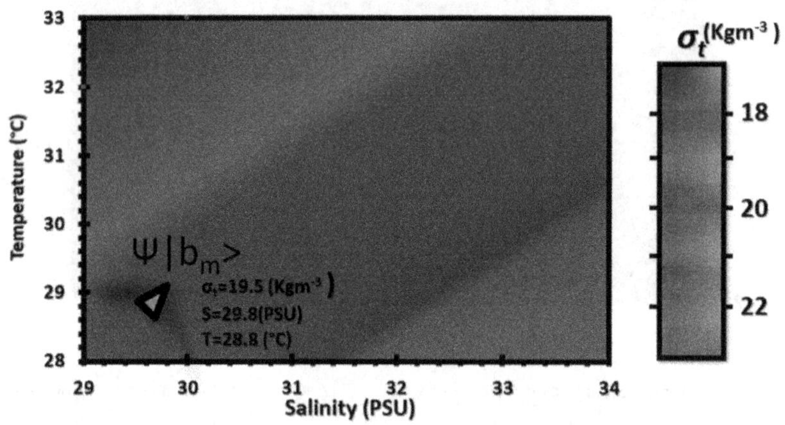

FIGURE 6.27 Characteristics of the core water mass as determined by QAE on March 2004. *QAE*, Quantum amplitude estimation.

of 19.5 kg/m^3, the temperature is 28.8°C, and the salinity is 29.8 psu (Fig. 6.27). Compared with what happened in February 1993, this scenario is very different (Maged Marghany & Mahmoud, 1994).

Therefore across the quantized Marghany turbulence boundary addressed in Chapter 5, the RADARSAT-1 S2 mode captured irregular patterns of the fronts (Fig. 6.28). These fronts are formed in waters with bathymetry ranging from 40 m to 50 m offshore. A pool of cold and heavy deep water mass of 27.2°C and 22.3 kg/m^3 that flows from offshore Terengganu and penetrates coastal waters (Fig. 6.29) is coupled to these fronts. The fronts also equate to filament, meandering currents (Fig. 6.30), and eddy. In this sense, filaments can develop as a consequence of meandering and deformation of the frontal boundaries, which are dynamic zones of convergence and mixing. These frontal zones can be strained and prolonged by ocean currents and other dynamic processes, resulting in filament development (Johannessen et al., 1991; Kim et al., 2017; Kuang et al., 2012; Marghany, 2021).

This unique combination of filament and meandering current creates a mesmerizing display of swirling patterns and intricate formations. The quantized Marghany front adds an element of precision and order to the otherwise dynamic and unpredictable nature of this phenomenon. In active imaging with radar, therefore, a bright signature of -4 dB at the ocean surface can indicate the presence of an ocean front. Ocean fronts often coincide with regions of convergence, where water masses with different properties (e.g., temperature, salinity) come together. The convergence of water masses can lead to changes in surface roughness and small-scale features, which can be detected by radar (Alpers & Holt, 1995; Marghany, 2011, 2012a, 2012b, 2021).

These fronts are aligned in Chapter 5 and have large internal wave groups that were also discussed in Chapter 5. This unusual feature appeared on March 31, 2005, when the northeast monsoon passed through Peninsular Malaysian coastal waters. It is crucial for understanding the mechanisms of this complex phenomenon. These fronts

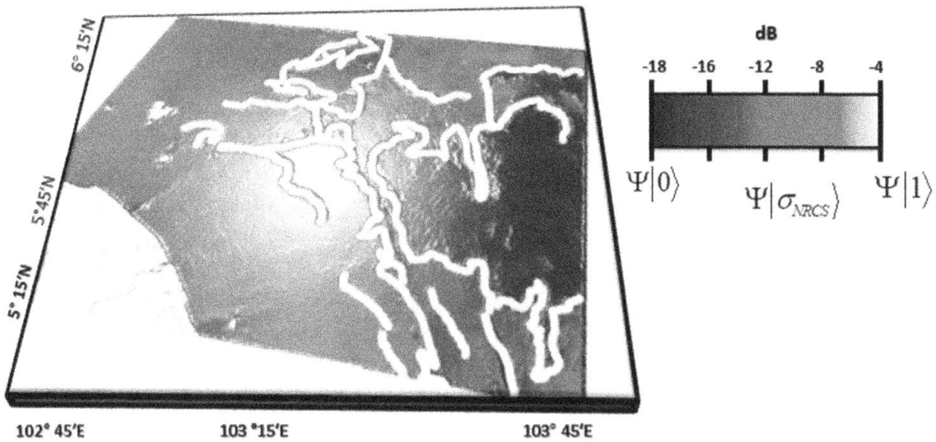

FIGURE 6.28 Automatic detection of quantized Marghany bound front state in RADARSAT-1 S2 mode on March 31, 2005.

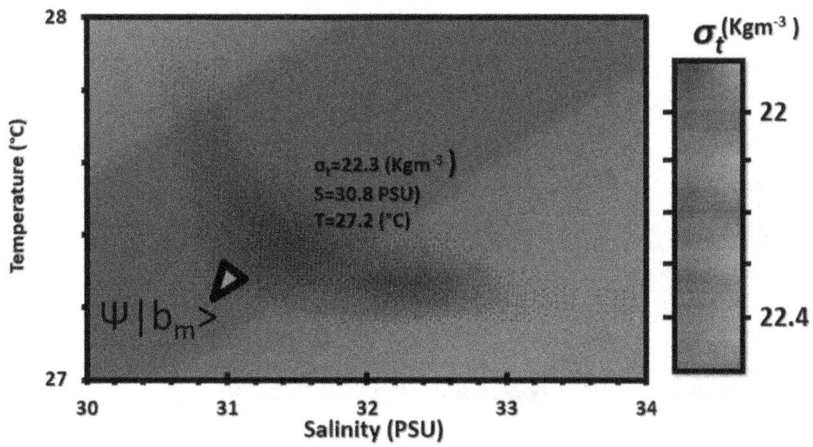

FIGURE 6.29 Characteristics of the cool core water mass as determined by QAE on March 2005. *QAE*, Quantum amplitude estimation.

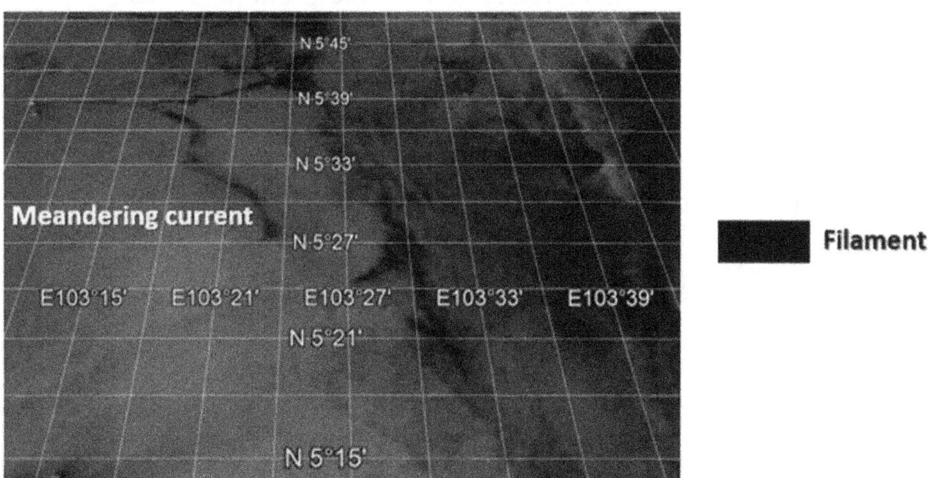

FIGURE 6.30 Filament and meandering current couple with quantized Marghany front.

belong to the shelf front, convergence front, and western boundary current front under the effect of the wind stress flow. These fronts are surrounded by a variety of quantum states that are entangled and superimposed simultaneously in the same period to create quantized Marghany turbulence boundaries from the perspective of quantum mechanics (Alpers & Holt, 1995; Maged Marghany & Mahmoud, 1994; Vogelzang et al., 1997).

Later, on October 27, 2009, the ALOSPALSAR satellite in fine mode and using the L_{HH}-band also had a chance to capture a long shelf front of about 35 km that ran parallel to the direction of the coastal waters of Kuala Terengganu. The front was dominated by the highest NRCS of -2 dB than the background (Fig. 6.31). The quantum edge detection algorithm effectively perceives the distinct train pattern of curvature fronts. This algorithm is particularly useful in detecting edges and boundaries in images with complex and intricate shapes (Fig. 6.32). It utilizes the principles of quantum mechanics to accurately identify and analyze the curvature fronts, providing precise results in edge detection tasks.

The cool core water mass is identified by the QEA as having a density of 22.4 kg/m^3, a temperature of 28,6°C, and a salinity of 33.8 psu (Fig. 6.33). The T-S diagram's shape shows that downwelling occurs along Kuala Terengganu's coastline in Malaysia. This implies that the convergence shelf front is the existence front. Downwelling is a process in the ocean where surface water sinks and moves downward into

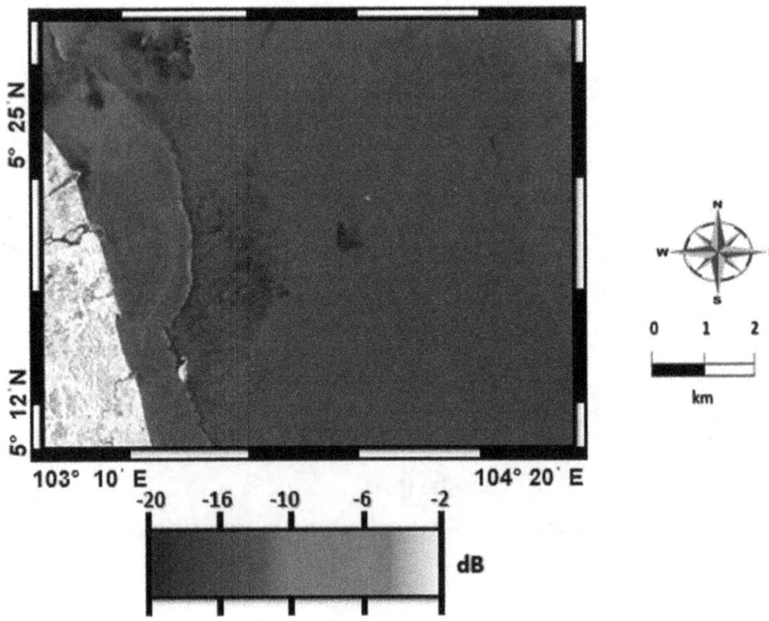

FIGURE 6.31 Backscatter variation across the front as seen by ALOS PALSAR fine mode data on October 27, 2009.

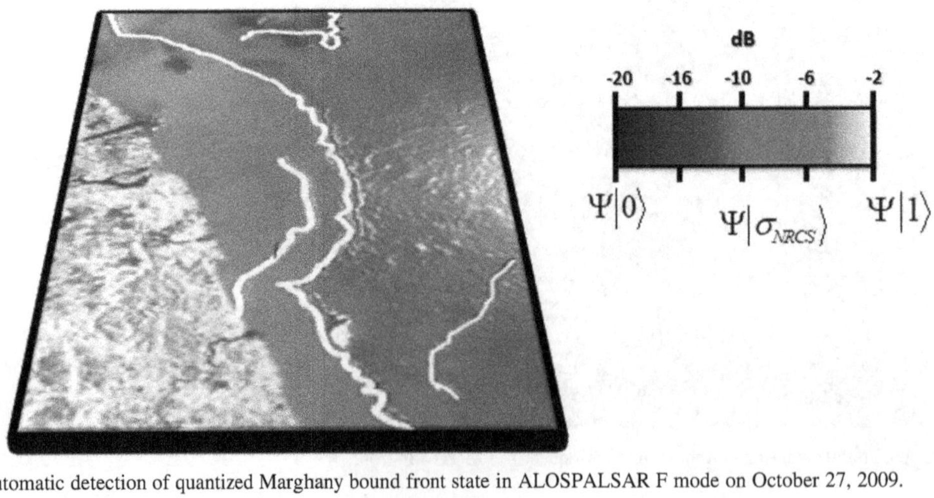

FIGURE 6.32 Automatic detection of quantized Marghany bound front state in ALOSPALSAR F mode on October 27, 2009.

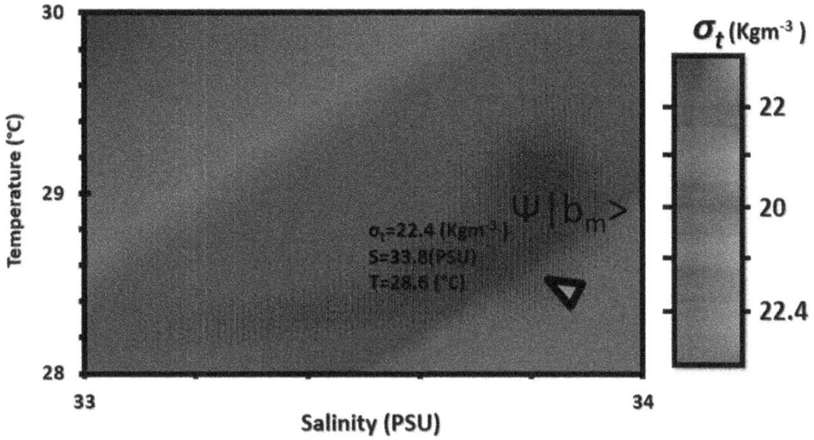

FIGURE 6.33 T-S diagram associated with ALOSPALSAR F mode on October 27, 2009.

deeper layers. This typically happens in areas where surface waters become denser, either due to cooling or salinity changes. Downwelling is associated with the transport of surface water toward deeper layers, which leads to the convergence of water masses. As the water sinks, it creates a net inflow of fluid particles in the deeper ocean layers, indicating convergence (Alpers & Holt, 1995; Kim et al., 2017; Marghany, 2021). This front is characterized by a rapid change in temperature and salinity, indicating a significant mixing of water masses, which is shown by slight differences in physical water properties from surface to bottom. In this perception, the temperature difference was 0.6°C, the salinity difference was 0.6 psu, and the water density difference was approximately 0.4 kg/m^3. This agreed with previous work done by Marghany (2012a, 2012b, 2021).

Temperature differences between the water masses on either side of an ocean front can result in variations in surface roughness. The presence of different temperatures may influence the distribution of surface-active substances such as surfactants, which can affect surface tension and roughness. These small differences in roughness can be detected by radar, leading to slight changes in radar backscatter. By studying these radar backscatter patterns, scientists can gain insights into the dynamics and properties of ocean fronts, such as the convergence of water masses, the sharpness of the front, and the associated surface roughness variations (Alpers & Holt, 1995; Hartmann et al., 2021; Marghany, 2021).

Recently, on December 22, 2022, the Sentinel-1 satellite with C_{VV}-band also can capture a convergence front along the east coast of peninsular Malaysia, mainly Kelentan coastal waters with the highest backscatter of -8dB than the background (Fig. 6.34). The author has a chance to collect physical water properties using SEA Bird CTD plus Rosette sampler bottle set instrument during the satellite overpass (Fig. 6.35).

During Sentinel-1 satellite overpass on December 22, 2022, the T-S diagram demonstrates penetration of deeper dense cool water mass of 26.4°C, 30.8 psu, and water density of 22.2 kg/m^3 across the coastal water of Kelantan, Malaysia (Fig. 6.36). Additionally, it suggests the presence of a shelf convergence front. The penetration of the deeper cool dense water mass indicates a potential mixing event in the coastal waters of Kelantan, Malaysia.

During the northeast monsoon season, strong northeasterly winds blow over the South China Sea and the surrounding region, including Vietnam and the eastern coast of Malaysia. These winds lead to the development of a strong current along the eastern coast of Vietnam of more than 0.5 m/s, pushing southward along the coastline toward Malaysia. As the strong current moves southward, it can induce upwelling along the eastern coast of Malaysia. Upwelling is a phenomenon where cold, nutrient-rich waters from deeper layers of the ocean rise to the surface. This upwelling process leads to the formation of a "tongue" of cool water along the east coast of Malaysia. The possibility of an upwelling occurring during the northeast monsoon rather than just the southwest monsoon will be discussed in the following chapters. Understanding the factors that contribute to upwellings during the northeast monsoon is crucial for comprehending the overall dynamics of ocean currents in this region (Johannessen et al., 1991; Maged Marghany & Mahmoud, 1994; O'Donnell et al., 1998; Vogelzang et al., 1997). Additionally, exploring the potential impacts of these upwellings on marine ecosystems and local weather patterns will provide valuable insights for future research and coastal management strategies.

The detection of frontal signatures on synthetic aperture radar (SAR) images can indeed raise valuable questions and provide insights into the dynamic processes occurring in the ocean. SAR imagery allows researchers and scientists to observe oceanographic features, such as

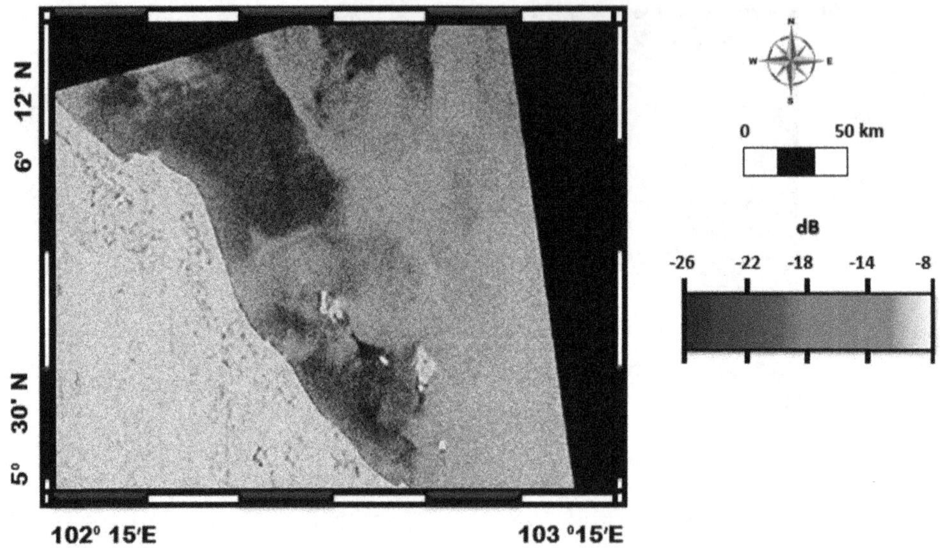

FIGURE 6.34 Shelf convergence front was imagined in the Sentinel-1 satellite with C_{VV}-band on December 22, 2022.

FIGURE 6.35 SEA Bird CTD and a set of Rosette sampler bottles were used for in-the-moment measurement during satellite overpasses by the author.

fronts, with a unique perspective, and it can serve as a valuable tool for studying various ocean phenomena. The appearance of frontal signatures near island headlands can be indicative of complex interactions between ocean currents, bathymetry (underwater topography), and coastal geography (Marghany, 2011, 2012a, 2012b, 2021). Islands and headlands can act as barriers to the flow of ocean currents, leading to the convergence of water masses and the formation of fronts. It is intriguing to initiate this phenomenon on December 6, 2022, in the C_{VV}-band of the Sentinel-1 satellite along the Tioman and Anambas Islands (Fig. 6.37). A quantum edge detection algorithm (Table 6.4) that yields an enhanced SAR image with a maximum NRCS of -12 compared with the background is effective at identifying the edge boundary of these fronts (Fig. 6.38).

Coastal features such as island headlands can act as natural boundaries that influence the behavior of ocean currents. The interaction between the currents and the headlands can create complex flow patterns and sharp gradients in water properties, which may manifest as frontal signatures in SAR images. Tidal currents around island headlands can also contribute to the formation of frontal structures. The changing water levels associated with tides can induce convergence or divergence of currents, leading to the development of fronts.

According to the aforementioned perspective, the mechanism of synthetic aperture radar (SAR) imaging of ocean fronts can be influenced by several factors, as outlined in the description (Alpers & Holt, 1995; Kim et al., 2017; Marghany, 2021):

Damping of Short Gravity Waves by Slicks: The presence of natural slicks aligned along the frontal boundary can dampen short gravity waves by reducing the surface roughness. Slicks are smooth areas on the ocean surface, often caused by the accumulation of

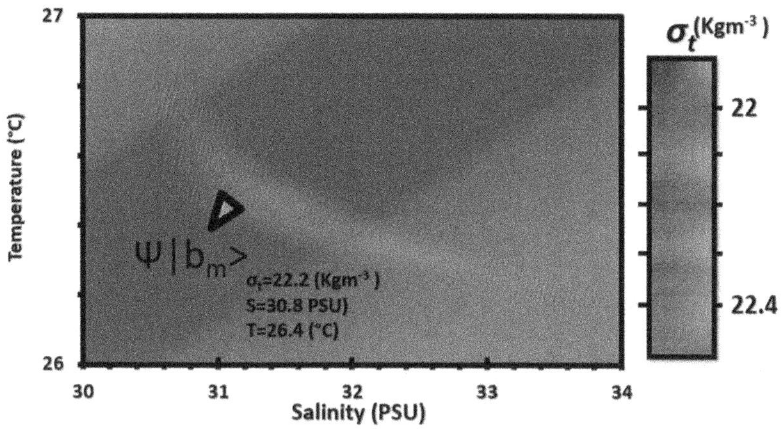

FIGURE 6.36 Dense cool water mass occurrence during Sentinel-1 satellite overpass on December 22, 2022.

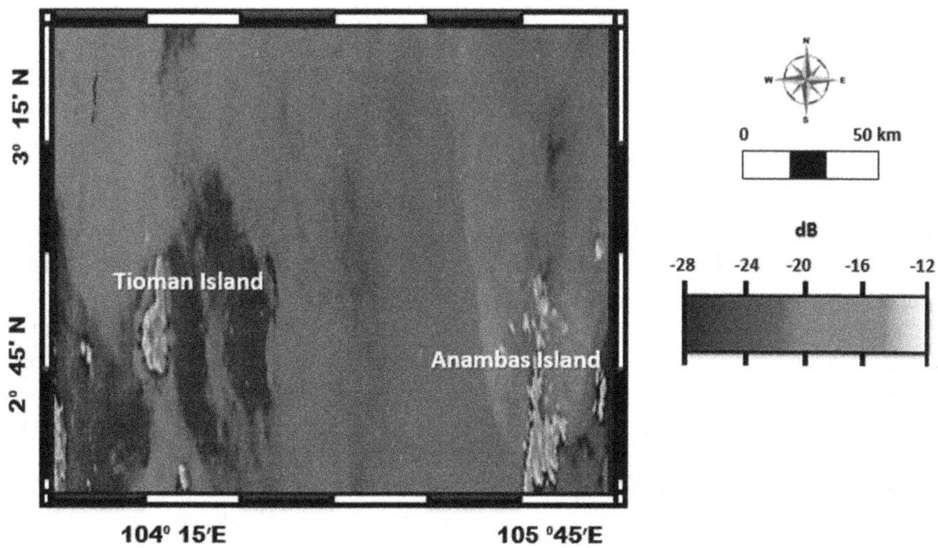

FIGURE 6.37 Front and current boundary triggered by Islands and headlands in Sentinel-1 satellite C_{VV}-band on December 6, 2022.

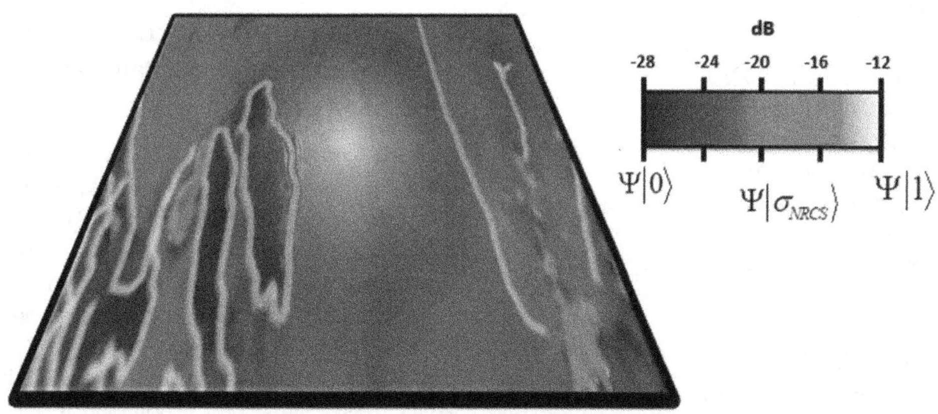

FIGURE 6.38 Quantum edge detection algorithm for well detection of front signatures close to islands and headlands.

organic materials such as oil or algae. When short gravity waves encounter these slicks, their energy is dissipated, leading to weaker radar backscatter signals in the SAR imagery.

Short Gravity-Wave/Current Interaction: Along shear and/or convergence zones within the front, there can be interactions between short gravity waves and ocean currents. These interactions can cause wave energy to be scattered or refracted, affecting the radar backscatter patterns observed in the SAR images. Variations in current velocities and gradients within the front can influence how the waves propagate and appear in the radar images.

Changes in Wind Stress and Sea Surface Temperature Gradients: Strong gradients in sea surface temperature can induce changes in wind stress patterns. These changes, in turn, affect the generation and propagation of ocean waves, including short gravity waves. The presence of fronts with different temperature properties can lead to alterations in wind stress, influencing the appearance of wave patterns in the SAR imagery.

Long-Gravity-Wave/Current Refraction: Long ocean waves (swell) can be refracted or bent when they encounter ocean currents, especially in regions with different current velocities. The refraction of long waves in the presence of fronts can alter their direction and propagation, affecting their appearance in SAR images. In the succeeding chapters, more information will be provided on this.

6.13 Spectral of quantized Marghany bound front state

The question of whether SAR images can assist in tracking the seasonal variation of the quantized Marghany bound front state spectral energy in large-scale areas has become crucial. More than hundreds of mosaic SAR data, that is, ERS-1, RADARSAT-1 S2, F1, ALOSPALSAR, TERRASAR-X spot mode, and Sentinel-1 satellite CVV-band, were fused with MODIS data from 1999 to 2023 to track the quantized Marghany bound front state along the western boundary of Malaysian coastal waters, that is, the east coast of peninsular Malaysia. Indeed, both the Aqua and the Moderate Resolution Imaging Spectroradiometer (MODIS) spacecraft are currently in use. On December 18, 1999, the Terra satellite was put into orbit; on May 4, 2002, the Aqua satellite.

6.13.1 Quantum image fusion for tracking quantized Marghany bound front state

According to this perspective, quantum image fusion involves the application of quantum operations on quantum states representing different images to create a fused quantum state that contains the combined information. The equation of quantum image fusion can be described using the principles of quantum mechanics, including quantum superposition and quantum entanglement. Let's denote the quantum states of the input images as $|\Psi_1\rangle$, $|\Psi_2\rangle$,., $|\Psi_n\rangle$, where n is the number of input images (Liu & Xiao, 2019; Majji et al., 2022). These quantum states represent the pixels of the images in a quantum representation. The⟩ fusion process can be represented using the following equation:

$$\Psi|I_F\rangle = \Sigma_i c_i|\Psi_i\rangle \tag{6.62}$$

In Eq. (6.62), $\Psi|I_F\rangle$ represents the fused quantum state that contains the combined information from all the input images. Σ_i, therefore, denotes the sum over all input images, where i varies from 1 to n. Consequently, c_i represents the fusion coefficients for each input image $|\Psi_i\rangle$(Li et al., 2023; Liu & Xiao, 2019). These coefficients determine the contribution of each input image to the fused state. The coefficients can be determined based on the specific fusion algorithm used. The pseudo-code of quantum image fusion can be demonstrated in Table 6.7.

According to Table 6.7, applying quantum operations to entangle quantum pixels from different images involves manipulating the quantum states of the pixels to create entangled states. The quantum states of two or more particles become highly correlated during entanglement, which makes it impossible to describe the states of individual particles separately. As a result, the entangled quantum pixels contain data from all the input images, enabling the fused state to quantum-enable the capture of pertinent details from each image. To produce a quantum superposition of the fused state, the entangled quantum pixels are superimposed (Liu & Xiao, 2019; Tawfik et al., 2021). The information from all the input images is combined and encoded by this superposition.

6.13.2 Tracking the fluctuations of quantized Marghany bound front state across different monsoon periods

For ease of description, we divide the north and south monsoon periods (November to March) into three periods: (1) initial time; (2) middle time; (3) end time. The initial period of the northeast monsoon is considered as November, as the quantized Marghany bound front state is well pronounced along the western boundaries and runs from 7°N to 4°N along 105°E to 107°E forms length of 600 km and a longitudinal length of 272 km, respectively (Fig. 6.39). This sharp bounded front state separates between two cool water masses of 27.5° and 28°C, respectively. In Malacca Straits, however, the warmer

TABLE 6.7 Pseudo-code for quantum image fusion.

```
# Pseudo-code for Quantum Image Fusion
# Step 1: Load and preprocess the input images def load_and_preprocess_images(): # Load and
preprocess the input images # Convert pixel values to quantum states (hypothetical quantum
representation) quantum_pixels_image1 = convert_to_quantum_states(image1)
quantum_pixels_imageN = convert_to_quantum_states(imageN) # . Additional images if available
# Step 2: Quantum Image Representation def quantum_image_representation(): # Represent
quantum pixels as quantum states (qubits) in a quantum register quantum_register_image1 =
create_quantum_register(quantum_pixels_image1) quantum_register_imageN =
create_quantum_register(quantum_pixels_imageN) # . Additional registers for other images
# Step 3: Quantum Entanglement def quantum_entanglement(): # Entangle quantum pixels from
different images using quantum gates entangled_state =
entangle_quantum_registers(quantum_register_image1, quantum_register_image2) # . Additional
entanglement if multiple images
# Step 4: Quantum Superposition def quantum_superposition(): # Superpose entangled quantum
pixels to create a quantum superposition of the fused state superposed_state =
superpose_quantum_state(entangled_state) # . Additional superposition if multiple images
# Step 5: Measurement and Fused Image Extraction def
measurement_and_fused_image_extraction(): # Perform quantum measurements on the quantum
superposition to obtain classical pixel values fused_classical_pixels =
measure_quantum_state(superposed_state) # Convert the fused quantum state back to classical
pixel values to create the fused image fused_image =
convert_to_classical_pixels(fused_classical_pixels)
# Main Function def main(): # Step 1: Load and preprocess the input images
load_and_preprocess_images()
 # Step 2: Quantum Image Representation quantum_image_representation()
 # Step 3: Quantum Entanglement quantum_entanglement()
 # Step 4: Quantum Superposition quantum_superposition()
 # Step 5: Measurement and Fused Image Extraction measurement_and_fused_image_extraction()
 # Display or save the fused image display_or_save_fused_image(fused_image)
# Run the main function main()
```

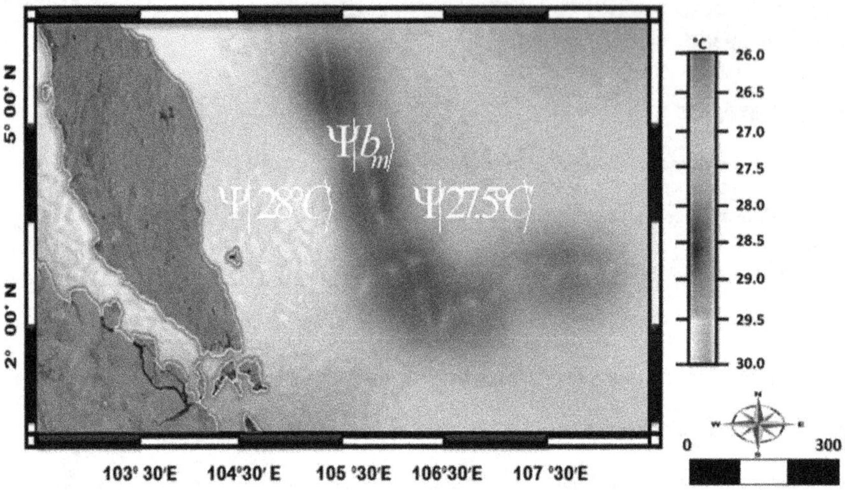

FIGURE 6.39 Formation of a quantized Marghany bound front state during the initial period of the northeast monsoon.

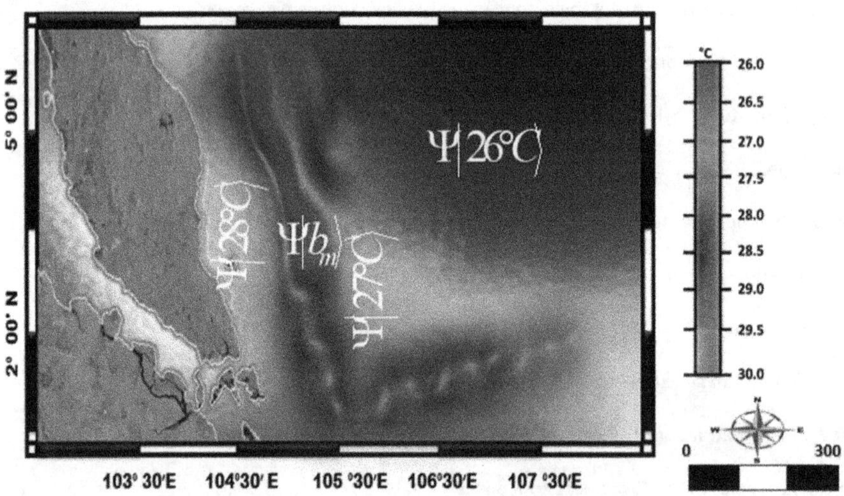

FIGURE 6.40 Formation of widest and largest quantized Marghany bound front state during December.

water mass of 30°C bounds the Andaman Sea water mass of 28°C in the Andaman Sea. It is worth noting that the tongue of the cool water of 27°C runs toward the western boundary of the east coast of Malaysia. In this view, the widest and largest front compared with November, which runs to 1°N, is formed by the tongue of 26°C cool water in December, which stretches from 7°N to 4°N along 106°E (Fig. 6.40) to form the length of 670 km along the western boundary. The tongue of the cool water of 28 C runs parallel to the western boundary and then turns into the Malacca Straits in December.

In the middle period of the northeast monsoon (January to February), the cool water of 26°C is a dominant large cluster along the western boundary as a well-mixed scenario (Maged Marghany & Mahmoud, 1994; Marghany et al., 1996). Therefore this cool water mass cluster creates the widest and largest front with a narrow water mass zone of 27°C along the western boundary. However, in February, there will be a noticeable start to the front's dissipation along 107°E of longitude at 2°N of latitude (Fig. 6.41). The cool front is extended along the western boundary with a slight temperature increase of 28°C at the end of the northeast monsoon period, which is presented in March. However, the mixing of 27°C water mass effectively replaces the cool water tongue's dissipation (Fig. 6.42).

The surface water mass tends to be warmer during the transitional period between the northeast monsoon and southwest monsoon periods, specifically in April. In this view, a pool of mixing water forms between the South China Sea and the Malacca Straits as the weak front zone of 29.0°C tends to move parallel to the western boundary and turn into it (Fig. 6.43). Consequently, at the beginning of the southwest monsoon period in May, the quantized

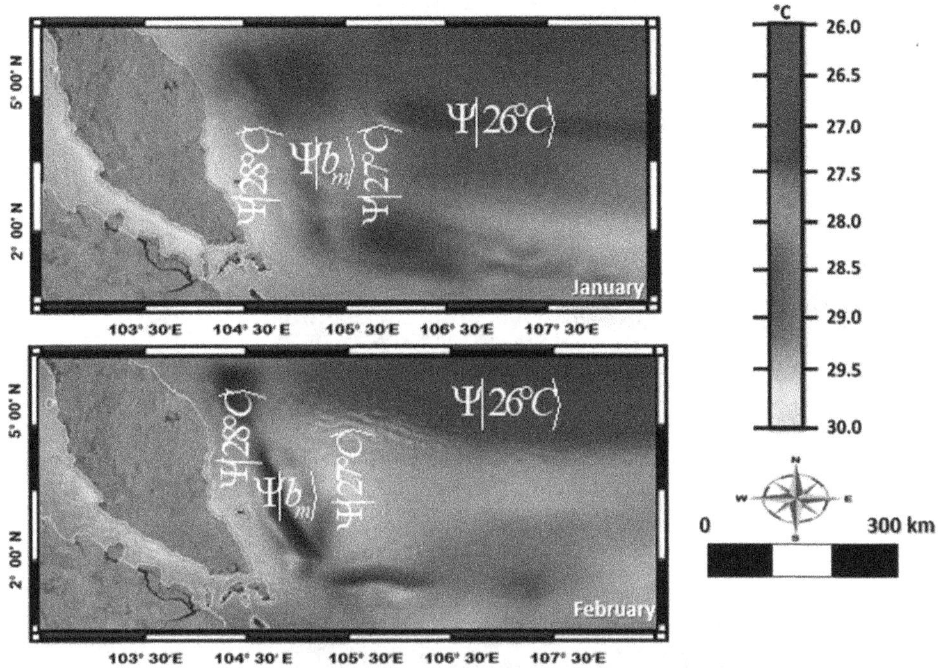

FIGURE 6.41 Fluctuations of quantized Marghany bound front state during the middle period of the northeast monsoon.

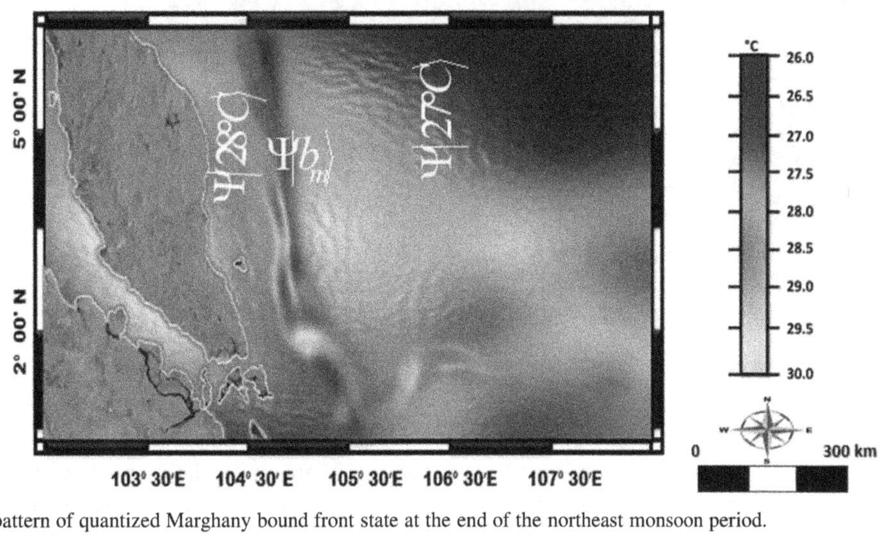

FIGURE 6.42 The pattern of quantized Marghany bound front state at the end of the northeast monsoon period.

Marghany bound front state became weaker because of the advection of cooler water flowing from the western boundary's southern domain (Fig. 6.44). Subsequently, in the middle of the southwest monsoon in June and July, the dominant water mass at 29°C spreads parallel to the western boundary and alliances with two cores of cold water intrusion at 2.3°N and 4.7°N between 104°E and 105°E(Fig. 6.45). The quantized Marghany bound front state appears to have formed at the end of the southwest monsoon period as a result of convergence between two water mass cores that are slightly different from one another and are curled from 2°N to 5°N and from 6°N to 7°N, respectively. There has been an average temperature difference of 0.5°C (Fig. 6.46).

Within the transitional period between the southwest monsoon period and the northeast monsoon period mainly in September and October, the water masses turn to be cooler of 28.5°C as demonstrated by well-mixed occurrences because of the strong turbulence (Marghany, 2012a, 2012b) (Fig. 6.47).

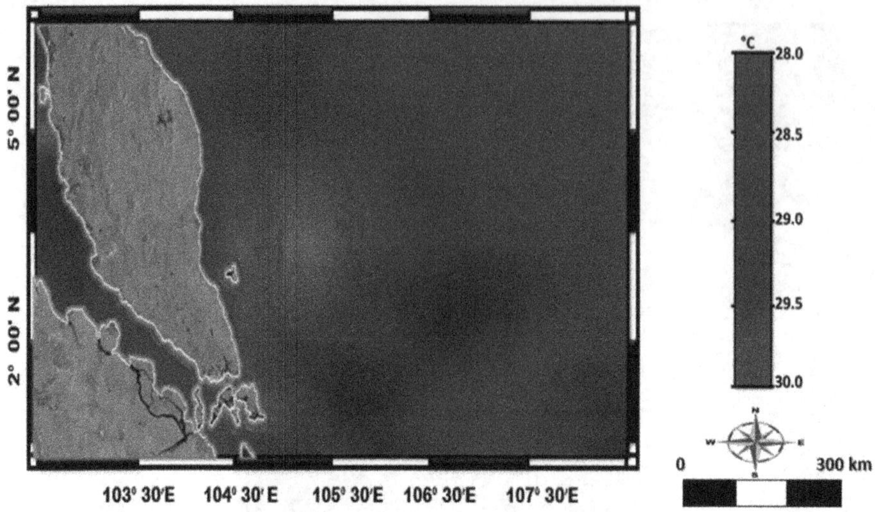

FIGURE 6.43 Weakest quantized Marghany bound front state in April.

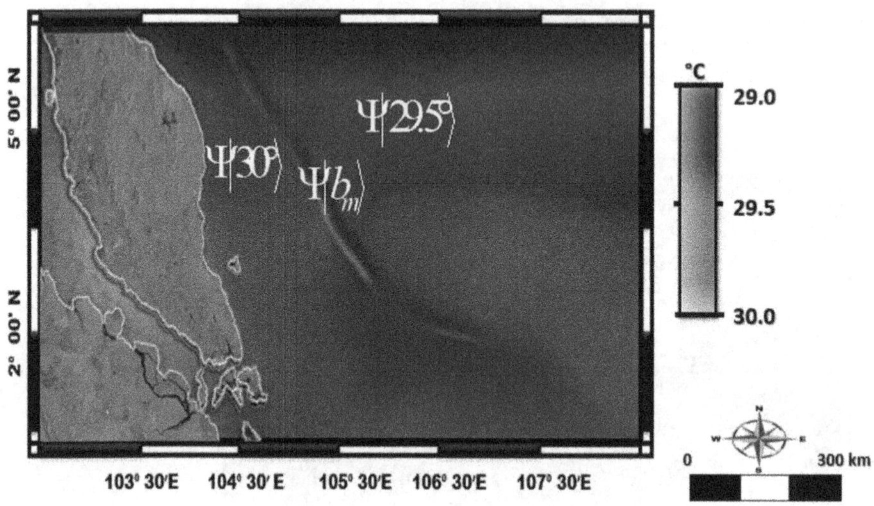

FIGURE 6.44 Weak quantized Marghany bound front state at the beginning of the southwest monsoon period.

The key query right now is how the warm and cool pools in the South China Sea form a thermal shelf front in the quantized Marghany turbulence boundary. In the real sense, the monsoonal winds have such a major impact on the South China Sea's (SCS) circulation patterns, resulting in distinct water currents that directly affect the water's temperature and salinity in various areas. The predominant atmospheric circulation patterns in the SCS are the northeast and southwest monsoons, which are also very pertinent to how the ocean currents are shaped.

Strong winds from the Asian continent blow toward the sea during the northeast monsoon (typically from November to March), driving a southward current in the northern part of the SCS. The equator and the southern coast of Vietnam are the destinations of this southward current, which carries cooler and saltier water from the SCS's northern regions (Saadon, Rojana-anawat Snidvongs, 1997; Zheng et al., 2016). Along the eastern coasts of Peninsular Malaysia and Borneo, where the wind blows the surface waters southward, deeper, cooler, and nutrient-rich waters upwell. This process supports healthy marine ecosystems.

Conversely, the predominant southwest winds change (typically from May to September), affecting the SCS in the opposite way. South of the SCS in the Karimata Strait, warmer and less salinized waters are forced northward by a boundary current driven by the southwest monsoon. The warmer waters from the equatorial regions are brought in by this northward current, which helps explain

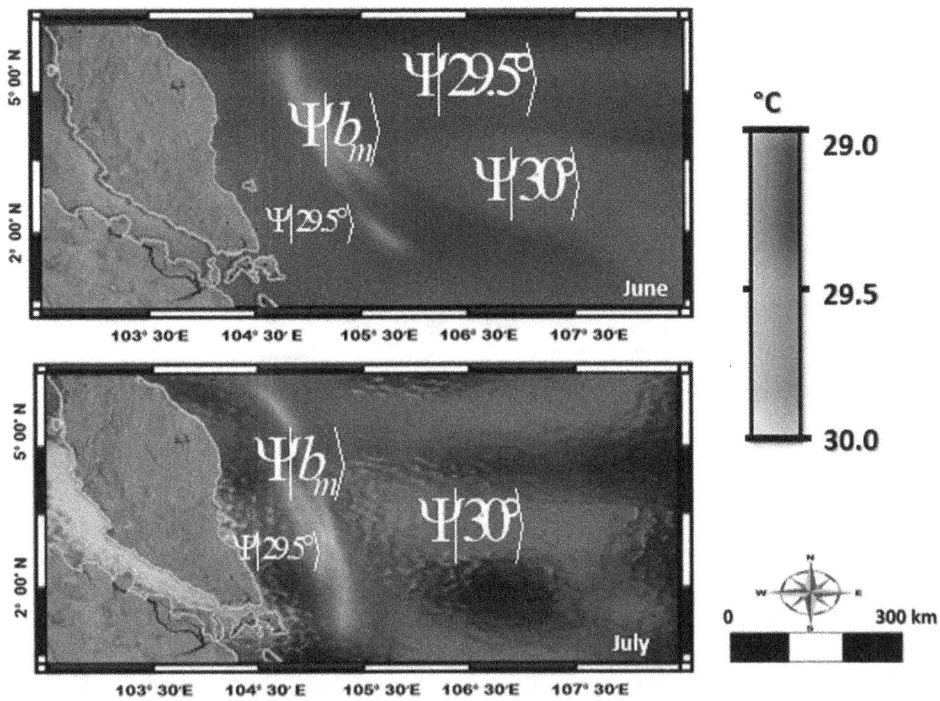

FIGURE 6.45 The pattern of the quantized Marghany bound front state at the middle of the southwest monsoon period.

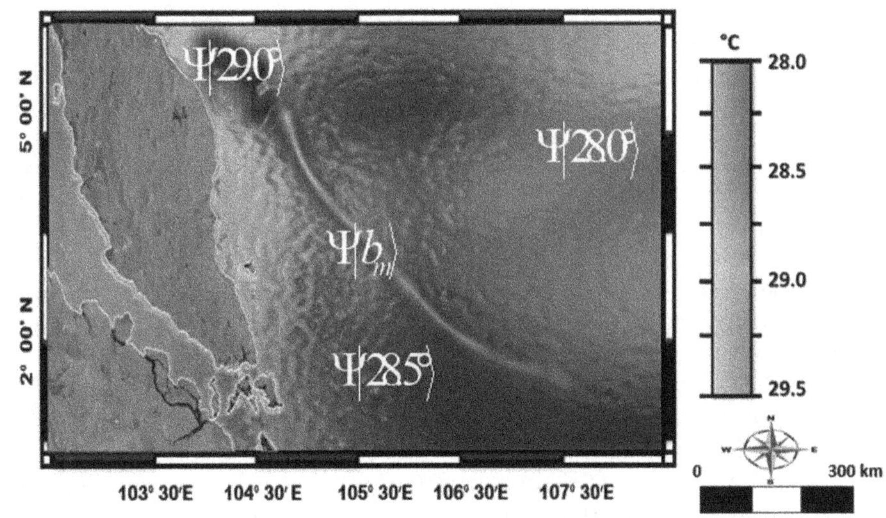

FIGURE 6.46 The pattern of quantized Marghany bound front state at the end of the southwest monsoon period.

why the northern parts of the SCS have higher sea surface temperatures at this time (Leenawarat et al., 2022).

The two monsoons' different circulation patterns have a significant impact on the South China Sea's coastal regions, weather patterns, and marine ecosystems. These variations in temperature and salinity, as well as the transportation of nutrients caused by the monsoonal currents, have an impact on the distribution of marine life, the occurrence of seasonal weather patterns, and the region's climate as a whole.

For the South China Sea to be forecasted and managed in terms of marine resources, fisheries, and environmental changes, it is crucial to comprehend these monsoonal-driven circulation patterns. The complex dynamics of the SCS have implications for regional climate variability and may affect weather patterns in Malaysia, Vietnam, the Philippines, and other neighboring coastal nations in Southeast Asia (Maged Marghany & Mahmoud, 1994; Marghany et al., 1996; Zheng et al., 2016).

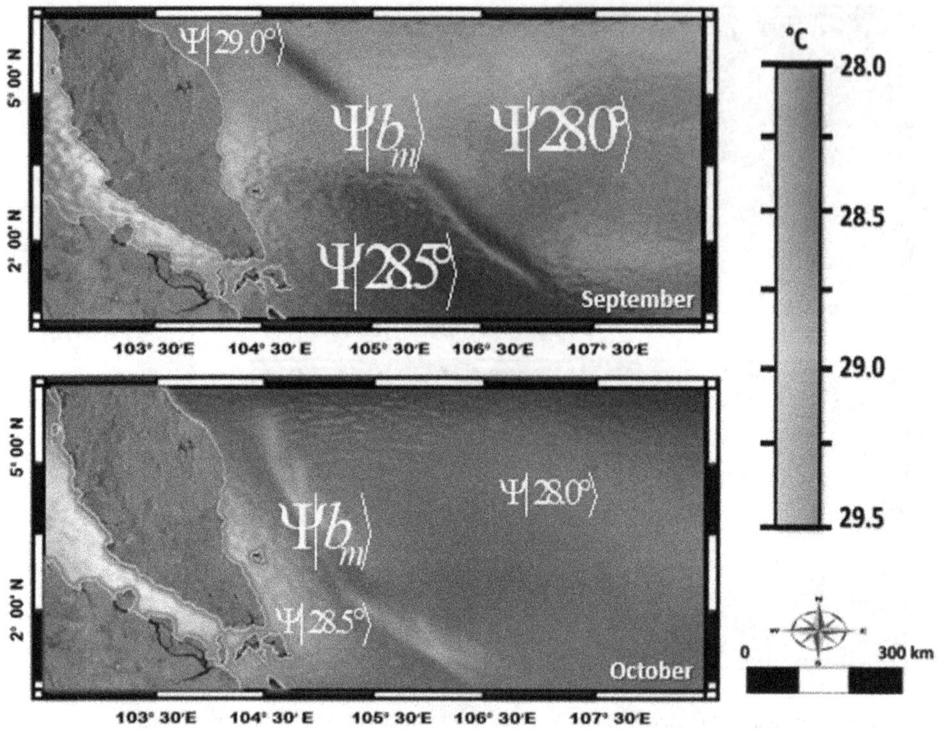

FIGURE 6.47 Dissipation of quantized Marghany bound front state during a transitional period to the northeast monsoon.

6.13.3 Can the quantized Marghany bound front state be used as a spectral turbulence incidence index?

The aforementioned viewpoint states that the annual cycle of QMBFS reaches its maximum peak during the northeast monsoon period, and dissipation is observed to have occurred during the transitional period, specifically in April. This scenario, which is the primary characteristic of QMBFS, is referred to as spectrum in the equilibrium range. The spectrum in the equilibrium range refers to the distribution of energy within the QMBFS system. It is during this period that the system experiences a balance between energy input and dissipation, resulting in a stable and predictable annual cycle. Equation 2.65 and the Hilimation of QMBFS, which is shown early in Eq. (6.38), therefore, can be used to construct the turbulence spectrum of the QMBFS. Keep in mind that $\eta =$ is the Kol mogorov microscale and $v = (\nu\varepsilon)^{1/4}$ is the Kolmogorov velocity (Fig. 6.48). Thus the equilibrium range also encapsulates the wave number dissipation range, similarly to how the boundary layer's wave layer encompasses the viscous sublayer. The spectrum of the dissipation, *D(k)*, can be demonstrated by:

$$D(k) = 2\nu k^2 E(k) \tag{6.63}$$

Eq. (6.36) shows that equilibrium occurs when $k\eta \geq 0.5$, which is well pronounced during northeast monsoon periods and transitional periods (September–October). However, a crucial factor should be the amount of energy that cascades down the spectrum. Since viscosity ultimately dissipates all energy, the total amount of energy transferred must be equal to the dissipation rate ε, and the viscosity itself should be the second important parameter (Johnson, 2016). Therefore the dissipation is well demonstrated during the transitional period (April) than in other monsoon periods. However, the intensity and characteristics of turbulence can vary depending on the environmental conditions, including seasonal changes. On the other hand, the quantized Marghany bound front state exhibits strong anisotropy as a result of $kl < 1.8$, and the strain rate of the mean flow maintains a steady orientation (Blevins, 1984; Pedlosky, 1979; Tritton, 2012). This anisotropy was further enhanced by the presence of vortices and eddies within the quantized Marghany bound front state. These dynamic features contribute to the overall complexity and nonuniformity of the flow field. During the transitional period, such as April in certain, the southwest monsoon season may be approaching, and atmospheric conditions can lead to specific turbulence patterns. These transitional periods are known for their changing weather patterns and shifts in wind and temperature gradients, which can influence the behavior of turbulence. In the context of monsoons, the transition from one season to another can lead to different atmospheric and

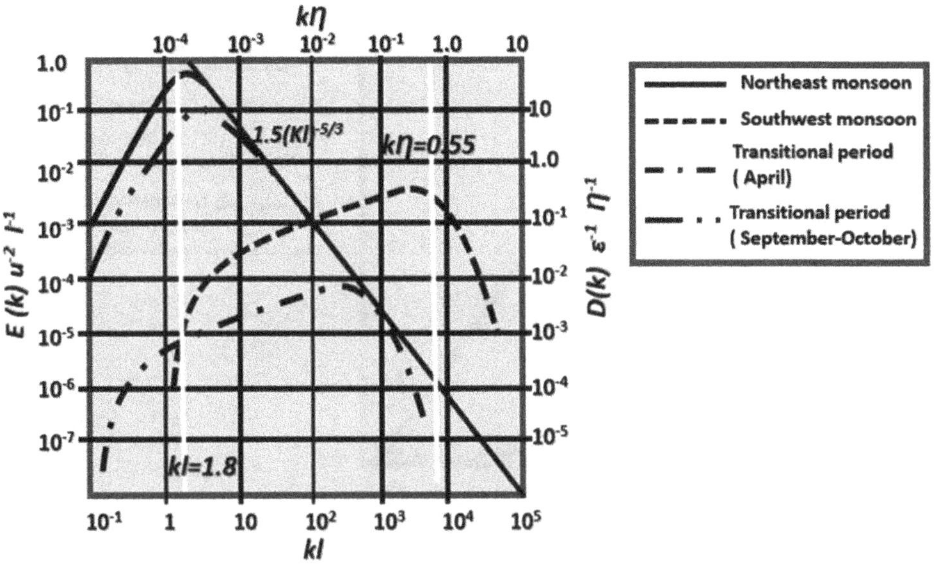

FIGURE 6.48 Normalized QMBFS energy and dissipation spectra. *QMBFS*, Quantized Marghany bound front state.

oceanic conditions, affecting the development of turbulence. For instance, the onset of the monsoon season may bring strong wind and rainfall, leading to enhanced turbulence in the atmosphere and ocean. This is very well illustrated by the Hamiltonian of the QMBFS curve. The highest peaks of the Hamiltonian of QMBFS occur during the northeast monsoon and transitional season (September to October), with values of 0.9 and 0.6, respectively.

Notably, the early northeast monsoon gains energy from the transitional season (September to October), which contributes to enhancing the spectral turbulence energy. Owing to less energy transfer from the end of the northeast monsoon periods, the transitional period (April) is dominated by a strong dissipation of 0.001 of the Hamiltonian of the QMBFS. Furthermore, the Hamiltonian of the QMBFS curve showed that dissipation persisted through the middle of the southwest monsoon (June–July). The 0.05 Hamiltonian of the QMBFS at the end of the southwest monsoon is a source of energy turbulence transfer during the transitional period (September–October). This indicates that the dissipation of the Hamiltonian of the QMBFS curve continues during the transitional period, resulting in a strong energy transfer (Fig. 6.49). Additionally, the presence of a significant source of energy turbulence transfer from the end of the southwest monsoon to the transitional period suggests sustained dissipation throughout this period (Busalacchi & O'Brien, 1980; Camerlengo & Demmler, 1997; Camerlengo, 1982; Hartmann et al., 2021; Johnson, 2016; Marghany, 2021). The transfer of energy turbulence during the transitional period is crucial for maintaining the overall stability and balance of the monsoon system. This allows for a smooth transition between different monsoon seasons and helps regulate weather patterns in the region.

Needless to say, QMBFS can act as a source of energy for the turbulent kinetic energy cascade. Turbulence is known for its ability to transfer energy from larger scales to smaller scales, leading to the dissipation of energy as heat. QMBFS can provide the necessary energy input to initiate and sustain turbulence in the surrounding fluid. In this regard, when a QMBFS moves over the boundary layer or oceanic surface layer, it is entangled with the lower portion of the ocean. This superposition can lead to the generation of turbulence owing to the complex interactions between different water masses.

The quantized Marghany bound front state (QMBFS) can potentially be used as a spectral turbulence incidence index. QMBFS is a mathematical model that captures turbulence behavior in various scenarios. It is a promising candidate for assessing turbulence incidence. Further research and validation are necessary to determine the effectiveness and reliability of this specific application.

6.14 Why can algorithms for quantum image fusion and quantum edge detection automatically identify the boundaries in different images?

Quantum image fusion and quantum edge detection algorithms offer the potential to automatically detect boundaries in images due to their inherent ability to process and analyze information in a highly parallel and superposed manner, which can lead to more efficient and accurate boundary detection compared with classical algorithms. In this perspective, in classical computing, algorithms process data sequentially, examining each pixel of an

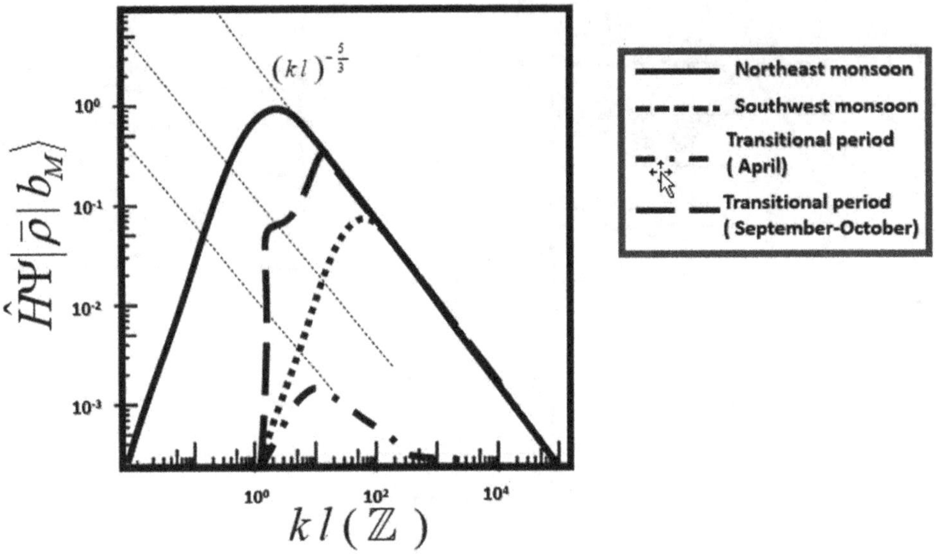

FIGURE 6.49 Hamiltonian of the QMBFS turbulence spectra. *QMBFS*, Quantized Marghany bound front state.

image one after the other. However, in quantum computing, quantum bits (qubits) can exist in a superposition of states, meaning they can represent multiple possibilities simultaneously. This allows quantum algorithms to explore all potential boundaries in an image at once, vastly speeding up the computation process. Quantum algorithms, therefore, can utilize entanglement, a unique property of qubits that allows them to be highly correlated even when separated. By exploiting entanglement, quantum edge detection algorithms can take advantage of relationships between neighboring pixels to detect edges more accurately, as neighboring pixels are often related to form boundaries. Quantum image fusion algorithm, consequently, can efficiently extract relevant features from different source images and combine them in a superposed state to create a fused image (Baym, 2018; Cavalieri & Maio, 2020; Chetia et al., 2021; Grinko et al., 2021; Majji et al., 2022). This process inherently focuses on preserving significant spatial and spectral details, making it easier to detect boundaries accurately.

The author's earlier work on "nonlinear ocean dynamics" laid the foundation for the extension of the quantized Marghany's front theory, as demonstrated in this chapter. The discovery of a shallow coastal front on the east coast of the Malaysian Peninsula is theorized in the context of nonlinear ocean dynamics. The author first identified in 1994 that subtle differences in water masses could lead to the emergence of separate states known as a "quantized Marghany bound front state." It is imperative to note that this remarkable phenomenon was initially discovered and studied extensively by the author in his master's thesis in 1994. Therefore the Marghany front is an exceptional oceanographic characteristic that has drawn unique phenomena alliances with the quantized Marghany turbulence boundary as was shown in Chapter 4. It maintains well-defined water mass states, even with minor variations in water mass gradient features. This remarkable phenomenon, known as the quantized Marghany bound front state (QMBFS), was initially grounded by the author in his master's thesis in 1994 (Maged Marghany & Mahmoud, 1994).

This chapter similarly utilized cutting-edge quantum image processing tools to automatically find the quantized Marghany bound front state (QMBFS) in a single SAR dataset. Quantum image fusion technology has also been developed to track thermal fronts using multiSAR and MODIS data. The energy transfer from the mean flow to turbulence and the energy transfer from large to small scales can both be quantized by the quantized Marghany bound front state. The turbulence receives its energy from the mean strain rate and transfers energy to small scales at a dissipation rate; therefore the quantization of the large-scale part of the spectrum should be based on the mean strain rate and dissipation rates. Needless to say, the Marghany front is a unique oceanographic feature that is characterized by its ability to maintain distinct water mass states despite the minimal variations in water mass gradient characteristics.

References

Adcroft, A., & Hallberg, R. (2006). On methods for solving the oceanic equations of motion in generalized vertical coordinates. *Ocean Modelling*, *11*(1–2), 224–233. Available from https://doi.org/10.1016/j.ocemod.2004.12.007.

Alpers W. Holt B. 1995 1 1995/01 International Geoscience and Remote Sensing Symposium (IGARSS), 1588–1590, IEEE Germany, Imaging of ocean features by SIR-C/X-SAR: An overview, 2

Baym, G. (2018). *Lectures on quantum mechanics* (pp. 1–594). United States: CRC Press. Available from http://www.tandfebooks.com/doi/book/10.1201/9780429499265.

Blevins R.D., Applied fluid dynamics handbook.. (1984).

Bowman, M. J. (2018). *Estuarine fronts. Hydrodynamics of estuaries: Volume I estuarine physics* (pp. 85–132). United States: CRC Press. Available from http://www.tandfebooks.com/doi/book/10.1201/9781351073257.

Brink, K. H. (2023). *Physical oceanography of continental shelves*. Princeton University Press.

Busalacchi, A. J., & O'Brien, J. J. (1980). The seasonal variability in a model of the tropical Pacific. *J. Phys. Oceanogr*, *10*(12), 1929–1951. Available from https://doi.org/10.1175/1520-0485(1980).

Camerlengo, A. (1982). Large-scale response of the pacific ocean subarctic front to momentum transfer: A numerical study. *Journal of Physical Oceanography*, *12*(10), 1106–1121. Available from https://doi.org/10.1175/1520-0485(1982), http://journals.ametsoc.org/loi/phoc.

Camerlengo A.L., Demmler M.I., On the dissipation of an upper oceanic front. Geoacta. (1997).,

Cavalieri G., Maio D., A quantum edge detection algorithm. arXiv, Italy arXiv. 23318422. (2020), Available from: https://arxiv.org.

Chetia, R., Boruah, S. M. B., & Sahu, P. P. (2021). Quantum image edge detection using improved Sobel mask based on NEQR. *Quantum Information Processing*, *20*(1). Available from https://doi.org/10.1007/s11128-020-02944-7, http://www.kluweronline.com/issn/1570-0755/contents.

Dijkstra, H. A. (2008). *Dynamical oceanography* (pp. 1–407). Netherlands: Springer Berlin Heidelberg. Available from http://www.springerlink.com/openurl.asp?genre = book&isbn = 978-3-540-76375-8.

Eckart, C., & Ferris, H. G. (1956). Equations of motion of the ocean and atmosphere. *Reviews of Modern Physics*, *28*(1), 48–52. Available from https://doi.org/10.1103/RevModPhys.28.48.

Fu X. Ding M. Sun Y. Chen S. 2009 12 2009/12 Proceedings of SPIE - The International Society for Optical Engineering, Volume 7497, 10.1117/12.832499, 0277786X, China. A new quantum edge detection algorithm for medical images.

Galindo A., Pascual P., Quantum mechanics I. Springer Science & Business Media, (2012).

Grinko, D., Gacon, J., Zoufal, C., & Woerner, S. (2021). Iterative quantum amplitude estimation. *Nature Research, Switzerland npj Quantum Information*, *7*(1). Available from https://doi.org/10.1038/s41534-021-00379-1, https://www.nature.com/npjqi/.

Hartmann A. Davari A. Seehaus T. Braun M. Maier A. Christlein V. 2021 1 2021/01 International Geoscience and Remote Sensing Symposium (IGARSS), 10.1109/IGARSS47720.2021.9554292 3479–3482, Institute of Electrical and Electronics Engineers Inc. Germany. Bayesian U-Net for segmenting glaciers in SAR imagery.

Huzzey, L. M., & Brubaker, J. M. (1988). The formation of longitudinal fronts in a coastal plain estuary. *Journal of Geophysical Research*, *93* (C2), 1329. Available from https://doi.org/10.1029/jc093ic02p01329.

Johannessen, J. A., Shuchman, R. A., Johannessen, O. M., Davidson, K. L., & Lyzenga, D. R. (1991). Synthetic aperture radar imaging of upper ocean circulation features and wind fronts. *Journal of Geophysical Research*, *96*(6), 10–422. Available from https://doi.org/10.1029/91jc00301.

Johnson, R. W. (2016). *Handbook of fluid dynamics*. CRC Press.

Kao, T. W. (1980). The dynamics of oceanic fronts. Part I: The Gulf Stream. *Journal of Physical Oceanography*, *10*(4), 483–492. Available from https://doi.org/10.1175/1520-0485(1980), 010 < 0483:tdoofp > 2.0.co;2.

Kim, T. S., Park, K. A., Li, X., Mouche, A. A., Chapron, B., & Lee, M. (2017). Observation of wind direction change on the sea surface temperature front using high-resolution full polarimetric SAR data. *Journal of Selected Topics in Applied Earth Observations and Remote Sensing*, *10*(6), 2599–2607. Available from https://doi.org/10.1109/JSTARS.2017.2660858, http://ieeexplore.ieee.org/xpl/RecentIssue.jsp?punumber = 4609443.

Konishi, K., & Paffuti, G. (2009). *Quantum mechanics: A new introduction*. OUP.

Kowalik, Z., & Murty, T. S. (1993). *Numerical modeling of ocean dynamics* (5). World Scientific.

Krogstad, H. E., & Schyberg, H. (1990). *On Hasselmann's nonlinear ocean-SAR transformation. Remote sensing: Global monitoring for earth management*. Espoo: IEEE.

Kuang, H. L., Perrie, W., Xie, T., Zhang, B., & Chen, W. (2012). Retrievals of sea surface temperature fronts from SAR imagery. *Geophysical Research Letters*, *39*(10). Available from https://doi.org/10.1029/2012GL051288, http://onlinelibrary.wiley.com/journal/10.1002/(ISSN)1944-8007/issues?year = 2012.

Largier, J. L. (1993). Estuarine fronts: How important are they? *Estuaries*, *16*(1), 1–11. Available from https://doi.org/10.2307/1352760.

Leenawarat, D., Luang-on, J., Buranapratheprat, A., & Ishizaka, J. (2022). Influences of tropical monsoon and El Niño Southern Oscillations on surface chlorophyll-a variability in the Gulf of Thailand. *Frontiers Media S.A., Japan Frontiers in Climate.*, *4*. Available from https://doi.org/10.3389/fclim.2022.936011, http://www.frontiersin.org/journals/climate.

Li, N., Yan, F., Venegas-Andraca, S. E., & Hirota, K. (2023). Quantum image scaling with applications to image steganography and fusion. *Signal Processing: Image Communication*, *117*, 117015. Available from https://doi.org/10.1016/j.image.2023.117015.

Liu, X., & Xiao, D. (2019). Multimodality mmage fusion based on quantum wavelet transform and sum-modified-Laplacian rule. *International Journal of Theoretical Physics*, *58*(3), 734–744. Available from https://doi.org/10.1007/s10773-018-3971-4, http://www.kluweronline.com/issn/0020-7748.

Macvean, M. K., & Woods, J. D. (1980). Redistribution of scalars during upper ocean frontogenesis: A numerical model. *Quarterly Journal of the Royal Meteorological Society*, *106*(448), 293–311. Available from https://doi.org/10.1002/qj.49710644805.

Maged Marghany., & Mahmoud. (1994). *Coastal circulation off Kuala Terengganu*.

Majji, S. R., Chalumuri, A., Kune, R., & Manoj, B. S. (2022). Quantum processing in fusion of SAR and optical images for deep learning: A data-centric approach. *IEEE Access*, *10*, 73743–73757. Available from https://doi.org/10.1109/ACCESS.2022.3189474, http://ieeexplore.ieee.org/xpl/RecentIssue.jsp?punumber = 6287639.

Mandl, F. (2013). *Quantum mechanics*. John Wiley & Sons, John Wiley & Sons.

Marghany M. 2012a 7 2012/07Lecture Notes in Computer Science (including subseries Lecture Notes in Artificial Intelligence and Lecture Notes in Bioinformatics), Volume 7335, 10.1007/978-3-642-31137-6_34, 16113349 3, 447–456, Malaysia. Three-dimensional coastal front visualization from RADARSAT-1 SAR satellite data.

Marghany, M. (2012b). Intermonsoon water mass characteristics along coastal waters off Kuala Terengganu. Malaysia. *International Journal of Physical Sciences.*, *7*(8), 1294–1299.

Marghany M., Nonlinear ocean dynamics: Synthetic aperture radar. Elsevier, (2021).

Marghany, M. (2011). Three-dimensional coastal water front reconstruction from RADARSAT-1 synthetic aperture radar (SAR) satellite data. *International Journal of Physical Sciences*, *6*(29), 6653–6659. Available from https://doi.org/10.5897/IJPS11.1239Malaysia, http://www.academic-journals.org/ijps/PDF/pdf2011/16Nov/Marghany%202.pdf.

Marghany M.M., Saadon M.S., Hussain M.L., Mohmed M.I., Seasonal thermohaline variation in \ncoastal waters off Kuala Terengganu National Conference on Climate Change. (1996), 152–167.

McLellan, H. J. (1965). *Elements of physical oceanography*. Pergamon Press.

Mellor, G., & Huang, R. X. (1997). Introduction to physical oceanography. *American Journal of Physics*, *65*(10), 1028–1029. Available from https://doi.org/10.1119/1.18716.

Miranda, F. P., Andutta., & Filho, B. K. (2017). *Fundamentals of estuarine physical oceanography* (8). Springer.

Muelbert, J. H., Acha, M., Mianzan, H., Guerrero, R., Reta, R., Braga, E. S., Garcia, V. M. T., Berasategui, A., Gomez-Erache, M., & Ramírez, F. (2008). Biological, physical and chemical properties at the Subtropical Shelf Front Zone in the SW Atlantic Continental Shelf. *Continental Shelf Research*, *28*(13), 1662–1673. Available from https://doi.org/10.1016/j.csr.2007.08.011.

Neumann, V. (2018). Mathematical foundations of quantum mechanics: New edition. *Princeton university press*, *53*.

O'Donnell, J., Marmorino, G. O., & Trump, C. L. (1998). Convergence and downwelling at a river plume front. *Journal of Physical Oceanography*, *28*(7), 1481–1495. Available from https://doi.org/10.1175/1520-0485(1998), http://journals.ametsoc.org/loi/phoc.

Palmen, C. W., & Newton. (1969). *Atmospheric circulation system* (602). Academic Press.

Pan, J., Jay, D. A., & Orton, P. M. (2007). Analyses of internal solitary waves generated at the Columbia River plume front using SAR imagery. *Journal of Geophysical Research: Oceans*, *112*(7). Available from https://doi.org/10.1029/2006JC003688, http://onlinelibrary.wiley.com/journal/10.1002/(ISSN)2169-9291.

Pedlosky, J. (1979). *Geophysical fluid dynamics*. Springer US.

Piola, A. R., Campos, E. J. D., Möller, O. O., Charo, M., & Martinez, C. (2000). Subtropical shelf front off eastern South America. *Journal of Geophysical Research: Oceans*, *105*(3), 6565–6578. Available from https://doi.org/10.1029/1999jc000300, http://agupubs.onlinelibrary.wiley.com/agu/jgr/journal/10.1002/(ISSN)2169-9291/.

Pond, G. L. P. (1978). *Introductory dynamic oceanography*. Pergamon Press.

Garvine, R. W. (1974). Dynamics of small-scale oceanic fronts. *Journal of Physical Oceanography*, *4*(4), 557–569. Available from https://doi.org/10.1175/1520-0485(1974).

Roden, G. I. (1976). On the structure and prediction of oceanic fronts. *Naval Research Reviews.*, *29*(3), 18–35.

Roden, G. I. (1972). Temperature and salinity fronts at the boundaries of the subarctic-subtropical transition zone in the western Pacific. *Journal of Geophysical Research*, *77*(36), 7175–7187. Available from https://doi.org/10.1029/jc077i036p07175.

Roden, G. I. (1974). Thermohaline structure, fronts, and sea-air energy exchange of the trade wind region east of Hawaii. *Journal of Physical Oceanography*, *4*(2), 168–182. Available from https://doi.org/10.1175/1520-0485(1974).

Saadon M. Rojana-anawat Snidvongs A. 1997 1997 Proceedings of the First Technical Seminar on Marine Fishery Resources Survey in the South China Sea, Area I: Gulf of Thailand and Peninsular Malaysia, 1–5. Physical characteristics of watermass in the South China Sea, Area I: Gulf of Thailand and east coast of Peninsular Malaysia.

Simpson, J. H. (1472). The shelf-sea fronts: Implications of their existence and behaviour. *Philosophical Transactions of the Royal Society of London. Series A, Mathematical and Physical Sciences*, *302*, 531–546.

Simpson, J. H., Allen, C. M., & Morris, N. C. G. (1978). Fronts on the continental shelf. *Journal of Geophysical Research*, *83*(C9), 4607. Available from https://doi.org/10.1029/jc083ic09p04607.

Sundani, D., Widiyanto, S., Karyanti, Y., & Wardani, D. T. (2019). Identification of image edge using quantum canny edge detection algorithm. *Journal of ICT Research and Applications*, *13*(2), 133–144. Available from https://doi.org/10.5614/itbj.ict.res.appl.2019.13.2.4, http://journals.itb.ac.id/index.php/jictra/article/download/9386/4334.

Tawfik, N., Elnemr, H. A., Fakhr, M., Dessouky, M. I., & El-Samie, F. E. A. (2021). Survey study of multimodality medical image fusion methods. *Multimedia Tools and Applications*, *80*(4), 6369–6396. Available from https://doi.org/10.1007/s11042-020-08834-5, https://link.springer.com/journal/11042.

Thompson, L. A. (2000). Ekman layers and two-dimensional frontogenesis in the upper ocean. *Journal of Geophysical Research: Oceans*, *105*(C3), 6437–6451. Available from https://doi.org/10.1029/1999jc900336.

Tritton D.J., Physical fluid dynamics. Springer Science & Business Media, Springer Science & Business Media, (2012).

Vogelzang, J., Wensink, G. J., Calkoen, C. J., & Van Der Kooij, M. W. A. (1997). Mapping submarine sand waves with multiband imaging radar 2. Experimental results and model comparison. *Journal of Geophysical Research: Oceans*, *102*(1), 1183–1192. Available from https://doi.org/10.1029/96jc02836, http://agupubs.onlinelibrary.wiley.com/agu/jgr/journal/10.1002/(ISSN)2169-9291/.

Webb P., Introduction to oceanography. (2021).

Yao, X. W., Wang, H., Liao, Z., Chen, M. C., Pan, J., Li, J., Zhang, K., Lin, X., Wang, Z., Luo, Z., Zheng, W., Li, J., Zhao, M., Peng, X., & Suter, D. (2017). Quantum image processing and its application to edge detection: Theory and experiment. *Physical Review X*, *7*(3). Available from https://doi.org/10.1103/PhysRevX.7.031041, https://journals.aps.org/prx.

Yu, L. (2015). Sea-surface salinity fronts and associated salinity-minimum zones in the tropical ocean. *Journal of Geophysical Research: Oceans*, *120*(6), 4205–4225. Available from https://doi.org/10.1002/2015JC010790, http://onlinelibrary.wiley.com/journal/10.1002/(ISSN)2169-9291.

Zettili, N., & Zahed, I. (2003). Quantum mechanics: Concepts and applications. *American Journal of Physics*, *71*(1), 93. Available from https://doi.org/10.1119/1.1522702, 93.

Zheng, Z. W., Zheng, Q., Kuo, Y. C., Gopalakrishnan, G., Lee, C. Y., Ho, C. R., Kuo, N. J., & Huang, S. J. (2016). Impacts of coastal upwelling off east Vietnam on the regional winds system: An air-sea-land interaction. *Dynamics of Atmospheres and Oceans*, *76*, 105–115. Available from https://doi.org/10.1016/j.dynatmoce.2016.10.002, http://www.elsevier.com/inca/publications/store/5/0/3/3/2/7/index.htt.

Zhou, J., Liu, S., Qian, H., Li, Y., Luo, H., Wen, S., Zhou, Z., Guo, G., Shi, B., & Liu, Z. (2020). Metasurface enabled quantum edge detection. *American Association for the Advancement of Science, China Science Advances*, *6*(51). Available from https://doi.org/10.1126/sciadv.abc4385, https://advances.sciencemag.org/content/advances/6/51/eabc4385.full.pdf.

Further Reading

Roden, G. I. (1977). Oceanic subarctic fronts of the Central Pacific: Structure of and response to atmospheric forcing. *Journal of Physical Oceanography*, *7*(6), 761–778. Available from https://doi.org/10.1175/1520-0485(1977).

Chapter 7

Texture and quantum entropy algorithms for turbulence and oceanic front structural detections in synthetic aperture radar images

Texture analysis holds a prominent position within radar imagery, particularly within the realm of remote sensing and image processing. Texture algorithms serve the purpose of scrutinizing the spatial fluctuations in pixel intensities within an image, thereby yielding invaluable insights for a myriad of radar imaging applications. These applications include front turbulence automatic identification, classification, and scene interpretation. Hence, radar imagery frequently grapples with the issue of speckle noise, potentially masking crucial details. Texture algorithms come to the rescue by mitigating speckle noise through image smoothing or filtering techniques, all the while safeguarding the integrity of essential texture characteristics. Nonetheless, the preceding chapters have omitted an examination of the influence of dynamic spatial variations in speckles on the interpretation and analysis of radar image texture. This chapter is dedicated to a comprehensive exploration of the extent to which speckles may pose hindrances to elucidating targets or oceanic features within radar imagery.

7.1 What is the significance of speckles in revealing turbulence front features in SAR data?

Speckle, often referred to as speckle pattern or speckle noise, is a granular noise texture that diminishes image quality due to interference occurring among wavefronts in coherent imaging systems. These systems encompass radar, synthetic aperture radar (SAR), medical ultrasound, and optical coherence tomography. Speckle should not be regarded as extraneous noise; instead, it represents an inherent variability stemming from diffuse reflections. This occurs due to nonidentical scatterers within each cell, with the coherent illumination wave exhibiting a high sensitivity to even minor fluctuations in phase changes. While the exploration of this phenomenon dates back to the era of Newton [citation needed], the prominence of speckles has surged notably with the advent of the laser. These distinctive reflections can manifest on various materials, including but not limited to paper, white paint, uneven surfaces, and in media characterized by a substantial concentration of scattering particles within space, such as airborne dust or cloudy liquids. Over time, speckles have found diverse applications in the realms of microscopy, imaging, and optical manipulation (Argenti & Alparone, 2002; Bromberg & Cao, 2014; Dong & Pan, 2017).

The overwhelming majority of surfaces, whether of synthetic or natural origin, exhibit pronounced roughness at scales comparable with the wavelength of the probing radiation. This phenomenon becomes apparent when we model the reflectivity function as an array of individual scatterers. Given the finite resolution inherent in the imaging process, we are, at any given moment, receiving signals emanating from a distribution of scatterers contained within the resolution cell. These scattered signals undergo coherent summation, resulting in constructive and destructive interference patterns, visually manifested as bright and dark spots within the image.

In conventional radar imaging, speckle contributes to an increase in the mean gray level within a local area. In the context of SAR, speckle is a prevalent concern that poses challenges in image interpretation. This phenomenon arises due to the coherent processing of backscattered signals originating from numerous distributed targets. In SAR turbulence front structures, for instance, speckle emerges from signals originating from elementary scatterers, such as gravity-capillary ripples, and takes the

Synthetic Aperture Radar Image Processing Algorithms for Nonlinear Oceanic Turbulence and Front Modeling. DOI: https://doi.org/10.1016/B978-0-443-19155-8.00007-6

form of a pedestal-like structure beneath the image depicting the sea waves (Asakura & Takai, 1981; Goodman, 2007; Liu et al., 2021).

It is noteworthy that speckles, while often considered a nuisance, can also carry valuable information, especially in connection with laser speckles and dynamic speckle phenomena. In cases where spatial speckle patterns evolve over time, they can serve as a means to measure the activity of a surface. This attribute proves particularly useful for tasks such as measuring displacement fields through digital image correlation techniques.

Needless to say, speckles, within the framework of SAR imaging of oceanic features, encapsulate a dualistic nature—simultaneously masking information and unveiling intricate insights into the ocean's finely detailed characteristics and dynamics. Consequently, researchers and oceanographers are tasked with the meticulous analysis and processing of SAR data, harnessing the informational potential harbored within speckle patterns, while also mitigating their disruptive influence on the interpretation of such imagery (Bromberg & Cao, 2014; Dong & Pan, 2017; Szalai & Dogossy, 2021).

The signal-to-noise ratio (SNR) is a critical concept in various fields, including electronics, communications, and signal processing. It represents the ratio of the desired signal's strength or power to the level of unwanted background noise or interference present in a given system or signal. SNR is typically expressed in decibels (dB) and can be calculated using the following formula:

$$\mathrm{SNR(dB)} = 10 \cdot \log_{10}(N^{-1}S) \tag{7.1}$$

where S represents the signal power or amplitude; and N represents the noise power or amplitude. In this sense, a high SNR indicates that the signal is much stronger than the noise, which is desirable in most applications because it implies a clear and reliable signal. Conversely, a low SNR suggests that the noise level is comparable with or stronger than the signal, making it difficult to extract meaningful information. In other words, a radar image with a high SNR is critical for detecting and tracking targets while minimizing false alarms. Let's say that after data processing, the calculated SNR is found to be 32 dB.

In this example, an SNR of 32 dB indicates that the power of the signal (ocean wave reflections) is 32 dB higher than the power of the noise present in the SAR data (Fig. 7.1). This suggests that the SAR image of the ocean surface features has a relatively strong signal-to-noise ratio, indicating that the quantized Marghany turbulence boundary feature is clearly detectable and distinguishable from the background noise. A higher SNR value would imply an even sharper and more reliable image for analyzing ocean surface features within incident angle ranges from 32.5 to 31 degrees (Fig. 7.2).

7.2 Formation of speckles

In the realm of physics, the term "interference" denotes a phenomenon wherein two coherent waves are superimposed by combining their intensities or displacements, while considering their relative phase disparity. This amalgamation

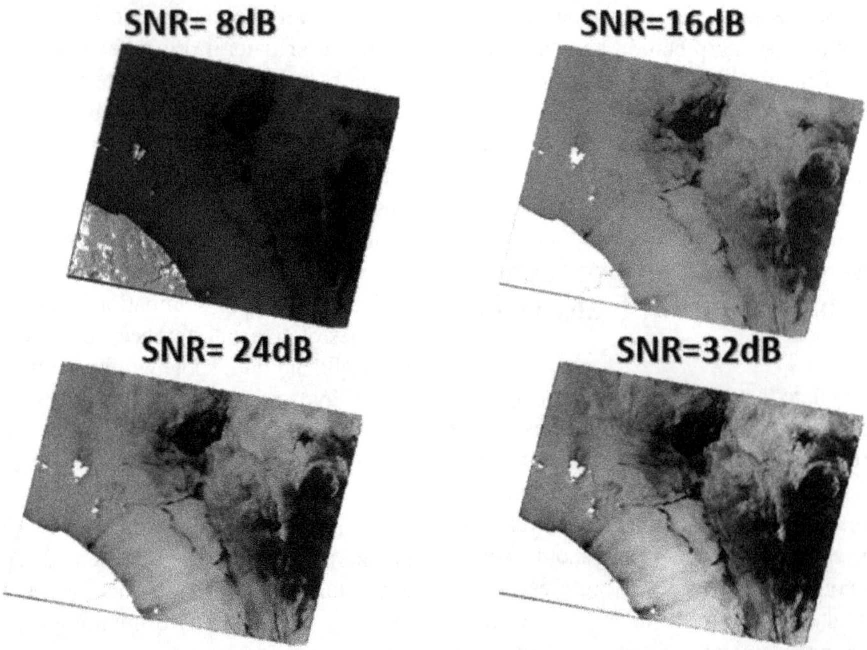

FIGURE 7.1 **Different SNR spatial variation impacts on quantized Marghany turbulence boundary feature in SAR image.**

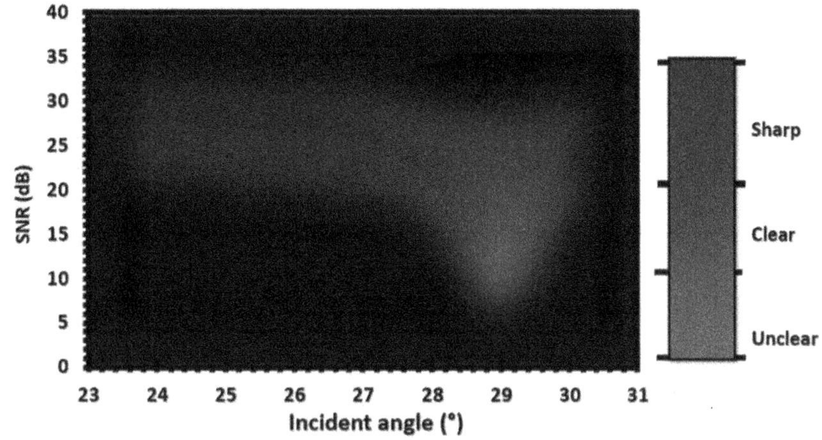

FIGURE 7.2 Spatial changes in incident angles as they traverse the quantized Marghany turbulence boundary feature coincide with the spatial fluctuations in signal-to-noise ratio.

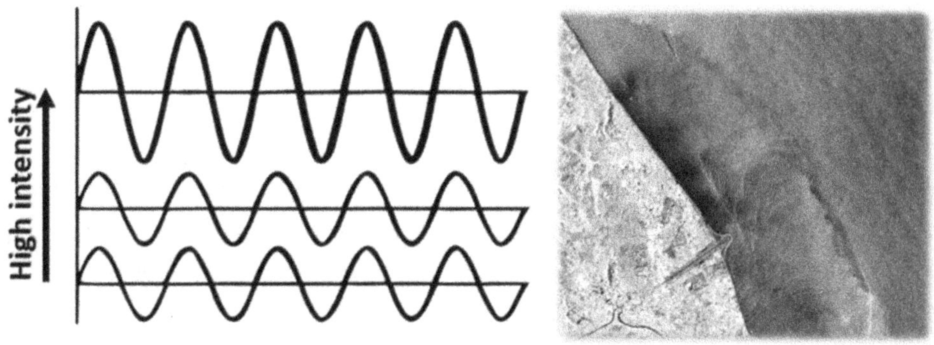

FIGURE 7.3 Constructive interference.

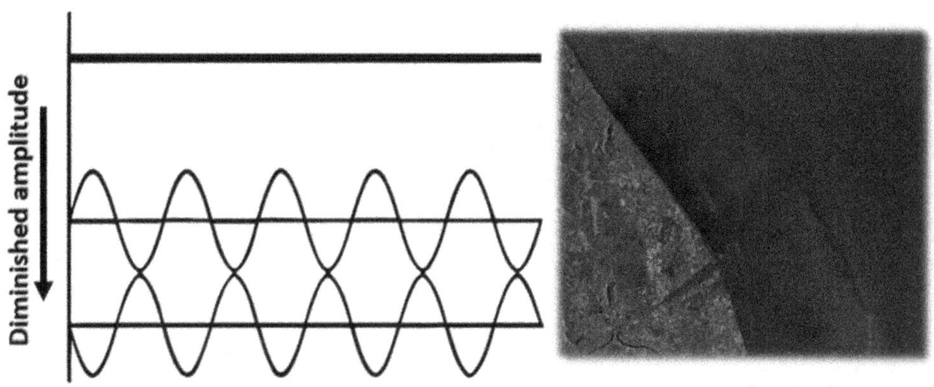

FIGURE 7.4 Destructive interference.

yields a resultant wave that can manifest either as increased intensity, termed "constructive interference," (Fig. 7.3) or diminished amplitude, known as "destructive interference (Fig. 7.4)." The outcome hinges on whether the two waves are in phase or out of phase, respectively. Interference phenomena are evident across a wide range of applications, including the quantized Marghany turbulence boundary features within SAR data. Notably, these phenomena are not limited to natural occurrences; within SAR pixels, bright pixels denote constructive interference, whereas dark pixels signify destructive interference (Asakura & Takai, 1981; Lopes et al., 1993; Okamoto & Asakura, 1995).

The interference of numerous waves' signals, therefore can be incorporated into the framework of quantum mechanics. The central concept that addresses this phenomenon is the theory of wave superposition. In this perspective, the fundamental concept of wave superposition posits that when multiple propagating waves of the same nature intersect at a common point, the resulting amplitude at that point corresponds to the vector summation of the amplitudes of the individual waves(Fig. 7.5). In cases where a crest of one wave coincides with a crest of another wave of identical frequency at the same spatial location, constructive interference occurs, resulting in the addition of individual amplitudes. Conversely, when the crest of one wave aligns with the trough of another wave, destructive interference emerges, leading to an amplitude equal to the difference between the individual amplitudes (Asakura & Takai, 1981; Raney & Wessels, 1988; Reu, 2015; Szalai & Dogossy, 2021).

As a result, the superposition of numerous wave signals within radar images gives rise to equivalent spatial variations in intensity, including both destructive and constructive aspects. Within this framework, the presence of destructive signals manifests as dark pixels, whereas constructive superposition wave signals are represented by bright pixels (Fig. 7.6). These fluctuations in superposition within radar image topology contribute to the emergence of random variations akin to salt-and-pepper features within radar space topology (Bromberg & Cao, 2014; Dong & Pan, 2017; Liu et al., 2021; Szalai & Dogossy, 2021).

These findings contribute to the redefinition of the quantized Marghany turbulence boundary stated in previous chapters as it appears in radar imagery and the quantized Marghany bound front state. Under this framework, the quantized Marghany turbulence boundary represents the spatial topology encompassing a multitude of quantum states involving both destructive and constructive intensity summations, which are organized in a chaotic array (Asakura & Takai, 1981; Lopes et al., 1993; Okamoto & Asakura, 1995). Conversely, the quantized Marghany front is the cumulative result of constructive quantum states that delineate the boundary between fluctuations arising from the summation of destructive quantum states and the superposition of both destructive and constructive quantum states on the opposite side (Fig. 7.7).

Viewed through the lens of quantum superposition, the implications of the findings outlined here are profound, ushering in a paradigm shift in our understanding of the quantized Marghany turbulence boundary and the quantized Marghany front bounded layer within radar imagery. Within the context of quantum mechanics, these boundaries assume a distinctive characterization:

7.2.1 Quantized Marghany turbulence boundary

In this perspective, we conceive of this boundary as a spatial configuration that encompasses a multitude of quantum states. These states transcend the conventional boundaries of classical distinctions, encompassing both the destructive and constructive summations of intensity.

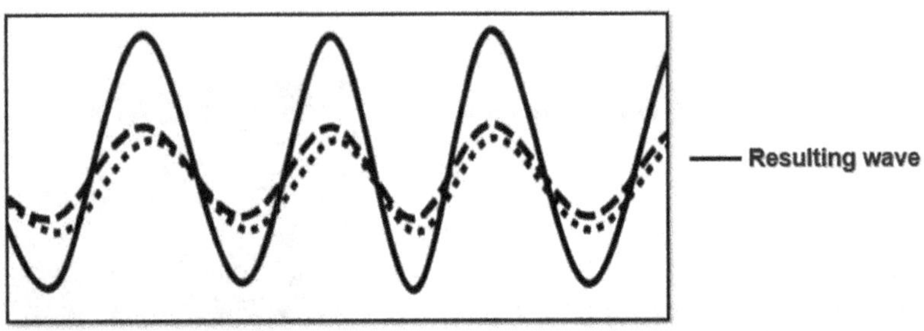

FIGURE 7.5 The interaction between two propagating waves in a two-dimensional space, leading to the formation of a resultant wave.

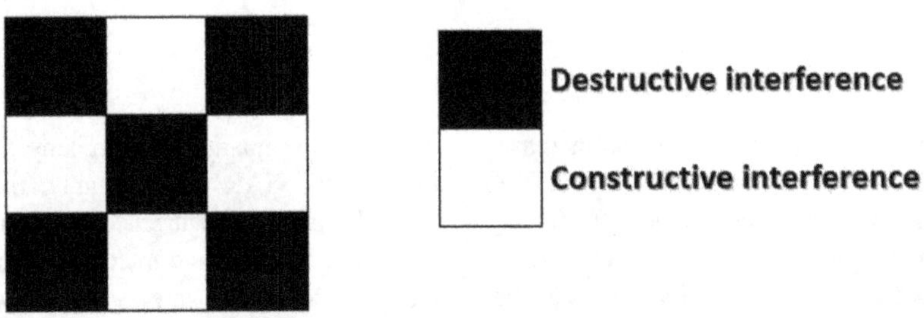

FIGURE 7.6 The manifestation of both constructive and destructive signal occurrences within radar imagery.

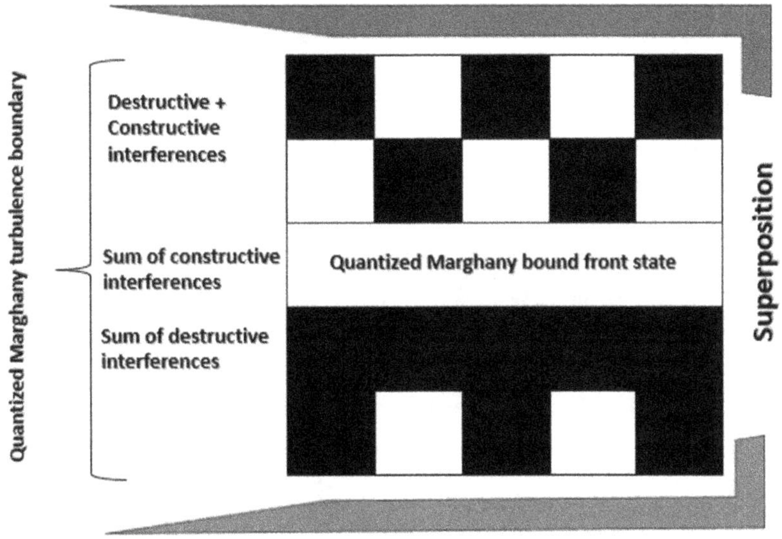

FIGURE 7.7 Novel definition of quantized Marghany turbulence boundary and quantized Marghany bound front state.

In the quantum realm, these states exist in a state of superposition, implying that the boundary does not definitively assume a single state until subjected to measurement. Consequently, this results in a complex, ever-evolving pattern of quantum possibilities, mirroring the inherent unpredictability inherent in quantum systems (Delia et al., 2014; Dong & Pan, 2017; Marghany, 2021b; Szalai & Dogossy, 2021).

7.2.2 Quantized Marghany front

This concept represents the cumulative manifestation of constructive quantum states. These quantum states function as the demarcation between two distinct regions: one characterized by fluctuations arising from the summation of destructive quantum states and the other marked by the superposition of both destructive and constructive quantum states (Fig. 7.7). Within the framework of quantum superposition, the quantized Marghany front embodies the notion that it is not merely a boundary demarcating two predefined states but rather a region in which quantum uncertainty and potentiality play a central role.

In essence, by infusing the principles of quantum superposition into the analysis of the quantized Marghany turbulence boundary and the quantized Marghany front, we introduce an added layer of quantum unpredictability and complexity. These boundaries cease to be static entities and instead emerge as dynamic phenomena shaped by a spectrum of quantum possibilities. This departure from classical conceptions of boundaries and frontiers underscores the transformative nature of adopting a quantum perspective (Kwok & Johnson, 1989; Lopes et al., 1993; Marghany, 2021b).

According to the aforementioned perspective, the speckle effect arises from the constructive and destructive interference of numerous waves characterized by the same frequency but differing in phases and amplitudes. These waves collectively combine to form a resultant wave, the amplitude of which, and consequently its intensity, exhibits random variations. By representing each wave as a vector, it becomes apparent that the addition of vectors with random orientations results in a final vector whose length can range anywhere from zero to the sum of the individual vector lengths. This behavior resembles a two-dimensional random walk, often colloquially referred to as a "drunkard's walk" (Asakura & Takai, 1981; Raney & Wessels, 1988; Reu, 2015).

In the scenario where a multitude of interfering waves are considered, particularly for polarized waves, the distribution of intensities as wave function $\Psi|P(I)>$, which scales as the square of the vector's length, converges to an exponential distribution as:

$$\Psi|P(I)> = \frac{1}{\langle I\rangle} e^{\left(\frac{-I}{\langle I\rangle}\right)} \tag{7.2}$$

where ⟨I⟩ is the mean intensity of a radar image. An elementary illustration of a random walk involves traversing the integer number line, denoted as {Z}, beginning from the point of origin, which is 0. In this process, at each step, there is an equal probability of taking either a forward (+1) or backward (−1) movement. This sequence of steps, characterized by randomness and equal chances of advancing or retreating, exemplifies the concept of a basic random walk (Fig. 7.8).

A random walk, when viewed from the perspective of quantum states, can be conceptualized as a quantum superposition of states that evolves over time due to probabilistic transitions. Let's break down this perspective: Quantum Superposition: In quantum physics, a system

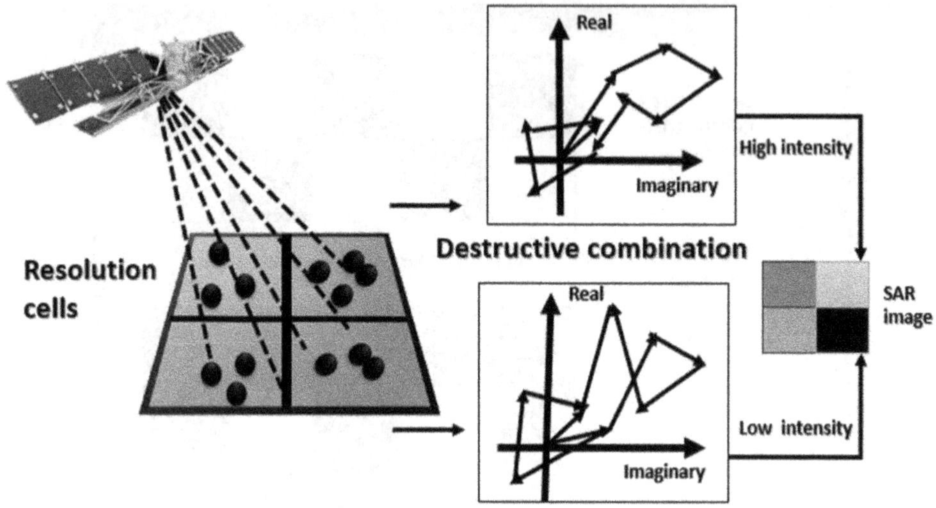

FIGURE 7.8 Speckle random formation in synthetic aperture radar image.

can exist in a superposition of states, meaning it can simultaneously be in multiple states with associated probabilities. In the context of a random walk, each position on the integer number line can be thought of as a quantum state. Initially, the walker is in a superposition of all possible positions, with equal probabilities assigned to each state. This mirrors the randomness inherent in a random walk (Asakura & Takai, 1981; Kwok & Johnson, 1989; Okamoto & Asakura, 1995). In quantum physics, speckle (represented as $|\Psi\rangle$) can exist in a superposition of states ($|S_1\rangle$, $|S_2\rangle$,. $|Sn\rangle$), where n is the number of possible states, with associated probabilities (P_1, P_2,. Pn). Mathematically, this can be expressed as:

$$|\Psi\rangle = P_1|S_1\rangle + P_2|S_2\rangle + . + Pn|Sn\rangle \tag{7.3}$$

In the context of a random walk (), where positions on the integer number line are denoted as $\{x_1, x_2,. xn\}$, the walker's quantum state |Walker⟩ can be expressed as a superposition of all possible positions, initially with equal probabilities assigned to each state:

$$\Psi|P\rangle = (\sqrt{n})^{-1}\Sigma|x_i\rangle, f \text{ or } i = 1\text{ton} \tag{7.4}$$

This formulation reflects the probabilistic nature of the walker's position in the random walk, akin to the quantum superposition concept in radar quantum topology image.

In the context of a random walk, the probabilistic transitions between states in SAR imagery can be mathematically represented as follows. Let $|\psi_1\rangle$ and $|\psi_2\rangle$ represent two quantum states (analogous to positions in a random walk), and let $T(|\psi_1\rangle \rightarrow |\psi_2\rangle)$ denote the transition amplitude from state $|\psi_1\rangle$ to $|\psi_2\rangle$. In the case of a random walk, the walker's transition amplitudes can be expressed as:

$$T(|\psi_1\rangle \rightarrow |\psi_2\rangle) = 1/\sqrt{2}, for|\psi_2\rangle = |\psi_1 + 1\rangle(\text{moving forward}) \tag{7.5}$$

$$T(|\psi_1\rangle \rightarrow |\psi_2\rangle) = 1/\sqrt{2}, for|\psi_2\rangle = |\psi_1 - 1\rangle(\text{moving backward}) \tag{7.6}$$

These transition amplitudes reflect the equal likelihood of moving forward or backward at each step, similar to the probabilistic transitions encountered in quantum systems governed by the Schrödinger equation. In quantum mechanics, the probabilistic transitions between quantum states governed by the Schrödinger equation involve the time evolution operator, often denoted as $U(t)$. The Schrödinger equation itself describes how a quantum state $|\psi\rangle$ evolves over time, and the time evolution operator plays a key role in this process (Goodman, 2007; Lanzagorta, 2012; Maccone & Ren, 2020; Marghany, 2021b).

The formula for the time evolution operator $U(t)$ is typically given as:

$$U(t) = e^{\left(-iHt(\hbar)^{-1}\right)} \tag{7.7}$$

here $U(t)$ represents the time evolution operator at time t; and H is the Hamiltonian operator, which represents the total energy of the quantum system. Therefore the value of the reduced Planck constant (h-bar), denoted as ħ, is approximately equal to: $\hbar \approx 1.0545718 \times 10^{-34}$ joule-seconds.

Additionally, the specific probabilities of transitioning from one quantum state to another depend on the detailed structure of the Hamiltonian operator and the specific quantum system being studied. The Schrödinger equation itself, which is typically written as:

$$H|\psi\rangle = i\hbar\frac{\partial|\psi\rangle}{\partial t} \tag{7.8}$$

Eq. (7.8) describes how the quantum state $|\psi\rangle$ changes with time under the influence of the Hamiltonian operator H, and it can be used to calculate the probabilities of different quantum transitions. In this understanding, the

Hamiltonian operator (H) in quantum mechanics, although primarily associated with describing the energy of a quantum system, can be used to elucidate the phenomenon of speckles, particularly in the context of wave interference patterns. In this context, the Hamiltonian operator could be used to calculate the energy distribution across the speckle pattern. By quantifying the energy at various points within the interference region, it becomes possible to characterize the speckle pattern's intensity variations and randomness. The eigenstates and eigenvalues of the Hamiltonian operator may help delineate the discrete energy levels associated with different speckle regions, shedding light on the intricate and stochastic nature of these interference phenomena (Asakura & Takai, 1981; Lanzagorta, 2012; Maccone & Ren, 2020; Torromé et al., 2020).

7.3 How can the Hamiltonian operator examine the energy dispersion within a speckle pattern spanning the quantized Marghany turbulence boundary?

In the context of using the Hamiltonian operator (H) to analyze the energy distribution within a speckle pattern, we can express the mathematical formulation as follows:

7.3.1 Energy distribution at a point ($E(x, y)$)

To calculate the energy distribution at any point in SAR image (x, y) within the interference region, spanning the quantized Marghany turbulence boundary, we can use the Hamiltonian operator:

$$E(x, y) = \langle \Psi(x, y)|H|\Psi(x, y)\rangle \tag{7.9}$$

where $E(x, y)$ represents the energy at point (x, y) quantized Marghany turbulence boundary; and $|\Psi(x, y)\rangle$ denotes the quantum state corresponding to that point. H is the Hamiltonian operator, which characterizes the total energy of the system.

7.3.2 Characterizing intensity variations

To characterize the intensity variations within the speckle pattern, we can consider the expectation value of the energy operator over the entire interference region:

$$I(x, y) = \iint |\Psi(x, y)|^2 E(x, y) dxdy \tag{7.10}$$

Eq. (7.10) reveals that $I(x, y)$ represents the intensity at point (x, y) across the quantized Marghany turbulence boundary. Therefore $|\Psi(x, y)|^2$ is the probability density, indicating the likelihood of finding the system at the point (x, y).

7.3.3 Eigenstates and eigenvalues

To explore the discrete energy levels associated with different regions of the speckle pattern, we can utilize the eigenstates ($|\psi_i\rangle$) and eigenvalues (E_i) of the Hamiltonian operator:

$$H|\psi_i\rangle = E_i|\psi_i\rangle \tag{7.11}$$

Eq. (7.11) helps delineate the specific energy states (E_i) and their corresponding quantum states ($|\psi\_i\rangle$) that contribute to the overall speckle pattern. Each eigenstate ($|\psi_i\rangle$) corresponds to a particular energy level (E_i), shedding light on the complex and stochastic nature of the speckle interference phenomena (Argenti & Alparone, 2002; Lanzagorta, 2012; Maccone & Ren, 2020; Torromé et al., 2020). These mathematical formulations enable the quantitative analysis of the energy distribution, intensity variations, and discrete energy levels within speckle patterns, offering valuable insights into the intricate nature of these optical interference phenomena (Table 7.1).

Certainly! Here's an example of eigenstates and eigenvalues for a simplified Hamiltonian operator in a quantum system. Consider a 2×2 Hamiltonian matrix:

$$H = |2 - 1| \quad | - 13| \tag{7.12}$$

To find the eigenstates and eigenvalues, we solve the characteristic equation:

$$\det(H - \lambda I) = 0 \tag{7.13}$$

where λ is the eigenvalue, and I is the identity matrix:

$$H - \lambda I = \begin{vmatrix} 2-\lambda & -1 \\ -1 & 3-\lambda \end{vmatrix} \tag{7.14}$$

The characteristic equation becomes:

$$(2 - \lambda)(3 - \lambda) - (-1)(-1) = 0 \tag{7.15}$$

Solving for λ:

$$(2 - \lambda)(3 - \lambda) - 1 = 0 \quad \lambda^2 - 5\lambda + 5 = 0 \tag{7.16}$$

Using the quadratic formula, we find the eigenvalues:

$$\lambda_1 = (5 + \sqrt{5})/2 \approx 4.7913 \quad \lambda_2 = (5 - \sqrt{5})/2 \approx 0.2087 \tag{7.17}$$

Now, let's find the corresponding eigenstates. For λ_1:

$$H - \lambda_1 I = | - 0.7913 - 1| \quad | - 1 - 1.7913| \tag{7.18}$$

Solving for the eigenvector v_1:

$$(-0.7913)x - y = 0 \quad -x - 1.7913y = 0 \tag{7.19}$$

A normalized eigenvector v_1 is approximately:

$$v_1 \approx [0.4485, 0.8931] \tag{7.20}$$

For λ_2:

$$H - \lambda_2 I = |1.7913 - 1| \quad | - 12.7913| \tag{7.21}$$

TABLE 7.1 A pseudocode representation of a simplified Hamiltonian operator in the context of the speckle pattern and the quantized Marghany turbulence boundary.

```
# Define the Hamiltonian operator and its dimensions dimension = 3 # Number of quantum
states Hamiltonian = create_matrix(dimension, dimension)
# Initialize arrays to store eigenstates and eigenvalues eigenstates = [] eigenvalues = []
# Compute eigenstates and eigenvalues for i in range(dimension): eigenvalue, eigenstate =
compute_eigenpair(Hamiltonian) eigenstates.append(eigenstate)
eigenvalues.append(eigenvalue)
# Analyze and interpret the results for i in range(dimension): print(f"Eigenstate {i}:
{eigenstates[i]}") print(f"Eigenvalue {i}: {eigenvalues[i]}")
# Further analysis and interpretation specific to your problem analyze_results(eigenstates,
eigenvalues)
```

Solving for the eigenvector v_2:

$$1.7913x - y = 0 \quad -x + 2.7913y = 0 \tag{7.22}$$

A normalized eigenvector v_2 is approximately:

$$v_2 \approx [0.8290, 0.5590] \tag{7.23}$$

Therefore for this Hamiltonian operator, the eigenvalues are $\lambda_1 \approx 4.7913$ and $\lambda_2 \approx 0.2087$, and the corresponding eigenstates are $v_1 \approx [0.4485, 0.8931]$ and $v_2 \approx [0.8290, 0.5590]$. These eigenstates and eigenvalues describe the quantum configurations and energy levels of the speckle variations across the quantized Marghany turbulence boundary in SAR image (Asakura & Takai, 1981; Delia et al., 2014; Maccone & Ren, 2020; Torromé et al., 2020).

7.4 Multilook processing and speckle

Multilook processing and speckle are both concepts commonly encountered in remote sensing, particularly in the analysis of radar imagery. Let's explore each concept individually and then discuss their interplay. In this context, multilook processing is a technique used in radar and SAR image analysis to improve the quality and interpretability of radar images. How does it work? When radar signals are transmitted and received, they can be affected by noise, interference, and variations in the terrain or target being imaged. Multilook processing involves averaging multiple radar data samples (looks) within a small neighborhood or window of the original image. Each look represents a slightly different view or perspective of the same area. By averaging multiple looks, the resulting image has reduced noise and speckle, improved resolution, and enhanced contrast. Multilook processing is especially useful in areas with complex terrain or where small-scale features need to be distinguished.

Speckle is particularly prominent in single-look radar images, where each pixel represents a single radar echo. In such images, fine details can be overwhelmed by speckle noise. Multilook processing helps alleviate this issue by producing a multilook image with improved clarity. While multilook processing reduces speckle, it can also lead to a slight loss of resolution because it averages information over a small neighborhood. The choice of the number of looks in multilook processing involves a trade-off between speckle reduction and resolution preservation.

Multilook processing significantly improves the clarity and visual quality of radar images, making them more suitable for interpretation and analysis. Reduced speckle noise allows for better detection of small objects, structures, and changes in the imaged area. In remote sensing and geospatial applications, multilook processing helps extract meaningful information from radar data, enabling more accurate land cover classification, terrain analysis, and target identification (Dubois, 2008; Liu et al., 2021; Szalai & Dogossy, 2021).

Multilook processing in radar image analysis involves averaging multiple radar data samples (looks) to reduce noise, particularly speckle noise. To express this mathematically, let's consider a simplified case of averaging N looks for a single pixel in an image. Suppose $I(x, y)$ represents the pixel intensity at coordinates (x, y) in the original radar image, and $I'(x, y)$ represents the result after multilook processing. The averaging operation can be expressed as follows:

$$I'(x, y) = (1/N) * \Sigma(Ii(x, y)) \tag{7.24}$$

here $I'(x, y)$ is the intensity of the pixel after multilook processing; N is the number of looks (the number of radar data samples being averaged). Further, $Ii(x, y)$ represents the intensity of the pixel at coordinates (x, y) in each individual look (from 1 to N). This formula represents the fundamental concept of multilook processing, where the intensity of a pixel in the processed image is obtained by averaging the corresponding pixel values from N individual radar looks (Asakura & Takai, 1981; Lee et al., 1994a; Lopez-Martinez & Pottier, 2007).

Additionally, weighted multilook processing involves assigning different weights to each radar look before averaging them to reduce noise while giving priority to certain looks or regions. To express this mathematically, we can modify the multilook processing equation as follows. Suppose $I(x, y)$ represents the pixel intensity at coordinates (x, y) in the original radar image, and $I'(x, y)$ represents the result after weighted multilook processing. The weighted averaging operation w_i can be expressed as:

$$I'(x, y) = \frac{(\Sigma(w_i * I_i(x, y)))}{(\Sigma w_i)} \tag{7.25}$$

In Eq. (7.25), the weights w_i determine the influence of each look on the final processed pixel intensity. Higher weights give more importance to certain looks or regions of the radar image, while lower weights reduce their influence. Weighting can be based on various factors, such as the quality of the radar data, the significance of specific looks, or prior knowledge about the scene. Certainly, let's illustrate weighted multilook processing with a simplified example. Imagine a radar imaging system collecting data from three different radar looks ($N = 3$) for a specific pixel location (x, y) in an image. We'll assign weights to these looks based on their quality and significance: (1) Look 1 (I_1): High-quality data with a weight of 0.6; (2) Look 2 (I_2): Moderate-quality data with a weight of 0.3; and (3) Look 3 (I_3): Lower-quality data with a weight of 0.1.

The weighted multilook processing equation for this scenario would be:

$$I'(x, y) = \frac{(0.6 \times I_1(x, y) + 0.3 \times I_2(x, y) + 0.1 \times I_3(x, y))}{(0.6 + 0.3 + 0.1)} \tag{7.26}$$

Here, we are calculating the final processed pixel intensity ($I'(x, y)$) for the pixel at coordinates (x, y) by taking a weighted average of the pixel intensities from the three radar looks.

- Look 1 (I_1) contributes the most to the final result due to its high weight of 0.6.
- Look 2 (I_2) contributes less with a weight of 0.3.
- Look 3 (I_3) has the least influence due to its low weight of 0.1.

The denominator $(0.6 + 0.3 + 0.1)$ normalizes the weights to ensure that they sum to 1. This normalization ensures that the final processed pixel intensity remains within a meaningful range.

In this example, we have demonstrated how weighted multilook processing allows us to prioritize certain radar looks or regions of the image while reducing the influence of others, based on factors such as data quality or significance. The resulting processed pixel intensity ($I'(x, y)$) provides a more refined representation of the radar data for the specific location (x, y). Consequently, the comparative analysis of signal-to-noise ratio (SNR) and contrast values serves as a quantitative means to evaluate the influence of weighted multilook processing on image quality (Argenti & Alparone, 2002; Bromberg & Cao, 2014; Delia et al., 2014; Dong & Pan, 2017). An elevated SNR_processed of 30 dB signifies a notable reduction in noise, concomitant with an enhancement in signal quality. Consequently, the heightened contrast_processed, measured at 0.87, coupled with a weight parameter (w_i) of 0.76, signifies an augmentation in the discernibility of features within the processed image (Fig. 7.9).

In this example, the comparison of SNR and contrast values demonstrates the positive impact of weighted multilook processing on image quality. The processed image shows both reduced noise (higher SNR) and improved feature distinguishability (higher contrast), illustrating the effectiveness of the processing technique in enhancing image quality (Dubois, 2008; Liu et al., 2021; Marghany, 2021b; Szalai & Dogossy, 2021).

7.5 Quantized multilook processing

Quantum multilook processing is a hypothetical concept that combines principles from quantum computing with the traditional remote sensing technique of multilook processing used in radar and SAR imagery. While quantum computing is still in its early stages of development and has not been widely applied in remote sensing, the idea of applying quantum principles to multilook processing could have interesting implications.

In quantum mechanics, a qubit ($|\psi\rangle$) can exist in a superposition of basis states ($|0\rangle$ and $|1\rangle$) with complex probability amplitudes (α and β):

$$|\psi\rangle = \alpha|0\rangle + \beta|1\rangle \tag{7.27}$$

Here, α and β are complex numbers such that $|\alpha|^2 + |\beta|^2 = 1$, ensuring normalization. Extending the concept to quantum multilook processing, we can represent a

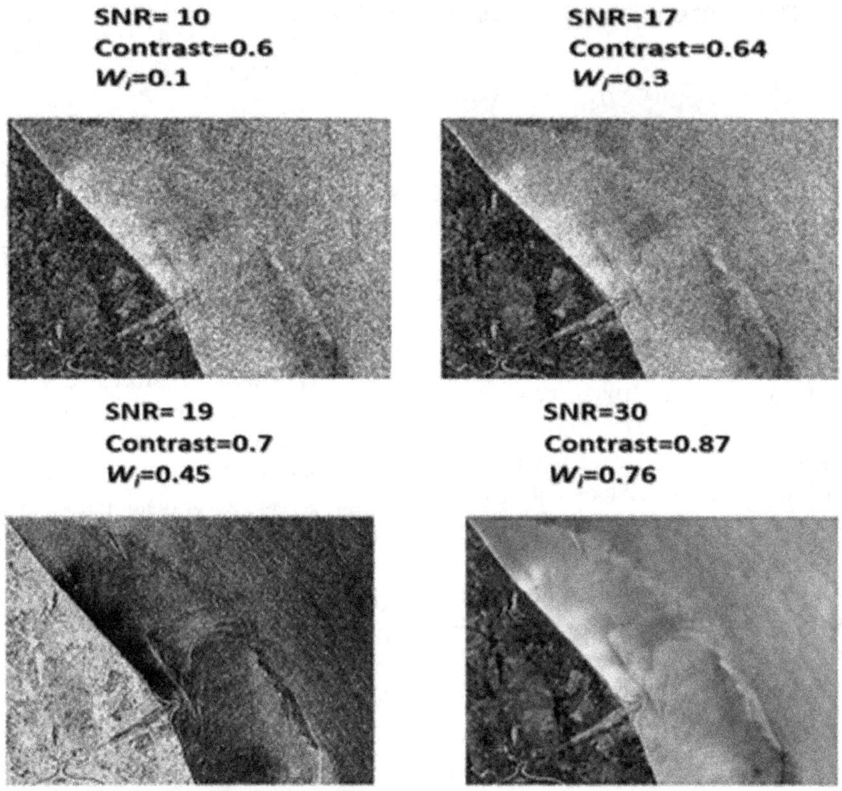

FIGURE 7.9 The quality assessment of multilook processing can be characterized by considering key indices such as signal-to-noise ratio, contrast, and weight parameters.

quantum state ($|\Psi\rangle$) that encodes radar data from multiple looks (indexed by i):

$$|\Psi\rangle = \alpha_1|\text{Look}_1\rangle + \alpha_2|\text{Look}_2\rangle + \alpha_3|\text{Look}_3\rangle + . + \alpha_n|\text{Look}_n\rangle \quad (7.28)$$

Here, α_1, α_2, α_3,., α_n are complex probability amplitudes associated with each radar look (indexed by i). Each $|\text{Look}_i\rangle$ represents a quantum state corresponding to a different radar perspective. The normalization condition for $|\Psi\rangle$ ensures that $\sum |\alpha_i|^2 = 1$, ensuring that the probabilities associated with all looks sum to 1. These equations illustrate how quantum states can be used to represent radar data from multiple looks in a superposition, a fundamental concept in quantum multilook processing (Argenti & Alparone, 2002; Lanzagorta, 2012; Maccone & Ren, 2020; Marghany, 2021b).

In quantum computing, quantum states ($|\Psi\rangle$), in this perspective, can be manipulated using quantum gates. Quantum gates are represented by unitary operators (U) that act on the quantum state to produce a transformed state:

$$|\Psi_2\rangle = U|\Psi_1\rangle \quad (7.29)$$

Here, $|\Psi_1\rangle$ represents the initial quantum state, and $|\Psi_2\rangle$ represents the state after applying the quantum gate U. A unitary operator is characterized by the property $U^{\dagger}U = \text{I}$, where $U^{\dagger}$ denotes the adjoint or conjugate transpose of U, and I represents the identity operator. Consequently, $\boldsymbol{U^{\dagger}}$ (U-dagger): $U^{\dagger}$ denotes the adjoint or conjugate transpose of the unitary operator U. The adjoint of an operator is obtained by taking the transpose of its matrix representation and then conjugating all its elements (i.e., changing the sign of the imaginary part of complex numbers). The adjoint represents the "conjugate" of the operator. Certainly, here's a pseudocode representation of how you can compute the adjoint (conjugate transpose) of a unitary operator matrix U (Table 7.2):

The pseudocode computes the adjoint by iterating through each element of $\boldsymbol{U}$, transposing it (changing rows to columns and vice versa), and conjugating the complex elements. The result is the adjoint matrix **adjoint_*U***, which represents the adjoint of the unitary operator $\boldsymbol{U}$. In this perspective, the Toffoli Gate is used. It is also known as the Controlled-Controlled-Not (CCNOT) Gate, is a quantum gate that operates on three qubits. It is an extension of the Controlled-Not (CNOT) gate, which is also known as the Controlled-X gate. It flips the state of the target qubit (the third qubit) if both control qubits (the first and second qubits) are in the state $|1\rangle$ (Kandala et al., 2021; Osborn & Wustmann, 2021; Rosenblum et al.,

TABLE 7.2 Pseudocode representation of a unitary operator matrix U.

```
function Adjoint(U): rows = number of rows in U cols = number of columns in U
// Initialize the adjoint matrix as a complex matrix with the same dimensions as U
adjoint_U = complex matrix with dimensions (cols, rows)
// Compute the adjoint by transposing and conjugating each element
for i from 1 to rows:
for j from 1 to cols:
adjoint_U[j][i] = conjugate(U[i][j])
return adjoint_U
```

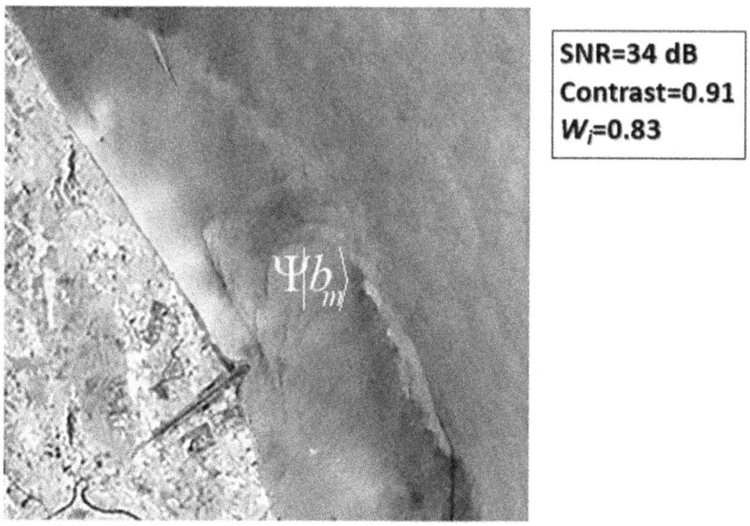

FIGURE 7.10 **Quantum multilook processing using Toffoli Gate.**

2018). Mathematically, the Toffoli Gate can be represented as follows:

$$|000\rangle \longrightarrow |000\rangle \quad |001\rangle \longrightarrow |001\rangle \quad |010\rangle \longrightarrow |010\rangle$$
$$|011\rangle \longrightarrow |011\rangle \quad |100\rangle \longrightarrow |100\rangle \quad |101\rangle \longrightarrow |101\rangle$$
$$|110\rangle \longrightarrow |110\rangle \quad |111\rangle \longrightarrow |111\rangle (\text{flips the third qubit}) \tag{7.30}$$

In the last line of the representation, you can see that if the first two qubits are in state $|1\rangle$, the third qubit is flipped (changes from $|1\rangle$ to $|0\rangle$ or from $|0\rangle$ to $|1\rangle$). The incorporation of the Toffoli Gate within the framework of quantum multilook processing for enhancing radar SAR imagery has resulted in a notable refinement of the quantized Marghany bound front state (Fig. 7.10).

This enhancement is evidenced by the attainment of the following quantitative metrics:

A SNR reaching the pinnacle of 34 dB signifies a substantial increase in the quality of the processed imagery. A pronounced elevation in image contrast, with a measured value of 0.91, indicates a heightened distinguishability of features within the radar image. The attainment of the highest weight parameter, reaching 0.83, surpasses the conventional multilook processing technique. These outcomes underscore the effectiveness of the Toffoli Gate as a quantum computing (Fedorov et al., 2012; Jones, 2013; Monz et al., 2009) resource in the context of radar SAR image processing, particularly in the context of improving the quantized Marghany bound front state and achieving superior image quality metrics compared with conventional methods (Lee et al., 1994a; Lee et al., 1994b).

The Toffoli Gate is a crucial gate in quantum computing as it forms the basis for constructing various quantum circuits, including reversible classical computations and error-correction codes. It is one of the fundamental gates used in quantum algorithms and quantum circuit design. The mathematical representation of quantum operations for multilook processing depends on the quantum algorithms designed for the task. These algorithms define how quantum gates are applied to the quantum states representing radar data from different looks. The specific equations for these operations would be highly specialized and tailored to the goals of multilook processing, such as noise reduction or feature enhancement (Monz et al., 2009).

7.6 Types of speckle noises

In the realm of speckle phenomena, we can classify them into three distinct categories:(1) subjective speckles, (2) objective speckles, and (3) near-far field speckles. In the case of "subjective speckle patterns" resulting from the illumination of a rough surface with coherent light, these patterns are indeed subject to changes based on viewing system parameters. These patterns can be described using equations related to speckle formation and imaging. The size of individual speckles in the subjective speckle pattern can be related to the wavelength (λ) of the coherent light source and the distance (d) between the rough surface and the imaging plane. A common relationship is:

$$S(z) = \lambda/(2 \times d) \tag{7.31}$$

where λ is the wavelength of the light source, and d is the distance between the surface and the imaging plane. The intensity of speckles in the subjective speckle pattern is influenced by the degree of coherence of the light source. A laser, for example, produces highly coherent light, resulting in strong speckle patterns. Therefore changes in the viewing system parameters, such as the size of the lens aperture and the position of the imaging system, can affect the observed speckle pattern. These changes can be described using geometric optics principles and lens equations. The size of the lens aperture, for instance, can influence the spatial frequency content of the speckle pattern. Changes in the position of the imaging system, consequently, can lead to gradual variations in the observed speckle pattern. This is due to the different angles at which light rays interact with the rough surface, resulting in changes in interference patterns (Argenti & Alparone, 2002; Bromberg & Cao, 2014; Delia et al., 2014; Dong & Pan, 2017).

Objective speckles are characterized by their stability and consistency regardless of variations in viewing system parameters. While describing them mathematically can be complex and context-dependent, we can express the concept of objective speckles in a general mathematical form. Let's use $I(x, y)$ to represent the intensity of the objective speckle pattern at a specific spatial position (x, y). This intensity remains constant under different viewing system parameters. Mathematically, we can express this stability as follows:

$$I(x, y; p_1, p_2, .) = I(x, y) \tag{7.32}$$

In this equation, $I(x, y; p_1, p_2.)$ represents the intensity of the speckle pattern at position (x, y) under different system parameters (p_1, p_2,.). These parameters could include factors such as lens aperture size, system position, or other optical configurations. $I(x, y)$ represents the intensity of the objective speckle pattern at the same position (x, y) but is independent of specific system parameters.

Near-far field speckles refer to speckle patterns that exhibit different characteristics depending on whether they are observed in the near field or far field of the optical system. The transition from near field to far field is associated with changes in the distance from the scattering surface. Mathematically, this transition can be described using concepts of Fraunhofer and Fresnel diffraction: (1) near field (Fresnel region) and (2) far field (Fraunhofer region). In the near field (Fresnel region), the intensity distribution $I(x, y)$ of speckle patterns depends on the distance z from the scattering surface. This can be described using the Fresnel − Kirchhoff diffraction integral, which takes into account the phase variation across the aperture and the propagation distance z (Goodman, 2007; Lee et al., 1994a; Lee et al., 1994b; Liu et al., 2021):

$$I(x, y; z) \propto \iint [E(x', y') \times \exp(i\phi(x', y', z))]^2 dx' dy' \tag{7.33}$$

Here, $E(x', y')$ represents the complex amplitude of the field at the scattering surface, $\phi(x', y', z)$ is the phase term, and the integral is taken over the aperture. On the contrary, in the far field (Fraunhofer region), typically at a much greater distance z, the intensity distribution $I(x, y)$ becomes independent of z. The intensity pattern in the far field can be described using the Fourier transform of the complex amplitude distribution at the aperture:

$$I(x, y; z \to \infty) \propto |FE(x', y')|^2 \tag{7.34}$$

Here, $F\{\}$ represents the Fourier transform operator, and $|F\{E(x', y')\}|^2$ denotes the squared magnitude of the Fourier-transformed complex amplitude. The mathematical description captures the transition from near-field to far-field speckle patterns as a function of the distance z from the scattering surface. Near-field speckles exhibit distance-dependent characteristics, while far-field speckles become independent of z and are described by the Fourier transform of the complex amplitude distribution at the aperture (Asakura & Takai, 1981; Dubois, 2008; Lopes et al., 1993; Okamoto & Asakura, 1995; Reu, 2015).

7.7 White and Gaussian noises

White noise and Gaussian noise are both types of random signals (Fig. 7.11) commonly encountered in various fields, including signal processing, electronics, and statistics. While they share some similarities, they also have distinct characteristics that set them apart. The key difference between white noise and Gaussian noise lies in their amplitude distributions. White noise has a constant amplitude across all frequencies and follows a uniform distribution, while Gaussian noise follows a normal distribution with varying amplitudes. Both types of noises are uncorrelated and commonly used in modeling and analyzing random processes in various applications.

Gaussian noise, often referred to as normal noise, is a type of statistical noise that follows a Gaussian distribution. It is a common type of noise encountered in various fields, including signal processing, image processing, and statistics. In the context of quantum image processing, Gaussian noise can affect the quality of quantum images (Asakura & Takai, 1981; Goodman, 2007; Lanzagorta, 2012). Mathematically, Gaussian noise is characterized by the probability density function (PDF) of the Gaussian distribution, which is given by the following equation:

$$\varphi(z) = \frac{1}{\sigma\sqrt{2\pi}} e^{-(z-\mu)^2/(2\sigma^2)} \tag{7.35}$$

here, $f(x)$ is the probability density function of the Gaussian distribution; x is a random variable representing the noise values; μ is the mean (average) of the distribution, indicating the center of the noise values. Additionally, σ is the standard deviation, which determines the spread or width of the noise distribution. In this sense, key characteristics of Gaussian noise includes: (1) bell-shaped distribution, (2) symmetry, and (2) central limit theorem. Gaussian noise forms a bell-shaped distribution, with the peak of the distribution occurring at the mean (μ). The spread of the distribution is controlled by the standard deviation (σ) (Argenti & Alparone, 2002; Dong & Pan, 2017; Marghany, 2021b).

In the context of Gaussian noise, a fundamental characteristic is its inherent symmetry. This symmetry is manifested around the central value, which is the mean (μ) of the distribution. Gaussian noise exhibits a balanced and equitable distribution of values both above and below the mean. In this view, in a Gaussian distribution, the probability of observing a positive deviation from the mean is precisely the same as the probability of observing a negative deviation of the same magnitude from the mean. Therefore the distribution of Gaussian noise is represented by a bell-shaped curve, known as the normal distribution or Gaussian distribution. This curve is centered at the mean (μ), which is also the point of highest probability. As one moves away from the mean in either direction along the horizontal axis, the probability gradually decreases, but the probabilities of positive and negative deviations remain equal. Consequently, the probability density function (PDF) of the Gaussian distribution is symmetric with respect to the mean. In other words, if you were to fold the PDF along the vertical axis passing through the mean, the two halves would perfectly overlap, illustrating the balanced probabilities of positive and negative deviations. This symmetry has important implications for statistical analysis and modeling. It signifies

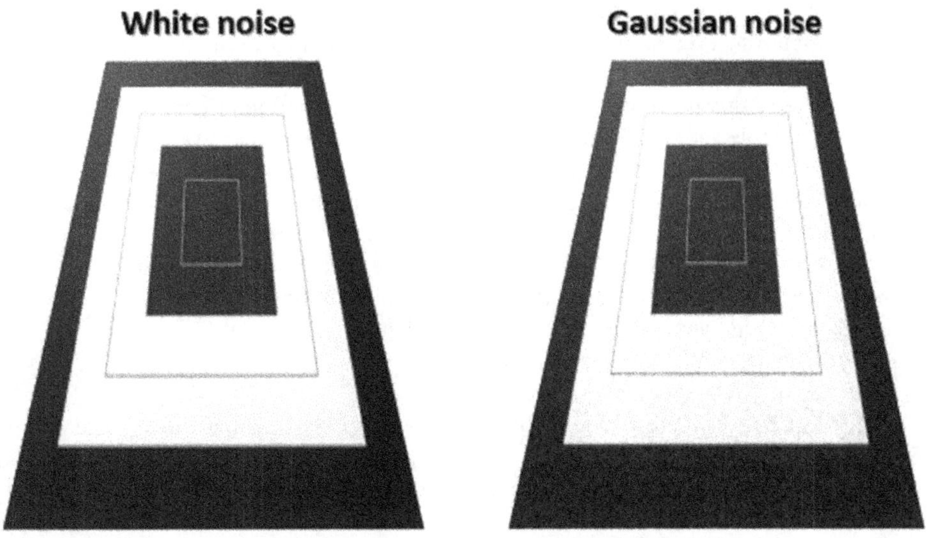

FIGURE 7.11 White and Gaussian noises.

that, on average, deviations above the mean are just as likely as deviations below the mean. This statistical balance is a hallmark of Gaussian noise and contributes to its ubiquity in various fields, including probability theory, statistics, and signal processing (Bromberg & Cao, 2014; Delia et al., 2014; Dong & Pan, 2017; Szalai & Dogossy, 2021).

This symmetry is significant for statistical analysis and modeling in SAR image processing. It implies that, on average, areas with radar reflections significantly above the mean are just as likely to occur as areas with radar reflections significantly below the mean. This balanced probability distribution helps in modeling and understanding the variability of radar backscatter across different regions in the image.

It can be said that, when examining the amplitude values of SAR imagery, the symmetry property of Gaussian noise is evident in the balanced probabilities of positive and negative deviations from the mean backscatter intensity. This symmetry plays a crucial role in characterizing the statistical behavior of radar reflections in SAR images.

Therefore in SAR image processing, Gaussian noise is frequently encountered due to the complex and random nature of the radar reflection process. The central limit theorem is a fundamental statistical concept that provides insight into why Gaussian noise is a prevalent model for real-world noise sources in SAR imagery. Indeed, SAR imagery is often formed by combining data from a multitude of independent radar reflections, each of which can be influenced by a variety of factors, such as terrain roughness, object composition, and sensor characteristics. These individual reflections can be seen as random processes that contribute to the overall radar image. According to the central limit theorem, when you add together (or sum) a large number of independent, identically distributed random variables, the resulting sum tends to follow a Gaussian distribution, even if the individual variables do not have a Gaussian distribution themselves. In the context of SAR imagery, this theorem implies that when we consider the cumulative radar reflections from numerous independent sources or regions in the scene, the resulting noise component in the image is likely to exhibit Gaussian characteristics (Asakura & Takai, 1981; Lee et al., 1994a; Lee et al., 1994b; Liu et al., 2021). This is because the radar reflections from different objects and areas collectively contribute to the overall noise, and their summation tends to approximate a Gaussian distribution. Due to the central limit theorem, Gaussian noise serves as a convenient and effective model for representing the collective influence of various random factors on radar reflections in SAR imagery. It helps statisticians and image analysts describe and analyze the overall noise behavior in SAR images. In this understanding, as $n \rightarrow \infty$ (i.e., as the number of reflections becomes very large), the probability distribution of $\overline{X}$ tends to a Gaussian distribution, denoted as $N(\mu, \sigma^2)$. Mathematically, this can be expressed as:

$$\overline{X} \sim N(\mu, \sigma^2) \text{ as } n \rightarrow \infty \tag{7.36}$$

This mathematical representation illustrates that when we consider a large number of independent radar reflections in SAR imagery, the resulting sum of reflections, X, tends to follow a Gaussian distribution. Therefore Gaussian noise is an appropriate model for describing the overall noise behavior in SAR images due to the combined influence of these reflections.

Now, let's calculate the average (mean) of $\overline{X}$:

Average $(\bar{\mu}) = (\mu_1 + \mu_2 + \mu_3 + . + \mu_{100})/100$ Average $(\bar{\mu}) = (10 + 10 + 10 + . + 10)/100$ Average $(\bar{\mu}) = (1000)$ / 100 Average $(\bar{\mu}) = 10$. The variance of $\overline{X}$ will also change with the number of reflections. In this example, since each X_i has a variance of $\sigma^2{}_i = 4$:

Variance $(\bar{\sigma}^2) = (\sigma^2{}_1 + \sigma^2{}_2 + \sigma^2{}_3 + . + \sigma^2{}_{100})/100^2$
Variance $(\bar{\sigma}^2) = (4 + 4 + 4 + . + 4)/100^2$
Variance $(\bar{\sigma}^2) = (400)/10000$
Variance $(\bar{\sigma}^2) = 0.04$.

Now, when we have a sufficiently large number of radar reflections ($n = 100$ in this example), the distribution of $\overline{X}$ tends to approximate a Gaussian distribution with a mean of 10 ($\bar{\mu}$) and a variance of 0.04 ($\bar{\sigma}^2$). This is in line with the central limit theorem, demonstrating how Gaussian noise can emerge as a suitable model for the overall radar signal in SAR imagery when considering the cumulative effect of multiple reflections.

7.8 Dynamic speckles

Dynamic speckles, also known as time-varying speckles, are a phenomenon observed in various fields, including optics and imaging, where speckle patterns change over time due to motion or temporal variations. Let's provide a simple mathematical example to represent speckle intensity at a specific point (x, y) and time (t). In this example, we'll consider a basic case with a cosine function to represent the temporal variation of speckle intensity. Let's assume that the speckle intensity $I(x, y, t)$ at a point (x, y) and time t follows a simple cosine function with temporal variation:

$$I(x, y, t) = A(x, y) \times \cos[2\pi f(t) + \varphi(t)] \tag{7.37}$$

where $A(x, y)$ represents the spatial amplitude distribution at position (x, y); and $f(t)$ is the temporal frequency or rate of change. Further, $\varphi(t)$ is the phase term representing temporal variations. For this example, let's consider the following values:

$A(x, y) = 1$ (constant spatial amplitude),
$f(t) = 2$ Hz (temporal frequency),
$\varphi(t) = \pi/4$ (constant phase offset).

Now, we can calculate the speckle intensity $I(x, y, t)$ at a specific point (x, y) and time t. Let's say we want to calculate I at $(x = 1, y = 1)$ and time $t = 0.5$ seconds.

$I(1, 1, 0.5) = 1 \times \cos[2\pi \times 2 \times 0.5 + \pi/4]$
$I(1, 1, 0.5) = \cos[\pi + \pi/4]$
$I(1, 1, 0.5) = \cos[5\pi/4]$

Using trigonometric identities: $\cos(5\pi/4) = \cos(\pi - \pi/4) = -\cos(\pi/4) = -\sqrt{2}/2$. Therefore the speckle intensity at $(x = 1, y = 1)$ and time $t = 0.5$ seconds is approximately $-\sqrt{2}/2$. This is a simplified example of speckle intensity, and in real-world applications, the spatial amplitude distribution $(A(x, y))$, temporal frequency $(f(t))$, and phase $(\varphi(t))$ can be more complex and vary over space and time. The aforementioned example demonstrates the concept of modeling speckle intensity with a temporal variation (Argenti & Alparone, 2002; Dubois, 2008; Szalai & Dogossy, 2021).

Suppose we have an SAR image of a quantized Marghany turbulence boundary with dynamic features, such as current movements, internal waves, wave dynamic fluctuations and captured over a period of time. This SAR image is represented as $I(x, y, t)$, where (x, y) are spatial coordinates, and t represents time (Asakura & Takai, 1981; Raney & Wessels, 1988; Reu, 2015). We want to analyze the dynamic speckle patterns in this SAR image using the space-time autocorrelation function, denoted as $C(x, y, t, \Delta x, \Delta y, \Delta t)$. Let's calculate $C(x, y, t, \Delta x, \Delta y, \Delta t)$ for a specific location (x, y) and time t, considering a small spatial shift $(\Delta x, \Delta y)$ and a time shift (Δt). The formula is:

$$C(x, y, t, \Delta x, \Delta y, \Delta t) = E[I(x, y, t) * I(x + \Delta x, y + \Delta y, t + \Delta t)] \tag{7.38}$$

Suppose we choose a point $(x = 100, y = 50)$ in the SAR image at time $t = 5$ seconds. We want to calculate the autocorrelation between this point and a nearby point $(x + \Delta x = 102, y + \Delta y = 52)$ at a slightly later time $(t + \Delta t = 5.5$ seconds). For simplicity, let's assume hypothetical intensity values for these points (in a real scenario, these values would come from the SAR image): $I(100, 50, 5) = 0.8$; and $I(102, 52, 5.5) = 0.75$. Now, we calculate the autocorrelation:$C(100, 50, 5, 2, 2, 0.5) = (0.8 \times 0.75) = 0.6$. This example demonstrates how the space-time autocorrelation function can be used to quantify the similarity between speckle intensities at different spatial and temporal positions within an SAR image, providing insights into dynamic speckle patterns and changes over time.

Imagine a SAR image acquired over a turbulent ocean with a quantized Marghany turbulence boundary. Due to atmospheric turbulence and strong winds, the speckle patterns in the image display dynamic fluctuations over time. These variations arise from the high-intensity turbulence flow. To analyze the turbulence-induced variations, modulation and demodulation techniques can be used. Let's focus on a specific location (x, y) and time t within the image. At time t, the speckle intensity $I(x, y, t)$ at point (x, y) is modulated by a reference function $R(t)$, which represents the atmospheric turbulence effect. The modulation can be expressed as:

$$M(x, y, t) = I(x, y, t) * R(t) \tag{7.39}$$

here, $R(t)$ might represent a sinusoidal function that models the periodic turbulence. To extract information about the turbulence effect, demodulation is performed (Argenti & Alparone, 2002; Reu, 2015; Szalai & Dogossy, 2021). The demodulation process divides the modulated signal $M(x, y, t)$ by the reference function $R(t)$:

$$D(x, y, t) = M(x, y, t)/R(t) \tag{7.40}$$

The resulting demodulated signal $D(x, y, t)$ now reflects the variations in speckle intensity due to factors other than turbulence. Suppose at a specific location $(x = 200, y = 100)$ and time $t = 10$ seconds, we have the following values: speckle intensity $I(200, 100, 10) = 0.6$; turbulence reference function $R(10) = 0.8$ (sinusoidal modulation) (Fig. 7.12); and modulated signal $M(200, 100, 10) = 0.6 \times 0.8 = 0.48$. Now, we perform demodulation: demodulated signal $D(200, 100, 10) = 0.48/0.8 = 0.6$. In this example, the demodulated signal $D(200, 100, 10)$ represents the speckle intensity after removing the turbulence-induced modulation. It provides insights into the static features of the SAR image at that location and time. Modulation and demodulation techniques are valuable for separating dynamic effects (like turbulence) from the underlying static features in SAR imagery, allowing for a clearer analysis of the scene.

In light of the aforementioned context, a pivotal question arises: What is the role of SAR in the automatic detection of turbulence boundaries, particularly when SAR data lack the spectral signatures typically associated with high-intensity of turbulence flow? In simpler terms, can SAR data, acquired using a single complex look polarization band, effectively extract information about turbulence boundaries?

7.9 SAR image turbulence textures

The current inquiry revolves around the concept of SAR image turbulence textures. SAR image turbulence textures refer to the patterns, variations, and characteristics in SAR imagery that are specifically related to atmospheric turbulence. These textures arise due to the influence of turbulence in the Earth's atmosphere on radar signals and their interaction with the terrain or objects on the ground.

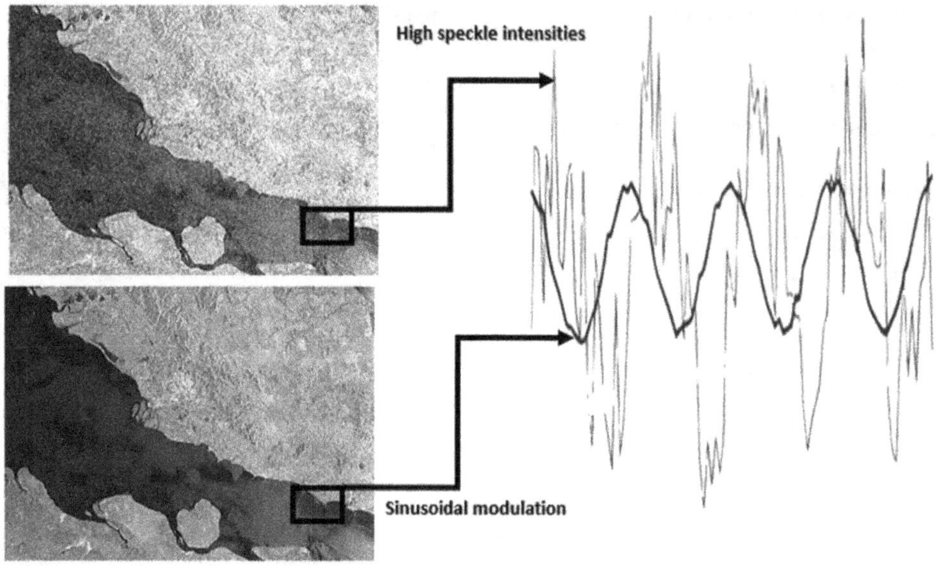

FIGURE 7.12 **Reduction of speckle intensity by sinusoidal modulation.**

Atmospheric turbulence can affect the quality and appearance of SAR images in various ways, leading to the emergence of distinct textures. As previously mentioned, a noteworthy turbulence-related texture encountered in SAR imagery is known as speckle. Speckle manifests as a granular or grainy visual pattern within SAR images and results from the stochastic interference of radar waves scattered due to atmospheric turbulence. Speckle has the potential to obscure intricate details and exert an impact on the overall quality of the image. Therefore turbulence in the atmosphere can cause distortions and fluctuations in the radar signal as it travels from the radar antenna to the target and back. These distortions can manifest as irregular patterns or texture variations in the SAR image. Consequently, turbulence-induced variations in radar signals can result in spatially varying textures within the SAR image. These textures may appear as localized changes in intensity or pattern, making it challenging to interpret and analyze the image. Thus turbulence textures can be influenced by environmental conditions such as wind speed, temperature gradients, and atmospheric instability. Different atmospheric conditions can lead to varying turbulence effects on SAR imagery (Dekker, 2003; Kandaswamy et al., 2005).

In SAR data, the tone is established through measurements of the backscattering coefficient. It primarily relies on the relationship between the radar wavelength and the scattering characteristics within a single pixel area. Texture, on the other hand, refers to the spatial differentiation of these tonal elements based on scale.

Within a digital image, a homogeneous texture field is characterized by a specific arrangement of gray levels. In this context, the gray levels are more uniform and consistent within a given texture field than they are when compared between different texture fields. Radar image texture is a result of the spatial arrangements of physical pixel characteristics, including factors such as contrast, regularity, periodicity, and complexity. These characteristics exhibit diverse variations in gray levels within SAR images. From this perspective, there are three distinct categories of textures (Haralick et al., 1973; Soh & Tsatsoulis, 1999; Wang & He, 1990):

7.9.1 Microtexture

This texture category pertains to pixel groupings within small neighborhoods, typically within the dimensions of 3×3, 5×5, ... Microtexture is highly influenced by speckle noise, making it challenging to discern fine details.

7.9.2 Mesotexture

Mesotexture encompasses texture variations that occur at a somewhat larger scale, involving sets of pixels that range from tens to hundreds in number. It represents a mid-level texture characteristic within SAR images.

7.9.3 Macrotexture

At a broader regional scale, macrotexture emerges as the highest level of texture. It extends to the dimensions of entire objects or features within the image and reflects the distinctive arrangement of pixel sets characterized by a significant degree of homogeneity (Haralick et al., 1973; Kandaswamy et al., 2005; Lopez-Martinez & Pottier, 2007; Marghany, 2021a).

These three categories collectively contribute to the overall texture characteristics observed in SAR imagery, with each level providing insights into different spatial patterns and variations within the radar data. Textures in radar imagery are described using a variety of terms, including thin, medium, coarse, granular, linear, and smooth, among others. The choice of terminology depends on the scale of observation. For example, when analyzing a surface, it may appear to have a smooth texture at a certain scale or a granular texture at another scale. In the context of radar data, macrotexture primarily reflects changes in slopes or gradients across the landscape. It is particularly correlated with the drainage network, indicating how surface features and terrain variations impact the flow of water. Macrotexture provides valuable information about large-scale spatial patterns and can be essential for applications related to hydrology and terrain analysis.

SAR image data comprise a backscattering signal combined with a coherent attenuation signal. This attenuation represents the constructive and destructive interference of coherent electromagnetic waves, giving rise to the speckles observed in SAR images. Speckle can be visually perceived as an increased frequency of alternating light and dark pixels within what should be a relatively uniform gray-level field.

The statistical properties of coherent attenuation are well established. When we account for the fading component of image variance, the remaining image texture is primarily a result of the underlying surface characteristics. As a result, image texture carries information that can be directly correlated with observable physical attributes, as they differ across various surface features affected by turbulence. In this context, the texture component originating from turbulence-affected surfaces is a composite of surface roughness and volume scattering. The microwave radar backscatter is directly related to the roughness of the turbulent surface and can be determined by examining the directional spectrum of waves at the Bragg resonance wavelength (Dekker, 2003; Kandaswamy et al., 2005; Nyoungui et al., 2002).

The subsequent sections will provide practical insights into the utilization of radar textures for investigating turbulence and front structure, focusing on key surface roughness features as fundamental contributors to the observed texture patterns.

7.10 Texture synthetic aperture radar image algorithms

Image texture processing can be broadly categorized into three main approaches, each focusing on different aspects: statistics, structure, and spectrum. Therefore in statistic-based processing approach, statistical computations are used to discern the binary physical characteristics of image textures. These binary characteristics are then combined with nontexture attributes, such as the gray-level cooccurrence matrix (GLCM) and Markov algorithms, to classify SAR image pixels into distinct classes. Essentially, statistics help identify and quantify the inherent texture patterns within the image. Consequently, texture analysis based on structure aims to understand the underlying rules governing the arrangement of texture primitives. This involves techniques such as mathematical morphology and topology, which examine how individual texture elements are organized and interact with each other. Structure-based analysis delves into the spatial relationships within the texture. Hence, texture analysis rooted in the spectrum focuses on multiscale and time − frequency analyses. Methods such as wavelet transform and fractal theory are used to study texture patterns across different scales and in various domains. Spectrum-based approaches provide insights into how textures evolve and vary over time and scale. These three approaches collectively offer a comprehensive framework for studying and interpreting image textures, allowing for a deeper understanding of the characteristics and properties of textures within SAR imagery (Haralick et al., 1973; Kandaswamy et al., 2005; Liu et al., 2021; Marghany, 2021a).

According to the aforementioned perspective, texture is a vital attribute used to discern objects or areas of interest within an image. Unlike spectral characteristics, which describe the overall tonal variations across various image bands, textural features provide insights into the spatial arrangement of tonal variations within a single band. One key tool for analyzing texture, therefore, is the gray-level cooccurrence matrix (GLCM), also known as the gray tone spatial dependency matrix. The GLCM serves as a record of how frequently different combinations of pixel brightness values (gray levels) manifest in an image. In this book, we use the GLCM for a series of "second-order" texture computations. These second-order measures take into account the relationships between pairs of pixels, typically neighboring ones, within the original image (Haralick et al., 1973; Lee et al., 1994a; Lee et al., 1994b; Lopez-Martinez & Pottier, 2007; Marghany, 2021a).

It's worth noting that there exist higher-order texture calculations, such as third-order and beyond, which consider the relationships among three or more pixels. While theoretically possible, these higher-order measures are not commonly utilized due to the computational time required and the complexity in interpreting the results. However, recent developments have aimed to streamline the calculation of third-order textures, making them more efficient for practical use.

7.11 What relationship exists between GLCM and the pixels within SAR images?

The question and answer provided shed light on the potential application of GLCM in identifying quantized Marghany turbulence boundary within SAR images

(Fig. 7.13). To delve deeper into this, let's explore how GLCM texture analysis captures the relationships between binary SAR pixels in a time domain, involving reference and neighboring pixels. In this context, the reference pixel is the focal point, and we examine its connection with the pixel to its immediate right, which serves as the neighbor pixel. This relationship is often denoted as a (1,0) relation, signifying that there is one pixel in the x-direction (horizontally, i.e., the row) and no pixels in the y-direction (vertically, i.e., the column). Now, consider a kernel window that encompasses the reference pixel. This window starts from the upper-left corner of the SAR image and extends diagonally to the lower-right corner. Pixels situated along the right edge of the image lack a right-hand neighbor, so they are excluded from this particular analysis.

In essence, GLCM texture analysis systematically examines how each pixel in the SAR image relates to its immediate right-hand neighbor, providing valuable insights into the spatial patterns and structures within the image (Haralick et al., 1973; Lopes et al., 1993; Lopez-Martinez & Pottier, 2007; Szalai & Dogossy, 2021).

7.12 How does GLCM function within the context of SAR turbulence image analysis?

As discussed earlier, the GLCM operates in a binary mode denoted as (1,0), signifying that the processing involves interactions between two pixels in SAR data. This means that the total number of pixel combinations within the GLCM, representing different possible combinations of gray levels, can be relatively small for a given kernel window size. The combinations of gray levels that are feasible for the test image, as well as their positions within the matrix, play a crucial role in this context. It's worth noting that the labels in the top row and left column of the GLCM matrix will only be utilized once during the process. For example, let's consider the first pair of reference and neighbor pixels, as indicated by the circled values 0 and 1 in Fig. 7.14.

Given that the reference pixel possesses a gray level value of 0 and the neighbor pixel has a value of 1, these values can be recorded as a single entry in a frequency table (Fig. 7.15). It's important to note that this process involves examining pairs of adjacent pixels. In essence, GLCM functions mathematically by considering two distinct pixels within the kernel window, typically positioned adjacent to each other along rows or columns (Haralick et al., 1973; Wang & He, 1990).

7.13 How the potential of GLCM transforms a SAR turbulence image into a symmetrical matrix?

The question at hand is how to create a GLCM from the entire SAR turbulence image, resulting in a new pattern that describes the statistical relationship between every pair of binary pixels. The fundamental question is: What does the SAR image look like after the GLCM transformation? In this context, the SAR image undergoes a transformation into a symmetrical matrix centered on the diagonal. To illustrate this, let's consider S as the SAR image matrix, and its transpose as S^T, which can be mathematically represented as:

$$S = S^T \tag{7.41}$$

This equation illustrates that the computation of SAR turbulence image texture necessitates symmetrical matrix. The subsequent step is to transform the gray-level cooccurrence matrix (GLCM) into this nonsymmetrical format. In this context, a symmetrical matrix implies that the same values appear in cells on opposite sides of the

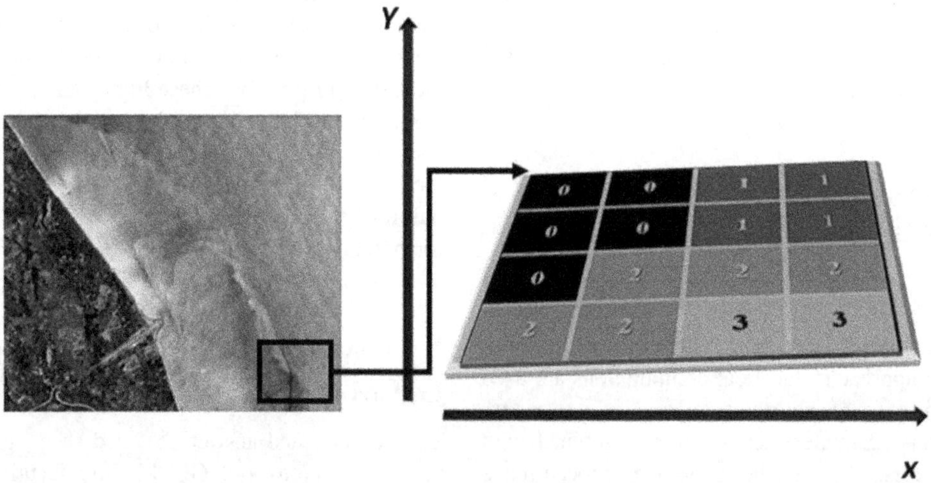

FIGURE 7.13 **Potential application of GLCM in identifying quantized Marghany turbulence boundary within SAR images.**

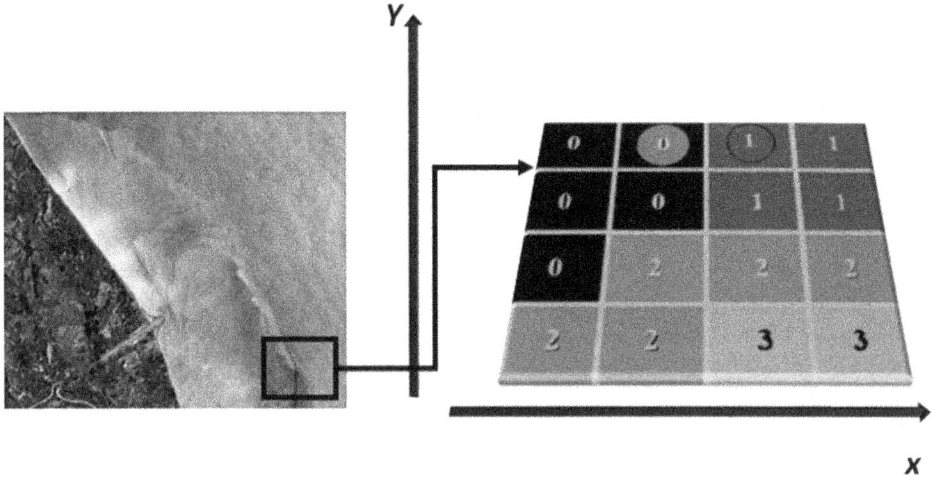

FIGURE 7.14 The spatial gap between two binary pixels across a turbulence boundary.

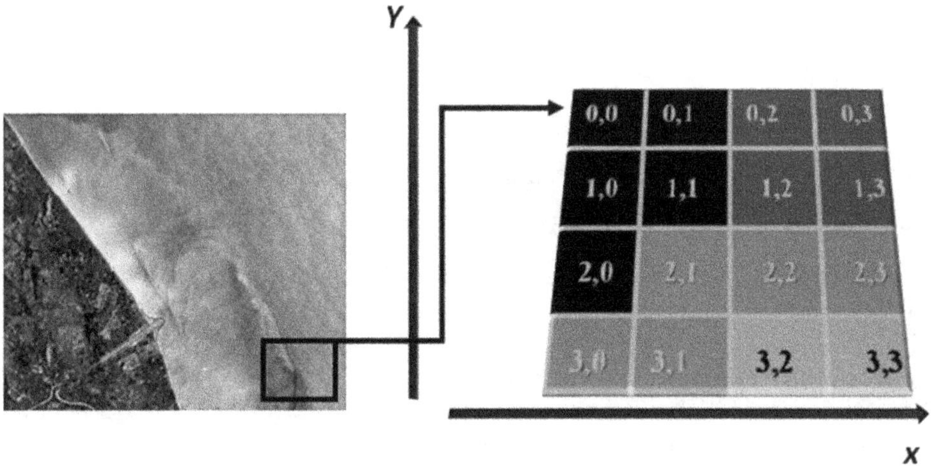

FIGURE 7.15 Table of frequency estimation across turbulence boundary.

diagonal (as shown in Fig. 7.16). For example, the value in cell 3,2 (i.e., 3) would be the same as the value in cell 2,3 (i.e., 3). The matrix on the right, calculated in the previous steps, is not symmetrical because it counts each reference pixel with the neighbor to its right. When counted in this manner, using only one direction, the number of times the combination 2,3 occurs is not equal to the number of times the combination 3,2 occurs (for instance, 3 may be to the right of 2,3 times, but to the left of two only once). However, symmetry can be achieved by counting every pair of binary pixels twice: once in the "forward" direction and once in the "backward" direction (by interchanging reference and neighbor pixels for the second count) (Haralick et al., 1973; Liu et al., 2021; Marghany, 2021a).

On the other hand, Fig. 7.17 is an example of a nonsymmetrical matrix across a turbulence boundary. In this matrix, the elements are not symmetrically arranged with respect to the main diagonal (from the top left to the bottom right). For instance, the element in row 1, column 2 (which is 2) is not mirrored by the element in row 2, column 1 (which is 4). Similarly, the element in row 1, column 3 (which is 3) is not mirrored by the element in row 3, column 1 (which is 7), and so on.

This lack of symmetry is evident when you compare the elements on one side of the main diagonal with the corresponding elements on the other side. In a symmetrical matrix, these pairs of elements would match, but in a nonsymmetrical matrix, they do not.

7.14 How to normalize symmetrical and nonsymmetrical SAR GLMC matrix?

Normalizing a GLCM (gray-level cooccurrence matrix) involves transforming the matrix so that the values represent probabilities or relative frequencies rather than raw

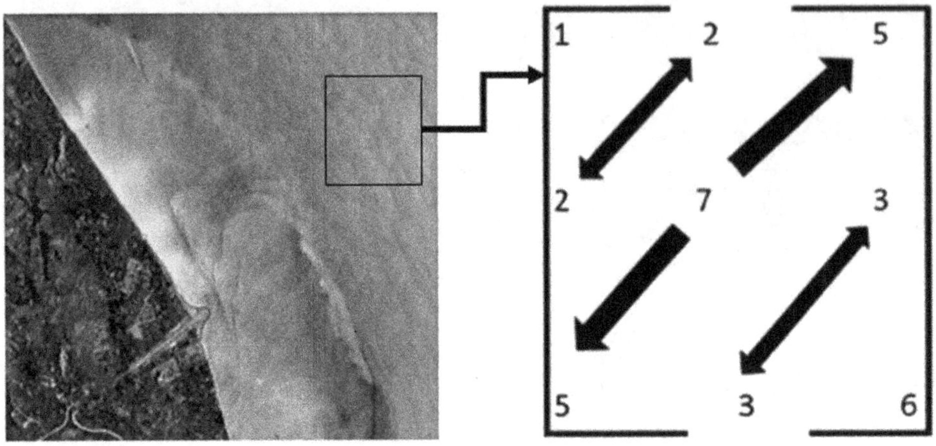

FIGURE 7.16 **Example of a symmetrical matrix in synthetic aperture radar image.**

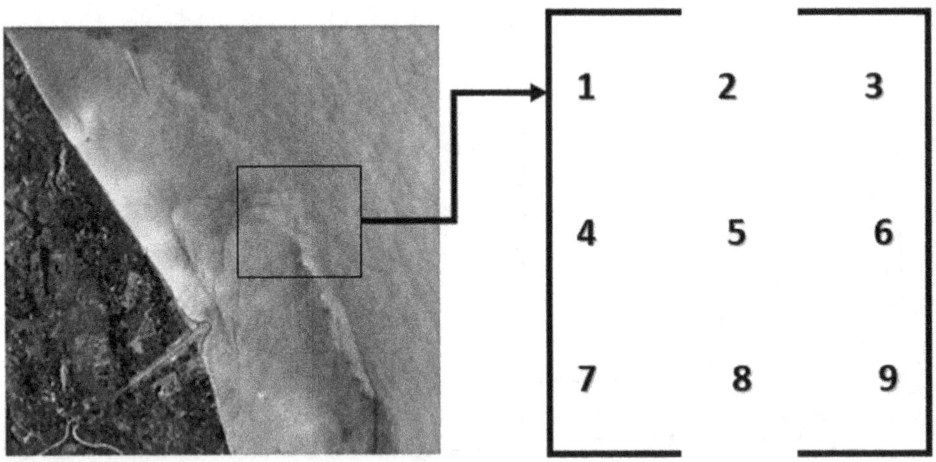

FIGURE 7.17 **Example of a nonsymmetrical matrix across a turbulence boundary.**

counts. This normalization step is important for making GLCMs comparable across different images or for extracting turbulence texture features that are independent of the image size and intensity range. First, one can calculate the GLCM, which results in a count matrix. This matrix contains the number of times specific pixel value pairs occur at a certain spatial relationship or offset within the image. For each row in the GLCM matrix (representing a specific pixel value), divide all the counts in that row by the sum of counts in that row. This step ensures that each row sums to 1, making the values represent probabilities. Similarly, you can choose to normalize by columns (relationship direction) in the matrix. This step ensures that each column sums to 1. Therefore if one wants to ensure that the entire matrix sums to 1, one can normalize by the total count (sum of all counts in the matrix). Symmetrical GLCMs have equal counts on either side of the main diagonal (i.e., pixel pairs (i, j) and (j, i) have the same count). You can choose to normalize them rowwise, columnwise, or based on the total count. The resulting matrix will remain symmetrical after normalization (Haralick et al., 1973; Kandaswamy et al., 2005).

Nonsymmetrical GLCMs, where (i, j) and (j, i) may have different counts, should be treated similarly. Normalizing them rowwise (Table 7.3) or columnwise will maintain their nonsymmetrical nature. If you want to create a symmetrical matrix for analysis purposes, one should normalize them by considering both directions (i.e., (i, j) and (j, i)).

This function takes the GLCM as input, normalizes it rowwise (each row sums to 1), and returns the normalized GLCM. One can apply similar logic for columnwise normalization or total count normalization as needed.

7.15 How can we generate an SAR turbulence texture image?

Normalization of the symmetrical GLCM plays a pivotal role in the creation of SAR texture images. This process involves applying a weighted average to the quantities of

TABLE 7.3 Normalize a GLCM (row-wise).

```
def normalize_glcm(glcm):
row_sums = glcm.sum(axis=1, keepdims=True)
normalized_glcm = glcm / row_sums
return normalized_glcm
```

dual pixels. Each pixel quantity is multiplied by a weight pixel value before being summed, and the result is divided by the total number of values. These weights are assigned to specify the relative significance of pixel values, similar to how exam scores may carry more weight than quiz scores in a student's final grade calculation (Haralick et al., 1973; Wang & He, 1990).

In essence, the outcome of a texture calculation is a single numerical value representing the entire kernel window. This value is placed at the position of the central pixel within the kernel window. The window is then shifted by one pixel, and the process is iterated to calculate a new GLCM and a new texture measure. This iterative approach is used to construct an entire image composed of texture values. However, it is important to note that each cell within a window must align with an occupied image cell. As a result, the central element of the window cannot correspond to a position within the image.

In cases where a window has dimensions of $N \times N$, a strip of unoccupied pixels, approximately $(N\text{-}1)/2$ pixels wide, surrounds the image. To handle this situation, these edge pixels are filled with the nearest texture calculation values. For example, in a 3×3 window (as illustrated in Fig. 7.18), the outer two rows and columns of the image receive texture values calculated from rows three to eight and columns three to eight, taking into account the dimensions (P and L) of the original image. In the illustrated image, both L and P are equal to 10.

7.16 Mathematical form of cooccurrence matrix

The mathematical form of a cooccurrence matrix (also known as a gray-level cooccurrence matrix or GLCM) is a square matrix that quantifies the joint occurrence of pixel values in a digital image. The cooccurrence matrix is typically denoted as P, and its elements $P(i,j)$ represent the frequency with which pixel values i and j occur together at a specific spatial relationship within the image. The mathematical form of the cooccurrence matrix P for a grayscale image is defined as follows:

$$P(i,j) = \sum_{x=1}^{M}\sum_{y=1}^{N} \delta(I(x,y) - i \text{ and } I(x + \Delta x, y + \Delta y) = j) \tag{7.42}$$

where $P(i,j)$ is the element at row i and column j of the cooccurrence matrix; M and N are the dimensions (width and height) of the image; and $I(x,y)$ represents the pixel value at position (x,y) in the image. Therefore Δx and Δy specify the spatial relationship or offset between the pixels being considered. Also, $\delta(\cdot)$ is the Kronecker delta function, which equals 1 if the condition inside it is true and 0 otherwise. In essence, the cooccurrence matrix P summarizes the statistics of pixel pairs in the image, taking into account their spatial relationships. It quantifies how often each pair of pixel values occurs within the specified spatial relationship (Marghany, 2021a).

The GLCM is a fundamental tool in texture analysis, image processing, and computer vision. It can be used to extract texture features that describe the spatial patterns and relationships within an image, which can be useful for various applications, including object recognition, classification, and segmentation.

7.16.1 Contrast

Distinguishing between various types of turbulence boundary features in SAR images poses a significant challenge. Therefore the focus is placed on analyzing the contrast variations in backscatter for each element, with respect to the diverse patterns of spatial texture variation. The contrast measure in the GLCM quantifies the local variations in pixel intensities within an image. It is defined as:

$$C_0 = \sum_{i,j} P(i,j)\cdot(i-j)^2 \tag{7.43}$$

where $P(i,j)$ represents the normalized GLCM, indicating the probability of pixel pairs with intensity values i and j. Therefore i and j represent the gray levels of pixel pairs in the GLCM (Table 7.4).

Identifying quantized Marghany turbulence boundaries in SAR (synthetic aperture radar) image, therefore can be a challenging task, especially when relying solely on contrast as the distinguishing factor. This difficulty arises because the contrast, which measures the difference in pixel intensities or backscatter values, may not always be a reliable indicator of these boundaries (Fig. 7.19). In SAR images, turbulence boundaries can manifest as regions with complex and variable backscatter patterns due to atmospheric conditions, surface properties, and other factors (Dekker, 2003; Kandaswamy et al., 2005).

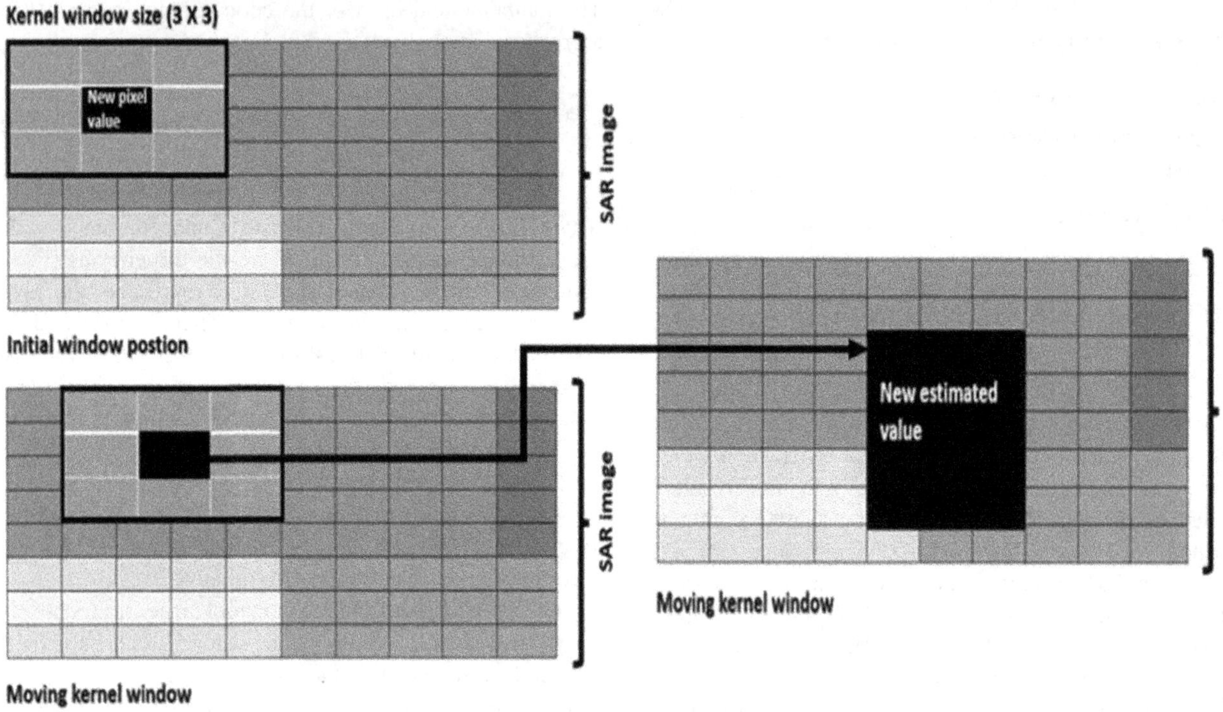

FIGURE 7.18 **Texture estimation based on the kernel window size of 3 × 3 pixels and lines.**

TABLE 7.4 Pseudocode example for calculating the contrast of an image region using a Gray-Level Cooccurrence Matrix (GLCM).

```
# Define the GLCM for the image region (example values) glcm = [ [0.1, 0.2, 0.1], [0.2, 0.3,
0.2], [0.1, 0.2, 0.1] ]
# Initialize the contrast variable contrast = 0.0
# Iterate through all gray level pairs (i, j) for i in range(len(glcm)): for j in range(len(glcm[i])):
# Calculate the contrast contribution for this pair contrast += glcm[i][j] * (i - j) ** 2
# The final contrast value print("Contrast:", contrast)
```

To effectively identify quantized Marghany turbulence boundaries, additional analysis and feature extraction techniques may be required. These techniques can include texture analysis, statistical measures, or the application of specialized algorithms designed to detect specific patterns associated with these boundaries. By considering multiple features and using advanced image processing methods, it becomes possible to improve the accuracy of turbulence boundary identification in SAR imagery.

7.16.2 Homogeneity

Homogeneity in a GLCM (gray-level cooccurrence matrix) represents a measure of how closely the gray levels of pixels in an image are distributed. It quantifies the similarity or uniformity of pixel pairs with respect to their intensity values. In a GLCM, each element (i, j) represents the frequency with which a particular pair of gray levels (i and j) occur in close proximity within the image. Homogeneity is calculated using these values and

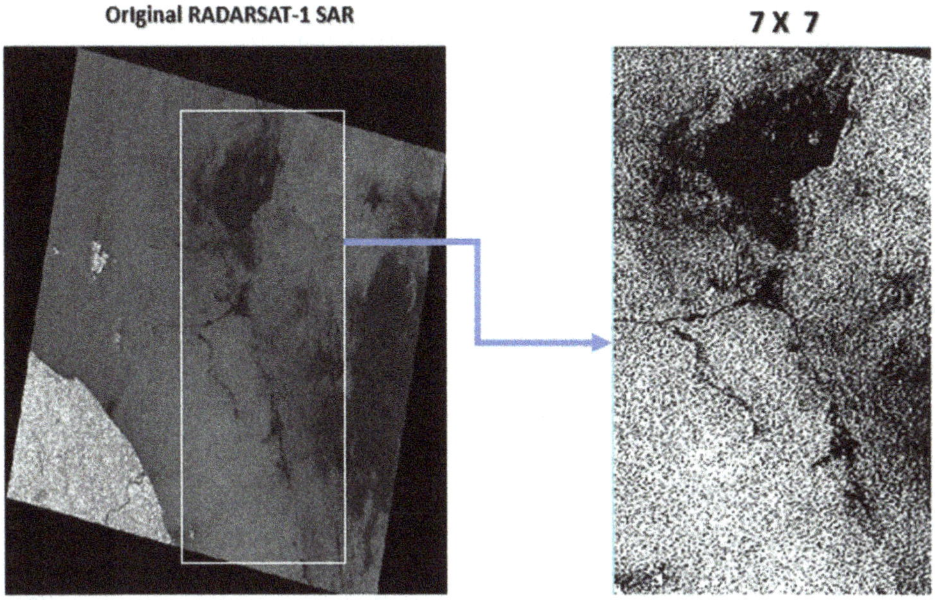

FIGURE 7.19 The outcome of the contrast analysis is obtained using a kernel window of dimensions 7 × 7 pixels and lines.

TABLE 7.5 Pseudocode example for calculating the homogeneity of an image region using a Gray-Level Cooccurrence Matrix (GLCM).

```
# Assuming you have the GLCM matrix 'glcm' already computed
# Initialize variables homogeneity = 0.0 rows, cols = glcm.shape
# Loop through each element in the GLCM for i in range(rows): for j in range(cols):
frequency = glcm[i][j] distance = abs(i - j) # Calculate the absolute difference between gray
levels
# Update homogeneity using the formula homogeneity += frequency / (1 + distance)**2
# Normalize the homogeneity value if needed (optional) # homogeneity /= (rows * cols)
# Print or return the homogeneity value print("Homogeneity:", homogeneity)
```

is designed to be high when most pixel pairs in the image have similar gray levels and low when there are significant variations in gray levels. Mathematically, homogeneity (H) in GLCM can be expressed as:

$$H = \sum (P(i,j)(1 + |i - j|))^{-2} \tag{7.44}$$

here $P(i, j)$ is the frequency of gray level pair (i, j) in the GLCM; and $|i-j|$ represents the absolute difference between the gray levels i and j. A pseudocode example to calculate homogeneity using a GLCM is demonstrated in Table 7.5. In simpler terms, the formula calculates the sum of squared reciprocal differences between gray levels for all pairs in the GLCM. When the difference between gray levels is small, the reciprocal term becomes larger, contributing to higher homogeneity values (Asakura & Takai, 1981; Rosenblum et al., 2018; Soh & Tsatsoulis, 1999; Szalai & Dogossy, 2021). Conversely, when there are significant differences in gray levels between neighboring pixels, the homogeneity value decreases.

Homogeneity is a useful metric for identifying quantized Marghany turbulent boundaries and bound front states in SAR (synthetic aperture radar) images (Fig. 7.20). In SAR image processing, homogeneity measures the uniformity or similarity of pixel values within a region or texture.

When quantized Marghany turbulent boundaries or bound front states are present in an SAR image, they often exhibit distinct textural patterns or variations compared with their surroundings. These patterns can result in regions of higher homogeneity within the image. Therefore by analyzing the homogeneity values across different regions of the SAR image, you can identify areas where turbulence boundaries or front states may be located.

In summary, higher homogeneity values in SAR images can indicate the presence of quantized Marghany turbulent boundaries or bound front states, making homogeneity a valuable tool for their identification and analysis in SAR image processing.

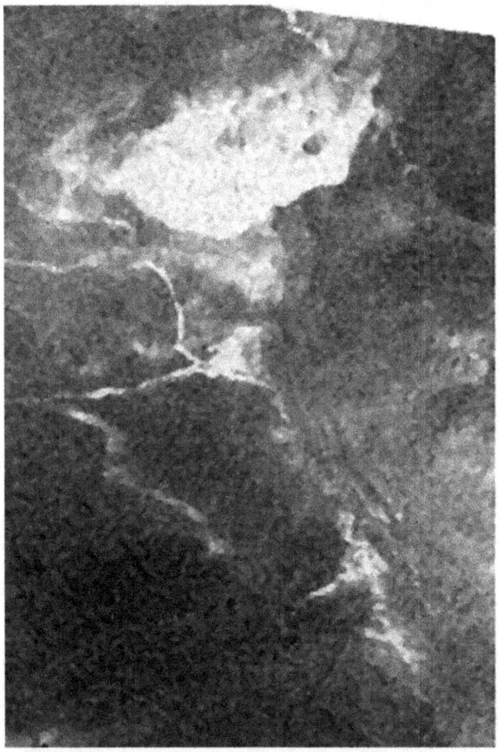

FIGURE 7.20 The outcome of the homogeneity analysis is obtained using a kernel window of dimensions 7 × 7 pixels and lines.

7.16.3 Dissimilarity

Dissimilarity in the context of GLCM (gray-level cooccurrence matrix) refers to a texture metric used to quantify the differences or variations in pixel values within a specified neighborhood or region of an image. GLCM is a valuable tool in image processing and analysis, particularly in characterizing texture patterns.

Dissimilarity is computed based on the GLCM, which captures the cooccurrence of gray-level values between pairs of pixels in an image. To calculate dissimilarity, you first create the GLCM for a specific neighborhood size and direction. The GLCM is a square matrix where each entry (i, j) represents the number of times a pixel with gray level i occurs in proximity to a pixel with gray level j in the specified direction (Haralick et al., 1973; Kandaswamy et al., 2005; Liu et al., 2021; Marghany, 2021a). In this sense, dissimilarity $D(i,j)$ can be mathematically defined as:

$$D(i,j) = \sum_{i=1}^{N} \sum_{j=1}^{N} P(i,j) \cdot |i - j| \tag{7.45}$$

here N represents the number of gray levels in the image; and $P(i,j)$ is the normalized GLCM entry at position (i, j), which indicates the probability of the pair (i, j) occurring

TABLE 7.6 Pseudocode example for calculating the dissimilarity of an image region using a Gray-Level Cooccurrence Matrix (GLCM).

```
# Define constants and variables N = number of gray levels GLCM = createGLCM(image) #
Create GLCM for the input image dissimilarity = 0
# Loop through each element in GLCM for i in range(N): for j in range(N): # Calculate the
dissimilarity contribution for the (i, j) pair dissimilarity += GLCM[i][j] * abs(i - j)
# Normalize dissimilarity dissimilarity /= N
# Output the dissimilarity metric print("Dissimilarity:", dissimilarity)
Return
End
```

in the image (Haralick et al., 1973; Kandaswamy et al., 2005; Marghany, 2001; Marghany & Genderen, 2014). Table 7.6 is a simplified pseudocode example to calculate the dissimilarity metric using the GLCM.

In the context of texture analysis using GLCM, dissimilarity is a metric that complements the contrast measure when identifying features within SAR (synthetic aperture radar) images, particularly in the context of quantized Marghany turbulence boundaries (Fig. 7.21).

Dissimilarity, as a metric, focuses on capturing variations or differences in gray-level pairs within the image. It emphasizes the contrast between different regions by considering how different the gray levels are between neighboring pixels. This makes it effective in highlighting areas with distinct textural variations, which can be indicative of certain features or boundaries. However, it's important to note that dissimilarity may not perform as effectively in identifying quantized Marghany bound front states. This is because these states might exhibit textural variations that are not adequately captured by dissimilarity alone. These variations could be subtle or less pronounced in terms of pixel intensity differences, making them less distinguishable using this metric.

Homogeneity, on the other hand, is better suited for identifying regions with pronounced differences in pixel intensities. It tends to highlight areas with sharp transitions or boundaries between materials or features within the image. Hence, when focusing on quantized Marghany turbulence boundaries, contrast can be a more suitable choice for highlighting these boundaries (Haralick et al., 1973; Raney & Wessels, 1988; Szalai & Dogossy, 2021).

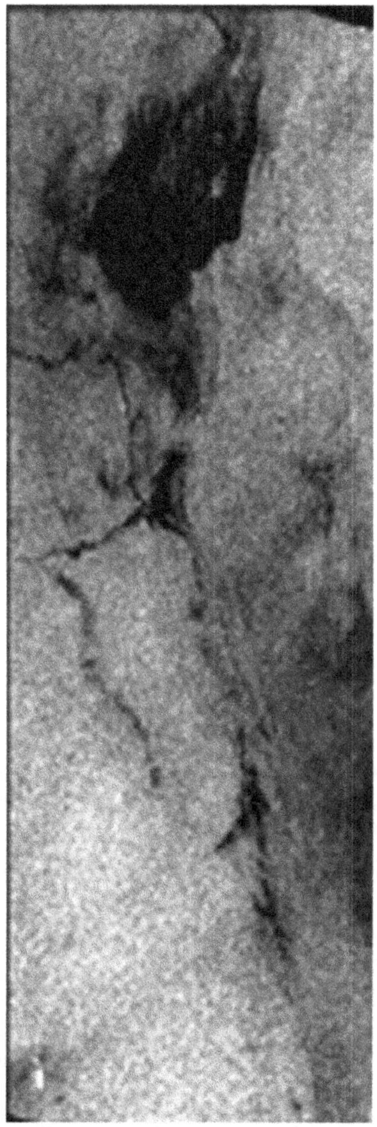

FIGURE 7.21 The outcome of the dissimilarity analysis is obtained using a kernel window of dimensions 7 × 7 pixels and lines.

7.16.4 Angular second moment and energy

Angular second moment (ASM) and energy are two important statistical measures derived from the GLCM in texture analysis. These measures provide insights into the homogeneity and complexity of textures within an image. Mathematically, ASM is calculated as the sum of the squared elements in the GLCM. It represents the probability of the occurrence of each gray-level pair squared, emphasizing highly recurrent patterns. ASM is defined as:

$$\text{ASM} = \sum_{i=0}^{N-1}\sum_{j=0}^{N-1} P(i,j)^2 \tag{7.46}$$

Therefore energy is closely related to ASM and also quantifies the texture's homogeneity. It represents the sum of the squared GLCM elements, similar to ASM. However, it is often used as a normalized version of ASM. Energy is calculated as:

$$\text{Energy} = \sqrt{\text{ASM}} \tag{7.47}$$

Both ASM and energy are used in texture analysis to characterize the regularity and homogeneity of textures within an image. Higher ASM and energy values indicate smoother and more uniform textures, while lower values indicate textures with greater variations and complexity. These measures are valuable for various image processing tasks, including pattern recognition, segmentation, and feature extraction (Argenti & Alparone, 2002; Kandaswamy et al., 2005; Marghany, 2001; Marghany & Hashim, 2010). Table 7.7 shows the pseudocode representation of how one might calculate the ASM and energy from a GLCM in a texture analysis algorithm.

TABLE 7.7 Pseudocode example for calculating the ASM and Energy of an image region using a Gray-Level Cooccurrence Matrix (GLCM).

```
# Assuming you have already computed the GLCM (represented as a 2D matrix)
# Initialize variables ASM = 0.0 Energy = 0.0
# Calculate the number of gray levels (N) based on the GLCM size N = len(GLCM)
# Calculate ASM for i in range(N): for j in range(N): P_ij = GLCM[i][j] # Get the GLCM
element at (i, j) ASM += P_ij ** 2
# Calculate Energy (square root of ASM) Energy = sqrt(ASM)
# Print or use the values of ASM and Energy as needed print("Angular Second Moment
(ASM):", ASM) print("Energy:", Energy)
```

The reason why the quantized Marghany turbulence boundary appears more discrete in ASM and energy than in homogeneity is rooted in the way these texture measures capture different aspects of the texture patterns within the image (Fig. 7.22). Indeed, ASM and energy are texture measures that focus on the orderliness and uniformity of pixel values. Specifically, ASM measures the overall orderliness of pixel pairs, and energy represents the uniformity of these values. When there are sharp transitions or discrete patterns, such as those seen at the quantized Marghany turbulence boundary, ASM and energy tend to pick up on these variations more effectively. The distinct transitions result in a higher ASM value and higher Energy, making the boundary appear more discrete in these measures.

Homogeneity in GLCM measures the closeness of gray-level values in an image. In other words, it quantifies how similar neighboring pixel values are. When an image region has relatively consistent and similar pixel values, homogeneity will be high. This means that it can smooth out the variations caused by the turbulence boundary, making it less distinct in homogeneity. The gradual transitions and subtle changes in pixel values at the boundary may not strongly affect the homogeneity measure (Marghany & Genderen, 2014; Nie et al., 2022).

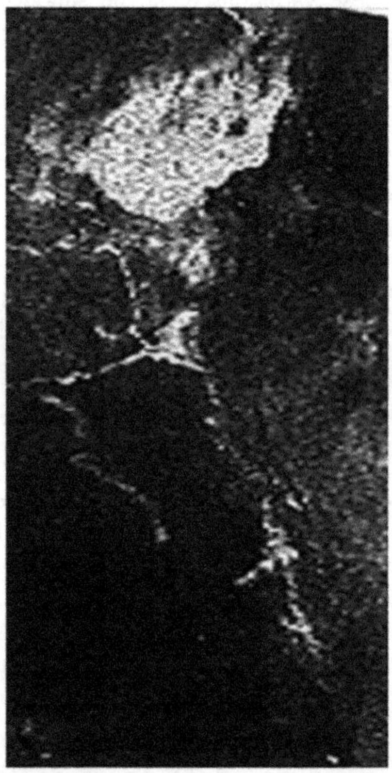

FIGURE 7.22 The outcome of the angular second moment value and higher energy is obtained using a kernel window of dimensions 7 × 7 pixels and lines.

7.17 Entropy

Entropy, in the context of texture analysis and image processing, is a fundamental concept that quantifies the randomness, uncertainty, or information content within an image or a specific region of interest. It plays a crucial role in characterizing the complexity and organization of pixel values or gray levels in an image. Entropy is often used as a texture feature to describe the degree of disorder or irregularity in a texture pattern (Argenti & Alparone, 2002; Kandaswamy et al., 2005; Marghany, 2001;

Marghany & Genderen, 2014; Nie et al., 2022). Mathematically, entropy $E(H)$ is calculated using the probability distribution of gray levels within the image or region. The formula for entropy is typically expressed as:

$$E(H) = -\sum_{i=1}^{N} P(i) \cdot \log_2(P(i)) \tag{7.48}$$

here N is the total number of gray levels in the image; and $P(i)$ is the probability of occurrence of the ith gray level. Table 7.8 shows a pseudocode example for calculating entropy from an image.

Entropy, in the context of image analysis, measures the amount of randomness or uncertainty in the pixel intensity values of an image. It's a statistical measure that quantifies the level of information content in the image. Higher entropy values indicate greater randomness, while lower entropy values suggest more uniform or homogeneous pixel intensities (Haralick et al., 1973; Marghany, 2001; Marghany & Genderen, 2014).

When comparing the ability of entropy and homogeneity to identify quantized Marghany turbulence boundaries, the choice between them depends on the characteristics of the boundaries and the specific goals of the analysis. Entropy is effective at identifying regions in an image with high variation or randomness in pixel intensities (Fig. 7.23). If the quantized Marghany turbulence boundary exhibits significant variations in pixel values, entropy may provide a good measure to distinguish it from homogeneous regions. However, if the boundary has both high and low-intensity areas, entropy alone might not provide a clear separation.

It can be said that the choice between entropy and homogeneity as a texture feature for identifying quantized Marghany turbulence boundaries depends on the specific characteristics of the boundaries and the desired sensitivity to different types of variations in the image. It's often beneficial to use multiple texture features and combine them in image analysis to achieve a more comprehensive understanding of the scene.

The ability of homogeneity to identify five of the quantized Marghany bound front states in Sentinel A data (Fig. 7.24), particularly due to the association of convergence and divergence zones, can be explained as follows:

TABLE 7.8 Pseudocode example for calculating entropy of an image region using a Gray-Level Cooccurrence Matrix (GLCM).

```
# Function to calculate entropy of an image function calculate_entropy(image): # Initialize
variables entropy = 0.0 total_pixels = 0 gray_level_counts = {}
# Step 1: Count the frequency of each gray level
for pixel in image:
gray_level = pixel.gray_value
if gray_level in gray_level_counts:
gray_level_counts[gray_level] += 1
else:
gray_level_counts[gray_level] = 1
total_pixels += 1
# Step 2: Calculate entropy using the probability distribution
for gray_level, count in gray_level_counts.items():
probability = count / total_pixels
entropy -= probability * log2(probability)
return entropy
# Example usage: image = load_image("sample_image.jpg") entropy_value =
calculate_entropy(image) print("Entropy:", entropy_value)
```

7.18 Why can homogeneity detect multiple phases of quantized Marghany bound front states?

The ability of homogeneity to identify five of the quantized Marghany bound front states in Sentinel A data, particularly due to the association of convergence and divergence zones, can be explained as follows:

7.18.1 Convergence and divergence zones

In SAR (synthetic aperture radar) imagery, convergence and divergence zones often correspond to regions where different air masses or wind patterns meet or separate. These zones can exhibit distinct radar backscatter characteristics due to variations in surface roughness, moisture content, or other factors. The radar signal interacts differently with these areas, leading to differences in pixel intensities.

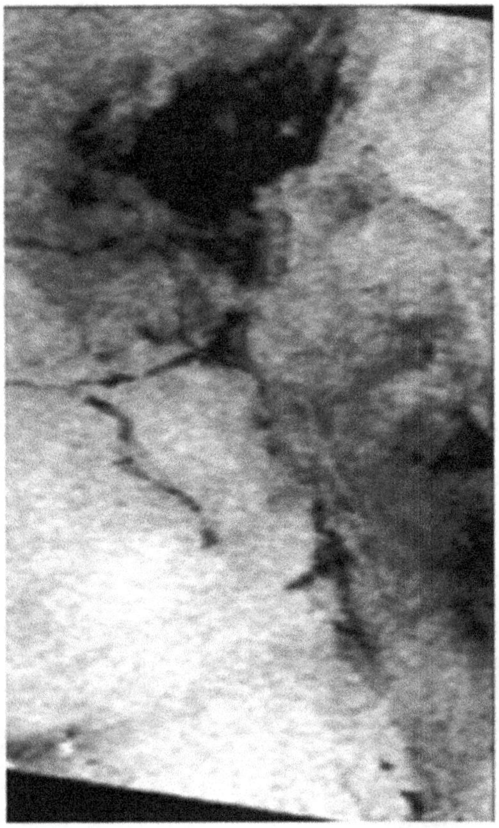

FIGURE 7.23 The outcome of the entropy analysis is obtained using a kernel window of dimensions 7 × 7 pixels and lines.

7.18.2 Homogeneity as a texture feature

Homogeneity, as a texture feature, measures the uniformity or similarity of pixel intensities within a local neighborhood. In areas with convergence or divergence zones, the radar backscatter characteristics are likely to be relatively uniform within these zones but differ significantly from the surrounding regions. This results in high homogeneity values within the zones.

7.18.3 Distinct boundaries

When convergence and divergence zones are well-defined and have clear boundaries, they create a sharp transition from homogeneous (inside the zones) to nonhomogeneous (outside the zones) regions. Homogeneity is effective at capturing these sharp transitions, making it a suitable texture feature for identifying such boundaries.

7.18.4 Quantized Marghany bound front states

The quantized Marghany bound front states likely correspond to different types or phases of convergence and divergence zones. These states may exhibit variations in their radar backscatter characteristics, but within each state, there is a level of uniformity or homogeneity.

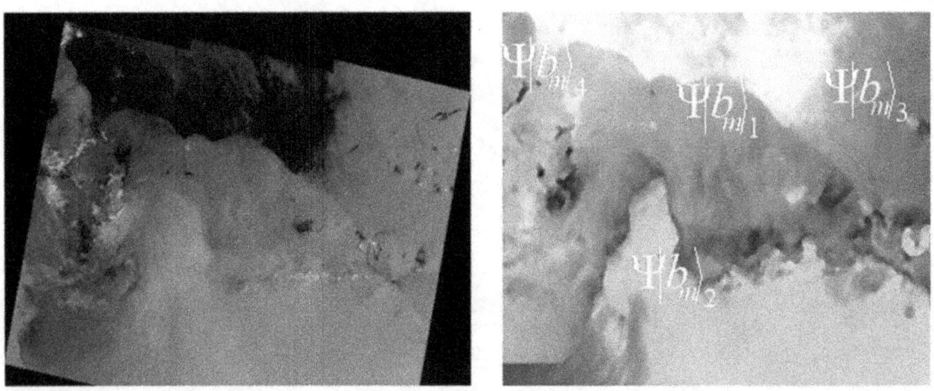

FIGURE 7.24 Identification of fifth of quantized Marghany bound front states in Sentinel A data using homogeneity.

7.18.5 Sensitivity to local patterns

Homogeneity focuses on local patterns within a given window or neighborhood. By analyzing these local patterns, it can detect areas with similar radar backscatter characteristics, which often correspond to specific features such as convergence and divergence zones.

In summary, homogeneity is effective at identifying the quantized Marghany bound front states associated with convergence and divergence zones because it captures the local uniformity and transitions in radar backscatter characteristics that are characteristic of these zones. When these zones have well-defined boundaries and distinct features, homogeneity can be a valuable texture feature for their identification in SAR imagery (Haralick et al., 1973; Marghany, 2001; Marghany, 2021a; Marghany, 2021b).

7.19 Quantum entropy

In the context of SAR (synthetic aperture radar) image processing, the probability $P(b_m)$ of observing a particular SAR image intensity across the quantized Marghany bound front state b_m can be expressed in terms of dual quantum states $|i\rangle$ and $\langle j|$ as follows:

$$P(b_m(\varepsilon)) = \sum_{i,j} \langle i|b_m(\varepsilon)|j\rangle \langle j|i\rangle \tag{7.49}$$

here $\langle i|$ and $\langle j|$ are the dual quantum states of quantized Marghany bound front state. Adationally, $|i\rangle$ and $|j\rangle$ are quantum states representing different possible outcomes or intensities in the SAR image. In this sense, b_m is an operator that corresponds to the quantized Marghany bound front states in SAR image intensity. This expression suggests that the probability of observing a specific intensity in the SAR image is determined by the quantum states $|i\rangle$ and $|\langle j|$ and their corresponding quantum amplitudes. It reflects the probabilistic nature of SAR image outcomes and their quantum representation (Marghany, 2021a).

In the realm of quantum science, quantum relative entropy serves as a metric for gauging the distinguishability between two quantum states. It operates as the quantum mechanical counterpart to relative entropy in classical information theory.

To put it into context, quantum entropy plays a pivotal role in quantifying the phenomenon of decoherence that occurs within quantum systems, particularly when considering the behavior of quantum particles interacting with their surrounding environment, such as quantize Marghany turbulence boundary involving quantized Marghany bound front states. The quantum entropy, when applied within the framework of quantum mechanics, utilizes the density matrix of a system. This density matrix is a mathematical construct that allows us to describe the quantum system's characteristics, accounting for a mixed state. A mixed state here implies that the quantum system may exhibit both coherence and decoherence, which are fundamental concepts in quantum physics. In essence, when we talk about quantum entropy interpreting SAR (synthetic aperture radar) data in the Hilbert space ($\mathscr{H}$), we are referring to the role it plays in capturing and quantifying the interplay of coherence and decoherence within the quantum system. This has applications in understanding the behavior of SAR data in a quantum context, shedding light on its unique properties and behavior at the quantum level (Hur, 2008; Lanzagorta, 2012; Maccone & Ren, 2020; Marghany, 2021a).

$$\mathscr{H}(b_m) = \sum_{i,j}^{N} \mathscr{H}(\psi|i > |j > ; \psi \in |b_m > \tag{7.50}$$

Formula (7.50) articulates that the quantum entropy image serves as a descriptor of the differences in quantized Marghany turbulence boundary diversities among different closses of the surrounding environment (Table 7.9). This makes it an excellent indicator, especially when assessing the possibilities of turbulence zone, particularly in cases where quantized Marghany bound front state diversities are evidents.

This can be demonstrated using ERS-1 imagery acquired along the Sumatra coastline on May 26, 1996 (Fig. 7.25).

More recently, Sentinel-A data have revealed the presence of various quantized Marghany fronts along the coastal waters of Sarawak on September 5, 2023 (Fig. 7.26). These fronts have emerged due to the existence of coral reefs, which, in turn, generate turbulent flows along the coastal waters of Sarawak.

Furthermore, recent Sentinel-A data have indicated the presence of large-scale quantized Marghany boundary turbulence, encompassing a variety of quantized Marghany bound front states along the coastal zone of Bali, Indonesia, on September 7, 2023 (Fig. 7.27). These states include mesoscale front vorticities (Fig. 7.28) and occurrences of internal waves (Fig. 7.29).

The internal waves generated at these fronts then radiate outward into the surrounding water. This radiation is similar to ripples spreading outward when you toss a stone into a pond. These internal waves can carry energy and mix water masses at different depths, which can have various ecological and oceanographic implications. When we observe oceanic fronts in SAR images, however, it is not uncommon to notice a connection between these fronts and the presence of internal waves. Internal waves are a type of ocean wave that occurs beneath the ocean surface, unlike surface waves that we see breaking at the shoreline. These waves typically form at the interface between water masses of different densities within the ocean, such as where a warm, less dense layer of water

TABLE 7.9 Pseudocode of quantum entropy algorithm.

1. Initialize a quantum state |ψ⟩ in a quantum register.

2. Apply quantum gates and operations to manipulate the quantum state |ψ⟩. These operations may include: - Quantum entanglement operations - Quantum measurements - Quantum state preparations

3. Calculate the quantum state's density matrix ρ from the resulting state |ψ⟩.

Initialize an empty density matrix ρ.

For each pair of basis vectors |i⟩ and |j⟩: Calculate the outer product |i⟩⟨j|, resulting in a matrix. If |i⟩ and |j⟩ are the same, multiply the matrix by the probability of observing state |i⟩ (e.g., p_i). Add the resulting matrix to ρ.

Output the density matrix ρ.

End

4. Calculate the von Neumann entropy of the quantum state: - H(ρ) = -Tr(ρ log2(ρ))

5. Output the calculated entropy value.

6. End

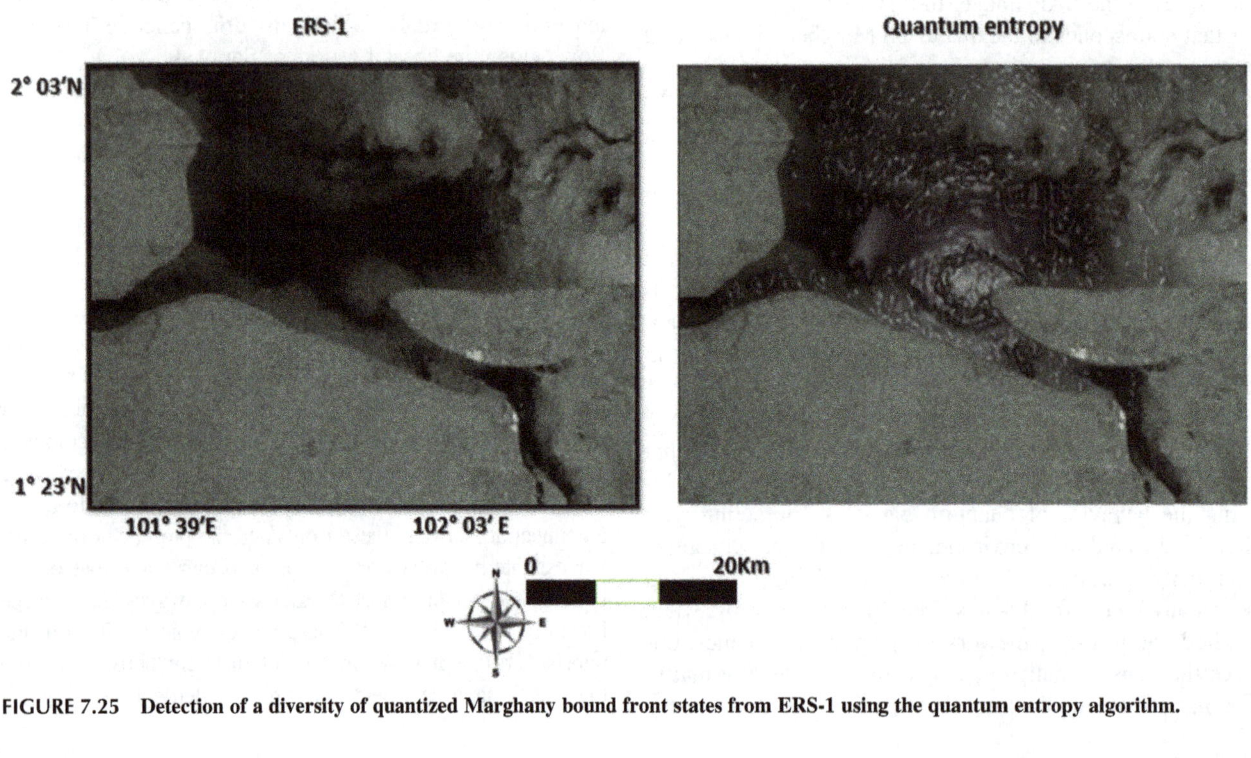

FIGURE 7.25 Detection of a diversity of quantized Marghany bound front states from ERS-1 using the quantum entropy algorithm.

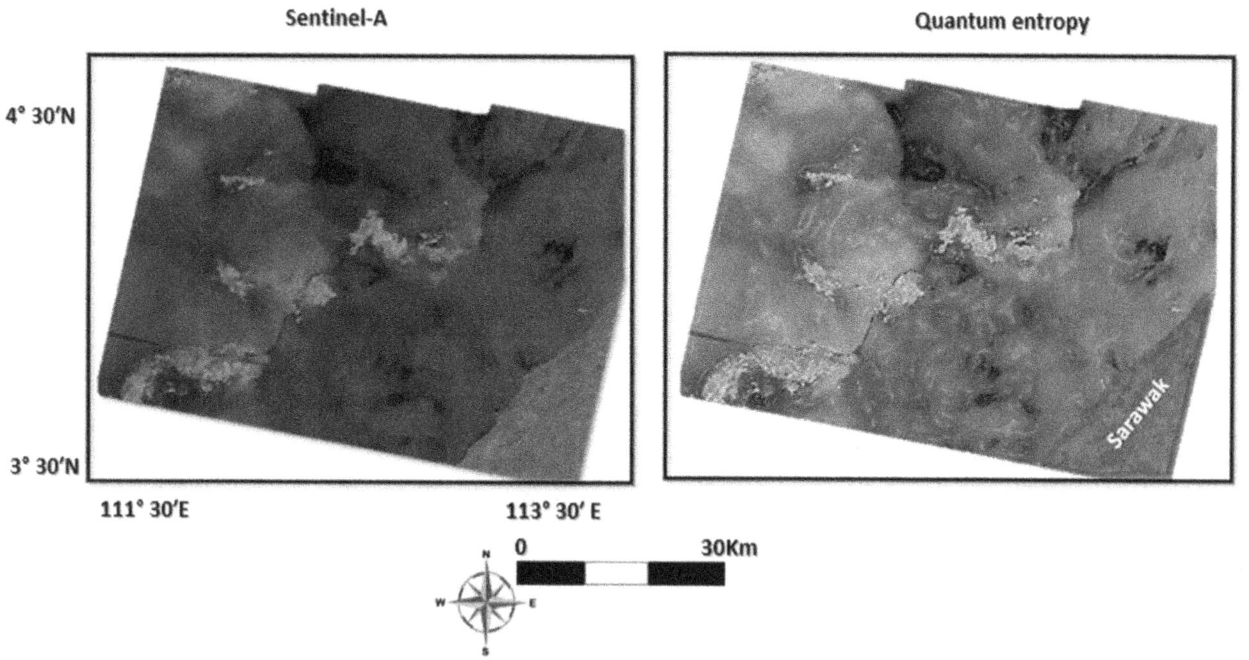

FIGURE 7.26 Detection of a diversity of quantized Marghany bound front states from Sentinel-A data using the quantum entropy algorithm.

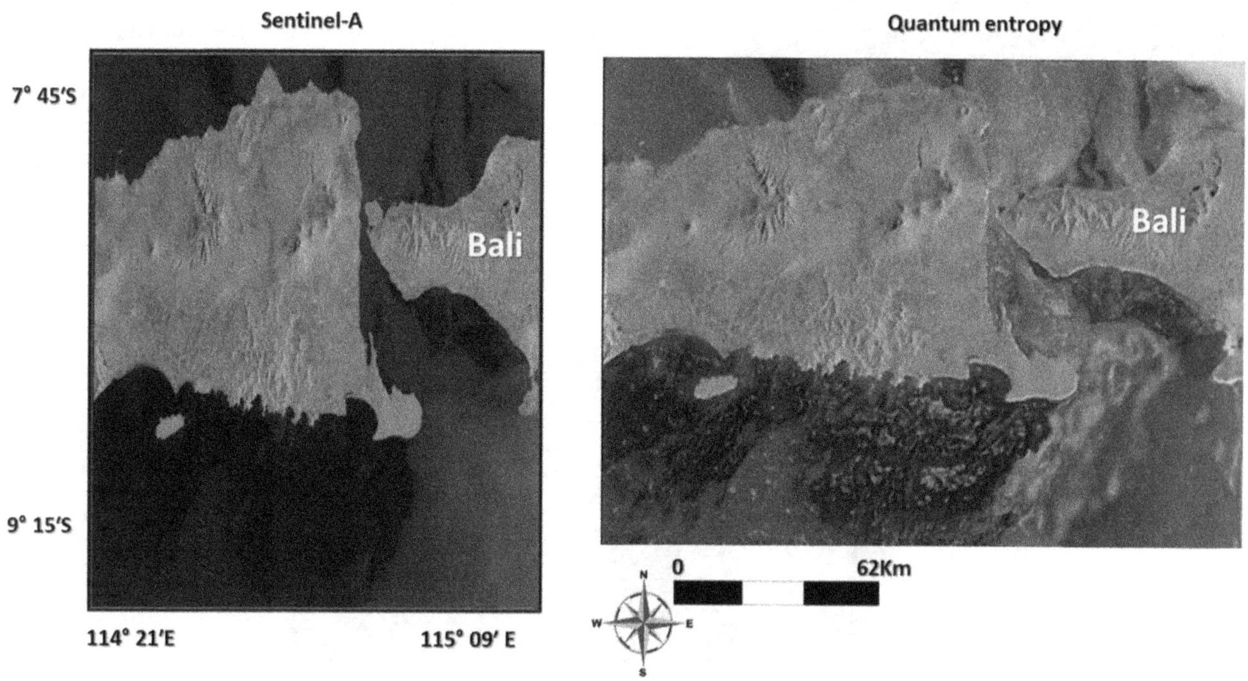

FIGURE 7.27 Detection of a diversity of quantized Marghany bound front states along Bali coastal waters from Sentinel-A data using the quantum entropy algorithm.

meets a colder, denser layer. In other word, the statement emphasizes that oceanic fronts are often associated with the generation of internal waves due to the convergence of water masses with different densities along the sloping isopycnal. These internal waves, once generated, propagate outward from the front and can be observed in SAR images, providing valuable insights into the dynamic processes occurring beneath the ocean's surface (Robinson, 2004; 2010a; 2010b).

From the perspective of Eq. (7.50), the density matrix, when considered within the quantum Hilbert space ($\mathscr{H}$), encompasses or contains information related to the quantized Marghany front. In other words, this equation provides a mathematical framework for understanding and quantifying the characteristics of the quantized Marghany front within the context of quantum entropy imaging. Put simply, this equation offers a mathematical structure that enables us to comprehend and measure the properties of the quantized Marghany front. It does so within the specific context of quantum entropy imaging. In essence, it provides a formalized way to analyze and describe the features and behaviors of the quantized Marghany front when using quantum entropy-based imaging techniques.

Consequently, quantum entropy algorithm also allows us to visualize the distinct structural characteristics of the quantized Marghany front within the topology of the quantized Marghany turbulence boundary. To illustrate, one of the most prominent features of the quantized Marghany bound front state is the presence of numerous small-scale vortices. These vortices exhibit well-defined morphological patterns, and this equation helps us capture and analyze their sharp structures. It provides a scientific framework to understand and study these intricate features in the context of quantum entropy imaging (Marghany, 2001; Marghany, 2021a; Marghany, 2021b).

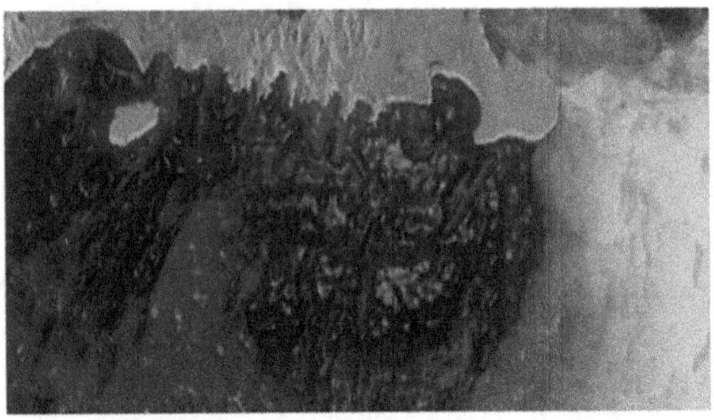

FIGURE 7.28 **Mesoscale front vorticities extracted from Fig. 7.27 along Gradjagan coastal waters.**

The quantized Marghany bound front states not only occur in Malaysian coastal waters but can also be found in other coastal regions such as the Sumatra, Indonesia, west coast of the Malacca Strait, Bali coastal water of Indian Ocean, and the North Sea in the Netherlands coastal waters (Fig. 7.30). This is emphasizing the broader geographical distribution.

It is worth noting the abundant occurrences of filaments in entire frontal zones. Filaments are commonly found near oceanic fronts, which are regions where water masses with different properties, such as temperature and salinity, converge. At fronts, there are significant density gradients, meaning there are abrupt changes in water density over a relatively short distance. These density gradients can be quite steep, and they often create conditions conducive to filament formation. Filaments, therefore are often associated with surface currents that transport water, nutrients, and organic material. These currents can create the characteristic elongated shape of filaments. As the currents move water masses and materials, they can enhance the convergence and concentration of organic material at the filament's core. Consequently, filaments

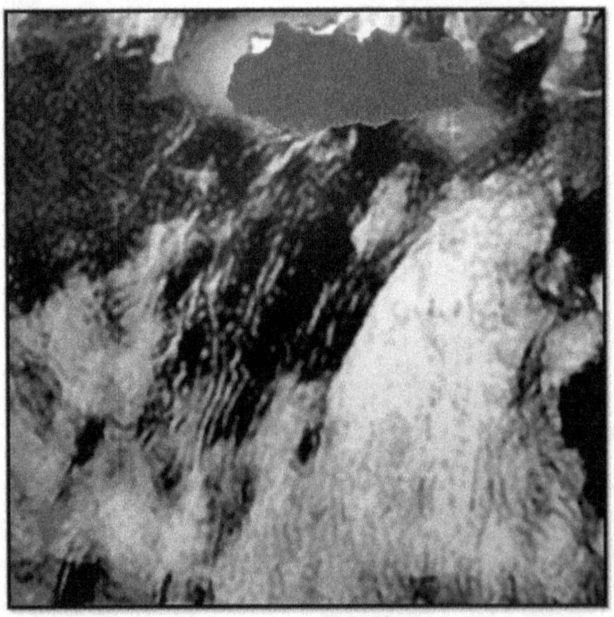

FIGURE 7.29 **Internal wave extracted from Fig. 7.27 across quantized Marghany turbulence boundary.**

FIGURE 7.30 Quantized Marghany bound front states were observed along the Netherlands' coastal waters in the North Sea using Sentinel-A data on September 6, 2023.

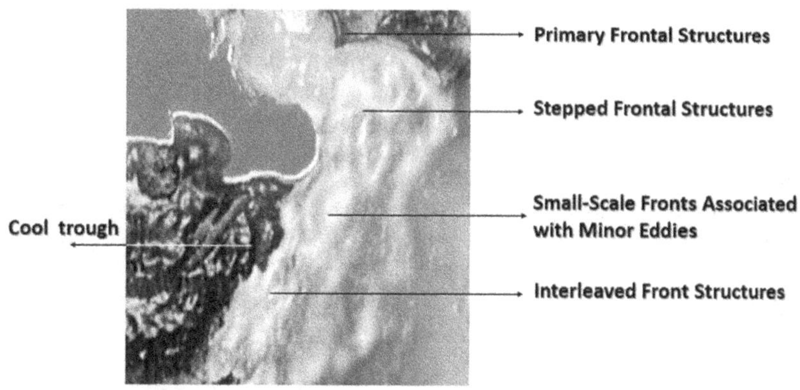

FIGURE 7.31 The structure of the quantized Marghany bound front state.

are associated with oceanic fronts because fronts create conditions that lead to the convergence and mixing of water masses with different properties. The concentration of organic material in these regions, coupled with the sensitivity of SAR to changes in surface roughness, causes filaments to appear as dark patches in SAR images (Robinson, 2004; Robinson, 2010a).

Certainly, it is evident that the quantized Marghany bound front states encompass distinctive stepped frontal structures, including primary fronts, interleaved front structures, and small-scale fronts associated with minor eddies. The term "quantized Marghany bound front states" refers to specific boundary conditions or features within a given area, often in the context of radar imagery. These boundary states exhibit certain structural characteristics that can be observed and analyzed in the imagery (Fig. 7.31).

7.19.1 Stepped frontal structures

One of the prominent features of these bound front states is the presence of stepped frontal structures. This means that within the boundary area, there are distinct and well-defined levels or steps in the arrangement of certain elements. These steps might represent variations in properties such as temperature, humidity, or other environmental factors.

7.19.2 Primary fronts

Within these boundary states, there are primary fronts. These are the main or central areas where significant changes or transitions occur. They can be thought of as the focal points within the boundary, where the most pronounced shifts in conditions are observed.

7.19.3 Interleaved front structures

In addition to primary fronts, there are interleaved front structures. These are secondary or intermediate regions within the boundary where different conditions interact or overlap. They create a complex and intricate pattern within the boundary area.

7.19.4 Small-scale fronts associated with minor eddies

Within these boundary states, you can also find small-scale fronts. These are typically associated with minor eddies or smaller-scale turbulent features. These fronts might not be as prominent as primary fronts but still contribute to the overall complexity of the boundary.

In summary, when analyzing quantized Marghany bound front states in radar imagery using quantum entropy algorithm, we observe a range of structural features. These include stepped frontal structures with distinct levels, primary fronts where significant transitions occur, interleaved front structures that create complexity, and small-scale fronts related to minor eddies. These features provide valuable insights into the boundary conditions and dynamics within the observed area (Robinson, 2010a).

This chapter presents a novel method for detecting quantized Marghany bound front states as part of the quantized Marghany turbulence boundary in multiSAR images using texture algorithms such as dissimilarity, entropy, and homogeneity. The chapter also discusses the impact of speckles on the topology of turbulence occurrence imaging in SAR images. It is worth noting that homogeneity performs better in identifying quantized Marghany turbulence boundaries than other texture algorithms. However, the quantum entropy algorithm can detect various quantized Marghany bound front states, primarily associated with the abundance of filaments in different geographical locations such as Sumatra, Bali, and the coastal waters of the Netherlands. The presence of filaments in SAR images, therefore, plays a pivotal role as both an indicator and a valuable tool for investigating spiral flows within the ocean. The forthcoming chapter will delve into a comprehensive examination of the interplay between filaments and spiral flows using SAR data.

References

Argenti, F., & Alparone, L. (2002). Speckle removal from SAR images in the undecimated wavelet domain. *IEEE Transactions on Geoscience and Remote Sensing.*, *40*(11), 2363–2374. Available from https://doi.org/10.1109/TGRS.2002.805083.

Asakura, T., & Takai, N. (1981). Dynamic laser speckles and their application to velocity measurements of the diffuse object. *Applied Physics*, *25* (3), 179–194. Available from https://doi.org/10.1007/BF00902971.

Bromberg, Y., & Cao, H. (2014). Generating non-rayleigh speckles with tailored intensity statistics. *Physical Review Letters*, *112*(21). Available from https://doi.org/10.1103/PhysRevLett.112.213904, http://harvest.aps.org/bagit/articles/10.1103/PhysRevLett.112.213904/apsxml.

Dekker, R. J. (2003). Texture analysis and classification of ERS SAR images for map updating of urban areas in the Netherlands. *IEEE Transactions on Geoscience and Remote Sensing.*, *41*(9), 1950–1958. Available from https://doi.org/10.1109/TGRS.2003.814628.

Delia, C., Ruscino, S., Abbate, M., Aiazzi, B., Baronti, S., & Alparone, L. (2014). SAR image classification through information-theoretic textural features, MRF segmentation, and object-oriented learning vector quantization. *IEEE Journal of Selected Topics in Applied Earth Observations and Remote Sensing*, *7*(4), 1116–1126. Available from https://doi.org/10.1109/JSTARS.2014.2304700.

Dong, Y. L., & Pan, B. (2017). A review of speckle pattern fabrication and assessment for digital image correlation. *Experimental Mechanics*, *57*(8), 1161–1181. Available from https://doi.org/10.1007/s11340-017-0283-1, http://www.springer.com/dal/home/generic/search/results?SGWID = 1-40109-70-36417762-0.

Dubois J.P. (2008). American conference on applied mathematics Poisson modulated stochastic model for partially developed multi-look speckle.

Fedorov, A., Steffen, L., Baur, M., Da Silva, M. P., & Wallraff, A. (2012). Implementation of a Toffoli gate with superconducting circuits. *Nature*, *481*(7380), 170–172. Available from https://doi.org/10.1038/nature10713.

Goodman, J. W. (2007). *Speckle phenomena in optics: theory and applications*. Roberts and Company Publishers.

Haralick, R. M., Dinstein, I., & Shanmugam, K. (1973). Textural features for image classification. *IEEE Transactions on Systems, Man, and Cybernetics*, *3*(6), 610–621. Available from https://doi.org/10.1109/TSMC.1973.4309314.

Hur, K. L. (2008). Entanglement entropy, decoherence, and quantum phase transitions of a dissipative two-level system. *Annals of Physics*, *323*(9), 2208–2240. Available from https://doi.org/10.1016/j.aop.2007.120.003.

Jones, C. (2013). Low-overhead constructions for the fault-tolerant Toffoli gate. *Physical Review. A*, *87*(2). Available from https://doi.org/10.1103/physreva.87.022328.

Kandala, A., Wei, K. X., Srinivasan, S., Magesan, E., Carnevale, S., Keefe, G. A., Klaus, D., Dial, O., & McKay, D. C. (2021). Demonstration of a high-fidelity cnot gate for fixed-frequency transmons with engineered suppression. *Physical Review Letters*, *127*(13). Available from https://doi.org/10.1103/PhysRevLett.127.130501, https://journals.aps.org/prl/abstract/10.1103/PhysRevLett.127.130501.

Kandaswamy, U., Adjeroh, D. A., & Lee, M. C. (2005). Efficient texture analysis of SAR imagery. *IEEE Transactions on Geoscience and Remote Sensing*, *43*(9), 2075–2083. Available from https://doi.org/10.1109/TGRS.2005.852768.

Kwok, R., & Johnson, W. T. K. (1989). Block adaptive quantization of Magellan SAR data. *IEEE Transactions on Geoscience and Remote Sensing*, *27*(4), 375–383. Available from https://doi.org/10.1109/36.29557.

Lanzagorta, M. (2012). *Quantum radar* (5). Morgan & Claypool Publishers.

Lee J.S. Schuler D.L. Lang R.H. Ranson K.J. (1994a) 12 1994/12. IEEE United States K-distribution for multi-look processed polarimetric SAR imagery 4. International geoscience and remote sensing symposium (IGARSS) 2179–2181.

Lee, J.-S., Hoppel, K. W., Mango, S. A., & M, A. R. (1994b). Intensity and phase statistics of multilook polarimetric and interferometric SAR imagery. *IEEE Transactions on Geoscience and Remote Sensing, 32*(5), 1017–1028. Available from https://doi.org/10.1109/36.312890.

Liu, S., Gao, L., Lei, Y., Wang, M., Hu, Q., Ma, X., & Zhang, Y. D. (2021). SAR speckle removal using hybrid frequency modulations. *IEEE Transactions on Geoscience and Remote Sensing, 59*(5), 3956–3966. Available from https://doi.org/10.1109/TGRS.2020.3014130, https://ieeexplore.ieee.org/servlet/opac?punumber = 36.

Lopes, A., Nezry, E., Touzi, R., & Laur, H. (1993). Structure detection and statistical adaptive speckle filtering in SAR images. *International Journal of Remote Sensing, 14*(9), 1735–1758. Available from https://doi.org/10.1080/01431169308953999.

Lopez-Martinez, C., & Pottier, E. (2007). On the extension of multidimensional speckle noise model from single-look to multilook SAR imagery. *IEEE Transactions on Geoscience and Remote Sensing, 45* (2), 305–320. Available from https://doi.org/10.1109/TGRS.2006.887012.

Maccone, L., & Ren, C. (2020). Quantum radar. *Physical Review Letters, 124*(20). Available from https://doi.org/10.1103/PhysRevLett.124.200503, http://harvest.aps.org/bagit/articles/10.1103/PhysRevLett.124.200503/apsxml.

Marghany, M. (2001). RADARSAT automatic algorithms for detecting coastal oil spill pollution. *International Journal of Applied Earth Observation and Geoinformation, 2001*(2), 191–196.

Marghany, M. (2021a). *Advanced Algorithms for Mineral and Hydrocarbon Exploration Using Synthetic Aperture Radar* (pp. 1–383). Malaysia: Elsevier. Available from https://www.sciencedirect.com/book/9780128217962, 10.1016/C2019-0-04105-2.

Marghany, M. (2021b). *Nonlinear ocean dynamics: synthetic aperture radar*. Elsevier.

Marghany, M., & Genderen, J. van (2014). Entropy algorithm for automatic detection of oil spill from radarsat-2 SAR data. *IOP conference series: Earth and environmental science, 18*(1), 012051. Available from https://doi.org/10.1088/1755-1315/18/1/012051.

Marghany, M., & Hashim, M. (2010). Texture entropy algorithm for automatic detection of oil spill from RADARSAT-1 SAR data. *International Journal of Physical Sciences, 5*(9), 1475–1480. Available from http://www.academicjournals.org/ijps/PDF/pdf2010/18%20Aug/Marghany%20and%20Hashim.pdf.

Monz, T., Kim, K., Hänsel, W., Riebe, M., Villar, A. S., Schindler, P., Chwalla, M., Hennrich, M., & Blatt, R. (2009). Realization of the quantum Toffoli gate with trapped ions. *Physical Review Letters, 102*(4). Available from https://doi.org/10.1103/PhysRevLett.102.040501Austria, http://oai.aps.org/oai?verb = GetRecord&Identifier = oai:aps.org:PhysRevLett.102.040501&metadataPrefix = oai_apsmeta_2.

Nie, H., Fu, Z., Tang, B.-H., Li, Z., Chen, S., & Wang, L. (2022). A dual-generator translation network fusing texture and structure features for SAR and optical image matching. *Remote Sensing, 14* (12), 2946. Available from https://doi.org/10.3390/rs14122946.

Nyoungui, A. N., Tonye, E., & Akono, A. (2002). Evaluation of speckle filtering and texture analysis methods for land cover classification from SAR images. *International Journal of Remote Sensing, 23*(9), 1895–1925. Available from https://doi.org/10.1080/01431160110036157, https://www.tandfonline.com/loi/tres20.

Okamoto, T., & Asakura, T. (1995). III: The statistics of dynamic speckles. *Progress in Optics, 34*(C), 183–248. Available from https://doi.org/10.1016/S0079-6638(08)70326-3.

Osborn, K. D., & Wustmann, W. (2021). Reversible fluxon logic with optimized CNOT gate components. *IEEE Transactions on Applied Superconductivity, 31*(2), 1–13. Available from https://doi.org/10.1109/tasc.2020.3035344.

Raney, R. K., & Wessels, G. J. (1988). Spatial considerations in SAR speckle consideration. *IEEE Transactions on Geoscience and Remote Sensing, 26*(5), 666–672. Available from https://doi.org/10.1109/36.7693.

Reu, P. (2015). All about speckles: Contrast. *Experimental Techniques, 39*(1), 1–2. Available from https://doi.org/10.1111/ext.12126, http://www.blackwellpublishing.com/aims.asp?ref = 0732-8818.

Robinson, I. S. (2004). *Measuring the oceans from space: The principles and methods of satellite oceanography*. Springer Science & Business Media.

Robinson, I. S. (2010a). *The methods of satellite oceanography* (pp. 7–67). Springer Science and Business Media LLC. Available from 10.1007/978-3-540-68322-3_2.

Robinson, I. S. (2010b). *Discovering the ocean from space: The unique applications of satellite oceanography*. Berlin Heidelberg: Springer. Available from 10.1007/978-3-540-68322-3.

Rosenblum, S., Gao, Y. Y., Reinhold, P., Wang, C., Axline, C. J., Frunzio, L., Girvin, S. M., Jiang, L., Mirrahimi, M., Devoret, M. H., & Schoelkopf, R. J. (2018). A CNOT gate between multiphoton qubits encoded in two cavities. *Nature Communications, 9*(1). Available from https://doi.org/10.1038/s41467-018-03059-5, http://www.nature.com/ncomms/index.html.

Soh, L. K., & Tsatsoulis, C. (1999). Texture analysis of sar sea ice imagery using gray level co-occurrence matrices. *IEEE Transactions on Geoscience and Remote Sensing, 37*(2 I), 780–795. Available from https://doi.org/10.1109/36.752194.

Szalai, S., & Dogossy, G. (2021). Speckle pattern optimization for DIC technologies. *Acta Technica Jaurinensis, 14*(3), 228–243. Available from https://doi.org/10.14513/actatechjaur.00573.

Torromé, R. G., Bekhti-Winkel, N. B., & Knott, P. (2020). Introduction to quantum radar, arXiv, Slovenia*arXiv*. Available from https://doi.org/10.48550/arxiv.2006.14238, https://arxiv.org.

Wang, L., & He, D. C. (1990). A new statistical approach for texture analysis. *Photogrammetric Engineering & Remote Sensing, 56*(1), 61–66.

Chapter 8

Developing a novel quantum cluster algorithm for the automatic detection of spiral patterns in SAR data

Spiral patterns in SAR images often result from atmospheric turbulence, which can affect the propagation of radar signals. When SAR sensors transmit and receive radar waves through a turbulent atmosphere, they encounter variations in the refractive index. These variations can introduce phase changes in the radar signals, causing interference patterns that manifest as spirals in the images.

This chapter, therefore, is dedicated to a comprehensive exploration of the intricate dynamics governing spiral sea patterns, particularly within the newly introduced concept of the quantized Marghany turbulence boundary. This boundary, a novel discovery, has its roots in the pioneering work of Marghany in 2021, where it was derived from the quantized Marghany bound front states. The focus of this investigation is on the coastal waters of Peninsular Malaysia, an area extensively elaborated upon in preceding chapters.

8.1 What is meant by spirals?

In mathematics, a spiral is a geometric curve that originates from a central point and extends outward while revolving around that point (Fig. 8.1). As it extends, it progressively moves farther away from the central point. Spirals belong to a category of patterns known as whorled patterns, which encompasses a wide range of shapes, including objects with concentric circles or other nested structures (Fig. 8.2). Therefore there are two major definitions of "spiral." These definitions succinctly describe the concept of a spiral in both a two-dimensional (planar) and three-dimensional context: (1) planar spiral and (2) three-dimensional spiral or helix. In this sense, a spiral is defined as a curve on a plane that revolves around a fixed center point. As it does so, the distance between the curve and the center point continuously increases or decreases. A spiral is also described as a three-dimensional curve that revolves around an axis while maintaining a constant or continuously varying distance from the axis. This three-dimensional spiral is commonly referred to as a helix. These definitions capture the essential characteristics of spirals and provide clarity about their properties in both two-dimensional and three-dimensional settings (Cook, 1903; Lindsley et al., 1995).

According to the aforementioned perspective, the term "Spiral on the Sea" typically refers to a complex and dynamic oceanic phenomenon characterized by the presence of spiral patterns or structures on the sea surface (Fig. 8.3). These patterns often manifest as swirling or helical formations of water, with visible spirals or vortices that can extend over various scales, from small eddies to large-scale oceanic features. Spiral seas can be attributed to a range of physical processes and environmental factors, including wind patterns, ocean currents, temperature gradients, and interactions with seafloor topography.

FIGURE 8.1 Spiral curvature pattern.

Synthetic Aperture Radar Image Processing Algorithms for Nonlinear Oceanic Turbulence and Front Modeling. DOI: https://doi.org/10.1016/B978-0-443-19155-8.00008-8

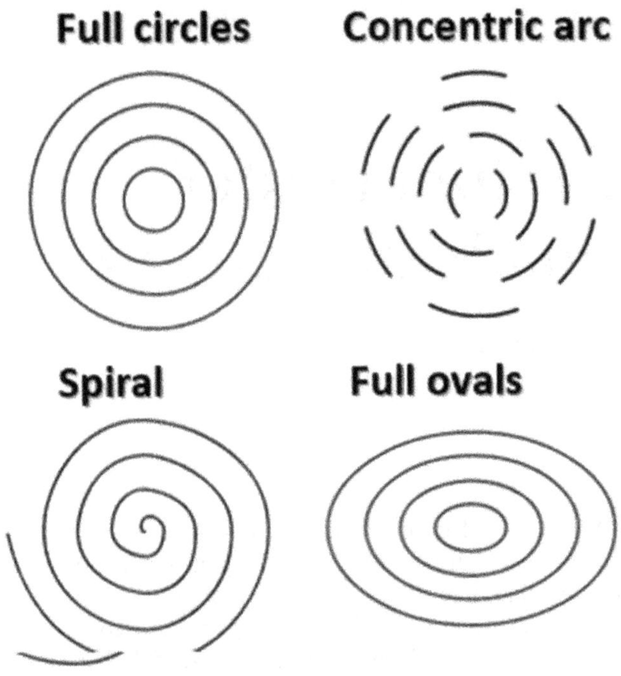

FIGURE 8.2 Whorled patterns.

FIGURE 8.3 Complex and dynamic spiral oceanic phenomenon.

These factors can create conditions conducive to the formation of these distinctive spiral patterns on the ocean's surface (Munk et al., 2000).

However, it is noteworthy that Malaysian researchers have not commonly employed the term "Spiral Sea" in their discourse. Their primary research focus has predominantly centered on fundamental inquiries within the realm of physical oceanography, with somewhat limited emphasis on comprehensively delving into the intricate nature of ocean dynamics. This preference can be attributed to a certain degree of oversight concerning the rigorous integration of physics and mathematics within their ocean dynamic investigations.

In this perspective, researchers and oceanographers often study spiral seas to gain insights into the underlying physical mechanisms driving these patterns and to better understand their impact on marine ecosystems, weather patterns, and other ocean-related phenomena. The study of spiral seas can also have practical applications in fields such as marine navigation, coastal engineering, and climate modeling.

8.2 Distinguishing between "spiral," "eddy," and "vorticity" is essential in fluid dynamics

The significant question is now: What are the differences between spirals, eddies, and vorticity?

"Spiral," "eddy," and "vorticity" are interconnected concepts within the field of fluid dynamics, specifically in the context of elucidating the motion and patterns exhibited by fluids, such as water and air (Fig. 8.4). Nonetheless, each of these terms pertains to distinct facets of fluid behavior. A spiral, fundamentally, denotes a particular geometric configuration characterized by a continuous and progressively widening curve that revolves around a central point. In the domain of fluid dynamics, the occurrence of a spiral pattern becomes evident when fluids, such as water or air, assume a twisting or helical trajectory as they flow. The genesis of spiral patterns can be attributed to diverse factors, including the Coriolis effect, which results from the Earth's rotation and imparts a deflection upon moving fluids. For instance, spirals can be discerned in oceanic contexts, manifesting in the trajectories of specific currents or the structural arrangements within hurricanes (Stuart, 1967).

Conversely, an eddy signifies a localized, self-contained motion occurring within a fluid. Typically, it manifests as the formation of swirling, rotating regions of fluid that stand apart from the primary flow. Eddies come in a spectrum of sizes, ranging from diminutive vortices encountered in streams to expansive eddies encountered in the atmosphere or oceans. They frequently adopt a circular or oval morphology and wield considerable influence over fluid transport, mixing processes, and turbulence. Eddies can emerge due to irregularities within the flow, obstacles impeding the fluid's course, or intricate interactions between disparate strata or currents (Doshi et al., 2021).

Vorticity, on the other hand, is a fundamental physical attribute that quantifies the local rotation or angular

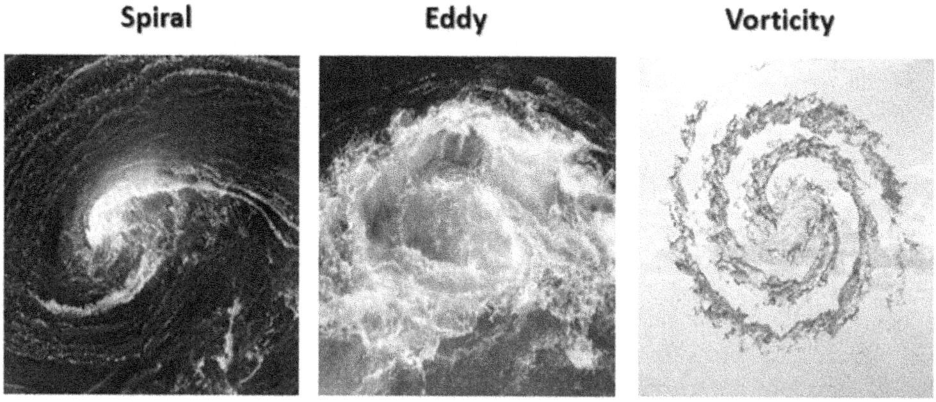

FIGURE 8.4 Distinguishing geometric patterns among spirals, eddies, and vorticity.

velocity exhibited by individual fluid particles as they traverse a fluid medium. It assumes the form of a vector field, providing a description of the proclivity of fluid elements to undergo rotation during their transit through the fluid. Notably, positive vorticity signifies counterclockwise rotation, while negative vorticity corresponds to clockwise rotation (in the Northern Hemisphere). Vorticity serves as a pivotal concept within fluid dynamics, affording valuable insights into the genesis and conduct of eddies, turbulence, and other intricate fluid phenomena. Its significance reverberates across various domains, prominently within meteorology and oceanography, where it constitutes an essential parameter underpinning the analysis of fluid motion (Zhang & Chen, 2022).

In summation, "spiral" denotes a geometric shape or pattern frequently encountered in fluid flows, "eddy" characterizes localized rotational motions within a fluid, and "vorticity" embodies a property that quantifies the local rotation of fluid elements within a flow. These concepts, while interconnected, assume distinct roles in the comprehension of fluid dynamics and the behavior of fluids across diverse natural settings, encompassing oceans, atmospheres, and river systems.

8.3 Formation mechanism of spirals

The formation mechanisms of spirals in the sea can be attributed to a variety of physical processes and environmental factors. In this sense, the Coriolis force (F_c) generating spiral patterns in the sea can be mathematically expressed as follows:

$$F_C = -2 \cdot m \cdot (\vec{V} \times \vec{\Omega}) \tag{8.1}$$

here F_c is the Coriolis force; m is the mass of the fluid element; $\vec{V}$ is the velocity vector of the fluid element; and $\vec{\Omega}$ is the Earth's angular velocity vector. The Earth's angular velocity vector ($\vec{\Omega}$) is indeed a vector with three components that represent the Earth's rotation in three-dimensional space. It is typically expressed as follows:

$$\vec{\Omega} = \begin{bmatrix} \Omega \cdot cos(\phi) \\ \Omega \cdot sin(\phi) \\ 0 \end{bmatrix} \tag{8.2}$$

where Ω is the magnitude of the Earth's angular velocity, and ϕ is the latitude of the fluid element. The components of this vector depend on the latitude (ϕ) of the fluid element. The first two components account for the variation in the Coriolis effect due to the change in latitude, and the third component is always zero, indicating that the Earth's rotation axis is perpendicular to the surface. With this expression for the Earth's angular velocity vector, one can use the earlier equation for the Coriolis force to determine the force responsible for generating spiral patterns in the sea. In this perspective, the Coriolis effect is a fundamental factor in the creation of spiral patterns in the sea. It results from the rotation of the Earth and causes moving fluids, such as ocean currents, to deflect to the right in the Northern Hemisphere and to the left in the Southern Hemisphere. This deflection leads to the development of rotating, spiral-like patterns in the ocean (Bassin et al., 2005; Lumpkin & Elipot, 2010; Ni et al., 2021).

The wind stress (τ) on the ocean's surface can generate currents with swirling or helical motion, particularly noticeable in regions with strong and persistent winds. The relationship between wind stress and the resulting ocean currents can be described using mathematical equations, primarily through the equations of motion in fluid dynamics. Wind stress, therefore, is the force exerted by the wind on the ocean's surface per unit area. It is typically described using the drag coefficient (C_d), the air density (ρ), and the wind speed ($\vec{u}$):

$$\tau = C_d \cdot \rho \cdot U^2 \cdot \vec{u} \tag{8.3}$$

The equations of motion describe the behavior of ocean currents in response to external forces such as wind stress. The relevant equation for this scenario is the momentum equation, which accounts for the acceleration of the fluid:

$$\rho \cdot Dt[DV]^{-1} = -\nabla p + \rho \cdot g + \underset{C}{\vec{F}} + \underset{Friction}{\vec{F}} + \underset{Wind}{\vec{F}} \tag{8.4}$$

the frictional force F_{Friction} within the ocean, which arises from the interaction between fluid layers and can be described using the concept of viscosity, is mathematically represented by the Navier – Stokes equations in fluid dynamics. The Navier – Stokes equations are a set of partial differential equations that govern fluid flow (McLellan, 1965). The frictional force component within these equations is typically expressed as:

$$F_{Friction} = -\mu \nabla^2 \vec{V} \tag{8.5}$$

The Laplacian operator (∇^2) represents the divergence of the gradient of velocity and accounts for the diffusion of momentum due to viscosity within the fluid. It describes how the velocity field changes over space and influences the frictional forces acting on the fluid. Let assume a fluid velocity profile like this:

$$\vec{V} = (0, 0, v_z(x)) \tag{8.6}$$

where $v_z(x)$ is the velocity component along the coastal water (in the z-direction), and it varies as a function of position x. Therefore the frictional force F_{Friction} experienced by the fluid is calculated by applying the Laplacian operator to the velocity profile (Dijkstra, 2005):

$$F_{Friction} = -\mu \frac{d^2 v_z}{dx^2} \tag{8.7}$$

Suppose we have an ocean current with the following characteristics:

1. Dynamic viscosity of seawater (μ) is approximately 1.0×10^{-3} Pa · s (pascal-seconds).
2. The current flows along a gradient of velocity, with a velocity profile given by $V(x) = 0.1x$ m/s, where x is the distance along the current in meters. To calculate the frictional force F_{Friction} experienced by the seawater due to turbulence, one can use the following equation for the shear stress (τ) at a given point within the current:

$$\tau = \mu \frac{dV}{dx} \tag{8.8}$$

Now, calculate the velocity gradient:

$$\frac{dV}{dx} = 0.1 m/sm^{-1} \tag{8.9}$$

Next, use this velocity gradient in the equation for shear stress to find the frictional force:

$$\tau = \left(1.0 \times 10^{-3} Pa \text{ cdotps}\right) \cdot \left(0.1 m/sm^{-1}\right) = 1.0 \times 10^{-4} Pa \tag{8.10}$$

Therefore the shear stress (τ) and the frictional force (F_{Friction}) experienced by the seawater due to turbulence at this particular point in the ocean current are both 1.0×10^{-4} pascals (Pa).

Let's explore how turbulence in the ocean can lead to the formation of small-scale eddies and spirals with a numerical example. Suppose we have a simplified scenario in a coastal region where two different water masses, one colder and denser (water mass A) and the other warmer and less dense (water mass B), are interacting. The interface between these two water masses is not smooth but irregular due to seabed topography. In this view, let water mass A have a temperature of 10°C and a salinity of 35 ppt (parts per thousand). Consequently, water mass B has a temperature of 20°C and a salinity of 32 ppt. The temperature and salinity differences between these two water masses create density gradients, which, when perturbed by turbulence, can lead to the formation of small-scale eddies and spirals.

Now, let's consider the simplified velocity field at the interface between these water masses, where turbulence is generated. We'll assume that the velocity (V) varies along the interface and has a sinusoidal pattern:

$$V(x) = 0.1 \sin\left(2 \frac{\pi x}{L}\right) ms^{-1} \tag{8.11}$$

$V(x)$ is the velocity of the water at a given position x; x is the distance along the interface in meters; and L is the characteristic length scale of the turbulence, let's say $L = 100$ meters. Now, let's calculate the Reynolds number (R_e), which is a dimensionless parameter that characterizes the flow regime. It's defined as:

$$R_e = [Viscousforces]^{-1}[Inertialforces] = [\mu]^{-1} \rho \cdot V \cdot L \tag{8.12}$$

Let's calculate R_e using the given values:
Average density (ρ) of seawater ≈ 1024 kg/m^3
Maximum velocity (V) ≈ 0.1 m/s
Characteristic length scale (L) = 100 m
Dynamic viscosity (μ) = 1.0×10^{-3} Pa · s

$$R_e = \frac{(1024 kg/m^3) \cdot (0.1 m/s) \cdot (100 m)}{1.0 \times 10^{-3} Pa \text{ cdotps}} \approx 10^8 \tag{8.13}$$

This Reynolds number (R_e) indicates that the flow at the interface between water mass A and water mass B is highly turbulent. Turbulence in this region can lead to the formation of small-scale eddies and spirals, which persist and contribute to the overall spiral pattern of ocean currents. These turbulent structures play a crucial role in mixing different water masses, distributing heat and nutrients, and shaping the complex flow patterns observed in the ocean (Pickard & Emery, 1961).

According to the perspective, the interface between these two water masses is subject to turbulence, which leads to the mixing of properties such as temperature and salinity. In this understanding, let's define a parameter called "Mixing Rate" (M), which represents the rate at which these two water masses mix in terms of temperature. For simplicity, we'll assume a linear mixing rate:$M = 0.02°C/m$. This means that for every meter of depth along the mixing interface, the temperature difference between water mass A and water mass B decreases by 0.02°C. Now, let's calculate the temperature (T) and salinity (S) profiles along a vertical section of the ocean, starting from the interface and extending downward:

Depth (d) = 10 meters, temperature (T) and salinity (S) can be given by the following formulas; respectively:

$$T = 20°C - (10m \cdot 0.02°C/m) = 19.8°C \tag{8.14}$$

$$S = 35ppt + (10m \cdot 0.01ppt/m) = 35.1ppt \tag{8.15}$$

We can continue this process for various depths, and at each depth, we calculate the temperature and salinity based on the linear mixing rate. As we move deeper into the ocean, the temperature and salinity profiles gradually converge toward a uniform value, reflecting the turbulent mixing between the two water masses. This mixing creates a layered structure where the properties of the water masses become more homogeneous with depth. The resulting layered structure and the gradual convergence of properties are aspects of turbulent mixing in the ocean, contributing to the complex dynamics and the formation of layered and spiral patterns in certain oceanographic contexts.

On the other hand, it's worth noting that seabed irregularities can give rise to spiral patterns within ocean currents. The ocean floor exhibits significant topographical variations, including the presence of underwater mountains, ridges, and valleys, all of which exert a pronounced influence on the dynamics of oceanic flows. When ocean currents intersect with these seabed features, they can undergo a transition to turbulent behavior, leading to intricate swirling or helical motions in the water masses.

Let's envision an underwater mountain ridge beneath the ocean's surface. This ridge extends prominently from the seabed over a defined span. In this simplified scenario, we'll specify the following parameters:

- Depth of the ridge (D): 200 meters
- Length of the ridge (L): 1,000 meters
- Width of the ridge (W): 200 meters
- Velocity of the ocean current (V): 0.5 meters per second (m/s)

Let's calculate the Reynolds number (R_e) using Eq. (8.13) to determine the flow regime.

$$R_e = \frac{(1024kg/m^3) \cdot (0.5m/s) \cdot (200m)}{1.0 \times 10 - 3Pa \text{ cdotps}} = 1024000 \tag{8.16}$$

The Reynolds number is quite large (greater than 1000), indicating that the flow around the ridge is turbulent. In turbulent flow, irregularities in the seabed, such as the underwater mountain ridge, can induce swirling and helical motion in the ocean current. These turbulent motions play a role in mixing water masses, redistributing nutrients, and influencing marine ecosystems in the vicinity of such features.

It is imperative to emphasize that the genesis of spiral patterns within ocean currents is a nuanced and intricate phenomenon influenced by a myriad of factors, with turbulence being one of the critical contributors. The precise attributes of these spiral patterns are contingent upon the intricate interplay of these multifarious factors, and as such, their manifestations can exhibit significant variability across diverse oceanographic scenarios (Gargett, 1989; Thorpe, 2004, 2007). To gain insights into the genesis of spiral patterns in the ocean, researchers frequently use numerical modeling techniques and empirical observations as integral components of their investigative methodologies.

8.4 Mathematical description of spirals

A mathematical description of a spiral provides a way to represent and understand the shape and characteristics of a spiral using equations. As it is mentioned at section 8.2, spirals are geometric shapes that emanate from a central point and extend outward in a continuous, often curving manner. There are various types of spirals, each with its own mathematical formula. Indeed, a two-dimensional or plane spiral can be most conveniently described using polar coordinates, where the radius (r) is a continuously changing function of the angle (φ). In polar coordinates, a point in the plane is represented by its radial distance (r) from a central point (usually the origin) and the angle (φ) between a reference line (often the positive x-axis) and the line connecting the central point to the point of interest.

$$r = r(\varphi). \tag{8.17}$$

When we say that the radius (r) is a "monotonic continuous function" of the angle (φ), it means that as you change the angle (φ), the radius (r) either consistently increases or decreases without abrupt jumps or discontinuities. This creates a smooth and continuous curve that characterizes the spiral. Therefore in x-y coordinates, the curve (spiral) can be represented parametrically, which means we express both the x-coordinate and the y-coordinate as functions of a parameter [usually denoted as (t)] (Habib & Sakai, 2005; Harary & Tal, 2011; Picado, 2009). This parametric representation allows us to describe the motion of a point on the curve as t varies. For a spiral, the parametric representation might look like this:

$$x(t) = f(t), y(t) = g(t) \tag{8.18}$$

here $x(t)$ represents the x-coordinate of a point on the curve as a function of the parameter t; $y(t)$ represents the y-coordinate of the same point as a function of t; and $f(t)$ and $g(t)$ are specific functions that define how the x and y coordinates change as t varies. In this perspective, the choice of functions $f(t)$ and $g(t)$ will depend on the type of spiral and its specific characteristics. Different types of spirals will have different parametric equations that govern their shapes and behaviors. In other words, in the x-y coordinate system, the curve is parametrically represented as follows:

$$x = r(\varphi)\cos\varphi, y = r(\varphi)\sin\varphi. \tag{8.19}$$

In Eq. (8.19), x and y represent the x and y coordinates of points on the curve; and $r(\varphi)$ is a function that determines how the distance from the origin (r) changes with the angle φ. Additionally, $\cos\varphi$ and $\sin\varphi$ are trigonometric functions that convert polar coordinates (r,φ) to Cartesian coordinates (x,y). These equations are commonly used to describe curves, including spirals, in the x-y plane when the curve's behavior is dependent on the parameter φ (Polezhaev, 2019; Siddique & Adeli, 2014).

Here are some common mathematical descriptions of spirals:

8.4.1 Archimedean spirals

An Archimedean spiral is a specific type of spiral with equations typically expressed in polar coordinates (Fig. 8.5). These equations describe the relationship between the distance from the origin (denoted as r) and the angle (denoted as θ) as follows:

$$r = a + b\theta \tag{8.20}$$

In this equation, r represents the distance from the origin to a point on the spiral; a is the initial radius, which signifies the distance from the origin to the starting point of the spiral. Therefore b is a constant that determines the rate of increase of the radius as the angle θ varies. As the angle θ increases, the value of r increases linearly with a slope determined by the constant b. This means that the spiral extends outward from the origin in a consistent and

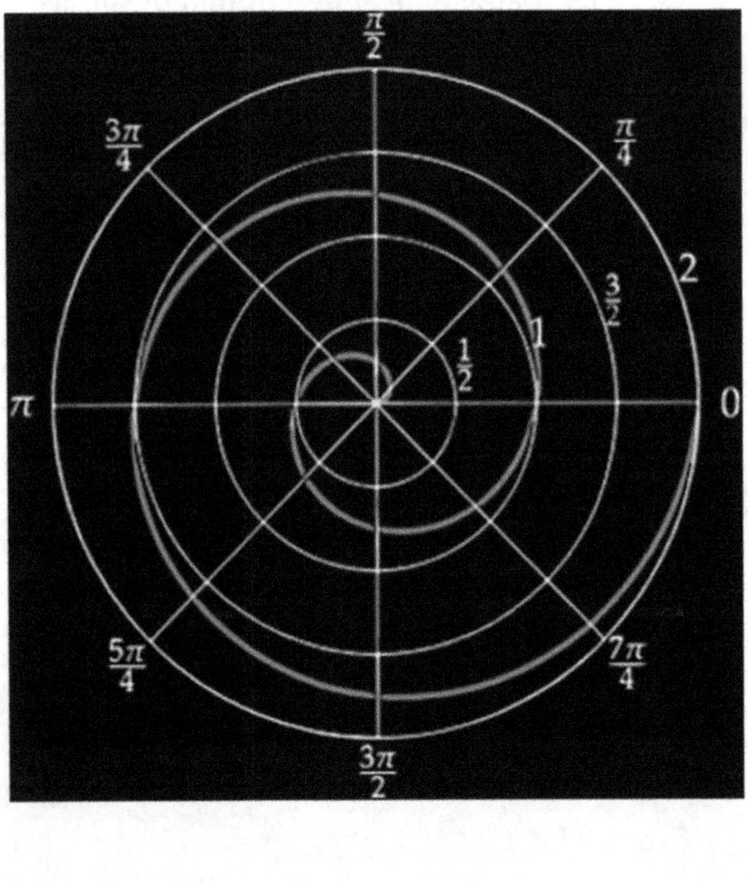

FIGURE 8.5 Archimedean spiral patterns.

evenly spaced manner, with the initial radius a determining how far from the origin the spiral begins. Indeed, Archimedean spirals are known for their uniform and predictable growth pattern, making them distinct from other types of spirals that exhibit exponential or logarithmic growth. They are commonly encountered in various natural and mathematical contexts, illustrating the concept of linear increase in radius as the spiral unfolds (Kelley & Huff, 2013; Rocha et al., 2022).

8.4.2 Logarithmic spirals

A logarithmic spiral (Fig. 8.6), often referred to as an equiangular spiral, is a specific type of spiral with equations typically expressed in polar coordinates. These equations describe the relationship between the distance from the origin (denoted as r) and the angle (denoted as θ) as follows:

$$r = a \cdot e^{b\theta} \tag{8.21}$$

Eq. (8.21) illustrates that r denotes the distance from the origin to a specific point along the spiral. Consequently, a serves as the initial radius, indicating the distance from the origin to the spiral's initial point. In simpler terms, b functions as a constant dictating the pace of expansion as the angle θ undergoes variation. This equation serves to delineate the defining characteristics of a logarithmic spiral:

Initial Radius (a): Within this context, the parameter a assumes the role of determining the distance from the origin to the spiral's point of origin. It essentially signifies the radius at $\theta = 0$ or the precise point from which the spiral commences.

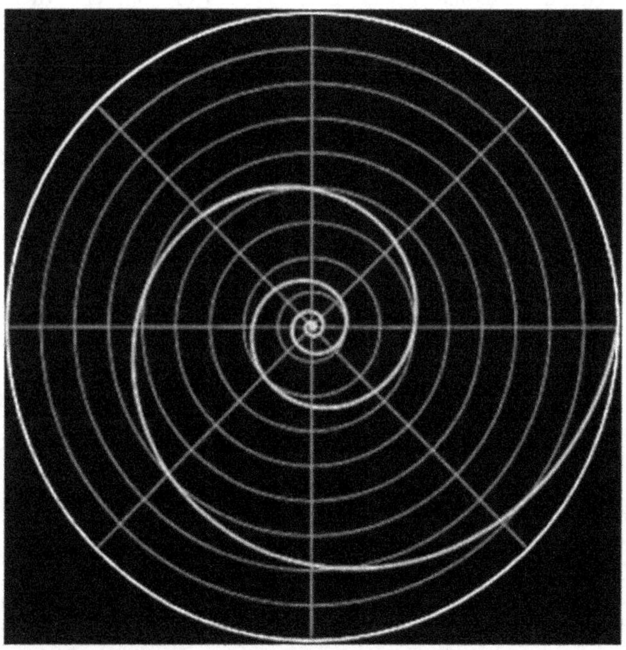

FIGURE 8.6 Logarithmic spiral patterns.

Rate of Expansion (b): The constant b exercises authority over the velocity at which the spiral extends with the progression of θ. A greater value of b begets an expedited expansion, whereas a diminutive b results in a more gradual unfurling.

Exponential Growth: The employment of the exponential function $e^b\theta$ signifies that the spiral's radius (r) undergoes exponential augmentation in tandem with θ augmentation. Consequently, this engenders a spiral that unfolds increasingly swiftly and expansively as it advances.

From the viewpoint presented earlier, logarithmic spirals are renowned for their distinctive growth pattern, where the distance from the origin expands in harmony with the angle of rotation (Jeong & Said, 2023; Skalak et al., 1982). This sets them apart from other spiral varieties such as Archimedean spirals, which display a straightforward linear increase in radius. Logarithmic spirals frequently emerge in natural phenomena, artistic expressions, and a wide array of mathematical and scientific domains, captivating minds with their captivating esthetics and intricate mathematical allure (Scherr et al., 2012; Shamloo et al., 2022).

8.4.3 Fermat's spiral

Fermat's spiral is a unique geometric curve that can be described using polar coordinates (Fig. 8.7). The equations for Fermat's spiral are as follows:

$$r = a \cdot \sqrt{\theta} \tag{8.22}$$

Eq. (8.22) shows that distance from origin (r) signifies how far a point on the spiral is from the central point or origin. As θ increases, r increases, creating a spiral pattern. The constant a, *therefore*, determines the size and shape of the spiral. Changing a will scale the spiral, making it larger or smaller. A larger a will result in a larger spiral, while a smaller a will produce a smaller one. The angle θ, consequently, defines the position of a point on the spiral relative to the positive x-axis. As θ varies, it determines the shape and direction of the spiral (Amin, 2014; Mellor & Huang, 1997; Sapkota et al., 2019; Yin et al., 2021). The square root function applied to θ leads to the unique growth pattern of Fermat's spiral. In this understanding, Fermat's spiral is distinctive because it grows outward from the origin as the angle θ increases (Weber et al., 2001). This spiral pattern is associated with the mathematical study of Fermat's last theorem and can be found in certain natural phenomena, but it is less common in comparison to other spiral types such as the Archimedean or logarithmic spirals (Xiao et al., 2022).

FIGURE 8.7 Fermat spiral patterns.

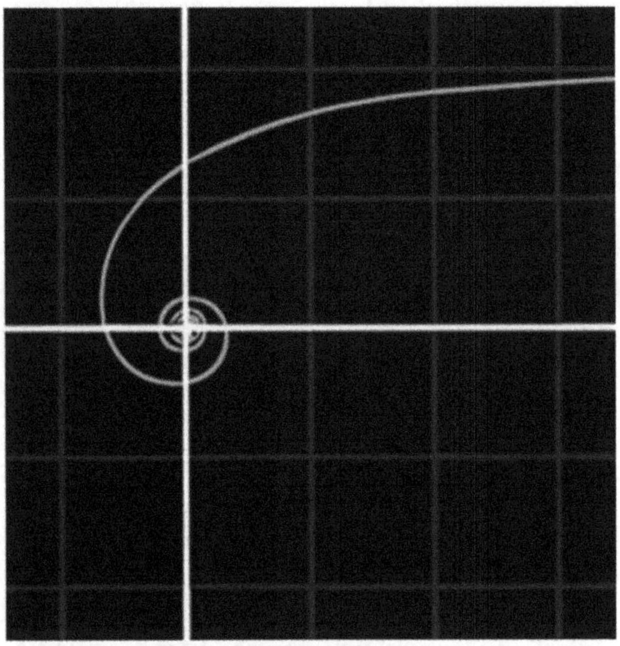

FIGURE 8.8 Hyperbolic spiral pattern.

8.4.4 Hyperbolic spiral

A hyperbolic spiral is a particular type of spiral that can be described using polar coordinates (Fig. 8.8). The equations for a hyperbolic spiral are as follows:

$$r = \frac{a}{\theta} \tag{8.23}$$

How a hyperbolic spiral behaves based on its equation? As θ approaches zero, r becomes extremely large, indicating that the spiral starts close to the origin and rapidly extends outward. As θ increases, r decreases, but it does so inversely proportional to θ. This means that the spiral continues to expand but at a decreasing rate as θ becomes larger. The shape of the hyperbolic spiral, consequently, is characterized by its tendency to stretch outward, with the distance from the origin increasing as the angle θ increases. Hyperbolic spirals are not as commonly encountered as some other spiral types such as Archimedean or logarithmic spirals (Wang et al., 2022). They exhibit unique mathematical properties that make them of interest in various scientific and mathematical contexts (Sohn, 2016; Xiao et al., 2005).

8.4.5 Golden spiral

The Golden spiral, also known as the Fibonacci spiral, is a special type of logarithmic spiral characterized by specific mathematical properties (Fig. 8.9). It is closely related to the golden ratio (ϕ), which is an irrational number approximately equal to 1.618. In this view, equation in polar coordinates is given by:

$$r = a \cdot e^{\phi\theta} \tag{8.24}$$

The defining feature of the golden spiral is its exponential growth in radius (r) as the angle (θ) increases. As θ increases, the value of r increases rapidly, causing the spiral to expand outward more and more with each rotation. The golden ratio (ϕ) is, therefore, closely associated with the Fibonacci sequence, a series of numbers where each number is the sum of the two preceding ones (e.g., 1, 1, 2, 3, 5, 8, 13). The ratios of successive Fibonacci numbers approximate the golden ratio, which is why this spiral is also called the Fibonacci spiral. The golden spiral exhibits a self-similar pattern, meaning that it retains its

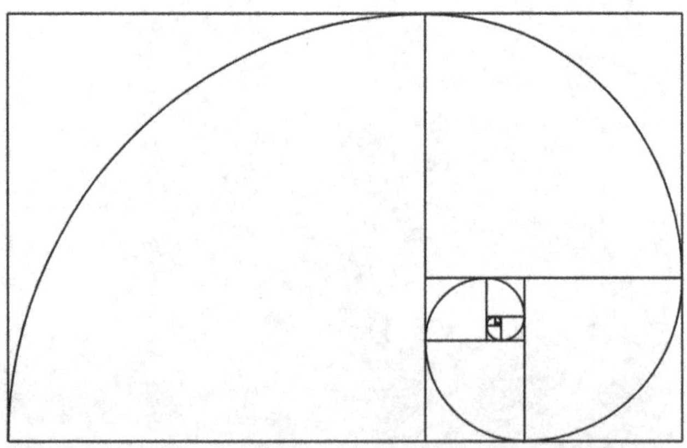

FIGURE 8.9 Golden spiral geometric pattern.

shape and proportions as it grows larger or smaller. This self-similarity is a fundamental characteristic of fractal geometry (Kommalapati et al., 2020; Kushal, 2023; Munteanu, 2010).

8.4.6 Spiral in Cartesian coordinates

A spiral in Cartesian coordinates refers to the representation of a spiral curve using the traditional x-y coordinate system. In contrast to polar coordinates, where the curve's position is defined by its distance from the origin (r) and the angle (θ), Cartesian coordinates use x and y coordinates to pinpoint the position of points along the spiral in a two-dimensional plane. In this view, the equations for a spiral in Cartesian coordinates typically describe how the x and y coordinates change as you move along the curve. Different types of spirals, such as Archimedean or logarithmic spirals, may have different equations, but they all follow a similar principle:

$$x(t) = a \cdot t \cdot \cos(t) \tag{8.25}$$

$$y(t) = a \cdot t \cdot \sin(t) \tag{8.26}$$

These equations describe how the x and y coordinates change with respect to the parameter "t." As "t" changes, it traces out the path of the spiral in the x-y plane. Different types of spirals have different equations that dictate their shapes and growth patterns. Spirals in Cartesian coordinates are particularly useful for visualizing and modeling spirals in a flat, two-dimensional space, making them applicable in various fields, including mathematics, engineering, and design (Delattre et al., 2010; Weiman & Chaikin, 1979; Wu et al., 2004).

These are just a few examples of mathematical descriptions of spirals. The choice of the appropriate equation depends on the specific characteristics and behavior of the spiral in question. Different spirals exhibit different rates of growth and geometries, making them suitable for various applications in mathematics, science, and art.

8.5 Geometric properties of spirals

The polar slope angle plays a pivotal role in defining the geometric characteristics of spirals. The polar slope angle, represented by α, is the angle between the tangent to a spiral and the corresponding polar circle. It is also known as the angle of the polar slope. The tangent represents the direction in which the spiral is curving at a particular point, and the polar circle is a reference circle in the polar coordinate system. The formula for calculating the polar slope, denoted as tanα, using vector calculus in polar coordinates is as follows:

$$\tan\alpha = [r]^{-1} r' \tag{8.27}$$

here tanα is the polar slope angle, or the angle between the tangent and the polar circle. Therefore r' is the derivative of the radial distance r with respect to the polar angle φ. In other words, it measures how quickly the distance from the origin is changing as the angle φ varies. For a specific type of spiral described by the equation $r = a\varphi^{\mathrm{n}}$, the formula for the polar slope becomes:

$$\tan\alpha = [\varphi]^{-1} n \tag{8.28}$$

This formula allows you to calculate the polar slope angle at any point on the spiral, providing insights into the direction and steepness of the spiral's curve at that location (Harary & Tal, 2011; Holcombe et al., 2016; Zhao et al., 2016). Consequently, the curvature (κ) of a curve described by a polar equation $r = r(\varphi)$ is given by the following formula:

$$\kappa = \frac{r^2 + 2(r')^2 - rr''}{(r^2 + (r')^2)^{3/2}}. \tag{8.29}$$

For a spiral described by $r = a\varphi n$, the curvature (κ) can be calculated using the following formula:

$$\kappa = \cdots = \frac{1}{a\varphi^{n-1}} \frac{\varphi^2 + n^2 + n}{(\varphi^2 + n^2)^{3/2}}. \tag{8.30}$$

In this equation, a is a constant that determines the scale or size of the spiral; φ is the polar angle; and n is a constant related to the growth pattern of the spiral. This formula allows you to compute the curvature at any point on the spiral, providing information about how sharply or gradually the spiral is curving at that specific location.

The crucial question is: can the spiral areas be determined?

The formula for calculating the area (A) of a sector of a curve described by the polar equation $r = r(\varphi)$ is given by:

$$A = \frac{1}{2} \int_{\varphi_1}^{\varphi_2} r(\varphi)^2 d\varphi. \tag{8.31}$$

In this equation, $r(\varphi)$ represents the radial distance from the origin to a point on the curve as a function of the polar angle φ. In this view, φ_1 and φ_2 are the limits of integration, representing the starting and ending angles of the sector. For a spiral with the equation $r = a\varphi^{\mathrm{n}}$, the formula for calculating the sector area becomes:

$$A = \frac{1}{2} \int_{\varphi_1}^{\varphi_2} a^2 \varphi^{2n} d\varphi = \frac{a^2}{2(2n+1)} \text{\big}(\varphi_2^{2n+1} - \varphi_1^{2n+1} \text{\big}), \text{if } n \neq -\frac{1}{2}, \tag{8.32}$$

$$A = \frac{1}{2} \int_{\varphi_1}^{\varphi_2} \frac{a^2}{\varphi} d\varphi = \frac{a^2}{2} \left(\ln\varphi_2 - \ln\varphi_1\right), \text{if } n = -\frac{1}{2}. \tag{8.33}$$

In Eq. (8.32), a is a constant related to the spiral's scale; n is a constant determining the spiral's growth pattern; and φ_1 and φ_2 are the starting and ending angles of the sector. On the other hand, k is a constant affecting the spiral's growth; and $r(\varphi_1)$ and $r(\varphi_2)$ are the radial distances corresponding to φ_1 and φ_2 in Eq. (8.33). These formulas allow to calculate the area of a sector enclosed by a spiral curve, providing insights into the extent of the region it covers in the polar coordinate system (Harary & Tal, 2010; Lekkas et al., 2013; Zhi et al., 2022).

Consequently, the formula for calculating the area (A) of a sector enclosed by a logarithmic spiral described by $r = ae^k\varphi$ is as follows:

$$A = \frac{r(\varphi_2)^2 - r(\varphi_1)^2}{4k}. \tag{8.34}$$

Hence, the formula for calculating the length (L) of an arc of a curve described by a polar equation $r = r(\varphi)$ is given by:

$$L = \int_{\varphi_1}^{\varphi_2} \sqrt{(r'(\varphi))^2 + r^2(\varphi)}\, d\varphi. \tag{8.35}$$

Eq. (8.35) illustrates that $r'(\varphi)$ represents the derivative of the radial distance r with respect to the polar angle φ. In this sense, $r(\varphi)$ is the radial distance from the origin to a point on the curve as a function of φ. Additionally, φ_1 and φ_2 are the limits of integration, representing the starting and ending angles of the arc. For a spiral described by $r = a\varphi^n$, the formula for calculating the arc length becomes:

$$L = \int_{\varphi_1}^{\varphi_2} \sqrt{\frac{n^2 r^2}{\varphi^2} + r^2 d\varphi} = a\int_{\varphi_1}^{\varphi_2} \varphi^{n-1}\sqrt{n^2 + \varphi^2}d\varphi. \tag{8.36}$$

Not all of these integrals can be readily solved using standard mathematical tables. For instance, when dealing with a Fermat's spiral, the integral can only be expressed using elliptic integrals. When considering a logarithmic spiral described by $r = ae^k\varphi$, the formula for calculating the arc length is given by:

$$L = \frac{\sqrt{k^2+1}}{k}\left(r(\varphi_2) - r(\varphi_1)\right). \tag{8.37}$$

In this scenario, a represents a constant associated with the scale of the spiral; and k is a constant that influences the spiral's growth. Thus $r(\varphi_2)$ denotes the radial distances corresponding to the polar angles φ_1 and φ_2; respectively. These equations provide the means to determine the length of an arc along a curve described by a polar equation, offering valuable information regarding the total distance covered by the curve between two specified angles (Baumgarten & Farin, 1997; Sánchez-Reyes, 2002; Xu et al., 2022).

Circle inversion, when expressed in polar coordinates, is a straightforward transformation:

$$(r, \varphi) \mapsto \left(\frac{1}{r}, \varphi\right) \tag{8.38}$$

Under this inversion, the image of a spiral described by $r = a\varphi^n$ becomes a spiral with the polar equation:

$$r = \frac{1}{a}\varphi^{-n} \tag{8.39}$$

For instance, the inverse of an Archimedean spiral results in a hyperbolic spiral. Similarly, a logarithmic spiral $r = ae^k\varphi$ is transformed into another logarithmic spiral:

$$r = \frac{1}{a}e^{-k\varphi}. \tag{8.40}$$

The concept of circle inversion unveils fascinating connections among various spiral forms, highlighting how they undergo geometric transformations. In simpler terms, it demonstrates how one type of spiral can morph into another through this inversion process. This reveals the intricate interplay and transformations within the realm of spiral geometry.

In the world of spirals, the function $r(\varphi)$ typically behaves in a particular way—it's strictly increasing, smooth, and stretches infinitely. This behavior is quite common for standard spirals, where $r(\varphi)$ can take the form of power functions or exponential functions, allowing the spiral to grow without limit. However, if we decide to be a bit more restrained and use a bounded function for $r(\varphi)$, the resulting spiral will also be bounded. One such function that neatly contains the spiral within specific limits is the arctan function:

The arctan function ensures that the spiral doesn't spiral out of control but remains within well-defined boundaries. This choice adds an interesting twist to the world of spirals, offering a more confined yet esthetically intriguing variation. Consider setting $r = a\arctan(k\varphi)$ with the following choices: $a = 4$, $k = 0.1$, and $\varphi \geq 0$. This mathematical configuration defines a unique spiral, characterized by its behavior. Let's break it down with an example:

Imagine a spiral that begins at the origin, just like an Archimedean spiral. As one moves along the spiral, the distance from the origin (r) is determined by the arctan function applied to φ. In this case, let us set a to be 4 and k to be 0.1. As one progresses along the spiral ($\varphi \geq 0$), one would notice a fascinating pattern. Initially, the spiral starts from the origin and gradually approaches a circle with a radius of $a\pi/2$. This circle serves as an asymptote or a boundary that the spiral approaches but never quite reaches. On contrary, the spiral starts at the center (0) and gracefully winds its way outward, getting closer and closer to the circle with a radius of $a\pi/2$. However, it never actually intersects with or surpasses this circle, creating a visually captivating and mathematically intriguing pattern. Needless to say, this choice of parameters and the use of the arctan function add a unique twist to the conventional spiral, resulting in an elegant and bounded spiral that evolves gracefully as it approaches its defined limit. In this understanding, bounded spirals refer to spiral patterns or curves that have clear and defined boundaries or limits. In these spirals, there is a finite range or constraint within which the spiral evolves or exists. This constraint can be imposed by various factors, such as physical boundaries, mathematical conditions, or specific parameters that limit the extent or growth of the spiral (Beckman et al., 2000; Hall, 1982; Singh et al., 1999).

These bounded spirals often exhibit unique and intriguing characteristics within their defined limits, making them a subject of interest in various fields, including mathematics, physics, and art. They provide a fascinating glimpse into how spirals can manifest within specific boundaries, leading to diverse and captivating patterns and phenomena.

8.6 Numerical solution of spirals on the sea

Numerical solutions of spirals in the context of ocean dynamics can involve simulating and analyzing various aspects of spiral patterns and their formation in the sea. Numerical models are commonly used in oceanography to study complex fluid flow phenomena. It references laboratory experiments, theoretical frameworks, and observational studies related to these spiral formations. Therefore the formation mechanisms of spiral features in fluid flows, particularly emphasizing their close association with the instability of shear flow. In this context, shear flow instability, often characterized by the development of waves or disturbances in adjacent fluid layers with different velocities, can be described mathematically using various approaches, including linear stability analysis. Let's consider a one-dimensional shear flow in the x-direction, and we'll denote the velocity of the flow as $U(y)$, where y is the vertical direction perpendicular to the flow (Gerkema et al., 2008; Pannatoni, 1983; Soltanian & Beigzadeh, 2023). The basic flow profile for the shear flow is $U(y)$. This represents how the velocity varies with respect to the vertical coordinate y. For instance, $U(y)$ might be a linear profile for simple shear flow:

$$U(y) = U_0\left(1 - \frac{y}{d}\right) \tag{8.41}$$

where U_0 is the maximum velocity at the center of the flow, and d is the height of the shear layer. To investigate shear flow instability, we often linearize the equations governing fluid motion around the basic flow profile. The key equation involved is the linearized Navier − Stokes equation:

$$\frac{\partial u'}{\partial t} = -\,U(y)\partial\frac{\vec{u}'}{\partial x} - \frac{1}{\rho}\nabla p' + \nu\nabla^2 u' \tag{8.42}$$

here u' is the velocity perturbation (deviation from the basic flow); p' is the pressure perturbation; ρ is the fluid density; and ν is the kinematic viscosity. Therefore the continuity equation for incompressible flow is linearized as:

$$\nabla\cdot\vec{u}' = 0 \tag{8.43}$$

This equation represents the continuity equation, ensuring that mass is conserved within the flow as explained early in Chapter 2.

Therefore fully nonlinear theories in fluid dynamics, including those developed by Stuart in 1967, often involve the study of complex phenomena such as the development of spiral patterns in fluid flows. In this sense, Stuart's contributions to fluid dynamics include his work on stability theory and the development of mathematical models to describe nonlinear behavior in fluid flows. His work is relevant to understanding the dynamics of flow instabilities and the emergence of coherent structures, which may include spiral patterns. Indeed, the Stuart formalism is a mathematical framework used to study vorticity dynamics and instability in fluid dynamics, particularly in the context of two-dimensional flows. It provides a way to analyze the evolution of vorticity patterns and understand the behavior of fluid flows, including the formation of spirals and vortices. Consequently, the Stuart formalism involves the following key equations:

8.6.1 Vorticity transport equation

The vorticity transport equation describes how vorticity evolves in a fluid flow. In two dimensions, it can be written as:

$$\frac{D\omega}{Dt} = \vec{\omega}\cdot\nabla\vec{v} + \nu\nabla^2\vec{\omega} \tag{8.44}$$

here $\frac{D\omega}{Dt}$ is the material derivative of vorticity; $\vec{\omega}$ is angular frequency; and ν is the kinematic viscosity.

8.6.2 Stream function equation

The stream function equation helps describe the flow in terms of streamlines. For a two-dimensional, incompressible, and irrotational flow, it is given by:

$$\nabla^2\psi = -\,\omega \tag{8.45}$$

here ψ is the stream function, and ω is the vorticity. The velocity component along the stagnation streamline depends on the flow's boundary conditions. In potential flow theory, the velocity is related to the stream function (ψ) or velocity potential (ϕ) as follows:

For two-dimensional flow:

$$V = \frac{\partial \psi}{\partial n} \tag{8.46}$$

$$V = \frac{\partial \phi}{\partial s} \tag{8.47}$$

Here, $\frac{\partial \psi}{\partial n}$ represents the normal derivative of the stream function, and $\frac{\partial \phi}{\partial s}$ represents the tangential derivative of the velocity potential. The exact form of these derivatives depends on the boundary conditions and flow geometry (Wu et al., 2007; Wu & Wu, 1998).

The crucial question is how to effectively incorporate the vorticity motion equation to accurately describe spiral motions? Modeling ocean current spirals involves complex fluid dynamics equations. One common phenomenon associated with spiraling ocean currents is the formation of ocean eddies, which can have spiral-like shapes. While these systems are described by the Navier – Stokes equations for fluid flow, a simple model that captures some aspects of spiral ocean currents is the two-dimensional vorticity equation. According to the aforementioned perspective, the two-dimensional vorticity equation describes the evolution of vorticity, which is a measure of local fluid rotation. In polar coordinates (r, θ), the vorticity equation can be written as:

$$\partial\omega/\partial t + (u/r)(\partial\omega/\partial\theta) = \nu(\nabla^2\omega) \tag{8.48}$$

Eq. (8.48) demonstrates that:

- ω is the vorticity (a scalar value).
- t is time.
- u is the radial velocity component.
- r is the radial distance from the center of the spiral.
- θ is the azimuthal angle.
- ν is the kinematic viscosity of the fluid.
- ∇^2 is the Laplacian operator.

To study spiral-like ocean currents, one would need to specify the initial conditions and boundary conditions for the problem.

8.6.3 Initial conditions

Let's assume an initial vorticity distribution in the form of a Gaussian-like spiral:

$$\omega(r,\theta,t) = \omega_0 e^{\left(\frac{r^2}{r_0^2}\right)} \sin(\theta) \tag{8.49}$$

where ω_0 is the maximum vorticity, and r_0 is the characteristic radius of the spiral.

8.6.4 Boundary conditions

We'll assume no-slip boundary conditions at $r = 0$ and $r = R$ (where R is the outer boundary of the ocean region of interest).

8.6.5 Velocity field

The radial velocity component u can be determined based on the vorticity distribution through the stream function ψ using the relation:

$$u = -r^{-1}\frac{\partial \psi}{\partial \theta} \tag{8.50}$$

Assume that the vorticity (ω) is concentrated in a circular region and decreases radially outward from the center of the spiral. Mathematically, one can represent this vorticity distribution as:

$$\vec{\omega} = \omega_0 e^{-\alpha r}\hat{\vec{\theta}} \tag{8.51}$$

here $\hat{\vec{\theta}}$ is the unit vector in the azimuthal direction, and α is a decay rate.

8.6.6 Velocity components (V_r and Vθ)

The radial component of velocity represents the speed at which the particle moves away from or toward the center of the spiral. It can be expressed as follows:

$$V_r = \frac{dr}{dt} \tag{8.52}$$

where $\frac{dr}{dt}$ is the rate of change of radial distance with respect to time (t). Therefore the tangential component of velocity represents the speed at which the particle moves around the center of the spiral. In a potential flow, this component depends on the angular velocity ω and the radial distance (r) from the center of the spiral:

$$V_\theta = \omega \cdot r \tag{8.53}$$

These equations describe the particle's motion in terms of its radial and tangential velocities as it follows a spiral trajectory in a two-dimensional flow field (Morton, 1984).

For an Archimedean spiral, the components can be derived from the parametric equations that define the spiral's motion:

$$r(t) = a + b\omega t \tag{8.54}$$

The value of b determines the tightness or looseness of the spiral. If b is positive, the spiral expands outward as θ increases, and if b is negative, the spiral contracts inward. Here are some variations of the Archimedean spiral:

1. Outward Spiraling (Expanding) Archimedean Spiral: $a > 0$, and $b > 0$.
2. Inward Spiraling (Contracting) Archimedean Spiral: $a > 0$, and $b < 0$.
3. Outward Spiraling Logarithmic Spiral: $a = 0$ (starting from the origin), and $b > 0$.
4. Inward Spiraling Logarithmic Spiral: $a = 0$ (starting from the origin), and $b < 0$.

These conditions describe different types of spirals with various shapes and characteristics. By adjusting the values of a and b, you can create spirals of different sizes and orientations. The rate of strain, therefore, measures how the velocity of a fluid element changes as it moves along a path. In the context of spiral motion, du $(ds)^{-1}$ represents the rate of change of velocity as the particle moves along the spiral path. To calculate du $(ds)^{-1}$, one would differentiate the velocity components V_r and $V\theta$ with respect to the path length s, which corresponds to the arc length along the spiral trajectory. This involves the chain rule for differentiation

$$\frac{du}{ds} = \frac{d}{ds}\sqrt{V_r^2 + V_\theta^2} \tag{8.55}$$

The specific mathematical expression for du $(ds)^{-1}$ would depend on the equations that describe the spiral's motion and how the velocity components evolve as a function of position along the spiral. Let's work through an example of finding the rate of strain $(du(ds)^{-1})$ for a particle moving along a spiral trajectory. Suppose we have a spiral defined by the following equations:

- Radius of the spiral: $r(t) = 2t$ meters,
- Angle of rotation: $\theta(t) = 4\pi t$ radians.

We want to find the rate of strain du $(ds)^{-1}$ for a quantity u that depends on the position along the spiral, given $u = 3r^2 + 2\theta$. 1- Find $\frac{du}{dt}$:

Start by finding $\frac{du}{dt}$ using the chain rule. Differentiate u with respect to t:

$$\frac{du}{dt} = \frac{d(3r^2 + 2\theta)}{dt} \tag{8.56}$$

Using the derivatives of $r(t)$ and $\theta(t)$, we have:

$$\frac{du}{dt} = 3 \cdot 2(2t) + 2 \cdot \frac{\pi}{4} = 12t + \frac{\pi}{2} \tag{8.57}$$

1. Find $\frac{dt}{ds}$:

   To find $\frac{dt}{ds}$, differentiate the parameterization of $s = s(t)$ with respect to t:

$$s(t) = \int_0^t \sqrt{\left(\frac{dr}{d\tau}\right)^2 + \left(\frac{r}{\tau}\right)^2} d\tau \tag{8.58}$$

Using $r(t) = 2t$, we can calculate $\frac{dr}{d\tau} = 2$, and we have:

$$s(t) = \int_0^t \sqrt{(2)^2 + \left(\frac{2r}{\tau}\right)^2} d\tau = \int_0^t 2\sqrt{2d\tau} = 2\sqrt{2t} \tag{8.59}$$

Now, differentiate $s(t)$ with respect to t to find $\frac{dt}{ds}$:

$$\frac{dt}{ds} = \frac{1}{\frac{ds}{dt}} = \frac{\sqrt{2}}{4} \tag{8.60}$$

2. Multiply the Results:

Finally, multiply the results from steps 1 and 2 together to find $\frac{du}{ds}$:

$$\frac{du}{ds} = \frac{du}{dt} \cdot \frac{dt}{ds} = \left(12t + \frac{\pi}{2}\right) \cdot \left(\frac{\sqrt{2}}{4}\right) = 3\sqrt{2t} + \pi\frac{\sqrt{2}}{8} \tag{8.61}$$

Therefore the rate of strain $\frac{du}{ds}$ for the given quantity u along the spiral trajectory is $3\sqrt{2t} + \pi\frac{\sqrt{2}}{8}$. In other word, $3\sqrt{2t}$ indicates that the rate of strain is changing linearly with time (t). Specifically, it's increasing at a rate of $3\sqrt{2}$ units per unit of time. This component represents the time-dependent stretching or compressing of u along the spiral trajectory. Additionally, the constant term $\pi\frac{\sqrt{2}}{8}$ represents a fixed rate of strain that is independent of time. It's a constant contribution to the rate of strain along the spiral path. By following these steps, you can find the rate of strain along the spiral trajectory, taking into account the changes in u with respect to the position s along the path and the parametric description of the spiral's motion.

8.7 How does SAR imagine spiral patterns

How synthetic aperture radar (SAR) images can exhibit complex streak patterns, including spirals, and it attributes these patterns to surface characteristics and the interaction between radar waves and the surface. Let's denote the Earth's surface elevation profile as $Z(x,y)$, where x and y are the spatial coordinates. The surface roughness can be characterized by the deviation of Z from a smooth surface Z_{smooth}. One can represent this deviation as $H(x,y)$:

$$H(x, y) = Z(x, y) - Z_{smooth}(x, y) \tag{8.62}$$

The radar wave interaction with the surface roughness can be described by the radar backscatter coefficient (σ_0), which depends on the characteristics of the surface. The radar equation relates the backscatter coefficient to the surface roughness:

$$\sigma_0(x, y) = f(H(x, y), \lambda) \tag{8.63}$$

here λ is the radar wavelength; f is a function that relates the backscatter to the surface roughness; and $\sigma_0(x,y)$ is the backscatter coefficient at location (x,y). In this view, the function f that relates the radar backscatter (σ_0) to the surface roughness (H) represents the mathematical relationship between these two variables. This function is often expressed using an empirical or semiempirical model that accounts for the scattering properties of the surface. The choice of the specific model depends on the type of surface, radar wavelength, and other factors. In this view, the actual form of $f(H)$ can vary depending on the context and the specific scattering model used. For example, one common model is the "Kirchhoff approximation" for radar backscatter from rough surfaces (Vlander et al., 2007; Zilman et al., 2015). In this model, the backscatter coefficient is related to the surface roughness through parameters such as the correlation length (L) and the root mean square (RMS) height of the roughness (σ_H):

$$\sigma_0 = f(H) = K \cdot \left(\frac{2\pi}{\lambda}\right)^2 \cdot \left(\frac{\sigma_H}{L}\right)^2 \cdot e^{\left(-2 \cdot \left(\frac{2\pi}{\lambda}\right)^2 \cdot \left(\frac{\sigma_H}{L}\right)^2\right)} \tag{8.64}$$

The Bragg resonance scattering condition occurs when the radar wavelength matches the characteristic wavelength of surface features. For a spiral structure, this can be expressed as:

$$2d\sin(\theta) = m\lambda \tag{8.65}$$

where d is the characteristic spacing or pitch of the spiral; θ is the incident angle; and m is an integer representing the order of resonance. In the case of an extremely turbulent sea surface, there is a phenomenon known as constructive interference between the incident radar wave and the scattered wave originating from the rough surface. This condition can be mathematically represented as follows:

$$2k_0\sin(\theta) = m \cdot \Delta k \tag{8.66}$$

here k_0 is the wavenumber of the incident radar wave, and Δk is the wavenumber mismatch, which depends on the surface's roughness and is calculated as $\Delta k = k_s - k_0$, where k_s is the wavenumber of the scattered wave. The mathematical formula for representing the scattering order "m" as an integer is: $m \in \mathbb{Z}$. It typically takes on values such as 1 for first-order scattering, 2 for second-order scattering, and so on, depending on the specific scattering order being considered. In the context of radar or SAR imaging, these scattering orders represent different levels of interactions between the incident radar wave and the scattering objects or surfaces. First-order scattering typically refers to the primary interaction, while higher-order scattering involves more complex interactions. In other words, the return signal received by the SAR antenna will vary depending on whether the Bragg resonance condition is met. When the condition is satisfied, constructive interference occurs, leading to enhanced backscatter. In contrast, when the condition is not met, destructive interference may occur, reducing the backscatter. The overall impact of the spiral on the return signal, consequently, is influenced by the radar wavelength, spiral geometry, and incident angle. Spirals that satisfy the Bragg resonance condition for the radar wavelength will appear more prominently in the SAR image due to increased backscatter (F. Xu et al., 2015).

The SAR imaging process involves collecting backscattered signals from various spatial locations. The image formation process integrates the radar signals received from different locations along the SAR platform's path. This integration leads to the formation of streak patterns. Let's denote the SAR image as $I(x,y)$, and the SAR imaging process can be represented mathematically as:

$$I(x,y) = \int \sigma_0(x', y')dx'dy' \tag{8.67}$$

Here, the integral integrates over the area covered by the SAR platform's path. Eq. (8.67) represents a double integral. It means that the SAR image intensity at point (x,y) is determined by integrating the radar backscatter coefficients (σ_0) over the entire scene. This integration is performed with respect to both the x' and y' coordinates, which effectively means summing up the contributions from all points in the scene to calculate the pixel value at (x,y) in the SAR image. Therefore the radar backscatter coefficient (σ_0) is influenced by the presence of these waves on the surface. The radar backscatter can be a function of the wave amplitudes, wavelengths, and other parameters. Let's denote the radar backscatter considering the effects of both capillary and short gravity waves as $\sigma_{0c,g}$:

$$\sigma_{0c,g}(x,y) = f\left(A_C(x,y), \lambda_C, A_g(x,y), \lambda_g, \lambda, \theta\right) \tag{8.68}$$

This equation relates the radar backscatter coefficient ($\sigma_{0c,g}$) at a specific point in the SAR image to various parameters associated with capillary waves (denoted with subscript c) and short gravity waves (denoted with subscript g). These parameters include the amplitudes ($A_c(x,y)$ and $A_g(x,y)$), wavelengths (λ_c and λ_g), the radar wavelength (λ), and the angle between the radar beam and the wave direction (θ). The appearance of spiral patterns in the SAR image occurs when there are systematic variations in surface roughness or elevation as the SAR platform moves along its path. These variations are captured in the function $H(x,y)$ (Bertacca et al., 2005; Tomiyasu, 1978).

When variations in $\sigma_0(x,y)$ occur along the SAR platform's path, they manifest as spatially modulated patterns in the SAR image. This modulation can be mathematically represented as a sinusoidal or helical function, which captures the systematic variation:

$$\sigma_0(x,y) = A\cos(kx + \phi) \tag{8.69}$$

Here, A represents the amplitude of the modulation, k is the spatial frequency or wave number, x is the spatial coordinate along the SAR platform's path, and ϕ is the phase of the modulation. To achieve a spiral pattern in the surface properties $H(x,y)$, we can introduce a helical or spiral structure into $H(x,y)$ as follows:

$$H(x,y) = H_0\cos\left(kx + \phi_s\right) \tag{8.70}$$

here, H_0 represents the amplitude of the surface property variation, k is the same spatial frequency as in the backscatter modulation, x is the spatial coordinate along the SAR platform's path, and ϕ_s is the phase of the spiral pattern. When one combines the variations in $\sigma_0(x,y)$ due to spatial modulation and the spiral pattern in surface properties $H(x,y)$, one obtains spiral patterns in the SAR image. The mathematical relationship between these variations can be expressed as:

$$\sigma_0(x,y) = A\cos\left(kx + \phi_s\right) = f\left(H_0\cos\left(kx + \phi_s\right)\right) \tag{8.71}$$

Needless to say, these mathematical equations illustrate how variations in radar backscatter σ_0 $(x,y))$ due to spatial modulation, coupled with a spiral pattern in surface properties $H(x,y))$, result in the appearance of spiral patterns in the SAR image. The modulation and spiral structures are captured by sinusoidal functions with the same spatial frequency, creating a visually distinct pattern in the SAR image.

The alteration of surface roughness resulting from the presence of these substances induces variations in the radar backscatter coefficient, consequently influencing the emergence of spiral patterns within SAR images. The interaction of surfactants and surface films with the surface roughness $H(x,y))$ introduces a source of decoherence $D(x,y))$ in the system. This decoherence process can be mathematically described as:

$$H'(x,y) = \sqrt{1 - D(x,y)} \cdot H(x,y) \tag{8.72}$$

here $\sqrt{1 - D(x,y)}$ represents the decoherence factor that quantifies the reduction in coherence or correlation between the original surface roughness $H(x,y))$ and the modified surface roughness $H'(x,y))$ due to the presence of surfactants and surface films. This decoherence process leads to changes in the radar backscatter coefficient $(\sigma_0'(x,y))$:

$$\sigma_0'(x,y) = f(H'(x,y)) \tag{8.73}$$

In this context, the concept of decoherence captures the loss of coherence or correlation between the original and modified spiral surface properties caused by surfactants and surface films, ultimately affecting the radar backscatter characteristics. It can be said that the modification of surface roughness due to these substances leads to changes in the radar backscatter coefficient, ultimately impacting the appearance of spiral patterns in SAR images.

Indeed, based on the perspective described earlier, it's clear that both the phase shift introduced by the spiral structure $(\varphi(\theta))$ and the scattering amplitude $(F(\theta))$ can be complex functions, and they depend on various factors including the spiral's geometry and orientation. These parameters play a crucial role in determining how radar waves interact with the spiral structure, and they have a significant impact on the spatial distribution of radar return signals. Mathematically, the radar return signal, which we can denote as $S(\theta)$, can be expressed as:

$$S(\theta) = A * F(\theta) * e^{(i\varphi(\theta))} \tag{8.74}$$

here A represents a constant related to the radar system's characteristics, and $F(\theta)$ represents the scattering amplitude, which can be a complex function depending on the spiral's geometry and orientation. Hence, $e^{(i}\varphi(\theta))$ represents the complex exponential term that accounts for the phase shift. These equations describe how radar waves interact with the spiral and how the complex functions F (θ) and $\varphi(\theta)$ influence the radar return signals at different angles θ (Kim & Johnson, 2002; Xu et al., 2015).

Needless to say, the interaction between spiral structures and SAR radar waves is a multifaceted phenomenon influenced by factors such as radar wavelength, spiral geometry, and incident angle. Spirals meeting the Bragg resonance condition can display heightened backscatter, resulting in unique patterns within SAR imagery. Grasping this intricate interplay is essential for the accurate interpretation of SAR images and the extraction of pertinent details regarding surface characteristics, including spiral formations.

8.8 Automatic detection of spirals in SAR images

The primary challenge in extracting spiral features from SAR (synthetic aperture radar) images lies in the presence of speckles, as discussed in Chapter 7. Speckles are random noise patterns that can obscure the desired features in SAR imagery, making their extraction a complex task. To address this challenge, advanced algorithms based on quantum computing are being explored to overcome these complexities. In this context, the quantum distance estimation algorithm is being leveraged as a promising solution to extract spiral features from the intricate and noisy surrounding environment in SAR images. This quantum algorithm offers the potential to efficiently process and analyze SAR data, even in the presence of speckles and other forms of noise. By utilizing quantum properties and computational advantages, the quantum distance estimation algorithm can enhance the accuracy and reliability of spiral feature extraction in SAR images.

To execute the quantum distance estimation algorithm, we must first encode the generated data points into a multiqubit quantum state. To achieve this, we employ two prevalent embedding techniques: amplitude embedding and angle embedding. We then assess the performance of each method individually.

Amplitude embedding is a method to represent classical data in a quantum state by encoding information in the amplitudes of quantum states. Suppose we have a set of data points {x1, x2,·, xn}. To perform amplitude embedding, we map these data points to quantum states using the following process: (1) normalize the data and (2) quantum superposition. Normalize the data points; therefore they lie within a specific range, often between |0> and |1> (Gu et al., 2009; Khan et al., 2019). Encode the normalized data points into quantum superposition states. Each data point corresponds to a coefficient (amplitude) in the superposition. Mathematically, the SAR data points $I_1, I_2, \cdot, I_n$, can represent the quantum state as:

$$|\psi\rangle = \alpha_1|0\rangle + \alpha_2|1\rangle + . + \alpha_n|n-1\rangle \tag{8.75}$$

Eq. (8.75) means that the quantum state $|\psi\rangle$ is described as a linear combination of the basis states $|0\rangle$, $|1\rangle$, $|2\rangle, \cdot |n-1\rangle$, with the coefficients $\alpha_1, \alpha_2, \cdot, \alpha_n$ determining the probabilities associated with each basis state along the spiral features in SAR data. This is a fundamental representation in quantum computing, where complex amplitudes allow for the representation of quantum superposition, entanglement, and other quantum phenomena. Mathematically, the quantum state $|\psi\rangle$ in angle embedding, therefore, can be represented as:

$$|\psi\rangle = |0\rangle \otimes e^{(i\theta_1)}|0\rangle + |0\rangle \otimes e^{(i\theta_2)}|1\rangle + . + |0\rangle \otimes e^{(i\theta_n)}|n-1\rangle \tag{8.76}$$

here $e^{(i\theta_1)}, e^{(i\theta_2)}, .., e^{(i\theta_n)}$ are complex numbers represented using the exponential notation. Each $e^{(i\theta_i)}$ is associated with a specific basis state of spiral features in the SAR image and represents a complex probability amplitude. This notation is often used in quantum computing to represent quantum states, and it allows for the description of quantum superposition and phase information associated with each basis state (Anand et al., 2022). In this sense, these embeddings allow classical data to be efficiently processed on a quantum computer, opening up possibilities for quantum data analysis and machine learning.

Let's create a new dissimilarity measure inspired by Marghany for measuring dissimilarity or similarity between two vectors $\vec{a}$ and $\vec{b}$ of spiral features in a SAR image. We'll call this measure "Spiral Dissimilarity Measure (SDM)." SDM is designed to capture the dissimilarity or similarity between two vectors representing spiral features in SAR images. It takes into account both the differences in feature values and their relative importance through element-wise weighting (Table 8.1). SDM is defined as follows:

$$MD(|a\rangle,|b\rangle,p) = \left(\sum_{i=1}^{n} W_i\delta_i\left(\left|\vec{a_i} - \vec{b_i}|\right)^{1+i}\right)^{p^{-1}} \tag{8.77}$$

here quantum states $|a\rangle$ and $|b\rangle$, where $|a\rangle = \sum_i \alpha_i|i\rangle$ and $|b\rangle = \sum_i \beta_i|i\rangle$ represent vectors a and b in a quantum superposition, respectively. Therefore α_i and β_i are complex probability amplitudes for the $|i\rangle$ basis state. $\delta_i(|\alpha_i|, |\beta_i|)$ is the dissimilarity measure between the complex probability amplitudes α_i and β_i. Additionally, p is the order parameter controlling the sensitivity to individual elements. Therefore in this pseudocode, we assume that one have functions or methods for creating quantum superposition states, obtaining amplitudes, calculating dissimilarity measures, and accessing element-wise weights. The **quantum_superposition_mdd** function iterates through elements in superposition states and calculates the MDD using the specified order parameter (**p**). The result is the Marghany distance in quantum superposition. Here W_i represents the weight assigned to the dissimilarity measure for the i-th element in SAR image. Therefore the spiral similarity score as the inverse of the weighted distance, normalized to the range [0, 1]:

$$S_{Spiral} = 1/(1+\sqrt{\left(\sum_i\left(W_i * \left(\vec{a_i} - \vec{b_i}\right)^2\right)\right)} \tag{8.78}$$

This ensures that higher values of S indicate greater similarity, while lower values indicate dissimilarity. This equation enables the calculation of the Marghany distance between two vectors, $\vec{a_i}$ and $\vec{b_i}$, using different distance metrics (Euclidean, Manhattan, or Minkowski) by adjusting the value of p. It offers a unified dissimilarity measure capable of accommodating various distance metrics based on the specified order. In this pseudocode, we use the **quantum_superposition_mdd** function to calculate the MDD between two quantum superposition states representing spiral features. The MDD is then normalized to the range [0, 1], and the spiral similarity score (SSS) is computed as 1 minus the normalized distance. This score quantifies the similarity between the spiral features in the quantum superposition states (Table 8.2).

Consequently, quantum element-wise weighted distance decomposition harnesses the remarkable quantum property of superposition (Wereszczynski et al., 2018). In classical computing, data elements are processed sequentially or in parallel, but quantum computing takes a

TABLE 8.1 Pseudocode for quantum superposition of Marghany dissimilarity distance.

```
function quantum_superposition_mdd(|a⟩, |b⟩, p):
# Initialize variables to store intermediate values result = 0.0 n =
number_of_elements_in(|a⟩) # Assuming |a⟩ and |b⟩ have the same number of elements
# Loop over each element in the superposition states for i in range(n):
# Get the complex probability amplitudes
αi = amplitude_of(|a⟩, |i⟩) βi = amplitude_of(|b⟩, |i⟩)
# Get the element-wise weight
Wi = get_weight_for_element(i)
# Calculate the dissimilarity measure for the element dissimilarity =
calculate_dissimilarity(|αi|, |βi|) # Apply the weight and order parameter
weighted_dissimilarity = Wi * (dissimilarity ** p)
# Add the weighted dissimilarity to the result result += weighted_dissimilarity
# Apply the p-th root to the sum of weighted dissimilarities MDD_result = result ** (1/p)
# Return the Marghany Distance in Quantum Superposition return MDD_result
# Example usage: |a⟩ = create_quantum_superposition_vector(a)
# Create a quantum superposition state for vector a |b⟩ =
create_quantum_superposition_vector(b)
# Create a quantum superposition state for vector b p = 2
# Choose the order parameter
distance = quantum_superposition_mdd(|a⟩, |b⟩, p) print("Marghany Distance in Quantum
Superposition:", distance)
function SpiralSimilarityScore(a, b, weights, metricType, p):
// Compute the Marghany Dissimilarity Measure (MDM) using a loop // Normalize MDM to
the range [0, 1]
minMDM = 0 // Minimum possible MDM
maxMDM = 1 // Maximum possible MDM
normalizedMDM = (MDM - minMDM) / (maxMDM - minMDM)
// Compute the Spiral Similarity Score as the inverse of normalized MDM spiralSimilarityScore
= 1 - normalizedMDM
return spiralSimilarityScore
```

different approach (Tomesh et al., 2021). With Q-EWDD, we delve into the quantum realm, where quantum bits, or qubits, can exist in a multitude of states simultaneously. This unique property forms the foundation of Q-EWDD and empowers us to explore an array of weighted combinations of elements in an incredibly efficient manner (Table 8.3).

In this pseudocode, we represent the quantum state $|\psi\rangle$, initialize the MDD to zero, define element-wise weights for vectors a and b, and loop through the elements to calculate the dissimilarity measure for each element. The dissimilarity measure is then weighted by the respective element-wise weights and added to the MDD. This process yields the quantum element-wise weighted

TABLE 8.2 Pseudocode representation of spiral similarity score using the quantum_superposition_mdd.

```
function spiral_similarity_score(|a⟩, |b⟩, p):
# Calculate Marghany Distance (MDD) using quantum_superposition_mdd function
MDD = quantum_superposition_mdd(|a⟩, |b⟩, p)
# Normalize MDD to the range [0, 1] max_possible_distance = 1.0
# Maximum possible MDD (can be determined empirically)
if MDD > max_possible_distance: normalized_distance = 1.0 else: normalized_distance =
MDD / max_possible_distance
# Calculate Spiral Similarity Score (SSS) SSS = 1.0 - normalized_distance
# Return the Spiral Similarity Score return SSS
# Example usage: |a⟩ = create_quantum_superposition_vector(a)
# Create a quantum superposition state for vector a |b⟩ =
create_quantum_superposition_vector(b)
# Create a quantum superposition state for vector b p = 2
# Choose the order parameter
similarity_score = spiral_similarity_score(|a⟩, |b⟩, p) print("Spiral Similarity Score:",
similarity_score)
```

distance decomposition (Q-EWDD) for the given data vectors a and b.

8.9 Identifying spiral feature pattern in SAR data

To articulate the process of analyzing decomposed distances to identify patterns linked with spiral features, we can formulate a pioneering mathematical equation that capitalizes on the quantum advantage to enable efficient exploration. This equation shall be referred to as the "Spiral Pattern Identification Index (SPII)."

$$SPII(MDD) = \sum_{i=1}^{n} \left(1 - \frac{MDD_i}{max(MDD)}\right) \cdot P_i \tag{8.79}$$

In this equation, SPII(D) represents the spiral pattern identification index, MDD is the set of decomposed distances obtained through Q-EWDD, n is the number of data vectors, MDD_i is the decomposed distance for the i-th data vector, max(MDD) is the maximum decomposed distance, and P_i is the pattern detection coefficient for the i-th data vector. To formulate a quantum mechanics formula for the spiral pattern identification index (SPII), we can start by defining the SPII as a quantum expectation value. Let's assume that we have a quantum state $|\psi\rangle$ that encodes the information about decomposed distances and pattern detection coefficients. The SPII can be represented as:

$$SPII = \langle\psi|\widehat{O}|\psi\rangle \tag{8.80}$$

In Eq. (8.80), $|\psi\rangle$ is the quantum state that encodes decomposed distances, and $\widehat{O}$ is the quantum operator corresponding to the SPII calculation. The quantum operator $\widehat{O}$ should be constructed based on the specific equation or method used for SPII calculation. This equation often involves weighted sums or integrations over the decomposed distances. The exact form of $\widehat{O}$ would depend on the mathematical representation of SPII.

For example, if SPII involves a weighted sum of decomposed distances, the quantum operator could be constructed as:

$$\widehat{O} = \sum_{i} W_i M\widehat{M}D_i \tag{8.81}$$

where W_i are the pattern detection coefficients, and $M\widehat{M}D_i$ are quantum operators corresponding to decomposed distances. This quantum mechanics formula allows you to calculate the SPII using quantum states and operators. The specific construction of $\widehat{O}$ and the quantum state $|\psi\rangle$ would depend on the mathematical expression of SPII and the details of the quantum computing system being used. In this sense, quantum computers perform

TABLE 8.3 Pseudocode for quantum element-wise weighted distance decomposition.

```
# Define the quantum state for data vectors a and b # |ψ⟩ = α1|00⟩ + α2|01⟩ + α3|10⟩ +
α4|11⟩ # Where α1, α2, α3, α4 are complex coefficients
# Initialize the Marghany Distance (MDD) to zero MDD = 0
# Define the element-wise weights for vectors a and b weights_a = [W1, W2, ., Wn] # Weight
vector for vector a weights_b = [W1, W2, ., Wn] # Weight vector for vector b
# Loop through the elements of vectors a and b for i in range(len(a)):
# Calculate the dissimilarity measure δi(ai, bi) between elements ai and bi dissimilarity =
calculate_dissimilarity(a[i], b[i])
# You can replace 'calculate_dissimilarity' with the desired distance metric
 # Multiply the dissimilarity measure by the element-wise weights
weighted_dissimilarity = dissimilarity * weights_a[i] * weights_b[i]
 # Add the weighted dissimilarity to the MDD
MDD += weighted_dissimilarity
# MDD now contains the Quantum Element-wise Weighted Distance Decomposition
```

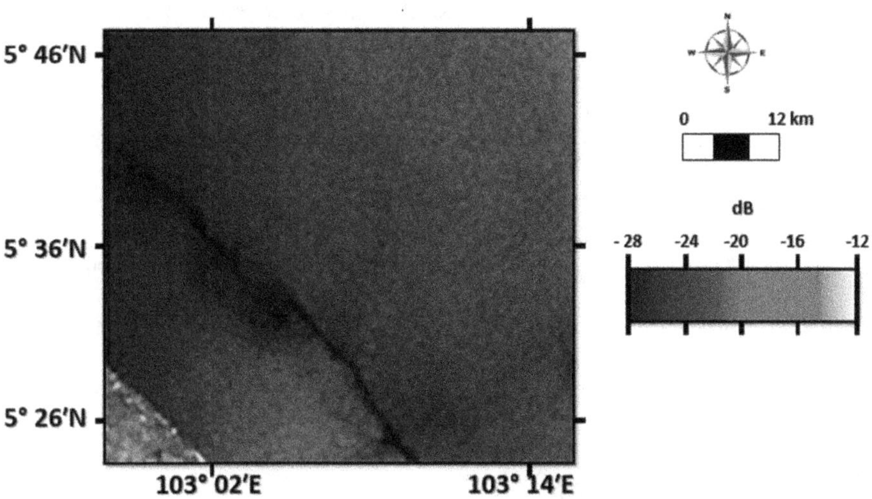

FIGURE 8.10 The occurrence of spiral flow in RADARSAT-1 SAR S1 mode.

operations in parallel, which dramatically speeds up the computation of distances for large datasets. SPII and Q-EWDD can handle high-dimensional data with ease. Q-EWDD uses a quantum circuit that leverages quantum parallelism to calculate the weighted distances in parallel. It utilizes controlled SWAP gate as discussed in Eq. (6.50), Chapter 6, and quantum operations to perform element-wise calculations efficiently.

Therefore a mathematical representation of the step where quantum clustering algorithms or techniques are utilized to group similar data points based on dissimilarity values from the distance matrix:

$$Minimize{:}J = i = \sum_{i=1}^{n}\sum_{j=1}^{k} X_i.MDD_{ij}^2 \tag{8.82}$$

Subject to the constraint that each data point i is assigned to exactly one cluster:

$$\sum_{j=1}^{k} X_i = 1, \forall_i \in 1, 2, \ldots, n \tag{8.83}$$

Possibly using a quantum optimization algorithm will yield the assignment of data points to clusters, effectively grouping similar data points based on the dissimilarity values in the distance matrix. Consequently, the quantum approximate optimization algorithm (QAOA) utilizes the quantum nature of qubits to explore multiple potential solutions simultaneously. This quantum parallelism allows them to search for the optimal configuration of cluster assignments more efficiently than classical optimization methods (Choi & Kim, 2019; Crooks, 2018; Shaydulin & Alexeev, 2019). The quantum optimization algorithm iteratively adjusts the values of the binary variables X_i to minimize the cost function J. These adjustments aim to group data points into clusters such that the overall dissimilarity, as represented by the cost function, is minimized. The result is an assignment of data points to clusters that best captures the inherent patterns in the data.

QAOA starts by preparing a quantum state $|\psi(\beta,\gamma)\rangle$, which depends on two sets of parameters β and γ. This state is created by applying quantum operators to an initial state, typically the uniform superposition state $|+\rangle = N^{-1}\sum_x |x\rangle$. The expectation value of the objective function is calculated with respect to the quantum state $|\psi(\beta,\gamma)\rangle$. This is done by measuring the Hamiltonian C in this state:

$$\langle\psi(\beta,\gamma)|C|\psi(\beta,\gamma)\rangle \tag{8.84}$$

Computing this expectation value requires running quantum circuits and measuring the results. QAOA is designed to optimize the parameters β and γ to find the values that minimize the expectation value of the objective function.

8.10 Tested SAR data

Multiple SAR images, each containing single-look complex data, are utilized to extract spiral pattern features along the quantized Marghany turbulence boundary within the coastal waters of Peninsular Malaysia. RADARSAT-1 SAR S1 mode data with incidence angle ranges between 19.1° and 26.9°, collected on March 19, 1998, revealed the presence of a spiral flow pattern parallel to the coastal waters of Kuala Terengganu, Malaysia, situated approximately 23 kilometers from the shoreline. This spiral pattern significantly reduced the backscatter signal, resulting in a low backscatter reading of -28 dB (Fig. 8.10). The attenuation in signal strength was attributed to the high concentrations of filamentous material in the water, which was indicative of elevated chlorophyll levels. The length of this spiral feature extended approximately 60 kilometers.

The classic K-means clustering algorithm delivers discontinuous patterns in spiral flow because it is based on centroid-based clustering (Fig. 8.11). K-means aims to partition data points into clusters by minimizing the distance between data points and the centroids of their respective clusters. However, spiral flow patterns are often continuous and exhibit gradual changes in flow characteristics. K-means clustering is not well-suited for capturing the continuous and intricate nature of spiral patterns (Ahmad & Dey, 2007). It tends to create clusters with compact and spherical shapes around centroids, which may not align with the spiral structure. Therefore it can lead to disjointed and discontinuous clusters that do not accurately represent the underlying spiral flow.

On the other hand, the newly developed quantum algorithm based on quantum superposition and Marghany distance decomposition (quantum_superposition_MDD) delivers sharper patterns of spiral flows compared with

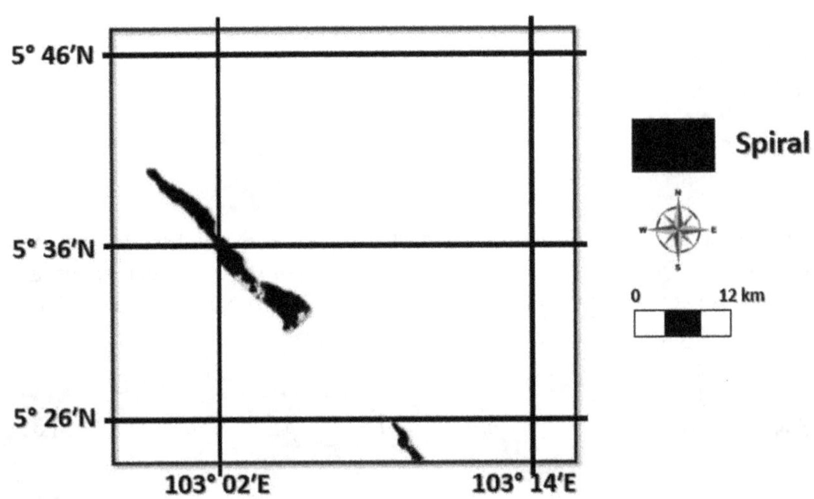

FIGURE 8.11 Result of classic K-means clustering algorithm.

those obtained using the classical K-means clustering algorithm (Fig. 8.12).

RADARSAT-1 SAR S2 mode data, acquired on February 2, 1999, with incidence angles of 23.7° and 31°, illustrate another instance of spiral flow moving parallel to the coastal line. This phenomenon results in a significant damping effect, with backscatter values reaching as low as -30dB (Fig. 8.13). The length of this spiral pattern measures approximately 76 kilometers. As a result, the quantum superposition Marghany distance decomposition (quantum_superposition_MDD) algorithm yields a similar outcome, automatically detecting sharp patterns of spiral flow (Fig. 8.14).

ERS-2 data in the Cvv-band, acquired over Bidong Island on May 14, 2000, captured a high concentration of rotated spiral flow within incidence angle ranging approximately from 23.7° and 31°. This shows high damping backscatter value by spiral of -20dB than previous examples (Fig. 8.15).

Consequently, the quantum superposition Marghany distance decomposition (quantum_superposition_MDD) algorithm outperforms raw ERS-2 data, automatically detecting sharper patterns of spiral flow (Fig. 8.16).

In the dataset of Radarsat-1 SAR standard mode 2, a groundbreaking discovery was made by the author on March 31, 2005, as elucidated in Chapter 4, Section 4.11. Interestingly, at that point in Chapter 4, the existence of spiral patterns had not yet been unveiled along the quantized Marghany turbulence boundary. The value of -25dB quantifies the reduction in signal strength compared with the surrounding area (Fig. 8.17). A lower backscatter signal suggests that the radar waves may be interacting differently with the features or conditions in the turbulence zone, and this could be a point of interest for further analysis, especially in applications such as oceanography and SAR imaging. In radar remote sensing, a decrease in signal backscatter can indicate various factors, including differences in surface roughness, composition, or wave

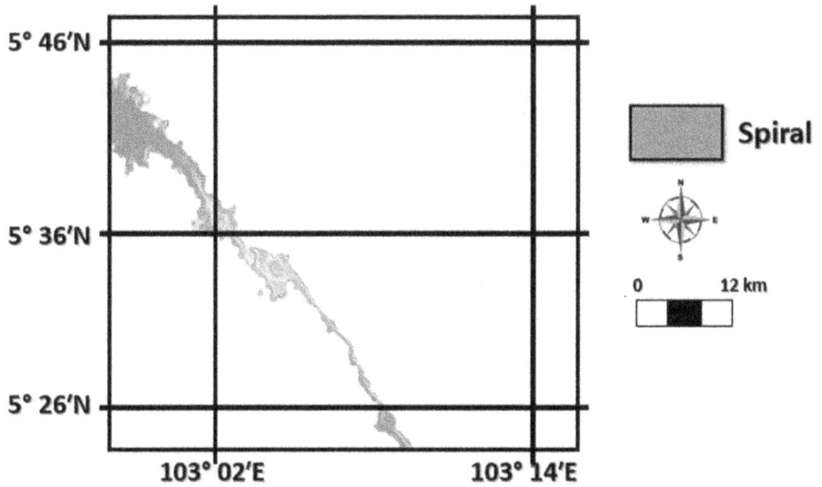

FIGURE 8.12 Automatic detection of hyperbolic spiral pattern in RADARSAT-1 SAR S1 mode data using quantum superposition Marghany distance decomposition.

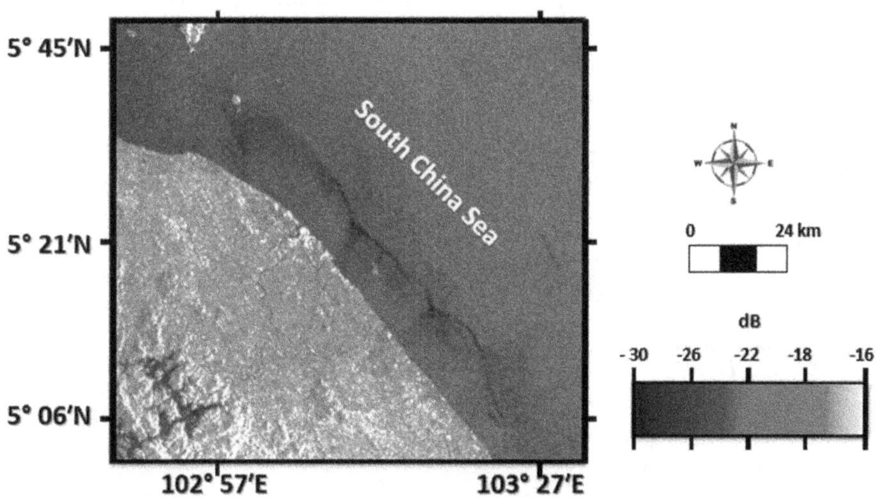

FIGURE 8.13 The occurrence of spiral flow in RADARSAT-1 SAR S2 mode.

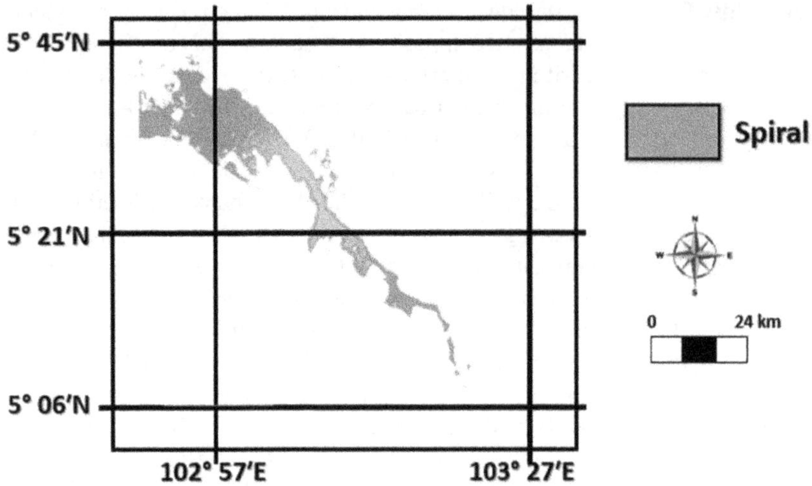

FIGURE 8.14 Automatic detection of hyperbolic spiral pattern in RADARSAT-1 SAR S2 mode data using quantum superposition Marghany distance decomposition.

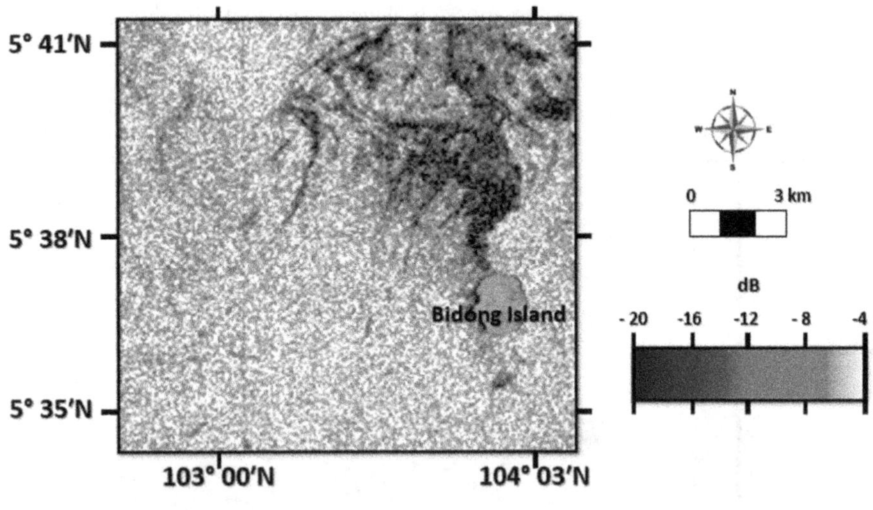

FIGURE 8.15 The occurrence of spiral flow in ERS-2 data.

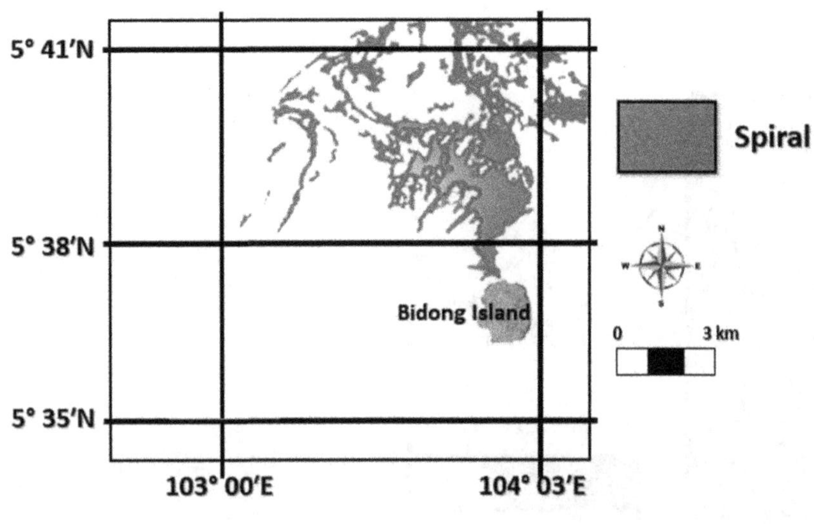

FIGURE 8.16 Automatic detection of rotated spiral patterns in ERS-2 data using quantum superposition Marghany distance decomposition.

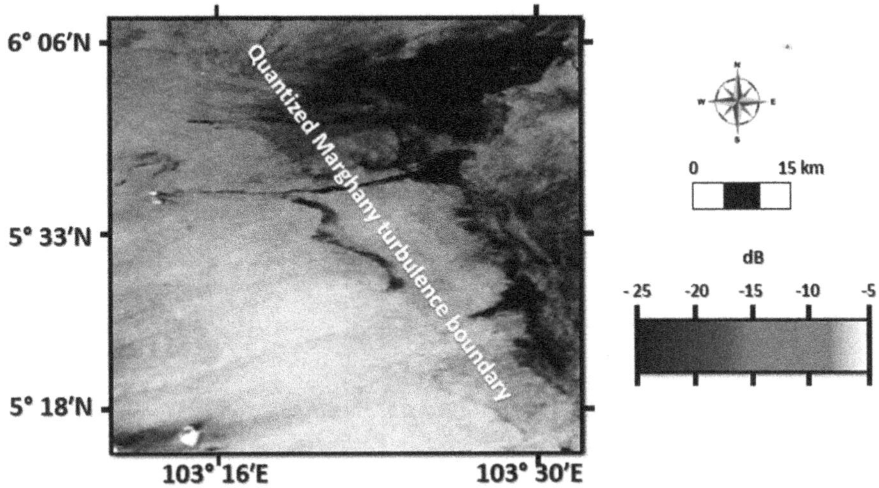

FIGURE 8.17 The emergence of extended spiral patterns along the quantized Marghany turbulence boundary was observed on March 25, 2005.

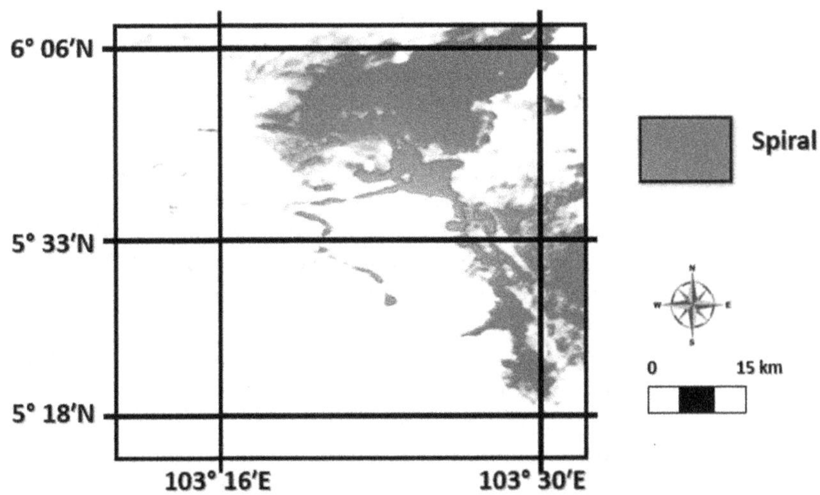

FIGURE 8.18 Automatic detection of complex spiral patterns in S2 data using quantum superposition Marghany distance decomposition.

characteristics within the observed area. The presence of turbulence in this zone might be one of the contributing factors to the lower backscatter, as turbulence can affect the way radar waves interact with the ocean surface (Hasselmann et al., 1985).

However, the introduction of a novel algorithm, the "Quantum Superposition Marghany Distance Decomposition," has proven to be a game-changer. This innovative algorithm enables the automatic detection of spiral pattern flows along the quantized Marghany turbulence boundary (Fig. 8.18). This particular spiral sector stretches across an impressive 42 kilometers along the western boundary of the East Coast of Peninsular Malaysia. The utilization of quantum principles in conjunction with the Marghany distance decomposition has opened up new avenues for understanding and identifying intricate spiral patterns in SAR data, further advancing our comprehension of coastal phenomena in the South China Sea (Marghany, 2021).

A fascinating discovery lies in the correlation between spiral patterns and the internal wave zone along the quantized Marghany turbulence boundary. In the S2 mode data, we observe the intricate interplay of spiral flows within the internal zone, resulting in a complex phenomenon. Remarkably, the quantum_superposition_MDD algorithm exhibits the capability to autonomously distinguish the spiral boundary from the internal wave propagation zone (Fig. 8.19).

The central inquiry revolves around the phenomenon of spirals becoming superimposed on internal wave zones. Internal waves, on the other hand, are subsurface waves that propagate within the ocean layers beneath the surface. They arise from the interactions between different water density layers, often due to temperature and salinity gradients.

The intriguing question arises: why do spirals appear superimposed on internal wave zones? The answer likely lies in the intricate interplay between these phenomena. It

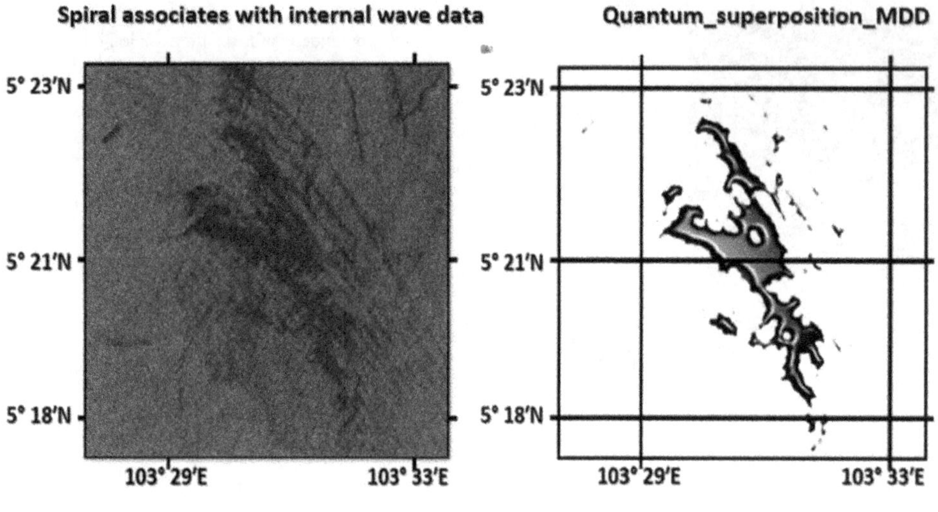

FIGURE 8.19 Automated identification of spiral patterns superimposed on internal waves through the utilization of the quantum superposition Marghany distance decomposition (quantum_superposition_MDD) algorithm.

is conceivable that certain conditions, such as the presence of specific underwater features or the convergence of ocean currents, create an environment where spirals and internal waves coincide. In this understanding, as internal waves travel, they interact with ocean currents. These currents can be horizontal or vertical and are often caused by tidal forces, winds, or other factors. The interaction between the waves and currents can result in the formation of spiral patterns. When internal waves encounter a shear zone (a region where there's a rapid change in velocity between adjacent water layers), consequently, they can become unstable (Gill, 1982). This instability can lead to the development of vortices or spirals. The Coriolis effect, caused by the Earth's rotation, can also influence the direction and rotation of these spirals (Timmermans & Marshall, 2020). These spiral patterns are often observed on the ocean's surface or through remote sensing techniques such as synthetic aperture radar (SAR) imaging. SAR can capture the surface manifestations of these internal processes, revealing the spiral patterns caused by the interaction of internal waves with surface currents.

Understanding this phenomenon requires advanced techniques and algorithms that can separate and analyze the contributions of spirals and internal waves within remote sensing data. This is where innovative approaches such as the quantum superposition Marghany distance decomposition (quantum_superposition_MDD) algorithm come into play. By harnessing the power of quantum mechanics, this algorithm can automatically distinguish and characterize these complex patterns, shedding light on their coexistence and interactions.

Recently, Sentinel-A data recorded the occurrence of spirals off the coastal waters of Peninsular Malaysia, specifically in the Kelantan coastal region, on August 13, 2023. Similarly, the quantum_superposition_MDD algorithm demonstrated its ability to automatically detect short spiral flows, spanning a length of approximately 1 kilometer (Fig. 8.20). These spirals were observed moving in a northeasterly direction toward the offshore area.

The ROC curve illustrates the superior performance of the quantum_superposition_MDD algorithm compared with the K-means clustering algorithm. Specifically, the quantum_superposition_MDD algorithm adeptly identified 95% of spiral flow patterns present in the SAR data (true positive rate) while also accurately recognizing 95% of zones where spiral patterns were absent (true negative rate). The ROC curve for the quantum_superposition_MDD algorithm depicting spiral patterns exhibits a steep ascent toward the upper-left corner of the graph, signifying its high true positive and true negative rates. Consequently, the AUC (area under the curve) in this scenario approaches 1, the highest attainable value, underscoring the superior performance of the quantum_superposition_MDD algorithm in comparison to K-means clustering (Fig. 8.21).

It is evident that the quantum superposition Marghany distance decomposition (quantum_superposition_MDD) algorithm can effectively recognize the spiral pattern as a hyperbolic spiral.

8.11 Theory of quantized spiral turbulence

The fundamental inquiry arising from the aforementioned presentation is whether this chapter can offer an innovative theory concerning spiral flow rooted in the concept of quantization. In the realm of fluid dynamics and the fascinating world of coastal waters, there exists a curious phenomenon that beckons us to delve into its mysteries. It is the enigmatic dance of spirals, those intricate patterns that seem to emerge from the very fabric of the fluid itself. But what if these spirals might hold the secrets of quantum mechanics within them? Imagine a coastal region where wind and tides conspire to create chaos in

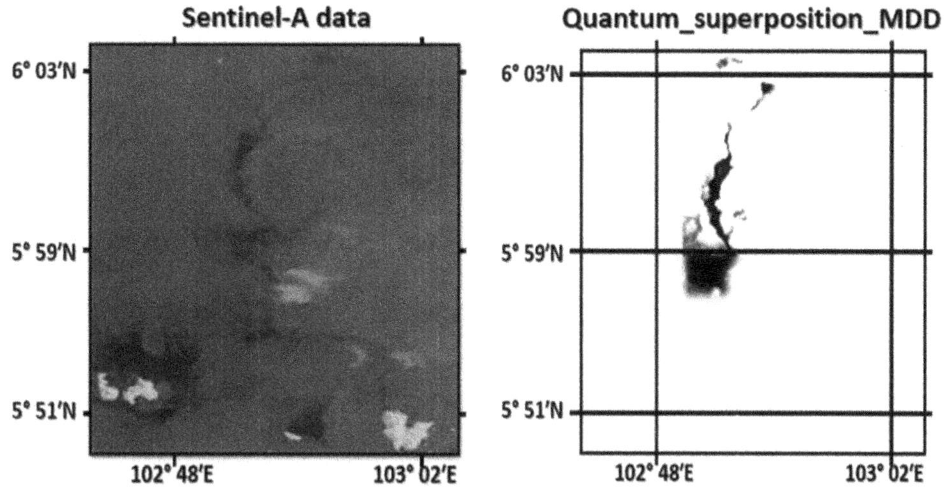

FIGURE 8.20 Recent automatic detection of spiral flow in Sentinel-A data.

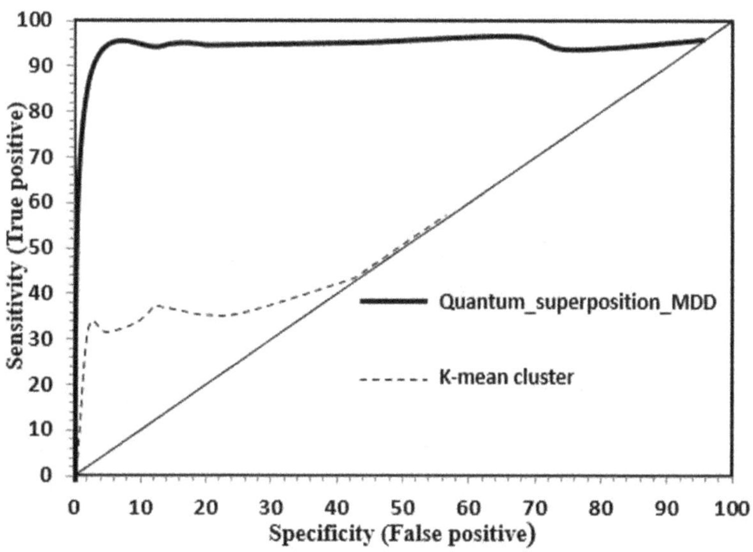

FIGURE 8.21 Accuracy of quantum_superposition_MDD algorithm and K-mean cluster algorithm.

the water. This is where our story unfolds. The Marghany turbulence boundary, as we demonstrate in previous chapters, is a critical juncture where turbulent forces and wave dynamics collide. But it's not a random collision; it's quantized. Fluid elements follow specific, interconnected pathways within this boundary, and that's where the spirals come into play.

These spirals are not just pretty patterns; they are markers, quantum markers. They signify that beneath the surface, the fluid dynamics are not behaving as classical physics would dictate. Instead, they hint at the existence of quantum-like phenomena, like quantized energy levels and wave-particle duality, right there in the turbulent flow. But what sets the stage for these spirals? It's the interplay of wind and tidal forces, the architects of this mesmerizing dance. The wind brings energy into the system, while the tides introduce their periodic rhythms. Within the Marghany turbulence boundary, these forces interact in a quantized manner, leading to the emergence of spiral patterns.

Now, why should we care about these spirals, you might ask? Well, they hold vital environmental clues. The presence and characteristics of spirals tell us about the health of coastal ecosystems, the presence of upwelling zones, and the concentrations of filaments. They are like nature's own indicators, guiding us to hidden treasures beneath the waves. But it doesn't end there. Inspired by this quantum dance of spirals, we've developed Quantum_Superposition_MDD, an algorithm that mimics these quantum principles. It automatically detects and analyzes spiral patterns in SAR imagery, offering us a deeper understanding of our coastal waters and their secrets.

Therefore as we gaze upon these spirals, let us remember that even in the seemingly chaotic world of fluid dynamics, there may be a touch of quantum magic, waiting to be unraveled by curious minds like ours. Needless to say, the "Theory of Quantized Spiral Turbulence (QST)" represents a novel framework for understanding the occurrence and characteristics of spiral patterns in turbulent fluid dynamics, particularly in coastal regions and internal wave zones. QST integrates principles from quantum mechanics, fluid dynamics, and environmental science to elucidate the intriguing phenomena associated with quantized spiral turbulence.

8.12 What makes Quantum_Superposition_MDD a valuable tool for automated spiral detection?

The difference in the sharpness of patterns between the newly developed quantum algorithm (quantum_superposition_MDD) and the classical K-means clustering algorithm can be attributed to several key factors:

1. Quantum Parallelism: Quantum computing harnesses the power of quantum parallelism, allowing it to process and analyze data in parallel across multiple quantum states. This capability enables quantum algorithms to explore a vast number of possibilities simultaneously, which can result in more detailed and precise pattern recognition.
2. Quantum Superposition: Quantum algorithms such as quantum_superposition_MDD leverage the concept of superposition, where quantum bits (qubits) can exist in multiple states at once. This property allows for the simultaneous examination of various combinations of data elements, leading to a more comprehensive understanding of patterns.
3. Quantum Parallelism: Quantum computing harnesses the power of quantum parallelism, allowing it to process and analyze data in parallel across multiple quantum states. This capability enables quantum algorithms to explore a vast number of possibilities simultaneously, which can result in more detailed and precise pattern recognition.
4. Marghany Distance Decomposition (MDD): The inclusion of Marghany distance decomposition in the quantum algorithm provides a unique way to assess the dissimilarity or similarity between data points. By assigning element-specific weights and considering individual contributions to the overall distance, MDD can highlight subtle variations in patterns that might be overlooked by classical clustering techniques.
5. Quantum SWAP gates and Operations: Quantum circuits in algorithms such as quantum_superposition_MDD involve intricate combinations of quantum SWAP gates and operations. These SWAP gates can perform complex transformations on data, enhancing the algorithm's ability to capture intricate details in patterns.
6. Data Dimensionality: Quantum algorithms are well-suited for handling high-dimensional data efficiently. In cases where data points have many dimensions or features, quantum algorithms can effectively reduce the computational complexity and provide clearer insights into patterns.
7. Noise and Data Variability: Quantum algorithms can be more resilient to noise and data variability, as they can adapt and make probabilistic assessments. This adaptability helps in distinguishing true patterns from noise, resulting in sharper and more accurate pattern recognition.
8. Optimization Techniques: Quantum algorithms often employ quantum optimization techniques, such as quantum approximate optimization algorithm (QAOA), to find optimal solutions. These techniques can help refine patterns and yield sharper results.

In summary, the combination of quantum computing principles, Marghany distance decomposition, and quantum-specific capabilities such as superposition and parallelism enables quantum_superposition_MDD to excel in pattern recognition tasks, delivering sharper and more detailed patterns compared with classical K-means clustering algorithms. This makes quantum algorithms particularly promising for applications where pattern recognition and data analysis are critical.

In this chapter, we have introduced a groundbreaking algorithm, Quantum_Superposition_MDD, for the automatic detection of spiral flow patterns in multiSAR data, focusing on the coastal waters of Peninsular Malaysia. This algorithm leverages the power of quantum computing principles, such as superposition and parallelism, to significantly enhance our ability to recognize and understand these intricate patterns.

The research has shown that Quantum_Superposition_MDD surpasses classical K-means clustering algorithms in delivering sharper and more detailed spiral flow patterns. The quantum advantage lies in its capability to efficiently handle high-dimensional data, withstand noise, and explore multiple data combinations simultaneously. This has opened up new horizons for pattern recognition in SAR imagery.

Furthermore, we have unveiled the fascinating association between spiral patterns and internal wave zones along the quantized Marghany turbulence boundary. The algorithm's ability to isolate spiral boundaries from internal wave propagation has contributed to our comprehension of this complex phenomenon. It has also been revealed that spiral patterns are promising indicators of turbulence features, often triggered by the interplay of wind and tidal forces. The prevalence of filament concentrations and surface currents in coastal waters underscores their significance as markers of upwelling phenomena.

In summary, Quantum_Superposition_MDD has emerged as a powerful tool for the automatic detection of spiral patterns in multiSAR data. This research contributes not only to the realm of pattern recognition but also to a deeper understanding of the intricate dynamics of coastal waters. As we move forward, the potential applications of this innovative algorithm in various fields, such as oceanography and environmental monitoring, are both exciting and promising.

References

Ahmad, A., & Dey, L. (2007). A k-mean clustering algorithm for mixed numeric and categorical data. *Data and Knowledge Engineering, 63*(2), 503–527. Available from https://doi.org/10.1016/j.datak.2007.03.016.

Amin, A. High throughput particle separation using differential fermat spiral microchannel with variable channel width (Doctoral dissertation). (2014).

Anand, A., Schleich, P., Alperin-Lea, S., Jensen, P. W. K., Sim, S., Díaz-Tinoco, M., Kottmann, J. S., Degroote, M., Izmaylov, A. F., & Aspuru-Guzik, A. (2022). A quantum computing view on unitary coupled cluster theory. *Chemical Society Reviews, 51*(5), 1659–1684. Available from https://doi.org/10.1039/d1cs00932j.

Bassin, C. J., Washburn, L., Brzezinski, M., & McPhee-Shaw, E. (2005). Sub-mesoscale coastal eddies observed by high frequency radar: A new mechanism for delivering nutrients to kelp forests in the Southern California Bight. *Geophysical Research Letters, 32*(12), 1–4. Available from https://doi.org/10.1029/2005GL023017.

Baumgarten, C., & Farin, G. (1997). Approximation of logarithmic spirals. *Computer Aided Geometric Design, 14*(6), 515–532. Available from https://doi.org/10.1016/S0167-8396(96)00043-X.

Beckman, J. E., Rozas, M., Zurita, A., Watson, R. A., & Knapen, J. H. (2000). Populations of high-luminosity density-bounded H [CSC]ii [/CSC] regions in spiral galaxies: Evidence and implications. *The Astronomical Journal, 119*(6), 2728–2744. Available from https://doi.org/10.1086/301380.

Bertacca, M., Berizzi, F., Mese, E. D. 2005 IEEE Transactions on Geoscience and Remote Sensing 01962892 11 2484–2493, Italy. A FARIMA-based technique for oil Slick and low-wind areas discrimination in sea SAR imagery, 43, Available from https://doi.org/10.1109/TGRS.2005.857622.

Choi, J., Kim, J. 2019 10 ICTC 2019 - 10th International Conference on ICT Convergence: ICT Convergence Leading the Autonomous Future, 10.1109/ICTC46691.2019.8939749 9781728108926, 138–142, Institute of Electrical and Electronics Engineers Inc. South Korea, A tutorial on quantum approximate optimization algorithm (QAOA): Fundamentals and applications, http://ieeexplore.ieee.org/xpl/mostRecentIssue.jsp?punumber = 8932631

Cook, T. A. (1903). Spirals in nature and art. *Nature, 68*(1761), 296. Available from https://doi.org/10.1038/068296a0.

Crooks, G. E. Performance of the quantum approximate optimization algorithm on the maximum cut problem. arXiv, United States arXiv. 23318422. (2018), https://arxiv.org, https://doi.org/10.48550/arxiv.1811.08419.

Delattre, B. M. A., Heidemann, R. M., Crowe, L. A., Vallée, J. P., & Hyacinthe, J. N. (2010). Spiral demystified. *Magnetic Resonance Imaging, 28*(6), 862–881. Available from https://doi.org/10.1016/j.mri.2010.03.036.

Dijkstra, H. A. (2005). *Nonlinear physical oceanography: a dynamical systems approach to the large scale ocean circulation and El Nino* (532). Springer.

Doshi, D., Jain, L., & Gala, K. (2021). Review of the spiral model and its applications. *International Journal of Engineering Applied Sciences and Technology, 5*(12). Available from https://doi.org/10.33564/ijeast.2021.v05i12.053.

Gargett, A. E. (1989). Ocean turbulence. *Annual Review of Fluid Mechanics, 21*(1), 419–451. Available from https://doi.org/10.1146/annurev.fl.21.010189.002223.

Gerkema, T., Zimmerman, J. T. F., Maas, L. R. M., & Van Haren, H. (2008). Geophysical and astrophysical fluid dynamics beyond the traditional approximation. *Reviews of Geophysics, 46*(2). Available from https://doi.org/10.1029/2006RG000220.

Gill, A. E. (1982). *Atmosphere-ocean dynamics* (30). Academic Press.

Gu, M., Weedbrook, C., Menicucci, N. C., Ralph, T. C., & van Loock, P. (2009). Quantum computing with continuous-variable clusters. *Physical Review A., 79*(6). Available from https://doi.org/10.1103/physreva.79.062318.

Habib, Z., & Sakai, M. (2005). Spiral transition curves and their applications. *Scientiae Mathematicae Japonicae., 61*(2), 195–206.

Hall, R. W. (1982). Efficient spiral search in bounded spaces. *IEEE Transactions on Pattern Analysis and Machine Intelligence, 4*(2), 208–215. Available from https://doi.org/10.1109/TPAMI.1982.4767228.

Harary G. Tal A. 2010 Israel 3D Euler spirals for 3D curve completion, Proceedings of the Annual Symposium on Computational Geometry, 393–402, Available from https://doi.org/10.1145/1810959.1811025.

Harary, G., & Tal, A. (2011). The natural 3D spiral. *Computer Graphics Forum, 30*, 2237. Available from https://doi.org/10.1111/j.1467-8659.2011.01855.x, 246, http://www.blackwell-synergy.com/loi/CGF.

Hasselmann, K., Raney, R. K., Plant, W. J., Alpers, W., Shuchman, R. A., Lyzenga, D. R., Rufenach, C. L., & Tucker, M. J. (1985). Theory of synthetic aperture radar ocean imaging: A MARSEN view. *Journal of Geophysical Research, 90*(C3), 4659. Available from https://doi.org/10.1029/jc090ic03p04659.

Holcombe, S. A., Wang, S. C., & Grotberg, J. B. (2016). Modeling female and male rib geometry with logarithmic spirals. *Journal of Biomechanics, 49*(13), 2995–3003. Available from https://doi.org/10.1016/j.jbiomech.2016.07.021, http://www.elsevier.com/locate/jbiomech.

Jeong, I. J., Said, A. R. Logarithmic spirals in 2d perfect fluids. arXiv, South Korea arXiv. 23318422. (2023), Available from: https://arxiv.org. doi: 10.48550/arXiv.2302.09447.

Kelley, M.S., Huff, G. H. (2013) Fluidic tuning of a frequency selective surface based on a four-arm Archimedean spiral IEEE Antennas and PropagationSociety, AP-S International Symposium (Digest) 15223965 468-469 United States https://doi.org/10.1109/APS.2013.6710895.

Khan, S. U., Awan, A. J., Vall-Llosera, G. K-Means clustering on noisy intermediate scale quantum computers. arXiv, Sweden arXiv. 23318422. (2019), https://arxiv.org.

Kim, H., & Johnson, J. T. (2002). Radar images of rough surface scattering: Comparison of numerical and analytical models. *IEEE Transactions on Antennas and Propagation, 50*(2), 94–100. Available from https://doi.org/10.1109/8.997977.

Kommalapati, S., Agrawal, A., & Duryodhan, V. S. (2020). Enhancing miscible fluid mixing by introducing pseudo turbulent flow in golden ratio spiral microchannel. *Industrial and Engineering Chemistry Research, 59*(9), 3784–3793. Available from https://doi.org/10.1021/acs.iecr.9b05276, http://pubs.acs.org/journal/iecred.

Kushal, M. (2023). Comparison of aerodynamic coefficients of tall tapered triangular building modified with golden spiral in plan. *Materials Today: Proceedings*, *72*, 3121–3132. Available from https://doi.org/10.1016/j.matpr.2022.09.516, https://www.sciencedirect.com/journal/materials-today-proceedings.

Lekkas, A. M., Dahl, A. R., Breivik, M., & Fossen, T. I. (2013). Continuous-curvature path generation using Fermat's spiral. *Modeling, Identification and Control: A Norwegian Research Bulletin*, *34*(4), 183–198. Available from https://doi.org/10.4173/mic.2013.4.3.

Lindsley, D. H., Brass, D. J., & Thomas, J. B. (1995). Efficacy-performing spirals: A multilevel perspective. *Academy of Management Review*, *20*(3), 645–678. Available from https://doi.org/10.5465/amr.1995.9508080333.

Lumpkin, R., & Elipot, S. (2010). Surface drifter pair spreading in the North Atlantic. *Journal of Geophysical Research: Oceans*, *115* (C12). Available from https://doi.org/10.1029/2010jc006338.

Marghany, M. (2021). *Nonlinear ocean dynamics: Synthetic aperture radar*. Elsevier.

McLellan, H. J. (1965). *Elements of physical oceanography*. Pergamon Press.

Mellor, G., & Huang, R. X. (1997). Introduction to physical oceanography. *American Journal of Physics*, *65*(10), 1028–1029. Available from https://doi.org/10.1119/1.18716.

Morton, B. R. (1984). The generation and decay of vorticity. *Geophysical & Astrophysical Fluid Dynamics*, *28*(3-4), 277–308. Available from https://doi.org/10.1080/03091928408230368.

Munk, W., Armi, L., Fischer, K., & Zachariasen, F. (2000). Spirals on the sea. *Proceedings of the Royal Society of London. Series A: Mathematical, Physical and Engineering Sciences*, *456*(1997), 1217–1280. Available from https://doi.org/10.1098/rspa.2000.0560.

Munteanu, M. I. (2010). From golden spirals to constant slope surfaces. *Journal of Mathematical Physics*, *51*(7). Available from https://doi.org/10.1063/1.3459064.

Ni, Q., Zhai, X., Wilson, C., Chen, C., & Chen, D. (2021). Submesoscale eddies in the South China Sea. *Geophysical Research Letters*, *48*(6). Available from https://doi.org/10.1029/2020GL091555, http://agupubs.onlinelibrary.wiley.com/hub/journal/10.1002/(ISSN)1944-8007/.

Pannatoni, R. F. (1983). On a general fluid dynamical theory of discrete unstable spiral modes in disk-shaped galaxies. *Geophysical & Astrophysical Fluid Dynamics*, *24*(3), 165–212. Available from https://doi.org/10.1080/03091928308209065.

Picado, J. (2009). Seashells: The plainness and beauty of their mathematical description. *Loci.*, *1*. Available from https://doi.org/10.4169/loci003294Portugal, http://mathdl.maa.org/mathDL/23/?pa = content &sa = viewDocument&nodeId = 3294.

Pickard, G. L., Emery, W. J. Descriptive physical oceanography: an introduction. (1961).

Polezhaev, A. (2019). Spirals, their types and peculiarities. *Russian Federation Frontiers Collection*, *1072*. Available from https://doi.org/10.1007/978-3-030-05798-5_4, http://www.springer.com/series/5342.

Rocha, L. A. M., Thorne, L., Wong, J. J., Cartwright, J. H. E., & Cardoso, S. S. S. (2022). Archimedean spirals form at low flow rates in confined chemical gardens. *Langmuir*, *38*(21), 6700–6710. Available from https://doi.org/10.1021/acs.langmuir.2c00633, http://pubs.acs.org/journal/langd5.

Sapkota, S., Bhattarai, A., Bashyal, H. P., Nepal, U. 2019 Design, CFD analysis and modelling of archimedean-spiral type wind turbine. Proceedings of IOE Graduate Conference 6 563-571.

Scherr, T., Quitadamo, C., Tesvich, P., Park, D. S. W., Tiersch, T., Hayes, D., Choi, J. W., Nandakumar, K., & Monroe, W. T. (2012). A planar microfluidic mixer based on logarithmic spirals. *Journal of Micromechanics and Microengineering*, *22*(5). Available from https://doi.org/10.1088/0960-1317/22/5/055019, http://iopscience.iop.org/0960-1317/22/5/055019/pdf/0960-1317_22_5_055019.pdf, United States.

Shamloo, A., Mozhdehbakhsh Mofrad, Y., Safari, M., & Naseri, T. (2022). Comparison of logarithmic, elliptic, and conical helical spiral for isolation of circulating tumor cells based on inertial method. *Physics of Fluids*, *34*(9). Available from https://doi.org/10.1063/5.0100079, http://scitation.aip.org/content/aip/journal/pof2.

Shaydulin R. Alexeev Y. 2019 10th International Green and Sustainable Computing Conference, IGSC 2019 9781728154169, Institute of Electrical and Electronics Engineers Inc. United States Evaluating Quantum Approximate Optimization Algorithm: A Case Study, http://ieeexplore.ieee.org/xpl/mostRecentIssue.jsp?punumber = 8954738 https://doi.org/10.1109/IGSC48788.2019.8957201.

Siddique, N., & Adeli, H. (2014). Spiral dynamics algorithm. *International Journal on Artificial Intelligence Tools*, *23*(06), 1430001. Available from https://doi.org/10.1142/s0218213014300014.

Singh, V., Ojha, S. P., & Singh, L. K. (1999). Modal behavior, cutoff condition, and dispersion characteristics of an optical waveguide with a core cross section bounded by two spirals. *Microwave and Optical Technology Letters*, *21*(2), 121–124. Available from https://doi.org/10.1002/(SICI)1098-2760(19990420)21:2 < 121::AID-MOP11 > 3.0.CO;2-J.

Skalak, R., Dasgupta, G., Moss, M., Otten, E., Dullemeijer, P., & Vilmann, H. (1982). Analytical description of growth. *Journal of Theoretical Biology*, *94*(3), 555–577. Available from https://doi.org/10.1016/0022-5193(82)90301-0.

Sohn, S. I. (2016). Self-similar roll-up of a vortex sheet driven by a shear flow: Hyperbolic double spiral. *Physics of Fluids*, *28*(6). Available from https://doi.org/10.1063/1.4953780, http://scitation.aip.org/content/aip/journal/pof2.

Soltanian, S., & Beigzadeh, R. (2023). Computational fluid dynamics and fuzzy logic for modeling conical spiral heat exchangers. *Chemical Engineering and Technology*, *46*(4), 747–755. Available from https://doi.org/10.1002/ceat.202200488, http://onlinelibrary.wiley.com/journal/10.1002/(ISSN)1521-4125.

Stuart, J. T. (1967). On finite amplitude oscillations in laminar mixing layers. *Journal of Fluid Mechanics*, *29*(3), 417–440. Available from https://doi.org/10.1017/S0022112067000941.

Sánchez-Reyes, J. (2002). p-Bézier curves, spirals, and sectrix curves. *Computer Aided Geometric Design*, *19*(6), 445–464. Available from https://doi.org/10.1016/S0167-8396(02)00130-9.

Thorpe, S. A. (2004). Recent developments in the study of ocean turbulence. *Annual Review of Earth and Planetary Sciences*, *32*, 91–109. Available from https://doi.org/10.1146/annurev.earth.32.071603.152635.

Thorpe, S. A. (2007). *An introduction to ocean turbulence An Introduction to Ocean Turbulence* (pp. 1–291). Cambridge University Press. Available from http://doi.org/10.1017/CBO9780511801198, 10.1017/CBO9780511801198.

Timmermans, M. L., & Marshall, J. (2020). Understanding Arctic Ocean circulation: A review of ocean dynamics in a changing climate.

Journal of Geophysical Research: Oceans, 125(4). Available from https://doi.org/10.1029/2018JC014378, http://agupubs.onlinelibrary.wiley.com/agu/jgr/journal/10.1002/(ISSN)2169-9291/.

Tomesh, T., Gokhale, P., Anschuetz, E. R., & Chong, F. T. (2021). Coreset clustering on small quantum computers. *Electronics, 10*(14), 1690. Available from https://doi.org/10.3390/electronics10141690.

Tomiyasu, K. (1978). Tutorial review of synthetic-aperture radar (SAR) with applications to imaging of the ocean surface. *Proceedings of the IEEE, 66*(5), 563–583. Available from https://doi.org/10.1109/PROC.1978.10961.

Vlander, L. M. H., Carlström, A., & Askne, J. (2007). EtTect of frost flowers, rough saline snow and slush on the ERS-l SAR backscatter of thin Arctic sea-ice. *International Journal of Remote Sensing, 16* (17), 3287–3305. Available from https://doi.org/10.1080/01431169508954631.

Wang, X., Liu, Z., Wang, B., Cai, Y., & Wan, Y. (2022). Vortices degradation and periodical variation in spiral micromixers with various spiral structures. *International Journal of Heat and Mass Transfer., 183*, 122168. Available from https://doi.org/10.1016/j.ijheatmasstransfer.2021.122168.

Weber M. Alexa M. Müller W. 2001 Proceedings of the IEEE Symposium on Information Visualization, 7-13, undefined, Visualizing time-series on spirals.

Weiman, C. F. R., & Chaikin, G. (1979). Logarithmic spiral grids for image processing and display. *Computer Graphics and Image Processing, 11*(3), 197–226. Available from https://doi.org/10.1016/0146-664X(79)90089-3.

Wereszczynski K. Michalczuk A. Josinski H. Polanski A. 2018 Quantum computing for clustering big datasets, Applications of Electromagnetics in Modern Techniques and Medicine, PTZE 2018 9781728111773 25-28. Institute of Electrical and Electronics Engineers Inc. Poland, http://ieeexplore.ieee.org/xpl/mostRecentIssue.jsp?punumber = 8491723 10.1109/PTZE.2018.8503109.

Wu, J. Z., Ma, H. Y., & Zhou, M. D. (2007). *Vorticity and vortex dynamics*. Springer Science & Business Media.

Wu, J. Z., & Wu, J. M. (1998). Boundary vorticity dynamics since Lighthill's 1963 article: Review and development. *Theoretical and Computational Fluid Dynamics, 10*(1-4), 459–474. Available from https://doi.org/10.1007/s001620050077, http://link.springer-ny.com/link/service/journals/00162/index.htm.

Wu Q. He X. Hintz T. 2004 Virtual spiral architecture. Proceedings of the International Conference on Parallel and Distributed Processing Techniques and Applications, PDPTA'04 1 399-405, Australia.

Xiao, F., Honma, Y., & Kono, T. (2005). A simple algebraic interface capturing scheme using hyperbolic tangent function. *International Journal for Numerical Methods in Fluids., 48*(9), 1023–1040. Available from https://doi.org/10.1002/fld.975.

Xiao, Y., Ribe, N. M., Zhang, Y., Pan, Y., Cao, Y., & Shum, H. C. (2022). Generation of Fermat's spiral patterns by solutal Marangoni-driven coiling in an aqueous two-phase system. *Nature Communications, 13*(1). Available from https://doi.org/10.1038/s41467-022-34368-5, https://www.nature.com/ncomms/.

Xu, F., Li, X., Wang, P., Yang, J., Pichel, W. G., & Jin, Y. Q. (2015). A backscattering model of rainfall over rough sea surface for synthetic aperture radar. *IEEE Transactions on Geoscience and Remote Sensing, 53*(6), 3042–3054. Available from https://doi.org/10.1109/TGRS.2014.2367654.

Xu, F., Zhang, D., Zhang, L., Bu, S., Huang, H., Ji, Y., Huo, G., Ding, H., & Xu, W. (2022). Influence of key geometry parameters on heat transfer and flow resistance properties of toothed internal spiral piece cracking furnace tube. *Applied Thermal Engineering, 207*, 118170. Available from https://doi.org/10.1016/j.applthermaleng.2022.118170.

Yin, B., Yue, W., Muhtasim Fuad Sohan, A. S. M., Zhou, T., Qian, C., & Wan, X. (2021). Micromixer with fine-tuned mathematical spiral structures. *ACS Omega, 6*(45), 30779–30789. Available from https://doi.org/10.1021/acsomega.1c05024.

Zhang, J., & Chen, N. Z. (2022). Vortex-induced vibration of flexible riser transporting high-speed spiral flow in deep-sea mining. *Ocean Engineering, 261*. Available from https://doi.org/10.1016/j.oceaneng.2022.112084, http://www.journals.elsevier.com/ocean-engineering/.

Zhao, H., Gu, F., Huang, Q.-X., Garcia, J., Chen, Y., Tu, C., Benes, B., Zhang, H., Cohen-Or, D., & Chen, B. (2016). Connected fermat spirals for layered fabrication. *ACM Transactions on Graphics, 35* (4), 1–10. Available from https://doi.org/10.1145/2897824.2925958.

Zhi, P., Wu, Y.-C., Yang, Q., Kong, X., & Xiao, J. (2022). Effect of spiral blade geometry on 3D-printed concrete rheological properties and extrudability using discrete element modeling. *Automation in Construction, 137*, 104199. Available from https://doi.org/10.1016/j.autcon.2022.104199.

Zilman, G., Zapolski, A., & Marom, M. (2015). On detectability of a ship's Kelvin wake in simulated SAR images of rough sea surface. *IEEE Transactions on Geoscience and Remote Sensing, 53*(2), 609–619. Available from https://doi.org/10.1109/TGRS.2014.2326519.

Chapter 9

Fractal dimension algorithm for automatic detection of oceanic front and turbulence structural in MultiSAR data

In the previous chapter, we explored the complexities surrounding spiral-induced turbulence in ocean waters, where the nonlinearity of interactions between the ocean surface and radar signals presented significant challenges. However, there is a promising avenue for advancing our understanding and automating the detection of these spiral patterns in synthetic aperture radar (SAR) images, thanks to a novel algorithm known as "quantum_superposition_MDD."

This chapter aims to delve deeper into the intricacies of coastal water structures, particularly focusing on understanding the dynamics of water fronts. Furthermore, we have delved into and expanded upon a pioneering theoretical framework known as the "quantized Marghany bound front state," initially introduced by Marghany in 2021. This framework holds immense promise in illuminating the intricate dynamics of ocean fronts, offering a fresh perspective on our understanding of these complex phenomena. One intriguing question that arises is whether these water fronts can encompass vortices, surface wake, eddies, and spiral patterns. This chapter seeks to address this question by introducing an innovative SAR image processing technique designed to detect and analyze the intricate structures of turbulence and fronts within coastal waters, especially when dealing with radar data affected by speckles using quantum fractal algorithm.

9.1 What is a magic of fractal art?

The term "fractal" was introduced by Benoit Mandelbrot in 1975, derived from the Latin word "fractus," meaning "broken" or "fractured." Mandelbrot pioneered a mathematical framework for describing irregular shapes, patterns, and phenomena that classical geometry could not adequately address. Fractals are readily observed in the natural world, including coastlines, clouds, mountain ranges, trees, leaves, snowflakes, crystals, lightning, sea shells, and more. In contemporary science, fractals serve as a fundamental concept in chaos theory and dynamical systems theory (DST). Physicists' interest in this field was greatly influenced by Benoit Mandelbrot's groundbreaking work, particularly his renowned book, "The Fractal Geometry of Nature" (1983) (Mandelbrot & Wheeler, 1983). In this context, Mandelbrot noted that "turbulence involves many fractal facets" and asserted that "Fractals. make it possible to apply the technique of self-similarity to the geometry of turbulence" (Raizer, 2021; Tarboton et al., 1988). Indeed, over the ensuing decades, it became evident that turbulence serves as the underlying source of numerous fractal, self-similar, and nonlinear scaling properties. The topological and geometrical aspects of turbulent flow, encompassing chaotic patterns, coherent structures, and attractors, can be effectively explored through the concept of fractal dimension.

Fractals are created using mathematical algorithms that generate self-replicating patterns across different scales. These algorithms typically rely on recursive equations, iterative processes, and self-similarity. The mathematical elegance of fractals is a captivating aspect appreciated by both artists and observers (Raizer, 2021).

One famous example of a fractal is the Mandelbrot set, defined by the following recursive equation:

$$z_{n+1} = z_n^2 + c \tag{9.1}$$

where z_n represents a complex number at iteration *n, and* c is a constant complex number. Therefore the iteration continues until $|zn|$ exceeds a predetermined threshold or a maximum number of iterations is reached. Indeed, Eq. (9.1) generates the intricate and visually stunning Mandelbrot set (Fig. 9.1), exhibiting self-replicating patterns at varying levels of magnification. To ascertain the inclusion of a point "C" within the Mandelbrot set, a recursive process is employed. This process initiates with an initial condition, Z(0) = 0, and proceeds with iterative

Synthetic Aperture Radar Image Processing Algorithms for Nonlinear Oceanic Turbulence and Front Modeling. DOI: https://doi.org/10.1016/B978-0-443-19155-8.00009-X

calculations using the Eq. (9.1). The critical criterion for membership in the Mandelbrot set is the behavior of the magnitude of Z(n) as the number of iterations "n" increases. If this magnitude remains finite, signifying that it does not undergo unbounded growth, then the point "C" is deemed a member of the Mandelbrot set. Conversely, if the magnitude of Z(n) becomes unbounded during the iterative process, the point "C" is classified as outside the Mandelbrot set.

Now, let's explore the concept of infinite complexity within this context. The Mandelbrot set exhibits self-similarity, which means that as one can zoom in on various regions of the set, one sees similar patterns repeated at different scales (Fig. 9.2).

Mathematically, this self-similarity arises from the recursive nature of the equation, where each point's fate (whether it belongs to the set or not) depends on the outcome of the equation applied to neighboring points. This recursive process creates the intricate patterns observed in the Mandelbrot set. In this understanding, recursive equations are mathematical equations that are defined in terms of themselves. They describe a process or relationship where a function or value is calculated based on one or more previous instances of the same function or value. Recursive equations are often used to solve problems that can be broken down into smaller, similar subproblems.

$$F(n) = F(n-1) \tag{9.2}$$

In this equation, F(n) represents the value or function at a given level n. It's defined in terms of F(n − 1), which is the value or function at the previous level or iteration. The recursion continues until it reaches a base case, which provides a direct value without further recursion (Fig. 9.3). A classic example of a recursive equation is the Fibonacci sequence (Fig. 9.4):

$$F(n) = F(n-1) + F(n-2) \tag{9.3}$$

In this case, each term in the sequence is defined in terms of the two previous terms, and the recursion continues until it reaches the base cases F(0) = 0 and F(1) = 1. Recursive equations can describe various phenomena, from fractal patterns to population growth, and they are a fundamental concept in both mathematics and computer science. The concept of infinite complexity, consequently, is reflected in the infinite depth of the Mandelbrot set. No matter how deeply you zoom into the set, you will continue to encounter new patterns and structures, each with its own level of complexity. The recursive nature of the equation

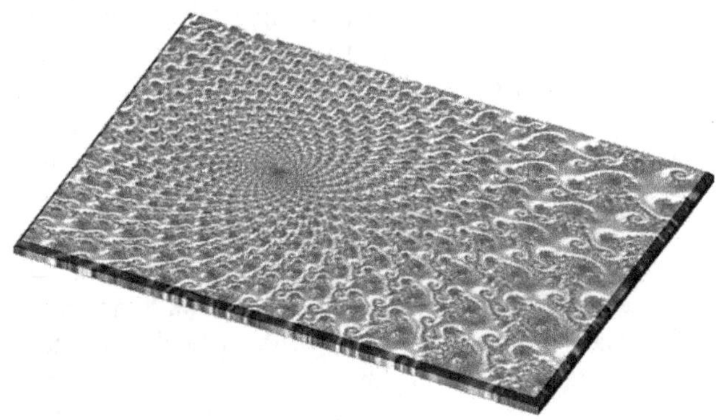

FIGURE 9.1 Mandelbrot set fractal.

FIGURE 9.3 Recursive fractals.

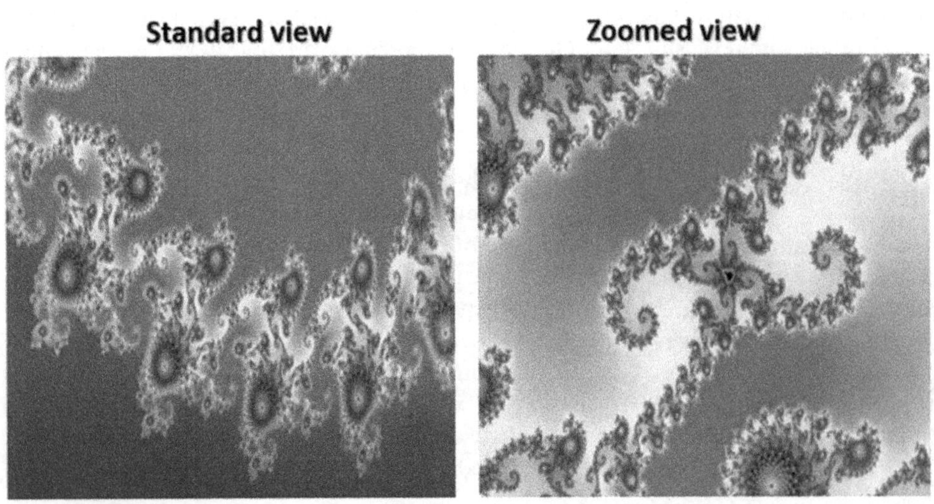

FIGURE 9.2 The Mandelbrot set exhibits self-similarity when zoomed in.

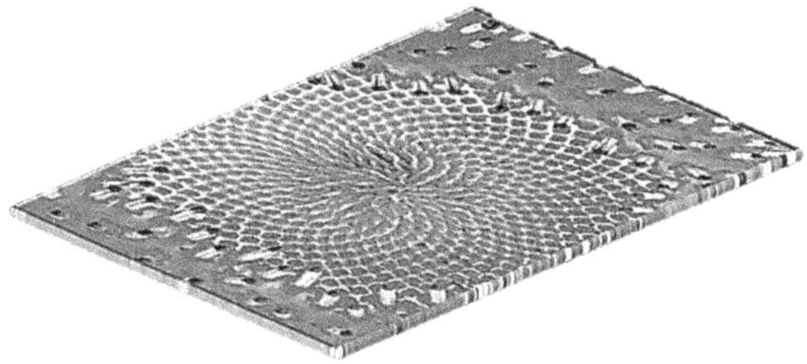

FIGURE 9.4 Fibonacci sequence in nature.

ensures that there is no limit to how deeply one can explore the set. Hence, at each level of magnification, the Mandelbrot set reveals new details and patterns. These emergent patterns are the result of the recursive iterations and the interactions between neighboring points. Mathematically, this emergence of new patterns is a consequence of the iterative process continuing indefinitely.

Fractal art often produces shapes and patterns that are not found in nature or traditional art forms. The unique esthetics of fractals include the fractal dimension, self-similarity, and the Mandelbrot set, among others. These characteristics make fractal art distinct and captivating. It can be said that the magic of fractal art lies in its mathematical foundation, infinite complexity, and its place in the realm of algorithmic art. It continues to captivate and inspire scientists with its blend of mathematics and artistic expression.

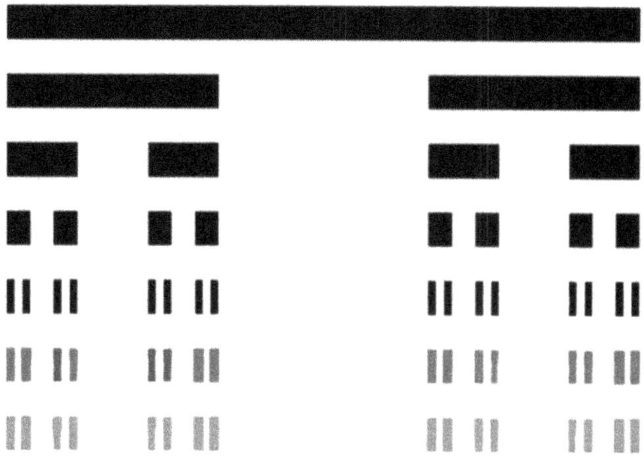

FIGURE 9.5 Cantor set fractal pattern.

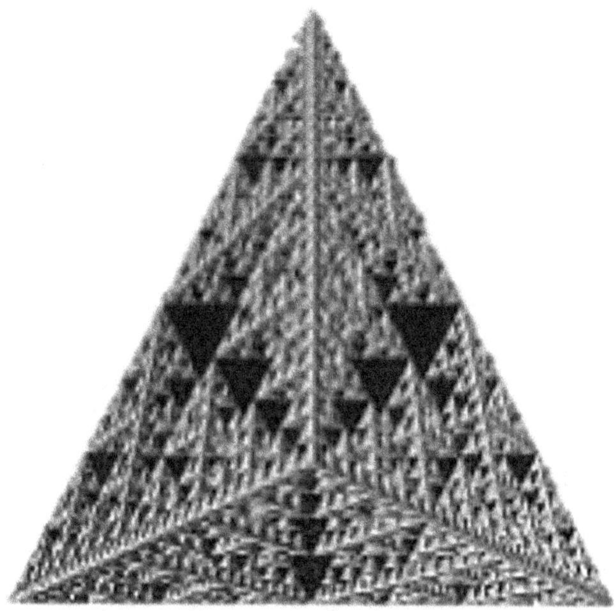

FIGURE 9.6 Sierpinski triangle fractal pattern.

9.2 Classical fractals

The concept of a fractal is universally recognized as a geometric form or pattern distinguished by a remarkable attribute referred to as self-similarity, which becomes apparent at various levels of scale. Put simply, when closely scrutinizing a fragment of a fractal, one can discern similarities to the entire fractal or specific portions of it. This innate self-replicating characteristic endures whether you zoom in or out on the fractal, unveiling analogous shapes or patterns across a spectrum of magnification levels.

Some of the most renowned examples of classical fractals include the Cantor set (Fig. 9.5), the Sierpinski Triangle (Fig. 9.6), the Koch Snowflake (Fig. 9.7), the Dragon Curve (Fig. 9.8), the Julia set (Fig. 9.9), and the Mandelbrot set. These geometrical fractals are constructed within a specific class using recursive equations.

The shape is repeatedly divided into numerous parts following the same scaling rule, with each part bearing a resemblance to the shape of the entire structure. This remarkable property is closely associated with self-similarity. Consequently, a fractal represents a geometrical entity that exhibits similarity to itself across all scales. It's worth noting that there are also "nonscaling" fractals within this classification. One notable example is the Julia set, which is named after the French mathematician Gaston Julia, known for his exploration of this set back in 1918. The Julia set is defined by a family of complex quadratic polynomials represented by Eq. (9.1). In the case of a Julia set with a significantly small value of c, the capacity dimension is (Turbulence, strange attractors and chaos, 1995):

$$D_f = 1 + \frac{|c|^2}{4\ln 2} + O\left(|c|^3\right) \tag{9.4}$$

Fractals, as described by Falconer (2005), exhibit several defining features. These include the ability to reveal intricate details even at minute scales, surpassing the limits of conventional Euclidean space with their irregularity. Fractals demonstrate self-similarity, whether with exact

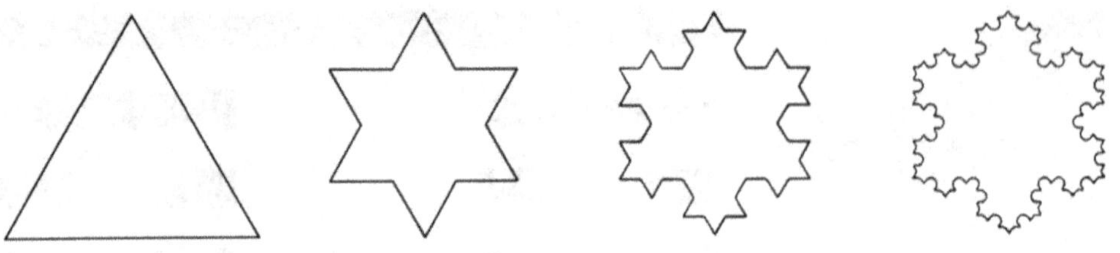

FIGURE 9.7 Koch Snowflake fractal pattern.

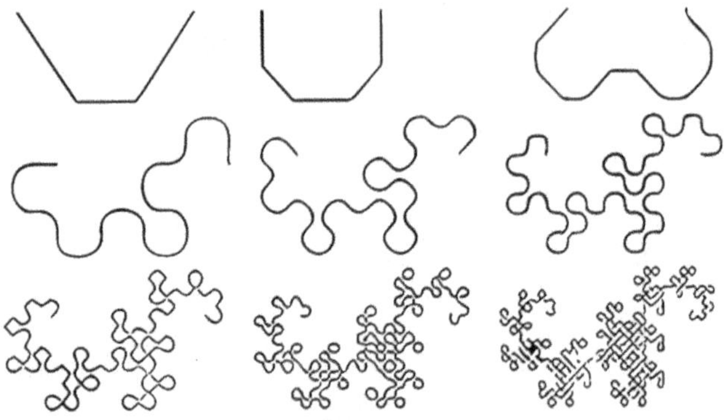

FIGURE 9.8 Dragon curve.

FIGURE 9.9 Julia set fractal pattern.

precision or stochastic variations, and they display scale invariance. Moreover, their Hausdorff dimension surpasses their topological dimension, all while being defined in a simple and recursive manner. Fractals, therefore, can be categorized into two groups: deterministic and stochastic (or random) fractals. Deterministic fractals, such as the Julia set and Mandelbrot set, are constructed using precise scaling transformations based on the same rules in each iteration. Random fractals, on the other hand, employ random scaling transformations and find applications in modeling and characterizing natural phenomena and stochastic processes (Raizer, 2021). One fundamental model of random fractals is an extension of Brownian motion known as fractional Brownian motion (Falconer, 2005).

9.3 Statistical self-similarity

Statistical self-similarity in fractals is typically described using statistical measures and probability distributions. There isn't a specific equation for statistical self-similarity, but it involves analyzing how statistical properties of a fractal object or dataset change as you zoom in or out. In this sense, probability density function (PDF) describes the probability distribution of data values in a fractal. It can be represented by the equation:

$$P(X = x) = lim_{n\to\infty} n_I n^{-1} \tag{9.5}$$

where $P(X = x)$ is the probability of finding a specific data value x, n_i is the number of data points with the value x, and n is the total number of data points. Statistical self-similarity, therefore, often involves scaling laws that describe how statistical properties change with scale. For example, if N is the number of data points in a given region of a fractal, and r is the size of that region, a scaling law might be:

$$N(r) \propto r^{-D_f} \tag{9.6}$$

where D_f is the fractal dimension. In this view, the fractal dimension is a measure of how a fractal object fills space at different scales. It can be calculated using various methods, such as box counting, information dimension, or correlation dimension. On the other hand, a stochastic process denoted as {X(t), t ≥ 0} is considered to be self-similar if its finite-dimensional distributions satisfy the scaling relation:

$$X(t_1,\ t_2,\ ..,\ t_n) \stackrel{d}{=} t_1^H X(t_1,\ t_2,\ ..,\ t_n), \tag{9.7}$$

where $X(t_1,t_2,..,t_n)$ represents the joint distribution of the process at time points $t_1,t_2,..,t_n$. Therefore t_1^H denotes the scaling factor with exponent H, where H is a parameter that characterizes the degree of self-similarity. Furthermore, $=^d$ indicates that the finite-dimensional distributions of the process on both sides of the equation are equivalent in a statistical sense. In essence, this definition states that a stochastic process is self-similar if, when you observe its behavior at different time points, one can scale the observations by a factor to obtain t_1^H, a distribution that is statistically equivalent to the original distribution. The parameter H quantifies the extent of self-similarity within the process.

Now, suppose we have fractional Brownian motion with $H = 0.7$. If we observe the process at two different time points, say $t_1 = 2$ and $t_2 = 4$, we can use the scaling relation to find the joint distribution:

$$X(2,4) \stackrel{d}{=} 2^{0.7} X(1,4), \tag{9.8}$$

This means that the joint distribution at $t_1 = 2$ and $t_2 = 4$ is statistically equivalent to $2^{0.7}$ times the joint distribution at $t_1 = 1$ and $t_2 = 4$. In this perspective, the Hurst exponent H quantifies how the process maintains its statistical properties as you scale the time interval. A value of $H = 0.5$ corresponds to standard Brownian motion, which is not self-similar. Values of H different from 0.5 indicate varying degrees of self-similarity, with higher H values indicating stronger self-similarity.

9.4 Fractal dimension

The fractal dimension of a geometric object is a mathematical measure that quantifies its complexity, irregularity, and self-similarity at different scales. There are several methods to compute the fractal dimension of an object, and the choice of method depends on the nature of the object and the data available. The box-counting dimension, often referred to as the Minkowski–Bouligand dimension, represents one of the most straightforward approaches for determining the fractal dimension of a 2D entity. This method entails overlaying the object with a grid of progressively smaller boxes and tallying the quantity of boxes required to encompass the object at each scale. The equation for this process is expressed as follows:

$$D(r) = \lim_{r \to 0} \frac{\log(N(r))}{\log(r^{-1})} \tag{9.9}$$

In Eq. (9.9), D(r) is the fractal dimension; r is the size of the boxes; and N(r) is the number of boxes needed to cover the object at scale ϵ. The definition provided in Eq. (9.9) implies that $N(\epsilon)$ scales as $r^{(-D(r))}$ for small values of r, and the limit emphasizes that the dimension D(r) remains finite. The fractal dimension, denoted as D (r), resides within the bounds of the entire Euclidean n-dimensional space and can manifest as an integer or non-integer value (fractional). This dimension D(r), as articulated in Eq. (9.9), is technically referred to as a "capacity" dimension and was initially introduced by Kolmogorov in 1958 (Falconer, 2005; Raizer, 2021). In this sense, the box-counting dimension is determined through a log − log diagram, which entails plotting log (N) against $\log(r^{-1})$ while varying cube sizes (r). Subsequently, the fractal dimension D(F) is computed as the slope of a linear regression on this plot. In practical or numerical experiments, the limit as r approaches 0 in Eq. (9.9) is utilized for this purpose (Raizer, 2021).

It's important to note that, by mathematical definition (Falconer, 2005), the Hausdorff dimension and the box-counting dimension are not generally equal. However, for strictly self-similar sets like classical fractals, these dimensions can be equivalent. The box-counting method, despite this distinction, remains a straightforward and valuable tool for analyzing fractals found in various natural phenomena, scaling processes, and quasistochastic datasets (Mandelbrot & Wheeler, 1983). This method is particularly practical for remote sensing data analysis. In this view, the distinction between the Hausdorff dimension (H) and the box-counting dimension (D(r)) can be represented mathematically as follows:

$$H = \lim_{\varepsilon \to 0} \left[\frac{\log(N(\varepsilon))}{\log(\varepsilon^{-1})} \right] \tag{9.10}$$

where N(ε) represents the minimum number of ε-diameter balls (or covers) required to cover a given set. ε and r are positive real numbers that denote the size of the balls or cubes, respectively. As ε or r approaches zero, the covers become finer and finer, and the dimensions are calculated based on how the number of covers changes. However, the limits as ε or r approaches zero ensure that the dimensions are calculated at the finest level of detail. Both the Hausdorff dimension and the box-counting dimension, therefore, describe the scaling properties of fractal sets, but they are not necessarily equal for all sets. The box-counting dimension is often used because it is more practical and easier to compute, especially for real-world datasets (Raizer et al., 1994).

Alternatively, when dealing with a single complex geometric object, such as an atmospheric cloud, a turbulent spot, or a vortex, the fractal dimension (Ds) can be determined using a concept known as the perimeter − area relation:

$$P\alpha\left(\sqrt{A}\right)^{D_s}, \tag{9.11}$$

In this equation, P represents the measured perimeter, and A represents the area of a single object with a distinct boundary (Mandelbrot & Wheeler, 1983). This relation is commonly employed in the fractal analysis of individual objects observed in images. For instance, both fractal algorithms Eqs. (9.9) and (9.10) have been put to the test and used for the geometric analysis of foam and whitecap coverage in ocean aerial photography (Raizer et al., 1994).

9.5 Multifractal systems

A multifractal system represents a fascinating extension of the concept of fractals. While traditional fractals are characterized by a single exponent, known as the fractal dimension, multifractals take complexity to the next level. In these systems, a solitary exponent is no longer sufficient to capture their intricate behavior. Instead, we require a continuous spectrum of exponents, which we aptly refer to as the "singularity spectrum."

To put it simply, multifractals are like the virtuosos of complexity. They don't rely on a single musical note; instead, they produce an entire symphony of exponents. Each exponent in this spectrum tells a unique story about the system's behavior at different scales. It's akin to having a diverse set of lenses to view the same landscape, revealing hidden details and nuances that a single lens could never capture. Multifractals, in all their complexity, offer a richer and more comprehensive understanding of the intricate dynamics they represent (Tarboton et al., 1988; Turbulence, strange attractors and chaos, 1995).

In a multifractal system, we introduce the concept of a local power law, denoted by the variable "s." This local power law signifies how the behavior around any point within the system is governed. For a given point "x" within the multifractal system, we express its local power law as:

$$f(x) \sim |x - x_0|^s \tag{9.12}$$

here $f(x)$ represents the quantity or measurement of interest at point "x"; x_0 is a reference point around which we are examining the local behavior; and $|x - x_0|$ is the distance between the point of interest "x" and the reference point x_0. s is, therefore, the exponent that characterizes the local power law behavior. This exponent can vary continuously across the multifractal landscape, forming the singularity spectrum.

The singularity spectrum, often denoted as $f(\alpha)$, provides insights into how these exponents "s" are distributed across the multifractal system. It describes the multifaceted nature of the system by associating each exponent "s" with a corresponding strength or measure, denoted as α. This relationship can be expressed as:

$$f(\alpha) = Strength(s - \alpha) \tag{9.13}$$

The key distinction that sets multifractals apart from their classical fractal counterparts lies in the presence of this unbroken continuum of exponents, elegantly encapsulated within what we term the "singularity spectrum." This nuanced feature serves as the gateway to unraveling the intricate and multifarious dynamics inherent to multifractal systems. On the other hand, singularity spectrum can be expressed as:

$$s(\vec{x} + \vec{a}) - s(\vec{x}) \sim a^{h(\vec{x})}. \tag{9.14}$$

The term "singularity exponent," denoted as $h(\vec{x})$, is a fitting descriptor for its role in characterizing the localized nature of singularity or regularity surrounding the point $\vec{x}$. In real-world applications, understanding the multifractal behavior of a physical system denoted as X doesn't typically involve a direct characterization of its singularity spectrum D(h). Instead, data analysis provides insights into the multiscaling exponents $\zeta(q)$, $q \in \mathbb{R}$, where q belongs to the real numbers. Multifractal signals, in general, adhere to a scale invariance property, resulting in power-law behaviors for multiresolution quantities, contingent upon their scale denoted as a. These multiresolution quantities, symbolized as $T_X(a)$, may encompass local averages within boxes of size a, gradients spanning a distance of a, wavelet coefficients at scale a, and more, depending on the specific object under investigation (Bunde et al., 1990; Falconer, 2005; Macé et al., 2019). For multifractal entities, a prevalent observation is a global power-law scaling pattern of the following form:

$$\langle T_X(a)^q \rangle \sim a^{\varsigma(q)} \tag{9.15}$$

Utilizing the multifractal formalism, it can be demonstrated that, under suitable assumptions, there exists a relationship between the singularity spectrum D(h) and the multiscaling exponents ζ(q) achieved through a Legendre transform. While determining D(h) necessitates intricate local data analysis, often involving challenging and numerically unstable calculations, estimating ζ(q) relies on statistical averages and linear regression techniques applied to log − log plots. Once ζ(q) is known, one can derive an estimate of D(h) using a straightforward Legendre transform (Mandelbrot & Wheeler, 1983; Tarboton et al., 1988). This can be represented as:

$$D(h) = \max[qh - \varsigma(q)] \tag{9.16}$$

In this equation, h represents the singularity exponent, and q is a scaling parameter. By maximizing qh − ζ(q) over various q values, we can obtain an estimate of D(h). The Legendre transform provides a method to relate the singularity spectrum (ζ(q)) and the fractal dimension (D(h)) in multifractal analysis. In multifractal systems, stochastic processes such as multiplicative cascades are commonly used for modeling. The ζ(q) values are statistically interpreted as they describe how the distributions of multiresolution quantities $T_X(a)$ change when we vary the scale parameter "a" from larger to smaller values. This variation is referred to as statistical intermittency, indicating a deviation from Gaussian models. Let's denote ζ(q) as the scaling exponent that characterizes the statistical properties of the evolution of $T_X(a)$ distributions, where "a" represents the scaling factor. In multifractal analysis, this scaling exponent is used to understand how the moments of these distributions change as we vary the scaling factor. It is defined as follows:

$$\zeta(q) = lim_{a \to 0}[a]^{-1} \log\left(\sum_{i=1}^{N} |T_X(a_i)|^q\right) \tag{9.17}$$

In this equation, ζ(q) represents the scaling exponent for the qth moment; "a" is the scaling factor, and the limit (\lim) indicates that "a" approaches 0; N is the total number of scaling levels considered; and $|T_X(a_i)|$ represents the multiresolution quantities at different scales "a_i." In this sense, $\sum_{i=1}^{N}$ calculates the sum of the qth powers of $|T_X(a_i)|$ for i ranging from 1 to N.

9.6 Calculating multifractal scaling using box counting

Multifractal spectra can be extracted through a process known as box counting applied to digital images. This procedure involves a systematic examination of how the pixels within the image are distributed. The key concept revolves around the probability (P) of observing a certain number of pixels (m) within a particular box (i), which changes with variations in the box size (ϵ). This variability is characterized by an exponent (α), and it undergoes changes across different regions of the image. It's important to note that for monofractals, the exponent remains relatively constant across the entire set.

$$P_{[i,\epsilon]} \propto \epsilon^{-\alpha_i} \therefore \alpha_i \propto \frac{\log P_{[i,\epsilon]}}{\log \epsilon^{-1}} \tag{9.18}$$

Let N_ϵ present total count of boxes that contain more than zero pixels, calculated for each specific box size (ϵ). Therefore the total mass or sum of pixels in all boxes for this particular box size (ϵ) is given by:

$$M_\epsilon = \sum_{i=1}^{N_\epsilon} m_{[i,\epsilon]} \tag{9.19}$$

where $m_{[i,\epsilon]}$ represents the number of pixels or mass in any box, i, at size ϵ. Hence, the likelihood of this mass $P_{[i,\epsilon]}$ at i relative to the total mass for a given box size is represented as:

$$P_{[i,\epsilon]} = \frac{m_{[i,\epsilon]}}{M_\epsilon} \tag{9.20}$$

Let's consider Q as a set of arbitrary values that it can be used as exponents to alter implemented SAR dataset. Mathematically, this can be expressed as:

$$I_{(Q)_{[\epsilon]}} = \sum_{i=1}^{N_\epsilon} P^{Q}_{[i,\epsilon]} \tag{9.21}$$

Eq. (9.21) represents the summation of all mass probabilities that have been altered by raising them to the power of Q for this specific box size. When Q is set to 1, Eq. (9.21) evaluates to 1, which is the typical sum of all probabilities. However, when Q is set to 0 ($Q = 0$), every term within the summation becomes equal to 1. Consequently, the sum becomes equivalent to the number of boxes that were counted, denoted as N_ϵ. This mathematical expression helps us understand how altering the probabilities by exponentiating them with Q influences the summation outcome, which varies depending on the value of Q.

Hence, the pivotal question that arises is: How does the mass probability, after being distorted within a specific box, compare with the distorted sum of all probabilities across all boxes at this particular size? This crucial comparison can be mathematically illustrated as follows:

$$\mu_{(Q)_{[i,\epsilon]}} = \frac{P^{Q}_{[i,\epsilon]}}{I_{(Q)_{[\epsilon]}}} \tag{9.22}$$

These distortion equations serve the purpose of examining the behavior of the set as it undergoes scaling, resolution, or division into a sequence of ε-sized segments, each subjected to the distortion by (Guarneri & Mantica, 1994; Macé et al., 2019; Raizer, 2021). This exploration aims to uncover various values for the dimension of the set, as expressed in the following equation:

$$I_{(Q)_{[\epsilon]}} \alpha \epsilon^{\tau(Q)} \tag{9.23}$$

A notable characteristic of Eq. (9.21) is its dependence on the scale raised to the exponent τ in. Hence, a range of values for τ(Q) can be derived by analyzing the slopes of the regression line, which relates the logarithm of Eq. (9.21) to the logarithm of ε for each Q, using the following formula:

$$\tau_{(Q)} = \lim_{\epsilon \to 0} \left[\frac{\log I_{(Q)_{[\epsilon]}}}{\log \epsilon} \right] \tag{9.24}$$

This equation, denoted as τ(Q), represents the limit as ε (epsilon) approaches zero of the logarithm of a specific expression involving Q and ε. The expression calculates the slope of the logarithmic relationship between a quantity denoted as I(Q)_ε and ε. This equation is used to characterize how the multifractal set behaves when it is scaled down into smaller and smaller pieces and then distorted by different values of Q. The resulting τ(Q) values provide insights into the dimensionality and scaling properties of the set under different distortions. When considering the generalized dimension:

$$D_{(Q)} = \lim_{\epsilon \to 0} \left[\frac{\log I_{(Q)_{[\epsilon]}}}{\log \epsilon^{-1}} \right] (1 - Q)^{-1} \tag{9.25}$$

where

$$D_{(Q)} = \frac{\tau_{(Q)}}{Q - 1}, \tag{9.25.1}$$

$$\tau_{(Q)} = D_{(Q)}(Q - 1), \tag{9.25.2}$$

$$\tau_{(Q)} = \alpha_{(Q)} Q - f_{(\alpha_{(Q)})}. \tag{9.25.3}$$

Therefore the estimation of $\alpha_{(Q)}$ involves determining the slope of the regression line for log A(ε, Q) versus log ε, is given by:

$$A_{\epsilon,Q} = \sum_{i=1}^{N_\epsilon} \mu_{i,\epsilon_Q} P_{i,\epsilon_Q} \tag{9.26}$$

In this equation, A(ε, Q) is the sum we want to calculate; μ_{i,ϵ_Q} represents a specific mass or number of pixels in the ith box at size ε, which has been distorted or modified by a factor Q. Additionally, P_{i,ϵ_Q} represents the probability associated with the mass in the ith box at size ε, which has also been distorted or modified by a factor Q. Certainly, the equation essentially involves adding together the altered masses μ_{i,ϵ_Q}, each multiplied by their respective adjusted probabilities P_{i,ϵ_Q}, across all boxes of size ε. This summation allows us to compute the value of A(ε, Q). The mean value τ(Q) is estimated by calculating the slope of the logarithmic regression line for $\tau(Q)_{[\varepsilon]}$ versus ε, where:

$$\tau(Q)_{[\epsilon]} = \frac{\sum_{i=1}^{N_\epsilon} P_{[i,\epsilon]}^{Q-1}}{N_\epsilon} \tag{9.27}$$

This expression entails summing up the Q^{-1} power of the probabilities ($P_{[i,\varepsilon]}$) for each box of size ε and dividing it by the total number of boxes, N_ε. This calculation allows us to determine the mean value τ(Q) as the slope of the regression line when plotted against ε (Mandelbrot & Wheeler, 1983; Tarboton et al., 1988; Turiel et al., 2006).

9.7 What is the relationship between turbulence and multifractals?

This crucial question can guide us in understanding how to implement multifractal algorithms to trace turbulence in SAR data. The significance of this question lies in its potential to shed light on the practical utilization of multifractal algorithms in the context of turbulence detection within SAR (synthetic aperture radar) data. It essentially encourages us to delve into the effective deployment of these algorithms for the identification and examination of turbulence patterns inherent in SAR datasets. By addressing this inquiry, we can devise methodologies and approaches that harness the principles of multifractals to enrich our comprehension of turbulence phenomena as observed through SAR technology. Consequently, this endeavor holds substantial ramifications across diverse domains, encompassing remote sensing, oceanography, and meteorology, all of which rely heavily on the accurate detection and characterization of turbulence for various research and application purposes.

In the realm of fluid dynamics, turbulent flow is distinguished by the irregular and chaotic motion of fluid particles. Turbulent flow represents a nonlinear dynamic system characterized by an exceedingly vast number of degrees of freedom. Unlike laminar flow, turbulent flow is marked by pronounced mixing, intermittency, and irregularities, rendering its behavior intricate and challenging to articulate mathematically.

The assertion that turbulence, particularly the velocity field, must exhibit fractal or multifractal properties arises from its stochastic and multiscale nature. However, from an empirical standpoint, conclusive experimental evidence supporting this assertion remains elusive. Nevertheless, there exist several widely recognized and pragmatic multifractal models that find utility in the statistical analysis and simulation of turbulent phenomena.

In this context, much attention is directed toward multifractal spectra, denoted as f(á), which are computed for time series or imaging data. Essentially, two fundamental multifractal concepts are contemplated: the first is rooted in self-similar solutions derived from the Navier–Stokes equations, while the second is grounded in a phenomenological multifractal approach.

The initial concept hinges on the presumption of scale invariance, often referred to as self-similarity, within the framework of the unsteady incompressible 2D or 3D Navier–Stokes equations governing isotropic turbulence. This approach enables us to explore turbulence through the lens of multifractality. The Navier–Stokes equations, which govern the behavior of turbulent fluid flow, exhibit a formal invariance under a specific scaling transformation. This scaling transformation is mathematically expressed as follows:

$$\frac{\delta \vec{u}}{\delta t} + (\nabla.\vec{u})\vec{u} = -\rho^{-1}\nabla P + v\nabla^2\vec{u} + \vec{F}(\vec{x},t), \quad \nabla.\vec{u} = 0 \tag{9.28}$$

$$\vec{x} \to \lambda\vec{x}, \vec{u} \to \lambda_u^{\vec{h}}, \quad t \to \lambda^{h-1}t, \quad v \to \lambda^{h+1}v, \rho \to \lambda^{h'}\rho, \varepsilon \to \lambda^{3h-1}\varepsilon \tag{9.29}$$

Here, λ represents the scaling factor applied to the spatial coordinates x. This invariance implies that the behavior of turbulent flow is self-similar across different scales, a fundamental assumption in this concept. Mathematically, the scaling transformation implies that if u(x, t) is a solution to the Navier–Stokes equations, then it follows that u(λx, t) is also a solution, where u represents the velocity field, x denotes spatial coordinates, and t represents time. This concept essentially explores the implications of this scale invariance in terms of multifractality and how it relates to the behavior of turbulent flow described by the Navier–Stokes equations.

Under the assumption that energy dissipation, denoted as "ε," is homogeneous in space and time, it implies that the scaling exponent "h" is not arbitrary but should be fixed to "h = h′ = 1/3," which corresponds to the K41 theory. In the context of turbulence theory, the K41 theory, or Kolmogorov's 1941 theory, is a fundamental theory that describes the scaling behavior of turbulence in the inertial subrange. According to this theory, the energy dissipation rate ε scales with the wave number k as $\varepsilon \propto k^{(4/3)}$. This relationship leads to the conclusion that the scaling exponent "h" is equal to 1/3. While the K41 theory assumes idealized conditions of homogeneous and isotropic turbulence, real-world turbulence often exhibits intermittency, which is a form of spatial variability not considered in K41. In such cases, multifractal models provide a more comprehensive description of turbulence by accounting for the varying degrees of fluctuations at different scales. There is a connection between the scaling laws of the K41 theory and multifractal models. Both involve scaling exponents that describe how turbulent quantities change with scale (Meneveau & Sreenivasan, 1987; Nelkin, 1990; Turbulence, strange attractors and chaos, 1995). However, multifractal models offer a broader framework for characterizing the multifractal spectra, which can capture more complex behaviors beyond the K41 predictions.

There exist approximate solutions, therefore, for the unsteady incompressible Navier–Stokes equations in the context of fully developed turbulence as the kinematic viscosity (ν) approaches zero (indicating very turbulent flow). Additionally, it notes that wavelet-based multifractal (MF) analysis has been developed and utilized for studying turbulence, with reference to the work by Massopust in 2016 (Massopust, 2016). This implies that in the regime of fully developed turbulence, traditional methods of solving the Navier–Stokes equations may not be practical due to the complexity of turbulence. Instead, researchers have turned to approximate solutions that offer insights into turbulent behavior as ν becomes very small.

9.8 Wavelet-based multifractal analysis for turbulence

In wavelet-based multifractal analysis for turbulence, several equations and mathematical relationships are used to characterize the multifractal properties of turbulent flows. In this view, the mathematical expression for the wavelet transform can be written as follows:

$$W(j,k) = \int f(t)\Psi_{j,k}(t)dt \tag{9.30}$$

This equation represents the wavelet transform, where *W(j,k)* is the wavelet coefficient at scale *j* and position *k*. It decomposes the velocity signal *f(t)* into different scales and positions using a wavelet basis function *Ψj,k(t)*. The statistical moments of the wavelet coefficients at scale *j* and position *k*, *therefore*, can be mathematically expressed as follows:

First moment:

$$\langle |f(j,k)| \rangle = N^{-1} \sum_{k=1}^{N} |W(j,k)| \tag{9.31}$$

Second moment:

$$\langle |f(j,k)|^2 \rangle = N^{-1} \sum_{k=1}^{N} |W(j,k)|^2 \tag{9.32}$$

Third moment;

$$\langle |f(j,k)|^3 \rangle = N^{-1} \sum_{k=1}^{N} |W(j,k)|^3 \tag{9.33}$$

Fourth moment:

$$\langle |f(j,k)|^4 \rangle = N^{-1} \sum_{k=1}^{N} |W(j,k)|^4 \tag{9.34}$$

In these equations, *f(j,k)* represents the wavelet coefficients at scale *j* and position *k; N* is the total number of wavelet coefficients; and *W(j,k)* is the wavelet coefficient value at scale *j* and position *k*. These moments provide statistical measures of the wavelet coefficients, capturing their behavior and distribution at a specific scale and position. They are useful for analyzing the multifractal properties of signals and data. In this sense, these moments provide information about the distribution of turbulence intensity at different scales (Arneodo et al., 2008; Goncalves et al., 1998; Kestener & Arneodo, 2003).

The scaling exponents can be expressed as follows:

$$h(j) = \frac{d\log\langle |W(j,k)| \rangle}{d\log\epsilon} \tag{9.35}$$

$$h(2j) = \frac{d\log\langle\langle |W(j,k)|^2 \rangle}{d\log\epsilon} \tag{9.36}$$

$$h(3j) = \frac{d\log\langle\langle |W(j,k)|^3 \rangle}{d\log\epsilon} \tag{9.37}$$

$$h(4j) = \frac{d\log\langle\langle |W(j,k)|^4 \rangle}{d\log\epsilon} \tag{9.38}$$

Equations from (9.35) to (9.38) represent how the statistical moments scale with respect to the scale parameter ϵ. The scaling exponents *h(j)* provide insights into the multifractal properties of turbulence. In this perspective, the multifractal spectrum can be mathematically expressed as:

$$f(\alpha) = \tau(q) - qh(q) - 1 \tag{9.39}$$

This equation relates the singularity strength (α) to the scaling exponent (*h(q)*). The multifractal spectrum *f*(α) characterizes how turbulence intensity is distributed across different scales. The multifractal spectrum is often visualized as a plot of the Hölder exponent (*h*) against the singularity strength (α) (Table 9.1). This visualization provides a clear representation of the multifractal properties of turbulence. The Hölder exponent calculation, therefore, is a crucial step in analyzing fractal turbulence. This calculation is performed using the following equation:

$$h = \frac{\log(|W(j,k)|)}{\log(2)} - \frac{\log(|W(j+1,k)|)}{\log(2)} \tag{9.40}$$

TABLE 9.1 Comprehensive pseudocode for a wavelet-based multifractal algorithm used in turbulence analysis.

```
# Load SAR data
# Define parameters max_scale = 5 wavelet_name = 'db2' scales = np.arange(1, max_scale +
1) holder_exponents = []
# Iterate through different scales for scale in scales: # Perform wavelet transform
wavelet_coefficients = pywavelets.wavedec(turbulence_data, wavelet_name, level=scale)
 # Calculate local singularity strengths singularity_strength =
CalculateSingularityStrength(wavelet_coefficients)
 # Store singularity strength in the holder exponent array
holder_exponents.append(singularity_strength)
# Analyze the holder exponents (e.g., determine the multifractal spectrum)
multifractal_spectrum = AnalyzeHolderExponents(holder_exponents)
# Visualize the results
```

Here's what this equation does: it computes the Hölder exponent (h), which serves as a measure of local regularity or singularity within the turbulence. This exponent is determined by examining the relative change in wavelet coefficients between two adjacent scales, denoted as j and $j+1$. The Hölder exponent provides insights into how the turbulence behaves at a specific location and scale, helping researchers understand its intricate structure and variations (Kestener & Arneodo, 2003, 2004; Macé et al., 2019; Massopust, 2016).

The second, phenomenological concept of multifractals involves the utilization of multifractal measures through the analysis of structure functions. Let's denote the Eulerian velocity field as $\vec{u}$, and consider a set of Lagrangian trajectories of fluid particles moving within this field. Lagrangian structure functions of higher order can be defined as follows:

$$S(n,r) = \langle |\delta u|^n \rangle \alpha C r^{\varsigma(n)} \tag{9.41}$$

here S(n, r) represents the Lagrangian structure function of order n (Table 9.2), which depends on the separation distance r between fluid particles, and ς(n) is a single scaling exponent. Therefore δu signifies the difference in the velocity field $u(x+r)$ and $u(x)$ along the fluid particle trajectory. The angle brackets ⟨⟩ denote an ensemble average over all possible fluid particle trajectories. In essence, this equation quantifies the statistical properties of velocity fluctuations at different orders (n) and separation distances (r) within a turbulent flow. It plays a crucial role in understanding the multifractal properties of turbulence, particularly in terms of its intermittency and scaling behavior.

In the realm of turbulent flow, we often delve into the fascinating world of structure functions to understand how velocity fluctuations behave across different scales. The second-order structure function, denoted as $S_2(r)$, provides a valuable insight. It exhibits a power-law behavior, specifically $S_2(r) \propto C_2\epsilon^{2/3}r^{2/3}$, where ϵ represents the turbulence energy dissipation rate, and r signifies the scale of interest. Now, what truly makes this intriguing is the scaling exponent, denoted as $\zeta(n)$, which characterizes how these structure functions evolve with various orders, represented by n. In the case of the second-order structure function, $2n = 2$. When $\zeta(n)$ equals $n/3$, we are in the realm of the famous Kolmogorov scaling law. This means that at this particular scaling, the behavior follows the Kolmogorov scaling law without any additional corrections (Bunde et al., 1990; Du et al., 1995; Lien & D'Asaro, 2002; Nelkin, 1990; Sawford & Yeung, 2011).

Now, here's where the real magic happens. This scaling exponent, $\zeta(n)$, has a profound connection with the fractal dimension, $D(F)$, of the turbulence. In fact, the relationship is quite elegant: $\zeta(n) = n/3 + (3 - D(F))\ (1 - n/3)$. This equation links the scaling exponent to the fractal dimension, revealing a deep connection between the statistical properties of turbulence and its underlying fractal structure.

Therefore in essence, when we specialize to $n = 2$, we're talking about the power-law energy spectrum of three-dimensional turbulence in the inertial range. This spectrum, as dictated by the Kolmogorov scaling law, plays a fundamental role in our understanding of turbulent flows, shedding light on the intricate dance of energy across different scales. Let's delve into the mathematical expression for the power-law energy spectrum of three-dimensional (3D) turbulence in the inertial range. When we specialize to $n = 2$, we are specifically looking at the

TABLE 9.2 Pseudocode for Lagrangian structure function of order n.

```
# Load SAR image data sar_image = LoadSARImageData()
# Define Lagrangian structure function order 'n' n = 2 # You can choose the desired order
# Define spatial lag distances lags = [1, 2, 4, 8, 16] # Choose appropriate lag distances for
your analysis
# Initialize an empty array to store structure function values structure_function = []
# Iterate through lag distances for lag in lags: # Calculate the structure function for each lag
sf_value = CalculateStructureFunction(sar_image, lag, n)
# Append the result to the structure function array structure_function.append(sf_value)
# Analyze the structure function values (e.g., calculate turbulence parameters)
turbulence_parameters = AnalyzeStructureFunction(structure_function)
# Print or store the turbulence parameters for further analysis
# Define functions for structure function calculation and analysis as needed def
CalculateStructureFunction(image, lag, order): # Implement your method for calculating
the structure function here # Use 'image,' 'lag,' and 'order' as input parameters # Return
the calculated structure function value
def AnalyzeStructureFunction(sf_values): # Implement your method for analyzing the
structure function values here # Use 'sf_values' as the input array # Return relevant
turbulence parameters or information
```

second-order structure function, denoted as E(k). Here's the power-law expression for E(k) in this context:

$$E(k) \propto C_l \in^{2/3} k_l^{-5/3} (k_l)^{3-D/3} \tag{9.42}$$

In the context of turbulence and the Kolmogorov scaling theory, the length-scale ℓ represents the size of the largest eddies within the Kolmogorov inertial interval. This interval is a specific range of scales within a turbulent flow where energy is transferred from large eddies to smaller ones. Now, when we mention the "K41 theory," we are referring to the Kolmogorov 1941 theory, which is a fundamental theory in the field of turbulence. In the context of this theory, when the fractal dimension D (F) (used to characterize the geometry of the turbulent structure) equals 3, it corresponds to the classical Kolmogorov theory. In this case, the energy spectrum $E(k)$, which describes how energy is distributed across different wavenumbers k in the turbulent flow, follows the well-known Kolmogorov's law: $E(k) \propto k^{-5/3}$.

However, the situation changes when the fractal dimension $D(F)$ is greater than 1 but less than 3, denoted as $1 < D(F) < 3$. In this scenario, the energy spectrum given by Eq. (9.42) in your text becomes steeper than the classic Kolmogorov $k^{-5/3}$ spectrum. This means that the distribution of energy across wavenumbers behaves differently than what Kolmogorov's theory predicts. The exact nature of this deviation depends on the specific value of $D(F)$ within the range $1 < D(F) < 3$. This reflects the diverse and complex nature of turbulence in various scenarios.

It is important to emphasize that the (multi)fractal measures described here rely on a phenomenological self-similar approach rather than exact solutions of the Navier–Stokes equations. This approach is considered a qualitative representation of turbulence dynamics. However, it's worth noting that direct experimental support for multifractal behavior in turbulence is relatively limited and has primarily been conducted in controlled environments such as laboratory wind tunnels.

In real-world observations, the detection and analysis of multifractal turbulent structures, whether in two-dimensional (2D) or three-dimensional (3D) geometries, are made possible through high-resolution imagery and topological processing. In earlier experiments, the focus was primarily on flow visualization. Extracted trajectories, boundaries, patterns, surfaces, and volumes are subjected to analysis using fractal-based methods, along with the consideration of the Lyapunov characteristic exponent for dynamical systems.

Multifractal measures serve as valuable tools for capturing the scaling and dynamics of turbulent flow. Therefore efficient criteria for identifying turbulent characteristics involve tracking variations in the fractal dimension and/or multifractal spectrum at different stages of turbulent flow. Turbulent coherent structures, often exhibiting attributes such as fractal-like vortexes, clusters, spots, intrusions, and multiphase flows, contribute to the complex tapestry of multifractal behavior in these systems.

9.9 Automatic determination of turbulence front types in MultiSAR data

The application of the wavelet-based multifractal algorithm to ERS-2 data, specifically acquired along the coastal waters of Koto Baharu, Kelentan, Malaysia, on September 22, 1999, reveals intriguing insights into the local current patterns and backscatter fluctuations. The observations depict an irregular flow pattern characterized by fluctuations in backscatter ranging from -24 dB to -16 dB. Notably, the dominant feature within this dataset is a distinct front exhibiting a notably high backscatter reading of -16 dB (Fig. 9.10).

In this view, the month of September signifies a transitional phase within the climatic cycle, bridging the gap between the waning southwest monsoon season and the impending onset of the northeast monsoon, which typically takes hold in November. During the specific date of ERS-2 data acquisition, notable meteorological features come into play, particularly in the form of curl wind stress. The curl wind stress recorded in September, therefore, exhibits a magnitude of approximately 4 (10^{-7} N/m^3) along the coastline of Koto Baharu, Kelentan, Malaysia. This stress vector predominantly assumes a southeasterly direction, indicating the prevailing wind patterns at the time of data capture by ERS-2. These wind dynamics play a significant role in shaping local environmental conditions and are instrumental in driving various natural processes, underlining the significance of their precise measurement and analysis. Farther away from the coastline of Kota Bharu, Kelantan, Malaysia, the curl wind stress diminishes to a level of -2 (10^{-7} N/m^3) (Fig. 9.11). It is important to note that positive wind stress curl contributes to the Ekman divergence effect during the summer season, while a negative wind stress curl induces convergence, particularly during the winter months. This phenomenon underscores the significance of wind stress curl in influencing local oceanic conditions and circulation patterns in the region.

To gain insights into the sea water temperature dynamics, a series of real-time, in situ measurements were carried out employing a CTD instrument. These measurements were systematically conducted at three distinct time intervals: just before the ERS-2 overpass, during the actual ERS-2 overpass, and immediately after it, with an hour of separation between each measurement point. Analysis of the obtained data has unveiled an intriguing isothermal pattern that extends from the water's surface to a depth of 17 meters. This pattern signifies a gradual temperature variation, amounting to a mere 0.6°C, within this specific depth range.

What is particularly noteworthy is the conspicuous intrusion of colder water into the region, manifesting as temperatures ranging from 29.2°C to 29.8°C. In stark contrast, the shallower waters closer to the shore maintain a relatively uniform temperature of 30°C (Fig. 9.12). The existence of a thermocline layer, typically indicative of a rapid temperature transition with increasing depth, appears to be notably subdued in this particular geographical area.

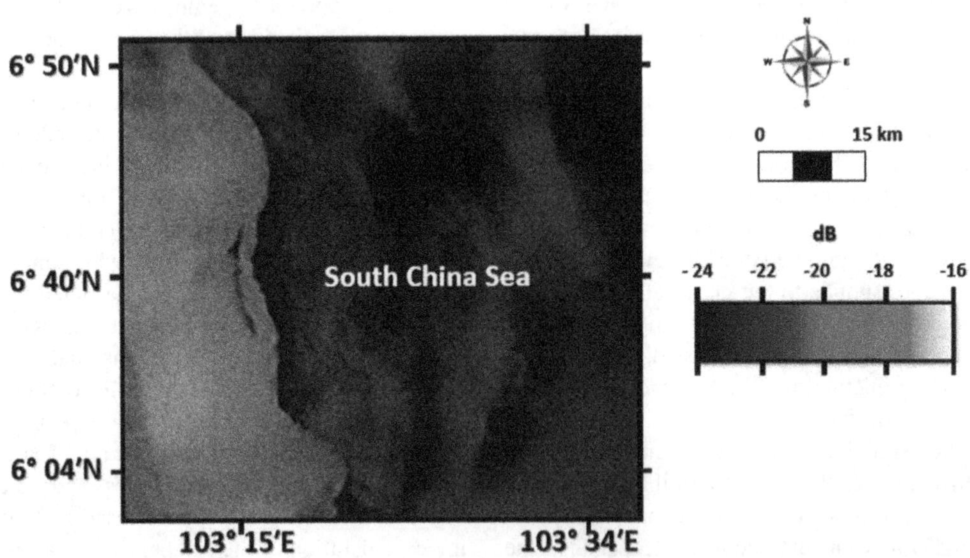

FIGURE 9.10 Detection of sharp front along coastal waters of Koto Baharu, Kelentan, Malaysia, on September 22, 1999, in ERS-2 image.

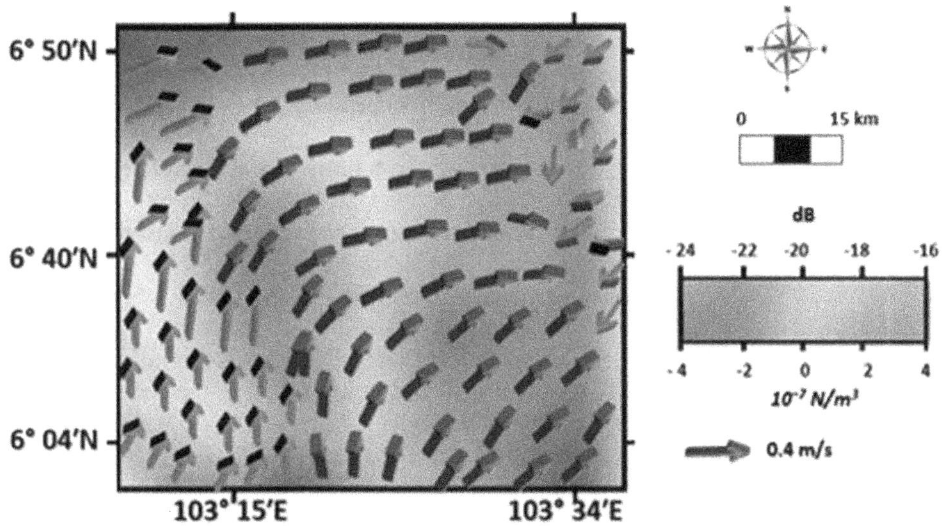

FIGURE 9.11 Curl wind stress observed across the front in ERS-2 data.

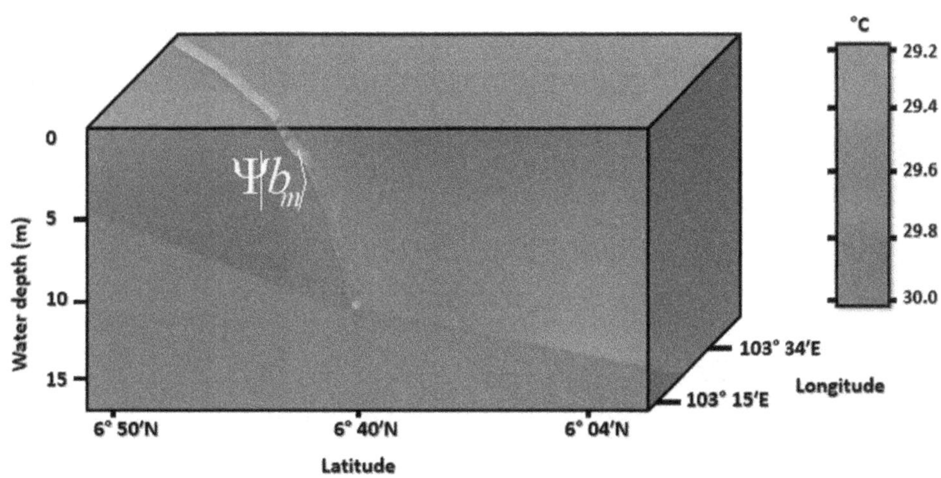

FIGURE 9.12 3-D isothermal pattern observed during the ERS-2 overpass.

This provides compelling evidence for the existence of quantized Marghany front-bound states in the coastal waters of Malaysia. This attenuation can plausibly be attributed to the presence of significant turbulence and thorough mixing phenomena, particularly evident within the shallow water depth. These observations align consistently with the geographical location of the ERS-2 data acquisition.

However, it is important to note that, at this stage, the ERS-2 data alone do not provide a comprehensive understanding of the precise nature or structural characteristics of this front. As discussed in previous chapters, the conventional analysis falls short in elucidating the finer details of such phenomena, primarily due to the inherent challenges posed by speckle impacts, which can obscure the underlying structure. Hence, a pivotal question emerges: Can the advanced wavelet-based multifractal algorithm, as demonstrated earlier, offer a more comprehensive and nuanced characterization of the front's type and underlying structure, even in the presence of the significant speckle effects? This inquiry underscores the potential of this algorithm to unveil hidden details within complex coastal dynamics, promising a deeper understanding of these intricate environmental processes.

The initial data processing in box counting involves dividing an image into smaller boxes to identify the locations of quantized Marghany front-bound states in SAR (synthetic aperture radar) images. This process helps in analyzing the distribution and characteristics of these states within the image (Fig. 9.13).

On the other hand, when it comes to fractal box counting, using a kernel window size of 10 x 10 pixels and lines can yield excellent detection of the quantized Marghany front-bound states $\Psi|b_m>$ compared with window sizes of 9 x 9 and 11 x 11 pixels and lines, respectively (Fig. 9.14). In simpler terms, a 10 x 10 window strikes a good balance between capturing intricate details and ensuring a smooth analysis. Smaller windows, such as 9 x 9, may capture more details but can be influenced by noise, making the results less reliable. Conversely, larger windows, such as 11 x 11, may yield blur results but could overlook subtle features.

However, the singularity spectrum (h) offers superior clustering compared with fractal box counting. Consequently, the singularity spectrum algorithm can effectively identify distinct quantized Marghany front-bound states in SAR data, with a maximum value of 0.7 (Fig. 9.15). This aligns with the maximum backscatter of -16dB depicted in Fig. 9.10.

In this context, the singularity spectrum (h) serves as a tool for discerning the characteristics of the quantized Marghany front-bound states. It reveals the presence of distinct features within these states, including secluded fronts, occluded fronts, and convection cells (Fig. 9.16). Notably, the singularity spectrum for convection cells exhibits a lower value of 0.4 compared with occluded and secluded fronts (Fig. 9.15). Conversely, the quantized Marghany front-bound states exhibit the highest singularity spectrum value, reaching 0.7.

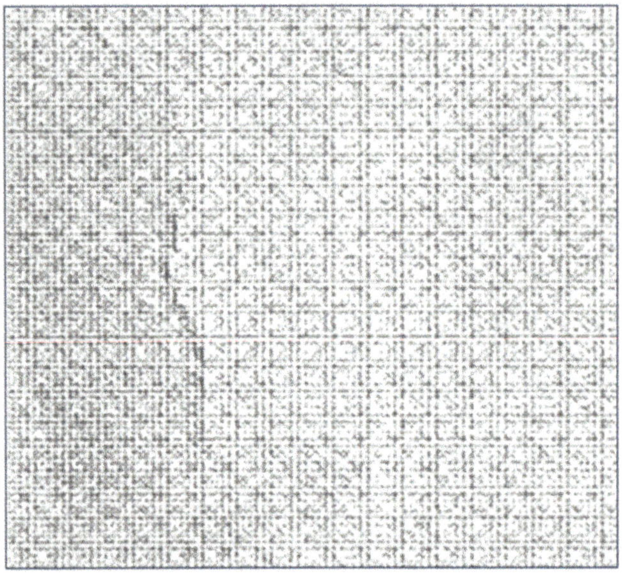

FIGURE 9.13 The initial data processing in box counting algorithm.

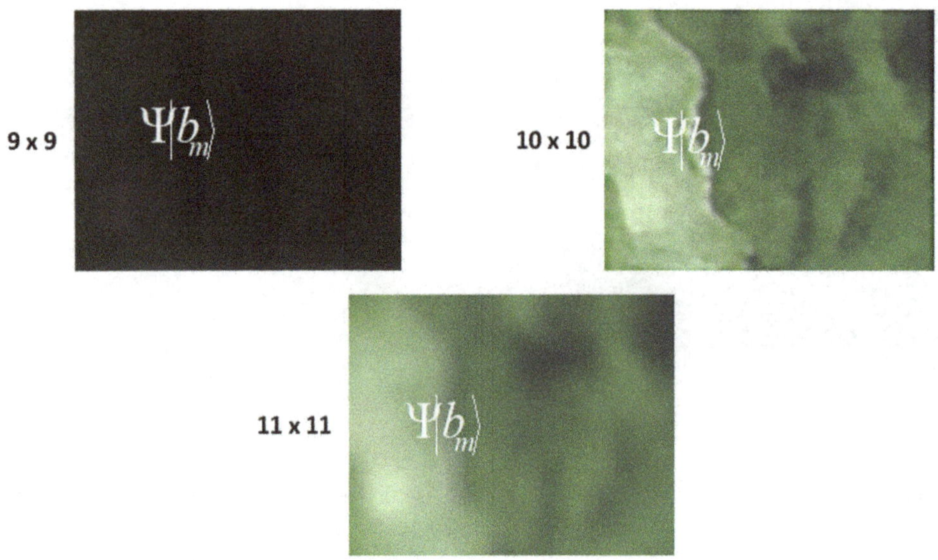

FIGURE 9.14 The fractal dimension box-counting results for ERS-2 image with different box-counting sizes.

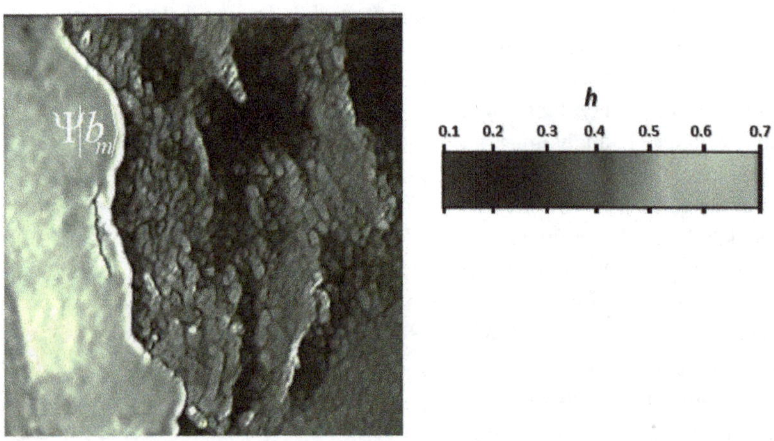

FIGURE 9.15 The singularity spectrum (h) algorithm results or ERS-2 image.

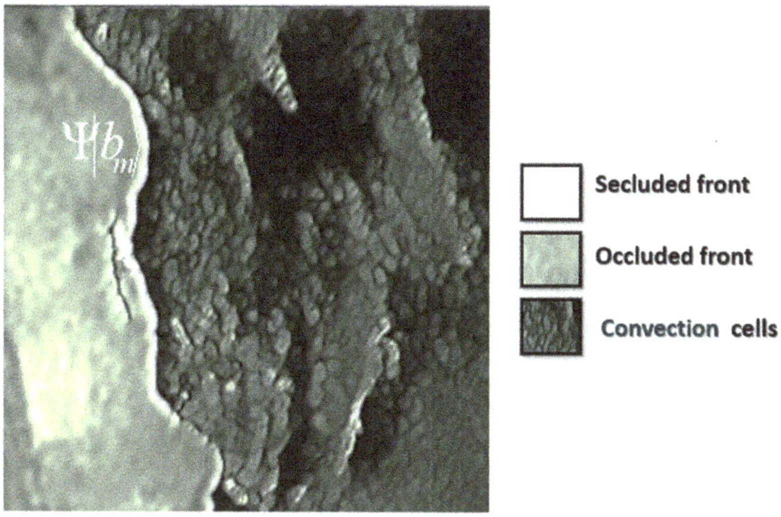

FIGURE 9.16 Deciphering the nature of quantized Marghany front-bound states within the ERS-2 dataset using the singularity spectrum.

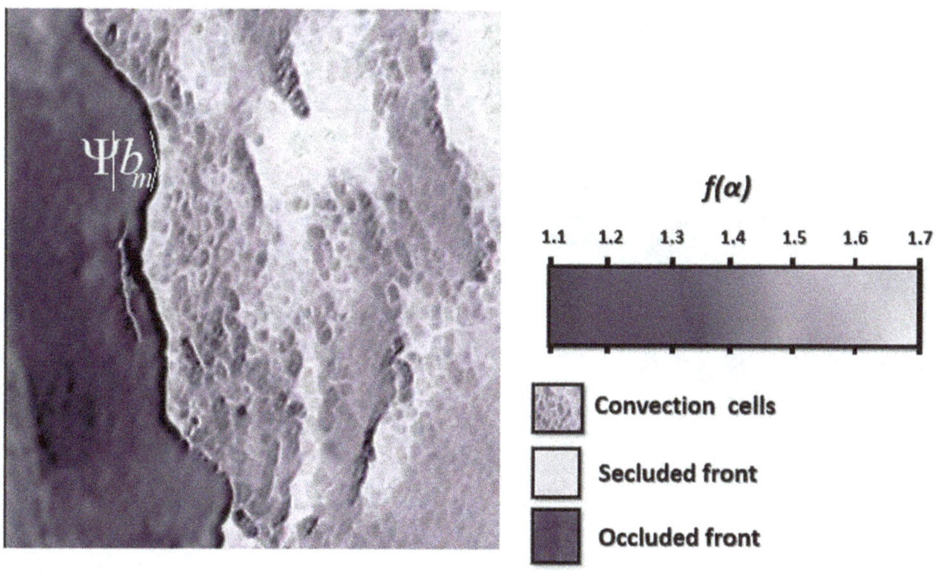

FIGURE 9.17 The singularity strength f(α) results within ERS-2 image.

In the context of multifractal analysis, the singularity strength f(α) is a measure of the multifractality of a signal or dataset. It represents the Hölder exponent, which characterizes the local regularity or singularity of the signal at different scales. In this context, it's noteworthy that the quantized Marghany front-bound states exhibit elevated fractal dimensions, aligning with the superposition of convection cells, as both possess higher multifractal values of 1.3 compared with other features within the ERS-2 image (Fig. 9.17). This means that convection cells correspond to stronger singularities more pronounced irregularities in the image at that scale than occluded front. On the contrary, the singularity strength f(α) value of 1.7 suggests that regions with weaker singularities are more regular as shown in region of occluded front.

The occurrence of an occluded front within a quantized Marghany bound front state along Malaysian coastal waters is a fascinating phenomenon influenced by a combination of atmospheric and oceanic conditions specific to the region. In this view, the coastal waters of Malaysia are strategically located at the confluence of different air masses and ocean currents. This geographic location exposes the region to a dynamic interplay of weather systems, making it susceptible to the formation of complex fronts, including occluded fronts.

From previous chapters, quantized Marghany bound front states are known for their distinct temperature gradients. In this case, warm and cold air masses converge. When a faster-moving cold air mass overtakes a slower-moving warm air mass along the front, it can lead to the formation of an occluded front. The coastal environment amplifies this interaction due to the presence of the sea and the land. The occurrence of quantized Marghany bound front states, therefore, and, by extension, occluded fronts, may be subject to seasonal variability. Different weather systems dominate at various times of the year, resulting in changes in front formation. This phenomenon showcases the intricate relationship between the oceans and the atmosphere in shaping the local climate and weather patterns (Marghany, 2021).

Similarly, the ALOSPALSAR data with L-band HH polarization, acquired on December 26, 2008, along the coastal waters of Perhentian Island, Malaysia, reveal a significant singularity value of 0.9, indicative of convection cell signatures (Fig. 9.18). This means that the singularity spectrum analysis reveals the existence of a specific oceanographic feature known as a "cold conveyor belt." This feature is identified by a sequence or series of vorticity flows within the ocean, indicating areas where cold water masses are transported and mixed with surrounding waters. The singularity spectrum analysis helps in detecting and understanding the patterns and dynamics associated with such oceanic phenomena.

Notably, the dominant characteristic of quantized Marghany front-bound states is the presence of strong vorticity, modeled at a speed of 1.8 m/s (Fig. 9.19) as shown in Section 4.10, chapter 4. Additionally, the progression vector plot, derived from an in situ current meter deployed north of Perhentian Island, illustrates a hyperbolic water flow originating from the northeast and extending toward the southeast in an irregular pattern, with a flow velocity of 1.77 m/s (Fig. 9.20).

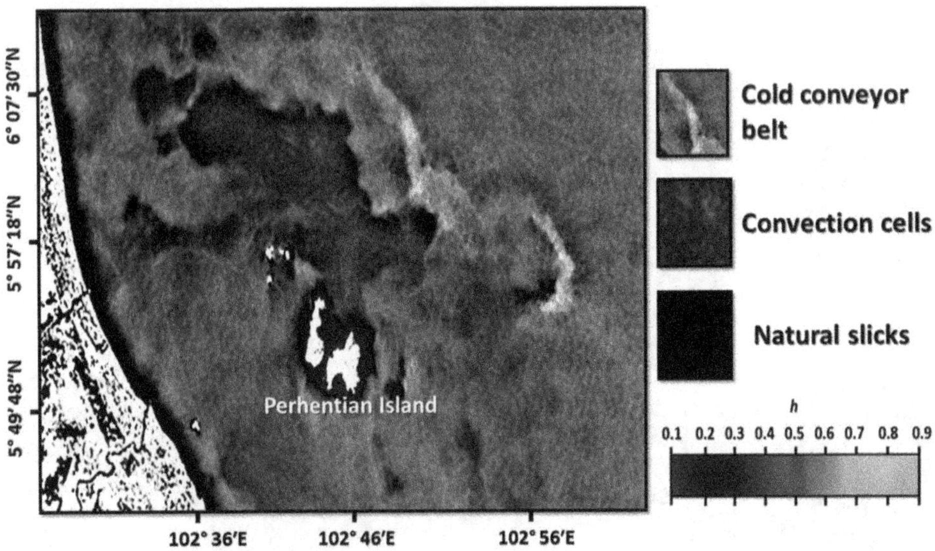

FIGURE 9.18 The singularity spectrum (h) algorithm cluster results for ALOSPALSAR image.

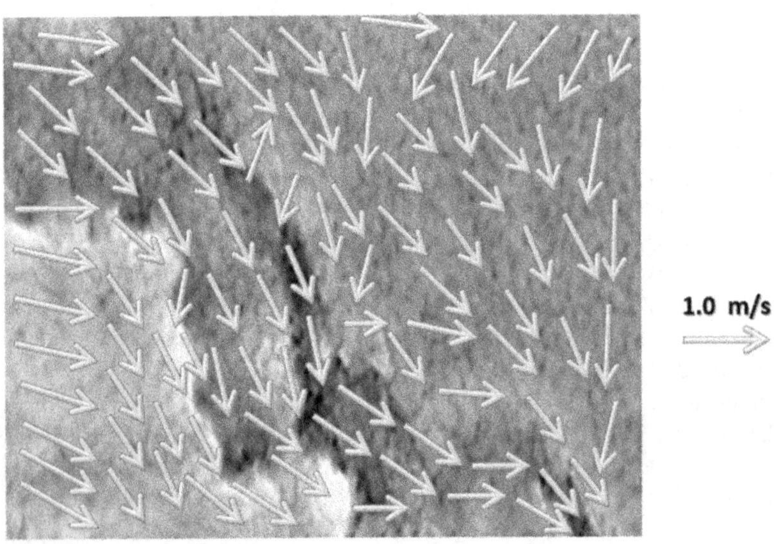

FIGURE 9.19 Vorticity flow speed retrieved from ALOSPALSAR image.

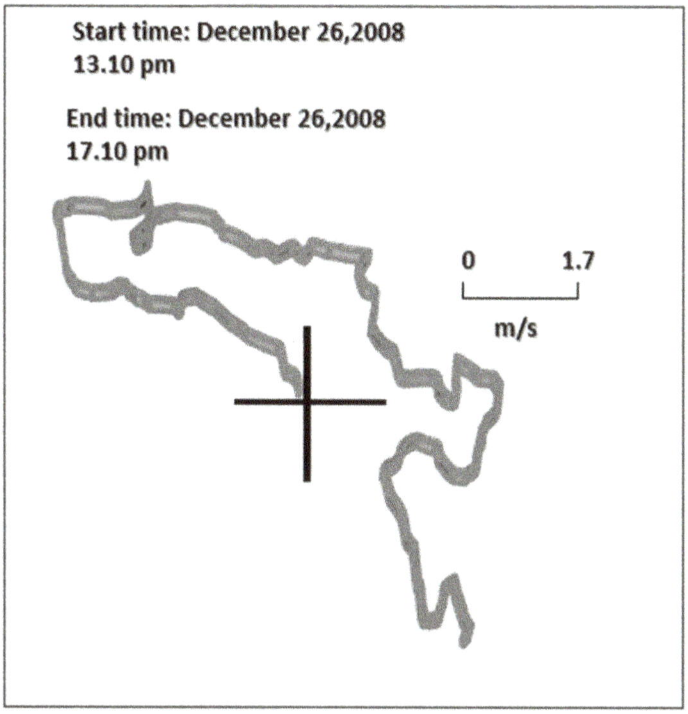

FIGURE 9.20 Progressive vector plot retrieved from in situ current meter measurements during ALOSPALSAR overpass.

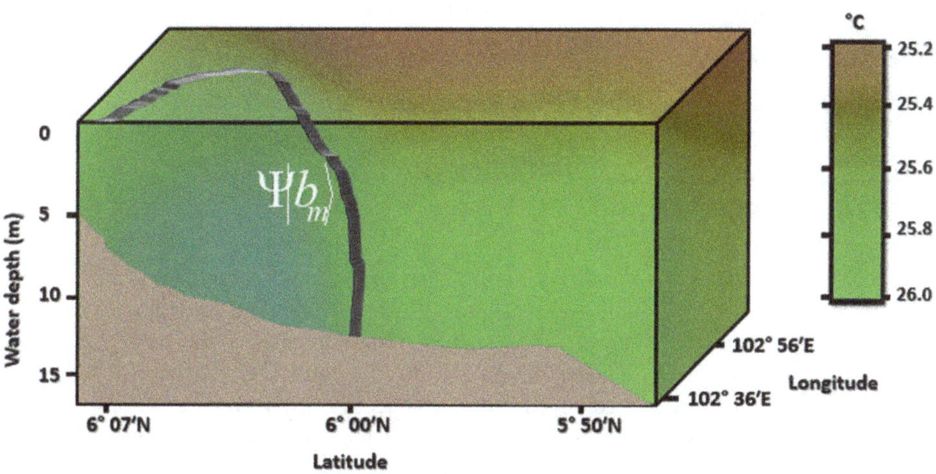

FIGURE 9.21 3-D isothermal pattern observed during the ALOSPALSAR overpass.

The isothermal patterns provide compelling evidence for the existence of a cold conveyor belt front, characterized by temperature ranges between 25.2°C and 26°C. This observation signifies the intrusion of a cold dome originating offshore and moving toward the inshore region. Indeed, the quantized Marghany front-bound state exhibits a distinct concave shape, serving as a demarcation between the temperature ranges of 25.2°C – 26°C. This boundary extends to a considerable water depth of 12 meters. Notably, the month of December corresponds to the northeast monsoon period, during which the water column experiences a decrease in temperature, reaching approximately 25°C at the bottom, while surface temperatures mostly hover around 26°C (Fig. 9.21).

Similar observations were made in ENVISAT ASAR data collected on June 3, 2010, along the coast of Tioman Island, Malaysia. ENVISAT ASAR operates in the C band, offering various operational modes and the ability to detect surface height changes with remarkable submillimeter precision. It played a pivotal role as a data relay for ERS-1 and ERS-2, enabling a wide array of functions, including the examination of different polarities of light, the amalgamation of diverse polarities, angles of incidence, and spatial resolutions (Table 9.3) (Li et al., 2011).

TABLE 9.3 ENVISAT different Modes.

Mode	Id	Polarization	Incidence	Resolution	Swath
Alternating polarization	AP	HH/VV, HH/HV, VV/VH	15–45°	30–150 m	58–110 km
Image	IM	HH, VV	15–45°	30–150 m	58–110 km
Wave	WV	HH, VV		400 m	5 × 5 km
Suivi global (ScanSAR)	GM	HH, VV		1 km	405 km
Wide Swath (ScanSAR)	WS	HH, VV		150 m	405 km

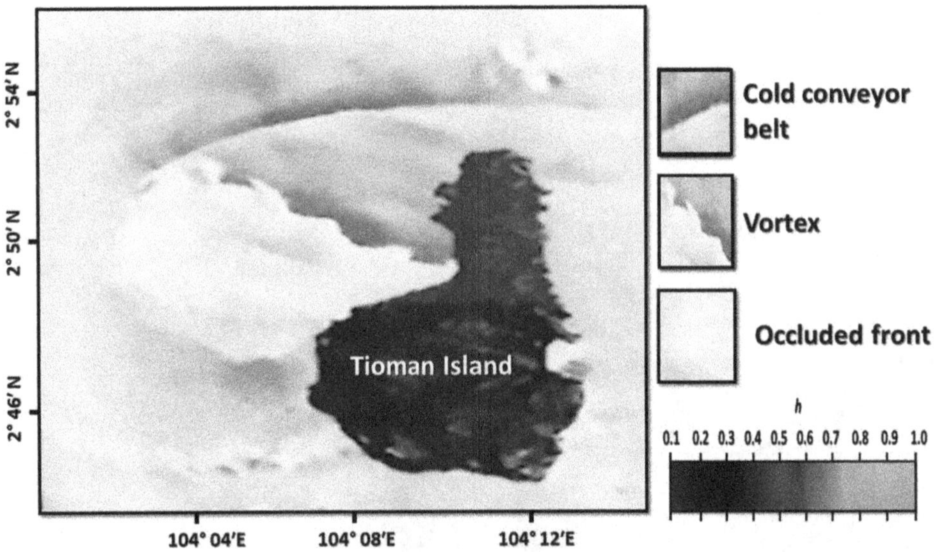

FIGURE 9.22 The singularity spectrum (h) algorithm cluster results for ENVISAT image.

In this context, the singularity spectrum provides supporting evidence for the presence of a conveyor belt phenomenon within the coastal waters of Malaysia. This phenomenon entails the convergence of a warm conveyor belt with an occluded front, resulting in distinctive series of vortex features within the western side of the island (Fig. 9.22). Notably, the northern part of the island exhibits a prevalence of occurrences related to cold fronts. In this sense, the isothermal flow patterns observed indicate an upward intrusion of colder water, with temperatures ranging from 28.5°C to 29.5°C. This colder water interacts with the warmer surface layer, which maintains temperatures around 30.0°C (Fig. 9.23). As a result of this interaction, a convergence zone forms along the island. This pattern strongly suggests the presence of an upwelling flow phenomenon in the vicinity of Tioman Island.

In essence, the isothermal flow data reveal that colder water from beneath is moving upward and mixing with the warmer surface waters, creating a convergence zone near the island. This is indicative of upwelling, a process where deeper, colder waters rise to the surface. Furthermore, the water column near the island's coastline exhibits well-mixed water stratification, extending to depths of approximately 14 meters. This suggests a thorough mixing of water layers in this region. The vorticity flow speed retrieved from ENVISAT ASAR image serves as additional confirmation of the presence of a sequence of cold vortices with temperatures of 28.5°C and a corresponding current velocity of 0.84 m/s (Fig. 9.24). This observation aligns closely with a singularity spectrum reading of 1.0, signifying instances of intense turbulence. This turbulence can be attributed to the rotational motion of the southeast current around the island. These findings agreed with the studies of (Chai et al., 2009; Chiang et al., 2008; Fang et al., 2012; Jilan, 2004; Shaw & Chao, 1994; Zeng et al., 2016).

Furthermore, the progression vector plot, generated from data collected by an in situ current meter deployed to the north of Perhentian Island, provides a visual representation of the complex flow patterns in the region. It reveals a series of vortices connected to closed eddy water flow. These vortices originate from the southwest and extend in

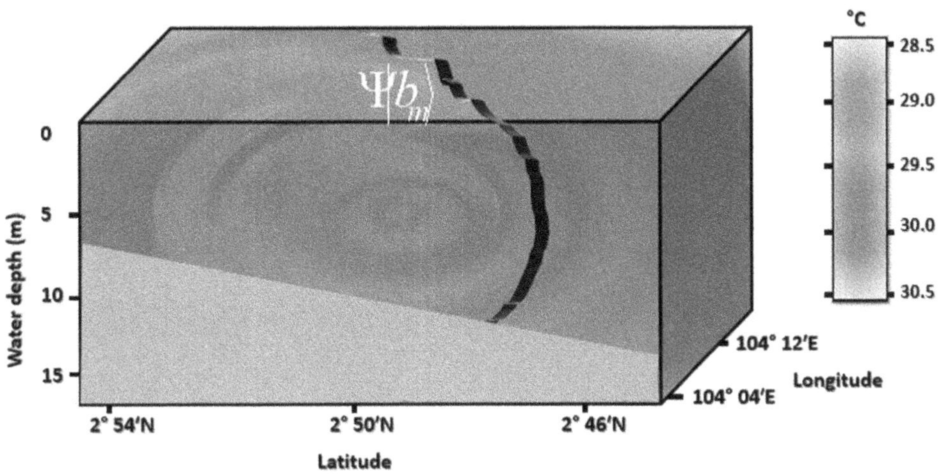

FIGURE 9.23 3-D isothermal pattern observed during the ENVISAT overpass.

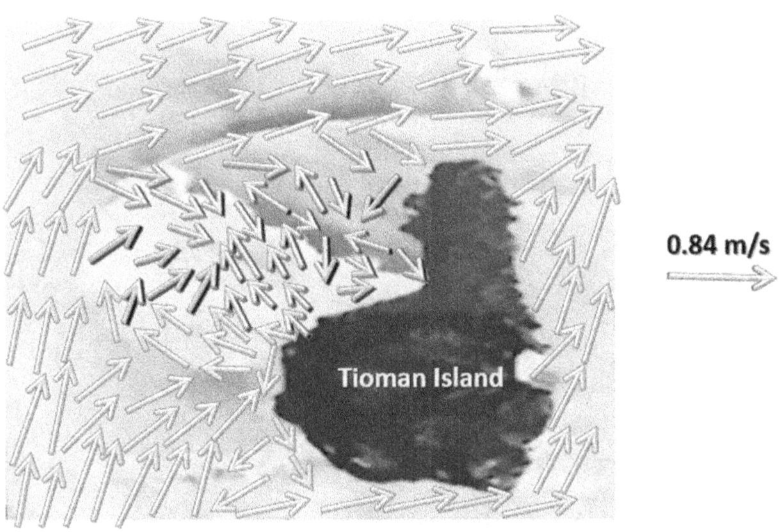

FIGURE 9.24 Vorticity flow speed retrieved from ENVISAT ASAR image.

an irregular pattern toward the northeast. The flow velocity associated with this phenomenon measures at 0.7 m/s (Fig. 9.25). Remarkably, these observations align closely with the findings from ENVISAT ASAR data.

In essence, the progression vector plot offers insights into the dynamic behavior of water currents near Perhentian Island. The presence of vortices and closed eddy water flow patterns, coupled with their specific origin and direction, is clearly depicted. This information complements the data obtained from ENVISAT ASAR, providing a comprehensive understanding of the coastal water dynamics in the region.

In simpler terms, the progressive vector plot generated from on-site data reinforces the idea that a series of cold vortices with specific temperature and velocity characteristics exists in the area. This observation is consistent with the singularity spectrum's reading of 1.0, which indicates the occurrence of significant turbulence resulting from the swirling motion of the southeast current around the island.

The pivotal question at hand is this: How can we leverage the singularity spectrum to pinpoint turbulence zones within SAR images?

9.10 Quantitative turbulence zones using singularity spectrum

Interpreting the singularity spectrum within the context of turbulent flows reveals the multifractal nature of turbulence. By studying how the singularity spectrum changes with variations in Reynolds number (*Re*), which characterizes flow conditions, valuable insights into the evolution of turbulence properties under different flow regimes can be gained. In the context of our analysis, the Sentinel-A satellite data collected on September 25, 2023, utilizing VV band and the IW beam mode along the eastern coastline of Malaysia, provide a valuable insight into the turbulent flow patterns observed, as indicated by the

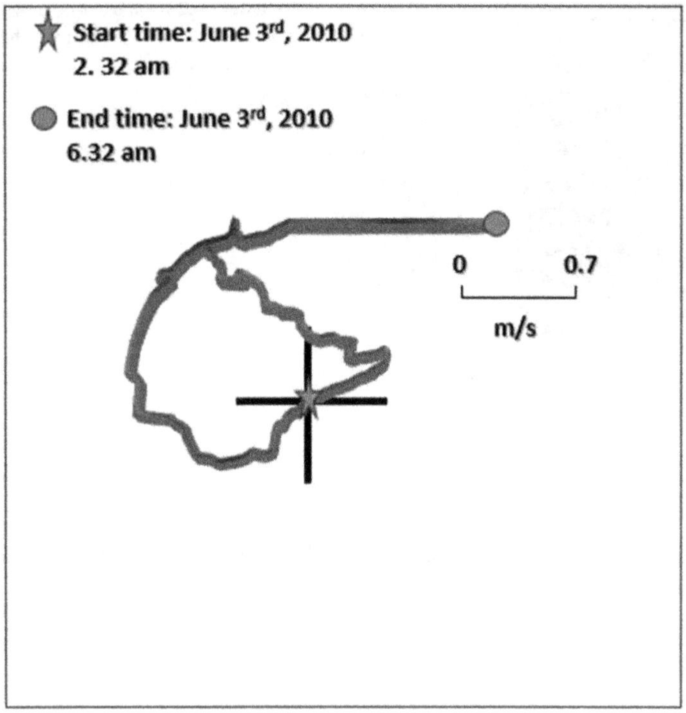

FIGURE 9.25 Progressive vector plot retrieved from in situ current meter measurements during ENVISAT ASAR overpass.

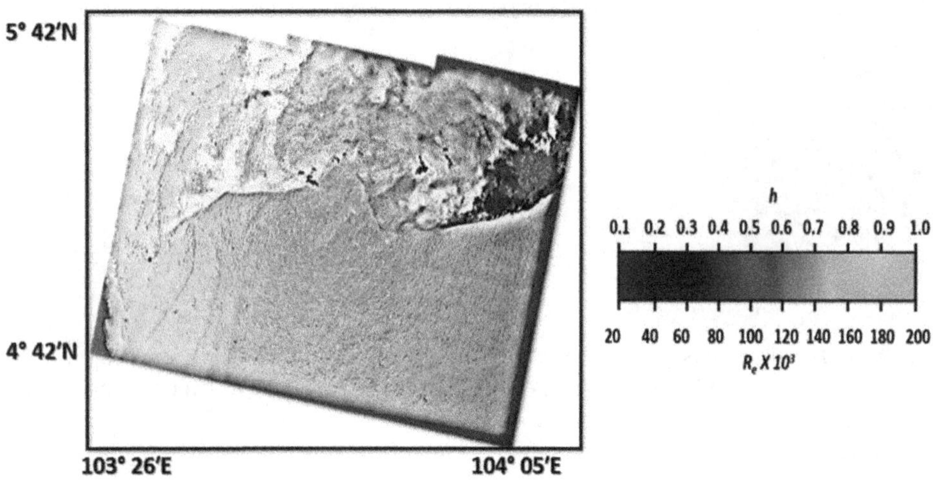

FIGURE 9.26 The alignment between the singularity spectrum and the Reynolds number in Sentinel-A satellite data.

singularity spectrum. The singularity spectrum is a critical measure of turbulence in this scenario.

Specifically, we observe that the highest singularity spectrum value, denoted by 1, corresponds to a Reynolds number of 200 x 10^3 (Fig. 9.26). This alignment between the singularity spectrum and the Reynolds number serves as a compelling indicator of the presence of strong turbulence in the eastern coastal region of Malaysia. This signifies a notable level of turbulence in the area, potentially of considerable interest for further study and analysis.

Therefore the Lagrangian structure function S(n,r) exhibits an inverse relationship with both the singularity spectrum and the multifractal dimension in the Sentinel data. In the turbulent flow captured by Sentinel data, the highest singularity corresponds to a Lagrangian structure function of approximately 0.55 and a multifractal value of around 0.5. As the turbulence follows the Kolmogorov scaling law, the Lagrangian structure function tends to approach 0.4, with a multifractal value of 0.85 (Fig. 9.27). In this sense, it illustrates that turbulence consists of regions with varying degrees of singularity. Regions with high singularity values (large h, small α) correspond to intense vortices or other singular features, while those with lower singularity values (small h, larger α) represent more regular flow regions. Additionally, in the context of turbulence, particularly in the study of turbulence in fluid dynamics, the

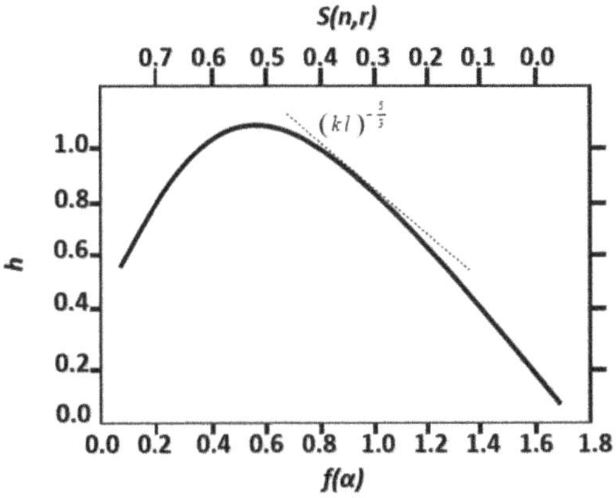

FIGURE 9.27 The Lagrangian structure function S(n, r) is determined based on the singularity spectrum in the Sentinel-A dataset.

singularity spectrum can provide insights into the scaling behavior and multifractal properties of turbulence.

According to the aforementioned perspective, the Lagrangian structure function measures the statistical properties of the velocity differences experienced by a fluid element as it moves through the turbulent flow. It quantifies how turbulent fluctuations vary at different spatial scales. In the context of your data, a Lagrangian structure function of 0.55 suggests a certain level of turbulence intensity. The singularity spectrum, therefore, characterizes the distribution of singularities in a dataset, revealing where the data exhibit sharp transitions or variations. In the context of turbulence, it helps identify the scales at which turbulent structures are most pronounced. A high singularity spectrum (value of 1) indicates strong turbulence features. Consequently, the multifractal dimension provides information about the scaling behavior of a system. In turbulence analysis, it helps understand the self-similar or multifractal nature of turbulent fluctuations. A multifractal value of 0.5 indicates a certain degree of multifractality, which means that turbulence exhibits different scaling properties at different scales (Bacry et al., 1993; Chhabra et al., 1989; Kuznetsov, 2004).

The inverse relationship between these metrics can be attributed to the intricate nature of turbulence. When turbulence is intense and displays sharp variations (high singularity spectrum), the Lagrangian structure function can exhibit a certain value. However, as turbulence follows the Kolmogorov scaling law, the Lagrangian structure function tends to approach 0.4, indicating a more consistent behavior at smaller scales. Similarly, the multifractal value of 0.85 suggests that the turbulence exhibits multifractal scaling properties, which may not be perfectly aligned with the Lagrangian structure function. This discrepancy reflects the complex, multifaceted nature of turbulence, where different scaling laws and behaviors come into play at various scales.

This chapter has presented a novel method for the automatic detection of oceanic fronts and turbulence structures. To achieve this, a multifractal dimension algorithm was employed in conjunction with a wavelet-based multifractal approach, and multifractal measures were derived by analyzing structure functions, particularly the Lagrangian structure function. The findings in this chapter contribute to a better understanding of ocean fronts and turbulence structures using MultiSAR data. Notably, while the algorithm was applied to different SAR sensors, the combination of the singularity spectrum and multifractal dimension algorithm yielded similar effective results in the automatic detection of ocean fronts and turbulence structures across all these diverse SAR sensors.

References

Arneodo, A., Audit, B., Kestener, P., & Roux, S. (2008). Wavelet-based multifractal analysis. *Scholarpedia*, *3*(3), 4103. Available from https://doi.org/10.4249/scholarpedia.4103.

Bacry, E., Muzy, J. F., & Arnéodo, A. (1993). Singularity spectrum of fractal signals from wavelet analysis: Exact results. *Journal of Statistical Physics*, *70*(3–4), 635–674. Available from https://doi.org/10.1007/BF01053588.

Bunde, A., Havlin, S., & Roman, H. E. (1990). Multifractal features of random walks on random fractals. *Physical Review. A*, *42*(10), 6274–6277. Available from https://doi.org/10.1103/PhysRevA.42.6274.

Chai, F., Liu, G., Xue, H., Shi, L., Chao, Y., Tseng, C. M., Chou, W. C., & Liu, K. K. (2009). Seasonal and interannual variability of carbon cycle in South China Sea: A three-dimensional physical-biogeochemical modeling study. *Journal of Oceanography*, *65*(5), 703–720. Available from https://doi.org/10.1007/s10872-009-0061-5.

Chhabra, A. B., Meneveau, C., Jensen, R. V., & Sreenivasan, K. R. (1989). Direct determination of the f (α) singularity spectrum and its application to fully developed turbulence. *Physical Review. A*, *40*(9).

Chiang, T. L., Wu, C. R., & Chao, S. Y. (2008). Physical and geographical origins of the South China Sea warm current. *Journal of Geophysical Research: Oceans*, *113*(8). Available from https://doi.org/10.1029/2007JC004794, http://onlinelibrary.wiley.com/journal/10.1002/(ISSN)2169-9291.

Du, S., Sawford, B. L., Wilson, J. D., & Wilson, D. J. (1995). Estimation of the Kolmogorov constant (C0) for the Lagrangian structure function, using a second-order Lagrangian model of grid turbulence. *Physics of Fluids*, *7*(12), 3083–3090. Available from https://doi.org/10.1063/1.868618.

Falconer, K. (2005). *Fractal geometry: Mathematical foundations and applications*. Wiley. Available from https://doi.org/10.1002/0470013850.

Fang, G., Wang, G., Fang, Y., & Fang, W. (2012). A review on the South China Sea western boundary current. *Acta Oceanologica Sinica*, *31* (5), 1–10. Available from https://doi.org/10.1007/s13131-012-0231-y.

Goncalves, P., Riedi, R., & Baraniuk, R. (1998). Simple statistical analysis of wavelet-based multifractal spectrum estimation. Conference Record of the Asilomar Conference on Signals, Systems and Computers. IEEE Comp Soc France 1:287-291.

Guarneri, I., & Mantica, G. (1994). Multifractal energy spectra and their dynamical implications. *Physical Review Letters*, *73*(25), 3379–3382. Available from https://doi.org/10.1103/PhysRevLett.73.3379.

Jilan, S. (2004). Overview of the South China Sea circulation and its influence on the coastal physical oceanography outside the Pearl River Estuary. *Continental Shelf Research*, *24*(16), 1745–1760. Available from https://doi.org/10.1016/j.csr.2004.06.005, http://www.journals.elsevier.com/continental-shelf-research/.

Kestener, P., & Arneodo, A. (2003). Three-dimensional wavelet-based multifractal method: The need for revisiting the multifractal description of turbulence dissipation data. *Physical Review Letters*, *91*(19). Available from https://doi.org/10.1103/PhysRevLett.91.194501.

Kestener, P., & Arneodo, A. (2004). Generalizing the wavelet-based multifractal formalism to random vector fields: Application to three-dimensional turbulence velocity and vorticity data. *Physical Review Letters*, *93*(4), 1–44501. Available from https://doi.org/10.1103/PhysRevLett.93.044501, https://journals.aps.org/prl/issues.

Kuznetsov, E. A. (2004). Turbulence spectra generated by singularities. *Journal of Experimental and Theoretical Physics Letters*, *80*(2), 83–89. Available from https://doi.org/10.1134/1.1804214.

Li, X. M., Lehner, S., & Bruns, T. (2011). Ocean wave integral parameter measurements using envisat ASAR wave mode data. *IEEE Transactions on Geoscience and Remote Sensing*, *49*(1), 155–174. Available from https://doi.org/10.1109/TGRS.2010.2052364.

Lien, R. C., & D'Asaro, E. A. (2002). The Kolmogorov constant for the Lagrangian velocity spectrum and structure function. *Physics of Fluids*, *14*(12), 4456–4459. Available from https://doi.org/10.1063/1.1518695, http://scitation.aip.org/content/aip/journal/pof2.

Macé, N., Alet, F., & Laflorencie, N. (2019). Multifractal scalings across the many-body localization transition. *Physical Review Letters*, *123* (18). Available from https://doi.org/10.1103/PhysRevLett.123.180601, http://harvest.aps.org/bagit/articles/10.1103/PhysRevLett.123.180601/apsxml.

Mandelbrot, B. B., & Wheeler, J. A. (1983). The fractal geometry of nature. *American Journal of Physics*, *51*(3), 286–287. Available from https://doi.org/10.1119/1.13295.

Marghany, M. (2021). *Nonlinear ocean dynamics: Synthetic aperture radar*. Elsevier.

Massopust, P. R. (2016). *Fractal functions, fractal surfaces, and wavelets* (Second Edition, pp. 1–405). Germany: Elsevier Inc. Available from http://www.sciencedirect.com/science/book/9780128044087.

Meneveau, C., & Sreenivasan, K. R. (1987). Simple multifractal cascade model for fully developed turbulence. *Physical Review Letters*, *59*(13), 1424–1427. Available from https://doi.org/10.1103/PhysRevLett.59.1424.

Nelkin, M. (1990). Multifractal scaling of velocity derivatives in turbulence. *Physical Review. A*, *42*(12), 7226–7229. Available from https://doi.org/10.1103/PhysRevA.42.7226.

Raizer, V. (2021). *Remote sensing of turbulence*. CRC Press.

Raizer, V. Y., Novikov, V. M., & Bocharova, T. Y. (1994). The geometrical and fractal properties of visible radiances associated with breaking waves in the ocean. *Annales Geophysicae*, *12*(12), 1229–1233. Available from https://doi.org/10.1007/s00585-994-1229-3.

Sawford, B. L., & Yeung, P. K. (2011). Kolmogorov similarity scaling for one-particle Lagrangian statistics. *Physics of Fluids*, *23*(9). Available from https://doi.org/10.1063/1.3643852, http://scitation.aip.org/content/aip/journal/pof2.

Shaw, P. T., & Chao, S. Y. (1994). Surface circulation in the South China Sea. *Deep Sea Research Part I: Oceanographic Research Papers*, *41*(11–12), 1663–1683. Available from https://doi.org/10.1016/0967-0637(94)90067-1.

Tarboton, D. G., Bras, R. L., & Rodriguez-Iturbe, I. (1988). The fractal nature of river networks. *Water Resources Research*, *24*(8), 1317–1322. Available from https://doi.org/10.1029/WR024i008p01317.

Turbulence, strange attractors and chaos. *World Scientific*, *16* (1995).

Turiel, A., Pérez-Vicente, C. J., & Grazzini, J. (2006). Numerical methods for the estimation of multifractal singularity spectra on sampled data: A comparative study. *Journal of Computational Physics*, *216*(1), 362–390. Available from https://doi.org/10.1016/j.jcp.2005.12.004, http://www.journals.elsevier.com/journal-of-computational-physics/.

Zeng, L., Wang, D., Chen, J., Wang, W., & Chen, R. (2016). SCSPOD14, a South China Sea physical oceanographic dataset derived from in situ measurements during 1919–2014. *Scientific Data*, *3*(1). Available from https://doi.org/10.1038/sdata.2016.29.

Chapter 10

Quantum finite automaton-based algorithm for identifying rain cell turbulences in synthetic aperture radar images

In the preceding chapters, various algorithms were showcased for tracking and mapping different types of fronts and turbulence in coastal waters, primarily attributed to the interactions between ocean currents and wind patterns. However, this chapter is dedicated to introducing a distinct form of turbulence stemming from heavy rainfall over the sea surface, characterized by specific footprint signatures discernible in synthetic aperture radar data. These rain cells represent a unique focus area in this chapter.

10.1 Rain cells

Rain cells, often referred to as rain showers (Fig. 10.1), are specific regions where rain precipitation occurs within a larger weather system. These cells are relatively small in scale compared with the overall area covered by the weather system. They can be thought of as isolated patches of rainfall within a broader meteorological context. In this perspective, rain cells come in different intensities, ranging from light drizzle (Fig. 10.2), where the raindrops are small and sparse, to heavy downpours (Fig. 10.3), characterized by a rapid and substantial volume of rain falling in a short period of time. The intensity of a rain cell can vary over time, with some cells producing light rain and others leading to more intense rainfall.

These cells, therefore, are often associated with convective processes in the atmosphere. Convective rain occurs when warm, moist air rises, cools, and condenses, forming clouds and precipitation (Fig. 10.4). As this moist air ascends, it may create localized areas of rain cells. These cells can develop relatively quickly and are usually short-lived, lasting for a few hours at most. Rain cells, consequently, can have a significant impact on local weather conditions, affecting visibility, road conditions, and outdoor activities. Weather radar and satellite technology play a crucial role in detecting and tracking the movement of these cells, helping meteorologists provide accurate and timely weather forecasts and warnings for the affected regions (Capsoni et al., 1987; Peleg & Morin, 2012).

10.2 Rain cell shapes and orientations

Rain cells, also known as rain showers or rain patches, can take on various shapes and orientations in the atmosphere. Their characteristics are influenced by a combination of meteorological factors, including wind patterns, atmospheric instability, and the underlying topography. In this context, round or circular cells are often observed in more stable atmospheric conditions. Circular rain cells can form when there's a uniform distribution of moisture and temperature in the air (Capsoni et al., 1987; von Hardenberg et al., 2003). They tend to produce steady, relatively light rainfall. The shape is influenced by the absence of strong wind shears or converging air masses. Therefore linear cells are elongated in shape, forming lines or bands of rainfall. Linear rain cells are often associated with weather features such as cold fronts, warm fronts, or troughs. The orientation of the line is typically perpendicular to the direction of the front or boundary, with rain occurring along the length of the line. The intensity of rainfall within linear cells can vary, and they are often linked to heavier precipitation.

Clustered or cellular cells, consequently, are characterized by a grouping or clustering of individual cell-like structures. They can take on various shapes and orientations, often appearing as a cluster of circular or irregular shapes. This formation is linked to the interaction of multiple atmospheric factors, such as the convergence of air

Synthetic Aperture Radar Image Processing Algorithms for Nonlinear Oceanic Turbulence and Front Modeling. DOI: https://doi.org/10.1016/B978-0-443-19155-8.00010-6

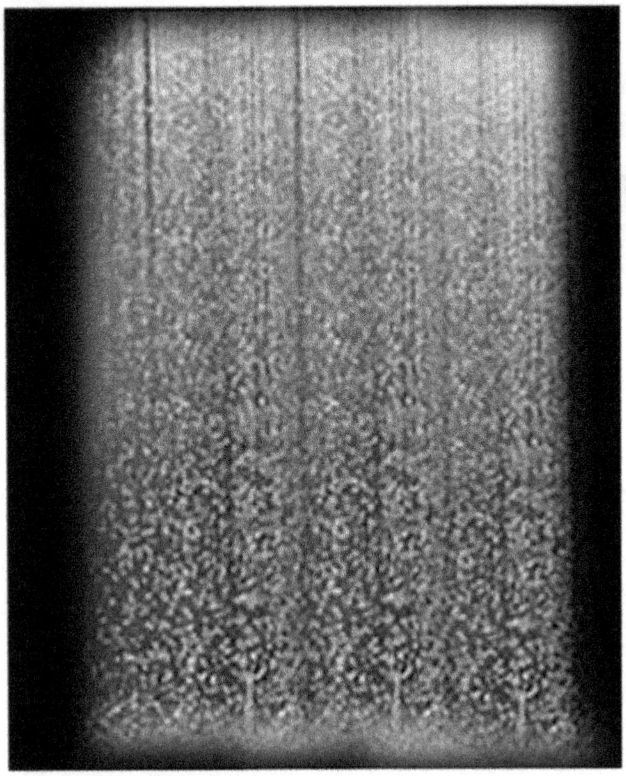

FIGURE 10.1 Example of rain showers.

FIGURE 10.2 Light drizzle.

masses and local topographic features. The intensity and duration of rainfall in clustered cells depend on the specific meteorological conditions. On the other hand, some rain cells have an oblong or elliptical shape. These cells are often associated with broader-scale weather systems such as low-pressure areas or tropical cyclones. The elongation of the cell is influenced by the wind field around the storm system. Rainfall within these cells can be persistent and sometimes heavy (Capsoni et al., 1987; Enjamio et al., 2005; Peleg et al., 2018).

FIGURE 10.3 Heavy downpours.

Additionally, in certain conditions, rain cells may appear irregular or fragmented in shape. This can occur when there are multiple interacting weather systems or when the atmospheric conditions are less organized. The irregular shape and orientation result from the complex interplay of various meteorological factors. The presence of mountains or other topographical features, subsequently, can significantly affect the shape and orientation of rain cells. When moist air is forced to rise over mountains, it can lead to the formation of orographic rain cells, which are typically elongated along the wind direction (Capsoni et al., 1987; Feral et al., 2000; Peleg et al., 2018). Needless to say, the shape and orientation of rain cells provide valuable information to meteorologists for understanding local weather conditions, forecasting precipitation patterns, and assessing the potential impact of rainfall on a given area.

10.3 Formation mechanisms of rain cells

Rain cells are formed through a complex interplay of atmospheric conditions. The process commences with the presence of moisture in the atmosphere. As warm, moist air rises, it cools due to the decreasing air pressure, reaching a point known as the dew point, where it can no

FIGURE 10.4 Occurrence of convection rain.

longer retain all its moisture. At this juncture, the moisture condenses into minuscule water droplets, which accumulate to form clouds. These clouds are essentially vast collections of these tiny water droplets suspended in the atmosphere.

Within a cloud, different regions may experience various vertical air movements, fostering the growth of raindrops. Some of these droplets continue to grow and coalesce, eventually forming larger raindrops. The growth of raindrops can also occur through the collision and coalescence of smaller and larger droplets. In colder regions, ice nucleation is another mechanism for raindrop formation, with tiny ice crystals forming in the upper regions of the cloud. As these ice crystals interact with supercooled water droplets, they grow large enough to fall as rain.

As a raindrop descends, it undergoes a fascinating process of growth and fragmentation. When two raindrops collide during their descent, they may merge, leading to an increase in the size of the resulting drop. However, there is a limit to how large a raindrop can become (Fig. 10.5). Once a raindrop reaches a critical size, typically around 4 millimeters or more, the forces of surface tension can no longer support its cohesion. At this point, the raindrop undergoes a transformation, breaking apart into smaller drops. The surface tension, which initially held the raindrop together, ultimately yields, and the once-considerable raindrop disintegrates, allowing its individual components to continue their journey toward the Earth's surface. This intriguing phenomenon illustrates the intricate dynamics that govern the behavior of raindrops during their descent (Capsoni et al., 1987; Chen et al., 2023).

When these raindrops become sufficiently large and heavy, they overcome the updrafts within the cloud and descend to the Earth's surface. The size and intensity of the raindrops determine the nature of the precipitation—whether it will be light drizzle, moderate rain, or heavy

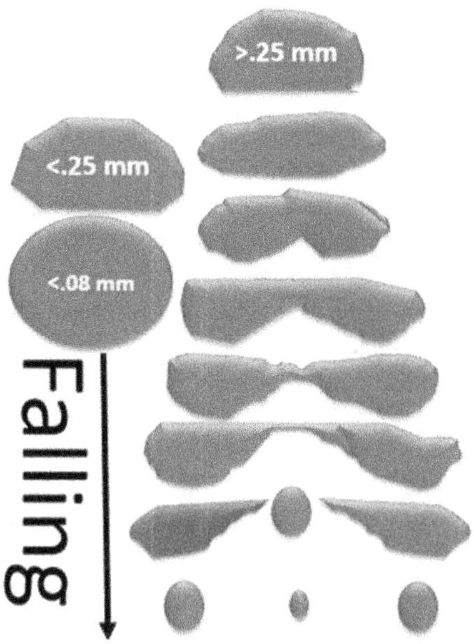

FIGURE 10.5 Falling of rain drops.

downpours. Rain cells are localized areas where these raindrops reach the ground. These cells can vary in size, shape, and intensity, influenced by local wind patterns, terrain, and larger weather systems. Additionally, atmospheric turbulence and dynamics play a role in their formation, with factors such as temperature gradients, convergence of air masses, and topography contributing to the development of these cells. Understanding the mechanisms behind rain cell formation is crucial for weather forecasting and assessing the impact of rainfall in various regions (Capsoni et al., 1987; Kumjian et al., 2015; Takahashi & Shimura, 2004; von Hardenberg et al., 2003).

In light of the aforementioned perspective, the question that arises is whether rainfall has the capacity to induce turbulence across the surface of the ocean.

10.4 How can rainfall generates turbulence across the surface of the ocean?

Let us consider that ocean turbulence is identified as *T*; the rate at which rainwater falls into the ocean surface is *R*, which is controlled by wind speed *U* and sea surface temperature *SST*. In this view, the basic equation of ocean turbulence due to the impact of rainfall can be simply expressed as:

$$T = f(R, U, SST) \tag{10.1}$$

This equation implies that ocean turbulence (T) is a function of rainfall rate (R), wind speed (U), and sea surface temperature (SST). The specific form of the function, as well as the units and coefficients, would require more detailed empirical research and modeling. In this simplified model, therefore, we assume that turbulence (T) is a function of rainfall intensity (I), the radius of raindrops (r), and wind speed (U). We'll use a power-law relationship to represent this:

$$T = k \cdot I^{\alpha} \cdot r^{\beta} \cdot U^{\gamma} \tag{10.2}$$

Eq. (10.2) suggests that turbulence is influenced by the combined effects of rainfall intensity, raindrop size, and wind speed. The specific values of the coefficients (k, α, β, and γ) would depend on the units used for each parameter and would require empirical analysis. In this view, k, α, β, and γ are coefficients that would need to be determined through empirical data and more sophisticated modeling.

Raindrops form through a process called coalescence. In this process, smaller cloud droplets collide and merge, forming larger raindrops. The rate of raindrop formation (*dN*/*dt*) can be represented as a function of the number of cloud droplets (*Nc*), collision efficiency (*E*), and collision frequency (f_R):

$$\frac{dN}{dt} = E.f_R.N_c^2 \tag{10.3}$$

Raindrops grow as they collide with smaller droplets and collect them. The growth rate (*dD*/*dt*) of an individual raindrop can be described by the collection equation:

$$\frac{dD}{dt} = \pi \cdot [N_O]^{-1} N_c \cdot r^2 \cdot v \cdot \rho \tag{10.4}$$

where N_O is the number of cloud droplets per unit volume; *r* is the radius of the raindrop; *v* is the relative velocity between the raindrop and cloud droplets; and ρ is the air density. Raindrops, therefore, can break up into smaller droplets when subjected to forces such as wind shear and turbulence. The critical size (*Dc*) at which raindrops break up depends on the shear stress (τ) and the viscosity of air (μ):

$$D_c = \frac{4\mu}{\tau} \tag{10.5}$$

Smaller raindrops with radii less than D_c are subjected to breakup. Raindrops, consequently, can impact the ocean surface and induce turbulence (Avissar & Liu, 1996; Feral et al., 2000; Frenger et al., 2013; Kumjian et al., 2015). Now, we can formulate an equation to describe the turbulent kinetic energy E(*K*) generated by raindrop impact:

$$E(k) = 0.5\rho v^2 \tag{10.6}$$

here v is the raindrop velocity (in meters per second); and ρ is the density of water (approximately 1000 kg/m^3). In this understanding, the induced turbulent velocity (*w*) can be approximated as:

$$w = \frac{R}{D} v \tag{10.7}$$

Eq. (10.7) represents the vertical component of the raindrop's motion penetrating into the water. In this sense, the raindrop impact can generate waves and ripple effects on the ocean surface, which can contribute to turbulence. This can be described using a wave-generation term w_g:

$$w_g = \frac{u}{R} \tag{10.8}$$

To quantify the overall turbulence intensity (*T*), we can use the ratio of the turbulent kinetic energy to the ocean surface wave velocity:

$$T = Ku^{-2} \tag{10.9}$$

This provides an estimate of the turbulence intensity resulting from raindrop impact on the ocean surface. The raindrop impact can also generate waves and ripple effects on the ocean surface, influenced by the local wind speed:

$$w_g = \int \left(\frac{4}{3}\pi R^3\right) \cdot \left(\frac{4}{3}\pi R^3 N(R) \frac{u}{R}\right) dR. \tag{10.10}$$

This model takes into account the raindrop size distribution and local wind speed, providing a more detailed representation of the impact of rain on ocean turbulence (Capsoni et al., 1987; Paluch & Lenschow, 1991; Takahashi & Shimura, 2004). It integrates over different raindrop sizes, acknowledging that larger raindrops may have a greater impact.

10.5 Quantized rain cells interaction with sea surface

The concept of "Quantized Rain Cells Interaction with Sea Surface" is a theoretical framework that combines the quantum world with macroscopic meteorological phenomena for creative and imaginative purposes. In this concept, we introduce the notion of "quantanized rain cells" to describe rain cells that exhibit quantum-like behavior when interacting with the sea surface. This concept is entirely fictional and not grounded in actual science, but it can be explored for creative and speculative purposes.

These hypothetical rain cells exhibit quantum properties, such as wave-particle duality, uncertainty, and entanglement. They are described as discrete units of rain with wave-like characteristics when in motion and particle-like behavior when interacting with the sea surface. In this sense, let us assume that Ψ represents the quantum wave function of a rain cell; x and t represent spatial position and time, respectively (Berezin, 1975; Gray & Neuhoff, 1998; Yang et al., 2019). Thus the hypothetical wave-particle duality equation for a rain cell can be formulated as:

$$\Psi(x,t) = Asin(kx - \omega t) + Be^{(ikx)} + Ce^{(ikx)} \tag{10.11}$$

In Eq. (10.11), A represents the amplitude of the wave-like component; *B* represents the probability amplitude of a rain cell behaving as a particle; and *C* represents the probability amplitude of entanglement with other rain cells. Additionally, *k* is the wave number; and ω is the angular frequency.

Eq. (10.11) symbolizes the quantum-like behavior of rain cells. When in the atmosphere, they exhibit a wave function described by the first term (wave-like characteristics). Upon reaching the sea surface, they transition into a particle-like state, which is described by the second term. The third term represents the potential for entanglement with other rain cells.

Therefore from the point of view of wave-particle duality, quantanized rain cells can behave as both waves and particles. When in the atmosphere, they exhibit wave-like behavior, characterized by interference patterns and diffraction. However, upon reaching the sea surface, they transition to a particle-like state. In other words, when these hypothetical quantanized rain cells are suspended in the atmosphere, they reveal their wave-like characteristics. At this stage, they are described by a wave function, Ψ, which manifests as undulating patterns in space and time. Interference phenomena, analogous to ripples on a pond, can be observed as these rain cells interact with one another. Their behavior is marked by the familiar phenomena of interference and diffraction (Berezin, 1974; Yang et al., 2019).

However, as quantanized rain cells descend and approach the sea surface, they undergo a mysterious transformation. At this critical juncture, they shift from their wave-like state to exhibit particle-like attributes. It's as though they condense into individual, localized entities. These rain cells, once spread out like waves, now appear as distinct particles with defined positions and trajectories. This transition in behavior is akin to the way quantum entities, such as electrons, exhibit both wave and particle properties depending on the conditions of observation. Just as quantum particles exist in superpositions of states, the quantanized rain cells exist in superpositions of wave-like and particle-like states (Ali & Engliš, 2005; Berazin, 2012; Yang et al., 2019).

The uncertainty in the impact of quantanized rain cells, as described by Heisenberg's uncertainty principle, can be elucidated as follows: When applied to the impact of quantanized rain cells on the sea surface, the uncertainty principle implies that we cannot precisely determine both their position and momentum simultaneously. In this context, the position uncertainty (Δx) pertains to the exact impact location, while the momentum uncertainty (Δp) concerns the rain cells' velocity and mass. The mathematics of the uncertainty principle introduces an element of unpredictability into the interactions between these quantanized rain cells and the sea surface. This suggests that, while we can acquire some knowledge about either their position or their momentum, inherent uncertainty always exists in predicting their impact characteristics.

Consequently, rain cells can become entangled with one another. This means that the behavior of one rain cell can instantaneously affect another, even when they are physically separated. In this scenario, let's denote two entangled rain cells as *A* and *B*. The behavior of one rain cell (*A*) affects the other (*B*), and vice versa, in an entangled state. In this understanding, we can represent the entangled state using the following equation:

$$\Psi = \Psi(A) \otimes \Psi(B) \tag{10.12}$$

Ψ represents the entangled state of the rain cells *A* and *B*, while $\Psi(A)$ and $\Psi(B)$ are the individual states of rain cells *A* and (B). This equation shows that the behavior of rain cell *A* is intertwined with that of rain cell *B*. Changes in one rain cell's properties, such as position or momentum, will instantaneously affect the properties of the other rain cell, no matter how far apart they are.

10.6 Quantanized Marghany rain interaction coefficient

The author introduces a novel concept to describe the rain cell interactions with the sea surface, which is important for identifying rain cell footprints in SAR data. This concept is named after the author as "Quantanized Marghany Rain Interaction Coefficient (QMRIC)." The QMRIC is introduced as a parameter to describe the extent of quantum behavior exhibited by these rain cells when they interact with the sea surface. It can be a measure of the degree of uncertainty or the potential for entanglement during raindrop impacts. In this context, the QMRIC can be defined mathematically as:

$$Q_{MR} = \Psi(U)[E(K)]^{-1} \tag{10.13}$$

In Eq. (10.13), $\Psi(U)$ represents the quantum potential energy as the rain cell particles can exhibit wave-like behavior as discussed in previous section. The energy, therefore, associated with this behavior is given by the de Broglie wavelength equation:

$$\Psi(U) = h\lambda^{-1} \tag{10.14}$$

here h is Planck's constant (46.626×10^{-34} J · s); and λ represents the raindrop's de Broglie wavelength. using Eqs. (10.6) and (10.14); then Eq. (10.13) is rearranged as:

$$Q_{MR} = 2h\left[\lambda m v^2\right]^{-1} \tag{10.15}$$

Now, we can incorporate this fictional quantum parameter into a classical equation describing the generation of waves and ripple effects on the ocean surface during a raindrop impact:

$$H(t) = Q_{MR} \cdot \sin(2\pi f t) \tag{10.16}$$

here $H(t)$ is the height of the surface disturbances over time; t is the time elapsed since the raindrop impact; and f represents the frequency of the waves generated by the impact. In more details, we can formulate the following equation to describe the height of the surface disturbances (H) as a function of time (t) and the parameters aforementioned:

$$H(t) = \frac{E(K)}{\rho g D}\left(1 - e^{\left(\frac{2gDt}{v^2}\right)}\right) \tag{10.17}$$

This equation represents the growth of surface disturbances over time due to the impact of raindrops. The term $\rho \cdot g \cdot D.E(K)$ accounts for the initial energy imparted by the raindrop impact. The expression $\left(1 - e^{\left(\frac{2gDt}{v^2}\right)}\right)$ models the time-dependent growth of these surface disturbances, which gradually reach a maximum height (Ali & Engliš, 2005; Berazin, 2012; Berezin, 1974, 1975; Gersho, 1977, 1978; Gray & Neuhoff, 1998; Yang et al., 2019).

10.7 Radar signature of rain cells on the sea surface

The radar signatures of rainfall within SAR images of the sea surface stem from various intricate physical processes. In this sense, radar backscattering is a phenomenon influenced by microwaves interacting with the sea's surface, particularly short surface waves known as "Bragg waves." The amplitudes of these waves undergo modification due to the presence of raindrops impacting the sea surface, resulting in alterations in radar backscatter patterns. In this view, let σ represent the radar backscattering amplitude, A_{Brag} be the amplitude of the incident radar waves, h denote the height of Bragg waves, and ρ represent the raindrop density. Thus the modification of the Bragg waves and their impacts on radar backscattering can be expressed as:

$$\sigma = \lambda \cdot \rho \cdot h \cdot \left(1 - e^{(-2 \cdot \rho \cdot h \cdot d^2)}\right). \tag{10.18}$$

here, Eq. (10.18) demonstrates how the radar backscattering amplitude (σ) is influenced by the amplitude of the incident radar waves (λ), the height of Bragg waves (h_{Bragg}), and the raindrop density (ρ). The term d represents the distance traveled by radar waves. Therefore Eq. (10.18) provides a mathematical description of how raindrop impacts on Bragg waves at the sea's surface can lead to modifications in radar backscatter patterns. Let σ_r denote the radar signal roughness, β_0 be the baseline roughness of the sea surface, U_w represent the wind speed associated with the rain cells, and ρ denote the size or intensity of the rain cells. The effect of rain cells and wind gusts on sea surface roughness can be expressed as:

$$\sigma_r = \beta_0 + f(U_W, \rho). \tag{10.19}$$

This equation provides a mathematical description of how rain cells, accompanied by wind gusts, introduce surface roughening effects and influence radar signatures. Therefore let $\sigma_r(\theta)$ represent the roughness factor, which depends on the radar incident angle θ, raindrop size or intensity ρ, and sea surface roughness due to rain can be expressed as:

$$\sigma_r(\theta) = \frac{k_3}{\rho}\int_0^{\rho} P(\rho)d\rho.\sin(k_4.\theta) \tag{10.20}$$

here, k_3 and k_4 are constants, and $P(\rho)$ is the raindrop size distribution. Consequently, the atmospheric attenuation $\mu(\lambda)$ due to the presence of raindrops depends on the radar wavelength λ, raindrop density $D(\rho)$, and the scattering cross section σ; which is given by:

$$\mu(\lambda) = k_5\int_0^{\infty} D(\rho)\sigma(\lambda)d\rho \tag{10.21}$$

here k_5 is constant. To develop a comprehensive model that combines these effects, we can express the total radar backscattering coefficient σ_T as:

$$\sigma_T(\theta, \lambda) = \sigma(\theta, \lambda) \cdot \sigma_r(\theta) \cdot e^{-\mu(\lambda)} \tag{10.22}$$

This equation represents the combined influence of Bragg waves, sea surface roughening due to rain, and atmospheric effects caused by raindrops on the radar backscattering coefficient. The attenuation factor $e^{-\mu(\lambda)}$ accounts for the attenuation of radar signals in the atmosphere due to raindrop interactions. In summary, the radar signatures of rain in SAR images of the sea surface are the result of complex interactions between radar waves, raindrops, the sea surface, and the atmosphere. These interactions can be observed and analyzed in SAR imagery to provide valuable information about rainfall and its effects on the sea surface (Melsheimer & Gade, 1998; Melsheimer, Romeiser, et al., 1998; Melsheimer, Alpers, et al., 1998, 2001; Zhang & Alpers, 2018).

Therefore involving Eq. (10.17) into Eq. (10.22), one can obtain:

$$\sigma = \sigma_0 + k_3 \cdot \rho \cdot D^3 \cdot t \cdot \left(1 - e^{\left(-V^2/(2 \cdot g \cdot D \cdot t)\right)}\right) + k_4 \cdot \tau_w \tag{10.23}$$

Eq. (10.23) demonstrates the impact of rain cells on the backscatter of the sea roughness under wind stress (τ_w) and raindrop size (D). Needless to say, this equation represents simplified relationships and is part of a more comprehensive model for radar backscattering due to rain cells on the sea surface.

10.8 Quantanization of rain cell impact on radar backscatter

Quantanizing the impact of rain cells on radar backscatter involves a quantum mechanical perspective, where rain cells exhibit both wave-like and particle-like behaviors. While this approach is typically applied to the quantum realm, it's a creative way to conceptualize the radar backscattering process in the presence of rain cells. In this view, the quantum radar backscattering equation is given as:

$$\sigma_{RCS} = \frac{\sigma_0}{\sqrt{2\pi\sigma}} e^{\left(\frac{-(h-h_0)^2}{2\sigma^2}\right)} \tag{10.24}$$

here σ_{RCS} is the radar backscattering cross section due to the impact of raindrops. Therefore h is the height of the radar system above the sea surface; and h_0 is the mean height of the sea surface. In this context, σ is the standard deviation representing the uncertainty in the radar backscatter. The quantum radar backscattering equation expresses how radar waves interact with the sea surface, which is perturbed by the presence of rain cells. This equation encapsulates the fundamental characteristics of radar backscattering and its uncertainty due to these rain cells. On other words, this equation is formulated based on a probabilistic framework, acknowledging the inherent uncertainties in radar backscattering. It allows us to calculate the probability of obtaining a specific radar backscattering amplitude (Alpers et al., 2016; Ye et al., 2016; Zhang & Alpers, 2018). This concept reflects the dual nature of radar waves interacting with the sea surface, much like particles in the quantum world.

Quantum radar backscattering probability perhaps is given by:

$$P(\sigma_{RSC}) = \int_{-\infty}^{\infty} |\Psi(\sigma_{RSC})|^2 d\sigma_{RSC} \tag{10.25}$$

where $P(\sigma_{RSC})$ is the probability of obtaining a radar backscattering cross section; and $\Psi(\sigma_{RSC})$ is the radar backscattering amplitude wave function. This representation treats radar backscattering due to rain cells as a quantum phenomenon, where uncertainties in both height (h) and radar backscatter (σ) are related by Heisenberg's uncertainty principle as follows:

$$\Delta h \cdot \Delta\sigma \geq \frac{\hbar}{2} \tag{10.26}$$

here Δh is the uncertainty in height; and $\Delta\sigma$ is the uncertainty in radar backscatter. It's important to note that this approach is a conceptual interpretation of radar backscattering influenced by rain cells, utilizing quantum mechanical principles. As we explore this quantum radar backscattering equation, we're essentially applying the quantum mechanics framework to understand how rain cells influence radar backscattering. This conceptual approach opens up exciting possibilities for a deeper comprehension of radar phenomena in the presence of rain cells (Melsheimer & Gade, 1998; Melsheimer, Romeiser, et al., 1998; Ye et al., 2016; Zhang & Alpers, 2018). While it can be considered as the new method for rain cells understanding imaging mechanism in radar image, it offers a unique perspective inspired by quantum mechanics.

10.9 A quantum finite automaton algorithm

Let's delve into the intriguing world of quantum automata and their interaction with rain cell pixels in SAR images. In this theoretical exploration, we're envisioning a scenario where a quantum automaton operates within a state space representing N possible rain cell pixels in SAR images. These states are akin to N-state qubits, denoted as |j, within the complex and real number space $\mathbb{R}$.

Within this SAR image space, rain cells are introduced as agents of change, altering the Bragg surface and inducing irregular pattern flows, often referred to as turbulence flow. This dynamic effect can be mathematically represented. Specifically, we describe the quantum state of N-state qubits associated with exploring rain cell pixels and their role in generating turbulence flow as $|\Psi| \in \mathbb{R}$. Now, in the quantum automaton framework, state transitions are at the heart of its operation. These transitions are elegantly captured through a set of N × N unitary matrices, denoted as Uα. Each of these matrices serves as a key component of the quantum automaton algorithm. The transition process in the quantum automaton algorithm involves a collection of N × N unitary matrices denoted as Uα. These unitary matrices describe the transition of the automaton from its current state |ψ|to its subsequent state |φ|. Mathematically, this transition can be represented as:

$$U_\alpha : |\psi| \rightarrow |\varphi| \tag{10.27}$$

Eq. (10.27) encapsulates the foundational dynamics of quantum automata and their state transitions. It is within these state transitions that the quantum automaton's algorithm unfolds, guiding its exploration of rain cell pixels and their influence on the emergence of turbulence flow patterns. This theoretical framework, inspired by quantum mechanics, provides a unique lens through which we can understand the intricate interplay between quantum automata and rain cells in SAR images. According to pervious sections, the rain cells can increase the sea surface roughness, which leads to high backscatter (Ambainis & Freivalds, 1998; Ambainis & Watrous, 2002; Birkan et al., 2021).

In the context of quantum radar backscattering, we introduce an important concept: the impact of rain cells on sea surface backscatter, denoted by σ. This impact is not a static process but is governed by a transition function, T, that operates on the space of sea surface backscatter signals, denoted as *S*. The transition function *T* describes how the presence of rain cells modifies the backscattered signals. The quantum semiautomaton, which plays a central role in modeling this process, can be mathematically expressed as:

$$T_{rp} = C\mathbb{R}^N; S; U_\alpha | \alpha \in S \tag{10.28}$$

In this equation, T_{rp} represents the quantum semiautomaton, where σ signifies the input of sea surface backscatter signals affected by rain cells; N represents the number of possible states, which are characteristic of the quantum automaton's behavior; and *S* denotes the space of sea surface backscatter signals. The set $\{U\alpha \mid \alpha \in S\}$, therefore, represents a collection of unitary matrices, each indexed by elements from the set S. These unitary matrices describe the transitions within the quantum semiautomaton. Essentially, this mathematical expression encapsulates the dynamic interplay between sea surface backscatter, rain cell impact, and the quantum automaton's state transitions (Ambainis & Freivalds, 1998; Birkan et al., 2021; Qiu, 2021). It provides a framework for understanding how rain cells influence backscattered radar signals in a quantum context.

Therefore when implementing the input strings s = (s_0, s_1, ⋯ , s^k;), the probability of the state machine can be modified as:

$$Pr(s) = k \times P\left(U_{\mathrm{sk}}/U_{\mathrm{s1}}U_{\mathrm{s0}}\right)/|D_{\mathrm{k}}|^2 \tag{10.29}$$

In this expression, *Pr*(*s*) represents the probability of the state machine based on the input strings *s*, while *k*, *P*, *U*, and D_k have their respective defined meanings within the context of the problem or system being described. This string array is built as function of the rain cell properties, that is, backscatter, size and shape, length, width, and coverage pixel areas in SAR data. The expression indicates a probability calculation that depends on the specific input sequence and associated parameters. In the realm of radar image backscatter mechanisms, a string typically represents a fundamental data structure. It acts as an array, a conglomeration of qubits specifically arranged to harbor a sequence of rain cell pixels. These pixels are intrinsically tied to rain cell pixels, encoded within the qubits (Ambainis & Freivalds, 1998; Ambainis & Watrous, 2002; Qiu et al., 2011; Say & Yakaryılmaz, 2014). The term "string," in this context, extends its definition to encompass more generalized arrays or sequences, serving as repositories of rain cells pixels derived from synthetic aperture radar (SAR) images. These arrays adhere to intricate structures housing rain cells pixels, as depicted in Fig. 10.6.

Consequently, the automaton's initial state can be adequately portrayed as the vector |Ψ>. In this context, the automaton exists in a state of readiness, poised to process the input string prior to its submission. Constructing the unit matrix predicated on the likelihood of the initial state's acceptance assumes the following form:

$$Pr(\emptyset) = |P||\Psi\rangle||^2 \tag{10.30}$$

Eq. (10.16) elucidates the essence of the empty string (Ø), presenting itself as the antithesis of the unit matrix. The identification of rain cell properties is orchestrated through the utilization of a Mealy machine. In this context, the SAR single look complex serves as the input, generating irregular backscatter and variation of rain cell texture properties as the output. Each transition edge within this process bears labels denoting the values of the input |O|B and the output |O'|B', as illustrated in Fig. 10.7 (Ambainis & Watrous, 2002; Qiu et al., 2011; Say & Yakaryılmaz, 2014).

Initiating from the state $\Psi|S_i>$, this machine operates as an edge detector. Consequently, it produces a "one" each time the input undergoes a switch and a "zero" otherwise. Certainly, Table 10.1 shows the pseudocode representation for a Mealy machine designed for the automatic detection of rain cells in SAR (synthetic aperture radar) images:

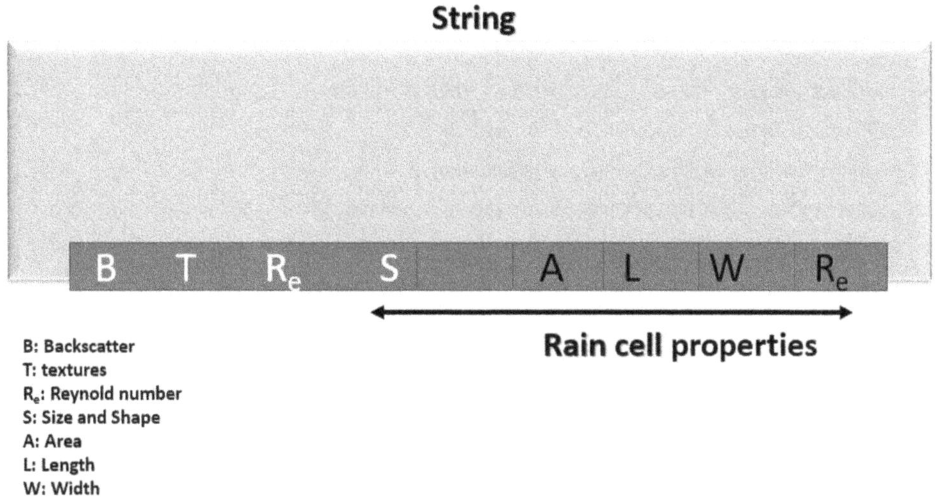

FIGURE 10.6 The string array encompasses the properties of rain cells within SAR data. *SAR*, Synthetic aperture radar.

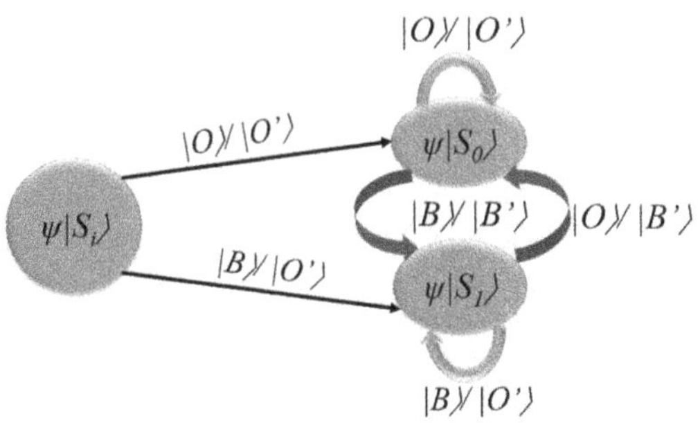

FIGURE 10.7 Mealy machine designed for the automatic detection of rain cells in SAR imagery. *SAR*, Synthetic aperture radar.

This pseudocode presents a basic structure for a Mealy machine that scans through each pixel in the SAR image. It operates based on defined threshold values and different states to detect rain cells within the image. The specific implementation and choice of threshold values will depend on the actual requirements and characteristics of the SAR imagery being analyzed.

10.10 Automatic detection of rain cells in synthetic aperture radar imagery

To develop a mathematical framework for the automatic detection of multiple clusters caused by rain cells using Hilbert space, one could represent the rain cell properties in terms of quantum states. This approach can involve the encoding of these properties into qubits within a Hilbert space framework. Let's consider a Hilbert space denoted by $\mathcal{H}$ and the rain cell properties encoded as quantum states within this space. Let $|c_1\rangle, |c_2\rangle, \ldots, |c_n\rangle$ represent the quantum states associated with different clusters of rain cell properties. The superposition of these states within the Hilbert space $\mathcal{H}$, therefore, can be expressed as:

$$|\psi\rangle = \alpha_1|c_1\rangle + \alpha_2|c_2\rangle + \cdots + \alpha_n|c_n\rangle \tag{10.31}$$

Here, $|\psi\rangle$ represents the overall state of the rain cell clusters, and $\alpha_1, \alpha_2, \ldots, \alpha_n$ are probability amplitudes corresponding to each cluster state $|c_i\rangle$. In this perspective, the quantum measurements applied after encoding the rain cell properties in qubits can involve operators corresponding to observables related to these properties (Debnath & Mikusinski, 2005; Skorohod, 2012). These operators could be represented as $M_1, M_2, \ldots, M_n$ in the Hilbert space. In this sense, the probability of detecting a specific rain cell cluster $|c_i\rangle$ after a measurement is given by the Born rule as:

$$P_i = \langle\psi|M_i^\dagger M_i|\psi\rangle \tag{10.32}$$

Eq. (10.32) represents the probability associated with measuring the state $|c_i\rangle$ after the action of the operator M_i on the state $|\psi\rangle$. In this understanding, the three components of the Hilbert space $\mathcal{H}$ could correspond to different aspects or parameters of rain cell properties, such as backscatter $|B\rangle$, width $|W\rangle$, and texture of turbulence $|T\rangle$. These components could be represented mathematically as subspaces within the larger Hilbert space, allowing for a multidimensional representation of the rain cell properties (Gross, 1967; Young, 1988). Mathematically, these components can be indicated as subspaces, providing a multidimensional representation of the rain cell properties as follows:

$$\mathcal{H} = \mathcal{H}_\beta \otimes \mathcal{H}_W \otimes \mathcal{H}_T \tag{10.33}$$

where $\mathcal{H}_\beta$ is the subspace corresponding to intensity, spanned by $|\beta\rangle$ and its associated basis; $\mathcal{H}_W$ is the subspace

TABLE 10.1 Mealy machine pseudocode for the automatic detection of rain cells in synthetic aperture radar data.

```
FUNCTION detectRainCells(SAR_image): STATE currentState = INITIAL_STATE BOOLEAN
rainDetected = FALSE FOR EACH pixel IN SAR_image: READ pixelIntensity =
GET_PIXEL_INTENSITY(pixel) CASE currentState OF STATE_1: IF pixelIntensity >=
THRESHOLD_VALUE_1: rainDetected = TRUE OUTPUT "Rain cell detected" currentState =
STATE_2 ELSE: currentState = INITIAL_STATE
 STATE_2: IF pixelIntensity < THRESHOLD_VALUE_2: rainDetected = FALSE OUTPUT "No rain
cell detected" currentState = INITIAL_STATE ELSE: currentState = STATE_1 END FOR
RETURN rainDetected END FUNCTION
```

TABLE 10.2 Pseudocode of Hilbert space for automatic detection of rain cell properties.

```
FUNCTION detectRainCellsInHilbertSpace(SAR_image): INITIALIZE Hilbert space H for rain
cell properties FOR EACH pixel IN SAR_image: intensity = GET_Backscatter(pixel) //
Calculate intensity value of the pixel size = GET_width (pixel) // Calculate Width value of
the pixel duration = GET_texture (pixel) // Calculate texture turbulence value of the pixel //
Create quantum states for backscatter, size, and texture turbulence quantumState =
CREATE_QUANTUM_STATE(backscatter, size, and texture turbulence) // Apply operators
on the Hilbert space for measurement or computation result =
APPLY_OPERATORS(quantumState, HilbertSpaceOperators) // Evaluate the result or take
further actions based on the computation END FOR END FUNCTION
```

for width and shape, spanned by $|W\rangle$ and its associated basis.; and $\mathcal{H}_T$ is the subspace for texture of turbulence, spanned by $|T\rangle$ and its associated basis. Each subspace represents one specific aspect of the rain cell properties. The tensor product ($\otimes$) combines these subspaces to create the multidimensional Hilbert space $\mathcal{H}$ that collectively captures the various parameters characterizing rain cells in SAR imagery. In this view, Table 10.2 provides a representation that demonstrates the concept using pseudocode.

This pseudocode outlines a conceptual implementation of using a Hilbert space in a quantum-like fashion for rain cell automatic detection in SAR imagery. It involves the creation of quantum states based on backscatter, size or width, and texture of turbulence, and the application of certain operators or computations within the Hilbert space. Thus the automatic detection of rain cells is given by:

$$\mathcal{H}_{accept} = span\{|q\rangle : |q\rangle \in S_{accept}\} \tag{10.34}$$

Eq. (10.34) vividly delineates the linear span within the subset of accepted rain cell clusters in SAR data. Conversely, the identification of disregarded, modified rain cell pixels in the SAR image follows a similar process. The resultant space within the SAR image denotes the nonhalting subspace. Each subpixel within SAR data is projected into three constituents: P_{acc}, P_{rej}, and P_{non}, aligning with the actual rain cell pixels as depicted by:

$$P_{accept}: \mathcal{H}_Q \rightarrow \mathcal{H}_{accept} \tag{10.35}$$

Therefore Table 10.3 shows a conceptual pseudocode representation of linear span for rain automatic detection in SAR images. This pseudocode outlines a conceptual representation of determining the linear span within the subset of accepted rain cell clusters in SAR data. It involves iterating through the SAR data, identifying pixels belonging to the accepted cluster, and then performing mathematical operations (such as calculating linear spans) on the collected subset.

TABLE 10.3 Linear span pseudocode representation of linear span for rain automatic detection in synthetic aperture radar data.

FUNCTION linearSpanAcceptedRainCells(SAR_data): ACCEPTED_RAIN_CELLS_SUBSET = EMPTY_SET // Initialize an empty set for accepted rain cells FOR EACH pixel IN SAR_data: IF pixel BELONGS_TO_ACCEPTED_CLUSTER: ADD pixel TO ACCEPTED_RAIN_CELLS_SUBSET // Add the pixel to the set of accepted rain cells // Perform linear algebra or mathematical operations to determine linear span LINEAR_SPAN = CALCULATE_LINEAR_SPAN(ACCEPTED_RAIN_CELLS_SUBSET) RETURN LINEAR_SPAN // Return the calculated linear span within the accepted rain cell subset END FUNCTION

In this scenario, the accurate probability of the actual rain cell clusters is given by:

$$Pr(\sigma)_{accept} = \left|\left|P_{accept}|\psi\rangle\right|\right|^2 \tag{10.36}$$

Eq. (10.36) posits that when the wave function collapses into the "accept" or "reject" clusters, the subsequent processing comes to a halt (Ambainis & Freivalds, 1998; Ambainis & Watrous, 2002; Birkan et al., 2021; Nielsen & Chuang, 2000; Qiu, 2021). Conversely, the processing continues onward when dealing with subsequent clusters derived from the input qubits, unfolding into what can be described as an eigenstate of P_{non}.

10.11 Quantum operator gate

The parameter inputs in the canonical algorithm of the quantum finite automaton are developed through the utilization of the controlled-NOT gate, often referred to as C-NOT or CNOT. According to Monroe et al. (1995), this quantum logic gate is intricately associated with each pixel and line of every individual qubit. The CNOT gate operates within a quantum register containing two qubits, as depicted in Fig. 10.8. Its function involves flipping the second qubit, defined as the target qubit, exclusively when the first qubit, known as the control qubit, assumes a state of |1> (Monroe et al., 1995; O'Brien et al., 2003; Okamoto et al., 2005).

The transformation of the quantum state by the CNOT gate allows for the general input to be a linear superposition of $\{|0\rangle, |1\rangle\}$. This transformation can be represented mathematically as:

$$CNOT(|\psi\rangle \otimes |\phi\rangle) = |\psi\rangle \otimes |\phi \oplus \psi\rangle \tag{10.37}$$

Here, $|\psi\rangle$ and $|\phi\rangle$ represent the quantum states of the two qubits involved, while the symbol $\oplus$ represents the XOR operation. This equation denotes the action of the

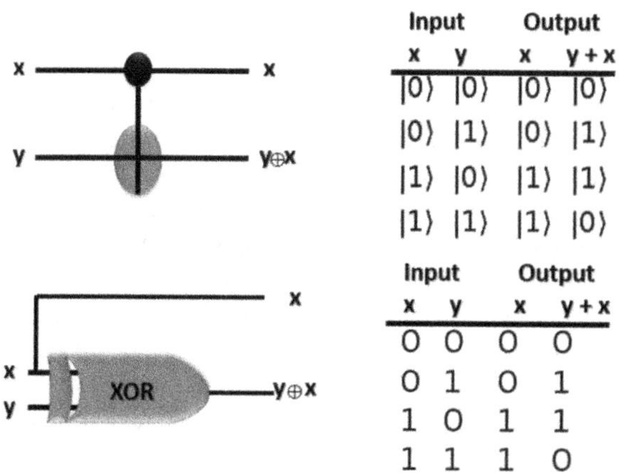

FIGURE 10.8 Reversible XOR gate serves as the classical equivalent of the CNOT gate.

CNOT gate on the input states, resulting in a specific transformation of the quantum state (Chen et al., 2008). Thus the action of the CNOT gate can be represented by the following matrix:

$$CNOT = \begin{bmatrix} 1 & 0 & 0 & 0 \\ 0 & 1 & 0 & 0 \\ 0 & 0 & 1 & 0 \\ 0 & 0 & 0 & 1 \end{bmatrix} \tag{10.38}$$

The CNOT gate, integral in quantum computing, operates within a two-qubit system where one qubit acts as the controller and the other as the recipient of change. This matrix serves as a graphical representation of how the gate influences the quantum states of these qubits. Essentially, the control qubit determines whether the transformation will occur, whereas the target qubit is subject to change based on the state of the control qubit. The matrix entries delineate this behavior: when the control qubit is in the $|0\rangle$ state, the target qubit remains unaffected, while a state change is triggered in the target qubit when the control qubit is in the |1⟩ state. In essence, this gate's functionality lies in its ability to entangle qubits, facilitating quantum superposition and parallelism, thus enabling the intricate operations essential for quantum algorithms and circuit design in quantum computing (Nielsen & Chuang, 2000).

10.12 In situ measurement and synthetic aperture radar satellite data

Measuring rain intensity while navigating the open sea from a boat demands a blend of ingenious methods. Implementing rain gauges or collection bins directly on the boat allows for the quantitative assessment of rain volume, providing a hands-on approach to gage the precipitation's strength. Employing marine radars equipped on the vessel grants the ability to scrutinize and interpret rainfall intensity in the surrounding marine expanse, revealing the varying degrees of rainfall. Additional weather instruments, including anemometers, hydrometers, and weather stations, offer supplementary data on atmospheric conditions, indirectly hinting at the potential intensity of rainfall. Remote sensing devices, along with satellite imagery, widen the horizon by providing a broader perspective on rain cells and precipitation rates, offering insights into the grand mosaic of rain formation over the sea. Integrating these diverse methods forms a robust strategy, combining direct measurements with comprehensive remote and satellite observations to decode the tapestry of rain intensity at sea from the boat's vantage point.

In this study, novel acoustic rain gages (ARGs) have been implemented to be securely positioned beneath the ocean surface (Fig. 10.9), capturing and registering the acoustic emissions generated by rain. Distinctive sound profiles produced by various types of rainfall enable researchers to leverage these recordings, enabling identification and quantification of raindrop dimensions, precipitation intensity, and other attributes of rain patterns above the ocean. The auditory data captured by the acoustic rain gage emanate from the specific sea surface area above the device, with the coverage expanding as the gage is submerged deeper in the ocean (Avanzato & Beritelli, 2020; Kumari et al., 2023; Ma, 2022). A comprehensive assessment of rainfall over an extended region proves more advantageous than isolated point measurements, as rainfall exhibits considerable variability within short proximities. This approach yields a more holistic understanding of rainfall behavior over the ocean expanse (Avanzato & Beritelli, 2020; Ma, 2022). In this instance, during the TerraSAR wide-scan SAR (WS) passage over the study area on October 7, 2022, along the coastal waters of Kelantan, Malaysia (Fig. 10.10). In this perspective, the device was deployed to unravel the mechanisms underlying how intense rain induces turbulence within the sea surface and water column.

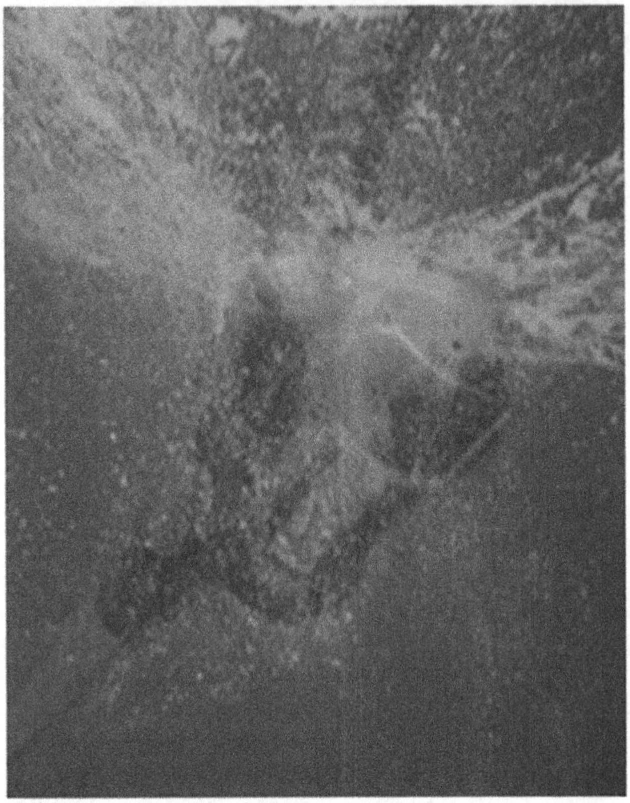

FIGURE 10.9 Novel acoustic rain gages (ARGs) used during TerraSAR wide-scan SAR (WS) transited over the study area. *SAR*, Synthetic aperture radar.

Utilizing wide-Scan SAR (WS) mode, an impressive swath width of up to 270 kilometers is attained by scanning six adjoining ground subswaths. However, due to the beam-switching process, solely bursts of SAR echoes are received, leading to a constrained bandwidth and subsequently, a decrease in azimuth resolution. This mode achieves a spatial resolution of 40 meters. For easier management of image file size, the standard scene size spans up to 270 kilometers by 200 kilometers in Wide-Scan SAR. Additionally, the acquisition length can reach up to 1500 kilometers in Wide-Scan SAR (Steinbrecher et al., 2014).

10.13 Automatic detection of rain cells in synthetic aperture radar image

Within the quantum finite automaton algorithm, the transformed physical properties of rain cells pass through gates, symbolizing a "1" in the digital data, and are then mapped onto corresponding SAR pixels. Meanwhile, the

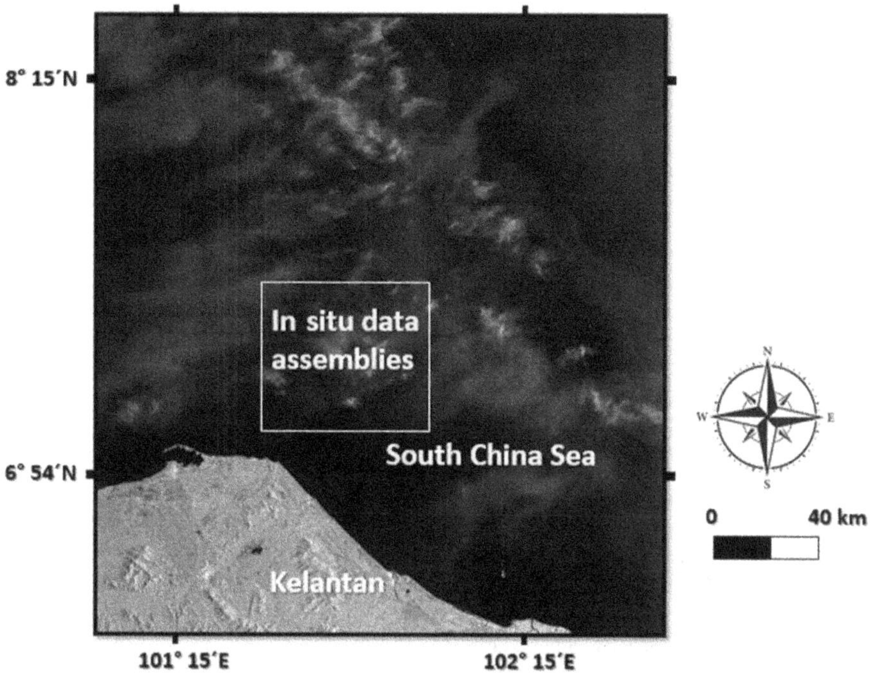

FIGURE 10.10 TerraSAR wide-scan SAR (WS) passed over the study area on October 7, 2022, along the coastal waters of Kelantan, Malaysia. *SAR*, Synthetic aperture radar.

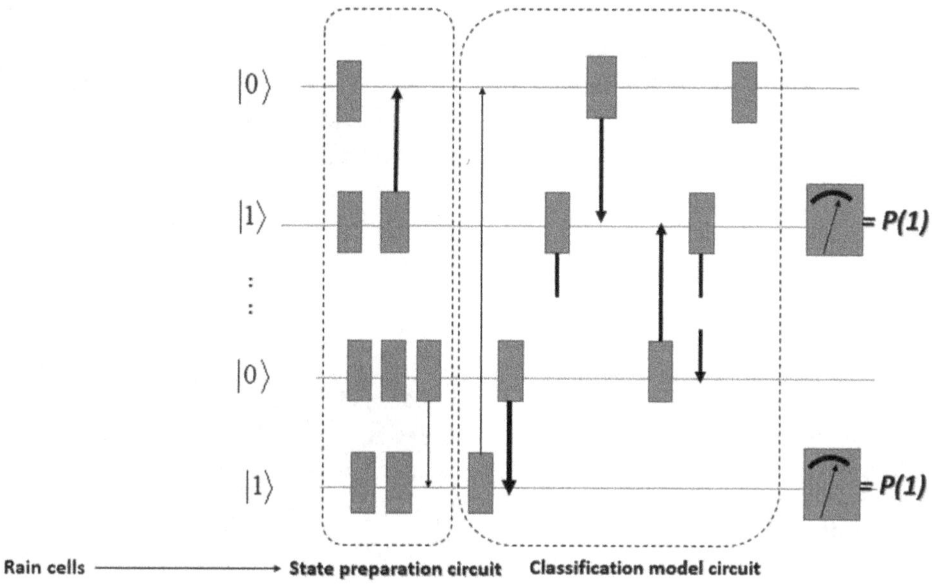

FIGURE 10.11 Quantum circuit used in rain cell automatic detections.

surrounding features are denoted as "0." In this context, these logic gates serve as foundational elements for processing rain cells information. Particularly, the AND gate plays a pivotal role in swiftly determining the agreement between in situ measurements and SAR imagery. Operating on two bits of information, it yields an output of "1" only if both incoming bits are registered as 1 s. Hence, when the algorithm detects conformity between in situ measurements and SAR image, signaling a "yes" (1), the AND gate outputs a corresponding "1." On the contrary, if there is disagreement, where one or both components diverge, indicating a "no" (0), the AND gate generates an output of "0" or a negative result (Fig. 10.11).

In this perspective, the qubits exhibiting both qudits' spin orientations, whether up or down, traverse across the image. Their journey aims to discern and attribute specific pixels to the unique signatures of rain cell backscatter and their texture of turbulence within the SAR image (Fig. 10.12).

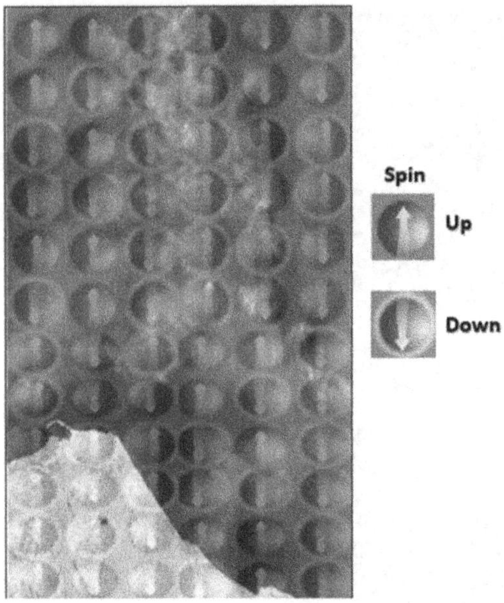

FIGURE 10.12 Qubits exhibit both spin-up and spin-down states to ascertain the physical characteristics of rain cells within SAR imagery. *SAR*, Synthetic aperture radar.

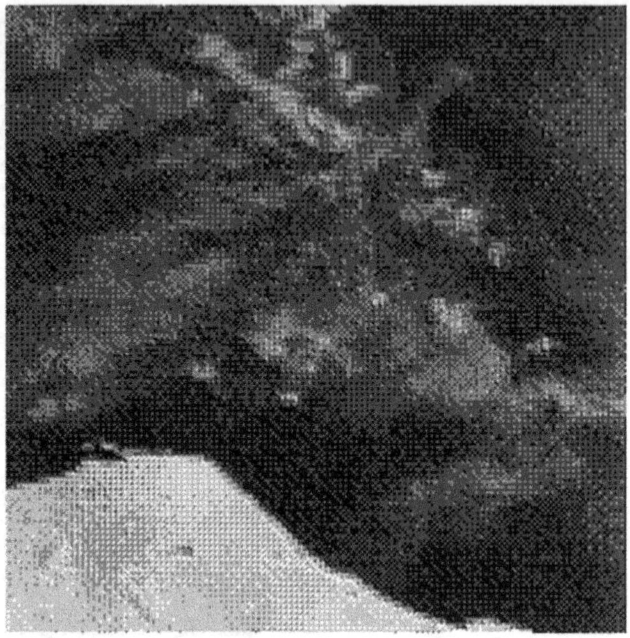

FIGURE 10.13 Rain cell probability partition function.

Upon convergence to the genuine rain cell backscatter and texture turbulence signatures in a quantum state during training, the loss function's value can serve as an indicator of the free energy contained within the actual physical properties' signature states. This free energy is directly proportional to the logarithm of the backscatter signature partition function, depicted by the following equation:

$$\mathcal{L}(\theta) = -\log Z_s \tag{10.39}$$

In this context, Z_s represents the spectral partition function, instrumental in formulating the quantum simulation task as a quantum-probabilistic variational learning assignment (as depicted in Fig. 10.13). Within this scenario, a quantum-probabilistic framework is established for the genuine rain cell physical property signature states, characterized as $\beta(\theta, \varphi)$, parameterized by (θ, φ). Consequently, the derivation of a variational cluster for the quantum probability states (as illustrated in Fig. 10.14) is attained by minimizing the free energy acting as the loss function.

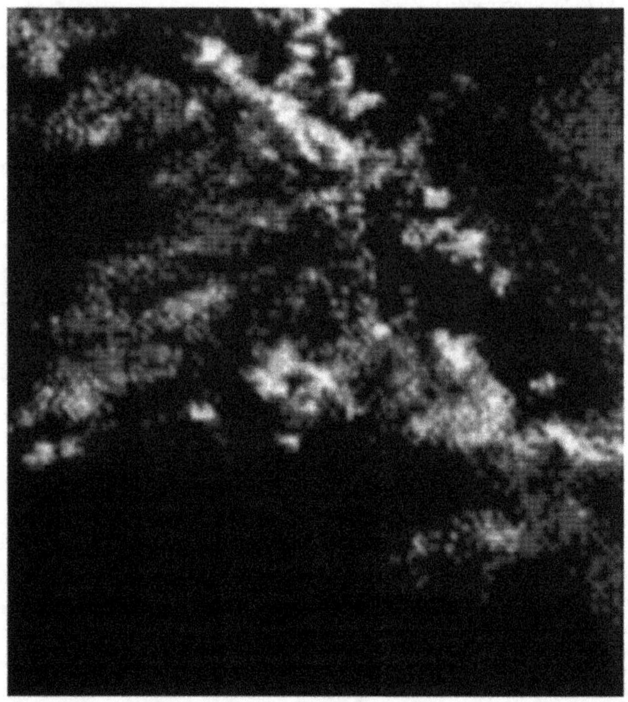

FIGURE 10.14 Minimizing the loss function of the quantum rain cell states.

$$\mathcal{L}(\theta, \varphi) = Z_s tr\left(\widehat{\beta}\ (\theta, \varphi)\hat{H}\right) - \hat{M}\left(\widehat{\beta}\ (\theta, \varphi)\right) \tag{10.40}$$

This leads to the restoration of the loss function for the ground state variational quantum eigensolver (VQE), thereby reviving the ground state variational principle, depicted as:

$$\mathcal{L}(\theta, \varphi) = \langle \Psi(\theta, \varphi) | \hat{H} | \Psi(\theta, \varphi) \rangle \tag{10.41}$$

here $\Psi(\theta,\varphi)$ represents the variational quantum state parameterized by θ and φ, and $\hat{H}$ denotes the Hamiltonian operator. Eq. (10.41), therefore, delves into the limited realm where a zero spectral signature prevails, denoting an absence of necessity for hidden space parameters due to the absence of information. Under these circumstances, the hidden state aligns unitarily with any initial basic pure state. In this context, the computational simulation of the system's dimension becomes viable on a classical computer. Essentially, the simulation of the wave function directly within the Hilbert space becomes imperative (Ambainis & Freivalds, 1998; Ambainis & Watrous, 2002; Birkan et al., 2021; Qiu, 2021). As a result, models based on quantum Hamiltonians function across a diverse spectrum of quantum rain cell states, as illustrated in Fig. 10.15. Simply put, it acts akin to a kernel window detector, unraveling the genuine partition associated with the sea surface turbulence induced by the footprint of rain cells, a discussion which follows in subsequent sections.

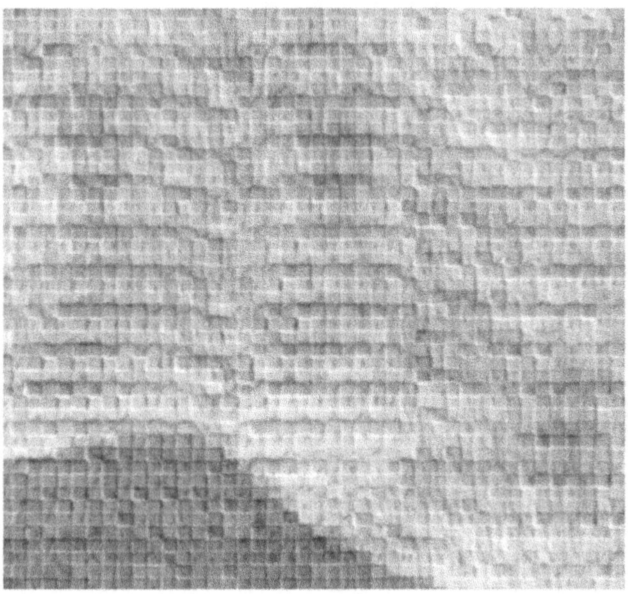

FIGURE 10.15 Quantum Hamiltonian for searching footprint signature of rain cells.

10.14 Rain cells' training score and the bias of the automated classifier

Both the training score and the bias of the classifier hinge upon the precise gradient estimation of the partition function within the authentic spectrum of gold feature indices. In this context, Hamiltonian-based models serve as a basis for formulating an analytical expression for the partition function. Consequently, by training solely on the Hamiltonian gradient, the construction of gradient variations between diverse rain cell patterns and the encompassing sea environment is initiated, as depicted in Fig. 10.16.

Fig. 10.16 illustrates the gradient boundary fluctuations stemming from the Hamiltonian gradient across distinct spectral feature categories. Subsequently, upon determining the boundary gradients among various spectral classes, the identification of class labels becomes more straightforward. Employing the Hamiltonian gradient for simulating the backscatters and turbulence textures of rain cells aids in mitigating bias among the different classes.

Hence, the precise rain cell footprint signatures, causing turbulence across the sea surface, must conform within the constrained interval of (-0.5, +0.5) to yield significant meaning, unveiling the probabilistic fluctuations for the development of quantum rain cell turbulence states with inherent bias. Notably, a substantial probability of 0.9 correlates with a minimal bias of -0.5, as illustrated in Fig. 10.17.

Fig. 10.18 illustrates various quantum states characterizing rain cell features and their significant impact on sea surface turbulence, utilizing the quantum finite automaton algorithm. Notably, the most pronounced turbulence with a Reynolds number of 2.8×10^5 correlates with a recorded rainfall of 560 mm using ARGs. Within this context, the observed wind speed, approximately 6 m/s, spanning 200 km along the offshore region of Kelantan's

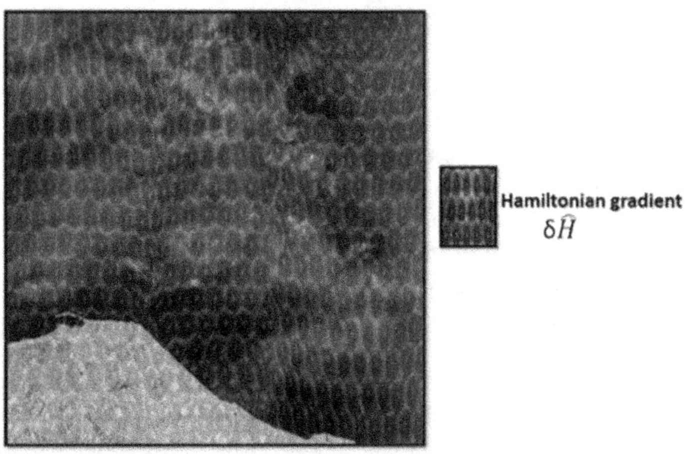

FIGURE 10.16 Most pronounced gradient boundaries observed within the spectrum of classes in the Hamiltonian-gradient mapping.

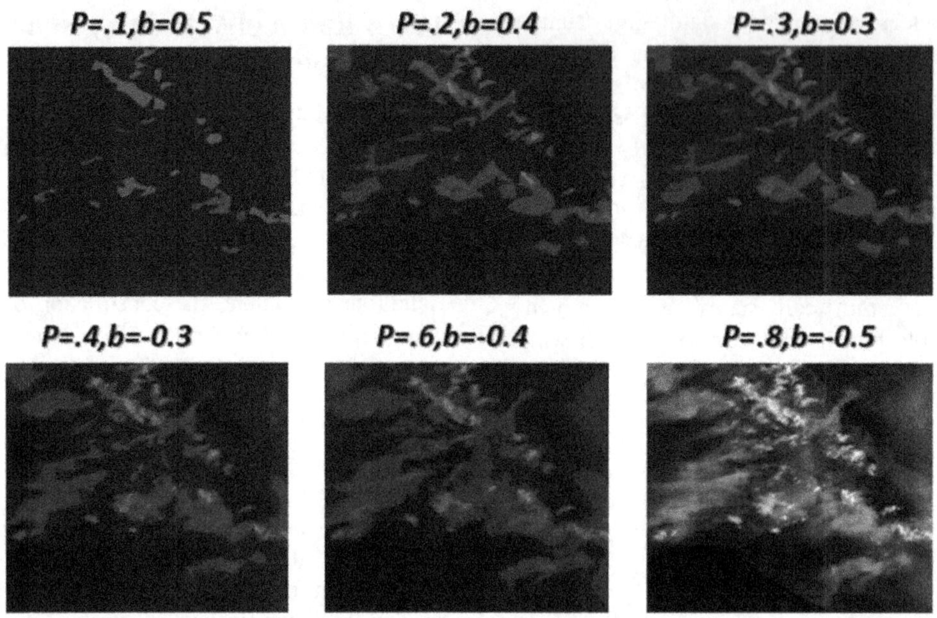

FIGURE 10.17 Construction of rain cell states: probability and bias.

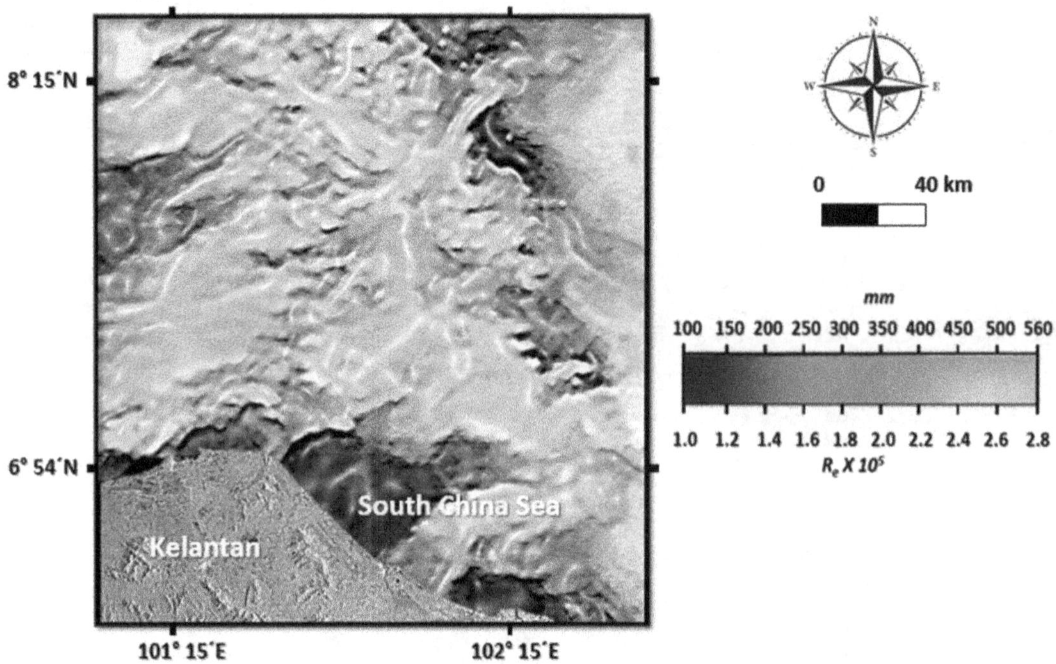

FIGURE 10.18 Automatic classification of rain cells and turbulence flow using the quantum finite automaton algorithm.

coastal waters, intensified the turbulence over this expanse. It is also noteworthy that the distinct boundaries of rain cells are effectively discerned through the application of the quantum finite automaton algorithm.

In a parallel fashion, the quantum finite automaton algorithm functions as a quantum supervised algorithm, enabling the automatic detection of rain cells on October 6, 2023, along the coastal waters of Kelantan, Malaysia, inducing a convergence in the coastal front within the Sentinel-1 strip map mode VH polarization. This event led to the creation of two converging fronts due to the reduction in sea surface temperature, as rainfall typically exhibits lower temperatures associated with cloud systems of lower temperatures (as depicted in Fig. 10.19).

This trajectory demonstrates the proficient identification of rain cell clusters and turbulence boundaries by employing the Hamiltonian-gradient algorithm within the framework of the quantum finite automaton algorithm.

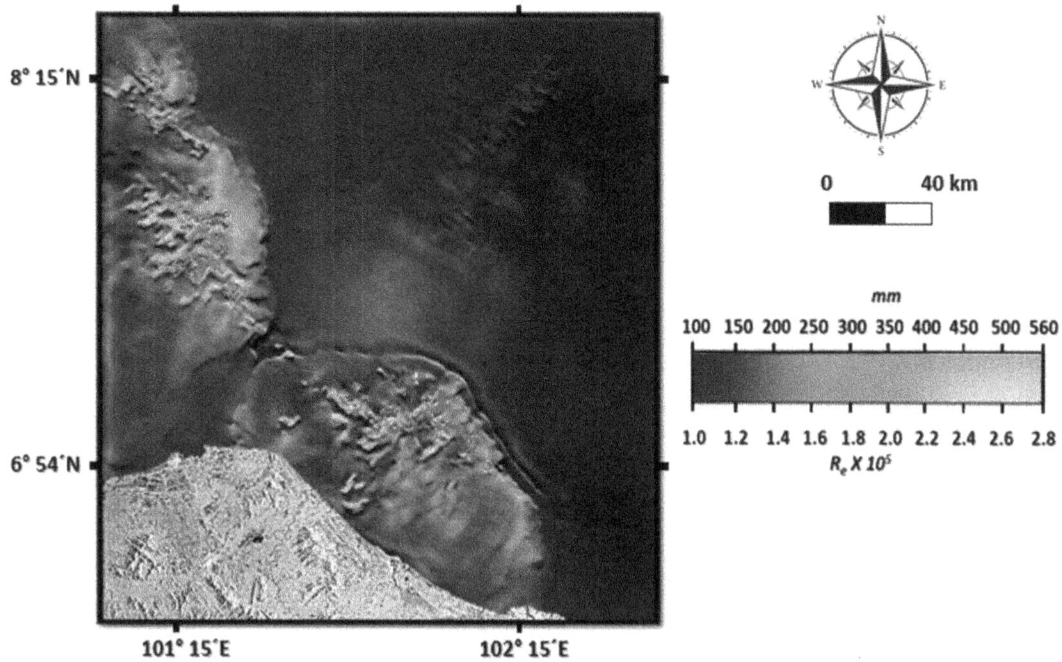

FIGURE 10.19 Automatic classification of rain cells and turbulence flow within the Sentinel-1 strip map mode using the quantum finite automaton algorithm.

Notably, these outcomes corroborate the methodologies elucidated in Fig. 10.17. The primary inquiry at hand revolves around the mechanism through which rainfall exerts its influence on the reduction of sea surface temperature. Rainfall influences the sea surface temperature through various mechanisms. First, as raindrops make contact with the ocean's surface, they initiate a process of evaporative cooling. This process involves the absorption of heat energy from the surrounding area as the rainwater evaporates, resulting in a cooling effect. Additionally, the rainwater, often cooler than the ocean surface, contributes to a dilution effect as it mixes with the warmer seawater, leading to an overall reduction in temperature. Moreover, rainfall alters the surface properties of the water, affecting its ability to absorb solar radiation. While clouds associated with rain can block sunlight and reduce heating, the rain itself can modify the reflective properties of the water, temporarily limiting its capacity to absorb solar energy. Collectively, these mechanisms associated with rainfall contribute to the lowering of the sea surface temperature (Alpers et al., 2016; Ye et al., 2016; Zhang & Alpers, 2018).

Additionally, rain can contribute to the formation of a sea surface convergence zone. When rain falls over the ocean, it impacts the characteristics of the water surface. Factors such as the temperature of the rainwater and its interaction with the warmer sea surface can influence the local sea surface conditions.

Rainfall often introduces cooler water into the warmer ocean surface, leading to a mixing of different temperature layers. This temperature difference and the subsequent mixing can create a convergence zone, where the contrasting temperatures and flow dynamics cause a gathering or convergence of water masses.

Additionally, the evaporation and cooling effect of rainwater, along with altered surface properties, can lead to changes in the local atmosphere. These changes in atmospheric conditions, including temperature, humidity, and pressure, can further contribute to the formation of convergence zones. Convergence zones can foster distinct weather patterns, including the development of clouds, precipitation, and even variations in wind patterns. Therefore rain can be a contributing factor in the creation or intensification of sea surface convergence zones (Avissar & Liu, 1996; Capsoni et al., 1987; Chen et al., 2023).

Certainly, the relationship between wind speed and both the energy spectrum and rain cell width dimension within SAR data showcases an interesting inverse correlation. Within the turbulent flow observed in SAR data, the most substantial energy spectrum, denoted as E(k), aligns with a wind speed of approximately 5.5 m/s and a rain cell width of roughly 5 km. Embracing the principles of the Kolmogorov scaling law, a convergence emerges where the wind speed tends toward 4 m/s, correspondingly displaying an energy spectrum E(k) of 0.85 (Fig. 10.20).

This presentation underscores the diverse nature of turbulence, reflected in its association with regions of varying rain cell widths. Smaller rain widths seemingly coincide with intensified vortices, while larger width segments depict more orderly flow regions marked by extensive

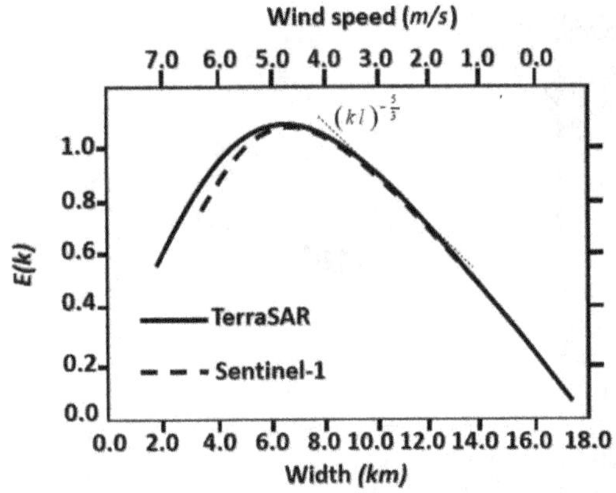

FIGURE 10.20 The spectrum of turbulence in rain cells under the influence of wind speed.

eddy fronts. This distinct categorization illuminates the multifaceted aspects within the turbulent flow field.

Moreover, concerning the study of turbulence in fluid dynamics, the energy spectrum plays a pivotal role in unraveling the scaling behavior and discerning the properties of rain cells amidst the turbulent milieu. Its analysis presents valuable insights into the underlying characteristics of turbulence, aiding in the comprehension of this complex and multifarious natural phenomenon.

According to the aforementioned perspective, rainfall upon the sea surface is not merely a passive interaction but an orchestrator of complex hydrodynamic performances. As raindrops disturb the tranquil demeanor of the ocean expanse, a symphony of eddies ensues. This dance is choreographed by the interplay of several forces—the initial impact of raindrops generating surface perturbations, the subsequent intermingling of rainwater's cooler density with the warmer sea, and the consequential onset of convective movements. Furthermore, the supporting role of wind, often accompanying rainstorms, contributes its force, sculpting the aqueous canvas with shear stress and setting the stage for these mesmerizing water whirls. The very rotation of our Earth, dictated by the Coriolis effect, bestows upon these eddies their characteristic spinning motion. In harmony, these forces orchestrate the elegant and intricate formation of eddies, intricate circular currents that waltz across the sea surface, influencing the redistribution of properties and imparting a dynamic rhythm to the ocean's intricately woven fabric (Chen et al., 2023; Kumjian et al., 2015).

The implementation of the quantum finite automaton offers an automated solution for analyzing rain cell footprint signatures and detecting turbulence flow within SAR imagery. At the onset of the algorithm, the quantum probability state serves as the foundation, establishing a precise trajectory to decipher the rain cell footprint signatures contained within each qubit. Operating within the domain of SAR imagery, this automaton navigates through a topological space housing an array of diverse quantum states. Unlike its topological counterpart, the quantum automaton deviates in its approach; rather than a binary outcome of inclusion or exclusion within the final set, it works with a probability measure, enhancing the precision and nuanced nature of its findings.

The quantum probability, derived from the square of the initial state projected onto a final state, serves as a pivotal element in the automated detection of rain cell footprint signatures and the identification of turbulence flow within SAR imagery. Augmenting this quantum exploration, the Hamiltonian-gradient algorithm works in tandem with the quantum automaton, enabling the discrimination between the distinct boundaries of rain cell footprint signatures and the identification of turbulence flow clusters within SAR data. This synergy leads to the automatic pinpointing of the precise locations where turbulence arises due to rain cell footprint signatures embedded in SAR images. The efficacy of the quantum automaton finds verification through the receiver operator curve (ROC), a tool illustrating its accuracy. The ROC curves unveil detailed insights into rain cell footprint signatures and the automated detection of turbulence flow, showcasing notably higher true-positive rates (TPRs), as exemplified in Fig. 10.21. Furthermore, concerning the ROC area, rain cell footprint signatures and turbulence flow manifest a remarkable clustering of approximately 98%, aligning closely with in situ measurements.

In conclusion, the author introduces an innovative concept to elucidate the interactions between rain cells and the sea surface, crucial for identifying rain cell imprints in SAR data. This concept, coined by the author as the "Quantanized Marghany Rain Interaction Coefficient (QMRIC)," serves as a parameter to characterize the level of quantum behavior displayed by these rain cells during their interaction with the sea surface. The QMRIC is presented as a metric to gage the extent of uncertainty or the potential for entanglement arising from raindrop impacts.

This chapter presents a groundbreaking algorithm designed for the automatic identification of rain cells that induce turbulence on the sea surface using SAR data. The convergence of advanced quantum automata with the nuanced capabilities of the Hamiltonian-gradient algorithm ushers in a new frontier in the precise identification of rain cell footprint signatures and the discernment of turbulence flow within SAR imagery. This dynamic fusion not only pinpoints the intricate boundaries of rain cell imprints but also elucidates the very spots where turbulence emerges due to these imprints in SAR images. The precision and efficacy of the quantum automaton find validation through the meticulous ROC analysis, showcasing the accuracy in identifying rain cell footprint signatures and the automatic detection of turbulence flow. The ROC curves, boasting remarkably high true-positive rates,

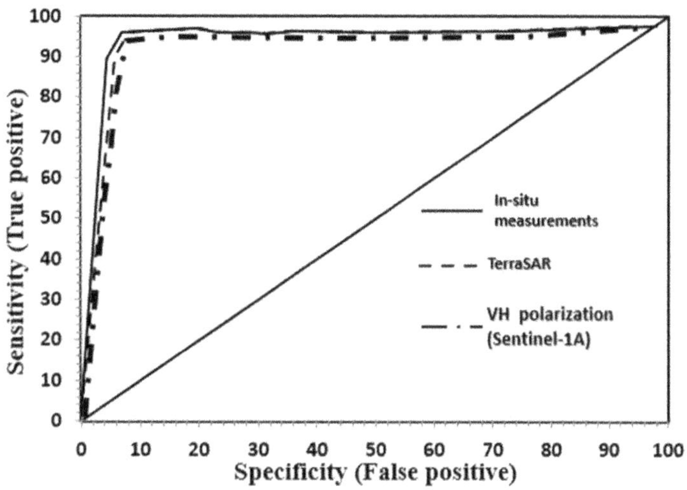

FIGURE 10.21 ROC for quantum automaton accuracy. *ROC*, Receiver operator curve.

serve as an emblem of the quantum automaton's precision. Moreover, with rain cell footprint signatures and turbulence flow exhibiting an approximate 98% concordance in terms of ROC area clustering, this synergy aligns closely with in situ measurements, underscoring the veracity and potential of quantum-assisted analyses in unraveling the complexities of SAR imagery. This heralds a significant stride in advancing our understanding of these intricate phenomena, promising extensive applications and insights into the intricate dynamics of rain cells and turbulence within SAR imagery.

References

Ali, S. T., & Engliš, M. (2005). Quantization methods: A guide for physicists and analysts. *Reviews in Mathematical Physics*, *17*(4), 391–490. Available from https://doi.org/10.1142/S0129055X05002376.

Alpers, W., Zhang, B., Mouche, A., Zeng, K., & Chan, P. W. (2016). Rain footprints on C-band synthetic aperture radar images of the ocean - Revisited. *Remote Sensing of Environment*, *187*, 169–185. Available from https://doi.org/10.1016/j.rse.2016.10.015, http://www.elsevier.com/inca/publications/store/5/0/5/7/3/3.

Ambainis A. Freivalds R. 1998 1-Way quantum finite automata: Strengths, weaknesses and generalizations, Annual Symposium on Foundations of Computer Science - Proceedings 332–341, IEEE Computer Society, United States.

Ambainis, A., & Watrous, J. (2002). Two-way finite automata with quantum and classical states. *Theoretical Computer Science*, *287*(1), 299–311. Available from https://doi.org/10.1016/S0304-3975(02)00138-X.

Avanzato, R., & Beritelli, F. (2020). An innovative acoustic rain gauge based on convolutional neural networks. *Information*, *11*(4), 183. Available from https://doi.org/10.3390/info11040183.

Avissar, R., & Liu, Y. (1996). Three-dimensional numerical study of shallow convective clouds and precipitation induced by land surface forcing. *Journal of Geophysical Research: Atmospheres*, *101*(D3), 7499–7518. Available from https://doi.org/10.1029/95jd03031.

Berazin, F. A. (2012). *The method of second quantization* (24). Elsevier.

Berezin, F. A. (1974). Quantization. *Mathematics of the USSR-Izvestiya*, *8*(5).

Berezin, F. A. (1975). General concept of quantization. *Communications in Mathematical Physics*, *40*(2), 153–174. Available from https://doi.org/10.1007/BF01609397.

Birkan, U., Salehi, Ö., Olejar, V., Nurlu, C., & Yakaryılmaz, A. (2021). *Implementing quantum finite automata algorithms on noisy devices, Lecture Notes in Computer Science (including subseries Lecture Notes in Artificial Intelligence and Lecture Notes in Bioinformatics)* (12747, pp. 3–16). Turkey: Springer Science and Business Media Deutschland GmbH. Available from https://www.springer.com/series/558, 10.1007/978-3-030-77980-1_1.

Capsoni, C., Fedi, F., Magistroni, C., Paraboni, A., & Pawlina, A. (1987). Data and theory for a new model of the horizontal structure of rain cells for propagation applications. *Radio Science*, *22* (3), 395–404. Available from https://doi.org/10.1029/RS022i003p00395.

Chen, F., Li, H., Hu, S., Jiang, S., Li, J., & Wu, R. (2023). Formation mechanisms of the "5· 31" record-breaking extreme heavy rainfall process in South China in 2021. *Atmosphere*, *14*(5).

Chen, J., Altepeter, J. B., Medic, M., Lee, K. F., Gokden, B., Hadfield, R. H., Nam, S. W., & Kumar, P. (2008). Demonstration of a quantum controlled-NOT gate in the telecommunications band. *Physical Review Letters*, *100*(13). Available from https://doi.org/10.1103/PhysRevLett.100.133603, http://oai.aps.org/oai?verb = GetRecord&Identifier = oai:aps.org:PhysRevLett.100.133603& metadataPrefix = oai_apsmeta_2, United States.

Debnath, L., & Mikusinski, P. (2005). *Introduction to Hilbert spaces with applications*. Academic Press.

Enjamio, C., Vilar, E., Redaño, A., Fontán, F. P., & Ndzi, D. (2005). Experimental analysis of microscale rain cells and their dynamic evolution. *Radio Science.*, *40*(3). Available from https://doi.org/10.1029/2004RS003119, RS3015-RS3033.

Feral, L., Mesnard, F., Sauvageot, H., Castanets, L., & Lemorton, J. (2000). Rain cells shape and orientation distribution in South-West of France. *Physics and Chemistry of the Earth, Part B: Hydrology, Oceans and Atmosphere*, *25*(10–12), 1073–1078. Available from https://doi.org/10.1016/S1464-1909(00)00155-6.

Frenger, I., Gruber, N., Knutti, R., & Münnich, M. (2013). Imprint of southern ocean eddies on winds, clouds and rainfall. *Nature Geoscience*, *6* (8), 608–612. Available from https://doi.org/10.1038/ngeo1863.

Gersho, A. (1977). Quantization. *IEEE Communications Society Magazine*, *15* (5), 16. Available from https://doi.org/10.1109/MCOM.1977.1089500, 16.

Gersho, A. (1978). Principles of quantization. *IEEE Transactions on Circuits and Systems*, *25*(7), 427–436. Available from https://doi.org/10.1109/TCS.1978.1084497.

Gray, R. M., & Neuhoff, D. L. (1998). Quantization. *IEEE Transactions on Information Theory*, *44*(6), 2325–2383. Available from https://doi.org/10.1109/18.720541.

Gross, L. (1967). Potential theory on Hilbert space. *Journal of Functional Analysis*, *1*(2), 123–181. Available from https://doi.org/10.1016/0022-1236(67)90030-4.

von Hardenberg, J., Ferraris, L., & Provenzale, A. (2003). The shape of convective rain cells. *Geophysical Research Letters*, *30*(24). Available from https://doi.org/10.1029/2003GL018539, http://onlinelibrary.wiley.com/journal/10.1002/(ISSN)1944-8007/issues?year = 2012.

Kumari, R., Sah, D. K., Cengiz, K., Ivković, N., Gehlot, A., & Salah, B. (2023). Acoustic signal-based indigenous real-time rainfall monitoring system for sustainable environment. *Sustainable Energy Technologies and Assessments*, *60*103398. Available from https://doi.org/10.1016/j.seta.2023.103398.

Kumjian, M. R., Lebo, Z. J., & Morrison, H. C. (2015). On the mechanisms of rain formation in an idealized supercell storm. *Monthly Weather Review*, *143*(7), 2754–2773. Available from https://doi.org/10.1175/MWR-D-14-00402.1, http://journals.ametsoc.org/doi/pdf/10.1175/MWR-D-14-00402.1.

Ma, B. B. (2022). Rainfall at sea: Using the underwater sounds of raindrops as a rain gauge for weather and climate. *Acoustics Today*, *18* (2), 62. Available from https://doi.org/10.1121/at.2022.18.2.62.

Melsheimer C. Gade M. 1998 Rain cells over the sea monitored by synthetic aperture and weather radars: A comparison, Oceans Conference Record (IEEE), 1445-1448, IEEE Germany 3.

Melsheimer C. Romeiser R. Alpers W. 1998 Modeling of radar signatures of rain cells over the sea, International Geoscience and Remote Sensing Symposium (IGARSS) 1680–1682 IEEE Germany 3, 10.1109/igarss.1998.692419.

Melsheimer, C., Alpers, W., & Gade, M. (1998). Investigation of multifrequency/multipolarization radar signatures of rain cells over the ocean using SIR-C/X-SAR data. *Journal of Geophysical Research: Oceans*, *103* (C9), 18867–18884. Available from https://doi.org/10.1029/98jc00779.

Melsheimer, C., Alpers, W., & Gade, M. (2001). Simultaneous observations of rain cells over the ocean by the synthetic aperture radar aboard the ERS satellites and by surface-based weather radars. *Journal of Geophysical Research: Oceans*, *106*(C3), 4665–4677. Available from https://doi.org/10.1029/2000jc000263.

Monroe, C., Meekhof, D. M., King, B. E., Itano, W. M., & Wineland, D. J. (1995). Demonstration of a fundamental quantum logic gate. *Physical Review Letters*, *75*(25), 4714–4717. Available from https://doi.org/10.1103/PhysRevLett.75.4714.

Nielsen, M. A., Chuang, I. (2000). Quantum computation and quantum information.

O'Brien, J. L., Pryde, G. J., White, A. G., Ralph, T. C., & Branning, D. (2003). Demonstration of an all-optical quantum controlled-NOT gate. *Nature.*, *426*(6964), 264–267. Available from https://doi.org/10.1038/nature02054.

Okamoto, R., Hofmann, H. F., Takeuchi, S., & Sasaki, K. (2005). Demonstration of an optical quantum controlled-NOT gate without path interference. *Physical Review Letters*, *95*(21). Available from https://doi.org/10.1103/PhysRevLett.95.210506Japan, http://oai.aps.org/oai/?verb = ListRecords&metadataPrefix = oai_apsmeta_2&set = journal:PRL:95.

Paluch, I. R., & Lenschow, D. H. (1991). Stratiform cloud formation in the marine boundary layer. *Journal of the Atmospheric Sciences*, *48*(19), 2141–2158. Available from https://doi.org/10.1175/1520-0469(1991)048 < 2141:SCFITM > 2.0.CO;2.

Peleg, N., Marra, F., Fatichi, S., Molnar, P., Morin, E., Sharma, A., & Burlando, P. (2018). Intensification of convective rain cells at warmer temperatures observed from high-resolution weather radar data. *Journal of Hydrometeorology*, *19*(4), 715–726. Available from https://doi.org/10.1175/JHM-D-17-0158.1, https://journals.ametsoc.org/doi/pdf/10.1175/JHM-D-17-0158.1.

Peleg, N., & Morin, E. (2012). Convective rain cells: Radar-derived spatiotemporal characteristics and synoptic patterns over the eastern Mediterranean. *Journal of Geophysical Research: Atmospheres*, *117* (D15). Available from https://doi.org/10.1029/2011jd017353.

Qiu, D. Learning quantum finite automata with queries. arXiv, China arXiv. 23318422. (2021), https://arxiv.org. doi: 10.48550/arxiv.2111.14041.

Qiu, D., Li, L., Zou, X., Mateus, P., & Gruska, J. (2011). Multi-letter quantum finite automata: Decidability of the equivalence and minimization of states. *Acta Informatica*, *48*(5), 271–290. Available from https://doi.org/10.1007/s00236-011-0139-6.

Say, A. C. C., & Yakaryılmaz, A. (2014). Quantum finite automata: A modern introduction. *Lecture Notes in Computer Science (including subseries Lecture Notes in Artificial Intelligence and Lecture Notes in Bioinformatics)*, *8808*, 208–222. Available from https://doi.org/10.1007/978-3-319-13350-8_16, http://springerlink.com/content/0302-9743/copyright/2005/.

Skorohod, A. V. (2012). *Integration in Hilbert space* (79). Springer Science & Business Media.

Steinbrecher, U., Kraus, T., Alfonzo, G. C., Grigorov, C., Schulze, D., Bräutigam, B. (2014).TerraSAR-X: Design of the new operational WideScanSAR mode, Proceedings of the European Conference on Synthetic Aperture Radar, EUSAR 21974403, 189-192, Germany, http://conference.vde.com/eusar/history/Pages/EUSARHistory.aspx

Takahashi, T., & Shimura, K. (2004). Tropical rain characteristics and microphysics in a three-dimensional cloud model. *Journal of the Atmospheric Sciences*, *61*(23), 2817–2845. Available from https://doi.org/10.1175/JAS-3294.1.

Yang J. Shen X. Xing J. Tian X. Li H. Deng B. Huang J. Hua X.S. 2019 IEEE Computer Society China. Quantization networks. Proceedings of the IEEE Computer Society Conference on Computer Vision and Pattern Recognition, 9781728132938, 7300-7308, 10.1109/CVPR.2019.00748.

Ye, X., Lin, M., Yuan, X., Ding, J., Xie, X., Zhang, Y., & Xu, Y. (2016). Satellite SAR observation of the sea surface wind field caused by rain cells. *Acta Oceanologica Sinica*, *35*(9), 80–85. Available from https://doi.org/10.1007/s13131-016-0936-4, http://www.springer.com/earth-+ sciences + & + geography/oceanography/journal/13131.

Young, N. (1988). *An introduction to Hilbert space*. Cambridge University Press.

Zhang, B., & Alpers, W. (2018). *The effect of rain on radar backscattering from the ocean* (pp. 325–338). Informa UK Limited. Available from 10.1201/9781351235822-20.

Chapter 11

Quantum cellular automata algorithm for automatic detection of upwelling in synthetic aperture radar data

The integration of quantum cellular automaton (QCA) into the analysis of synthetic aperture radar (SAR) imagery represents a significant response to the challenges arising from the limited availability of thermal bands in SAR, as compared with optical data. This limitation impedes our capacity to effectively identify crucial indicators essential for comprehending intricate oceanic processes, especially those linked to upwelling dynamics. The proposed methodology signifies a paradigm shift, harnessing QCA's groundbreaking capabilities to autonomously pinpoint upwelling zones. Rooted in principles inspired by quantum mechanics, QCA introduces a transformative approach to spatial pattern analysis, offering a novel avenue for the automatic detection of upwelling. By exploiting QCA's distinctive ability to enhance decoherence in SAR images, the method highlights areas indicative of upwelling, discerning natural slicks, and the emergence of cool waters brought to the surface. This paves the way for a comprehensive exploration of the interplay between SAR imagery and QCA, unveiling a promising frontier in advancing our understanding of intricate marine phenomena. In this chapter, we embark on an insightful journey into the realms of upwelling, with a focus on the mesmerizing phenomenon of Island Wakes. Casting our gaze upon the maritime tapestry, we unravel the occurrences of upwelling through the lens of Tioman, Tinggi, and Aur Islands in the enchanting expanse of Malaysia. Brace yourself for an exploration that transcends the ordinary, as we delve into the nuanced dance of oceanic forces and the captivating narratives etched within the wake of these islands.

11.1 What is upwelling?

Upwelling, a vital oceanic process, involves the vertical movement of deep, chilled water to the ocean surface (Fig. 11.1). This phenomenon is primarily a result of the complex interplay between wind patterns and the Earth's rotation. The Earth's eastward rotation on its axis plays a pivotal role in the Coriolis effect, directing wind movements (see Chapter 2; Section 2.14). In the Northern Hemisphere, winds veer to the right, and in the Southern Hemisphere, they veer to the left due to this effect. This atmospheric dynamic significantly influences the occurrence of upwelling in coastal regions.

Notably, the Coriolis effect is instrumental in generating upwelling even in open ocean areas near the Equator. Here, the trade winds, blowing parallel to the Equator, displace surface water both north and south, fostering the upwelling of deeper water to the surface, shaping the ocean's intricate dynamics.

The intricate wind patterns produced in the wake of sluggish cyclones possess the capability to displace surface water, resulting in upwelling directly beneath the cyclone's eye. The infusion of cooler water in this process ultimately contributes to the gradual attenuation of the cyclonic system (Hidaka, 1972; Pitcher et al., 2010).

11.2 Impacts of upwelling

The ramifications of upwelling on ecosystems are profound, especially concerning biodiversity and productivity. Through the elevation of nutrient-rich deep water to the surface, upwelling fosters the vibrant growth of seaweed and plankton. These foundational organisms, in turn, establish the fundamental link in the food chain for a diverse array of marine life, including fish, marine mammals, and avian species.

Regions influenced by upwelling serve as home to some of the most biologically diverse and productive ecosystems on a global scale. For example, a specific expanse spanning 25,900 square kilometers off the coast of Peru consistently experiences coastal upwelling, rendering it one of the planet's most fertile fishing areas.

Synthetic Aperture Radar Image Processing Algorithms for Nonlinear Oceanic Turbulence and Front Modeling. DOI: https://doi.org/10.1016/B978-0-443-19155-8.00011-8

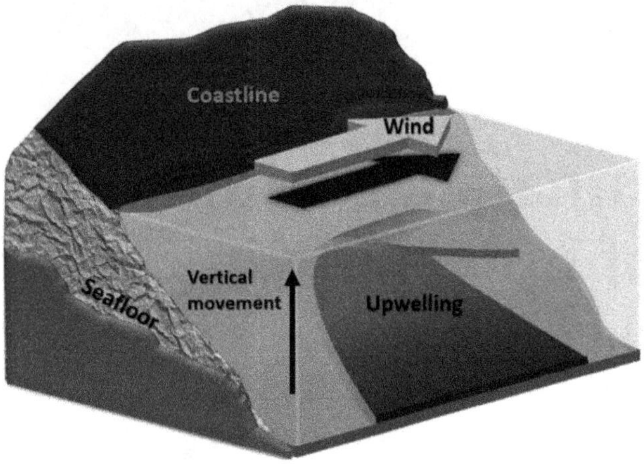

FIGURE 11.1 The simplification of the upwelling concept.

Despite encompassing merely 1% of the world's oceanic realm, coastal upwelling regions significantly contribute, constituting approximately 50% of the total fish catch hauled in by global fisheries. This underscores the critical role of upwelling in sustaining thriving marine environments and supporting substantial fishery yields on a worldwide scale.

Upwelling, additionally, exerts a significant influence on the mobility of animal life within the affected area. Small larvae, representing the developmental stages of numerous fish and invertebrates, often navigate ocean currents for extensive durations. During a robust upwelling occurrence, these larvae can be carried far from the coast, posing a risk to their survival in the offshore environment.

The process of upwelling, where the ascent of colder water occurs from deeper ocean layers to the surface, further, has a notable impact on the local atmospheric conditions. When this cold water emerges to the surface, it significantly cools the surrounding air. This cooling effect fosters specific atmospheric phenomena, such as the development of sea fog.

A prominent example of the influence of upwelling-induced fog can be observed in the city of San Francisco, California. This city is renowned for its characteristic chilly and foggy summers, which are markedly shaped by the seasonal upwelling events occurring in the nearby oceanic region. As the cold, nutrient-rich water surfaces along the California coast, the chilled air above it encounters a sharp decrease in temperature. This temperature disparity between the cold water and the warmer air above it leads to the formation of fog. San Francisco experiences this maritime-influenced climate during the summer months, where cool and dense fog frequently engulfs the city due to the upwelling-induced cooling effect in the adjacent coastal waters (Ekman, 1905; Hsuesh & O'Brien, 1971; Pitcher et al., 2010).

11.3 Artificial upwelling

The pivotal query that emerges is whether artificial generation of upwelling is achievable. Artificial upwelling, a pioneering concept being explored by scientists and industries, aims to engineer zones where "artificial upwelling" systems are established to facilitate the vertical movement of cold, nutrient-rich water from the deeper layers of the ocean to the surface. The overarching goal behind this innovative approach is to potentially boost fish populations and enhance crop yields in areas spanning from the Gulf of Mexico to the southwestern coast of Australia.

The process of artificial upwelling involves employing sophisticated technology that harnesses the kinetic energy derived from the motion of waves to stimulate the ascent of cold, nutrient-enriched water from the ocean's depths to the surface. By utilizing this advanced technology, researchers and businesses seek to replicate the beneficial effects of natural upwelling, known for its ability to promote the proliferation of marine life due to the influx of nutrient-rich water.

Initial experiments exploring artificial upwelling have been conducted in the Pacific Ocean, particularly in the vicinity of the Hawaiian Islands. These experiments serve as an initial step in comprehending and harnessing the potential of artificial upwelling systems. Researchers aim to assess the feasibility and effectiveness of this innovative method in influencing the marine ecosystem, potentially offering new strategies for bolstering fish populations and optimizing fisheries in various regions across the globe (Liu & Jin, 1995; Pan et al., 2016).

11.4 Why Coriolis effect is keystone of upwelling?

The Coriolis effect serves as a keystone in the process of upwelling due to its influence on the movement of ocean currents. This effect, caused by the Earth's rotation, alters the direction of moving objects (such as winds and ocean currents) on the planet.

In the context of upwelling, prevailing winds interact with the ocean's surface. When these winds blow parallel to a coastline, they set surface water in motion (Fig. 11.2). As a result, the water starts to move away from the shore due to the pressure created by the wind. This movement creates a vacuum or void near the coast, prompting deeper water to rise up and replace the surface water that has moved offshore.

Here's where the Coriolis effect comes into play. The rotation of the Earth causes a deflection in the movement of this newly rising deeper water. In the Northern Hemisphere, water moving away from the coast gets deflected to the right due to the Coriolis effect, while in

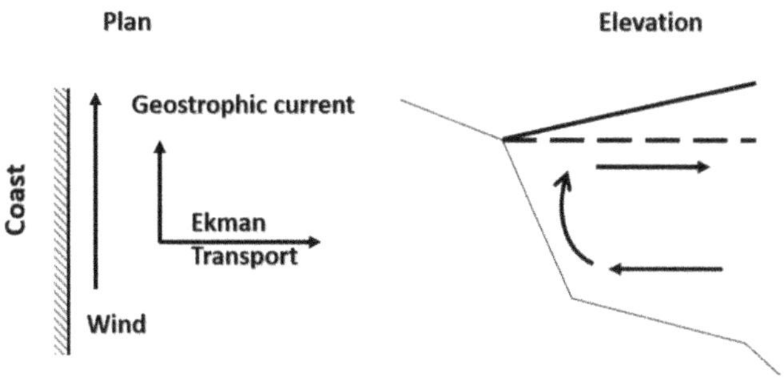

FIGURE 11.2 How parallel winds blowing along the coastline cause upwelling.

the Southern Hemisphere, it deflects to the left. This deflection guides the deeper water upward along the coast, leading to the phenomenon of upwelling.

Therefore the Coriolis effect is crucial for the redirection and movement of water that characterizes upwelling. It works in tandem with the wind patterns along the coastline, playing a pivotal role in fostering the rise of nutrient-rich deeper water to the surface, ultimately influencing the development and sustenance of this oceanic process.

11.5 Quantization of Coriolis effects

The attempt to simulate the Coriolis force using the Schrödinger equation and applying it to incompressible fluid flow appears to present a complex theoretical endeavor. The Schrödinger equation, a fundamental equation in quantum mechanics, usually describes the behavior of quantum particles and systems in terms of wave functions.

In this description, the attempt seems to modify the Schrödinger equation to include the influence of forces resembling the Coriolis effect experienced by a particle in a rotating frame. The Coriolis force, often associated with the rotation of the Earth, causes apparent deflection of moving objects. This effort, however, attempts to introduce similar effects in the context of quantum mechanics and incompressible fluid dynamics. The interpretation involves describing the Coriolis force as arising from the frictional force affecting a moving particle within a rotating frame, resulting in the generation of an additional force term (Anandan & Suzuki, 2004). This modified equation aims to incorporate this effect into the Schrödinger equation to account for the particle's behavior in this altered frame of reference, which is expressed mathematically by:

$$\overleftarrow{E'}\,\psi(u,\partial t) = \frac{(\overleftarrow{p} - m\vec{V})^2}{2m}\psi(u,\partial t) \tag{11.1}$$

here $\overleftarrow{E'}$ is kinetic energy that is generated due to the water particle mass transfers from frame F_0 to transferred frame F_T with resultant velocity $\vec{V}$, and it can be expressed by:

$$\hat{E'} = i\left(\frac{\partial}{\partial t} + \vec{V}.\nabla\right) \tag{11.2}$$

Using the nonrelativistic limit of the Lorentz transformation for the energy, one can obtain measured energy $\hat{E'}$ as:

$$\hat{E'} = E - \vec{V}.p + 0.5m\vec{V}^2 \tag{11.3}$$

Therefore the nonrelativistic Lorentz boost express the Galilei boost with a velocity as:

$$U = e^{\left(it\vec{V}.\overleftarrow{p'} - im\vec{V}.\left(x - \vec{V}t\right)\right)} \tag{11.4}$$

Therefore the velocity $\vec{V}$ is given by $\Omega \times x$, and the nonrelativistic Lorentz boost U can be reconstructed as:

$$U = e^{\left(it\left(\Omega x\left(x - \vec{V}t\right);\overleftarrow{p}\right)\right)} = e^{\left(it\Omega.\left(x - \vec{V}t\right).\overleftarrow{p}\right)} \tag{11.5}$$

Eq. (11.5) seems to suggest an equality of orbital angular momentum between the initial and transfer frames. However, these deductions are evidently flawed, even from a classical perspective. The nonrelativistic Lorentz boost is responsible for transitioning from the initial frame to the transfer frame, not directly to another frame as noted in references.

Additionally, a singular boost transformation cannot fully capture the scenario as it relies more on a sequence of successive transformations. Within this context, the mathematical representation of the gage field related to the rotational frame owing to the Coriolis force is articulated (Anandan & Suzuki, 2004; Rizzi & Ruggiero, 2004). The specifics of this expression would be contingent on the unique nuances of the problem and the underlying theory being employed.

$$A^{\mu}\left(-0.5(\Omega \times x)^2, \Omega \times x\right). \tag{11.6}$$

The Hamiltonian representing a water particle in a state of relaxation within the rotating frame is formulated as follows:

$$H = \left(\frac{0.5}{m}(\overleftarrow{p'} - m\Omega \times x)^2 - 0.5m(\Omega \times x)^2\right), \tag{11.7}$$

Hence, as the water particle transitions from the initial frame to frame under the influence of the Coriolis force, the particle can exhibit a spinning motion $\overleftarrow{S}$. Within this conceptual framework, Eq. (11.7) necessitates the inclusion of the spin interaction term, which can be represented as:

$$H = \frac{0.5}{m}(\overleftarrow{p} - m\Omega \times x)^2 - 0.5m(\Omega \times x)^2 - \Omega\vec{\hat{S}} \tag{11.8}$$

Simply put, angular momentum, often referred to as spin, is inherent in rotating objects such as a spinning golf ball or the rotating Earth. This property signifies the total rotational momentum of a body. To compute this angular momentum, one typically integrates the contributions from the motion of each minuscule mass constituting the body. This process results in the determination of the total angular momentum or spin $\vec{\hat{S}}$ of the system. The recognized formula for computing this total angular momentum is given by:

$$\vec{\hat{S}} = I\omega \tag{11.9}$$

here I is the moment of inertia of the body, and ω is its angular velocity. In this context, the spin vector $\vec{\hat{S}}$ is oriented along the axis of rotation (Anandan & Suzuki, 2004; Marghany, 2021b) (Fig. 11.3) and can be described using the right-hand rule (Fig. 11.4). Applying this rule involves curling the fingers of the right hand in the direction of rotation, and the resulting direction in which the thumb points indicates the orientation of the spin vector $\vec{\hat{S}}$.

The effect of the spin coupling to the rotation can be represented by an operator acting on the initial wave function. Mathematically, this can be denoted by the operator:

$$\overline{\Phi}_S = \hat{T} e^{\left[\int dt \vec{S'} . \Omega\right]}, \tag{11.10}$$

where $\hat{T}$ represents the time ordering operator. Eq. (11.10) can be added to the continuity equation to obtain:

$$\frac{\partial}{\partial t} q + \nabla . \left(q\vec{u} \right) + \hat{\Phi}_S = 0 \tag{11.11}$$

Eq. (11.11) illustrates that water movement is influenced by the spinning motion of water particles induced by the uniform rotation attributed to the Coriolis force. This insight suggests that the continuity and motion within the water must be governed by the spinning behavior of water particles.

In this context, the Schrödinger equation characterizes a field with a single complex value. To precisely describe the spinning term in Eq. (11.11), it becomes imperative to represent it as a vector comprising four complex numbers. Accomplishing this necessitates the implementation of a relativistic wave equation, such as the Dirac equation. The Dirac equation, known for its applicability in describing particles with spin and relativistic effects, becomes essential to accurately represent the spinning behavior of the water particles within this theoretical framework. The expression for the relativistic phases of the rotating frame in quantum mechanics utilizing the Dirac equation with a spin connection is mathematically represented as:

$$i\gamma^{\mu} D_{\mu} \Psi = m\Psi \tag{11.12}$$

where $i\gamma^{\mu} D_{\mu}$ denotes the Dirac operator coupled with the spin connection, m is the mass of the particle; and Ψ represents the wave function. This equation encapsulates the relativistic aspects within a rotating frame in the realm of quantum mechanics by employing the Dirac equation along with the spin connection. The mathematical representation of the low-energy limit of the Dirac equation is given by:

$$\begin{aligned} &\left[\gamma^0 (m + \overline{p}_0 - mA_0 - 0.5A^i \overline{p}_i)\right] \\ &= -\left[\gamma^i \left(\overline{p}_i - 0.5mA_i - \frac{i}{2} \nabla A_0\right) - m\right] \Psi, \end{aligned} \tag{11.13}$$

The formulation of the Hamiltonian for the upper constituent and dual spinors in the low-energy boundary, obtained through the common splitting of the four spinors

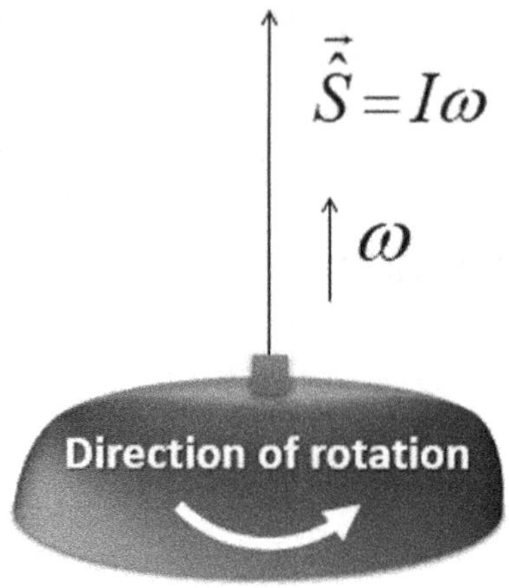

FIGURE 11.3 A simple explanation of angular momentum.

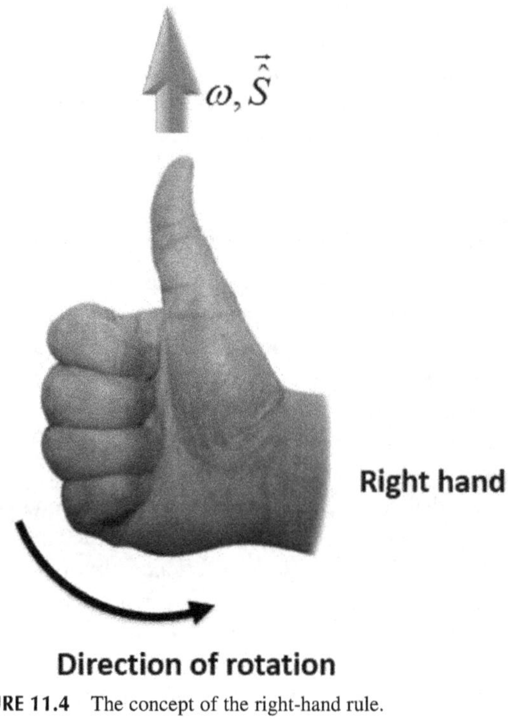

FIGURE 11.4 The concept of the right-hand rule.

into upper and lower constituents, can be mathematically represented as:

$$H = \frac{0.5}{m}(\hat{p}_0 - mA - (I.\omega) \times \nabla A_0)^2 + mA_0 - \Omega(I.\omega). \tag{11.14}$$

Eq. (11.14) elucidates that the low-energy approximation is derived within the rotating coordinate system, distinct from an inertial coordinate system. This distinction implies that the calculations or approximations represented in this equation are specifically applicable and relevant within a rotating frame of reference, rather than within an inertial (nonaccelerating) coordinate system.

According to the aforementioned perspective, the Hamiltonian equation in the context of the Coriolis force acting on an object within a rotating reference frame is typically not directly formulated using the Hamiltonian in classical mechanics. The Hamiltonian primarily describes the total energy of a system in classical mechanics or the quantum mechanical operator representing the total energy of a particle.

The specific formulation of the Coriolis force using the Hamiltonian typically involves the total energy or potential energy contributions associated with the rotating reference frame. This may include terms considering the rotational kinetic energy or the effects of the rotating frame on the dynamics of the system. The incorporation of the Coriolis force within the framework of the Hamiltonian typically involves complex derivations and depends on the specific context of the problem or system being analyzed in classical mechanics. The Hamiltonian itself may not directly feature the Coriolis force, but rather the effects of the rotating reference frame on the total energy or dynamics of the system.

The Hamiltonian formulation related to the Coriolis force might include terms related to the rotational dynamics but is a complex derivation, depending on the specifics of the problem and the context in classical mechanics.

In the context of ocean dynamics and upwelling, the Coriolis effect plays a crucial role in redirecting water movement along coastlines due to prevailing winds. This redirection, quantified by the Coriolis force, influences the behavior of ocean currents, including the upward movement of deeper water in the upwelling process. To quantify the Coriolis effect, the Earth's angular velocity is a key factor. The Earth completes a full rotation approximately every 24 hours. This velocity, denoted as ω, is approximately 7.292×10^{-5} radians per second.

Quantifying the Coriolis effect, therefore, aids in understanding and predicting the behavior of oceanic systems, providing a foundation for mathematical models used in oceanography, meteorology, and climatology. These models incorporate the Coriolis effect to simulate and predict the intricate dynamics of ocean currents, including phenomena such as upwelling, within a quantized framework.

11.6 Mechanism of upwelling

The complex interplay driving upwelling involves three key elements: wind, the Coriolis effect, and Ekman transport. Their collaborative action differs across various upwelling types, yet collectively, they induce similar effects. In the broader upwelling process, winds traverse the ocean's surface in a specific direction, sparking an interaction with the water. As the wind exerts its influence, the water moves at a net angle of approximately 90 degrees from the wind's path, influenced by the Coriolis forces and Ekman transport. Ekman transport prompts the surface water layer to drift at roughly a 45-degree angle from the wind's direction (Fig. 11.5). Friction between this surface layer and the layer beneath drives subsequent layers to move in unison. This coordinated movement forms a spiraling motion throughout the water column.

Subsequently, the direction of water movement is guided by the Coriolis forces; in the Northern Hemisphere, water is propelled to the right of the wind's direction, while in the Southern Hemisphere, it is directed to the left. If this collective water movement diverges, it triggers upwelling, where deep water ascends to replenish the displaced water. This intricate dance of wind, Coriolis forces, and Ekman transport orchestrates the cyclical replenishment of surface water, enriching it with nutrients and fostering thriving marine ecosystems (Mazzini & Barth, 2013; Roberts & Nieuwenhuys, 2016).

11.6.1 Ekman transport

Ekman transport is an integral facet of Ekman motion theory, initially explored in 1902 by Vagn Walfrid Ekman. Winds serve as the primary force propelling ocean circulation, and Ekman transport constitutes a vital element in

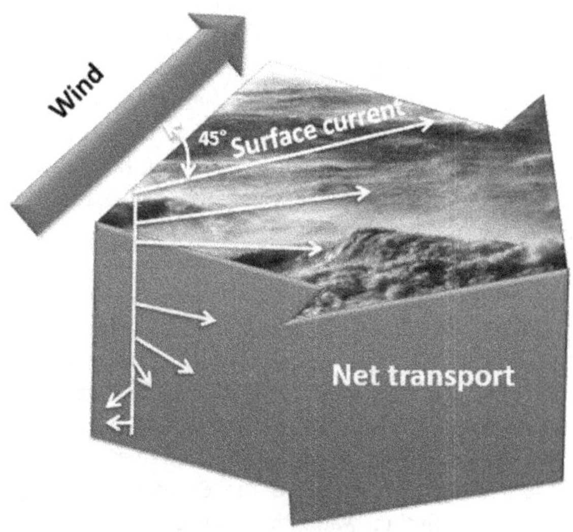

FIGURE 11.5 Ekman spiral and net transport.

wind-driven ocean currents. This phenomenon materializes when the frictional force exerted by the wind influences the upper layers of the ocean, typically spanning depths of 10–100 meters (Austin & Lentz, 2002; Ekman, 1905).

As the wind interacts with the ocean surface, it imparts a frictional force that drags the surface water. However, owing to the Coriolis effect, the water moves perpendicular to the direction of the wind, approximately at a 90° angle. The direction of this transport varies according to the hemisphere: in the northern hemisphere, it moves 90° clockwise from the wind direction, while in the southern hemisphere, it moves 90° anticlockwise. The initial observation of this phenomenon dates back to Fridtjof Nansen's Arctic expedition in the 1890s when he noted ice movement occurring at an angle to the wind direction.

Ekman transport significantly influences the biogeochemical characteristics of the Earth's oceans due to its role in generating upwelling (Ekman suction) and downwelling (Ekman pumping), adhering to the laws of mass conservation. Mass conservation, as it relates to Ekman transfer, mandates that any water displaced within an area must be replenished. This balancing act is achieved through either Ekman suction or Ekman pumping, contingent upon wind patterns, to ensure the preservation of mass within the oceanic systems (Enriquez, Mariño-Tapia, 2014).

11.6.2 Ekman suction

Ekman suction is an integral part of the Ekman transport phenomenon, generating areas of upwelling by causing water divergence. As per the principle of mass conservation, the water displaced by Ekman transport must be replenished. As water diverges, it creates a void, acting as a suction force, pulling up deep-sea water into the euphotic zone—an occurrence known as upwelling.

The repercussions of Ekman suction are profound for biogeochemical processes in the area due to the resulting upwelling. Upwelling brings nutrient-rich, cold deep-sea water to the euphotic zone, fostering phytoplankton blooms and creating a highly productive environment. This environment becomes a hotspot for fisheries, with nearly half of the world's fish catch originating from these upwelling regions.

Ekman suction is observed both along coastlines and in the open ocean, including occurrences along the equator. Along the Pacific coastlines of California, Central America, and Peru, as well as the Atlantic coastline of Africa, upwelling is generated due to Ekman suction, driven by equator-bound currents. The Coriolis effect causes surface water to deflect 90° to the left (in the Southern Hemisphere, toward the equator) of the wind direction, prompting water divergence from the coastal boundary, leading to Ekman suction.

Moreover, upwelling due to Ekman suction occurs at locations where the Polar Easterlies meet the Westerlies in the subpolar regions north of the subtropics and where the Northeast Trade Winds meet the Southeast Trade Winds along the Equator. The Coriolis effect, deflecting surface water 90° to the left (in the Southern Hemisphere), leads to the divergence of surface water along these boundaries, resulting in upwelling to maintain mass conservation.

Mathematically, Ekman suction and transport can be represented by equations describing the wind stress on the ocean surface, the Coriolis effect, and the resulting water movement. The Ekman spiral equation, derived from these physical principles, illustrates the depth-dependent structure of the wind-driven current caused by the balance between wind stress, Coriolis force, and friction within the ocean. Additionally, the equations describing the vertical and horizontal components of the Ekman transport offer a quantitative understanding of the movement and upwelling of water influenced by Ekman suction. In this understanding, the Ekman spiral equation is a fundamental concept in physical oceanography that describes the structure of the wind-driven current in the ocean. It reveals the depth-dependent behavior of water movement caused by the combined effects of wind stress, the Coriolis force, and friction within the ocean (Hart, 1995; Stevenson & Janowitz, 1977; Tan et al., 2019; Van Senden & Imberger, 1990).

The equation mathematically represents the velocity of ocean currents at different depths, yielding an insight into the spiral-shaped pattern created by the Ekman transport. It can be, therefore, expressed as:

$$V(z) = \frac{\tau}{\rho f} \times e^{\frac{-z}{D}} \tag{11.15}$$

In Eq. (11.15), V(z) represents the velocity of the water at depth z below the surface; τ symbolizes the wind stress on the ocean's surface; ρ denotes the density of water; and f stands for the Coriolis parameter, which depends on the latitude. In this sense, D is the Ekman depth, representing the depth at which the velocity of the water decreases to e^{-1} or approximately 37% of the surface velocity (Hart, 1995; Stevenson & Janowitz, 1977; Tan et al., 2019; Van Senden & Imberger, 1990).

This equation showcases an exponential decay of water velocity with depth, forming a spiral pattern of water movement. As depth increases, the velocity diminishes exponentially, creating a spiral flow pattern known as the Ekman spiral. This pattern characterizes the way in which water moves under the influence of wind stress and the Coriolis force.

The Ekman spiral equation plays a crucial role in understanding the vertical structure of ocean currents and is essential in various oceanographic studies, particularly in comprehending the distribution of nutrients and biological activity in the ocean's depths, as well as in climate modeling and ocean circulation studies.

11.6.3 Ekman pumping

Ekman pumping, a facet of Ekman transport, results in areas of downwelling by causing water convergence. As per the principle of mass conservation, the accumulation of surface water necessitates a downward push. This accumulation of warm, nutrient-poor surface water is vertically driven down the water column, leading to downwelling (Gaube et al., 2015; Julien et al., 2016).

The implications of Ekman pumping on the surrounding environments are significant. Downwelling caused by this process results in nutrient-poor waters, diminishing the biological productivity of the area. Moreover, it facilitates the vertical transport of heat and dissolved oxygen as warm, oxygen-rich surface water is pushed toward the deeper ocean water. Ekman pumping can be observed along coastlines and in the open ocean. Along the Pacific Coast in the Southern Hemisphere, northerly winds move parallel to the coastline. Due to the Coriolis effect, surface water is pulled 90° to the left of the wind current, leading to water convergence along the coast boundary, consequently causing Ekman pumping. In the open ocean, Ekman pumping occurs within gyres. Specifically, in the subtropical regions between 20°N and 50°N, Ekman pumping occurs as the trade winds transition to westerlies, resulting in an accumulation of surface water (Ekman, 1905; Wang et al., 2013).

Mathematically, Ekman pumping can be described using equations that consider the wind stress on the ocean surface, the Coriolis effect, and the resulting convergence of water. These equations capture the vertical movement of water caused by Ekman pumping, especially in regions where convergence of water occurs, such as along coastlines or within gyres. The Ekman pumping process, therefore, showcases the dynamic interactions between wind, the Coriolis effect, and ocean currents, profoundly impacting the vertical distribution of nutrients, heat, and dissolved oxygen within the oceanic environment. Understanding Ekman pumping is crucial in comprehending the complexities of oceanic circulation, its impact on marine ecosystems, and its role in the larger context of global climate dynamics (Abney & Epstein, 1996).

To capture the convergence or divergence of water leading to Ekman pumping, the continuity equation can be employed. This equation deals with mass conservation and describes the vertical movement of water.

$$\frac{\partial V}{\partial z} = -\nabla . V \tag{11.16}$$

Eq. (11.16) shows that the divergence of the velocity field $\nabla \cdot V$ through the water depth z. In this sense, V represents the vertical velocity of the water. The Sverdrup balance equation, therefore, describes the balance between the wind stress and the vertical motion of the water column in a rotating ocean. It's fundamental in explaining the causes of upwelling and downwelling. The Sverdrup balance equation, consequently, can be written as:

$$\nabla \times \left(f^{-1} \nabla \times \tau\right) = -\nabla \times \left(\rho^{-1} \nabla \times \vec{F}\right) \tag{11.17}$$

Where $\vec{F}$ is the vertically integrated vertical velocity.

11.7 Ekman spiral

In deriving the properties of an Ekman spiral, we begin with a uniform, horizontal geostrophic interior flow within a homogeneous fluid. This flow is represented by a vector $u = (u^{-}, v^{-})$, where both components remain constant due to the uniformity of the flow. As a consequence of this uniformity, the horizontal gradients are zero. Therefore the continuity equation results in $\partial z \partial w = 0$. It's important to note that the interior flow under consideration is purely horizontal, implying that $w = 0$ at all depths, including within the boundary layers.

Under these conditions, the Navier – Stokes momentum equations, which govern geophysical motion, can be simplified to focus on the behavior of the flow in a two-dimensional horizontal plane. This reduction in dimensionality accounts for the uniformity of the flow and the absence of vertical motion ($w = 0$), leading to a set of equations that describe the geostrophic, horizontal flow within the fluid. These equations primarily deal with the dynamics of the horizontal components of the flow, omitting the vertical component due to its absence in the simplified scenario.

$$-fv = -\frac{1}{\rho_0}\frac{\partial p}{\partial x} + \nu_E \frac{\partial^2 u}{\partial z^2}, \tag{11.18}$$

$$-fu = -\frac{1}{\rho_0}\frac{\partial p}{\partial y} + \nu_E \frac{\partial^2 v}{\partial z^2}, \tag{11.19}$$

$$0 = -\frac{1}{\rho_0}\frac{\partial p}{\partial z}, \tag{11.20}$$

In the given context, the Coriolis parameter (f), the fluid density (ρ_0), and the eddy viscosity (ν_E) are assumed to be constant for simplicity. These parameters are considered relatively constant within the scale of an Ekman spiral, allowing for this approximation to hold true.

A uniform flow necessitates a pressure gradient that uniformly varies. When the components of the interior flow, $u = u^{-}$ and $v = v^{-}$, are substituted into the equations governing the geostrophic motion, the following outcomes emerge. These results are indicative of the behavior of the flow within this uniform, horizontal context.

$$-f\overline{v} = -\frac{1}{\rho_0}\frac{\partial p}{\partial x} = constant \tag{11.21}$$

$$-f\overline{u} = -\frac{1}{\rho_0}\frac{\partial p}{\partial y} = constant \tag{11.22}$$

The last of the three equations in the context of this uniform, horizontal flow, when solved, indicates that the pressure within this flow is independent of depth. This outcome arises from the specific characteristics of the flow, leading to a constant pressure field throughout the fluid's depth. This independence of pressure from depth is a notable property of this uniform, geostrophic, horizontal flow within a homogeneous fluid.

$$-f(v-\overline{v}) = \nu_E\frac{\delta^2 u}{\partial z^2} \tag{11.23}$$

$$-f(u-\overline{u}) = \nu_E\frac{\delta^2 v}{\partial z^2} \tag{11.24}$$

Given the differential equations, the proposed solutions for the flow components are expressed as:

$$u = \overline{u} + Ae^{\lambda z} \tag{11.25}$$

$$v = \overline{v} + Be^{\lambda z} \tag{11.26}$$

Substituting these possible solutions into the governing equations yields the relation:

$$\nu_E^2\lambda^4 + f^2 = 0 \tag{11.27}$$

From this relation, the potential values for λ are as follows:

If $\nu_E^2\lambda^4 + f^2 = 0$, considering the equality holds, it implies that:

$$\lambda^4 = -\frac{f^2}{\nu_E^2} \tag{11.28}$$

λ^4 can be solved by taking the fourth root of $-\frac{f^2}{\nu_E^2}$. However, there is a fundamental issue here because $\frac{f^2}{\nu_E^2}$ would be a nonnegative real number, whereas the equation implies the need for an imaginary root. This signifies that no real λ can satisfy the equation within the real number system. The complex root of λ, therefore, can be derived from the fact that the equation results in the square of an imaginary number when $\frac{f^2}{\nu_E^2}$ is taken negatively. Consequently, the value of λ will be a complex number. The complexity of λ denotes that the solutions $\overline{u}$ and $\overline{v}$ in the proposed form contain an oscillatory behavior in the vertical direction. This behavior characterizes the nature of an Ekman spiral, where the flow exhibits a decay in amplitude with increasing depth. The presence of a complex λ indicates this decaying behavior in the vertical direction.

The no-slip condition at the bottom of the fluid and the assumption of a constant interior flow for $z \gg d$ allow the determination of coefficients A and B within the proposed solutions for u and v. However, under the no-slip condition at the bottom ($z = 0$), the velocity of the fluid must be zero. Considering the proposed solutions for $\overline{u}$ and $\overline{v}$:

$$u = \overline{u}\left[1 - e^{-z/d}\cos(z/d)\right] \overline{} \overline{v}e^{-z/d}\sin(z/d), \tag{11.29}$$

$$u = \overline{v}e^{-z/d}\sin(z/d) + \overline{v}\left[1 - e^{-z/d}\cos(z/d)\right], \tag{11.30}$$

In this context, the value of the parameter d is defined as: $d = \sqrt{\frac{\nu_E^2}{f}}$. This parameter d signifies the thickness of the Ekman layer. When z reaches a value on the order of d, the velocity vector will approach the values of the interior flow. This defines the transition where the influence of the Ekman spiral diminishes, and the flow aligns more with the constant interior flow.

The definition of d as the thickness of the Ekman layer is crucial in characterizing the behavior of the Ekman spiral. Several important properties of the Ekman spiral can be derived from this solution and the defined thickness of the Ekman layer. This includes the transition from the Ekman spiral to the interior flow as z approaches the order of d, signifying the limit of the Ekman layer's influence on the velocity profile.

Several key properties of the Ekman spiral stem from this solution:

As z approaches 0, the flow exhibits a transverse component concerning the interior flow. This component is oriented 45 degrees to the left in the Northern Hemisphere ($f > 0$) and 45 degrees to the right in the Southern Hemisphere ($f < 0$). This angle between the Ekman spiral flow and the interior flow is at its maximum initially and decreases with increasing z.

When dz reaches the value of π, the resulting flow aligns with the interior flow but is increased by a factor of $e^{-}\pi$ concerning the interior flow.

For higher values of dz, a minimal transverse component in the opposite direction emerges, similar to the initial orientation. The exponential term decreases toward zero for $z \gg d$, eventually leading to the velocity vector approaching the constant values of the interior flow $\left(\vec{u} = (\overline{u}, \overline{v})\right)$.

As a result of these properties, the velocity vector of the flow as a function of depth takes on the characteristic spiral shape known as the Ekman spiral. This spiral structure represents the transition from the influence of the Ekman layer to the constant interior flow, displaying a gradual change in the direction and magnitude of the flow as depth increases.

In the scenario where wind stress $\vec{\tau} = (\tau_x, \tau_y)$ acts along the air − water interface, causing shear stress on the flow, Ekman spirals can form. This situation occurs due to the exertion of wind stress on the water surface, influencing the flow beneath.

Assuming a uniform, geostrophic, and homogeneous fluid, the equations governing the motion for this scenario resemble those described in the previous section pertaining to the bottom Ekman spiral. These equations consider the flow below the air − water interface and its interaction with the wind stress exerted upon it.

The equations describing this geostrophic flow under the influence of wind stress will have similar characteristics to those presented earlier for the bottom Ekman spiral. These equations outline the behavior of the flow beneath the air − water interface and its response to the applied wind stress, leading to the formation of Ekman spirals. The fundamental concepts of the Ekman spiral persist in this scenario due to the shear stress induced by the wind at the air − water interface, resulting in a similar but distinct formation of Ekman spirals within the fluid flow beneath as shown in Eqs. (11.23 and 11.24). In this understanding, the boundary conditions for the situation involving wind stress acting along the air − water interface, influencing the flow beneath, are crucial in defining the behavior of the fluid motion. These conditions delineate the constraints under which the flow operates and interacts with the air − water interface due to the applied wind stress.

$$Surface(z=0){:}\rho_0\nu_E\frac{\partial v}{\partial z}=\tau_y \tag{11.31}$$

$$Towards\ interior(z\rightarrow -\infty){:}u=\overline{u}\ and\ v=\overline{v} \tag{11.32}$$

The solution for the flow in this scenario influenced by wind stress acting along the air − water interface can be expressed by the following equations:

$$u=\overline{u}+\frac{\sqrt{2}}{\rho_0 fd}e^{z/d}\left[\tau_x\cos\left(\frac{z}{d}-\frac{\pi}{4}\right)-\tau_y\sin\left(\frac{z}{d}-\frac{\pi}{4}\right)\right] \tag{11.33}$$

$$v=\overline{v}+\frac{\sqrt{2}}{\rho_0 fd}e^{z/d}\left[\tau_x sin\left(\frac{z}{d}-\frac{\pi}{4}\right)+\tau_y\cos\left(\frac{z}{d}-\frac{\pi}{4}\right)\right] \tag{11.34}$$

These equations depict the flow u and v as functions of depth z beneath the air − water interface, influenced by the wind stress components τ_x and τ_y acting on the water surface. The solution illustrates the Ekman spiral forming due to the interaction between the wind stress and the fluid flow beneath the air − water interface, showcasing the distinct behavior of the flow with increasing depth (Dritschel et al., 2020; Ellison, 1955; Gnanadesikan & Weller, 1995; Hieber & Stannat, 2013; Stacey et al., 1986; Wirth, 2010).

11.8 Ekman number

The Ekman number (Ek) is a dimensionless parameter in fluid dynamics that plays a crucial role in characterizing the balance between viscous forces and Coriolis forces, particularly in geophysical phenomena within the oceans and atmosphere. Named after the Swedish oceanographer Vagn Walfrid Ekman, this dimensionless number helps quantify the impact of planetary rotation on fluid motion.

In scenarios where the Ekman number is small, disturbances exhibit the ability to propagate over considerable distances before succumbing to the effects of low friction. This characteristic emphasizes the importance of Coriolis forces in governing the behavior of the fluid. The Ekman number (*Ek*) is defined by the following equation:

$$\text{Ek}=\frac{\nu}{2D^2\Omega\sin\varphi} \tag{11.35}$$

Here, the variables are defined as follows:

- D is a characteristic (usually vertical) length scale of a phenomenon,
- ν is the kinematic eddy viscosity,
- Ω is the angular velocity of planetary rotation,
- φ is the latitude.

The term $2\Omega\sin\varphi$ represents the Coriolis frequency. This dimensionless number provides insight into the ratio of viscous forces to Coriolis forces in a system, considering the influence of planetary rotation and the characteristic length scale of the phenomenon. The Ekman number is particularly useful in geophysical contexts, such as describing oceanic and atmospheric dynamics, where the balance between these forces is crucial to understanding fluid motion.

Now, consider the typical scales in the Ekman layer: u and v are of order U, L is the characteristic length scale, and ∇ is of order $1/L$. In this perspective, the Ekman number (*Ek*) is expressed as:

$$\text{Ek}=\frac{\nu}{\Omega L^2}. \tag{11.36}$$

Here, the parameters are defined as follows:

- ν is the kinematic viscosity,
- Ω is the angular velocity of planetary rotation,
- L is a characteristic length scale.

The Ekman number (*Ek*), therefore, can also be expressed as:

$$\text{Ek}=\sqrt{\frac{\nu}{2\Omega L^2}}=\sqrt{\frac{\text{Ro}}{\text{Re}}}. \tag{11.37}$$

In this formulation:

- ν is the kinematic viscosity,
- Ω is the angular velocity of planetary rotation,

- L is a characteristic length scale,
- Ro is the Rossby number,
- Re is the Reynolds number.

The Rossby number (Ro), named in honor of Carl-Gustav Arvid Rossby, serves as a dimensionless parameter essential for elucidating fluid dynamics. It meticulously captures the balance between inertial force and Coriolis force, symbolized by terms such as $|\mathbf{v}\cdot\nabla\mathbf{v}|\sim U^2/L$ and $\Omega\times\mathbf{v}\sim U\Omega$ in the Navier–Stokes equations. Widely applied in geophysical contexts, particularly in the oceans and atmosphere, it sheds light on the profound influence of Coriolis accelerations arising from planetary rotation. Another moniker for this parameter is the Kibel number. Expressed as $Ro = LfU$, the Rossby number entails characteristic velocity (U) and length (L) scales, with $f = 2\Omega\sin\phi$ denoting the Coriolis frequency. Here, Ω represents the angular frequency of planetary rotation, and ϕ signifies the latitude. In essence, the Rossby number encapsulates the intricate interplay between forces in the dynamic fluid realm, offering profound insights into geophysical phenomena.

A diminutive Rossby number indicates a system strongly influenced by Coriolis forces, while a substantial Rossby number signifies a system where inertial and centrifugal forces hold sway. In specific scenarios, such as tornadoes with a large Rossby number (approximately 10^3), the Coriolis force becomes negligible, and the equilibrium rests between pressure and centrifugal forces, a state known as cyclostrophic balance. This balance is also prevalent in the inner core of tropical cyclones. In contrast, low-pressure systems exhibit a low Rossby number (around 0.1–1), where centrifugal force is minimal, and balance is maintained between Coriolis and pressure forces, a condition termed geostrophic balance. In oceanic systems, all three forces are comparable, referred to as cyclogeostrophic balance.

The magnitude of the Rossby number varies across different phenomena, spanning orders of magnitude (approximately $10^{-2}-10^2$). In tornadoes, the Coriolis force is negligible, while in low-pressure systems, centrifugal force is inconsequential. In oceanic contexts, the Rossby number hovers around unity, but its range can extend over several orders of magnitude ($10^{-2}-10^2$), highlighting the diverse dynamics involved. The Rossby number's significance lies in delineating the impact of planetary rotation: a large Rossby number implies a system where rotational effects are minor, whereas a small Rossby number underscores the dominance of planetary rotation, allowing for the geostrophic approximation (Noir et al., 2001; Rotvig & Jones, 2002; Stellmach & Hansen, 2004; Takahashi et al., 2008; Valle-Levinson, 2008).

11.9 Ekman velocity

In the realm of oceanography, Ekman velocity, also known as a form of residual ageostrophic velocity deviating from geostrophy, constitutes a component of the overall horizontal velocity (u) in the upper layer of open ocean water. This particular velocity arises due to the influence of winds sweeping across the ocean surface, resulting in a balance between the Coriolis force acting on this layer and the force exerted by the wind.

The interaction between winds and the Ekman layer involves the direct and frictional impact of winds through vertical eddy viscosity, typically extending to the upper 50–100 meters of the ocean. The surface flow resulting from friction (u) deviates at an angle to the right of the wind in the northern hemisphere and to the left in the southern hemisphere (approximately 45 degrees if viscosity is uniformly distributed in the vertical z-direction). This surface flow, in turn, induces modifications in the underlying flow, causing it to veer slightly more to the right. The flow vectors progressively weaken with depth in an exponential manner, reaching their lowest intensity at around 50–100 meters and culminating in the formation of an Ekman spiral.

The angle of each subsequent layer within the spiral is contingent upon the strength and vertical distribution of the vertical eddy viscosity. Upon aggregating the contributions from all vertical layers—integrating the velocity over depth, from the bottom to the top of the Ekman layer—the overall "Ekman transport" aligns precisely at a 90-degree angle to the right of the wind direction in the Northern Hemisphere and to the left in the Southern Hemisphere, as elaborated in previous sections.

In the scenario where geostrophic balance is attained within the Ekman layer, and wind stress is exerted at the water surface, the governing equation is expressed as:

$$f\hat{z}\times\boldsymbol{u} = -\nabla\phi + \frac{\partial\tau}{\partial z}, \tag{11.38}$$

In the provided Eq. (11.38), ϕ represents the modified pressure, calculated as $p/\rho0$, τ is the applied stress divided by the mean density of water in the Ekman layer ($\rho0$), $\hat{z}$ is the unit vector in the vertical direction, opposing the direction of gravity, and ρ is the local density, normalized by the mean density of water in the Ekman layer ($\rho0$).

The Ekman velocity ($\mathbf{u}_E$) is defined as the difference between the total horizontal velocity ($\mathbf{u}$) and the geostrophic velocity ($\mathbf{u}_g$). Mathematically, this can be expressed as:

$$\boldsymbol{u}_E = \boldsymbol{u} - \boldsymbol{u}_g \tag{11.39}$$

As the geostrophic velocity (**ug**) is defined as the horizontal velocity that would result from the balance between the Coriolis force and the pressure-gradient force in the geostrophic flow. It can be expressed mathematically as:

$$u_g = \frac{1}{f}\hat{z} \times \nabla\tau, \tag{11.40}$$

$\nabla \times \tau$ represents the curl of the wind stress vector. In this sense,

$$f\hat{z} \times u_E = \frac{\partial \tau}{\partial z} \tag{11.41}$$

Or

$$u_E = -\hat{z} \times \left(\frac{1}{f}\frac{\partial \tau}{\partial z}\right). \tag{11.42}$$

The Ekman transport is obtained by integrating the Ekman velocity from the bottom level ($z = zb$)—at which the Ekman velocity vanishes—to the surface ($z = zt$).

$$U_E = \int_{z_b}^{z_t} u_E dz = -\hat{z} \times \left(\frac{\tau_t - \tau_b}{f}\right). \tag{11.43}$$

To quantify this transport, the Ekman velocity ($\mathbf{u}_E$) is introduced, representing the difference between the total horizontal velocity ($\mathbf{u}$) and the geostrophic velocity ($\mathbf{u}_g$). The geostrophic velocity itself is the balance between the Coriolis and pressure-gradient forces.

The Ekman transport ($\mathbf{u}_E$) is determined by integrating the Ekman velocity over the Ekman layer's depth, from the ocean bottom (z_b) where the Ekman velocity vanishes, to the ocean surface (z_t). Mathematically, this integration yields the Ekman transport as a cross product between the unit vector in the vertical direction ($\hat{z}$) and the difference in wind stress ($\tau_t - \tau_b$) divided by the Coriolis parameter (f). This formulation captures the cumulative effect of the wind-induced horizontal flow throughout the Ekman layer.

Let's consider a hypothetical numerical example to illustrate the concept of Ekman transport:

Assumptions:

- Wind stress at the ocean surface, τ, is ($1\ N/m^2, 0\ N/m^2$) (eastward direction) in the Northern Hemisphere.
- Depth of the Ekman layer, $z_t - z_b$, is 100 m.
- Coriolis parameter, f, is $10^{-4} s^{-1}$ at the given latitude.
- Density of water, ρ_0, is 1025 kg/m^3.

Now, let's calculate the Ekman transport (**Ue**) using the formula:

$$U_E = -\hat{\mathbf{z}} \times \left(10^4 m\right) \tag{11.44}$$

Assuming that the unit vector $\hat{\mathbf{z}}$ is in the vertical upward direction, the cross product results in an Ekman transport of (0 m/s, 10^4m/s). This means that, due to the wind stress, there is a net eastward transport of water in the upper ocean with a magnitude of 10^4 m/s over the depth of the Ekman layer.

11.10 Upwelling index

The upwelling index is a measure used in oceanography to quantify the strength and intensity of upwelling, a process where deep, nutrient-rich waters rise to the surface. This phenomenon is crucial for marine ecosystems as it brings nutrients to the surface layers, supporting the growth of phytoplankton and subsequently influencing the entire food chain.

The upwelling index is often defined based on physical and oceanographic parameters such as changes in sea surface temperature, wind patterns, and Ekman transport. Here, I'll provide a simplified conceptual explanation of how the upwelling index might be formulated:

Sea Surface Temperature (SST): Upwelling is associated with cooler waters rising from the depths. A decrease in sea surface temperature in a particular region might indicate the presence of upwelling.

Wind Patterns: Winds play a crucial role in driving upwelling. Strong, persistent winds blowing parallel to the coast can push surface waters away, allowing cold, nutrient-rich waters to replace them. The strength and duration of these winds are considered in the index.

Ekman Transport: As mentioned earlier, Ekman transport is the net movement of surface water perpendicular to the direction of the wind. The Ekman transport associated with a particular wind pattern can contribute to the upwelling index.

A hypothetical formula for the upwelling index might be something like:

$$\begin{aligned} \textit{Upwelling Index (UI)} &= k \times (\textit{Change in SST}) \\ &+ m \times (\textit{Wind Strength}) \\ &+ n \times (\textit{Ekman Transport}) \end{aligned} \tag{11.45}$$

here, k, m, and n are coefficients that weigh the importance of each factor in contributing to the overall upwelling index. The actual formulation would be more complex and would likely involve regional variations and other oceanographic parameters. Assume that we are examining a coastal region over a period of 1 month, and we have the following changes:

Change in Sea Surface Temperature (SST): The average SST decreases by 2 degrees Celsius over the month, indicating a cooling trend.

Wind Strength: The prevailing winds along the coast have an average speed of 10 meters per second during the month.

Ekman Transport: Through observational data and modeling, it's determined that the Ekman transport is contributing a net northward movement of 0.5 meters per second.

Now, let's assign some arbitrary coefficients to these factors based on their perceived influence:

$$\begin{aligned}\textit{Upwelling Index (UI)} &= 0.8\times(\textit{Change in SST})\\ &\quad+0.5\times(\textit{Wind Strength})\\ &\quad+1.2\times(\textit{Ekman Transport})\end{aligned} \tag{11.46}$$

Substitute the numerical values:

$$\begin{aligned}\textit{Upwelling Index (UI)} &= 0.8\times(-2)+0.5\times(10)\\ &\quad+1.2\times(0.5)=4\end{aligned} \tag{11.47}$$

In this simplified example, the upwelling index is 4. This numerical value, though arbitrary, indicates a moderate level of upwelling activity in the considered coastal region for the given month. On the other hand, the upwelling index (*UI*) as a function of the Ekman transport (V(z)) is given by the formula:

$$\begin{aligned}UI\left(m^2/s\right) &= V(z)\cos(a-\alpha)\\ &= \frac{\tau}{\rho_0\left|\left(\frac{4\pi}{T_{Earth}}\right)\sin\varphi\right|}\cos(a-\alpha)\end{aligned} \tag{11.48}$$

where Tearth = 86,400 s is the period of the Earth's rotation; ρ_0 is 1026 kg/m^3 is the average; and sinφ is geographical latitude. Therefore *a* is the angle of the coastline with respect to a reference direction, and α is another angle or phase, indicating the particular orientation of the Ekman transport relative to the coastline. This formula calculates the upwelling index based on the Ekman transport, coastline angle (*a*), and the phase angle (α). It represents the offshore Ekman transport's influence on upwelling along the coastline. In this view, the upwelling index (*UI*), determined by the offshore Ekman transport, serves as a direct measure of the vertical volume transport associated with upwelling. This dimensional parameter, *UI*, is commonly recognized as the upwelling index. Essentially, the upwelling index represents the offshore volume transport per unit length of the coastline. In some cases, this parameter is expressed in kilometers, and its calculation is typically referenced to a distance of 1 km.

According to the aforementioned perspective, the utilization of the upwelling index (*UI*) with coastal wind data offers a convenient means to examine and identify coastal upwelling occurrences within a specific region. When the *UI* is negative, it signifies an onshore Ekman drift, driving seawater toward the coast and initiating the downward movement of density interfaces, a process termed coastal downwelling. This interplay often results in a rhythmic pattern of upwelling and downwelling sequences, influenced by the prevailing coastal winds and the synoptic variability in weather patterns occurring over 3 − 10 days (Lent, Tougaw, & Porod, 1993; Tatro & Mollo-Christensen, 1967; Wijffels et al., 1994).

11.11 Quantized Ekman spiral

Wind energy transfers to the sea surface, initiating a frictional force that kick-starts motion at the surface. The energy's fluctuations within the water column lead to a shift in the quantum state, extending from the surface down to the ocean floor. This quantum state evolution throughout the water column results in the formation of swirling water particles below the sea's surface. Differences in the phases of spinor components along the water column create a distinctive spiral of spin.

Let's consider the existence of numerous quantum states stemming from the wind's influence on the sea surface and extending to the ocean's depth. The quantity of these quantum states is directly related to the water depth, meaning that as the water gets deeper, the number of quantum states increases. Shallow waters exhibit fewer quantum states compared with deeper oceanic regions.

Hence, if we represent Ψ as a collection of spinor components – $\Psi = [\psi_1, \psi_2, \psi_3,., \psi_N]$, where ψ_N signifies an N-component spinor from the sea surface to the water's depth—the scientific explanation of the water's spiral pattern can be mathematically articulated.

$$\begin{bmatrix}\psi_1\\ \psi_2\\ \vdots\\ \psi_N\end{bmatrix} = \begin{bmatrix}e^{in\theta+\phi_i}\cdot\varphi_1(r,\omega t\\ e^{in\theta+\phi_i}\cdot\varphi_2(r,\omega t\\ \vdots\\ e^{in\theta+\phi_i}\cdot\varphi_N(r,\omega t\end{bmatrix} \tag{11.49}$$

It appears that Eq. (11.48) provides information about the number of spiral arms (n) observed within the water column in the context of an Ekman spiral pattern. The Ekman spiral's geometric configuration might not be well-defined due to the influence of kinetic energy phase alterations, denoted by φi(r), stemming from the stratification of the fluid due to baroclinity. Baroclinity pertains to the variation of water density that induces nonuniformity in the water column, leading to complexities in the flow patterns.

The nonlinearity arising from the changes in kinetic energy, particularly the phase alterations occurring through the water column due to the interaction with stratified fluid and water density friction, is mathematically formulated as follows:

$$\phi_i'' + \left(r^{-1}+i\varphi_j'\right)\phi_j' = \left[\left(-\omega+n^2\left(r^{-2}\right)+\varphi_j'\right)^2\right]\varphi_j \tag{11.50}$$

The expression represented by Eq. (11.49) elucidates the radial modulation of amplitudes ϕj(r) within the context of the Ekman spiral pattern. This modulation of amplitudes is instrumental in shaping the spiral, and it is particularly influenced by the baroclinic term denoted as βj. The baroclinic term plays a significant role in governing the alterations in φi(r) across the water depth. This term, βj, is primarily responsible for inducing changes in the kinetic energy phase (φi(r)) at various radial distances as the spiral extends through the water column:

$$\beta_j = \frac{\partial^2 \varphi}{\partial z} + r^{-1}\left(\frac{\partial \varphi_j}{\partial z}\right). \tag{11.51}$$

The presence of the spiral pattern is contingent on the nonzero nature of βj. When βj equals zero, this quantum spiral adopts characteristics akin to a "point vortex" that exhibits zero vorticity except for the central phase singularity. Eq. (11.50) underlines that the effect induced by the quantum baroclinicity, encapsulated in βj, represents non-Hermitian terms, dynamically fluctuating to orchestrate the effective motion of the spiral.

In Fig. 11.6, the Ekman quantum spiral, observed in the waters off the east coast of Malaysia, unveils an intriguing feature. Notably, variations in water density from the surface to a depth of 40 meters curtail the rotational speed of water particles due to friction. As friction intensifies, so does the density. The maximum water density, recorded at 22 kg/m^3, diminishes the kinetic energy of water particle rotation, resulting in the gradual decay of the spiral near the 40-meter water bottom. From a quantum mechanics standpoint, Ekman's theory finds an innovative explanation. It hinges on the alteration of the quantum state of kinetic energy transferred by wind stress to the water body, thereby instigating the rotation of water column particles. The distinctive energy phase in the spinning of water particles is contingent upon the quantum baroclinicity of the water body.

A noteworthy concept emerges in this context: an entanglement is established among the state of the water spiral, water density, and the state of wind stress. Consequently, water particle spirals endeavor to reach equilibrium or a state of uniform energy distribution. This equilibrium, achieved through quantum entanglement with their surroundings, necessitates an infinite amount of time.

According to the aforementioned perspective, the concept presented here suggests that wind stress triggers an entangled connection between wind particles and seawater particles, allowing actions on one to influence the other, despite their physical separation (refer to). This phenomenon, often referred to as "spooky action at a distance," underscores the entanglement between wind and seawater particles.

The Ekman spiral's formation is attributed to the transfer of kinetic energy from wind particles to seawater particles, generating intertwined combinations of wind kinetic energy and kinetic energy associated with density fluctuations (refer to Fig. 11.7). The extent of this entanglement is directly proportional to the water depth. Within this span of

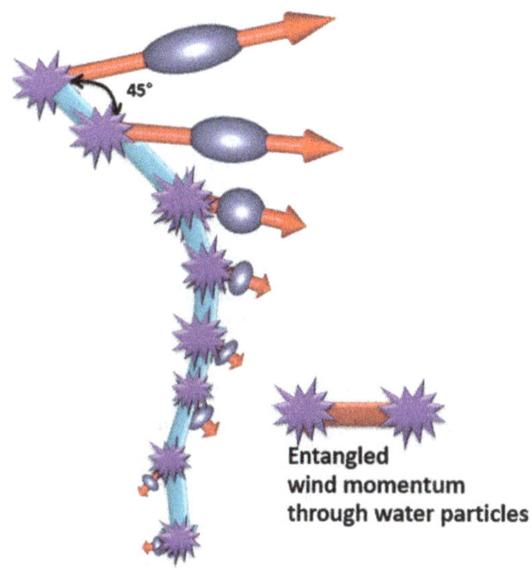

FIGURE 11.7 Quantized Ekman spiral.

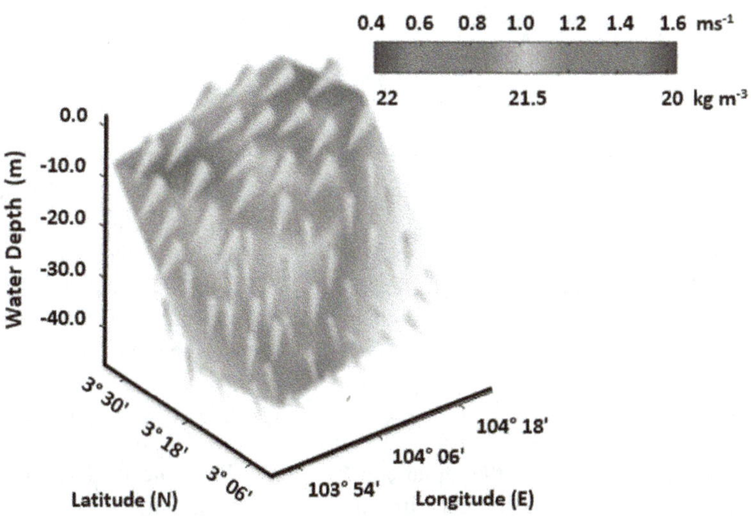

FIGURE 11.6 Ekman quantum spiral through the water column.

entanglement, the spiraling water experiences both spatial and temporal vibrations. Consequently, the Ekman spiral can be likened to a wave-particle duality, manifesting as water parcels or particles vibrating downward in both space and time. These vibrations occur in a spiraling manner, resembling wave-like motions until the spiral pattern reaches a state of gradual decay.

The water spiral pattern is described as a manifestation of the quantum state changes in kinetic energy transferred by wind stress to the water body, causing spinning of water column particles. The spinning of water particles exhibits different energy phases, determined by the quantum baroclinic nature of the water body. Entanglement occurs between the water spiral state, water density, and the state of wind stress. In summary, the quantum view of the Ekman spiral provides a novel perspective, suggesting that the classical understanding of fluid dynamics can be enriched by considering the quantum nature of the processes involved (Bewley et al., 2006; Bogoliubov et al., 1960; Fu & Xie, 2005; Schwinger, 1951; Snyder, 1947).

11.12 Introducing a novel theory: quantum-enhanced coastal upwelling imaging

Coastal upwellings, a crucial phenomenon for marine ecosystems, can be reexamined through a novel theoretical framework incorporating principles from quantum physics. This innovative perspective aims to shed light on the observed dark patterns in synthetic aperture radar (SAR) images, offering a quantum-enhanced explanation for the complexities involved in coastal upwelling detection.

In this theory, the conventional understanding of SAR, operating at microwave frequencies, encounters a quantum twist. The interaction of SAR with ocean surfaces is influenced by the quantum nature of the underlying processes. Specifically, the damping of Bragg waves, a key aspect in SAR imaging, is attributed to the quantum behavior of low-temperature water affected by coastal upwelling. Two quantum-enhanced mechanisms are proposed to elucidate the dark patterns imaged by SAR during coastal upwelling events. First, the upward movement of cooler water induces significant changes in marine boundary layers, reducing wind stress and contributing to lower sea surface roughness. This quantum-induced alteration in surface characteristics plays a crucial role in SAR backscatter signatures.

Second, the upwelling water carries biogenic slicks, introducing a quantum-damped effect on Bragg waves in the ocean surface. The quantum entanglement between the water's biogenic content and SAR signals contributes to the observed dark patterns. This novel theory suggests that the quantum properties of the marine environment, influenced by coastal upwelling, play a decisive role in SAR image interpretation.

In the quantum-enhanced theory of coastal upwelling imaging, therefore, the concept of decoherence plays a pivotal role in understanding the observed dark patterns in synthetic aperture radar (SAR) images. Decoherence refers to the process by which a quantum system loses its coherence and becomes entangled with its environment, leading to the emergence of classical behavior. Applying this concept to coastal upwelling, we can explore how quantum states of water molecules and biogenic slicks undergo decoherence, influencing SAR observations.

11.12.1 Quantum states of water molecules

In the inception of an upwelling phenomenon, water molecules manifest quantum characteristics, operating in a state of coherence. The ascent of cooler water disrupts this coherence, initiating a process of quantum decoherence. Consequently, the once collective quantum behavior of water molecules undergoes a transition into classical attributes. This decoherence phenomenon plays a pivotal role in altering marine boundary layers, thereby influencing wind stress and sea surface roughness—critical factors that significantly affect the backscatter observed in synthetic aperture radar (SAR) imaging.

Mathematically, this can be expressed as follows:

$$\frac{d\rho}{dt} = -\frac{i}{\hbar}[\mathcal{H}, \rho] - \mathfrak{D}(\rho) \tag{11.52}$$

here $\mathfrak{D}(\rho)$ is the decoherence term that encapsulates the environmental interactions leading to the loss of quantum coherence; ρ is the density matrix describing the quantum state; and $\mathcal{H}$ is the Hamiltonian operator, which is governing the quantum dynamics, and includes terms for water molecules and their interactions. In other words, the decoherence term represents the dissipation of quantum coherence due to environmental interactions. Mathematically, it can be expressed as:

$$\mathcal{H}(\rho) = \frac{1}{2}\sum_{k} \gamma_k \left(2A_k \rho A_k^{\dagger} - A_k^{\dagger} A_k \rho - \rho A_k^{\dagger} A_k\right) \tag{11.53}$$

Here, A_k are operators describing the environmental interactions, and γ_k represents the coupling strength. Now, let's introduce a novel entanglement term $E(\rho)$ to represent the intricate interdependence between the quantum coherence loss, Bragg scatter, and radar backscatter. This term signifies the entangled state resulting from the complex interactions involved. The equation evolves into:

$$\frac{d\rho}{dt} = -\frac{i}{\hbar}[\mathcal{H}, \rho] - \mathfrak{D}(\rho) + E(\rho) \tag{11.54}$$

Here, the entanglement term $E(\rho)$ encapsulates the correlations and interdependencies arising due to the quantum-to-classical transition, the interaction with Bragg waves, and the subsequent radar backscatter. In this sense, radar backscatter, influenced by the quantum state, Bragg scatter, and other environmental factors, can be expressed as the expectation value ($\langle \mathfrak{O}_{SAR} \rangle$). The entanglement term, therefore, captures the correlations emerging from the simultaneous influence of decoherence, Bragg scatter, and radar backscatter. It is a representation of the nontrivial connections between these phenomena (Marghany, 2019a, 2019b).

$$E(\rho) = \alpha(B(\rho) - \langle \mathfrak{O}_{SAR} \rangle) \tag{11.55}$$

Here, α is a parameter representing the strength of entanglement, and $B(\rho)$ is a term representing the Bragg scatter's influence on the quantum state. Needless to say, novel Eq. (11.55) derived by author shows that the entanglement term $E(\rho)$ serves as a bridge between the quantum realm, Bragg scatter, and radar backscatter, providing a unified understanding of their intricate relationships in the context of upwelling scenarios (Marghany 2019a,b).

11.12.2 Biogenic slicks and quantum entanglement

In the ethereal dance of upwelling phenomena, biogenic slicks, those enigmatic trails of life in the ocean, emerge as quantum entities entangled with synthetic aperture radar (SAR) signals. The intricate interplay between these slicks and SAR waves unfolds a quantum-damped ballet, where the very essence of entangled quantum states succumbs to the mesmerizing influence of radar interactions.

Mathematically, we can express this entanglement and decoherence process as follows:

Let $|\Psi\rangle$symbolize the quantum state associated with biogenic slicks. The entanglement term E(ρ) is a function that quantifies the intricate web of connections between the quantum states and SAR signals (Marghany, 2019a, 2019b). It evolves over time (t) due to the interaction with radar waves, leading to a dynamic entanglement state:

$$\frac{dE(\Psi)}{dt} = f(|\Psi\rangle, |I\rangle) \tag{11.56}$$

Here, the function f encapsulates the complex dynamics of entanglement, where the quantum states are entwined with SAR signals $|I\rangle$. As SAR waves permeate the quantum realm of biogenic slicks, a quantum-damped effect unfolds. The coherence of quantum states (Ψ) begins to wane, driven by the interaction term $I(\Psi, SAR)$:

$$\frac{d\Psi}{dt} = -i\mathcal{H}\Psi + I(\Psi, SAR) \tag{11.57}$$

In this equation, $\mathcal{H}$ represents the quantum Hamiltonian, and I embodies the intricate interaction term. Decoherence, the elegantly orchestrated unraveling of quantum coherence, is encapsulated by the density matrix ρ evolving over time:

$$\frac{d\rho}{dt} = \mathcal{L}(\rho) \tag{11.58}$$

here $\mathcal{L}$ is the Lindblad superoperator, which embodies the quantum dynamics, steering the system toward classical features. The modified SAR backscatter β_{mod} arising from the entangled quantum states is a manifestation of this enthralling quantum choreography:

$$\beta_{mod} = g(E(\rho), \beta_{SAR}) \tag{11.59}$$

Here, g captures the transformation of entanglement into observable SAR signatures. In essence, the entanglement of biogenic slicks with quantum states orchestrates a symphony of quantum-damped SAR responses, painting the ocean's canvas with enigmatic dark patterns witnessed by SAR sensors (Marghany, 2019a, 2019b).

Embarking on a celestial voyage, we cast our gaze upon coastal upwelling, interpreting it not merely as a marine dance but as a quantum ballet undergoing the delicate choreography of decoherence. This perception allows us to construct an ethereal bridge, seamlessly connecting the quantum elegance with classical narratives of the marine expanse. In this cosmic odyssey, the perception of coastal upwelling as a quantum symphony, adorned with entanglement and decoherence, enriches our understanding. It allows us to traverse the realms of quantum and classical descriptions, enveloped in the embrace of SAR signals and the enigmatic dance of coastal dynamics (Alpers & Bignami, 2020; Alpers & Zeng, 2021; Alpers, 2016, 2018).

In summary, the infusion of decoherence into the quantum tapestry of coastal upwelling imaging offers a transcendental framework, orchestrating a seamless quantum-to-classical transition. This perspective unravels the mysteries encoded in SAR images, shedding light on the quantum nuances lost in the coastal interplay.

11.13 Quantum cellular automata for automatic upwelling in synthetic aperture radar data

In the realm of quantum cellular automata (QCA), a groundbreaking theory unfolds, leveraging the quantum nature of particles to detect upwelling phenomena. The theory posits that the intricate interplay of electrons within QCA cells can be harnessed as a sensitive indicator of upwelling in synthetic aperture radar (SAR) images. The foundation of the theory, therefore, lies in the quantum states of electrons within the QCA cells, particularly their polarization configurations. As upwelling processes influence the marine environment, altering boundary layers and sea surface properties, these changes reverberate into the quantum domain of QCA cells.

The theory proposes that the quantum states, reflected in the polarization of QCA cells, respond dynamically to the variations induced by upwelling. The altered electron configurations within the QCA cells serve as a quantum signature, offering a unique and sensitive indicator of the upwelling activity in SAR images. To implement this theory, algorithms are devised to analyze SAR images, mapping the changes in QCA cell polarization configurations. The detection of specific patterns, influenced by upwelling-induced alterations, becomes a distinctive quantum fingerprint in the SAR data. This innovative approach not only bridges the gap between quantum information processing and environmental sensing but also opens new avenues for the exploration of quantum-enhanced technologies in Earth observation (Freedman et al., 2022; Sellapillay et al., 2022; Shirley et al., 2022).

In the realm of quantum cellular automata (QCA), the mathematical expression for cell polarization (CL) unfolds as a delicate balance of electron configurations. Symbolized by the equation:

$$CL(P) = \frac{(\alpha_2 + \alpha_4) - (\alpha_1 + \alpha_3)}{\alpha_1 + \alpha_2 + \alpha_3 + \alpha_4} \tag{11.60}$$

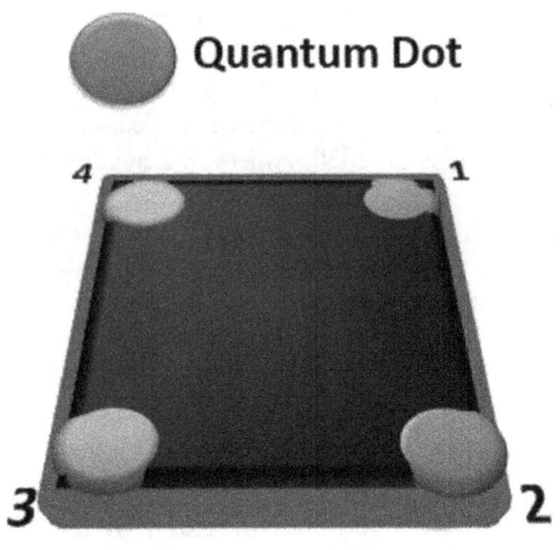

FIGURE 11.8 Schematic of the basic cell constructed from four quantum dots.

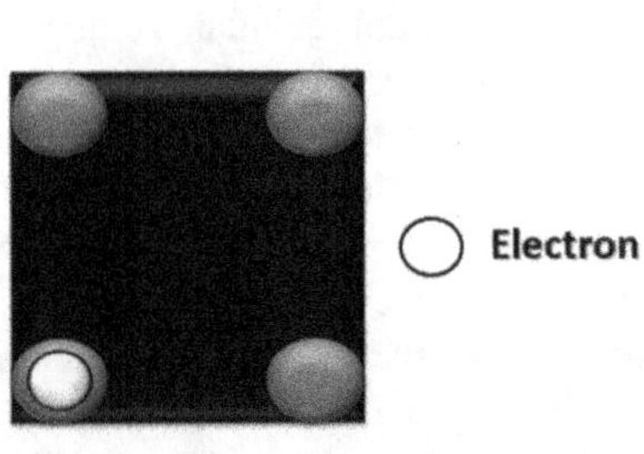

FIGURE 11.9 Dual bistable states consequence in cell polarization of P = -1(Binary 0).

Here, $\alpha_1, \alpha_2, \alpha_3, and \alpha_4$ represent parameters characterizing the quantum states within the QCA cell. The numerator, comprising the sum of $\alpha_2, and \alpha_4$, reflects the contribution of certain electron configurations, while the denominator, encompassing the sum of $\alpha_1, and \alpha_3$, encapsulates another set of quantum states (Fig. 11.8). This formula encapsulates the nuanced relationship between these quantum parameters, providing a quantitative measure of cell polarization (Fig. 11.9). In the context of the proposed upwelling detection theory based on QCA, variations in these parameters due to upwelling-induced environmental changes become pivotal. The computation of cell polarization serves as a mathematical lens through which the quantum signatures of upwelling phenomena are deciphered within the intricate framework of QCA.

11.14 Quantum-inspired adder circuit for automated upwelling detection

In the pursuit of automated upwelling detection in synthetic aperture radar (SAR) data, we introduce an advanced quantum-inspired adder circuit comprising five gates. This circuit leverages the principles of quantum cellular automata (QCA) to enhance the efficiency and accuracy of upwelling zone identification. The integral gates, therefore, encompass (Fig. 11.10):

11.14.1 Quantized Marghany turbulence boundary gate ($\Psi|E(I^{-1}(\mathbb{Z}))\rangle$)

This gate is dedicated to managing the intricate characteristics associated with turbulence boundaries in SAR data. Leveraging quantized Marghany turbulence principles (Chapter 4), it ensures a nuanced understanding and effective handling of turbulent regions within the upwelling zone in radar imagery.

11.14.2 Quantized Marghany bounded front state gate $\hat{H}\Psi|\bar{\rho}|b_M>$

Addressing the complexities of bounded front states within SAR data, this gate utilizes quantized Marghany

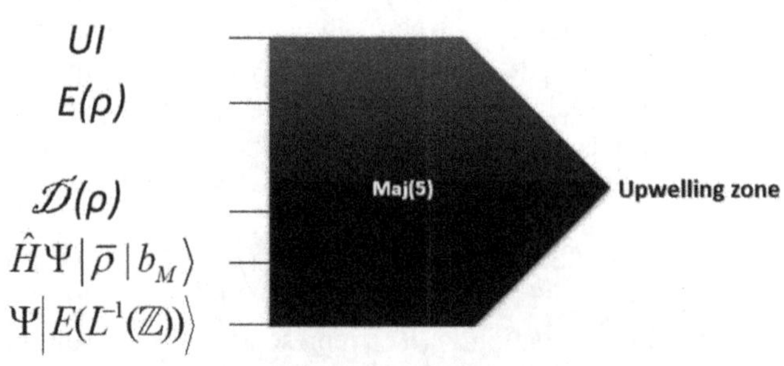

FIGURE 11.10 The symbolic representation for the majority gate capturing specific indices of upwelling features.

principles to navigate and interpret the specific conditions at the boundaries. It plays a pivotal role in discerning features related to bounded fronts, contributing to the comprehensive analysis of upwelling dynamics.

11.14.3 Decoherence gate(D(ρ))

The decoherence gate is designed to manage the intricate process of decoherence within SAR signals. Acknowledging the quantum-inspired nature of the adder circuit, this gate addresses the loss of coherence in quantum states due to interactions with radar waves. Its role is crucial in capturing the transition from quantum to classical behavior in SAR data.

11.14.4 Quantum entanglement and threshold logic gate($E(\rho)$)

This gate harnesses the principles of quantum entanglement and threshold logic to facilitate simultaneous consideration of multiple scenarios. By incorporating these quantum features, the gate aids in identifying regions surpassing predefined criteria, a vital step in pinpointing potential upwelling zones.

11.14.5 Upwelling index gate

Focused on quantifying the presence of upwelling phenomena, the UI gate utilizes the upwelling index as a key metric. This gate plays a central role in assessing and categorizing regions exhibiting characteristics indicative of upwelling, providing a quantitative measure to support the overall detection process.

In this configuration, the majority gate is designed to make decisions based on the given structure. The schematic symbol for a five-pin majority gate is depicted in Fig. 11.10. To illustrate, let's examine the functionality of a three-input AND gate and a three-input OR gate using the provided formula:

$$\begin{aligned} &\mathrm{H}(\Psi|E(l^{-1}(\mathbb{Z}))\rangle, \Psi|\overline{\rho}|b_M>, (\mathfrak{D}(\rho)), 0, 0) \\ &= \Psi|E(l^{-1}(\mathbb{Z}))\rangle \times \Psi|\overline{\rho}|b_M> \times (\mathfrak{D}(\rho)) \end{aligned} \tag{11.61}$$

$$\begin{aligned} \mathrm{H}\big(\Psi|E(l^{-1}(\mathbb{Z}))\rangle, 1, (\mathfrak{D}(\rho)), 0, UI\big) = \Psi|E(l^{-1}(\mathbb{Z}))\rangle \\ + (\mathfrak{D}(\rho)) + UI \end{aligned} \tag{11.62}$$

Let's delve into the realm of 1-bit full adders, where the intricacies of upwelling zones and their ambient conditions take center stage. In this scenario, our input features embody the quantized Marghany turbulence boundary and the quantized Marghany bound front state, while the carry bit elegantly represents the phenomenon of decoherence (Table 11.1). The unveiled design of quantum cellular automata (QCA) stands as the blueprint for the automated detection of upwelling zones within SAR data. This avant-garde design incorporates three majority gates and two inverters.

Notably, the QCA inverter showcases dual-standard cells in a diagonal orientation, mirroring the geometric elegance of dual rotated cells in a horizontal alignment (Fig. 11.11).

11.15 Quantum cellular automata for automated detection of upwelling zones

Let's denote Bmod as the set of SAR feature integers. Consequently, the 2D cellular space of SAR images can be represented as a 4-tuple, that is, (I_{SAR}, B_{mod}, ψ, N, T):

$$g(F_i) = \{F_1 + \delta_1, ., F_n + \delta_n\} \tag{11.63}$$

Eq. (11.63) indicates the selection of upwelling feature indices (F), where F ∊ B_{mod}.

$$\begin{aligned} H'(F) = \psi'(F + (0,0)), \psi'(F + (0,1)), \psi'(F + (1,0)), \\ \psi'(F + (0,-1)), \psi'(F + (-1,0))) \end{aligned} \tag{11.64}$$

Eq. (11.64) demonstrates that $\psi'(F + (0,0))$,

$$f\big(H^t(F)\big) = \psi^{t+1}(F) \tag{11.65}$$

Eq. (11.65) defines all possible pairs of the form ($H^t(F)$, $\psi^{t+1}(F)$), where F refers to the 2-D CA rule, as illustrated in Fig. 11.12.

The quiescent state is harnessed to limit the quantum computing states to a finite set, generating new SAR pixel states effectively within a finite range. This essential approach prevents an infinite computing product of unitaries, ensuring a robust identification of QCA.

From the perspective of Schrödinger quantum mechanics, the system's state at some time t is represented by a $|\Psi_t\rangle$ state vector in a Hilbert space $H|\Psi_t\rangle$ (Fig. 11.13).

The state vector evolves unitarily, maintaining its coherence over time. This controlled evolution within a finite set of states is a fundamental aspect of our quantum cellular automata (QCA) design, contributing to the stability and reliability of the quantum computing process. In the quantum mechanical framework, the evolution of the state vector is described by the unitary operator. Mathematically, this can be expressed as:

$$|\psi(t)\rangle = U(t)|\psi(0)\rangle \tag{11.66}$$

where $|\psi(t)\rangle$ is the state vector at time $U(t)$ is the unitary operator representing the evolution, and $|\psi(0)\rangle$ is the initial state vector. The automated identification of an upwelling zone in a synthetic aperture radar (SAR) image is realized through the utilization of block-partitioned quantum cellular automata (QCA), as illustrated in Fig. 11.14. In this context,

TABLE 11.1 Pseudocode of quantum cellular automata with decoherence, entanglement, and upwelling index.

```
# QCA with Decoherence, Entanglement, and Upwelling Index Pseudocode
class QuantumCellularAutomaton: def__init__(self, initial_state): self.state = initial_state
self.time_step = 0
def evolve(self): # Quantum evolution rules with decoherence and entanglement for cell in
self.state: neighbors = self.get_neighbors(cell) decoherence, entanglement =
self.calculate_quantum_effects(neighbors) cell.apply_decoherence(decoherence)
cell.apply_entanglement(entanglement)
# Calculate the upwelling index after quantum evolution upwelling_index =
self.calculate_upwelling_index()
# Update the state after quantum evolution self.time_step += 1
return upwelling_index
def get_neighbors(self, cell): # Retrieve neighboring cells of the given cell # .
def calculate_quantum_effects(self, neighbors): # Calculate decoherence and
entanglement based on the states of neighboring cells # .
return decoherence, entanglement
def calculate_upwelling_index(self): # Calculate the upwelling index based on the current
state of the QCA # Example: Upwelling index can be a function of decoherence and
entanglement upwelling_index = calculate_function_of_decoherence_and_entanglement()
return upwelling_index
# Example Usage: initial_state = initialize_qca_state() qca =
QuantumCellularAutomaton(initial_state)
# Evolve the QCA over multiple time steps for_ in range(num_time_steps): upwelling_index
= qca.evolve()
# Access the final state of the QCA final_state = qca.state
```

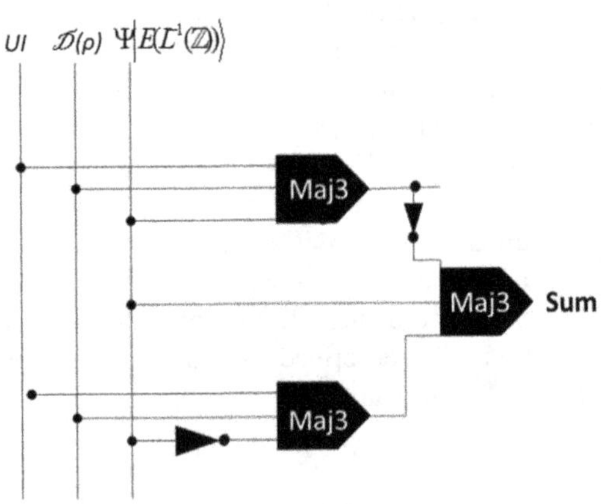

FIGURE 11.11 One-bit quantum cellular automata (QCA) adder with five gates for automated upwelling zone detection.

block-partitioned QCA can be represented as a 4-tuple. For a system with nearest-neighbor interactions, the most basic unitary QCA rule with a radius $r = 1$, describing a unitary operator applied over a three-cell neighborhood $j-1$, j, and $j+1$, is expressed as:

$$M(u_{00}, u_{01}, u_{10}, u_{11}) = |00\rangle\langle 00| \otimes u_{00} + |01\rangle\langle 01| \otimes u_{01} + |10\rangle\langle 10| \otimes u_{10} + |11\rangle\langle 11| \otimes u_{11} \quad (11.67)$$

In this quantum cellular automata (QCA) framework, the expression $M(u_{00}, u_{01}, u_{10}, u_{11})$ signifies a unitary operator. Let's break down its significance:

The terms $|00\rangle\langle 00||00\rangle\langle 00|$, $|01\rangle\langle 01||01\rangle\langle 01|$, $|10\rangle\langle 10||10\rangle\langle 10|$, and $|11\rangle\langle 11||11\rangle\langle 11|$ represent quantum states in a Hilbert space. The coefficients u_{00}, u_{01}, u_{10}, and u_{11} are parameters that influence the evolution of these quantum states. The overall expression $M(u_{00}, u_{01}, u_{10}, u_{11})$ essentially describes the unitary transformation applied to the quantum states. This transformation involves a linear combination of the initial states with the specified coefficients. In simpler terms, this equation captures the quantum evolution

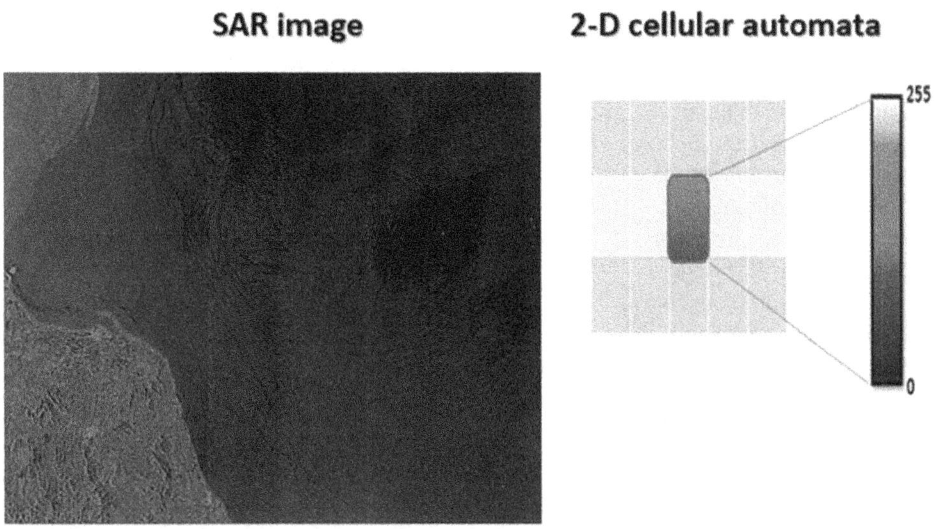

FIGURE 11.12 2-D cellular automata of SAR image with 256 states. *SAR*, Synthetic aperture radar.

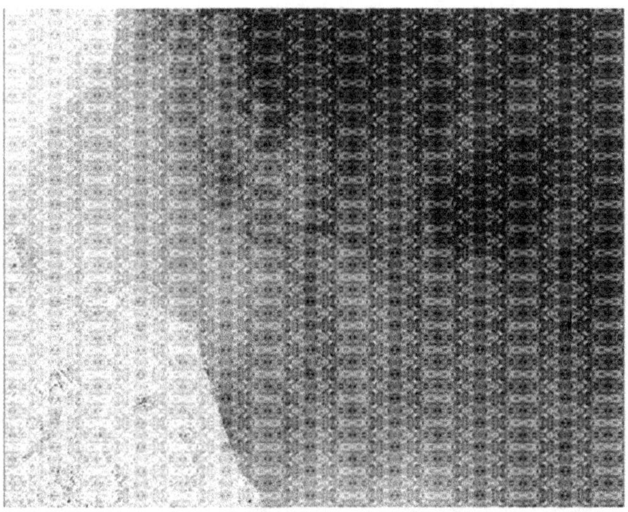

FIGURE 11.13 Cellular configuration of SAR images through quantum cellular automata (QCA). *SAR*, Synthetic aperture radar.

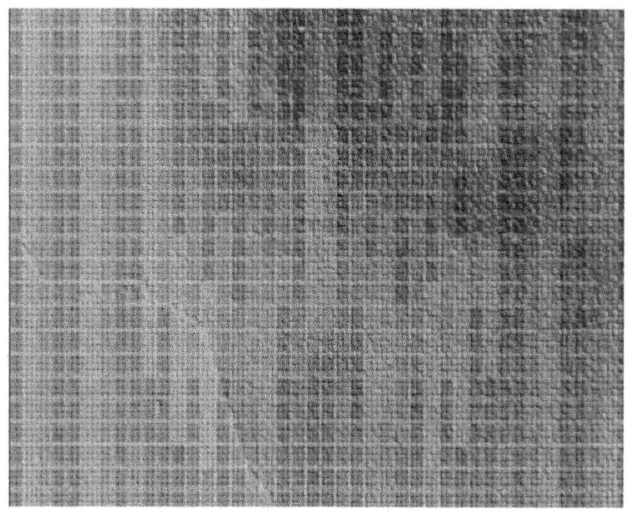

FIGURE 11.14 Quantum cellular automata (QCA) applied to SAR image with block partitioning. *SAR*, Synthetic aperture radar.

of a system governed by a unitary operator, with the coefficients determining the influence of each quantum state in the evolution process (Brennen & Williams, 2003; Freedman et al., 2022; Kent et al., 2023).

The expression $|01\rangle\langle 01| \otimes u_{01}$ signifies a qubit update at site j with the unitary u_{01} when the qubit at the site $j+1$ is in the state $|1\rangle$. In this context, the states $|1\rangle$ and $|0\rangle$ represent hydrocarbon zones correlated with the underlying drainage network. The transformation M operates with its unique 2-site conversion. Consequently, the partitioning of the SAR image for upwelling zones involves simultaneously updating all even qubits with the rule M before updating all odd qubits with the same rule M. Periodic boundaries are assumed in this scenario (Marghany, 2021a).

In the context of a nearest-neighbor one-dimensional block-partitioned quantum cellular automata (QCA), we can represent each quantum system r using the density operator formalism, specifying it by a probability distribution function given as:

$$\rho_r = \sum_{i,j=0}^{1} M_{ij}|i\rangle|j\rangle \tag{11.68}$$

Here, $|i\rangle$ and $|j\rangle$ are basis states, and M_{ij} represents the matrix elements of the density operator. This formalism captures the quantum state of the system in terms of its probabilities and quantum coherence. In the pursuit of this endeavor, a completely positive map $S(r)$ acting on state r is expressed through a set of Krauss operators K_m. The representation is given by:

$$S(r) = \sum_{m} K_m r K_m^{\dagger} \tag{11.69}$$

Here, $K_m^{\dagger}$ denotes the adjoint (conjugate transpose) of the Krauss operator K_m. This formulation encapsulates the evolution of the quantum state r under the influence of the completely positive map $S(r)$, capturing the dynamical aspects of the quantum system. The realization of a block-partitioned nonunitary quantum cellular automaton (QCA) is proposed within the framework of an even-ordered pixel configuration, composed of an irregular array featuring dual distinguishable upwelling index characteristics (Lent, Tougaw, & Porod, 1993; Lent, Tougaw, Porod, et al., 1993; Zhang et al., 2005). These characteristics are distinctly defined and interact through the Ising interaction (Wiesner, 2018).

To quantify the dynamics of entanglement, it would be beneficial to have a singular parameter that measures the degree of multiparticle entanglement present in a state at each time step (Lent, Tougaw, & Porod, 1993; Lent, Tougaw, Porod, et al., 1993). An ideal entanglement metric should capture the nonlocal aspects of quantum correlations among the spins and, consequently, should be a function of the state that, on average, decreases under local operations and classical communication. In this context, the extent of multispin entanglement in terms of pure states of n qubits can be expressed as follows:

$$R(|\psi\rangle) = 2\left[1 - n^{-1}\sum_{j=-1}^{n-1} Tr\left(\rho_j^2\right)\right]. \tag{11.70}$$

Eq. (11.70) elegantly demonstrates the linear relationship between R and the purity of individual qubits, averaged across the pixels. This relationship adheres to two fundamental properties. First, $0 \le R(|j\rangle) \le 1$, where $|j\rangle$ signifies the drainage network system, and $R(|j\rangle) = 1$ for entangled upwelling states. Secondly, the second term $R(|j\rangle)$ remains invariant under local unitary transformations Uj (Brennen & Williams, 2003).

11.16 Incorporating quantum cellular automata for the automated identification of upwelling zones through in situ measurements

In the investigation, we systematically gathered crucial physical oceanography data, including temperature, salinity, and current profiles. This comprehensive dataset spans from the ocean surface to the seabed, with measurements taken at regular 5-meter depth intervals. The meticulous collection occurred during the respective Sentinel-A satellite overpasses in the months of June 26 and August 31, 2023 (Fig. 11.15). In the pursuit of temperature and salinity data at identical geographical locations, the esteemed researcher concurrently gathered data pertaining to the current flow throughout the water column (Fig. 11.16).

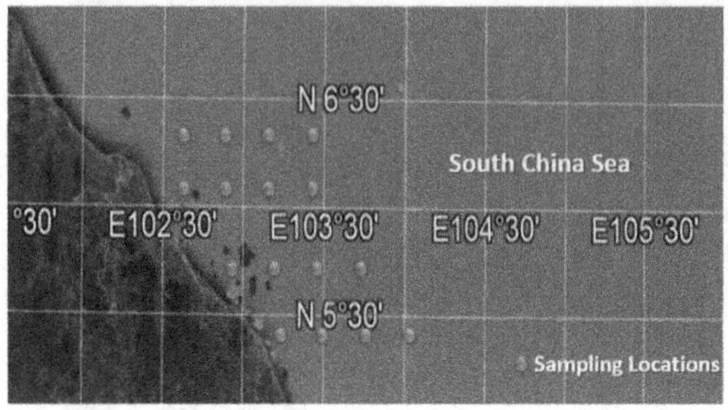

FIGURE 11.15 Geographical sample locations during Sentinel-A satellite overpasses.

FIGURE 11.16 During in situ data collection, the author made adjustments to the current meter.

The primary objective behind this rigorous data collection is to substantiate the presence of upwelling phenomena in SAR (synthetic aperture radar) images. By correlating these physical parameters with SAR data, we aim to enhance our understanding of the oceanic dynamics and validate the observed features in the satellite imagery. This multidimensional approach provides a holistic perspective, combining satellite observations with in situ measurements, to unravel the intricate dynamics of upwelling events.

Evidently, the quantum cellular automaton (QCA) algorithm exhibits a remarkable capability to discern the decoherence patterns on the sea surface induced by filament occurrences associated with upwelling events within the ambient environment (Fig. 11.17). Within the algorithm's

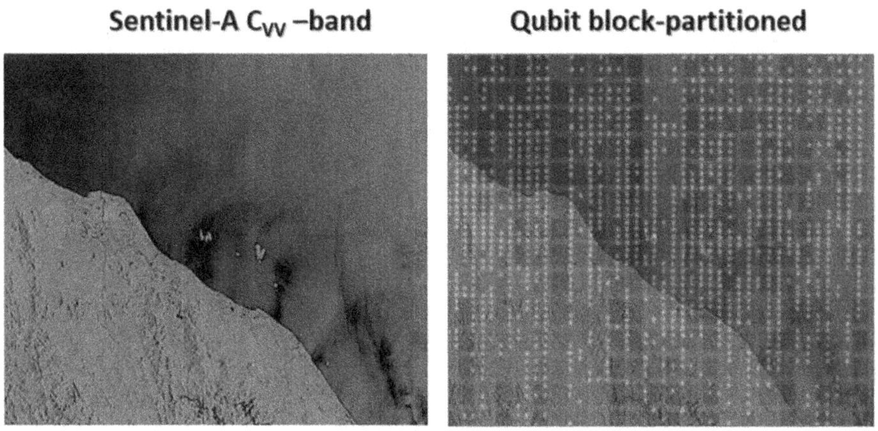

FIGURE 11.17 Qubit block-partitioned for SAR image. *SAR*, Synthetic aperture radar.

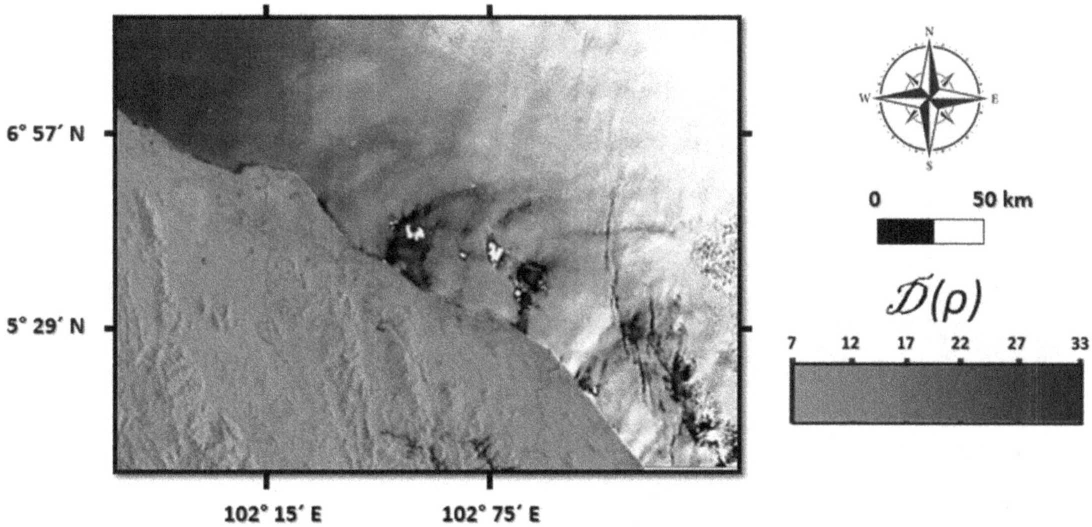

FIGURE 11.18 Modeling of the radar decoherence using QCA algorithm in SAR image. *QCA*, Quantum cellular automata; *SAR*, Synthetic aperture radar.

output, the distinctive features of decoherence are adeptly delineated, particularly when applied to Sentinel C-band and VV-band data.

In accordance with our observations, the decoherence occurring along the natural slick associated with upwelling events attains a maximum value of 33 (Fig. 11.18), surpassing that of the surrounding environment. Remarkably, regions with reduced decoherence manage to retain significant information regarding the roughness of the sea surface, contrasting with areas where decoherence is more pronounced, particularly in filament zones where the radar signal is deflected away from the antenna.

On June 26, 2023, during the Sentinel-A satellite overpasses, the wind curl stress along the coastal waters of the East coast of the Peninsular Malaysia reached its peak, registering a value of 0.04 x 10^{-7} N/m^3 (Fig. 11.19). This particular month corresponds to the southwest monsoon period, characterized by the prevailing wind direction from the south toward the offshore east coast of the Peninsular Malaysia. The pivotal question that arises pertains to whether this heightened wind curl stress could potentially instigate the emergence of filament zones characterized by pronounced decoherence, serving as an indicative signal of upwelling zones in the region.

Compelling evidence of upwelling occurrences is vividly illustrated through the analysis of normalized entanglement with sea surface temperature data acquired during Sentinel-A satellite overpasses. Notably, the coastal region exhibiting lower temperatures within the range of 28.5 − 29.0°C correlated with a robust entanglement value of 0.8 and a decoherence value of 33 (Fig. 11.20). This robust correlation serves as a promising indicator for the presence of upwelling phenomena in close proximity to the coastal waters. Furthermore, the quantum cellular automaton (QCA) algorithm demonstrates its proficiency in delineating the quantized Marghany turbulence boundary, closely associated with the quantized Marghany bound front state.

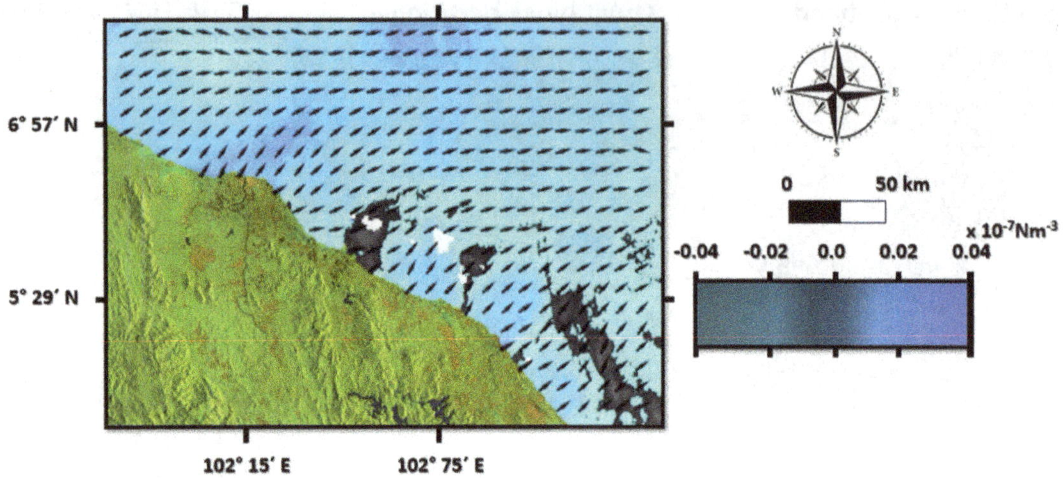

FIGURE 11.19 Wind curl stress modeled using QCA algorithm from Sentinel-A satellite on June 26, 2023. *QCA*, Quantum cellular automata.

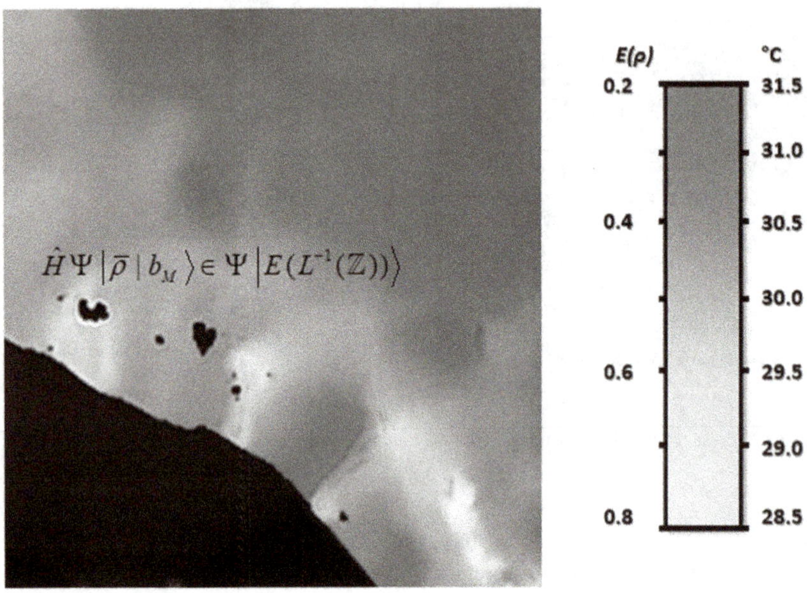

FIGURE 11.20 Normalized entanglement with sea surface temperature derived from SAR image and in situ sea surface temperature measurement using QCA. *QCA*, Quantum cellular automata; *SAR*, Synthetic aperture radar.

In light of the aforementioned perspective, the upwelling index (UI) derived from Sentinel-1A SAR data provides a compelling basis for asserting the occurrence of upwelling events. This assertion is substantiated by the highest recorded value of 0.5 m^2/s, aligning coherently with a low sea surface temperature of 28.5°C, a pronounced decoherence value of 33, and a normalized entanglement reaching its pinnacle at 0.8 (Fig. 11.21). These collective indicators fortify the argument for the reasonable occurrence of upwelling phenomena in the examined context. This finding agrees with recent of study of Kok et al. (2022).

The robust evidence supporting the reasonable occurrence of upwelling events, as derived from the upwelling index (UI) based on Sentinel-1A SAR data, can be comprehensively explained by considering the interconnected dynamics of various environmental parameters.

Firstly, the UI attaining its highest value of 0.5 m^2/s serves as a quantitative measure of the strength and magnitude of the upwelling process. This index reflects the upward movement of nutrient-rich, colder waters from the deeper ocean layers to the surface, a characteristic hallmark of upwelling events. This result validates the work established by Aziz et al. (2019) and Pitcher et al. (2010).

The correlation with a low sea surface temperature of 28.5°C further bolsters this assertion. Upwelling is known to bring colder waters to the surface, resulting in a decrease in sea surface temperature. The observed temperature aligns with the anticipated cooling effect associated with the upwelling phenomenon. Another compelling

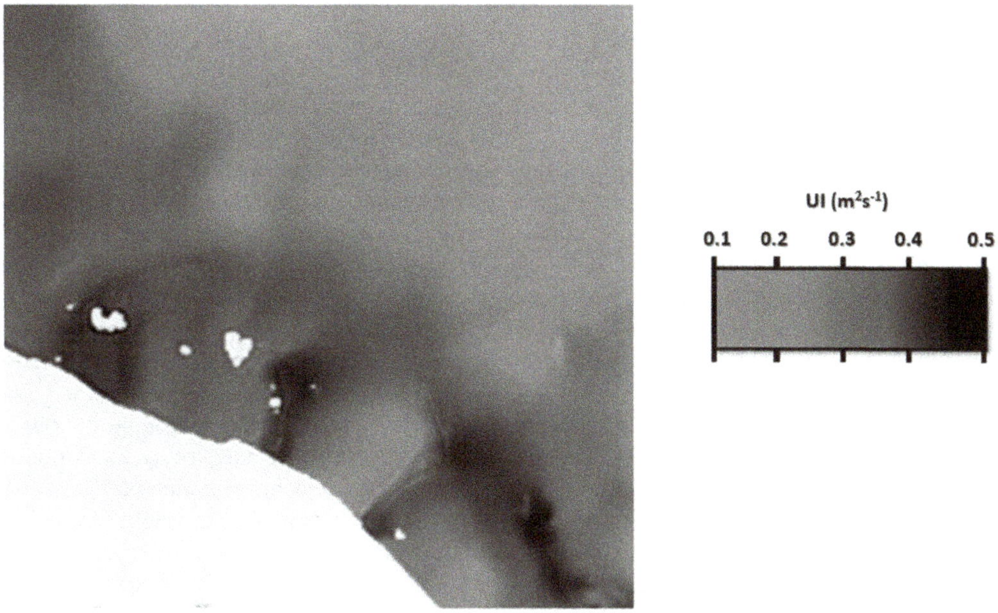

FIGURE 11.21 Upwelling index (UI) derived from Sentinel-1A SAR data using QCA. *QCA*, Quantum cellular automata; *SAR*, Synthetic aperture radar.

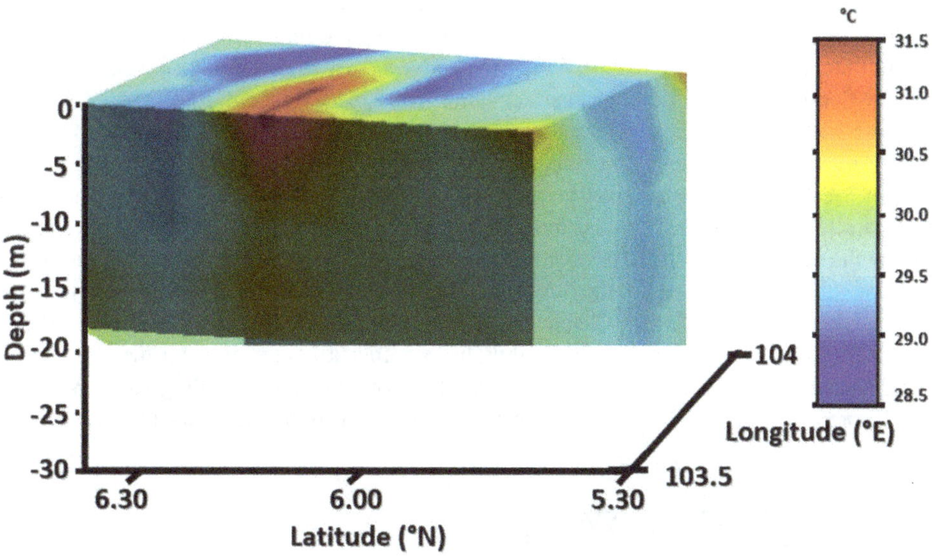

FIGURE 11.22 3-D isothermal pattern during Sentinel-1A SAR overpasses on June 26, 2023. *SAR*, Synthetic aperture radar.

observation manifests in the three-dimensional isothermal contours, depicting the cold upward movement originating from offshore regions with temperatures ranging from 28.5°C to 29.5°C (Fig. 11.22). Simultaneously, a distinct warm tongue of surface water, characterized by temperatures between 30°C and 31.5°C, dynamically progresses from onshore to offshore. This near-real-time depiction aligns with the Sentinel-A satellite overpass, showcasing the intricate dynamics of ocean temperature variations and the concurrent movement of contrasting thermal features.

Moreover, the simultaneous highest decoherence value of 33 and normalized entanglement reaching 0.8 contribute to the overall narrative. Decoherence is indicative of disruptions in the sea surface, often associated with the presence of filaments and turbulence, both of which are characteristic features of upwelling zones. Normalized entanglement provides additional insight into the complexity of interactions, emphasizing the intricate relationships between variables.

In essence, the coherent alignment of these multiple indicators—upwelling index, sea surface temperature, decoherence, and normalized entanglement—paints a comprehensive picture, compellingly supporting the conclusion that the observed SAR data are indeed reflective of a reasonable occurrence of upwelling in the studied area. The synergistic interpretation of these parameters underscores the interdisciplinary nature of the investigation, combining insights from oceanography, remote sensing, and environmental dynamics.

In contrast to the observations on June 26, 2023, the coastal waters of Kelantan on August 31, 2023, exhibited a prevailing influence of intense decoherence, notably reaching a peak value of 36. This heightened level of coherence is closely linked to the presence of filaments in close proximity to the coastal waters of Kelantan. Intriguingly, the lowest recorded decoherence value of 6 is distinctly associated with the quantized Marghany turbulence boundary and the quantized Marghany bound front state. This dichotomy in decoherence values underscores the diverse and dynamic nature of the coastal environment, where distinct phenomena such as filaments and quantized turbulence boundaries contribute to varying levels of coherence. Therefore the documented decoherence of the eddy within the coastal waters, correlated with a subtle natural slick, is measured at approximately 18 (Fig. 11.23). This ascertained value serves as a quantification of the degree of disruption or incoherence inherent in the eddy, offering valuable insights into the nuanced dynamics of this particular coastal turbulence features.

On August 31, 2023, during the Sentinel-A satellite overpasses, the wind curl stress along the coastal waters of the East coast of Kelantan reached its zenith, recording a value of 0.04 x 10^{-7} N/m^3 with an irregular pattern. In the northern part, the wind flow is parallel to the sea surface, while in the southern part, the wind takes an oblique direction. The convergence of these two distinct wind patterns results in the formation of turbulent flows, giving rise to an eddy. This specific timeframe aligns with the southwest monsoon period, characterized by prevailing winds from the south toward the offshore east coast of the Peninsular Malaysia. This result validates the studies done by Aziz et al., (2019), Daud et al. (2019), Kok et al. (2022).

The critical inquiry that arises revolves around whether this heightened wind curl stress has the potential to initiate the development of filament zones marked by conspicuous decoherence. Such filament zones could potentially serve as indicative signals of upwelling zones in the region. The intricate interplay between wind dynamics, turbulence formation, and potential upwelling phenomena during this specific period warrants comprehensive investigation to unravel the underlying mechanisms governing these coastal processes.

In stark contrast to the conditions observed on June 26, 2023, there is a noteworthy escalation in normalized entanglement, reaching a value of 1 on a date subsequent to that. This elevation in entanglement coincides with the attainment of the lowest sea surface temperature at 28.2°C, spanning an approximate distance of 172 km away from the coastal waters of Kelantan (Fig. 11.24). This notably low temperature is intricately associated with the quantized Marghany bound front state, signifying the presence of a cool front that distinctly separates it from a warmer front registering at 29.4°C.

Within the eddy, there is a discernible fluctuation in the sea surface temperature, ranging from 28.2°C to 29°C. This temperature variability is indicative of the complex interplay within the eddy, reflecting the combined influence of the convergence of westerly and southeasterly winds. The resultant turbulence features embody a fusion of a cool front and a cool eddy, both of which are embedded within the quantized Marghany turbulence boundary, as elaborated upon in Chapter 4. In this sense, the generation of mesoscale eddies by upwelling has significant implications for the characteristics and spatial distribution of the upwelling zone. When an upwelling jet is initiated, it brings nutrient-rich deep waters

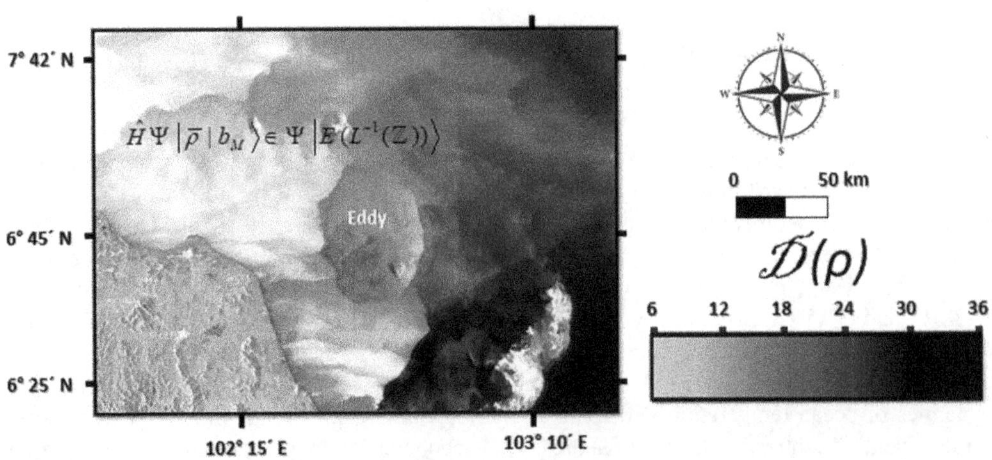

FIGURE 11.23 Highest decoherence on August 31, 2023 modeled from SAR image using QCA. *QCA*, Quantum cellular automata; *SAR*, Synthetic aperture radar.

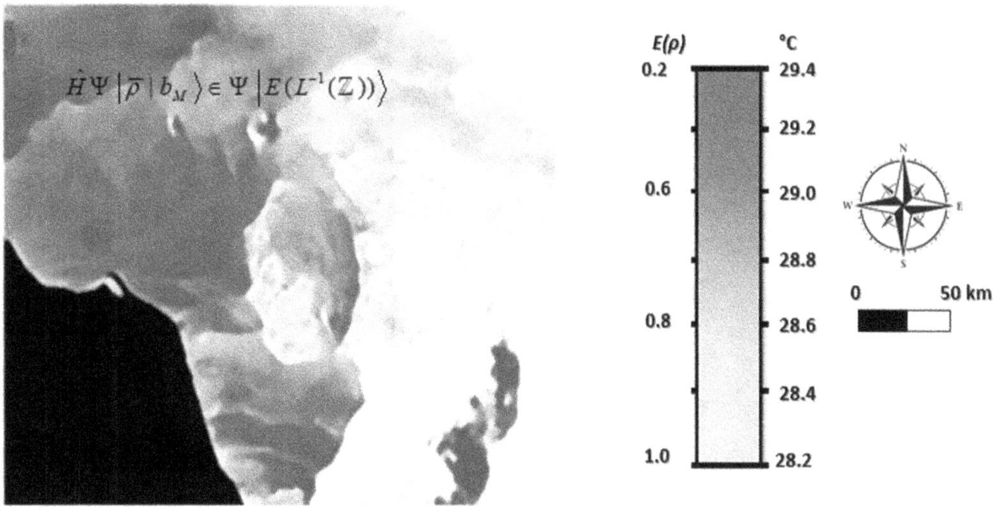

FIGURE 11.24 Normalized entanglement of SAR-derived sea surface temperature on August 31, 2023, using QCA. *QCA*, Quantum cellular automata; *SAR*, Synthetic aperture radar.

to the surface, creating favorable conditions for the development of mesoscale eddies. These eddies are rotational currents of water that circulate around a central point, influencing the surrounding oceanic environment (Klaus, 1987; Wyrtki, 1961, 1962).

One key effect of these mesoscale eddies, particularly those generated by the upwelling jet, is the widening of the upwelling zone as distance increases from the center of upwelling. As the eddies propagate outward, they contribute to the lateral expansion of nutrient-rich waters, enhancing the extent of the region influenced by upwelling. This widening effect is crucial for the dispersion of nutrients and the promotion of biological productivity over a larger area. However, the dynamics can be modified under specific circumstances (Abney & Epstein, 1996; Alpers & Bignami, 2020; Alpers & Zeng, 2021; Alpers, 2016, 2018). If there are natural physical features, such as headlands or preexisting eddy systems, that confine the upwelled water, a trophic succession can occur in the same area. In this context, the confinement of upwelled water within the boundaries set by these physical features allows for the establishment of a trophic succession—a sequence of changes in the composition of marine life and ecosystem structure. This succession can take place within the confines of the existing eddies or coastal features, leading to a more localized impact on marine life (Marghany & Hussien, 2003; Marghany, 1994).

This intricate scenario underscores the profound impact of wind dynamics and their convergence on the coastal environment, giving rise to distinct turbulence features characterized by varied sea surface temperatures. The alignment of these observations with the quantized Marghany turbulence boundary further emphasizes the need for a comprehensive understanding of the multifaceted interactions shaping the coastal dynamics in the specified region.

In contrast to the conditions observed on June 26, 2023, the upwelling index on August 31, 2023, reveals the most pronounced occurrence of upwelling at 1.0 m^2/s (Fig. 11.25). This aligns coherently with elevated normalized entanglement, the lowest sea surface temperature, and the highest decoherence, all modeled through quantum cellular automaton (QCA).

The confirmation of upwelling on August 31, 2023, is intricately tied to the observable vertical ascent of a three-dimensional isothermal layer characterized by cooler waters registering at 28.2°C (Fig. 11.26). This phenomenon aligns with the typical signature of upwelling, where colder, nutrient-rich waters from deeper ocean layers rise to the surface. The three-dimensional representation provides a nuanced understanding of the spatial distribution of this cool water ascent, reinforcing the credibility of the identified upwelling event.

Moreover, the demonstration of a cool eddy with a temperature of 29.4°C (Fig. 11.26) within the three-dimensional isothermal contours adds another layer to the narrative. Eddies are swirling, circular currents often associated with complex oceanic processes, and the presence of a cool eddy suggests intricate interactions in the marine environment. The coexistence of the cool eddy with the upwelling further underscores the dynamic nature of the oceanic system on that specific date (Robinson, 2010a).

The observed intensification in the vigor of the Ekman spiral on both June 26, 2023, and August 31, 2023, as deduced from in situ current measurements, can be ascribed to the dynamic shifts occurring in oceanic conditions during these specific timeframes. The Ekman spiral, a consequence of the Coriolis effect and wind-induced surface currents, is conventionally linked to upwelling phenomena. The surpassing of the June 26, 2023, measurement of 0.63 m/s (Fig. 11.27) implies an amplification

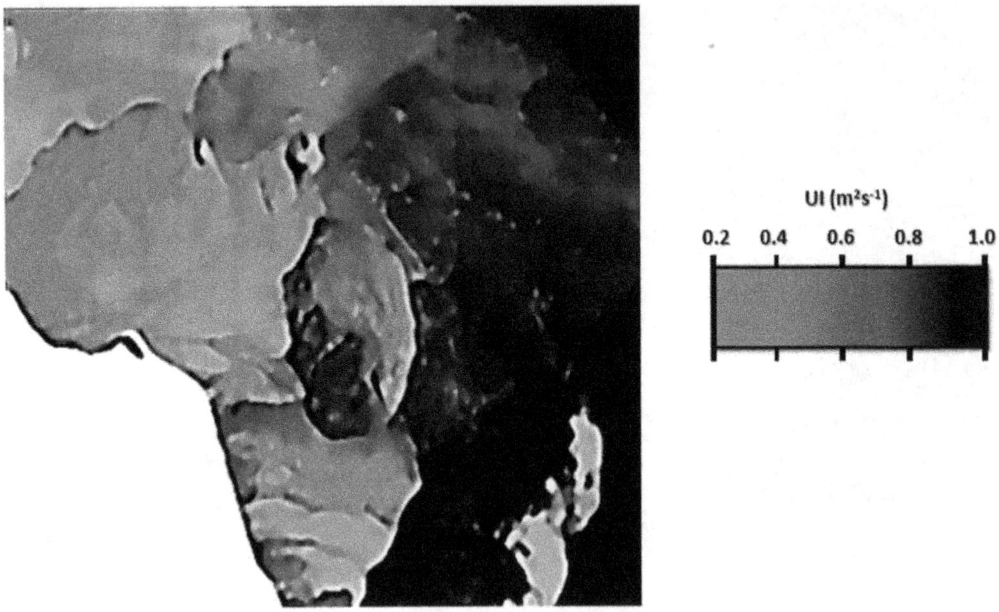

FIGURE 11.25 Upwelling index (UI) derived from Sentinel-1A SAR data using QCA on August 31, 2023. *QCA*, Quantum cellular automata; *SAR*, Synthetic aperture radar.

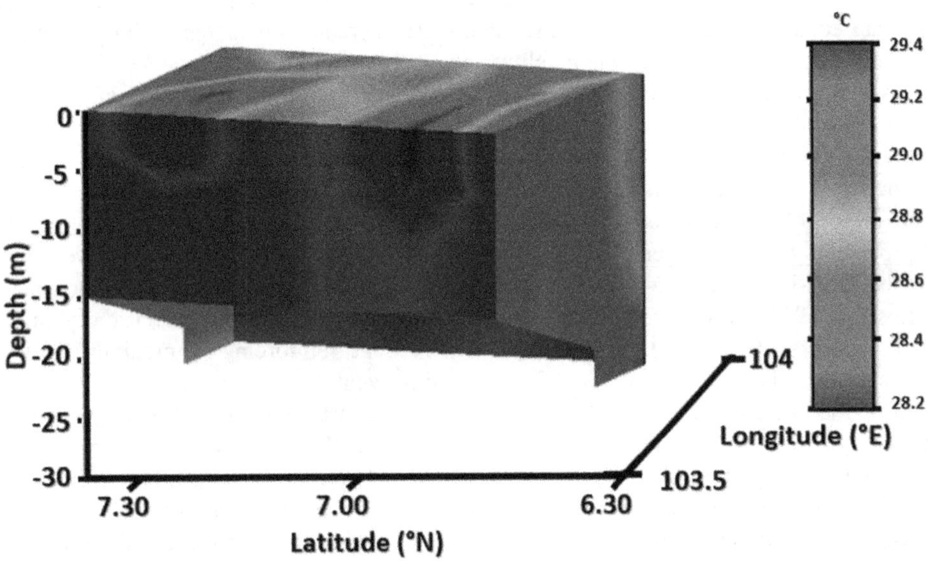

FIGURE 11.26 3-D isothermal pattern during Sentinel-1A SAR overpasses on August 31, 2023. *SAR*, Synthetic aperture radar.

in the vigor of the upwelling process on August 31, 2023. The recorded velocity of 0.8 m/s (Fig. 11.28) indicates a more forceful and discernible spiraling motion through the water column, denoting an enhanced displacement of deeper, nutrient-rich waters toward the surface. This reinforced Ekman spiral aligns with the concept of heightened upwelling activity during this later observation period and concurrently correlates with an increased decoherence in SAR images, reflecting the complex interplay of oceanic dynamics during this specific timeframe.

The occurrence of upwelling in the East Coast of Malaysia, particularly emphasizing the involvement of the quantized Marghany bound front state within the quantized Marghany turbulence boundary. The East Coast of Malaysia experiences upwelling primarily due to the influence of prevailing winds, notably the southwest monsoon. These sustained winds drive the surface waters away from the coast, allowing nutrient-rich, colder water from deeper layers to rise to the surface—a characteristic hallmark of upwelling (Austin & Lentz, 2002; Dritschel

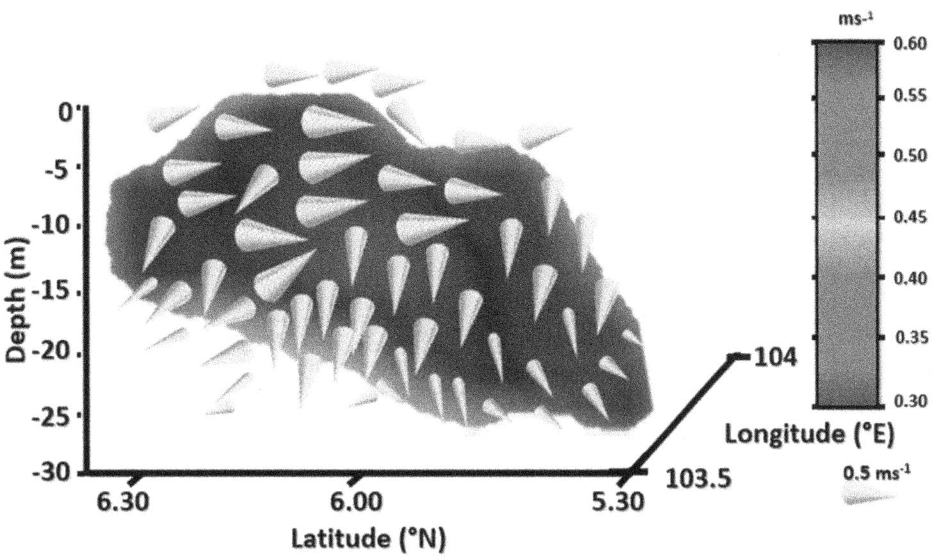

FIGURE 11.27 3-D current speed flow pattern during Sentinel-1A SAR overpasses on June 26, 2023. *SAR*, Synthetic aperture radar.

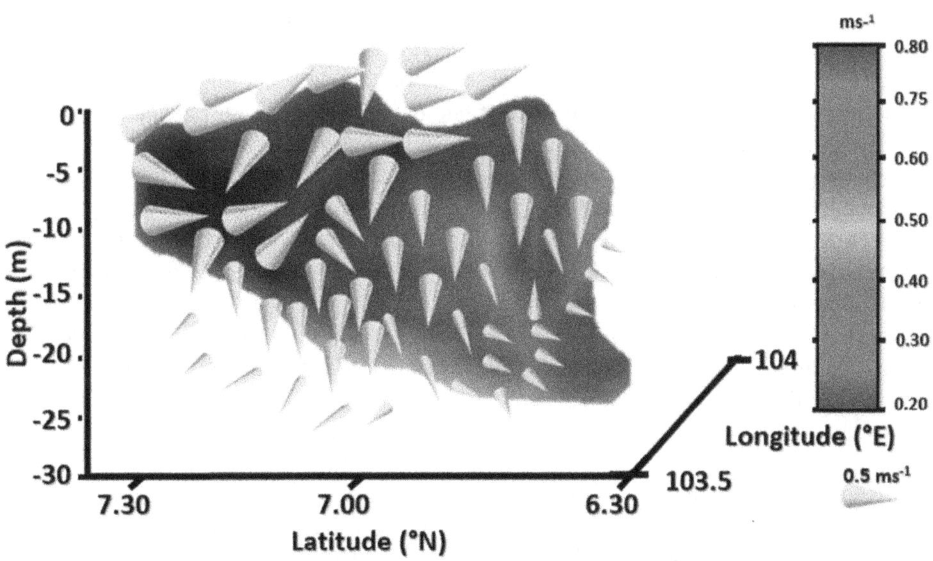

FIGURE 11.28 3-D current speed flow pattern during Sentinel-1A SAR overpasses on August 31, 2023. *SAR*, Synthetic aperture radar.

et al., 2020; Hu & Wang, 2016; Marghany, 1994, 2021b). The presence of the quantized Marghany bound front state within the quantized Marghany turbulence boundary, therefore, adds an additional layer of complexity to this phenomenon. The quantized Marghany bound front state represents a distinct boundary between cooler and warmer waters, often associated with unique oceanographic features. In the context of upwelling, this front state may indicate the separation between upwelled, colder water and the warmer surrounding waters.

The observed change can be attributed to the inflow of deep water from the South China Sea, facilitated by persistent westerly winds transporting warmer shallow water offshore. The sustained westerly winds play a pivotal role in displacing the relatively warmer surface waters, allowing the upwelling of cooler, nutrient-rich deep water from the South China Sea. This mechanism, driven by the dynamic interplay between wind patterns and ocean currents, contributes significantly to the alteration in sea surface temperature and the initiation of upwelling phenomena in the studied region (Chen et al., 2012; Gan et al., 2009; Wang et al., 2013).

The quantized Marghany turbulence boundary, as a broader concept, encompasses the turbulent interactions occurring at the interface between different water masses. In the East Coast of Malaysia, the convergence of these water masses, possibly marked by the quantized Marghany bound front state, contributes to the development of turbulence. This turbulence further facilitates the upwelling process, creating favorable conditions for nutrient transport and enhancing biological productivity in the region (Marghany & Hashim, 2010).

In summary, the occurrence of upwelling on the East Coast of Malaysia is intricately linked to the influence of monsoonal winds, driving the upwelling process(Aziz et al., 2019; Daud et al., 2019; Kok et al., 2022). The inclusion of the quantized Marghany bound front state within the quantized Marghany turbulence boundary reflects the complex interactions and boundary features that contribute to the nuanced dynamics of upwelling in this specific coastal region.

11.17 Island wake

In the realm of scientific exploration, the distinguished researcher stands as a trailblazer, unraveling the mysteries concealed within the undulating waters of Malaysian coastal realms. Remarkably, amid the abundance of facilities and esteemed marine research institutions, Malaysian scholars ardently engage in fundamental inquiries into the intricate domains of physical and dynamic oceanography. In the annals of scholarly pursuit, the phenomenon known as the island wake has hitherto eluded scrutiny along the coastal waters of Malaysia. This marks a groundbreaking juncture, as the erudite author spearheads the inaugural investigation into this enigmatic occurrence within the Malaysian maritime expanse (Alpers, 2016, 2018). Within the realm of remote sensing, a captivating illustration of location-tethered phenomena emerges in the form of island wakes. Satellite imagery, a lens into the hidden narratives of our oceans, occasionally unveils the distinctive signatures that cling to isolated islands or modest island clusters. These signatures, resembling wakes, gracefully extend over the adjacent ocean, propelled by a prevailing current.

The genesis of such maritime artistry unfolds through three distinct avenues. First, the ethereal ballet is choreographed by atmospheric nuances. Islands crowned with lofty volcanic peaks become conductors orchestrating disturbances in the wind flow. These disturbances, akin to poetic cloud streets trailing downwind or demarcated wind shadow zones in synthetic aperture radar (SAR) images, unveil themselves as regions of low backscatter—zones where the wind stress wanes in the islands' lee. Typically, these disturbances traverse a few hundred kilometers beyond the island, tracing the prevailing wind flow direction. However, on occasion, the influence of a lofty peak extends far into the ocean, shaping the wind flow on a grand scale (Robinson, 1983, 2004, 2006, 2010a, 2010b).

A prime exemplar of this atmospheric symphony unfolds in Hawaii. Here, the islands, crowned with majestic mountains, wield their influence as architects of wind patterns. The westward flow of easterly trade winds is harnessed and molded by the islands, creating a 3000-kilometer shadow behind the mountains. This atmospheric ballet begets the emergence of a narrow, easterly ocean surface current—a seemingly distant manifestation, intricately woven by the unseen hand of the islands themselves.

The second intricacy unfurls when the wake cast by high mountains on an island ushers forth a local oceanic response. This response, akin to a nuanced dance, manifests in various layers. On the surface, a diurnal thermocline gracefully emerges within the wind shadow, crafting warm patches of sea surface temperature (SST) that allure the gaze when captured through the lens of thermal infrared imagery.

Yet, in the realm of persistent winds fashioning enduring wind shadow realms, a more elaborate choreography ensues. The shear, the unseen force distinguishing the shadowed from the sunlit, orchestrates a mesmerizing ballet of Ekman convergence and divergence. This dynamic interplay births zones of upwelling and downwelling, with the former imparting a cooler hue to SST images. A nuanced symphony of biological significance accompanies this ballet, as upwelling delivers a bounty of nutrients, nurturing heightened primary production that echoes in the ocean's color palette. The intricate tapestry of temperature gradients and vibrant hues unfolds on the downwind canvas of island chains, exemplified by the Tioman Island in Malaysia. In the pursuit of understanding, the esteemed Sentinel-A satellite, casting its gaze upon the canvas of the ocean on the illustrious date of August 8, 2023, unfolds a narrative of unparalleled significance. The acquired data, a symphony of pixels capturing the intricate dance of oceanic forces, paint a vivid tableau dominated by the regal presence of the highest upwelling index, standing proudly at 1.0 m^2/s(Fig. 11.29).

This celestial alliance of data points extends its embrace to the zenith of knowledge, where the normalized entanglement attains its pinnacle. A harmonious convergence unfolds as these metrics intertwine, giving rise to the manifestation of a natural slick or filament (Alpers & Bignami, 2020; Alpers, 2016, 2018), a testament to the orchestrations transpiring behind the veiled curtains of the Timon Island. In this celestial choreography, the Sentinel-A satellite, as a virtuoso observer, unveils the nuanced nuances of upwelling occurrences, etching a poetic narrative on the canvas of scientific exploration.

In unraveling this phenomenon, the marriage of synthetic aperture radar (SAR) images, unveiling the intricacies of wind shadow and wind shear zones, with the poetic rendition of SST imagery, bestows profound insights into the ocean's response to the mountainous orchestrators of its dance.

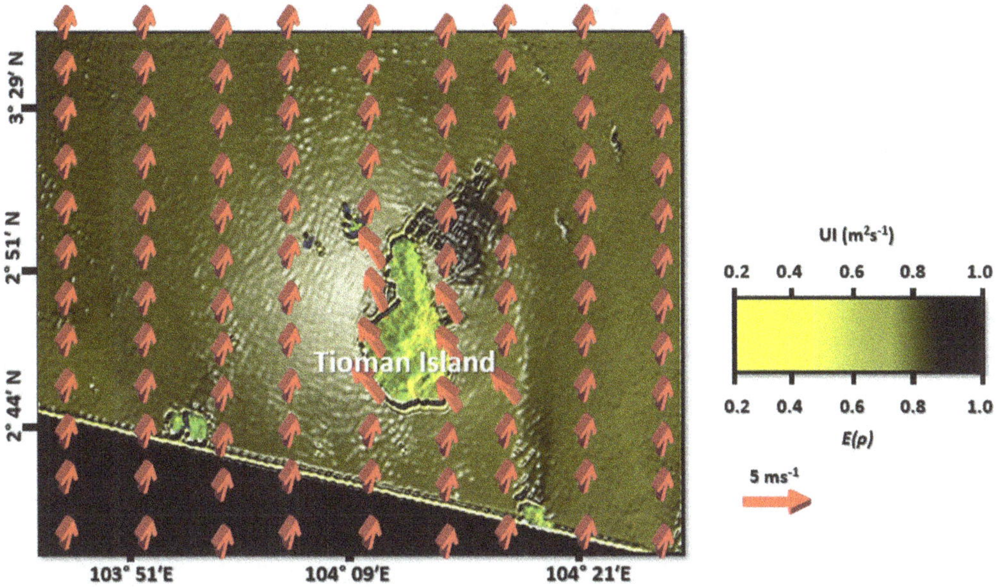

FIGURE 11.29 Island wakes around Tioman island, Malaysia.

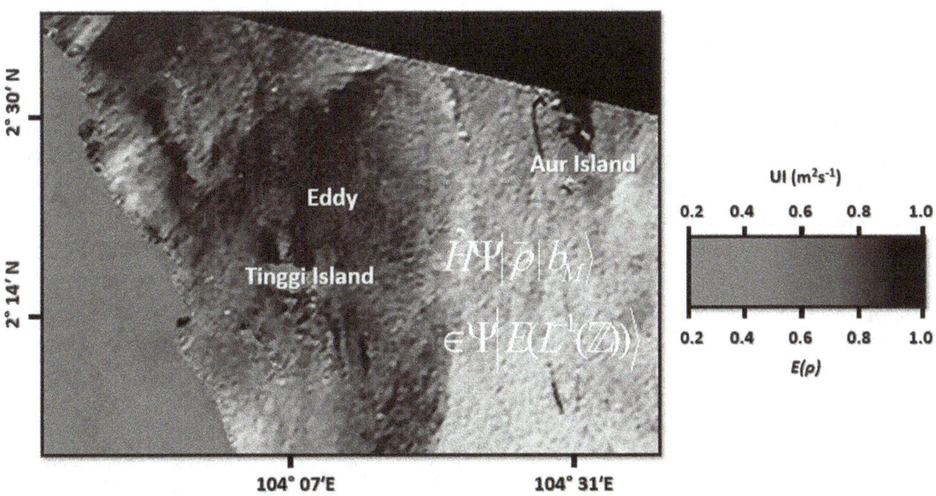

FIGURE 11.30 Eddy generates behind the island due to strong upwelling flow.

The third intricacy unfolds as a mesmerizing ballet of oceanic eddies, gracefully shedding their existence in the wake of isolated volcanic islands that boldly ascend from the ocean's depths. This captivating phenomenon, akin to a choreographed dance, has been artfully modeled as a classic fluid-dynamic puzzle. Within this intricate tableau, the ocean's depth, the island's diameter, and the Coriolis parameter stand as the maestros, orchestrating whether these eddies gracefully depart from the island's embrace or if a recirculation realm is ensnared in its midst. In proximity to Tinggi Island and Aur Islands, data acquired by the Sentinel Satellite on August 8, 2023, unequivocally illustrate a robust upwelling phenomenon. This is evidenced by the highest recorded upwelling index, reaching an impressive magnitude of 1.0 m^2/s, accompanied by the pinnacle value of normalized entanglement. The manifestation of this potent upwelling imparts its influence on the aquatic canvas, giving rise to an eddy flow trailing behind Tinggi Island. The dimensions of this eddy exhibit an approximate radius of 16 km and a length extending to 33 km, delineating the dynamic intricacies of the oceanic choreography at play (Fig. 11.30).

In the poetic language of fluid dynamics, the motion of these eddies becomes a manifestation of vertical water transport. A symphony of vertical currents emerges, breathing life into the dance of upwelling within the cocoon of the island's wake. This ethereal ballet not only paints a canvas of esthetic beauty but also unveils the potential for heightened production, a consequence of the nourishing embrace bestowed upon the ocean's depths.

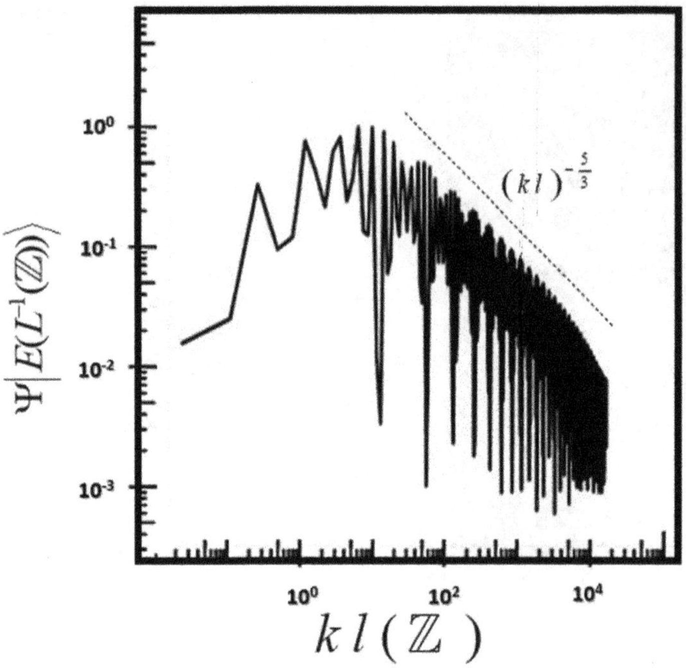

FIGURE 11.31 Quantum turbulence spectra of the upwelling occurrence along the quantized Marghany turbulence boundary layer.

The quantized Marghany turbulence boundary layer, as modeled, reveals intriguing insights into the intricacies of upwelling-induced turbulence. Notably, the spectra exhibit peaks of considerable magnitude, reaching the zenith at 0.98 around a wave number of 10°. The dissipation of this upwelling-induced turbulence spectra manifests itself in the lowest spectra energy values, notably at 10^{-2} (Fig. 11.31).

The discernible pattern within the upwelling turbulence spectra aligns harmoniously with the Kolmogorov spectra, adding an additional layer of complexity to the turbulence dynamics under consideration. This distinctive turbulence regime, characterized by its substantial size, unequivocally signifies a flow scenario where inertial forces hold sway over viscous forces, delineating the dominance of larger-scale dynamics within this intricate oceanic phenomenon (Robinson, 2004; Robinson, 2010a).

Island wakes emerge as a robust index for detecting upwelling in synthetic aperture radar (SAR) data owing to several key factors. First, the distinctive signature left by island wakes in SAR images serves as a telltale indicator of underlying oceanographic processes, including upwelling events. This distinct pattern arises from the interaction of ocean currents with the presence of islands. Second, the wind shadow effect induced by islands contributes to changes in sea surface roughness, a feature sensitively captured by SAR. This sensitivity allows for the identification of regions influenced by upwelling dynamics. Third, the formation of eddies in the vicinity of islands, detectable through SAR, provides insights into the dynamic processes associated with upwelling, as eddies are often linked to vertical movements of water. The high spatial resolution inherent in SAR data is a crucial asset, enabling the detailed observation of small-scale features such as the patterns associated with island wakes (Dong & McWilliams, 2007; Dong et al., 2007; Tomczak, 1988; Wolanski et al., 1984).

Finally, the synergy of SAR data with information from other sensors, such as those measuring sea surface temperature or ocean color, enhances the comprehensive understanding of oceanographic conditions, reinforcing the efficacy of island wakes as a valuable index for detecting and studying upwelling phenomena.

11.18 Why quantum cellular automaton can detect potential zone of upwelling automatically?

The author emphasizes the appropriateness of quantum cellular automaton (QCA) for the automatic detection of upwelling from synthetic aperture radar (SAR) images. QCA's utilization in this context is founded on its remarkable computational prowess, inspired by principles from quantum mechanics.

The intricacies of SAR images, particularly those associated with upwelling phenomena, demand a tool capable of deciphering complex spatial and temporal patterns. QCA's grid-based computational model excels in this regard, allowing for a nuanced analysis of neighboring cells and enabling the identification of subtle features within SAR imagery. In the specific case of upwelling detection, QCA proves invaluable due to its proficiency in quantifying decoherence—a measure of disruption or incoherence in observed patterns. The characteristic signatures of upwelling zones often present as regions of heightened decoherence, a quality that aligns seamlessly with QCA's analytical capabilities (Brennen & Williams, 2003; Freedman et al., 2022; Kent et al., 2023).

Furthermore, QCA's adaptability to the multifaceted nature of SAR images, which may encompass variables such as sea surface temperature, wind patterns, and turbulence, underscores its suitability for automatic upwelling identification. The algorithm's capacity to integrate diverse parameters and dissect their collective impact contributes to its efficacy in navigating the complexity inherent in SAR datasets.

In essence, the choice of QCA for automatic upwelling detection in SAR images is not merely a pragmatic one; it is a testament to the algorithm's innovative approach and its ability to address the intricate challenges posed by the analysis of oceanographic phenomena through remote sensing technologies.

This chapter introduces an innovative technique for the automatic detection of upwelling from synthetic aperture radar (SAR) images. SAR imagery encounters challenges due to the scarcity of thermal bands compared with optical data, limiting the ability to detect indicators of cool water and thermal fronts associated with upwelling. However, this chapter presents a groundbreaking approach utilizing quantum cellular automaton (QCA) to automatically identify upwelling zones. The method relies on enhancing the decoherence of SAR images resulting from the presence of natural slicks and cool waters brought to the surface during upwelling events. The occurrence of natural slicks leads to lower backscatter due to radar signal reflection away from the antenna. Consequently, entanglement serves as a confirmation of upwelling, with the highest entanglement values coinciding with elevated decoherence and lower temperatures. Ultimately, the upwelling index aligns with the highest normalized entanglement and decoherence values, serving as an excellent indicator for the occurrence of upwelling and showcasing the superior performance of QCA in the automatic detection of upwelling in SAR data.

The findings articulated in this chapter provide compelling evidence affirming the existence of the quantized Marghany turbulence boundary and the occurrence of the quantized Marghany bound front state. These revelations, extensively elucidated in Chapters 4 and 5, contribute significantly to our understanding of complex oceanic dynamics and their representation in remote sensing data. The abstract nature of these phenomena underscores their intricate interplay within the marine environment, emphasizing the nuanced insights gained through the exploration of chapters dedicated to their comprehensive analysis.

References

Abney, M., & Epstein, R. I. (1996). Ekman pumping in compact astrophysical bodies. *Journal of Fluid Mechanics*, *312*, 327–340. Available from https://doi.org/10.1017/S0022112096002030, http://journals.cambridge.org/action/displayJournal?jid = FLM.

Alpers W. 2016 International Geoscience and Remote Sensing Symposium (IGARSS), 10.1109/IGARSS.2016.7729568, 2201–2204, Institute of Electrical and Electronics Engineers Inc. Germany, SAR oceanography applied to southeast Asian Waters.

Alpers W. 2018 International Geoscience and Remote Sensing Symposium (IGARSS), 10.1109/IGARSS.2018.8517686, 6041-6044, Institute of Electrical and Electronics Engineers Inc. Germany, On the use of SAR in studies of upwelling.

Alpers W. Bignami F. 2020 International Geoscience and Remote Sensing Symposium (IGARSS), 10.1109/IGARSS39084.2020.9324140, 3537–3540, Institute of Electrical and Electronics Engineers Inc. Germany, Small-scale and sub-mesoscale phenomena associated with upwelling studied by SAR.

Alpers, W., & Zeng, K. (2021). On radar signatures of upwelling. *Journal of Geodesy and Geoinformation Science*, *4*(1).

Anandan, J., & Suzuki, J. (2004). *Quantum mechanics in a rotating frame* (pp. 361–370). Springer Science and Business Media LLC.

Austin, J. A., & Lentz, S. J. (2002). The inner shelf response to wind-driven upwelling and downwelling. American Meteorological Society, United States. *Journal of Physical Oceanography*, *32*(7), 2171–2193. Available from https://doi.org/10.1175/1520-0485(2002), http://journals.ametsoc.org/loi/phoc, 032 < 2171:TISRTW > 2.0.CO;2.

Aziz, A. A., Suratman, S., Kok, P. H., & Akhir, M. F. (2019). Distribution of nutrients concentration in the upwelling area off the east coast of Peninsular Malaysia during the Southwest Monsoon. *Malaysia Malaysian Journal of Analytical Sciences*, *23*(6), 1030–1043. Available from https://doi.org/10.17576/mjas-2019-2306-11, http://www.ukm.my/mjas/v23_n6/pdf/Azyyati_23_6_11.pdf.

Bewley, G. P., Lathrop, D. P., & Sreenivasan, K. R. (2006). Visualization of quantized vortices. *Nature*, *441*(7093), 588. Available from https://doi.org/10.1038/441588a, 588.

Bogoliubov, N. N., Shirkov, D. V., & Henley, E. M. (1960). Introduction to the theory of quantized fields. *Physics Today*, *13*(7), 40–42. Available from https://doi.org/10.1063/1.3057034.

Brennen, G. K., & Williams, J. E. (2003). Entanglement dynamics in one-dimensional quantum cellular automata. *Physical Review A.*, *68* (4). Available from https://doi.org/10.1103/physreva.68.042311.

Chen, C., Lai, Z., Beardsley, R. C., Xu, Q., Lin, H., & Viet, N. T. (2012). Current separation and upwelling over the southeast shelf of Vietnam in the South China Sea. *Journal of Geophysical Research: Oceans*, *117*(3). Available from https://doi.org/10.1029/2011JC007150, http://onlinelibrary.wiley.com/journal/10.1002/(ISSN)2169-9291.

Daud, N. R., Akhir, M. F., & Muslim, A. M. (2019). Dynamic of ENSO towards upwelling and thermal front zone in the east coast of Peninsular Malaysia. *Acta Oceanologica Sinica*, *38*(1), 48–60. Available from https://doi.org/10.1007/s13131-019-1369-7, http://www.springer.com/earth + sciences + & + geography/oceanography/journal/13131.

Dong, C., & McWilliams, J. C. (2007). A numerical study of island wakes in the Southern California Bight. *Continental Shelf Research*, *27*(9), 1233–1248. Available from https://doi.org/10.1016/j.csr.2007.01.016.

Dong, C., McWilliams, J. C., & Shchepetkin, A. F. (2007). Island wakes in deep water. *Journal of Physical Oceanography*, *37*(4), 962–981. Available from https://doi.org/10.1175/JPO3047.1.

Dritschel, D. G., Paldor, N., & Constantin, A. (2020). The Ekman spiral for piecewise-uniform viscosity. *Ocean Science*, *16*(5), 1089–1093. Available from https://doi.org/10.5194/os-16-1089-2020, http://www.ocean-sci.net/volumes_and_issues.html.

Ekman, V. W. (1905). On the influence of the Earth's rotation on ocean currents. *Archiv der Mathematik, Astronomy, and Physics*, *2*, 1–52.

Ellison, T. H. (1955). The Ekman spiral. *Quarterly Journal of the Royal Meteorological Society*, *81*(350), 637–638. Available from https://doi.org/10.1002/qj.49708135025.

Enriquez C. Mariño-Tapia I.S. M.A. E.L. 2014 2014 Proceedings of the 17th Physics of Estuaries and Coastal Seas (PECS) Conference, 17–20, Mechanisms driving a coastal dynamic upwelling.

Freedman, M., Haah, J., & Hastings, M. B. (2022). The group structure of quantum cellular automata. *Communications in Mathematical Physics*, *389*(3), 1277–1302. Available from https://doi.org/10.1007/s00220-022-04316-x, https://link.springer.com/journal/220.

Fu, M., & Xie, L. (2005). The sector bound approach to quantized feedback control. *IEEE Transactions on Automatic Control*, *50*(11), 1698–1711. Available from https://doi.org/10.1109/TAC.2005.858689.

Gan, J., Cheung, A., Guo, X., & Li, L. (2009). Intensified upwelling over a widened shelf in the northeastern South China Sea. *Journal of Geophysical Research: Oceans*, *114*(9). Available from https://doi.org/10.1029/2007JC004660, http://onlinelibrary.wiley.com/journal/10.1002/(ISSN)2169-9291.

Gaube, P., Chelton, D. B., Samelson, R. M., Schlax, M. G., & O'Neill, L. W. (2015). Satellite observations of mesoscale eddy-induced Ekman pumping. *Journal of Physical Oceanography*, *45*(1), 104–132. Available from https://doi.org/10.1175/JPO-D-14-0032.1, http://journals.ametsoc.org/doi/pdf/10.1175/JPO-D-14-0032.1.

Gnanadesikan, A., & Weller, R. A. (1995). Structure and instability of the Ekman spiral in the presence of surface gravity waves. *Journal of Physical Oceanography*, *25*(12), 3148–3171. Available from https://doi.org/10.1175/1520-0485(1995), 025 < 3148:SAIOTE > 2.0.CO;2.

Hart, J. E. (1995). Nonlinear Ekman suction and ageostrophic effects in rapidly rotating flows. *Geophysical & Astrophysical Fluid Dynamics*, *79*(1-4), 201–222. Available from https://doi.org/10.1080/03091929508228997.

Hidaka, K. (1972). Physical oceanography of upwelling. *Geoforum*, *3*(3), 9–21. Available from https://doi.org/10.1016/0016-7185(72)90082-6.

Hieber, M., & Stannat, W. (2013). Stochastic stability of the Ekman spiral. Scuola Normale Superiore. *Germany Annali della Scuola Normale - Classe di Scienze*, *12*(1), 189–208. Available from http://www.sns.it/en/edizioni/riviste/annaliscienze/.

Hsuesh, Y., & O'Brien, J. J. (1971). Steady coastal upwelling induced by an along-shore current. *Journal of Physical Oceanography*, *1*(3), 180–186. Available from https://doi.org/10.1175/1520-0485(1971), 001 < 0180:scuiba > 2.0.co;2.

Hu, J., & Wang, X. H. (2016). Progress on upwelling studies in the China seas. *Reviews of Geophysics*, *54*(3), 653–673. Available from https://doi.org/10.1002/2015RG000505, http://agupubs.onlinelibrary.wiley.com/agu/journal/10.1002/(ISSN)1944-9208/.

Julien, K., Aurnou, J. M., Calkins, M. A., Knobloch, E., Marti, P., Stellmach, S., & Vasil, G. M. (2016). A nonlinear model for rotationally constrained convection with Ekman pumping. *Journal of Fluid Mechanics*, *798*, 50–87. Available from https://doi.org/10.1017/jfm.2016.225, http://journals.cambridge.org/action/displayJournal?jid = FLM.

Kent, B., Racz, S., & Shashi, S. (2023). Scrambling in quantum cellular automata. *Physical Review B*, *107*(14). Available from https://doi.org/10.1103/PhysRevB.107.144306, https://journals.aps.org/prb/abstract/10.1103/PhysRevB.107.144306.

Klaus, W. (1987). Indonesian through flow and the associated pressure gradient. *Journal of Geophysical Research: Oceans*, *92*(C12), 12941–12946. Available from https://doi.org/10.1029/jc092ic12p12941.

Kok, P. H., Wijeratne, S., Akhir, M. F., Pattiaratchi, C., Chung, J. X., Roseli, N. H., & Daud, N. R. (2022). Modeling approaches in the investigation of upwelling along the east coast of Peninsular Malaysia: Its driven mechanisms. *Regional Studies in Marine Science*, *55*. Available from https://doi.org/10.1016/j.rsma.2022.102562, http://www.journals.elsevier.com/regional-studies-in-marine-science/.

Lent, C. S., Tougaw, P. D., & Porod, W. (1993). Bistable saturation in coupled quantum dots for quantum cellular automata. *Applied Physics Letters*, *62*(7), 714–716. Available from https://doi.org/10.1063/1.108848.

Lent, C. S., Tougaw, P. D., Porod, W., & Bernstein, G. H. (1993). Quantum cellular automata. *Nanotechnology*, *4*(1), 49–57. Available from https://doi.org/10.1088/0957-4484/4/1/004.

Liu, C. C. K., & Jin, Q. (1995). Artificial upwelling in regular and random waves. *Ocean Engineering*, *22*(4), 337–350. Available from https://doi.org/10.1016/0029-8018(94)00019-4.

Marghany, M. (2019a). *Automatic detection algorithms of oil spill in radar images*. CRC Press.

Marghany, M. (2019b). *Synthetic aperture radar imaging mechanism for oil spills* (pp. 1–322). Elsevier. Available from http://www.sciencedirect.com/science/book/9780128181119.

Marghany, M. (2021a). *Nonlinear ocean dynamics: Synthetic aperture radar*. Elsevier.

Marghany, M. (2021b). *Advanced algorithms for mineral and hydrocarbon exploration using synthetic aperture radar* (pp. 1–383). Malaysia: Elsevier. Available from https://www.sciencedirect.com/book/9780128217962.

M.M. Marghany, Coastal circulation off Kuala Terengganu (Master dissertation). (1994).

Marghany, M., & Hashim, M. (2010). MODIS satellite data for modeling chlorophyll-a concentrations in Malaysian coastal waters. *International Journal of Physical Sciences*, *5*(10), 1489–1495. Available from http://www.academicjournals.org/ijps/PDF/pdf2010/4%20Sept/Marghany%20and%20Hashim.pdf.

Marghany, M., & Hussien, M. L. (2003). Upwelling and downwelling along the coastal waters of Kuala Terengganu. *Ultra Scientist of Physical Sciences*, *15*(2), 201–208.

Mazzini, P. L. F., & Barth, J. A. (2013). A comparison of mechanisms generating vertical transport in the Brazilian coastal upwelling regions. Blackwell Publishing Ltd, United States. *Journal of Geophysical Research: Oceans*, *118*(11), 5977–5993. Available from https://doi.org/10.1002/2013JC008924, http://onlinelibrary.wiley.com/journal/10.1002/(ISSN)2169-9291.

Noir, J., Jault, D., & Cardin, P. (2001). Numerical study of the motions within a slowly precessing sphere at low Ekman number. *Journal of*

Fluid Mechanics, *437*, 283–299. Available from https://doi.org/10.1017/S0022112001004311.

Pan, Y. W., Fan, W., Zhang, D. H., Chen, J. W., Huang, H. C., Liu, S. X., Jiang, Z. P., Di, Y. N., Tong, M. M., & Chen, Y. (2016). Research progress in artificial upwelling and its potential environmental effects. *Science China Earth Sciences*, *59*(2), 236–248. Available from https://doi.org/10.1007/s11430-015-5195-2, http://www.springer.com/earth + sciences/journal/11430.

Pitcher, G. C., Figueiras, F. G., Hickey, B. M., & Moita, M. T. (2010). The physical oceanography of upwelling systems and the development of harmful algal blooms. *Progress in Oceanography*, *85*(1-2), 5–32. Available from https://doi.org/10.1016/j.pocean.2010.02.002.

Rizzi, G., & Ruggiero, M. L. (2004). *The relativistic Sagnac effect: Two derivations* (pp. 179–220). Springer Science and Business Media LLC.

Roberts, M. J., & Nieuwenhuys, C. (2016). Observations and mechanisms of upwelling in the northern KwaZulu-Natal Bight, South Africa. *African Journal of Marine Science*, *38*(supl), S43–S63. Available from https://doi.org/10.2989/1814232x.2016.1194319.

Robinson, I. (2006). *Satellite measurements for operational ocean models. Ocean Weather Forecasting: An Integrated View of Oceanography* (pp. 147–189). United Kingdom: Springer Netherlands. Available from http://www.springerlink.com/openurl.asp?genre = book&isbn = 978-1-4020-3981-2.

I.S. Robinson, Satellite observations of ocean colour.. The study of the ocean and the land surface from satellites. (1983), 415–432.

Robinson, I. S. (2004). *Measuring the oceans from space: the principles and methods of satellite oceanography*. Springer Science & Business Media.

Robinson, I. S. (2010a). *Discovering the ocean from space: The unique applications of satellite oceanography*. Berlin Heidelberg: Springer.

Robinson, I. S. (2010b). *The methods of satellite oceanography* (pp. 7–67). Springer Science and Business Media LLC.

Rotvig, J., & Jones, C. A. (2002). Rotating convection-driven dynamos at low Ekman number. *Physical Review E*, *66*(5). Available from https://doi.org/10.1103/physreve.66.056308, 1063-651X.

Schwinger, J. (1951). The theory of quantized fields. I. *Physical Review*, *82*(6).

Sellapillay, K., Verga, A. D., & Di Molfetta, G. (2022). Entanglement dynamics and ergodicity breaking in a quantum cellular automaton. *Physical Review B.*, *106*(10). Available from https://doi.org/10.1103/physrevb.106.104309.

Shirley, W., Chen, Y. A., Dua, A., Ellison, T. D., Tantivasadakarn, N., & Williamson, D. J. (2022). Three-dimensional quantum cellular automata from chiral semion surface topological order and beyond. *PRX Quantum*, *3*(3). Available from https://doi.org/10.1103/PRXQuantum.3.030326, https://journals.aps.org/prxquantum/abstract/10.1103/PRXQuantum.3.030326.

Snyder, H. S. (1947). Quantized space-time. *Physical Review*, *71*(1), 38–41. Available from https://doi.org/10.1103/PhysRev.71.38.

Stellmach, S., & Hansen, U. (2004). Cartesian convection driven dynamos at low Ekman number. *Physical Review E*, *70*(5). Available from https://doi.org/10.1103/physreve.70.056312.

Stevenson, J. W., & Janowitz, G. S. (1977). The effect of ekman suction on a flow over a shallow topography in the beta plane. *Dynamics of Atmospheres and Oceans*, *1*(3), 225–239. Available from https://doi.org/10.1016/0377-0265(77)90008-2.

Takahashi, F., Matsushima, M., & Honkura, Y. (2008). Scale variability in convection-driven MHD dynamos at low Ekman number. *Physics of the Earth and Planetary Interiors*, *167*(3-4), 168–178. Available from https://doi.org/10.1016/j.pepi.2008.03.005.

Tan, Dp, Li, L., Zhu, Yl, Zheng, S., Yin, Zc, & Li, Df (2019). Critical penetration condition and Ekman suction-extraction mechanism of a sink vortex. *Journal of Zhejiang University: Science A.*, *20*(1), 61–72. Available from https://doi.org/10.1631/jzus.A1800260, http://www.springerlink.com/content/1673-565X.

Tatro, P. R., & Mollo-Christensen, E. L. (1967). Experiments on Ekman layer instability. *Journal of Fluid Mechanics*, *28*(3), 531–543. Available from https://doi.org/10.1017/S0022112067002289, 14697645.

Tomczak, M. (1988). Island wakes in deep and shallow water. *Journal of Geophysical Research: Oceans*, *93*(C5), 5153–5154. Available from https://doi.org/10.1029/jc093ic05p05153.

Valle-Levinson, A. (2008). Density-driven exchange flow in terms of the Kelvin and Ekman numbers. *Journal of Geophysical Research: Oceans*, *113*(4). Available from https://doi.org/10.1029/2007JC004144, http://onlinelibrary.wiley.com/journal/10.1002/(ISSN)2169-9291.

Van Senden, D. C., & Imberger, J. (1990). Effects of initial conditions and Ekman suction on tidal outflows from inlets. *Journal of Geophysical Research: Oceans*, *95*(C8), 13373–13391. Available from https://doi.org/10.1029/jc095ic08p13373.

Wang, D., Wang, H., Li, M., Liu, G., & Wu, X. (2013). Role of Ekman transport versus Ekman pumping in driving summer upwelling in the South China Sea. *Journal of Ocean University of China*, *12*(3), 355–365. Available from https://doi.org/10.1007/s11802-013-1904-7.

Wiesner, K. (2018). Quantum cellular automata. *Cellular Automata: A Volume in the Encyclopedia of Complexity and Systems Science*, *93e104*.

Wijffels, S., Firing, E., & Bryden, H. (1994). Direct observations of the Ekman balance at 10°N in the Pacific. *Journal of Physical Oceanography*, *24*(7), 1666–1679. Available from https://doi.org/10.1175/1520-0485(1994), 024 < 1666:DOOTEB > 2.0.CO;2.

Wirth, A. (2010). On the Ekman spiral with an anisotropic eddy viscosity. *Boundary-Layer Meteorology*, *137*(2), 327–331. Available from https://doi.org/10.1007/s10546-010-9527-7.

Wolanski, E., Imberger, J., & Heron, M. L. (1984). Island wakes in shallow coastal waters. *Journal of Geophysical Research: Oceans*, *89* (C6), 10553–10569. Available from https://doi.org/10.1029/jc089ic06p10553.

Wyrtki, K. (1961). *Physical oceanography of the Southeast Asian waters, 2*.

Wyrtki, K. (1962). The upwelling in the region between java and Australia during the south-east monsoon. *Marine and Freshwater Research*, *13*(3), 217–225. Available from https://doi.org/10.1071/MF9620217.

Zhang R. Gupta P. Jha N.K. 2005 Proceedings of the IEEE International Conference on VLSI Design, 10639667, 229–234, United States, Synthesis of majority and minority networks and its applications to QCA, TPL and SET based nanotechnologies.

Further Reading

Estrade, P., Marchesiello, P., De Verdière, A. C., & Roy, C. (2008). Cross-shelf structure of coastal upwelling: A two-dimensional extension of Ekman's theory and a mechanism for inner shelf upwelling shut down. *Journal of Marine Research*, *66*(5), 589–616. Available from https://doi.org/10.1357/002224008787536790, http://www.ingentaconnect.com/content/jmr/jmr.

W.M. Graham, The physical oceanography and ecology of upwelling shadows. (1994).

Jürchott, M., Oschlies, A., & Koeve, W. (2023). Artificial upwelling—A refined narrative. *Geophysical Research Letters*, *50*(4).

Price, J. F., & Sundermeyer, M. A. (1999). Stratified Ekman layers. *Journal of Geophysical Research: Oceans*, *104*(9), 20467–20494. Available from https://doi.org/10.1029/1999jc900164, http://agupubs.onlinelibrary.wiley.com/agu/jgr/journal/10.1002/(ISSN)2169-9291/.

Stacey, M. W., Pond, S., & LeBlond, P. H. (1986). A wind-forced Ekman spiral as a good statistical fit to low-frequency currents in a coastal strait. *Science*, *233*(4762), 470–472. Available from https://doi.org/10.1126/science.233.4762.470.

van den Bremer, T. S., & Breivik, Ø. (2018). Stokes drift. *Philosophical Transactions of the Royal Society A: Mathematical, Physical and Engineering Sciences*, *376*(2111), 20170104. Available from https://doi.org/10.1098/rsta.2017.0104, 1364-503X.

Chapter 12

Quantum multiobjective algorithm for detecting turbulence vorticity and eddies in synthetic aperture radar satellite data

In the preceding chapters, various quantum computing algorithms have been showcased for their efficacy in scrutinizing turbulence flows across diverse SAR datasets. However, a critical examination of the fundamental attributes of turbulence flows, specifically vorticity and eddy, has been noticeably absent. Consequently, this chapter endeavors to delve into the profound significance of harnessing quantum multiobjective algorithm optimization. The objective is twofold: first, to model the degree of turbulence flows as a function of Reynolds number; and second, to autonomously detect both eddy and vorticity turbulence flows within SAR data. This strategic shift aims to unravel the intricate complexities inherent in turbulence phenomena and propel the understanding of quantum-inspired algorithms toward a more holistic exploration of SAR data intricacies.

12.1 What is meant by vortex?

In the realm of oceanography, the term "ocean vortex" denotes a hydrodynamic phenomenon characterized by the intricate, swirling motion of seawater (Fig. 12.1). This phenomenon emerges as a consequence of the complex interplay between various oceanic forces, resulting in dynamic patterns of water circulation known as vortices. At its core, an ocean vortex typically arises from the convergence and divergence of ocean currents. These vortices can manifest in diverse forms, ranging from expansive gyres (Fig. 12.2) in open oceanic expanses to more localized eddies along coastal regions. Their formation and behavior adhere to the fundamental principles of fluid dynamics, governing the movement and distribution of water masses (Truesdell, 1953; Wallace & Foss, 1995; Wu et al., 2007).

Ocean vortices play a pivotal role in influencing the transport of heat, nutrients, and marine life within the oceanic environment. They act as conduits of energy and momentum, shaping the intricate dynamics of marine ecosystems. Understanding the physics behind these vortices is essential for unraveling the complexities of oceanic circulation and its impact on the broader Earth system.

In essence, the study of ocean vortices delves into the scientific intricacies of fluid mechanics, where these swirling eddies serve as fundamental elements in deciphering the nuanced language of oceanic currents. Their presence and behavior contribute significantly to the broader

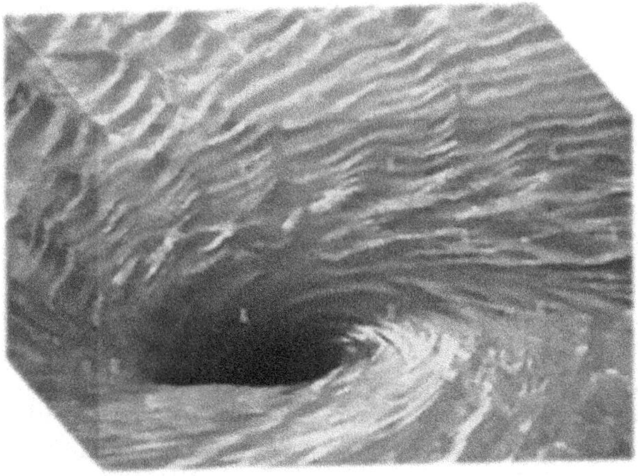

FIGURE 12.1 Swirling motion of seawater.

FIGURE 12.2 Ocean gyre.

Synthetic Aperture Radar Image Processing Algorithms for Nonlinear Oceanic Turbulence and Front Modeling. DOI: https://doi.org/10.1016/B978-0-443-19155-8.00012-X

narrative of Earth's hydrodynamic processes, illustrating the interconnectedness of the oceans and their vital role in maintaining global environmental equilibrium.

12.2 What are different between eddy and vortex?

In the realm of fluid dynamics, particularly in the context of oceans and atmospheres, the terms "eddy" and "vortex" are often used interchangeably, but they can carry nuanced distinctions based on context.

12.2.1 Size and scale

An eddy, in the context of fluid dynamics, represents a localized and relatively diminutive swirling current within water, air, or other fluids. These miniature circulations exhibit rotational motion, yet they tend to be on a smaller scale compared with more extensive vortices. Eddies are characterized by their relatively modest size and a degree of organizational simplicity. Conversely, the term "vortex" spans a wider spectrum of scales. While it can encompass smaller-scale phenomena akin to eddies, it extends its reach to encompass larger, more intricately organized swirling motions. Examples of such larger-scale vortices include tornadoes or hurricanes (Fig. 12.3). The term "vortex" implies a more extensive and potentially more structured rotational flow, covering a range from compact eddies to expansive, organized atmospheric or oceanic phenomena. In essence, the distinction lies in the scale and organization of the swirling motion: eddies being smaller and less organized, while vortices encapsulate a broader spectrum that includes both smaller, more chaotic motions and larger, more structured phenomena (Saffman, 1981; Wallace & Foss, 1995; Wu & Wu, 1996).

Mathematically comparing eddies and vortices involves delving into the equations that describe fluid dynamics, particularly the characteristics of rotational motion. Following is a conceptual representation, using generic symbols and variables:

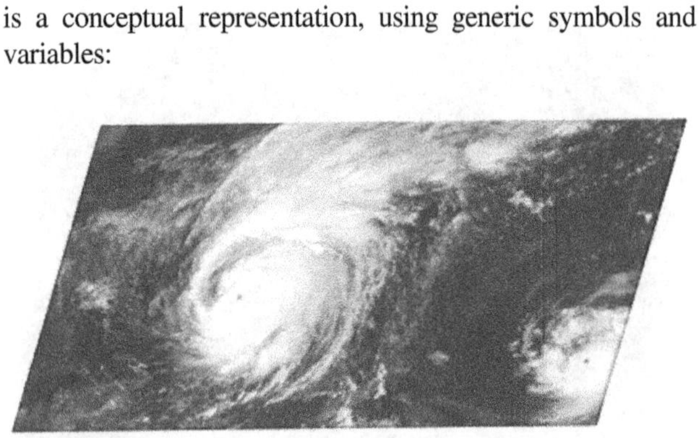

FIGURE 12.3 Hurricane generation.

Let R represent the characteristic radius or size of the rotational motion, and let ω denote the angular velocity. Therefore both eddy and vorticity can be described as function of their radius or size as follows:

$$KE = 0.5m.\omega^2.R^2 \tag{12.1}$$

In this representation, the comparison involves examining the characteristic radius and angular velocity for both eddies and vortices. The kinetic energy provides insight into the intensity of the rotational motion, factoring in both size and angular velocity. In this perspective, eddy may have lower angular velocities, reflecting the less organized nature of the motion. On the contrary, vortex can have higher angular velocities, especially in more organized and intense swirling phenomena (Saffman, 1981; Truesdell, 2018). Therefore if you want to estimate the characteristic radius of the vortex or eddy, you can rearrange Eq. (12.1) as follows:

$$R = \sqrt{\frac{2KE}{m\omega}} \tag{12.2}$$

This formula allows you to estimate the characteristic radius of the vortex or eddy based on the known values of kinetic energy, mass, and angular velocity.

12.2.2 Eddy and vortex formations

Eddies, those transient phenomena in fluid realms, materialize from the intricate interplay of turbulence, the nuances of irregularities in fluidic flow, and the subtle dance of interactions between diverse currents. Their existence is a fleeting waltz, marked by variations in intensity and duration. Conversely, vortices, the more stable and harmonious kin of fluid motion, emerge through a symphony of mechanisms. The conservation of angular momentum orchestrates their creation, a central point akin to a tornado's eye becomes their nucleus, and the interplay of different fluid layers contributes to their enduring stability and coherence.

In the intricate realm of fluid dynamics, an eddy manifests as a mesmerizing dance of swirling fluid and the creation of a countercurrent within the tumultuous embrace of turbulent flow. As the fluid courses through its journey, it gives birth to a unique space on the downstream side of an obstacle—a void where downstream-flowing fluid is notably absent.

The fluid, ever persistent, rushes in from behind the obstacle to fill this void, orchestrating a graceful swirl on each edge of the impediment. Following this captivating display, a brief interlude occurs—a reverse flow of fluid emerges behind the obstacle, defying the typical downstream course and moving upstream, toward the rear of the obstacle. This enchanting phenomenon gracefully unfolds, often observed as a natural spectacle behind substantial emergent rocks in swiftly flowing rivers (Maddison et al., 2015; Marshall et al., 1999, 2012; Shutts, 1983).

In essence, an eddy represents a deviation, a fluidic pirouette, from the broader flow of the fluid, showcasing the intricate and captivating dynamics that govern the movements of liquids in their ever-changing journey. In the intricate dance of fluid dynamics, the formation of eddies and vortices involves underlying physical principles. Let's delve into the intricate equations governing their creations. Therefore the formation mechanisms of eddies and vortices can be expressed in mathematical terms as follows:

$$E_{Eddy} = \frac{1}{2}mv^2 + \frac{R_{ec}\mu}{\rho d} + 2\mu_t S_{i,j} - \frac{2}{3}\rho\kappa\delta_{i,j} \tag{12.3}$$

where $\frac{1}{2}mv^2$ accounts for the kinetic energy of the fluid within the eddy, where m is the mass, and v is the velocity. In this perspective, $\frac{R_{ec}\mu}{\rho d}$ involves the Reynolds number (R_{ec}), viscosity (μ), density (ρ), and characteristic length (d). It characterizes the effects of turbulence and irregularities in fluid flow. Therefore $\frac{2}{3}\rho\kappa\delta_{i,j}$ contributes to the pressure gradients and turbulence effects as κ is the turbulence kinetic energy, and δ_{ij} is the Kronecker delta. Additionally, $2\mu_t S_{i,j}$ incorporates the turbulent viscosity (μ_t) and the rate of strain tensor (S_{ij}), representing the interactions and effects of turbulence.

Certainly, let's derive the equation expressing vortex formation energy (E_{vortex}) as the sum of angular momentum conservation energy, central point energy, and fluid interactions energy.

$$E_{Vortex} = \frac{1}{2}mv^2 + \frac{1}{2}I\omega^2 - \Delta p + \mu\Delta^2\vec{u} + \rho g \tag{12.4}$$

Like eddy, $\frac{1}{2}mv^2$ represents the kinetic energy of the fluid within the vortex, where m is the mass and v is the velocity. On the contrary to eddy, $\frac{1}{2}I\omega^2$ accounts for the rotational kinetic energy associated with the vortex as based on the moment of inertia I and the angular velocity ω. Unlike eddy, $-\nabla p$ involves the pressure gradient (∇p) and represents the work done against pressure forces. Therefore $\mu\nabla^2 u$ accounts for viscous dissipation within the vortex. In this view, the term ρg involves the density (ρ), gravitational acceleration (g), and represents the potential energy associated with the vortex's position in a gravitational field. Needless to say, Eq. (12.4) combines kinetic energy, rotational kinetic energy, work done against pressure, viscous dissipation, and potential energy to quantify the total energy associated with the formation of a vortex (Maddison et al., 2015; Truesdell, 2018).

12.2.3 Spatial distribution

In the realm of fluid dynamics, the spatial distribution of phenomena such as eddies and vortices unfolds with intricacy and diversity. Eddies, those dynamic swirls of fluid, manifest in regions where the otherwise continuous flow of fluid confronts obstacles or undergoes alterations in speed. These distinctive fluid circulations are commonly encountered in boundary layers and areas characterized by complex terrain. Imagine a swift-flowing river encountering substantial rocks—behind these obstacles, the fluid forms eddies, creating local circulations that dance around the impediments (Table 12.1).

On the other hand, vortices, a broader category encompassing a spectrum of scales, find their existence in a multitude of environments. They can emerge as microscopic entities within turbulent flows, showcasing the mesmerizing dance of fluid elements on a minuscule scale. Simultaneously, vortices also take on grand proportions in the atmospheric and oceanic realms. Picture the colossal vortices such as hurricanes or whirlpools, dynamic giants that command attention on a large scale, shaping the landscapes of the sky and sea (Truesdell, 1953; Wallace & Foss, 1995; Wu et al., 2007).

Thus the spatial tapestry of eddies and vortices unveils a captivating interplay of fluid dynamics, where the intricate details of local circulations and the majestic sweep of

TABLE 12.1 Comparison between eddy and vortex.

Aspect	Eddy	Vortex
Size and Scale	Generally smaller in scale	Can range in scale from small to large, may include both small eddies and larger, more organized structures
Formation Mechanism	Often a result of turbulence, irregularities, or local interactions	Can form through various mechanisms, including conservation of angular momentum, central point presence, or interaction between fluid layers
Spatial Distribution	Commonly found in regions with obstacles, changes in speed, or complex terrain	Can exist on different scales and in various environments, ranging from microscopic to large-scale atmospheric or oceanic phenomena

large-scale phenomena contribute to the rich and varied canvas of fluid motion.

12.2.4 Impact on flow

Eddies, in their fluidic intricacy, play a crucial role in the intricacies of mixing and the dispersion of properties within a fluid. However, their influence might not wield a dominant impact on the broader, larger-scale flow patterns.

In contrast, the significance of vortices extends beyond the local dynamics. Vortices possess the potential to exert a substantial influence on the surrounding flow, with repercussions that extend to impactful scales. This influence can be witnessed in phenomena such as weather patterns, ocean currents, and other substantial atmospheric and oceanic occurrences. Vortices, with their intricate dance, become key players in shaping the grand tapestry of large-scale fluid dynamics (Maddison et al., 2015; Marshall et al., 1999; Shutts, 1983).

It can be said, while both terms refer to swirling motions in fluids, "eddy" is commonly associated with smaller, more localized disturbances, while "vortex" can encompass a broader range of scales and often implies more organized and impactful rotational motion (Table 12.1).

12.3 Geometry of eddies and vortices: a comprehensive exploration of fluidic phenomena

Eddies, commonly observed in fluid dynamics, manifest distinctive features in their shapes. One prevalent form is the circular or spiral pattern, particularly prominent in rotating fluid systems where the inherent rotation induces swirling motion, giving rise to these characteristic shapes. However, eddies are not confined to uniform circularity; they can also adopt irregular shapes, especially in turbulent flows characterized by intricate interactions and irregularities in the fluid dynamics. The diversity in their shapes is a consequence of the complex and dynamic nature of turbulent flow fields. Moreover, eddies exhibit a considerable range in size, spanning from small-scale vortices to larger, more organized structures, highlighting the versatility and adaptability of these fluid phenomena across various scales (Browand & Troutt, 1985; Escher & Wunsch, 2012; Escher et al., 2010).

Conversely, vortices, with their intriguing shapes and dynamic nature, display characteristics that vary significantly based on scale and formation mechanisms. In this sense, the shape of vortices is intricately tied to their scale and the specific processes that give rise to them. From large atmospheric or oceanic phenomena to microscopic vortices, the scale plays a crucial role in determining their appearance. In larger atmospheric or oceanic vortices, such as tornadoes or hurricanes, therefore, the shape often resembles a funnel. These vortices exhibit a distinct central core and outward-spiraling winds, creating the characteristic tornado-like appearance. Consequently, on a smaller scale, microscopic vortices within turbulent flows can manifest as intricate swirls and spirals. These intricate patterns mirror the smaller eddies found in fluid systems. Needless to say, the geometry of these fluidic phenomena is a complex interplay of various factors, including the fluid properties, environmental conditions, and the underlying forces driving the flow (Browand & Troutt, 1985; Escher et al., 2010; Maddison et al., 2015; Truesdell, 2018).

12.4 Mathematical formulation of eddies and vorticities: a comprehensive description

Certainly, let's delve into the mathematical expressions describing eddy and vorticity in the context of ocean dynamics and meteorology. Eddy is often associated with the turbulent and swirling motion of fluid elements. In the realm of physics, particularly in fluid dynamics, the Reynolds-averaged Navier – Stokes equations (Chapter 2) can be utilized to model the behavior of turbulent eddies. These equations, derived from the fundamental principles of fluid mechanics, encapsulate the conservation of mass and momentum within a fluid. In this sense, large eddy simulation (LES) is an advanced computational approach used to simulate turbulent flows, capturing the large-scale structures of turbulence while modeling the smaller scales (Hunt, 1987; Majda, 1986; Mansfield et al., 1998; Meitz & Fasel, 2000; Shutts, 1983; Wu et al., 2007). In LES, the Navier – Stokes equations are filtered to separate the resolved (large) scales from the subgrid (small) scales. The filtered equations provide a system that can be solved numerically to simulate the behavior of the resolved scales. Let's derive the filtered Navier – Stokes equations for incompressible flow, which form the basis for LES:

The filtered velocity field is denoted as $\overline{u}(\vec{x},t)$, and the filtering operation is typically represented by a spatial filter $G(\vec{x})$ to separate large and small scales:

$$\vec{\overline{u}}\,(\vec{x},t) = \int \vec{u}\,(\vec{x} + \vec{r},t)G(\vec{r})d\vec{r} \qquad (12.5)$$

Therefore the velocity field is decomposed into the filtered part $(\vec{\overline{u}})$ and the fluctuating part $(\vec{u}')$:

$$\vec{u}\,(\vec{x},t) = \vec{\overline{u}}\,(\vec{x},t) + \vec{u}'\,(\vec{x},t) \qquad (12.6)$$

As well as the continuity equation is applied to the filtered velocity field that is, $\nabla.\vec{\overline{u}} = 0$, the filtered Navier – Stokes equation for incompressible flow is derived by applying the filtering operation to the original Navier – Stokes equations:

$$\frac{\partial \overline{u_i}}{\partial t} + \overline{u_j}\frac{\partial \overline{u_i}}{\partial x_j} = -\frac{1}{\rho}\frac{\partial \overline{p}}{\partial x_i} + \nu\frac{\partial^2 \overline{u_i}}{\partial x_j \partial x_j} - \frac{\partial(\overline{u_i u_j} - \overline{u}_i\overline{u}_j)}{\partial x_j}. \tag{12.7}$$

Here $\frac{\partial \overline{u_i}}{\partial t}$ represents how the mean velocity ($\overline{u}_i$) changes with time at a given point in the fluid. It accounts for the acceleration or deceleration of the fluid particles in the direction *i, and* $\overline{u_j}\frac{\partial \overline{u_i}}{\partial x_j}$ describes the advection or transport of the mean velocity in the *i*-th direction by the mean velocity itself. It captures the effect of the flow carrying momentum along with it. $-\frac{1}{\rho}\frac{\partial \overline{p}}{\partial x_i}$; therefore represents the pressure gradient force acting on the fluid. It accounts for how variations in pressure with respect to spatial coordinates affect the motion of the fluid. $\nu\frac{\partial^2 \overline{u_i}}{\partial x_j \partial x_j}$ consequently, represents the diffusion of momentum due to viscosity. It reflects how the mean velocity field smoothens out or diffuses over space due to the viscous effects, while $\frac{\partial(\overline{u_i u_j} - \overline{u}_i\overline{u}_j)}{\partial x_j}$ is subgrid-scale stress tensor. It accounts for the effects of unresolved turbulent fluctuations or eddies on the resolved scales. This term represents the additional stress due to turbulent motions that are not explicitly resolved in the simulation. Needless to say, the equation is a filtered representation of the Navier – Stokes equations, with the last term capturing the influence of turbulent eddies on the mean flow in the *i*-th direction. This is a crucial aspect of large eddy simulation, where the smaller turbulent scales are not resolved explicitly, but their impact is modeled through additional terms in the equations (Bauer et al., 2016; Escher et al., 2010; Maddison et al., 2015).

Now, turning our attention to vorticity, which characterizes the local rotation of fluid elements, the vorticity vector $\vec{\omega}$ can be defined as the curl of the velocity vector $\vec{u}$:

$$\vec{\omega} = \nabla \times \vec{u} \tag{12.8}$$

In the elegant realm of mathematical description, the vorticity of a three-dimensional fluid flow unveils itself as a pseudovector field, gracefully represented by the symbol ω. This illustrious pseudovector field is meticulously defined as the curl of the velocity field *v*, an exquisite portrayal of the continuous motion within the fluid medium. In the exquisite Cartesian coordinates, this mathematical ballet unfolds as follows:

$$\vec{\omega} = \nabla \times \vec{v} = \left(\frac{\partial v_z}{\partial y} - \frac{\partial v_y}{\partial z}\frac{\partial v_x}{\partial z} - \frac{\partial v_z}{\partial x}\frac{\partial v_y}{\partial x} - \frac{\partial vx}{\partial y}\right). \tag{12.9}$$

In simpler terms, vorticity reveals how the velocity vector alters when one shifts by an extremely small distance in a direction perpendicular to it. In a two-dimensional flow where velocity doesn't depend on the *z*-coordinate and lacks a *z*-component, the vorticity vector consistently aligns parallel to the *z*-axis. This alignment allows expressing the vorticity as a scalar field multiplied by a constant unit vector $\hat{z}$.

$$\vec{\omega} = \Delta \times \vec{v} = \left(\frac{\partial v_y}{\partial x} - \frac{\partial v_x}{\partial y}\right)\hat{z}. \tag{12.10}$$

Eq. (12.10) eloquently illustrates that within a rotating fluid, any discrete volume of fluid, referred to as a parcel, exhibits a nonzero vorticity (as depicted in Fig. 12.4). Conversely, a fluid parcel devoid of rotation, adhering to an irrotational state, manifests zero vorticity (as illustrated in Fig. 12.5).

It is worth noting that circulation and vorticity emerge as pivotal dual entities characterizing rotation within a fluid. Circulation, a scalar integral quantity, serves as a macroscopic metric, capturing the rotational essence over a finite area of the fluid. On the other hand, vorticity, a vector field, serves as a microscopic descriptor, delineating the rotation at each point within the fluid. This duality in representation provides a comprehensive understanding of rotation dynamics in fluid mechanics (Escher & Wunsch, 2012; Hunt, 1987; Majda, 1986; Mansfield et al., 1998; Truesdell, 2018).

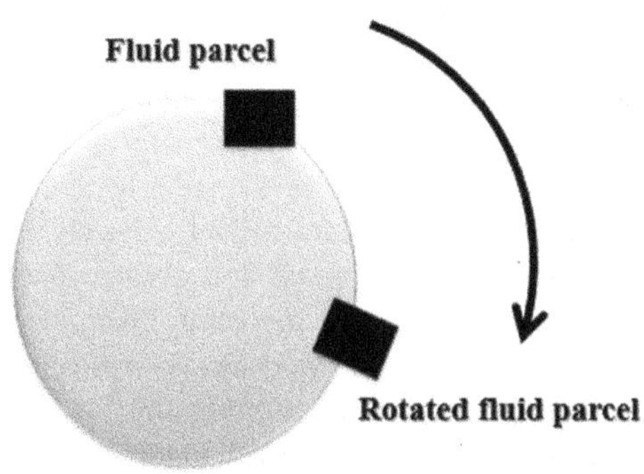

FIGURE 12.4 Rotation-induced vorticity in a fluid parcel.

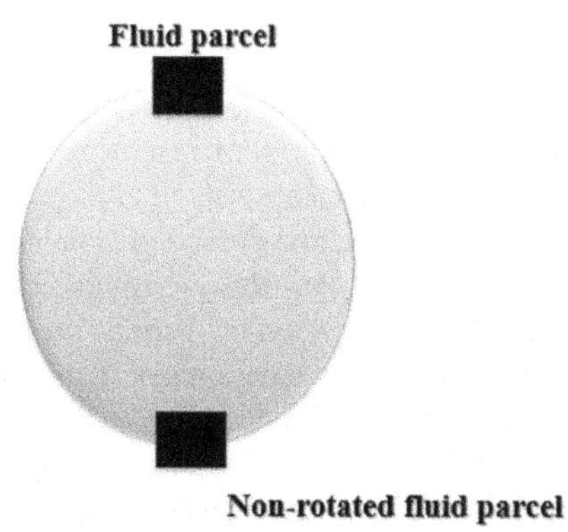

FIGURE 12.5 Vorticity absence caused by the nonrotation of a fluid parcel.

12.5 Unveiling abstract mechanisms underlying eddy and vorticity in synthetic aperture radar data

In this insightful perspective, mesoscale eddy or vorticity emerges as a common occurrence in the atmospheric layer adjacent to Earth's surface, driven by microscale gradients in velocity, temperature, and moisture. Its role is to diminish the corresponding physical gradients, leading to the elimination of microscale instability and a return to normal atmospheric conditions—an intriguing phenomenon known as convective collapse.

This reduction, stemming from eddy and vorticity energy fluxes, is integral to preventing the molecular diffusion of buoyancy, moisture, and momentum between Earth's intricate surface and the prevailing atmosphere. In contrast to turbulent diffusion, characterized by scales several orders of magnitude smaller, large-scale movements are now dominated by turbulent flux. The pivotal question arises: How does the SAR sensor capture ocean eddy and vorticity?

When an SAR sensor illuminates the ocean surface at moderate incident angles (between 20° and 60°), the ratio of backscattered energy is influenced by ocean surface roughness at the radar wavelength, known as the "Bragg" wavelength as demonstrated in previous chapters. Wind propagation induces slight waves in the longitudinal direction, and the intensity of backscattered energy from the turbulent surface peaks when the local wind aligns with the radar's line of sight. Conversely, when the wind opposes the radar's line of sight, there is a smaller local extreme in backscattered energy. SAR effectively detects the constraints imposed by atmospheric and oceanic phenomena on the centimeter-scale wave spectrum. Precipitation may occasionally interfere with radar signals, revealing dominant atmospheric conditions or presenting as incoherent patterns indicative of constrained cell dynamics.

Spaceborne SAR data, with its 10–100 m resolution and swath width of 100–1000 km, provide a comprehensive view of roughness patterns caused by sea surface tension. These data allow the inference of dynamic and thermodynamic environments, distinguishing between statically unstable and statically stable and baroclinic and barotropic conditions. Visible traces of macroscale, mesoscale, and microscale meteorological events aid in understanding the prevailing atmospheric and oceanic dynamics.

In this perspective, buoyancy-driven/shear-organized rolls, spiral circulations triggered by thermodynamic instability in a vertical wind shear environment, come into focus. Increased wind shear transforms randomly organized speckles into a distinct field of linearly arranged elongated speckles, leading to uniform development and a significant reduction in sea surface roughness (Cassianides et al., 2021; Karimova & Gade, 2016). The SAR thickness pattern accurately represents the frequent appearance of irregular dark and bright speckled lines. Determining the orientation of these roller surface tracks, and hence their SAR footprint, is achieved through matching methods and techniques, providing valuable insights into the complex dynamics of atmospheric and oceanic processes (Marghany 2021).

12.6 Developing a backscatter energy model for synthetic aperture radar imaging mechanisms involving vorticity and eddy roughness

Certainly, let's develop mathematical equations to describe how synthetic aperture radar (SAR) captures eddy and vorticity information under varying wind speeds (U) and incidence angles (θ). Assuming a simplified scenario where the ocean surface is characterized by small-scale surface perturbations (waves) and z-direction is normal to the Earth's surface, we can use the following equations:

12.6.1 Eddy generation due to wind shear

The eddy generation (ω) influenced by wind shear can be expressed as follows:

$$\omega = \alpha U \cos(\theta) \tag{12.11}$$

The constant α represents the sensitivity of eddy generation to wind shear.

12.6.2 Vorticity imprint on backscattered energy

The vorticity (ω) imprint on backscattered energy is influenced by the wind speed, incidence angle, and the surface roughness characteristics. Let's denote the radar wavelength as λ:

$$\vec{\omega} = \beta U \cos(\theta) \hat{z} \tag{12.12}$$

The term $\hat{z}$ represents the unit vector in the z-direction. The constant β incorporates the influence of wind speed and incidence angle on the vorticity. In the realm of synthetic aperture radar (SAR), the detection of backscattered energy ($E(\sigma)$) involves a profound understanding of the intricate interplay among various factors, including vorticity (ω), surface roughness (h), and the radar wavelength (λ). The backscattered energy ($E(\sigma)$) represents the radar signal that is reflected back toward the sensor after interacting with the ocean rough surface. The complexity of this phenomenon is encapsulated in the following expression:

$$E(\sigma) = \gamma e^{\left(-2\pi i \frac{\vec{\omega} \cdot \vec{h}}{\lambda}\right)} \tag{12.13}$$

TABLE 12.2 Pseudocode for modeling backscattered energy from vorticity and surface eddy roughness.

```
FUNCTION calculateBackscatteredEnergy(vorticity, surfaceRoughness, radarWavelength,
constantGamma): // Calculate the dot product of vorticity and surface roughness
vorticityDotSurfaceRoughness = dotProduct(vorticity, surfaceRoughness)
// Calculate the argument of the exponential term exponentArgument = -2 * π * i *
(vorticityDotSurfaceRoughness / radarWavelength)
// Calculate the exponential term exponentialTerm = exp(exponentArgument)
// Calculate the backscattered energy using the provided constants backscatteredEnergy =
constantGamma * exponentialTerm
// Return the result RETURN backscatteredEnergy END FUNCTION
```

The constant γ incorporates various factors, including system-specific parameters, calibration, and the overall efficiency of the SAR system. It ensures that the backscattered energy is appropriately scaled. The exponential term, $e^{\left(-2\pi i\frac{\vec{\omega}\cdot\vec{h}}{\lambda}\right)}$, accounts for the interference and phase relationships between the incident and reflected waves. This interference pattern is influenced by the alignment of vorticity, surface roughness, and the radar wavelength (Table 12.2).

This equation essentially encapsulates the quantum of information carried by the backscattered energy, revealing details about the rotational dynamics of the Earth's surface. The role of vorticity and surface roughness, entwined with the radar wavelength, is elegantly expressed through this mathematical formulation. Understanding and manipulating these relationships contribute to advancements in SAR technology and its applications in Earth observation and remote sensing (Cassianides et al., 2021; Ivanov & Ginzburg, 2002; Wang et al., 2019).

12.7 Automated detection of turbulent flows in synthetic aperture radar images using quantum algorithms

The process of accurately segmenting eddy and vorticity flows using partition clustering poses a significant challenge, which is acknowledged as a combinatorial optimization task. In conventional approaches, global optimization techniques, such as conventional genetic algorithms (CGAs) (Deb, 2001; Kim et al., 2006; Marghany, 2013), are commonly employed for this task. However, these methods often face drawbacks, including being time-consuming, exhibiting slow convergence rates, and being prone to getting trapped in local optima.

Traditional algorithms are inherently limited in their ability to operate in only one independent state at a time. This limitation implies that they can only complete phases sequentially, which can impede their efficiency in addressing complex optimization tasks.

In the domain of quantum physics, the fundamental particles constituting turbulent flow and ocean waves exhibit a unique characteristic—they are inherently unfixed and can exist in multiple phases simultaneously. This phenomenon is referred to as superposition. When these particles are distributed in synthetic aperture radar (SAR) images, they undergo a process of decomposition, settling into a specific state.

The concept of superposition, central to quantum mechanics, allows particles to exist in multiple states concurrently. In the context of SAR images, this implies that the constituent elements of eddy and vorticity flows can contribute to multiple aspects of the image simultaneously. The decomposition of these particles in SAR images represents a dynamic and simultaneous representation of various phases, contributing to a nuanced and comprehensive understanding of the underlying eddy and vorticity flows.

In essence, the utilization of quantum principles in this context facilitates a more holistic and parallel approach to representing eddy and vorticity flows, overcoming the limitations of traditional algorithms. It introduces a level of flexibility and complexity that aligns more closely with the inherently dynamic and multifaceted nature of eddy and vorticity flows, ultimately enhancing the accuracy and efficiency of the segmentation process in SAR images.

In the realm of quantum physics, the fundamental particles constituting turbulent flow and ocean waves possess a unique characteristic—they are inherently unfixed and can exist in multiple phases simultaneously, a phenomenon known as superposition. When these particles are distributed in synthetic aperture radar (SAR) images, they undergo a process of decomposition, settling into a specific state. This quantum-inspired approach introduces a dynamic and simultaneous representation of various phases in the SAR images, providing a more nuanced understanding of the underlying turbulent flows.

In this quantum context, concepts such as fuzziness, coherent superposition, interference, and implicit parallelism illustrate quantum computing's local foundations in quantum mechanics. These attributes contribute to increased diversity and consistency in exploring and employing standard evolutionary algorithms. Renowned scientists, exemplified by Han and Kim (2004), have leveraged quantum codes in evolutionary computation, progressively developing quantum-inspired algorithms for combinatorial optimization based on quantum computing principles and coding. Similarly, Jiao et al. (2008) explored a quantum-inspired immune clonal technique for global optimization in this prospective scenario.

Despite the potential benefits, challenges persist in the automated detection of turbulent boundary flow in SAR images. Limitations of single-factor clustering techniques and potential errors in image processing segmentation, particularly in distinguishing visible brightness and darkness patches, are obstacles that must be addressed to enhance the accuracy of automated detection in SAR images.

12.8 Quantum-enhanced synthetic aperture radar image processing for automated eddy and vorticity detection

In classical image processing, information is typically represented using bits, which are binary units of data that can exist in one of two states, 0 or 1. Classical image processing algorithms manipulate these bits to perform various operations on images. In quantum image processing, the key distinction lies in the utilization of qubits instead of classical bits. Qubits are quantum counterparts to classical bits but exist in a superposition of states. Unlike classical bits, which can only exist in a state of 0 or 1, qubits can exist in a linear combination of these states as demonstrated in previous chapters. This unique property is a consequence of quantum mechanics and is often described using a mathematical framework called Hilbert space.

$$|\psi\rangle = \alpha|0\rangle + \beta|1\rangle \quad (12.14)$$

Here, $|\psi\rangle$ represents a quantum state, and α and β are complex coefficients. The state $|0\rangle$ corresponds to the classical bit state 0, and $|1\rangle$ corresponds to the classical bit state 1. Therefore linear superposition of quantum states in SAR image can be given by:

$$|\Phi\rangle = \sum_i c_i|\psi i\rangle \quad (12.15)$$

In this equation, $|\Phi\rangle$ is a quantum state formed by the linear superposition of multiple quantum states $|\psi_i\rangle$, each with corresponding coefficients c_i. Now, how do radar signals relate to quantum states in both equations?

Radar signals can be represented in quantum states by encoding information about the radar echoes into the quantum states of qubits. The sensitivity and resolution of radar systems can be enhanced by leveraging the unique properties of quantum superposition. Quantum states can encode more information than classical states, allowing for more efficient and complex representation of radar signals. The utilization of quantum states in radar systems holds the potential to revolutionize signal processing capabilities, enabling more advanced and precise sensing technologies.

Therefore incorporating coherent superposition involves manipulating probability amplitudes coherently, introducing a degree of fuzziness to adapt to the dynamic nature of SAR data:

$$\Psi|C_f\rangle = \sum_i f_i c_i|\phi_i\rangle \quad (12.16)$$

Where f_i represents the fuzziness factor, and C_f is coherent fuzziness due to eddy and vorticity occurrences in SAR data. Consequently, the quantum interference is expressed through interference terms, introducing implicit parallelism:

$$\Psi|I_{Interfered}\rangle = \sum_{ij} I_{ij} f_i c_i|\phi_i\rangle + I_{ij} f_j c_j|\phi_j\rangle \quad (12.17)$$

To express the interference factor I_{ij} between phases i and j mathematically, you can use the inner product of the corresponding quantum states. Given two quantum states $|\psi_i\rangle$ and $|\psi_j\rangle$, the interference factor I_{ij} can be defined as follows:

$$I_{ij} = \langle\psi_i|\psi_j\rangle \quad (12.18)$$

Here, $\langle\psi_i|$ represents the conjugate transpose (also known as the adjoint or Hermitian conjugate) of the quantum state $|\psi_i\rangle$. If $|\psi_i\rangle$ and $|\psi_j\rangle$ are represented as column vectors, the interference factor can be calculated using the dot product:

$$I_{ij} = [\alpha_i^* \beta_i^*]\begin{bmatrix}\alpha_j \\ \beta_j\end{bmatrix} \quad (12.19)$$

Here, α_i and β_i are the complex coefficients corresponding to the quantum state $|\psi_i\rangle$, and α_j and β_j are the complex coefficients corresponding to the quantum state $|\psi_j\rangle$.

The asterisk (*) denotes the complex conjugate. This equation quantifies the degree of overlap or interference between the quantum states $|\psi_i\rangle$ and $|\psi_j\rangle$, providing a measure of their correlation in the context of the quantum-enhanced SAR image processing algorithm. Therefore the quantum state equations can distinguish between strongly scattered signals due to rough eddy and vorticity occurrences on the sea surface and weak ones that are scattered from a smooth sea surface state. This is because quantum states can provide more information about the radar signal, allowing for better differentiation between different types of scattering. Additionally, quantum sensors can be used to detect very weak radar signals that would otherwise be undetectable with traditional radar systems.

Eq. (12.17) illustrates that the qubit is in a linear superposition of these base states, achieved by varying the values of α and β with rotation angles θ and ϕ. This process can be mathematically explored using the Bloch sphere, as depicted in (), and is expressed as follows:

$$|\psi\rangle = \cos\left(\frac{\theta}{2}\right)|0\rangle + e^{i\phi}\sin\left(\frac{\theta}{2}\right)|1\rangle \tag{12.20}$$

Here, θ and ϕ are rotation angles, and the Bloch sphere provides a geometric representation of the qubit state (Fig. 12.6). The superposition of base states is achieved by adjusting θ and ϕ, allowing for a versatile representation of quantum information. This mathematical formulation and geometric interpretation showcase the fundamental principles of qubits and their manipulation in quantum computing and quantum information processing. The Bloch sphere provides a concise visual representation of the qubit's state and its evolution through quantum operations (Fig. 12.7).

According to the aforementioned perspective, the Bloch sphere can greatly assist us to distinguish between qubit and bit. In this sense, the qubit spins up and down

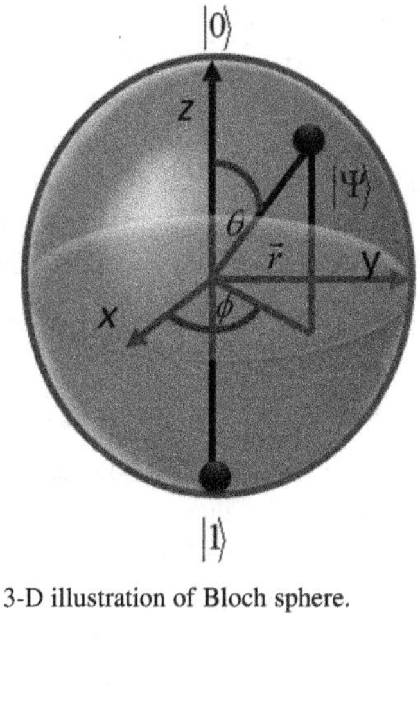

FIGURE 12.6 3-D illustration of Bloch sphere.

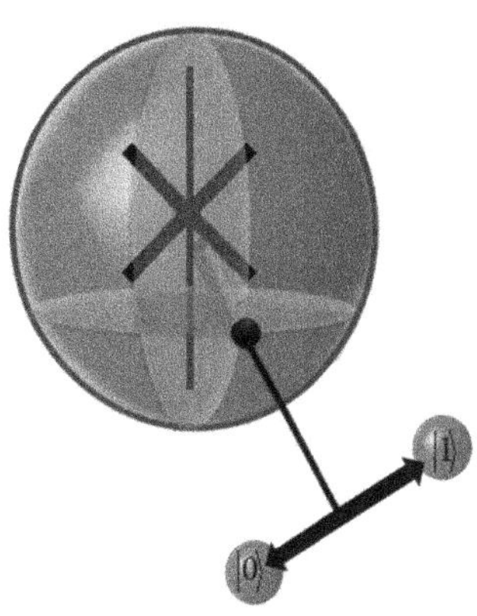

FIGURE 12.7 Sketch of a qubit.

and superposition between in the same time in less than a second (Fig. 12.2). Nevertheless, the conventional bit is either 0 or 1, which leads to a longer time of computation than a qubit (Hey, 1999).

The remaining question is now: How can qubit explore multiobjective evolution to create a novel quantum multiobjective evolution (QME) for automatic detection of turbulent based on eddy and vorticity patterns in synthetic aperture radar (SAR) images?

12.9 Quantum multiobjective evolutionary algorithm

In the context outlined earlier, the collapse of the quantum system occurs when a projection onto a specific basis is observed. This event is often referred to as measurement or decoherence, and it corresponds to the collapse of the wave function ψ in Eq. (12.20). Specifically, if a measurement is made in a particular basis, the quantum state is considered collapsed onto one of the basis states (Hey, 1999; Kaye et al., 2007; Williams, 2011). In this instance, the probability of the qubit collapsing onto the $|0\rangle$ state would approach $\cos^2(\theta/2)$, and onto the $|1\rangle$ state would approach $\sin^2(\theta/2)$. With this understanding, a novel qubit chromosome representation is employed in quantum multiobjective evolution (QME) based on the principles of quantum computing, encompassing qubits and linear superposition (Kaye et al., 2007). This representation is distinctive for its capacity to depict any linear superposition of solutions. A quantum bit (qubit) is the smallest unit of data in a dual-state quantum computer, capable of existing in the "1" or "0" states or any superposition of both (Bardin et al., 2020). This can be used to represent the state of a qubit:

$$|\psi\rangle = P|0\rangle + \sqrt{1-P^2}e^{i\phi}|1\rangle \tag{12.21}$$

Here, P represents the probability amplitude associated with the $|0\rangle$ state, and $e^{i}\phi$ introduces the phase factor associated with the $|1\rangle$ state. In the case of turbulent flow patterns in SAR data, P would correspond to the wave function of the turbulent flow pattern includes eddy and vorticity. Accordingly, m patterns of the eddy and vorticity features are assumed as a training set D in the SAR data, which is described as: $D = \{\Psi(P_m)\}$.

According to this perspective, the coherence state of the SAR features $|\Psi_{SAR}\rangle$ is mathematically described as follows:

$$C_{SAR} = |\langle\Psi_{SAR}|\Psi_{SAR}\rangle|^2 \tag{12.22}$$

Here, $\langle\Psi_{SAR}|\Psi_{SAR}\rangle$ represents the inner product of the quantum state $|\Psi_{SAR}\rangle|\Psi_{SAR}\rangle$ with itself. The modulus squared of this inner product provides the coherence state (C_{SAR}), offering insight into the correlation or similarity between the quantum states associated with the SAR features. Certainly, let's develop a mathematical framework for the Grover algorithm within the context of SAR features. The Grover algorithm involves three operations, leading to the coherent state of the SAR features ($|\Psi_{SAR}\rangle|\Psi_{SAR}\rangle$). Therefore the M matrix in Grover's algorithm is constructed based on the number of qubits $(2n + 1)$ and the number of patterns (m). The matrix is designed to mark the correct states that satisfy the search condition. The M matrix can be represented as follows:

$$M = I - 2|\Psi_{eddy}\rangle\langle\Psi_{vorticity}| \tag{12.23}$$

where I is the identity matrix, and $|\Psi_{eddy}\rangle|\Psi_{vorticity}\rangle$ are the turbulence feature states representing the accurate identification. In the context of SAR features, the M matrix is designed to mark the correct states related to the desired patterns. The specific form of $|\Psi_{eddy}\rangle|\Psi_{vorticity}\rangle$ would depend on the nature of the SAR data and the patterns being sought. The application of this M matrix in Grover's algorithm, therefore, contributes to the amplitude amplification process, enhancing the probability of measuring the correct state in subsequent iterations.

In the context of Grover's algorithm, the Oracle operation plays a pivotal role in marking the correct states that satisfy the search condition. This operation is crucial for the algorithm's success, as it allows the quantum computer to identify the states corresponding to the desired patterns. The Oracle operation is mathematically represented by the Oracle matrix O. The Oracle operation can be expressed as:

$$|\Psi_{eddy}\rangle\langle\Psi_{vorticity}| = O|\Psi_{SAR}\rangle \tag{12.24}$$

Here O is the Oracle matrix; and $|\Psi_{SAR}\rangle$ is the quantum state vector representing the SAR features. The Oracle matrix is designed to flip the sign of the amplitude of the states that satisfy the search condition. In the context of SAR image processing, this means identifying and marking the states that correspond to the patterns of interest in the data. The Oracle operation effectively guides the quantum computer toward the correct solutions during the iterative process of Grover's algorithm. In the mathematical framework of Grover's algorithm, we delve into the profound principles of amplitude amplification, a key concept that underlies the algorithm's efficiency (Wang et al., 2005; Williams, 2011; Zhang & Korepin, 2020). Amplitude amplification in Grover's algorithm is a process designed to enhance the probability of measuring the correct state, and it is ingeniously executed through the application of the Grover diffusion operator. Mathematically, this operation can be elegantly expressed as:

$$|\Psi_{SAR}\rangle = U_s M U_s^{\dagger} O|\Psi_{SAR}\rangle \tag{12.25}$$

Here, each element holds a profound significance in the quantum spinning. Indeed, U_s symbolizes the Grover diffusion operator, orchestrating the amplification of amplitudes. Additionally, $U_s^{\dagger}$ denotes the adjoint (conjugate transpose) of the Grover diffusion operator. In the dance of quantum states, U_s guides the amplitude amplification, gracefully tuning the probabilities of the states (Diao et al., 2002; Lavor et al., 2003; Wang et al., 2005). The Grover diffusion operator plays a central role in steering the quantum system toward the correct solutions, embodying the elegance and efficiency inherent in author's vision of quantum computation. Through this mathematical framework, we capture the essence of Grover's algorithm (Table 12.3) as it delicately manipulates probability amplitudes, showcasing the beauty of quantum mechanics in information processing.

Certainly, let's elegantly incorporate the provided information into the expression for the pattern state detection of turbulent flow using the M matrix. If we denote the state of turbulent flow as $|\psi^{\circ}\rangle$ and the state of the surrounding environment as $\alpha_{j0}|0\rangle + \alpha_{j1}|1\rangle$, then the pattern state detection using the M matrix can be gracefully expressed as:

$$|\Psi_{eddy}\rangle\langle\Psi_{vorticity}| = M \cdot |\psi^0\rangle \otimes (\alpha_{j0}|0\rangle + \alpha_{j1}|1\rangle) \tag{12.26}$$

Here, the symbol $\otimes$ gracefully represents the tensor product, delicately combining the state of turbulent flow due to eddy and vorticity with the nuanced state of the surrounding environment. If M is a matrix that delicately acts on the turbulent flow state, this operation can be further expanded using the matrix − vector product. Let the matrix elements of M be denoted as M_{ij}. Consequently, the pattern state detection can be expressed in a refined manner as:

$$|\Psi_{eddy}\rangle\langle\Psi_{vorticity}| = \sum_i M_{ij} \cdot |\psi_i^0\rangle \otimes (\alpha_{j0}|0\rangle + \alpha_{j1}|1\rangle) \tag{12.27}$$

Here, ψ_i^0 symbolizes the individual components of the turbulent flow state $|\psi^{\circ}\rangle$. The outcome manifests as an exquisite superposition of states, each gracefully corresponding to a distinct component of the turbulent flow

TABLE 12.3 Pseudocode of Grover algorithm for detection of eddy and vorticity states in synthetic aperture radar data.

```
# Import necessary libraries import numpy as np

# Define the SAR data and parameters sar_data = load_sar_data() # Replace with your SAR data

loading function num_qubits = calculate_qubits_required(sar_data) # Adjust based on data

complexity

# Initialize quantum state quantum_state = initialize_state(num_qubits)

# Define Oracle for eddy and vorticity detection oracle = create_oracle(sar_data)

# Define Grover diffusion operator diffusion_operator = create_diffusion_operator()

# Set the number of Grover iterations num_iterations = determine_iterations(num_qubits)

# Apply Grover algorithm for_ in range(num_iterations): # Apply Oracle quantum_state =

apply_oracle(quantum_state, oracle)

# Apply Diffusion Operator quantum_state = apply_diffusion_operator(quantum_state,

diffusion_operator)

# Measure quantum state to obtain result result = measure_quantum_state(quantum_state)

# Post-process and interpret the result detected_features = process_result(result, sar_data)

# Display or use the detected features display_detected_features(detected_features)
```

state and its harmonious interaction with the surrounding environment. Certainly, let's articulate the 3-qubit operation based on the transformation Ψ^{00} in a clearer manner. Assuming that $|\psi_1\rangle$, $|\psi_2\rangle$, and $|\psi_3\rangle$ signify the levels of turbulent flow (strong, medium, and light), with $|\psi_1\rangle$ and $|\psi_2\rangle$ in the $|0\rangle$ state and $|\psi_3\rangle$ in the $|0\rangle + |1\rangle\beta|0\rangle + \alpha|1\rangle$ state, the mathematical description of this operation can be elegantly expressed as:

$$\Psi^{00}(\Psi_1, \Psi_2, \Psi_3) = \begin{bmatrix} \alpha \\ \beta \\ 0 \\ 0 \\ 0 \\ 0 \\ 0 \\ 0 \end{bmatrix} \tag{12.28}$$

Eq. (12.28) represents the different levels of the turbulent pattern flows$|\psi_1\rangle$, $|\psi_2\rangle$, and $|\psi_3\rangle$, respectively, in the SAR image. In line with Wang et al. (2005), evolutionary computing with qubit representation has a better characteristic of population diversity than other representations since it can represent a linear superposition of state probabilities. In binary representation, at least eight strings (000), (001), (010), (011), (100), (101), (110), and (111) are required, whereas the above single qubit is enough to depict eight states (Fig. 12.8).

In this view, the quantum evolutionary algorithm (QMEA) and quantum genetic algorithms (QGAs) operate

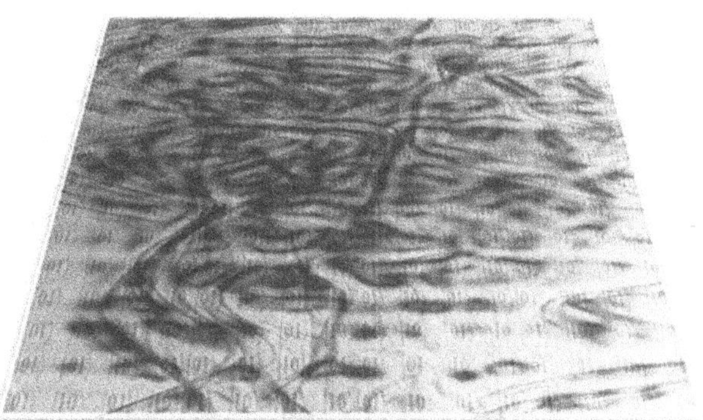

FIGURE 12.8 Qubits of eddy and vortex turbulent generations in SAR image. *SAR*, Synthetic aperture radar.

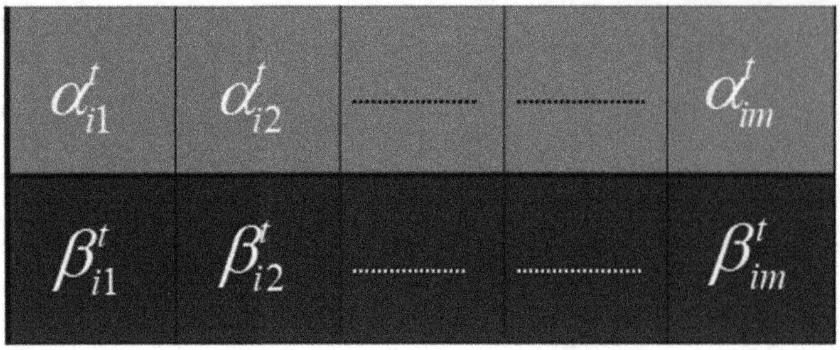

FIGURE 12.9 An array of one qubit populations.

efficiently on probability amplitudes of fundamental quantum states, whereas in the latter two (QEA), active crossovers and identified mutations are properly used to maintain population diversity (Hey, 1999; Kim et al., 2006).

The initial stage of QMEA is to generate accurate individual qubit populations of the turbulence flow levels in an SAR image, which can be signified as $|Q(t)> = \{|q^t_1>, |q^t_2>, ., |q^t_n>\}$. In this sense, let us consider that $|q^t_1> = \begin{pmatrix} \{|\alpha^t_{i1}>, |\alpha^t_{i2}>, ., ||\alpha^t_{im}>>\} \\ \{|\beta^t_{i1}>, |\beta^t_{i2}>, ., ||\beta^t_{im}>>\} \end{pmatrix}$ where i = 1,2,. . ..n is a qubit individual, n is the size of the population, and m is the span of the individual (Fig. 12.9).

12.10 Population pattern of quantum eddy and vorticity turbulent flow generations

In light of contemporary viewpoints, let us consider the constraint of the random qubit generation of the eddy and vorticity turbulence flows as $Q|N_0> \in |0,1>$. In this, the random number $|N_0>$ is confirmed as a turbulent flow if $Q|N_0>>|\alpha^t_i|^2$ in which the equivalent qubit equals 1, and is dissimilar when q^t_n is equivalent to 0, and turbulence does not exist in the SAR data. Consequently, the clustering center point of the turbulence is depicted by qubits |1> individually (Balicki et al., 2010; Bardin et al., 2020; Marghany, 2021).

12.10.1 Quantum fitness

The qubit chromosome form needs to be coded with the boundaries of the turbulence footprint. Every qubit in the qubit chromosome$|q_m>$ in this issue corresponds to a coefficient in the north-order surface fitting polynomial, which is given by:

$$f|i,j>_q = |q_0> + |q_1 i> + |q_2 j> + |q_3 i^2> + |q_4 ij> + |q_5 j^2> + . + |q_m j^n> \tag{12.29}$$

The connectedness metric is then employed to assess fitness. It is conceivable to mathematically express the level of the qubit chromosomes that are signed from adjacent pixels in the SAR image and clustered as the turbulent flow as follows:

$$conn|q^t> = \sum_{i=1}^{m}\left[\sum_{j=1}^{L}|q^t_{i,j}>\right] \tag{12.30}$$

Eq. (12.11) explains how to distinguish turbulent clusters in the environment of several qubit chromosomes L of m size. In this particular circumstance, each 10-qubit chromosome for every pixel closest to pixel i and its adjoining j-th is distinguished as the turbulent flows. In contrast, the absence of turbulent occurrence in SAR images and features belonging to the surrounding sea surface are recognized as well $|q^t_{ij}> = |j^{-1}>$.

12.10.2 Quantum mutation

The quantum mutation is carried out through the quantum rotation gate G(θ) with the assistance of the existing outstanding individual qubit, which behaves as a mutation operator in QEA (Fig. 12.10):

$$G(\theta) = \begin{bmatrix} \cos(\theta) & -\sin(\theta) \\ \sin(\theta) & \cos(\theta) \end{bmatrix} \tag{12.31}$$

Eq. (12.31) demonstrates the quantum rotation gate for searching in the quantum pixel in an SAR image (Fig. 12.6) as a function of rotation angle, which is cast as:

$$\theta = k \times f|\alpha_i, \beta_i> \tag{12.32}$$

In this case, the value of the coefficient k affects how quickly convergence arises. The number k must be selected wisely. The search grid of the algorithm would be large if k is too large, which would cause the solutions to diverge or prematurely converge to a local optimum. Conversely, if k is too small, the search grid would also be small, which would induce the algorithm to potentially become stuck. Additionally, $f|\alpha_i, \beta_i>$ maintains the direction of convergence of the search to a global optimum (Wang et al., 2005). The quantum gate determines whether the clustering of turbulence

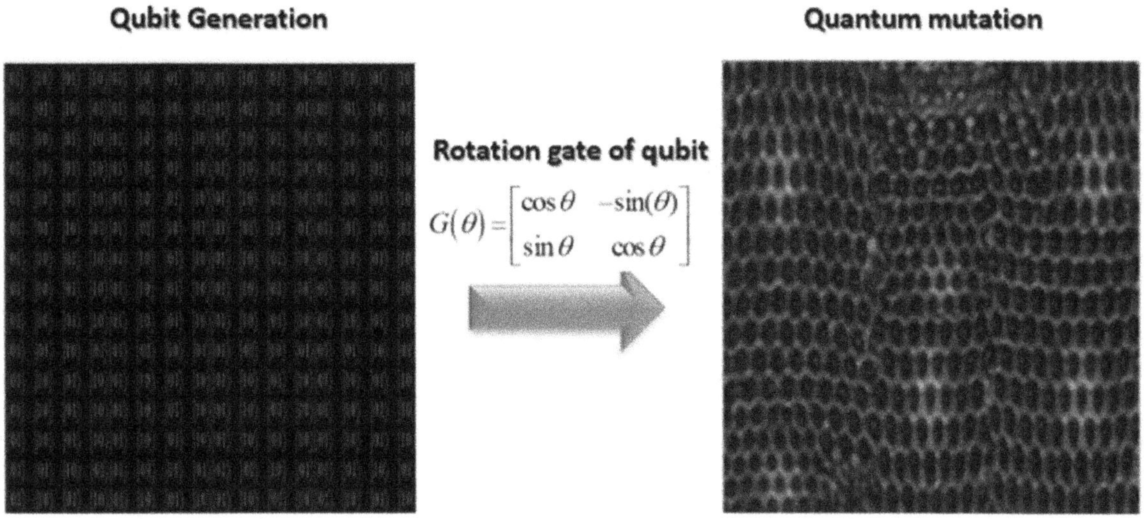

FIGURE 12.10 Quantum mutation of SAR image for eddy and vorticity turbulent detections. *SAR*, Synthetic aperture radar.

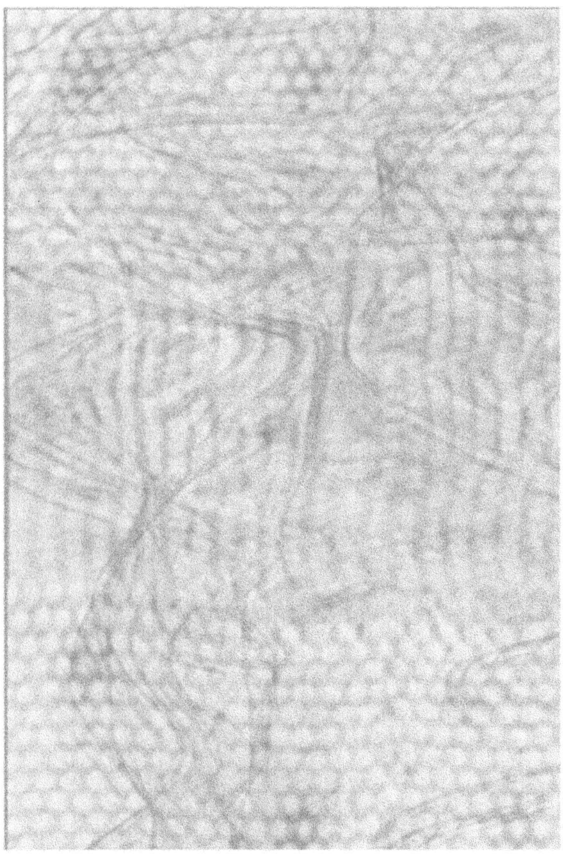

FIGURE 12.11 Quantum rotation gate for eddy and vorticity turbulent detections in SAR data. *SAR*, Synthetic aperture radar.

detection is a false alarm or a true clustering. Fig. 12.11 reveals how a quantum rotation gate can be applied to exhibit the turbulent detection as bright patches for the initial moment.

The rules in the opposite direction can be provided by Balicki et al. (2010), Marghany and Hakami (2019), Zhang and Korepin (2020):

$$|q_i^{t+1}> = G(t) \times |q_j^t> \tag{12.33}$$

In this sense, the probability amplitude of the qubit is signified by an individual population at t^{-th} creation as t is the evolutionary initiation at the quantum gate $G(t)$ (Beige et al., 2000; Cuomo et al., 2020; Marghany, 2021).

12.11 Pareto optimal solution for quantum nondominated sort and elitism

In brief, the enhanced iteration of the "Nondominated Sorting Genetic Algorithm" (NSGA-II), initially formulated by Prof. Kalyanmoy Deb to tackle nonconvex and nonsmooth single as well as multiobjective optimization challenges, has been further refined by Marghany (2021) to yield the evolved QNSGA-II. Let us denote the qubit chromosome as $Q|q_1^t, ..., q_n^t>$ for a population size of N individuals. To delve into the mathematical underpinnings of this model, we may express the QNSGA-II algorithm with an evolved set of equations. Consider the following representation:

$$|Q^{t+1}> = |Q^t> + |\Delta Q^t> \tag{12.34}$$

where $|Q^{t+1}>$ represents the qubit chromosome in the next iteration, $|Q^t>$ denotes the current qubit chromosome, and $|\Delta Q^t>$ signifies the change introduced in the chromosome at time step t. This dynamic equation encapsulates the quantum-inspired nature of the QNSGA-II algorithm, facilitating a nuanced exploration of solution spaces for nonconvex and nonsmooth optimization problems.

In the pursuit of advancing quantum multiobjective optimization solutions, let's formulate a quantum-inspired approach that can be juxtaposed against conventional methods for assessing the level of domination in the context of multiobjective problems.

Consider a quantum state representation for an individual in a population, denoted as $|Q_i\rangle$, where i corresponds to the index of an individual. In a quantum multiobjective optimization setting, we can define a quantum superposition state for a population P as follows:

$$|\Phi_P = \frac{1}{\sqrt{N}}\sum_{i=1}^{N}|q_i \tag{12.35}$$

Here, N represents the population size, and $|\Phi_P\rangle$ captures the collective quantum state of the population. This quantum superposition embodies the simultaneous consideration of multiple solutions in the search space. To assess the level of domination, a quantum-based dominance measure $\Delta_Q(q_i, q_j)$ can be introduced, incorporating quantum principles into the comparison of solutions. This measure might involve the quantum analog of dominance criteria, considering the interference effects inherent in quantum systems.

$$\Delta_Q(q_i, q_j) = \langle q_i|\hat{D}|q_j\rangle \tag{12.36}$$

Here $\hat{D}$ is a quantum operator representing the dominance criterion. The inner product $\langle q_i|\hat{D}|q_j\rangle$ captures the

TABLE 12.4 Pseudocode-based Pareto front for tracking eddy and vorticity in synthetic aperture radar image using quantum computing.

```
# Define the SAR image processing algorithm incorporating quantum computing concepts
# Initialization def initialize_population(Population_Size, Chromosome_Length): P =
generate_random_population(Population_Size, Chromosome_Length) return P
# Quantum Superposition def quantum_superposition(P): Q = apply_quantum_gate(P) return Q
# Pareto Front Determination def pareto_front(P): Pareto_Front = identify_pareto_front(P) return
Pareto_Front
# Quantum Evolution def quantum_evolution(Q, Pareto_Front): Updated_Q =
apply_quantum_operations(Q, Pareto_Front) return Updated_Q
# SAR Image Analysis def analyze_sar_image(P): # Perform SAR image analysis using classical methods
# Extract relevant features related to eddies and vorticity Features = extract_features(P) return
Features
# Main Quantum SAR Processing Loop def quantum_sar_processing(Population_Size,
Chromosome_Length, Max_Iterations): P = initialize_population(Population_Size,
Chromosome_Length)
 for iteration in range(Max_Iterations): Q = quantum_superposition(P) Pareto_Front = pareto_front(P)
Updated_Q = quantum_evolution(Q, Pareto_Front)
 # Convert quantum state to classical representation Updated_P =
measure_quantum_state(Updated_Q)
 # Analyze SAR image using classical methods Features = analyze_sar_image(Updated_P)
 # Further processing or visualization based on extracted features
 return Features
```

interference between the quantum states $\langle q_i|, and|q_j$ concerning the dominance criterion. To further refine this quantum multiobjective solution, one could explore the incorporation of quantum gates and operators that evolve the quantum state of the population over successive iterations, allowing for quantum parallelism and entanglement to enhance the exploration of Pareto fronts (Deb et al., 2000; Deb, 2001) (Table 12.4).

It is wort noting that the pseudocode assumes the existence of functions such as **generate_random_population**, **apply_quantum_gate**, **identify_pareto_front**, **apply_quantum_operations**, **measure_quantum_state**, and **extract_features**, which would need to be defined based on coherence of eddy and vorticity features in SAR image. The quantum operations and gates used in the pseudocode should be adapted to the quantum computing framework one is working with.

12.12 Boundary condition of quantum probability in tracking eddy and vorticity turbulence flows in synthetic aperture radar data

Certainly, let's express the mathematical boundary conditions that govern the quantum probability of the maximum sorting of the qubit chromosome, denoted as Pmax $|q_i\rangle$. This probability is associated with the quantum qubit chromosomes representing the probability amplitude of eddy and vorticity footprint in SAR data. The boundary conditions are as follows:

12.12.1 Normalization condition

$$\langle q_i|q_j\rangle = 1 \tag{12.37}$$

This condition ensures that the quantum state vector $|q_{ij}\rangle$ is properly normalized, representing a valid probability distribution.

12.12.2 Maximum sorting probability

$$P_{\max} = \max(|\langle q_i|q_j\rangle|^2) \tag{12.38}$$

Here, q_i represents individual qubit chromosomes in the quantum superposition. Pmax signifies the maximum sorting probability among these qubit chromosomes. In a multiobjective optimization context, the Pareto front represents a set of solutions that are nondominated, meaning no solution in the front is better than another in all objectives. Each solution in the Pareto front has a certain level of dominance. The maximum sorting probability ($P_{\max}$) refers to the highest probability among the qubit chromosomes in the quantum superposition. In the context of Pareto front, this can be associated with the quantum states representing optimal or nondominated solutions.

12.12.3 Eddy and vorticity probability amplitude

Let's express the eddy and vorticity probability amplitude in the context of quantum entanglement. In the language of quantum entanglement, the probability amplitude can be related to the entanglement between the quantum state $|\psi\rangle$ and the state representing eddy and vorticity features $|\text{eddy} - \text{vorticity state}\rangle|\text{eddy} - \text{vorticity state}\rangle$. The entanglement can be expressed using the Schmidt decomposition:

$$|\psi\rangle = \sum_i c_i|a_i\rangle \otimes |b_i\rangle \tag{12.39}$$

$$\psi|EV\rangle = \sum_j d_j|a_j^{'}\rangle \otimes |b_j^{'}\rangle \tag{12.40}$$

Here, $|a_i\rangle$ and $|b_i\rangle$ are basis states for the quantum system, and c_i are the corresponding coefficients. Similarly, $|a_j^{'}$ and $|b_j^{'}$ are basis states for the eddy – vorticity system, and $d_j^{'}$ are the coefficients. Therefore the entanglement between $|\psi\rangle$ and $|\text{eddy} - \text{vorticity state}\rangle$, that is, $|\text{EV}\rangle$ can then be quantified by the overlap of their coefficients:

$$P_{EV} = ||\sum_{i,j} c_i d_j \langle a_i|a_j^{'}\rangle\langle b_i|b_j^{'}\rangle||^2 \tag{12.41}$$

This expression captures the entanglement between the quantum state $|\psi\rangle$ and the state representing eddy and vorticity features $|\text{eddy} - \text{vorticity state}\rangle|\text{eddy} - \text{vorticity state}\rangle$, providing a quantum perspective on the probability amplitude associated with these features in the SAR data.

12.12.4 Consistency constraint

The condition $P_{\max} \geq P_{\text{eddy-vorticity}}$ establishes a crucial consistency between the quantum sorting process (represented by $P_{\max}$) and the presence of eddies and vorticity in the SAR data (represented by $P_{\text{eddy-vorticity}}$). On the other hand, if $P_{\max}$ is greater than or equal to $P_{\text{eddy-vorticity}}$, it implies that the quantum sorting mechanism is favoring qubit chromosomes with a high probability of representing eddy and vorticity features. This consistency ensures that the quantum algorithm is effectively identifying and prioritizing solutions that correspond to the desired features in the SAR data. In this perspective, the quantum superposition of states allows for the simultaneous consideration of multiple solutions. In the context of multiobjective optimization and Pareto front, this means considering a diverse set of nondominated solutions. By ensuring that the maximum sorting probability aligns with or exceeds the probability amplitude of the eddy and vorticity state, the quantum algorithm is effectively leveraging its ability to explore and exploit the solution space, emphasizing solutions that are not only Pareto optimal but also likely to exhibit eddy and vorticity characteristics in the SAR data.

It can be said that the condition $P_{\max} \geq P_{\text{eddy-vorticity}}$ establishes a coherent link between the quantum sorting process, Pareto optimality, and the representation of specific features (eddies and vorticity) in SAR data, ensuring that the quantum algorithm is guided toward solutions that align with the desired characteristics in a consistent manner. The pseudocode devised for formulating the boundary conditions of quantum probability, grounded in the principles of Pareto Front, is expounded in Table 12.5.

In this pseudocode, the **calculate_sorting_probability** and **calculate_eddy_vorticity_probability_amplitude** functions represent the quantum algorithms for determining the sorting probability and the probability amplitude for the eddy and vorticity state, respectively. The consistency condition is checked for each qubit chromosome in the Pareto Front, ensuring that the quantum algorithm prioritizes solutions that align with the desired eddy and vorticity features in a coherent manner. The **process_qubit_chromosome** function represents the subsequent steps you may want to take based on this consistency condition.

In this comprehension, the quantity of points constituting the Pareto front, contingent upon the optimized criteria for autonomously discerning turbulent flow within

TABLE 12.5 Pseudocode of modeling quantum probability for automatic detection of eddy and vorticity in synthetic aperture radar data.

```
# Assume there's a function to calculate sorting probabilities for each qubit chromosome def
calculate_sorting_probability(qubit_chromosome): # Implement the quantum algorithm to calculate
sorting probability # This could involve quantum gates, operations, and measurements
sorting_probability = quantum_sorting_algorithm(qubit_chromosome) return sorting_probability
# Assume there's a function to calculate the probability amplitude for eddy and vorticity state def
calculate_eddy_vorticity_probability_amplitude(qubit_chromosome): # Implement the quantum
algorithm to calculate probability amplitude # for the eddy and vorticity state
eddy_vorticity_amplitude = quantum_eddy_vorticity_algorithm(qubit_chromosome) return
eddy_vorticity_amplitude
# Main Quantum SAR Processing Loop def quantum_sar_processing(Population_Size,
Chromosome_Length, Max_Iterations): P = initialize_population(Population_Size,
Chromosome_Length)
 for iteration in range(Max_Iterations): Q = quantum_superposition(P) Pareto_Front = pareto_front(Q)
for qubit_chromosome in Pareto_Front: # Calculate sorting probability sorting_probability =
calculate_sorting_probability(qubit_chromosome) # Calculate probability amplitude for eddy and
vorticity state eddy_vorticity_amplitude =
calculate_eddy_vorticity_probability_amplitude(qubit_chromosome)
 # Check consistency condition if sorting_probability >= eddy_vorticity_amplitude: # Process the qubit
chromosome further or record relevant information process_qubit_chromosome(qubit_chromosome)
 return result
```

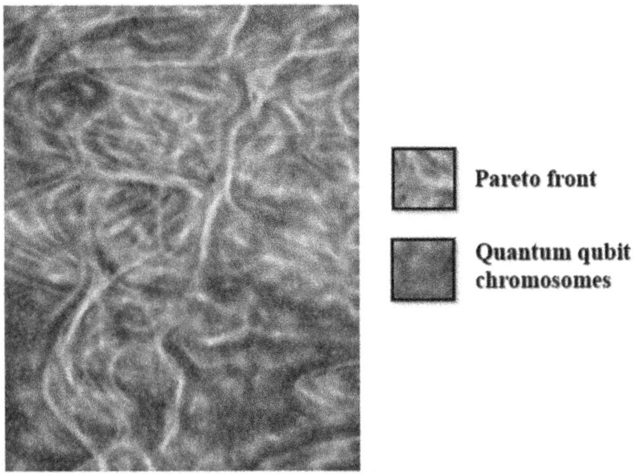

FIGURE 12.12 Pareto front curves elicited by quantum qubit chromosomes in SAR imagery. *SAR*, Synthetic aperture radar.

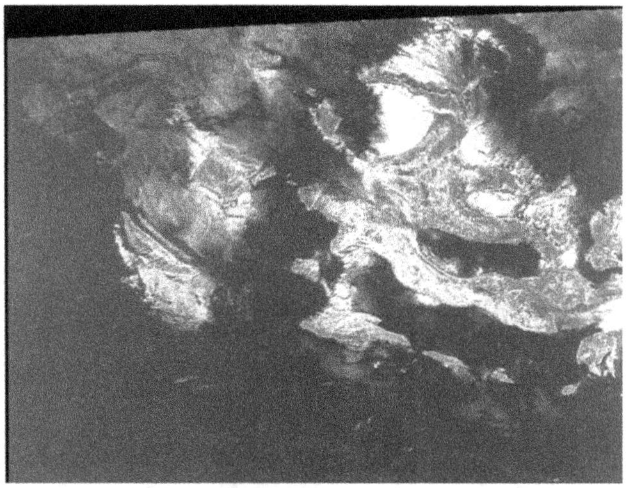

FIGURE 12.13 ALOS data acquired in this study. *ALOS*, Advanced Land Observing Satellite.

SAR data (as depicted in Fig. 12.12), adeptly traces the coherence of eddy and vorticity pixels in SAR images. Significantly, these entities are unequivocally precluded from the prospect of encountering quantum qubit chromosomes within the turbulence flow. In the realm of multiobjective optimization quandaries, the Pareto front embodies a collection of optimal solutions strategically employed for delineating trade-offs among diverse objectives, emphasizing prioritization over mere accuracy assessment. Researchers can judiciously select the most optimal solution, grounded in a profound comprehension of the Pareto front.

12.13 Automated identification of turbulent eddy and vorticity patterns in synthetic aperture radar imagery

The Advanced Land Observing Satellite (ALOS), which was launched on January 24, 2006, by the Japan Aerospace and Exploration Agency (JAXA), and which operated until May 12, 2011, provides the SAR data. ALOS data used here were acquired on March 22, 2007, around the Farasan Islands in Saudi Arabia's Red Sea coastal waters (Fig. 12.13). More precisely, the white box shown in Fig. 12.14 represents the ALOS data acquisition coverage area in the Red Sea.

ALOS (which carried PALSAR) was put into a sun-synchronous orbit and circumnavigated the earth every 100 minutes or 14 times throughout the day. Every 46 days, ALOS PALSAR made a return to the initial course (repeat cycle). At the equator, the interorbital distance was roughly 59.7 kilometers. The PALSAR was an active microwave sensor that could be operated day or night and was unaffected by the weather. The synthetic aperture radar (SAR) on Japan's first earth observation satellite served as its foundation (JERS-1). The PALSAR was a right-looking SAR that operated in the L-band and also had HH polarization and a cross-track aiming range of $8° - 60°$.

The coral reef − induced turbulent flow is investigated here using ALOS PALSAR data in the fine mode. The greatest ground resolution for fine mode data, which became 7 m, is one of the highest for an SAR (in contrast, the JERS-1 SAR had an approximately 18-meter resolution). With a look angle of 35°, a swath width of 70 km, a noise equivalent backscattering coefficient of − 25 dB, and a signal-to-ambiguity (S/A) ratio of 25 dB, the fine resolution mode provides a spatial resolution of 10 m in both range and azimuth for HH or VV with swath width of 70 km. However, the spatial resolution for full polarization HH + HV + VV + VH is 30 m with a swath width of 30 km.

The normalized radar backscatter cross section NRSC variation in the ALOS PALSAR image is demonstrated in Fig. 12.15. Over the Farasan Islands, the greatest NRSC of − 5dB occurs. The Farasan Islands contain sandy coasts, mangroves, and algae flats that reach their maximum extent near the coastline (Andréfouët et al., 2002), while reefs are only found offshore (Al-Anbaawy, 1993), which contributes to the maximum peak of NRSC. The Farasan Islands' coral reefs can generate an inconsistent pattern of water flow because of friction (Abu-Zied et al., 2011), which forces sea surface currents to modulate surface waves and create a very rough sea surface that produces the greatest SAR backscatter.

The ALOS PALSAR fine mode instrument has a reduced noise floor range of − 21dB, allowing for improved sea surface imaging. In particular, for the HH backscatter from turbulent flow cover, the instrument noise floor is only reached at the extreme edge of the swath.

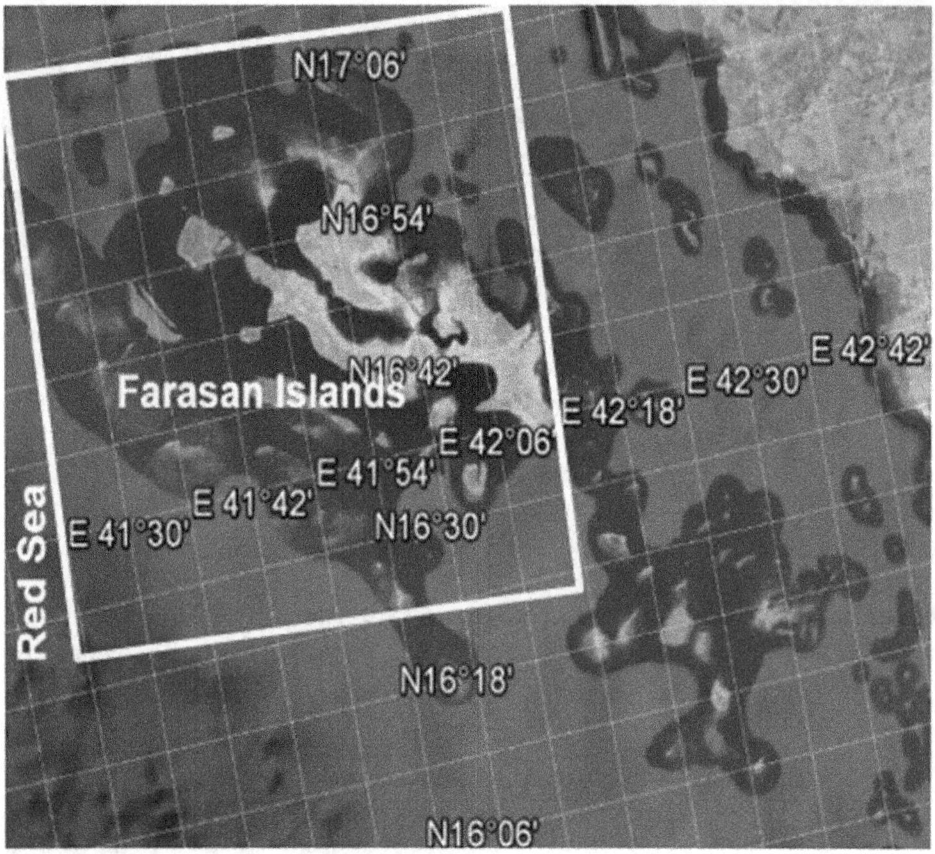

FIGURE 12.14 Google Earth map of Farasan Islands, Red Sea, and ALOS acquisition area (white box). *ALOS*, Advanced Land Observing Satellite.

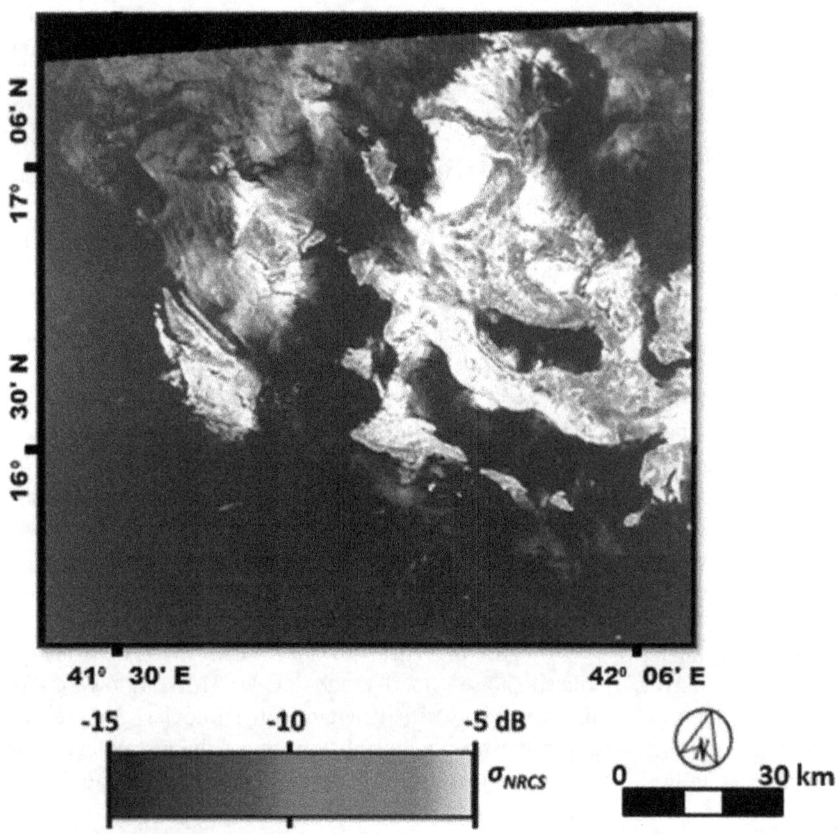

FIGURE 12.15 Normalized radar backscatter cross section (NRCS) variation in ALOS SAR image. *ALOS*, Advanced Land Observing Satellite; *SAR*, synthetic aperture radar.

This could also aid in determining the radar cross section of water utilizing L-band radar, when look-alikes such as alga blooms, low-wind zone, and high concentrations of chlorophyll are dampening the surface waves. Regarding parameter settings, certain quantities are used, including quantum population, encoding length, and maximum iteration number. To guarantee a precise consequence, we set N equal to 100, encoding length equal to 20, and iteration number equal to 100 (Marghany, 2013, 2021).

Fig. 12.16 depicts the results achieved using QMEA (Marghany, 2014a, 2014b). In this sense, QMEA can automatically distinguish between Bragg scattering and non-Bragg scattering (Marghany 2021). The dynamic variation between Bragg and non-Bragg scattering is precisely calculated. Bragg scattering tends to capture the uneven vorticity turbulent flow between the Farasan Islands (Marghany, 2014a, 2014b, 2015). The chaotic structure of the vorticity turbulent flow is distinctly twisted by QME in this regard (Fig. 12.11). This result agreed with previous work by Marghany and Hakami (2019), which used optical satellite data to track turbulent eddy flow. Indeed, PALSAR's unique characteristics would enable it to capture images of vorticity turbulence in areas where other sensors cannot. This would improve the accuracy of automatic detection systems and provide valuable data for weather forecasting and aviation safety (Singh et al., 2023).

In contrast to the antecedent studies conducted by Marghany (2015, 2021), the refined QNSGA-II expounded in this chapter exhibits proficiency in delineating vorticity chains and discerning their respective velocities, thereby assessing their propensity to induce turbulent flow. Remarkably, the vorticity velocities span from 0.2 to 0.6 m/s. Notably, these velocities, as corroborated by Reynolds number computations ranging from 3.0 to 9.0 $\times 10^5$, substantiate the capability of generating turbulence flow. This substantiation underscores the considerable potential inherent in the induction of turbulent phenomena.

Additionally, QNSGA-II can provide a variety of clusters in ALOS, such as spinning mesoscale eddies that envelop the sea surface (Fig. 12.17). As well as the iteration number increased to 100, N remains at 80 individual qubits, the encoding length is set to 20, and a crisp chaotic turbulent morphological flow is perfectly detected in ALOS data (Fig. 12.12).

Precise clusters of the turbulent morphological flow are obtained by increasing the number of variables. In other words, these clusters of turbulent morphology flow have demonstrated the range of turbulent morphology flow characteristics, from strong to mild flow (Fig. 12.18). In actuality, the water flow, which has a maximum Reynold number of 1.0 $\times 10^5$, creates erratic patterns as it strikes northern-western part of the Farasan

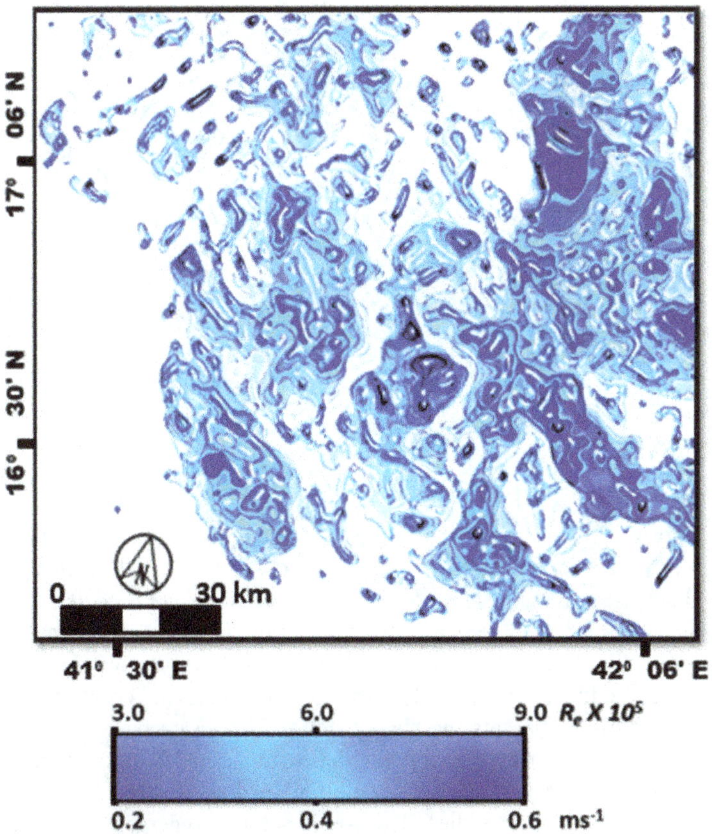

FIGURE 12.16 Vorticity chain flows between the Farasan Islands.

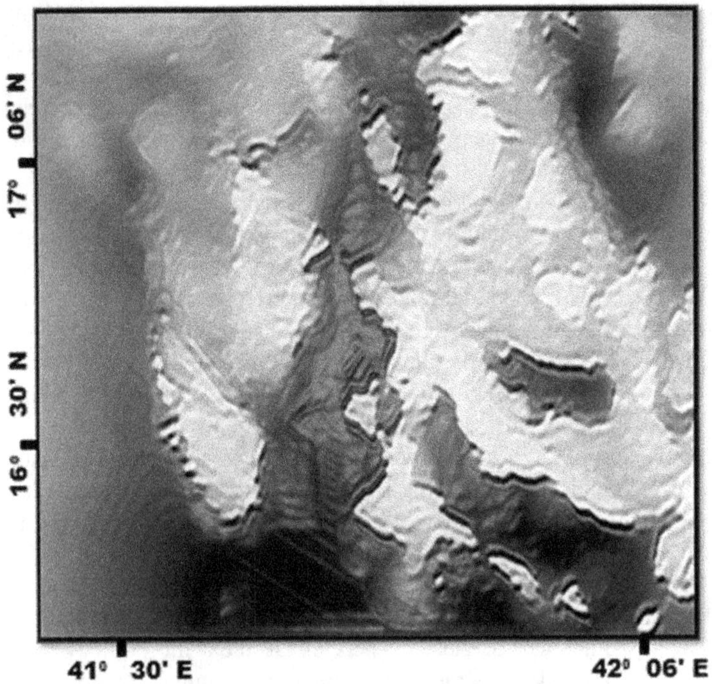

FIGURE 12.17 Mesoscale eddies detected using QNSGA-II.

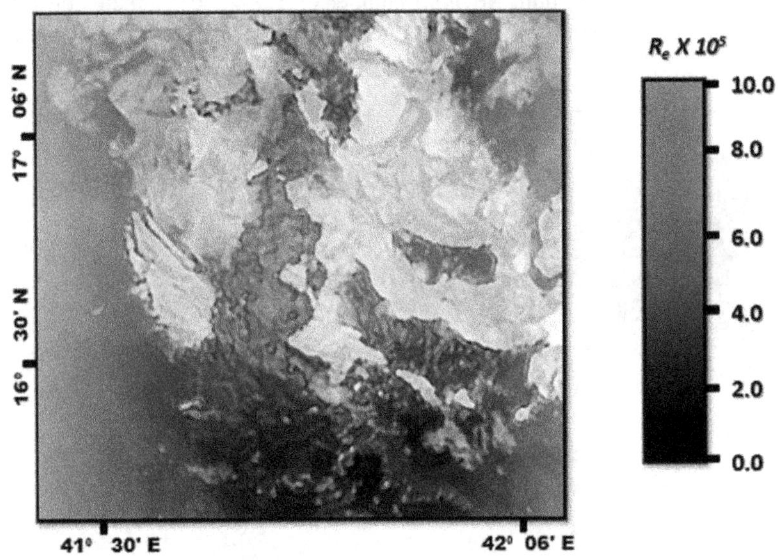

FIGURE 12.18 Simulation of eddy turbulence flow exploiting QNSGA-II.

Islands. Marghany and Hakami (2019) claim that the molecular diffusion that occurs at the coral – water interface results in a fluid viscosity that reduces turbulent motions and vertical transport. As noted by Marghany (2021), the synthesis or consumption of dissolved elements by the coral surface results in a steep gradient that forms a diffusive boundary layer. Henceforth, the obscurity region situated in the southern expanse of the island is governed by the influence of Wake Island, facilitating the dissipation of turbulent flows.

In the intricate dance of oceanic dynamics, the phenomenon of eddy flow around shallow coral reefs is a captivating subject of inquiry. Shallow coral reefs, with their submerged structures and intricate topography, induce distinctive hydrodynamic patterns that give rise to eddy formations. The irregular geometry of coral formations disrupts the steady flow of ocean currents. As oceanic currents encounter these submerged structures, they undergo complex interactions, leading to the creation of vortices or eddies. The corals act as submerged obstacles, prompting the fluid dynamics to adapt, causing the flow to curve around the coral formations. Moreover, the shallow depth characteristic of coral reef environments plays a pivotal role. In shallower waters, the influence of the seafloor on water movement becomes more pronounced. This interaction between the seafloor and the flowing water contributes to the generation of eddies. The shallow depths near coral reefs intensify these effects, fostering the development of eddy flows.

The intricate dance of eddies around shallow coral reefs is not merely a consequence of physical obstruction but a manifestation of the delicate balance between the intricate geometry of the reef structures and the dynamic forces exerted by the ocean currents. Understanding these processes is not only a testament to the beauty of natural phenomena but also a key to unraveling the complexities of fluid dynamics in coral reef ecosystems.

In essence, ALOS PALSAR detects variations in the roughness of a turbulent flow's surface and, for vigorous streams, variations in the conductivity of its surface layer. Because the vast majority of the radar energy that strikes a smooth surface is refracted away from the radar antenna, ALOS PALSAR "seems to see" substantial turbulence as a rough (radar-bright) patch against the smoother (radar-dark) ocean surface. The antenna was able to distinguish between strong and weak flows for ocean turbulence using a radar system because of the great sensitivity of the fine mode and other aspects. It is preferred to use HH polarization since its considerable tilt makes it susceptible to scattering from turbulent sea surfaces. Strong, moderate, and weak turbulent level flows are all analyzed quantitatively by QNSGA-II (Fig. 12.13).

In 100 iteration runs, QNSGA-II can therefore obtain the best solutions more frequently and with the highest clustering efficiency. The QNSGA-II runs stably and optimally. The consistent result of QNSGA-II, which shows a greater number of distinct clusters with turbulent flows tied to coral reefs, seems instantly to be precisely classified (Fig. 12.14).

The QNSGA-II algorithm reveals the possible interaction between water currents and coral reefs. Additionally, QNSGA-II is capable of reliably distinguishing between the island, coral reefs, and turbulence flow. It is possible to accurately confirm the profound influence of boundary movement on the generation of turbulent flow intensity, such as eddy, vorticity, and steady flows with QNSGA-II (Fig. 12.19). Properly compared with the irregular surface of the flowing sea, the turbulent flow typically retains a more marked degree of coherence.

According to the aforementioned perspective, irregular reefs are a common feature at shallow water depths between 5 and 20 m along the Farasan Islands. In this sense, the main surface current flow with a maximum velocity of 1 m/s along the Farasan Islands coastal waters modifies the tide. Therefore the coral reefs, sand bars, low islands, and shoals can cause turbulent flow with a maximum Reynold number of 10×10^5 when a tidal current passes through. This eddy turbulent flow can cause the development of chaotic eddies in shallow water areas and thus has a significant effect on the hydrodynamics of the Farasan Islands' coastal waters. Consequently, irregular reefs in shallow water can significantly alter the hydrodynamic patterns and cause also vorticity chain flows chaotic (Fig. 12.19). This finding agrees with the study of Marghany and Hakami (2019).

In the realm of advanced optimization algorithms, the QNSGA-II continues to demonstrate its prowess, exhibiting a capacity for analogous cluster detection in the intricate dynamics of eddy and vorticity phenomena along the Mozambique Channel. This discernment is notably drawn from an analysis of Sentinel HH polarization data acquired on the precise date of August 27, 2023.

Evidently, the QNSGA-II exhibits proficiency in delineating not only the boundary peripheries of mesoscale

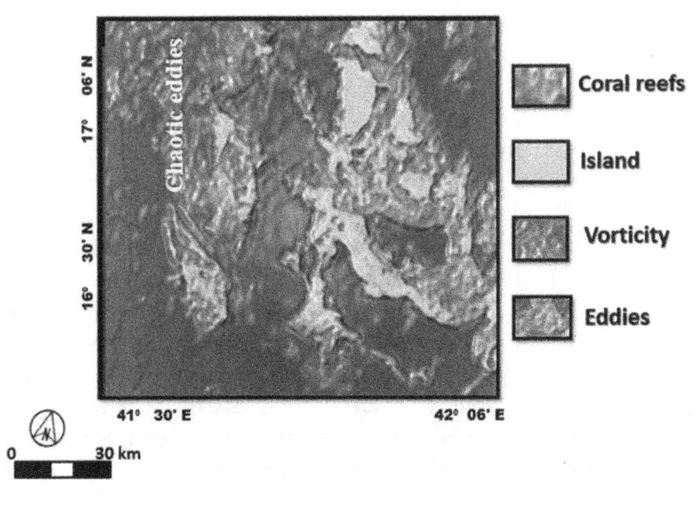

FIGURE 12.19 QNSGA-II automatically detects turbulent pattern flows caused by coral reefs.

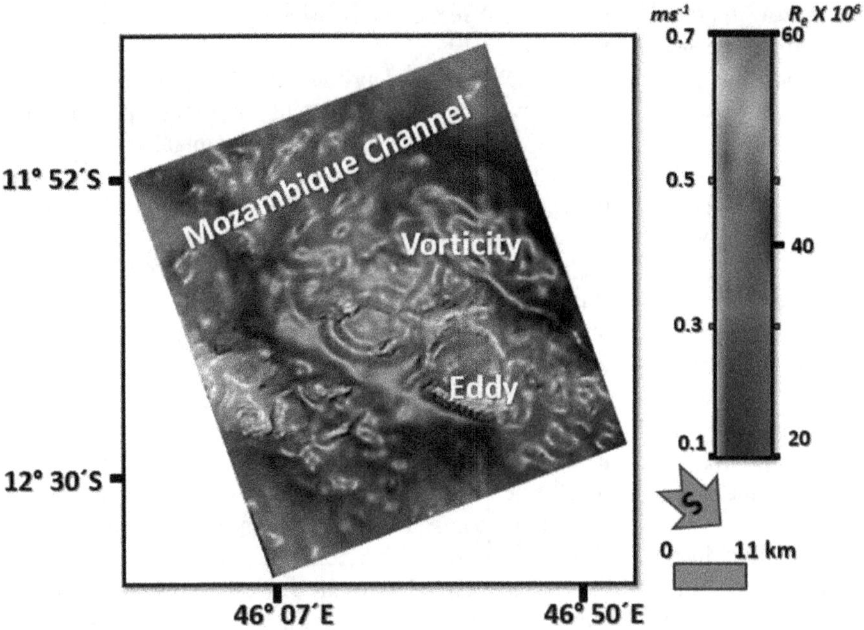

FIGURE 12.20 Automated identification of turbulent eddy and vorticity patterns in Sentinel-A imagery through QNSGA-II.

eddies but also the intricate flows within vorticity chains. These flows, propelled by a commendable velocity of 0.7 m/s, engender turbulence patterns, as exemplified by the calculated maximum Reynolds number reaching an impressive magnitude of 60.0×10^6. This intriguing interplay unfolds across a spatial expanse spanning 88 km, concurrently giving rise to mesoscale eddies characterized by a discerned radius of 11 km (Fig. 12.20).

The discernment and quantification achieved through the QNSGA-II algorithm shed light on the nuanced dynamics inherent in the Mozambique Channel, offering a valuable lens for comprehending the intricate interplay of eddies, vorticity, and turbulence within this maritime domain.

12.14 The significance of Pareto optimization in QNSGA-II

In the realm of optimization, the QNSGA-II algorithm stands as a remarkable fusion of local scrutiny and global exploration, affording a heightened precision in addressing optimal solutions when contrasted with traditional evolutionary algorithms. This algorithm, adept in both function optimization and multiuser detection within code division multiple access, finds applicability through its integration with quantum computing, enhancing accuracy through a genetic algorithm (Marghany, 2014a, 2014b, 2015, 2021). The culmination of these advancements is particularly noteworthy in the domain of quantum image processing, where QNSGA-II plays a pivotal role in advancing SAR image segmentation and processing.

QNSGA-II's prowess extends to the nuanced identification of morphological boundaries within SAR data, distinguishing turbulent flow magnitudes across varying segmentation layers, ranging from formidable to subtle. Notably, QNSGA-II doesn't yield a singular solution but rather a set of compromised solutions known as Pareto optimal solutions. This selection of alternatives allows decision-makers to make informed choices based on specific criteria (Beige et al., 2000; Deb et al., 2000; Marghany, 2014a). This methodology, classified as an a posteriori approach, underscores that decisions are made post the completion of the search process (Deb, 2001; Marghany, 2015; Marghany & Hakami, 2019).

Crucially, the quantum Pareto-optimization approach deployed by QNSGA-II eliminates the need for a priori preference decisions amid conflicting objectives, such as eddy and vorticity turbulent flow magnitudes, coral reefs, and delineations between land and sea. Quantum Pareto optimal points coalesce into a Pareto front within the multi-objective function of SAR feature space, as evidenced in Figs. 12.21 and 12.22, facilitating a comprehensive and nuanced understanding of the intricate SAR landscape.

The depiction of nondominated solutions across various algorithms is elucidated in Figs. 12.20 and 12.21. A discerning analysis reveals the notable superiority of QNSGA-II's solution (Fig. 12.21) in comparison to that of QMEA (Fig. 12.16). Noteworthy is the concentration of QMEA's solutions around the central axis of the Pareto front, signifying a tendency to focus on a specific region within the front. Conversely, QNSGA-II, in its pursuit of optimal solutions for turbulent flow magnitudes, coral reefs, and the demarcation of land and sea boundaries in

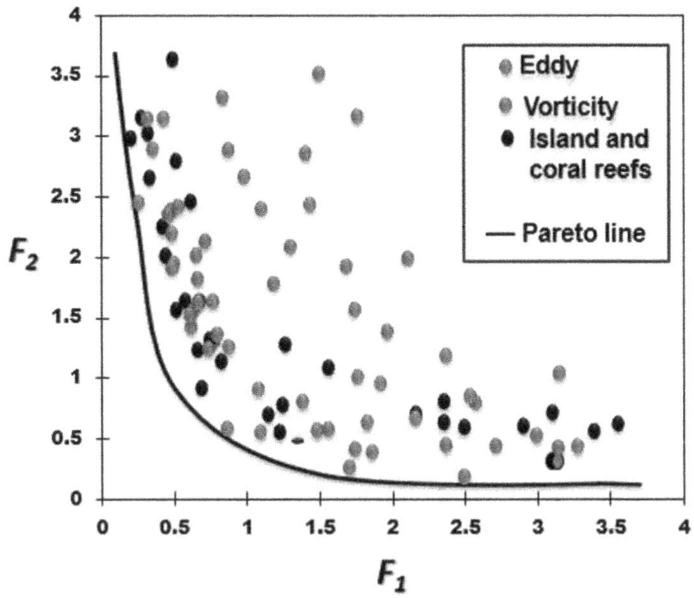

FIGURE 12.21 Nondominated solutions provided by QMEA.

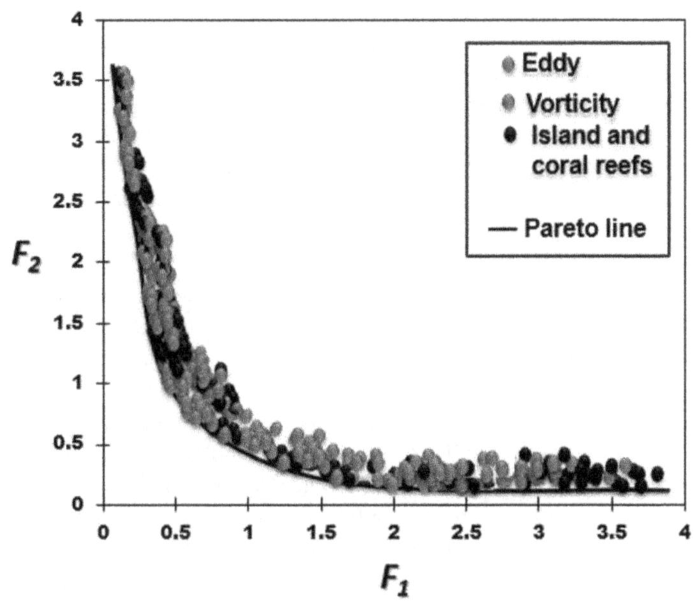

FIGURE 12.22 Nondominated solutions delivered by QNSGA-II.

SAR data, maintains a commendable degree of diversity in its solutions.

The capability of QNSGA-II to efficiently distribute its population along the acquired Pareto front eclipses that of QMEA. This is emblematic of QNSGA-II's prowess in leveraging quantum computing principles, including uncertainty, superposition, and interference, to enhance the quality and diversity of the nondominated set in multiobjective problems involving polarimetric SAR features. QNSGA-II's adeptness in automatic detection of eddy and vorticity turbulent flow magnitudes, coral reefs, and surrounding sea footprint boundaries is further highlighted by its accurate positioning in proximity to the Pareto optimal front (Fig. 12.21).

In essence, QNSGA-II not only refines the accuracy of the Pareto optimal front but also preserves a comprehensive diversity by embodying the compensatory mechanisms intrinsic to a quantum-inspired evolutionary algorithm (QEA). This nuanced approach aims at improving the proximity to the Pareto optimal front, thereby facilitating the identification of superior solutions, deemed virtuous individuals through a fitness function.

12.15 Quantum coherence in synthetic aperture radar imaging: exploring turbulent flow magnitudes with QNSGA-II

In the intricate dance of eddy and vorticity turbulent flow magnitudes, a realm of robust coherence unveils itself. Within the realm of quantum mechanics, this coherence manifests as an amplification in the strength of phase angles among the constituents of eddy and vorticity turbulent flows and the encompassing sea surface—a phenomenon akin to a quantum superposition.

This enigmatic coherence carries profound implications, defying conventional or probabilistically additive behaviors. Quantum coherence introduces the intriguing concept of wave function collapse, where the interplay involves both Bragg scattering and non-Bragg scattering. Essentially, it signifies the evolution of the physical attributes of sea roughness backscatter into a singular manifestation of eddy and vorticity turbulent flow magnitudes, akin to coherent pixels painted across the canvas of SAR images.

In this esoteric realm, the speculation of coherence in eddy and vorticity turbulent flow magnitudes within SAR data serves as the mechanism dictating the intricate boundaries between the quantum and classical domains—specifically, the domain of Bragg scattering. This coherence materializes as turbulent flow interacts with its surroundings in a peculiarly delectrical and irreversible manner, characteristic of a substance exhibiting a meager conductivity for electric current.

Within this context, the robust turbulence, possessing higher delectrical properties than its surroundings, establishes a formidable shield against the interplay of Bragg and non-Bragg scattering within the quantum superposition of the entire SAR scene's wave function. It's a symphony of interference, or rather the lack thereof, where the strong and eddy and vorticity turbulent flows play the lead role in orchestrating the dance of quantum coherence on the stage of SAR imagery.

In the realm of SAR data, the manifestation of a higher strength of backscatter within eddy and vorticity turbulent zones in comparison to their surroundings finds its roots in the intricate dance of coherence. This elevated strength, a testament to the quantum nature of eddy and vorticity turbulent flows, orchestrates a unique interplay where components of the wave function harmoniously amalgamate from the coherent sea surface and acquire phases from their immediate surroundings.

This nuanced interplay results in a total superposition of the global or universal wave function, unwaveringly coherent at the expansive level of sea roughness. The consequence is an immensely accurate depiction of eddy and vorticity turbulent flows within the SAR data, unraveling the complexities with precision. QNSGA-II emerges as a pivotal player in this landscape, contributing significantly to the deciphering of modulated hydrodynamics information embedded in eddy and vorticity turbulent flow pixels.

Beyond its prowess in deciphering turbulent flow, QNSGA-II addresses computational gridlock challenges faced by conventional spectral clustering algorithms in large-scale SAR image segmentation. Through an innovative encoding technique, QNSGA-II adeptly denotes eddy and vorticity turbulent flows representative of pixels and boundaries, optimizing their selection for enhanced clarity. The calculated affinity function, governing the clustering of turbulent and surrounding environments, efficiently reduces computation periods, enabling the handling of vast amounts of SAR data.

QNSGA-II's unique approach extends to the meticulous counting of a substantial number of eigenvalues, with a value of one, facilitating precise determination of cluster numbers across SAR data. This quantum-inspired algorithm, with its qubit distribution, achieves exceptional separation between diverse clusters, accurately identifying boundaries between eddy and vorticity turbulent flow clusters. The orchestrated decline in eigenvalues, particularly the pronounced gap between the second and third eigenvalues, illuminates the ease with which three or two classes of eddy and vorticity turbulent flows can be distinctly discerned. In essence, the presence of qubit distribution within QNSGA-II unfolds as a beacon, illuminating the path to unparalleled precision in turbulent flow analysis.

This chapter convincingly demonstrates a novel technique for the automatic detection of the level of eddy and vorticity flows using SAR data such as ALOS PALSAR and Sentinel HH polarization data. To this end, the quantum multiobjective algorithm is developed and examined in an SAR image, which was acquired over the Farasan Islands in the Red Sea and Mozambique Channel. In short, SAR backscatter and texture information is instantly turned into qubit chromosomes. In conclusion, quantum multiobjective evolution (QMEA) employs a novel qubit chromosomal depiction based on the modern notion and conceptual frameworks of quantum computing, including qubits and nonlinear superposition. The quantum multiobjective evolutionary (QMEA) can efficiently deliver a distinguishable classification map of turbulent flow without distinct boundaries and edges.

In this practical sense, quantum nondominated sort and elitism (QNSGA-II) is properly used to boost the efficiency of QMEA for automatic detection of the particular cause for turbulent flow in the Farasan Islands in ALOS PALSAR data. The efficient Pareto Front algorithm is implemented to accurately determine the most effective optimal solution for the automatic detection of turbulence and surrounding features. Accurately, the boundaries of the eddy and vorticity turbulent footprints are more accurately and automatically clustered and identified than QMEA. The surrounding sea features coral reefs and the local land, Wake island, and low-wind zone, and these are perfectly distinguished from the surrounding gradient variation of the eddy and vorticity turbulent flows.

Therefore the QNSGA-II can also better detect the varying level of eddy and vorticity turbulence flows as a function of Reynold number, than QMEA. In addition, QNSGA-II can show that the coral reefs are the leading cause of eddy and vorticity turbulent flow around the Farasan Islands. This accuracy is well shown by the Pareto front of value 3.7 in both dimensions of the front one (F_1) and front two (F_2) axes. In conclusion, QNSGA-II represents a promising tool for the automatic detection of eddy and vorticity turbulent flows in shallow coastal island waters such as the Farasan Islands and also in Mozambique Channel. This approach can be further extended to other coastal regions for efficient monitoring and management of water resources. However, the effectiveness of QNSGA-II may be influenced by the availability and quality of input data.

References

Abu-Zied, R. H., Bantan, R. A., Basaham, A. S., El Mamoney, M. H., & Al-Washmi, H. A. (2011). Composition, distribution, and taphonomy of nearshore benthic foraminifera of the Farasan Islands, Southern Red Sea, Saudi Arabia. *The Journal of Foraminiferal Research.*, *41* (4), 349–362. Available from https://doi.org/10.2113/gsjfr.41.4.349.

Al-Anbaawy, M. I. H. (1993). Reefal facies and diagenesis of pleistocene sediments in Kamaran Island, southern Red Sea, Republic of Yemen. *Egyptian Journal of Geology*, *37*(2), 137–151.

Andréfouët, S., Mumby, P. J., McField, M., Hu, C., & Muller-Karger, F. E. (2002). Revisiting coral reef connectivity. *Coral Reefs.*, *21*(1), 43–48. Available from https://doi.org/10.1007/s00338-001-0199-0, http://link.springer.de/link/service/journals/00338/index.htm.

Balicki, J. M., Balicka, H. T., Masiejczyk, J., Zacniewski, A. (2010). European Conference of Communications, ECCOM'10, ECCS'10 328–333 Poland Multi-criterion decision making in distrbiuted systems by quantum evolutionary algorithms. European Conference of Systems. ECS'10, European Conference of Circuits Technology and Devices, ECCTD'10.

Bardin, J. C., Sank, D., Naaman, O., & Jeffrey, E. (2020). Quantum computing: An introduction for microwave engineers. *IEEE Microwave Magazine.*, *21*(8), 24–44. Available from https://doi.org/10.1109/MMM.2020.2993475, https://ieeexplore.ieee.org/servlet/opac?punumber = 6668.

Bauer, M., Kolev, B., & Preston, S. C. (2016). Geometric investigations of a vorticity model equation. *Journal of Differential Equations.*, *260*(1), 478–516. Available from https://doi.org/10.1016/j.jde.2015.09.030, http://www.elsevier.com/inca/publications/store/6/2/2/8/6/8/index.htt.

Beige, A., Braun, D., Tregenna, B., & Knight, P. L. (2000). Quantum computing using dissipation to remain in a decoherence-free subspace. *Physical Review Letters*, *85*(8), 1762–1765. Available from https://doi.org/10.1103/PhysRevLett.85.1762.

Browand, F. K., & Troutt, T. R. (1985). The turbulent mixing layer: Geometry of large vortices. *Journal of Fluid Mechanics*, *158*, 489–509. Available from https://doi.org/10.1017/s0022112085002737.

Cassianides, A., Lique, C., & Korosov, A. (2021). Ocean eddy signature on SAR-derived sea ice drift and vorticity. *Geophysical Research Letters*, *48*(6). Available from https://doi.org/10.1029/2020GL092066, http://agupubs.onlinelibrary.wiley.com/hub/journal/10.1002/(ISSN)1944-8007/.

Cuomo, D., Caleffi, M., & Cacciapuoti, A. S. (2020). Towards a distributed quantum computing ecosystem. *IET Quantum Comm.*, *1*, 3–8.

Deb, K. (2001). Nonlinear goal programming using multi-objective genetic algorithms. *Journal of the Operational Research Society.*, *52* (3), 291–302. Available from https://doi.org/10.1057/palgrave.jors.2601089.

Deb, K., Agrawal, S., Pratap, A., & Meyarivan, T. (2000). A fast elitist non-dominated sorting genetic algorithm for multi-objective optimization: NSGA-II. *Lecture Notes in Computer Science*, *1917*, 849–858. Available from https://doi.org/10.1007/3-540-45356-3_83.

Diao, Z., Zubairy, M. S., & Chen, G. (2002). A quantum circuit design for Grover's algorithm. *Zeitschrift für Naturforschung A.*, *57*(8), 701–708. Available from https://doi.org/10.1515/zna-2002-0810.

Escher, J., & Wunsch, M. (2012). Restrictions on the geometry of the periodic vorticity equation. *Communications in Contemporary Mathematics.*, *14*(03), 1250016. Available from https://doi.org/10.1142/s0219199712500162.

Escher J., Kolev B., Wunsch M., The geometry of a vorticity model equation. (2010).

Han, K. H., & Kim, J. H. (2004). Quantum-inspired evolutionary algorithms with a new termination criterion, H/sub/spl epsi//gate, and two-phase scheme. *IEEE Transactions on Evolutionary Computation*, *8*(2), 156–169.

Hey, T. (1999). Quantum computing: An introduction. *Computing and Control Engineering Journal*, *10*(3), 105–112. Available from https://doi.org/10.1049/cce:19990303.

Hunt, J. C. R. (1987). Vorticity and vortex dynamics in complex turbulent flows. *Transactions of the Canadian Society for Mechanical Engineering.*, *11*(1), 21–35. Available from https://doi.org/10.1139/tcsme-1987-0004.

Ivanov, A. Y., & Ginzburg, A. I. (2002). Oceanic eddies in synthetic aperture radar images. *Journal of Earth System Science.*, *111*(3), 281–295. Available from https://doi.org/10.1007/bf02701974.

Jiao, L., Li, Y., Gong, M., & Zhang, X. (2008). Quantum-inspired immune clonal algorithm for global optimization. *IEEE Transactions on Systems, Man, and Cybernetics, Part B (Cybernetics)*, *38*(5), 1234–1253.

Karimova, S., & Gade, M. (2016). Improved statistics of sub-mesoscale eddies in the Baltic Sea retrieved from SAR imagery. *International Journal of Remote Sensing*, *37*(10), 2394–2414. Available from https://doi.org/10.1080/01431161.2016.1145367, https://www.tandfonline.com/loi/tres20.

Kaye, P., Laflamme, R., & Mosca, M. A. (2007). *Introduction to quantum computing*. Oxford University Press.

Kim Y., Kim J. H., Han K. H. 2006 EEE Congress on Evolutionary Computation, CEC 2006 2601–2606 South Korea Quantum-inspired multiobjective evolutionary algorithm for multiobjective 0/1 knapsack problems.

Lavor C., Manssur L. R. U., Portugal R., Grover's algorithm: Quantum database search. (2003).

Maddison, J. R., Marshall, D. P., & Shipton, J. (2015). On the dynamical influence of ocean eddy potential vorticity fluxes. *Ocean*

Modelling, *92*, 169–182. Available from https://doi.org/10.1016/j.ocemod.2015.06.003, http://www.elsevier.com/inca/publications/store/6/0/1/3/7/6/index.htt.

Majda, A. (1986). Vorticity and the mathematical theory of incompressible fluid flow. *Communications on Pure and Applied Mathematics.*, *39*(S1), S187–S220. Available from https://doi.org/10.1002/cpa.3160390711.

Mansfield, J. R., Knio, O. M., & Meneveau, C. (1998). A dynamic LES scheme for the vorticity transport equation: Formulation and a priori tests. *Journal of Computational Physics*, *145*(2), 693–730. Available from https://doi.org/10.1006/jcph.1998.6051, http://www.journals.elsevier.com/journal-of-computational-physics/.

Marghany, M. (2013). Genetic algorithm for oil spill automatic detection from envisat satellite data. *Lecture Notes in Computer Science*, *7972*, 587–598. Available from https://doi.org/10.1007/978-3-642-39643-4_42.

Marghany, M. (2014a). Multi-objective evolutionary algorithm for oil spill detection from COSMO-SkeyMed satellite. *Lecture Notes in Computer Science*, *8584*, 6355–6371. Available from https://doi.org/10.1007/978-3-319-09153-2_27, http://springerlink.com/content/0302-9743/copyright/2005/.

Marghany, M. (2014b). Utilization of a genetic algorithm for the automatic detection of oil spill from RADARSAT-2 SAR satellite data. *Marine Pollution Bulletin*, *89*(1–2), 20–29. Available from https://doi.org/10.1016/j.marpolbul.2014.10.041, http://www.elsevier.com/locate/marpolbul.

Marghany, M. (2015). Proceedings Asian Association on Remote Sensing Malaysia Flock 1 data Multi-Objective Evolutionary Algorithm for turbulent flow detection. 1 ACRS 2015 - 36th Asian Conference on Remote Sensing: Fostering Resilient Growth in Asia.

Marghany, M. (2021). *Nonlinear ocean dynamics: Synthetic aperture radar*. Elsevier.

Marghany, M., & Hakami, M. (2019). *Automatic detection of coral reef induced turbulent boundary flow in the Red Sea from Flock-1 satellite data* (pp. 105–122). Springer Science and Business Media LLC. Available from https://doi.org/10.1007/978-3-319-99417-8_6.

Marshall, D. P., Maddison, J. R., & Berloff, P. S. (2012). A framework for parameterizing eddy potential vorticity fluxes. *Journal of Physical Oceanography*, *42*(4), 539–557. Available from https://doi.org/10.1175/JPO-D-11-048.1, http://journals.ametsoc.org/doi/pdf/10.1175/JPO-D-11-048.1.

Marshall, D. P., Williams, R. G., & Lee, M. M. (1999). The relation between eddy-induced transport and isopycnic gradients of potential vorticity. *Journal of Physical Oceanography*, *29*(7), 1571–1578. Available from http://journals.ametsoc.org/loi/phoc, https://doi.org/10.1175/1520-0485(1999)029 < 1571:TRBEIT > 2.0.CO;2.

Meitz, H. L., & Fasel, H. F. (2000). A compact-difference scheme for the navier-stokes equations in vorticity-velocity formulation. *Journal of Computational Physics*, *157*(1), 371–403. Available from https://doi.org/10.1006/jcph.1999.6387, http://www.journals.elsevier.com/journal-of-computational-physics/.

Saffman, P. G. (1981). Dynamics of vorticity. *Journal of Fluid Mechanics*, *106*, 49–58. Available from https://doi.org/10.1017/S0022112081001511.

Shutts, G. J. (1983). The propagation of eddies in diffluent jetstreams: Eddy vorticity forcing of 'blocking' flow fields. *Quarterly Journal of the Royal Meteorological Society*, *109*(462), 737–761. Available from https://doi.org/10.1002/qj.49710946204.

Singh, A., Kushwaha, S. K. P., Nandy, S., Padalia, H., Ghosh, S., Srivastava, A., & Kumari, N. (2023). Aboveground forest biomass estimation by the integration of TLS and ALOS PALSAR data using machine learning. *MDPI, Czech Republic Remote Sensing*, *15*(4). Available from https://doi.org/10.3390/rs15041143, http://www.mdpi.com/journal/remotesensing.

Truesdell, C. (1953). Two measures of vorticity. *Indiana University Mathematics Journal.*, *2*(2), 173–217. Available from https://doi.org/10.1512/iumj.1953.2.52009.

Truesdell, C. (2018). *The kinematics of vorticity*. Courier Dover Publications.

Wallace, J. M., & Foss, J. F. (1995). The measurement of vorticity in turbulent flows. *Annual Review of Fluid Mechanics*, *27*(1), 469–514. Available from https://doi.org/10.1146/annurev.fl.27.010195.002345, http://arjournals.annualreviews.org/loi/fluid.

Wang, L., Tang, F., & Wu, H. (2005). Hybrid genetic algorithm based on quantum computing for numerical optimization and parameter estimation. *Applied Mathematics and Computation*, *171*(2), 1141–1156. Available from https://doi.org/10.1016/j.amc.2005.01.115.

Wang, Y., Yang, M., & Chong, J. (2019). Simulation and analysis of SAR images of oceanic shear-wave-generated eddies. *Sensors*, *19*(7), 1529. Available from https://doi.org/10.3390/s19071529.

Williams, C. P. (2011). *Quantum cryptography* (pp. 507–563). Springer Science and Business Media LLC. Available from https://doi.org/10.1007/978-1-84628-887-6_13.

Wu, J. Z., Ma, H. Y., & Zhou, M. D. (2007). *Vorticity and vortex dynamics*. Springer Science & Business Media.

Wu, J. Z., & Wu, J. M. (1996). Vorticity dynamics on boundaries. *Advances in Applied Mechanics*, *32*(C), 119–275. Available from https://doi.org/10.1016/S0065-2156(08)70077-2, http://www.elsevier.com/wps/find/bookdescription.cws_home/704246/description#description.

Zhang, K., & Korepin, V. E. (2020). Depth optimization of quantum search algorithms beyond Grover's algorithm. *Physical Review. A*, *101*(3). Available from https://doi.org/10.1103/PhysRevA.101.032346, http://harvest.aps.org/bagit/articles/10.1103/PhysRevA.101.032346/apsxml.

Chapter 13

Four-dimensional quantum holographic interferometry radar for nonlinear shear-wave chaotic flow detection

Turbulence within a chaotic system is an intricate and unpredictable occurrence characterized by irregular and disorderly fluid movement. In SUCH systems of turbulence flows that are ruled by chaos, the behavior displays a heightened understanding of primary circumstances, wherein slight alterations in the starting state can result in significantly divergent outcomes over time, which stands as a distinctive feature of chaotic systems. Within the realm of fluid dynamics, turbulence emerges when a fluid, be it air or water, undergoes irregular and chaotic motion. This phenomenon can arise from various factors, including the Reynolds number surpassing a critical threshold, prompting the laminar flow to be exchanged to turbulent flow as discussed early in Chapter 2.

Turbulence in chaotic systems manifests through vortices, eddies, and fluctuations across multiple scales. These structures interact complexly, instigating large-scale energy cascades to smaller scales. The inherent chaos of turbulence makes precise flow evolution predictions challenging, even with meticulous initial conditions. Exploring turbulence in chaotic systems often necessitates the utilization of mathematical models and simulations. Direct experimental observations prove challenging due to the intricate and rapidly changing nature of turbulent flows. Researchers employ methodologies rooted in chaos theory and statistical mechanics to glean insights into the fundamental dynamics governing turbulent systems.

In rundown, turbulence in chaotic systems, particularly within fluid dynamics, embodies a state of irregular and unpredictable motion marked by a consideration of preliminary conditions, the presence of vortices and fluctuations across various scales, and a complex interplay of forces culminating in chaotic behavior. The comprehension and prediction of turbulence in chaotic systems persist as active areas of investigation across diverse scientific disciplines.

This dedicated chapter is an exploration of several intricate topics. Initially, it delves into the mechanisms through which shear waves induce chaotic phenomena. Subsequently, it scrutinizes the transformative capabilities of hologram interferometry in envisioning the outcomes of this chaos in four-dimensional space. This visionary process is facilitated through the utilization of synthetic aperture radar satellite data, marking a paradigm shift in our ability to comprehend and portray the complexities inherent in shear-wave dynamics.

13.1 What constitutes chaos, and is turbulence synonymous with chaotic behavior?

Chaos, in scientific terms, refers to a state of dynamic unpredictability characterized by extreme sensitivity to preliminary circumstances. Turbulence is indeed considered a manifestation of chaotic behavior, exemplified by the intricate and irregular fluid motion it entails. The complex interactions and sensitivity to primary states within turbulent systems align with the fundamental characteristics of chaos in scientific discourse.

Turbulence, particularly in fluid dynamics, is a state of irregular and chaotic motion. It arises when a fluid, such as air or water, undergoes complex, irregular flow patterns. The behavior of turbulent flows is highly perceptive to preliminary laminar flow conditions, and small disturbances in the flow can amplify and result in intricate and seemingly random fluid motion.

In turbulent systems, the interactions between fluid elements occur at an extensive array of scales. These interactions lead to the formation of vortices, eddies, and fluctuations, creating a cascade of energy from larger to smaller scales. The complexity in turbulent flows arises not only from the nonlinearity of the fundamental computations but also from the multitude of interacting components within the system.

Synthetic Aperture Radar Image Processing Algorithms for Nonlinear Oceanic Turbulence and Front Modeling. DOI: https://doi.org/10.1016/B978-0-443-19155-8.00013-1

The common fundamental trait between turbulent systems and chaotic systems is their susceptibility to minute variations in the initial circumstances. This sensitivity is a consequence of the nonlinear character of the fluid dynamics equations governing turbulence. It means that, even with precise initial conditions, predicting the exact evolution of unsteady flow over time becomes challenging due to the amplification of small disturbances.

In the realms of chaos speculation and ocean dynamics, chaotic water mixing (Fig. 13.1) denotes a phenomenon where flow tracers evolve into intricate fractals propelled by the forces of ocean dynamics. This process is distinguished by the exponential expansion of fluid filaments. Notably, even seemingly uncomplicated movements, such as the irregular vortex or finitely determined wind propagation, can generate remarkably intricate patterns from initially straightforward tracer fields. Despite its prominence, the intricacies of this phenomenon remain elusive, constituting a focal point of ongoing and extensive research within the scientific community.

Needless to say, turbulence is indeed considered a form of chaos in the context of fluid dynamics. The intricate and seemingly random motion observed in turbulent flows is a result of the complex, nonlinear interactions between fluid elements, leading to a state of dynamic unpredictability akin to chaos in scientific systems.

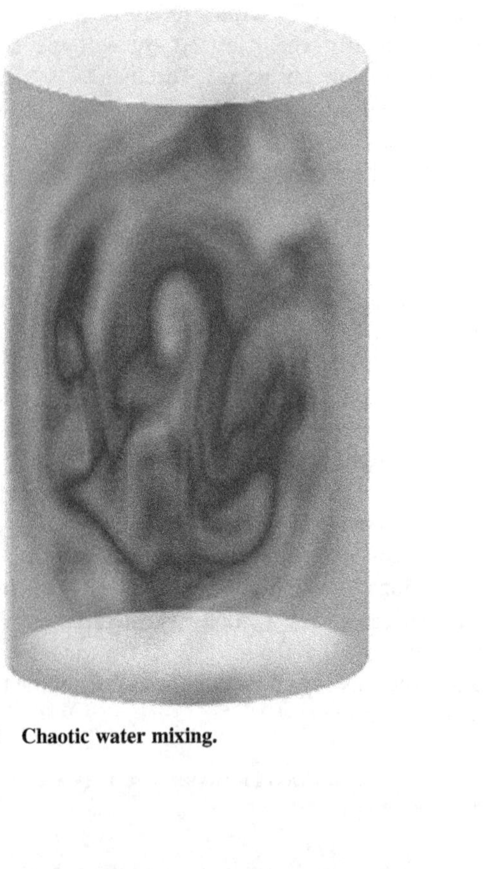

FIGURE 13.1 Chaotic water mixing.

13.2 Exploring chaotic turbulence in the ocean: a contextual perspective

Delving into the complex dynamics of chaotic turbulence in the ocean involves an exploration of complex fluid motion that defies simple predictability, which is recognized as chaotic advection, and is characterized by the intricate interplay of fluid elements leading to seemingly random and unpredictable patterns. Even within the vastness of the ocean, seemingly straightforward currents can give rise to remarkably intricate and chaotic behaviors.

It's crucial to emphasize that the study of chaotic ocean turbulence is an active area of research, marked by the pursuit of a deeper understanding of the underlying mechanisms governing these turbulent dynamics. The ocean, with its immense scale and complexity, presents a unique and challenging environment for such investigations.

By adopting a contextual perspective, we aim to comprehend the nuanced interactions within oceanic turbulent flows. The intricate dance of fluid filaments and the exponential growth of complexity in seemingly simple flows, such as those around vortices, captivate our scientific curiosity. The blinking vortex and other phenomena challenge our conventional notions, demanding a closer examination of the underlying principles governing chaotic advection in the ocean.

While our understanding of chaotic ocean turbulence is advancing, it remains a complex and multifaceted puzzle, prompting ongoing research endeavors. The exploration of this chaotic dance within ocean dynamics not only deepens our scientific knowledge but also holds implications for various fields, from climate science to marine ecology.

In Chapter 2, we elucidated that fluid flow can be conceptualized as a dynamical system—a collection of ordinary differential equations dictating the evolution of a Lagrangian trajectory. These equations, termed advection equations, can be expressed as:

$$\frac{d\vec{x}}{dt} = \vec{v}(\vec{x}, t) \tag{13.1}$$

where $\vec{v} = (u,v,w)$ represents the components of the velocity field, presumed to be determined by the solution of governing equations such as the Navier – Stokes equations. Additionally, $\vec{x} = (x,y,z)$ denotes the physical position. In this sense, if the dynamical system governing trajectories exhibits chaos, the integration of a trajectory becomes exceptionally sensitive to initial conditions. This sensitivity results in neighboring points undergoing exponential separation over time. This intriguing and complex phenomenon is recognized as chaotic advection. In essence, within a chaotic dynamical system, minor deviations in the starting conditions lead to vastly divergent trajectories, exemplifying the inherent unpredictability and complexity embedded in chaotic advection.

The facilitation of mixing in a flow is driven by the separation of adjacent fluid particles. This separation is a consequence of velocity gradients, a phenomenon referred to as shearing. Consider two neighboring fluid particles represented by $\vec{x_1}$ and $\vec{x_2}$, initially separated by $\partial\vec{x} = \vec{x_1} - \vec{x_2}$ at time t. As these particles are advected by a flow $\vec{v}$, the approximate separation at time $t + \delta t$ can be determined through Taylor expansion, as elaborated in Chapter 2:

$$\frac{d}{dt}(\vec{x} + \delta\vec{x}) \approx \vec{v} + \nabla\vec{v} \cdot \delta\vec{x} \tag{13.2}$$

This expression signifies the impact of the flow velocity and velocity gradients on the evolving separation between neighboring fluid particles, providing valuable insights into the mixing dynamics induced by shearing in the fluid flow.

Therefore

$$\delta x(t + \delta t) \approx \delta x(t) + \delta t(\delta x \cdot \nabla)\vec{v} \tag{13.3}$$

Consequently,

$$\frac{d}{dt}\delta\vec{x} \approx \nabla\vec{v} \cdot \delta\vec{x} \tag{13.4}$$

The growth rate of the separation is determined by the gradient of the velocity field along the direction of the separation. An illustrative example of this is the plane shear flow, which represents a straightforward instance of a large-scale stationary flow. In this scenario, fluid elements undergo deformation due to a uniform shear within the flow.

13.3 Initiating turbulence via shear waves: mechanisms and implications

Ocean wave shear flows refer to the horizontal displacements of water particles within the ocean induced by the passage of waves. In the context of fluid dynamics, shear flow represents the variation in velocity across different layers of a fluid. When applied to ocean waves, shear flows describe the differential movement of water particles along the horizontal plane (Johannessen et al., 1996; Lyzenga & Wackerman, 1997).

As ocean waves propagate, they transfer energy to the water particles, causing them to move in a circular or elliptical pattern. These motions result in shear flows where adjacent layers of water move at different velocities horizontally. The shear flow is particularly prominent near the surface of the ocean where the influence of wind and wave interactions is significant. Therefore the shear-wave mechanism, as discussed in literature (Johannessen et al., 1996; Karimova, 2012; Karimova, Lavrova, et al., 2012), is linked to interactions between waves and currents, particularly within zones characterized by current shear. The manifestation of eddies becomes apparent in the form of chaotic regions.

The term "shear" in shear flows signifies the change in velocity or the gradient of velocity across spatial dimensions. In the case of ocean waves, shear flows are essential components of the dynamic nature of wave motion. Understanding shear flows is crucial for studying the distribution of energy within the ocean, as well as for comprehending various oceanic processes such as turbulence and mixing.

Describing the characteristics of eddies generated by shear waves reveals a prevalent presentation in the form of spiral vortices, a topic extensively explored in preceding chapters. Consequently, the selection of an appropriate hydrodynamic vortex model becomes crucial for delineating the current field associated with these spiraled features.

Several vortex models have been considered for this purpose, including the Rankine vortex (Rankine, 1872), Oseen vortex (Oseen, 1911), Burgers – Rott vortex (Rott, 1958), and Sullivan vortex (Sullivan, 1959). In this view, the Rankine vortex model considers a discontinuity in core vorticity, a feature incongruent with shear-wave-generated eddies. While the Oseen vortex model, conceptualized as planar, does not fully capture the inherently three-dimensional nature of shear-wave-generated eddies. In response, two three-dimensional vortex models, incorporating an extra axial velocity, have been proposed: the Burgers – Rott and the Sullivan vortex models. The Burgers – Rott vortex model, characterized by a spiral vortex rather than the two-cell vortex of the Sullivan model, aligns more closely with the observed spiral features in SAR images. Given the prominent spiral patterns exhibited by shear-wave-induced eddies in SAR data, it can be accurately investigated using the Burgers – Rott vortex model. Indeed, it emerges as the most suitable for establishing the current field associated with these eddies.

An exact solution to the Navier – Stokes equations regulating viscous flow, the Burgers vortex, also known as the Burgers – Rott vortex, is a concept in fluid dynamics that was developed thanks to the work of Jan Burgers and Nicholas Rott. This vortex model describes a similar, stable flow with a toward-one circumferential motion that centers vorticity in a fine feature about the symmetry axis. At the same time, axial elongation contributes to an increase in vorticity. In the meantime, the vorticity continues to be dispersed due to intrinsic viscous diffusion. The stationary Burgers vortex results from the balance of these three processes.

In addition to its theoretical importance in explaining the vortex elongation technique, the Burgers vortex can be used to explain real-world events such as tornadoes. In situations such as tornadoes, the vorticity is provided by continual convection-driven vortex stretching, which is consistent with the features captured by the Burgers vortex model.

The Burgers vortex's flow is expressed in cylindrical coordinates (r,θ,z). Here, the flow field related to the axisymmetric stagnation point flow is taken into consideration, assuming axial symmetry (no dependence):

$$v_r = -\alpha r, \tag{13.5}$$

$$v_z = 2\alpha z, \tag{13.6}$$

$$v_\theta = \frac{\Gamma}{2\pi r} g(r), \tag{13.7}$$

In the specified equations, where 0α > 0 represents the strain rate and Γ > 0 is the circulation, both being constants, the flow adheres to the continuity equation as indicated by the first two equations provided. The azimuthal momentum equation within the Navier–Stokes equations can then be simplified to:

$$r\frac{d^2g}{dr^2} + \left(\frac{\alpha r^2}{\nu} - 1\right)\frac{dg}{dr} = 0 \tag{13.8}$$

The equation is expressing a relationship between the second derivative of g with respect to r, the first derivative of g with respect to r, and the function g itself, multiplied by certain coefficients. The vorticity equation, therefore, yields a nontrivial component solely in the z-direction, as expressed by:

$$\omega_z = \frac{\alpha\Gamma}{2\pi\nu} e^{\left(-\frac{\alpha r^2}{2\nu}\right)}. \tag{13.9}$$

The equation essentially articulates how the vorticity (ω_z) is influenced by the constants α, Γ, and ν in relation to the radial distance r. The exponential term $e^{\left(-\frac{\alpha r^2}{2\nu}\right)}$ indicates a rapid decrease in vorticity with increasing r. This kind of functional form is often encountered in physical systems where vorticity is concentrated near the center and diminishes as you move away radially (Drazin & Riley, 2006). A more intuitive grasp of the flow can be achieved by scrutinizing the three terms encapsulated in the vorticity equation for ω_z:

$$-\alpha r\frac{d\omega_z}{dr} = 2\alpha\omega_z + \frac{\nu}{r}\frac{d}{dr}\left(r\frac{d\omega_z}{dr}\right). \tag{13.10}$$

In essence, this equation describes the relationship between the derivatives of the vorticity component ω_z with respect to r, incorporating parameters such as α, ν, and the radial distance r. The terms signify different contributions to the overall behavior of the vorticity in the z-direction within the context of the fluid flow being analyzed (Robinson & Saffman, 1984).

In 1984 Tsutomu Kambe, therefore, presented a precise solution to the time-dependent Navier – Stokes equations for an arbitrary function $\alpha = \alpha(t)$. Specifically, when α is constant, the vorticity field $\omega(r,t)$ with an arbitrary initial distribution $\Omega(r) = \omega(r,0)$ is expressed as follows:

$$\omega(r,t) = \frac{\alpha}{2\pi\nu(1-e^{-2\alpha t})}\iint \Omega\left(\sqrt{\xi_1^2+\eta_1^2}\right)$$

$$\exp\left[\frac{-(x-\xi_1 e^{-\alpha t})^2 - (y-\eta_1 e^{-\alpha t})^2}{(2\nu/\alpha)(1-e^{-2\alpha t})}\right] d\xi_1 d\eta_1. \tag{13.11}$$

Consequently, if $t \to \infty$, the asymptotic behavior is described by:

$$\omega(r,t) = \frac{\alpha}{2\pi\nu}\exp\left(-\frac{\alpha r^2}{2\nu}\right) = \left[\Gamma + e^{-2\alpha t}\int_0^\infty\right.$$

$$\left.\Omega(S)\left(\frac{\alpha r^2}{2\nu} - 1\right)\left(\frac{\alpha r^2}{2\nu} - 1\right)2\pi s ds + O(e^{-4\alpha t})\right],$$

$$\Gamma = \int_0^\infty \Omega(S) 2\pi s ds \tag{13.12}$$

Therefore in the presence of a nonzero circulation ($\Gamma \neq 0$), any arbitrary vorticity distribution converges toward the Burgers' vortex. Conversely, if $\Gamma = 0$, as is the scenario when the initial condition comprises two equal and opposite vortices, the first term becomes zero, and the second term indicates that vorticity diminishes to zero as $t \to \infty$.

According to the aforementioned perspective, the Burgers vortex layer, often called the Burgers vortex sheet, represents a stretched shear layer and serves as a two-dimensional counterpart to the Burgers vortex, which is an exact solution to the Navier–Stokes equations. In this sense, the velocity field (vx, vy, vz) expressed in Cartesian coordinates accordingly:

$$v_x = -\alpha x, \tag{13.13}$$

$$v_z = \alpha z, \tag{13.14}$$

Therefore

$$v_y = U erf\left(\frac{\sqrt{\alpha}x}{\sqrt{2\nu}}\right), \tag{13.15}$$

In this context, where α > 0 represents the strain rate, $v_y(+\infty) = U$, and $v_y(-\infty) = -U$, the quantity $2U$ is considered as the strength of the vortex sheet. The vorticity equation only yields a significant component in the z-direction, expressed as:

$$\omega_z = \frac{2U}{\sqrt{2\pi}}\sqrt{\frac{\alpha}{\nu}}\exp\left(-\frac{\alpha x^2}{2\nu}\right). \tag{13.16}$$

In this scenario, where the strain rate (α > 0) plays a crucial role, the velocity field v_y is influenced. The conditions $vy(+\infty) = U$ and $vy(-\infty) = -U$ signify that, at both positive and negative infinity along the y-axis, the velocity in the y-direction tends toward U and $-U$, respectively. The magnitude $2U$ is identified as the strength characterizing the vortex sheet. In this understanding, the vorticity Eq. (13.16) associated with this configuration indicates that the vorticity predominantly manifests in the z-direction. This suggests significant rotation around the z-axis, and the intensity of this rotation is determined by the strain rate (α) and the specified conditions at positive and negative infinity along the y-axis.

On the other hand, nonaxisymmetric Burgers' vortices manifest in flows with nonaxisymmetric strains. For small vortex Reynolds numbers $Re = \Gamma[2\pi\nu]^{-1}$ and also achievable for $1Re \gg 1$, the development of nonaxisymmetric Burgers' vortices, suitable for various values of the vortex Reynolds number, can be investigated using numerical integrations. The velocity field is defined by the following expression:

$$v_x = -\alpha x + u(x, y), \quad (13.17)$$

hence,

$$v_y = -\beta y + v(x, y), \quad (13.18)$$

Therefore

$$v_z = \gamma z \quad (13.19)$$

Subject to the condition $\gamma = \alpha + \beta$, where $\alpha > 0$ and $\gamma > 0$ without any loss of generality, it is assumed that the vortex cross section is situated in the xy plane, contributing to a nonzero vorticity component in the z-direction.

$$\omega_z = \frac{\partial u}{\partial y} - \frac{\partial v}{\partial x}. \quad (13.20)$$

The axisymmetric Burgers' vortex is regained when $2\alpha = \beta = 2\gamma$, while the Burgers' vortex layer is regained when $\alpha = \gamma$ and $\beta = 0$. In this view, axisymmetric Burgers' vortex is achieved as well as $2\alpha = \beta = 2\gamma$. This configuration is achieved when the strain rates (α and β) are equal and half of the total strain rate (γ). It results in an axisymmetric vortex. Therefore Burgers' vortex layer is established under the circumstance of $\alpha = \gamma$ and $\beta = 0/$ in this sense, the strain rate in the xy plane (α) is equal to the total strain rate (γ), and there is no strain in the z-direction ($\beta = 0$). This leads to the formation of a vortex layer. These conditions define the specific characteristics of each vortex configuration in the context of the given parameters.

13.4 Exploring the dynamics of chaotic advection: a comprehensive characterization

In the presence of chaotic flow, minor initial discrepancies $\partial\vec{x}$ in a trajectory will exhibit exponential divergence. Our focus is on assessing stability, specifically determining how rapidly nearby trajectories deviate. The Jacobi matrix of the velocity field $\partial\vec{v}$ offers insights into the local rate of divergence of neighboring trajectories or the local stretching rate of Lagrangian space.

To quantify this, we introduce the matrix M as follows:

$$\frac{\mathrm{d}}{\mathrm{d}t}M \equiv \nabla\vec{v} \cdot M, M(t=0) = \boldsymbol{I} \quad (13.21)$$

Here, I denotes the identity matrix. Consequently:

$$0\delta\vec{x}(t) \approx M \cdot \delta\vec{x}_0 \quad (13.22)$$

The finite-time Lyapunov exponents are determined by averaging the natural logarithms of the magnitudes of the principal components of the vector M over a duration of time t:

$$M^T \cdot M \cdot \delta\vec{x}_{0i} = h_i \cdot \delta\vec{x}_{0i} \quad (13.23)$$

and

$$\lambda_i(\vec{x}, t) \equiv \frac{1}{2t}\ln L_i(\vec{x}, t). \quad (13.24)$$

where $\lambda_i(\vec{x}, t) \geq \lambda_{i+1}(\vec{x}, t)$ represents the ith Lyapunov exponent of the system, and $\delta\vec{x}_{0i}$ is the ith principal component of the matrix M, tending toward zero. If we initiate a set of orthonormal initial error vectors, denoted as $\{\delta\vec{x}_{0i}\}$, the matrix H will transform them into a set of final orthogonal error vectors with lengths represented by $\{\sqrt{L_i(\vec{x}, t)}\}$ (Ott, 2002). The system's operation converts an infinitesimal sphere of initial points into an ellipsoid, where the major axis is determined by $\sqrt{L_1(\vec{x}, t)}$ and the minor axis is determined by $\sqrt{L_N(\vec{x}, t)}$, with N being the number of dimensions (Mills, 2004).

This formulation of Lyapunov exponents is not only more elegant but also more suitable for real-world continuous-time dynamical systems compared with the common definition based on discrete function maps. Therefore the positive Lyapunov exponent is an excellent index of the existence of chaos.

In a chaotic system, we refer to the Lyapunov exponent as the asymptotic value of the greatest eigenvalue of M:

$$\lambda = \lim_{t \to \infty} \lambda_1(\vec{x}, t) \quad (13.25)$$

If there is a noticeable distinction between the Lyapunov exponents, as an error vector progresses in time, any deviation in the route of the largest vortex growth will tend to be amplified. Hence:

$$|\delta\vec{x}| \approx |\delta\vec{x}_0|e^{\lambda_1 t}. \quad (13.26)$$

In this view, the Lyapunov exponent of a flow is a distinctive quantity characterizing the asymptotic separation of fluid particles in a given flow. It is commonly utilized as a metric for the efficiency of mixing, measuring how rapidly vortex trajectories diverge due to chaotic advection.

13.5 Expansion of filaments versus the development of tracer gradients

In our investigation nitty gritty in Part Eleven, we dove into the viability of fibers as a dependable marker for recognizing upwelling twistings and the making of vortex structures. Moving our concentration, we are currently determined to reveal the inborn capability of fibers as unmistakable markers for communications with shear waves inside the mind boggling landscape of disturbance saw in SAR information. The major instrument driving fiber elements comes from the impact of normally happening surface movies that control gravity-slim waves. These movies change ocean surface strain, bringing about the smoothing of waves and a resulting decrease in NRCS (standardized radar cross segment). Vortexes produced by these movies, set apart by their articulated effect on sea surface harshness, are regularly effectively recognizable.

On the other side, whirlpools delivered by shear waves, appearing in SAR pictures through the balance of surface unpleasantness from wave-current communication, frequently display a more unobtrusive appearance. The qualities of shear-wave-created swirls are complicatedly impacted by different variables, including wind, flow, ocean bottom geology, SAR boundaries, and that's only the tip of the iceberg. These qualities ordinarily emerge from the mind boggling transaction of numerous variables, in some cases making their recognizable proof a difficult undertaking. Additionally, the imaging states of genuine SAR pictures remain covered in vulnerability and are challenging to control, adding further intricacy to the immediate translation of swirl highlight designs.

In exploring through these complexities, the fuse of turbulent hypothesis arises as vital, offering important bits of knowledge into how fiber development adds to the unpredictable elements of shear-wave-choppiness connections.

The phrase "Filament growth versus evolution of the tracer gradient" suggests a comparison between two phenomena: the expansion or growth of filaments and the changes in the gradient of a tracer substance within a system. Filaments typically refer to elongated structures, and their growth implies an increase in length or volume over time. On the other hand, the evolution of the tracer gradient involves examining how the concentration or distribution of a tracer substance changes within the system. In a dynamic system, such as fluid flow or any process involving the movement of substances, filaments may represent regions of increased concentration or specific patterns. The tracer gradient development, on the other hand, provides insights into how the concentration of a particular substance varies spatially. This comparison suggests an analysis of the competing or complementary aspects of these two phenomena within a given context, possibly in the context of fluid dynamics, chemical reactions, or similar processes. In this understanding, the mathematical expression can be given by:

$$\frac{\mathrm{d}\nabla q}{\mathrm{d}t} = -\nabla q \cdot \nabla \vec{v} \tag{13.27}$$

Concurrently with defining the Lyapunov exponent, we establish the matrix M' using the following formulation:

$$\frac{\mathrm{dM}'}{\mathrm{d}t} = -\nabla M' \cdot \nabla \vec{v} M'(t=0) = \boldsymbol{I} \tag{13.28}$$

It can be easily demonstrated that:

$$M' = M^{-1} \tag{13.29}$$

If we designate $\{Li'\}$ as the squared lengths of the principal components of the tracer gradient matrix M′, then:

$$L_i' = \frac{1}{L} \tag{13.30}$$

The $\{Li'\}$'s are arranged, as previously, in descending order. Hence, an increase in the error vector's magnitude will lead to a decrease in the tracer gradient, and vice versa. This can be intuitively understood by considering two adjacent points: with the difference in tracer concentration remaining constant, the sole factor influencing the gradient difference between them is their separation.

One strategy to distinguish the development of the fiber development under shear wave is carrying out form shift in weather conditions. To be sure, shape shift in weather conditions is a Lagrangian technique (Part 2) of recreating the development of at least one forms or isolines of a tracer as it is mixed by a moving liquid. Consider a mass of color infused into a waterway or stream: to initially arrange it very well may be demonstrated by following just the movement of its layouts. It is a brilliant strategy for concentrating on turbulent blending: in any event, when advected by smooth or limitedly settled speed fields, through a nonstop course of extending and collapsing, these shapes frequently form into complicated fractals. The tracer is ordinarily uninvolved as in yet may similarly be dynamic and addressing a dynamical property of the liquid like vorticity (Dritschel, 1988; Waugh & Plumb, 1994). As of now, shift in weather conditions of shapes is restricted to two aspects; however, speculations to three aspects are conceivable.

In this viewpoint, shape shift in weather conditions is one more helpful strategy for describing turbulent blending. In turbulent streams, advected shapes will develop dramatically over the long run. The given figure shows the casing by-outline development of a form advected north of a few days. The figure to the right shows the length of this shape as a component of time. The correlation between the intriguing evolution of shape and the presence of positive Lyapunov exponents is easily discernible (Mills, 2004; Ott, 2002;

Pattanayak, 2001). The rate of shape evolution is expressed as:

$$\frac{dL}{dt} = \int |\nabla\vec{v}\cdot d\vec{s}| \tag{13.31}$$

where $d\vec{s}$ represents the path, and the integral is carried out along the length of the contour. The growth rates of the contour will approximate the average of the prominent Lyapunov exponents:

$$L \approx L_0 \exp(\overline{\lambda}_1 t) \tag{13.32}$$

This mathematical expression represents an approximation where L is approximately equal to L_0 times the exponential function raised to the power of $\lambda_1 t$. Here, L and L_0 are variables representing lengths or scales, λ_1 is the first Lyapunov exponent, t is time, and exp denotes the exponential function. The expression suggests a growth or increase in the scale L over time t based on the initial scale L_0 and the value of the first Lyapunov exponent. The Lyapunov exponent characterizes the rate of divergence or stretching of nearby trajectories in a chaotic system.

In the perspective of chaotic advection, a fluid particle meanders across an extensive area, encountering other particles that were originally distant. It is plausible to assert that a particle becomes intermingled with others traversing the same region. Nevertheless, the trajectory's coverage doesn't invariably encompass the entire fluid domain. Poincaré sections serve the purpose of discerning areas characterized by either effective or ineffective mixing.

In the dominion of dynamical systems, the Poincaré map is an intriguing transformation denoted as M_P, acting on the state space $x_i(t_i)$ and mapping it to the subsequent state $x_i + 1$ $(t_i + 1 = t_i + T, x_i)$. The essence of M_P lies in its ability to evolve a point-like particle to its position after a specific time interval T. Particularly captivating is its application in the context of time-periodic flows, where iteratively applying the map unravels the successive positions of the particle over each period T. The construction of a Poincaré section involves initiating the process with various initial conditions and charting the corresponding iterates, essentially plotting trajectories stroboscopically at intervals of T.

In the pursuit of achieving fluid homogenization, the goal often revolves around unifying a species represented by its concentration field q. This species is commonly treated as a passive tracer, implying it exerts no influence on the flow dynamics. This species could manifest as, for instance, a dye intended for blending. The dynamics of the concentration field q adhere to the advection-diffusion equation, also known as the convection-diffusion equation:

$$\frac{\partial q}{\partial t} = \nabla \cdot D \nabla q - \vec{v} \cdot \nabla q. \tag{13.33}$$

In contrast to the straightforward diffusion equation, the component proportional to the velocity field $\vec{v}$ embodies the influence of advection. During the mixing of a tracer spot, the initial stages of the mixing process are primarily governed by the advection term in the evolution of the concentration field. Chaotic advection induces a transformation of the spot into a collection of slender filaments. The width of a dye filament undergoes exponential reduction over time, reaching an equilibrium scale where the impact of diffusion becomes significant. This particular scale is referred to as the Batchelor scale, defined as the square root of the ratio between the diffusion coefficient and the Lyapunov exponent, that is, $w_B = \sqrt{\frac{D}{\lambda}}$. In this context, where λ represents the Lyapunov exponent, and D is the diffusion coefficient, the Batchelor scale serves as a metric for the delicate equilibrium between filament stretching and diffusion in shaping the concentration field's evolution. In the realm of fluid and molecular dynamics, the Batchelor scale delineates the dimensions of a fluid droplet that undergoes diffusion within the same timeframe it takes for the energy in an eddy of size η to dissipate. The Batchelor scale, denoted as λ_B, can be expressed as:

$$\lambda_B = \frac{\eta}{\sqrt{Sc}} = \left(\frac{\nu D^2}{\varepsilon}\right)^{\frac{1}{4}} \tag{13.34}$$

Here, the parameter $\eta = (\nu^3/\varepsilon)^{1/4}$ represents the Kolmogorov length scale (see Chapter 2). Moreover,

Sc stands for the Schmidt number. In other word, $Sc = \frac{\nu}{D} = \frac{\mu}{\rho D} = \frac{\text{viscousdiffusionrate}}{\text{molecular(mass)diffusionrate}}$. Therefore η denotes the kinematic viscosity; D corresponds to the mass diffusivity; and ε signifies the rate of dissipation of turbulence kinetic energy per unit mass.

Much like the Kolmogorov microscales, which characterize the smallest turbulence scales before viscosity takes over, the Batchelor scale illuminates the minimum length scales of fluctuations in scalar concentration before succumbing to molecular diffusion dominance. It's noteworthy that for $Sc > 1$, a common scenario in many liquid flows, the Batchelor scale is considerably smaller when juxtaposed with the Kolmogorov microscales. This implies that scalar transport operates at scales even smaller than the smallest eddy size (Dritschel, 1988; Pattanayak, 2001; Paul et al., 2004).

Filament width, therefore, experiences a reduction due to stretching, countered by an increase facilitated by diffusion. Notably, the Batchelor scale sets a lower limit on observable length scales within the concentration field, as diffusion swiftly smudges out finer details. As most dye filaments approach the Batchelor scale, diffusion becomes a prominent force in diminishing the concentration contrast between the filament and its surrounding domain. The moment a filament attains the Batchelor scale is aptly named its mixing time. Solving the advection-diffusion equation reveals that beyond a filament's mixing time, the decline in concentration fluctuation induced by diffusion follows an exponential pattern, leading to rapid homogenization with the surrounding fluid.

The pivotal inquiry that emerges is as follows: Is it feasible to simulate chaotic turbulence in a four-dimensional space, diverging from the conventional two or three-dimensional paradigms within SAR images? Furthermore, could quantum hologram interferometry serve as a tool for visualizing the intricate 4-D shear-wave chaotic flow embedded in satellite SAR data?

13.6 Types of radar interferometry: an exploration

InSAR constitutes radar technology grounded in the interference of at least two coherence pulses, as illustrated in Fig. 13.2. The interference of electromagnetic waves can be mathematically represented using the superposition principle. The general equation for the electric field (S_{12}) of two interfering waves S_1 and S_2 can be expressed as:

$$S_{12} = S_1 + S_2 + 2\sqrt{S_1 + S_2}\cos(\phi) \quad (13.35)$$

This results in the backscattering of the phase difference between these pulses, enabling the detection of millimeter-scale deformations over time periods ranging from days to years. To delve into the phase mechanism, consider a real variable, such as time (t). The phase of the periodic function F is akin to an angle, representing the number of periods covered by that variable and denoted by $\phi(t)$:

$$\phi(t) = 2\pi\left[\frac{t - t_0}{f^{-1}}\right]. \quad (13.36)$$

Eq. (13.35) indicates that the phase difference is dependent on the rate of time changes per the inverse of pulse frequency, denoted by f. When the frequencies are not the same, the growth of phase deviation $\delta(t)$ is linear with the argument t. From this perspective, the periodic deviations resulting from reinforcement and confrontation lead to a phenomenon called beating. In practical terms, beating is an interference pattern between two signals with slightly different frequencies, perceived as a periodic variation in volume, where the rate is determined by the difference between the two frequencies.

In instances of dissimilar frequencies, the temporal growth of phase deviation (t) follows a linear trajectory with the variable t. From this perspective, the periodic oscillations between reinforcement and confrontation give rise to "beating." Essentially, beating is an interference pattern arising when two signals with slightly different frequencies interact, presenting as a periodic fluctuation in intensity, with the rate being the variation between the dual SAR frequencies.

It's worth noting that radar interferometry is a straightforward extension of Young's interferometry concept. The two radar (SAR) antennas function as coherent point sources in this context. Due to the relatively long wavelengths of radar signals, the signal can be easily digitized and coherently processed, directly measuring phase information (Lyzenga & Wackerman, 1997; Marghany, 2013a; 2013b). When imaging a surface, the phase fronts from the two sources interfere, and the measured phase differences delineate topographic information, with surface topography slicing through the interference pattern.

The two fundamental categories of interferometric radars distinguish themselves based on the geometric arrangement of the baseline vector, where the baseline represents the vector between the antenna locations within the interferometric pair. Hence, radar interferometry is classified into: (1) cross-track interferometers (Fig. 13.3) and (2) along-track interferometers (Fig. 13.4). Cross-track interferometers find application in topographic and surface deformation measurements, with antennas typically positioned apart in the cross-track direction.

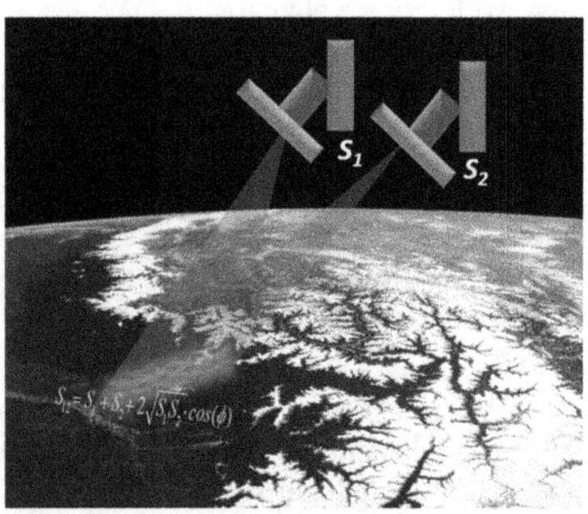

FIGURE 13.2 **Interference of dual SAR pulses**. *SAR*, Synthetic aperture radar.

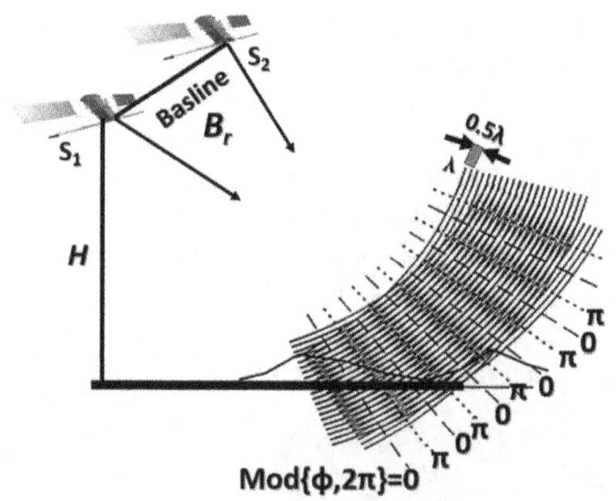

FIGURE 13.3 **Interferometric radar signal analysis for topographic mapping.**

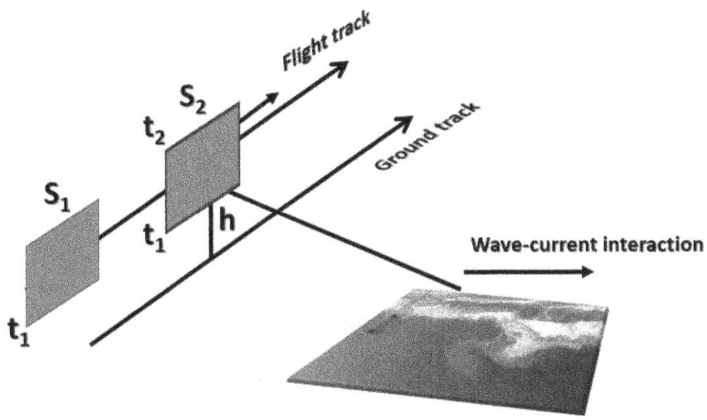

FIGURE 13.4 **Along-track interferometry.**

In contrast, along-track interferometers are utilized for measuring radial velocity, with antennas separated along the track direction. Both types of radar interferometry employ dual-antenna single-pass interferometers. However, cross-track interferometry employs single-antenna repeat-pass interferometers, enabling the measurement of topography and surface deformation. Conversely, along-track interferometry relies on an along-track separation geometry configuration conducive to measuring radial velocity (Woodhouse, 2017).

13.7 Challenges in InSAR data processing: an exploration

Certainly, as illustrated earlier in many studies (Bhattacharya & Mukherjee, 2017; Li et al., 2022; Macchiarulo et al., 2023), the phase information from an individual image lacks direct usability. However, significant phase changes between two or more disparate acquisitions are crucial, constituting the primary challenge in InSAR. By computing phase differences between acquisitions at distinct locations, generating a digital elevation model (DEM) is possible. Simultaneously, ground deformation can be detected at a millimeter scale by analyzing phase variations between acquisitions at different times.

In this context, examining phase variations between acquisitions at different times provides valuable insights into ground distortion with exceptional precision. The significant advantage of InSAR lies in its capability to cover extensive and challenging-to-access areas consistently, regardless of day or night and weather conditions. Nevertheless, InSAR faces constraints that must be consistently considered when interpreting InSAR products.

The first challenge is related to shifts in the direction of line of sight (LOS), dependent on the satellite's flying orientation and the radar beam's incidence angle. Fig. 13.5 illustrates a ground zone imaged at two different times, t_0 and t_1. The steeper the incidence angle, the more accurate the sensitivity to vertical displacements. For instance, when looking toward the west, a descending orbit predominantly covers the west-facing slope without distortion, while an ascending orbit primarily covers east-facing slopes (Fig. 13.6). Sensitivity is low when the actual surface displacement vector is nearly vertical to the LOS. Due to the north – south orbit direction, sensitivity to a surface shift in this plane is close to zero (Pinel et al., 2011; Woodhouse, 2017).

The second obstacle involves the issue of phase ambiguity due to the cyclic nature of the wave. This occurs when the variation between two phases of the two different acquisitions exceeds half of the effective wavelength, that is, 0.5λ. In simpler terms, if a sensor with a wavelength of 5.55 cm, like RADARSAT-2, has a revisit time of 24 days, and a moving area travels at a speed greater than 2.8 cm in 24 days, it introduces phase ambiguity. In such cases, InSAR fails to provide meaningful information about ground or object distortion. The third challenge in the InSAR process is a signal delay caused by atmospheric effects. In other words, changes in the Earth's surface pattern between acquisition times introduce an interferometric phase term, which functions on the changes of two slant ranges ∂R_d.

The objective of this chapter is to exclusively employ along-track interferometry, as it is the preferred technique for imaging sea surface dynamics, in contrast to cross-track interferometry. Assuming Ur represents the velocity of the scatterer's movement with a line of sight (LOS) to the target, O, the phase changes due to the target and scatterer movement are expressed as:

$$\frac{\partial\phi}{\partial t} \approx U_r.4\pi\mathrm{O}(\lambda)^{-1} \tag{13.37}$$

Eq. (13.37) unveils that changes in phase due to variations in lag time can be harnessed to deduce the movement of surface currents. In this context, the signal is sent out, scattered, and returned to the same location. It is crucial that the spatial interferometric baseline for ATI data is zero, indicating a purely temporal baseline. Ideally, one would employ a set of large stationary antennas to capture the phase $\partial\phi$ of an image, but this is impractical.

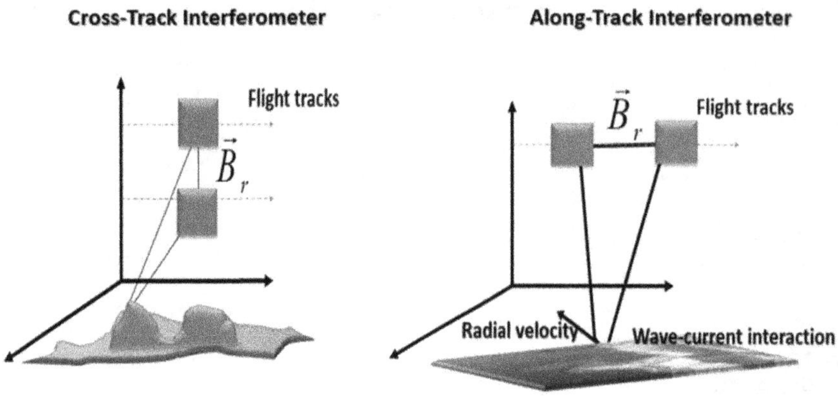

FIGURE 13.5 Interferometric radar configurations: cross-track and along-track interferometers.

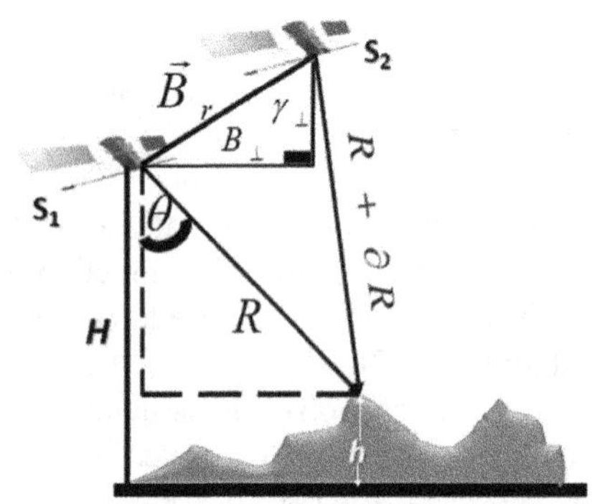

FIGURE 13.6 An ascending orbit predominantly captures slopes facing east.

Alternatively, a pair of SAR antennas can be displaced in the direction of the moving platform. If the platform travels at a speed V_S and the phase centers of the antennas move a distance d, then the time interval between the dual SAR images formed using the two antennas is $\partial t = (V_S)^{-1}d$. In this scenario, the radar transmits sequentially from each antenna and receives the signal with the same antenna used for transmission, known as "ping-pong" mode (Imel et al., 2002; Woodhouse, 2017).

Consequently, for measuring the current, it is essential to have a time lag that is sufficiently long to capture significant phase variations from the desired current differences. However, it should also be short enough to avoid phase ambiguities and prevent decorrelation of the backscattered signal. The decorrelation time is dependent on radar frequency and prevailing wind/wave conditions (Romeiser et al., 2010).

$$U_r = \frac{\phi_{ATI}}{4\pi \cdot B_{ATI} \cdot \sin\theta \cdot (\lambda V_S)^{-1}}. \tag{13.38}$$

Eq. (13.38) highlights the limitations associated with the low sensitivity when observing and measuring fast-moving objects in InSAR data. Specifically, in scenarios where there are dynamic ocean surface fluctuations or vehicles in free-flowing traffic, the interferogram phase undergoes significant spatial averaging, a process often referred to as multilooking. This spatial averaging diminishes the ability to capture fine details or rapid changes in the phase, particularly for objects or surfaces exhibiting high velocities (Romeiser et al., 2010; Suchandt & Runge, 2012). As a result, the interferogram phase undergoes spatial averaging, leading to a reduction in the ability to precisely detect and measure the movements of swiftly moving objects in the InSAR data.

The statement suggests that the chapter is oriented toward addressing challenges in InSAR data processing and proposes the use of a novel model to overcome these issues. One successful approach that has been mentioned is hologram interferometry. Hologram interferometry is a technique that involves using holographic principles in interferometric applications. It is likely that the chapter explores how hologram interferometry can offer solutions to the challenges inherent in InSAR, possibly providing more accurate and effective results in the context of the discussed issues.

13.8 Speculation for quantization of hologram interferometry

Professor Susskind is one of the most prominent figures in the world of holograms, having pioneered this field with his research and theories. Professor Susskind's influence in the field of holograms cannot be overstated. Professor Susskind has played an integral role in the advancement of holograms, pioneering research on the subject and producing groundbreaking results.

The quantum algorithm is a complex and interesting system, allowing computers to solve problems exponentially faster than traditional algorithms. The potential implications of this are immense, allowing for vast amounts of data to be analyzed quickly and efficiently. The quantum algorithm has opened up a world of possibilities in fields ranging from physics and chemistry to artificial intelligence and cryptography. By using the principles of quantum mechanics, the quantum algorithm takes advantage of the behavior of particles on a microscopic level to create solutions that are exponentially faster than traditional algorithms, especially in solving the uncertainty of InSAR procedures. With the quantum algorithm, complex problems can now be solved much more quickly, opening up new possibilities for research and development in a variety of fields.

Marghany (2021a) has innovated speculation quantization of hologram interferometry. In this scenario, Young's interferometry theory is a straightforward expansion of the classical version of wave interference known as InSAR (Marghany, 2021b). The dual SAR antenna wave function is obtained as a specific solution of the Schrödinger wave equation as the superposition of waves produced from dual SAR antennas from the point of view of quantum mechanics (Marghany, 2021c). The NOON state, which is widely acknowledged as the state of a radar interferometer, is represented by the following wave function in this understanding (Marghany, 2021a):

$$|\Psi_{NOON}^{\phi}> = \frac{1}{\sqrt{2N!}}\left(|N>_1|0>_2 + e^{iN\phi}|0>_0|N>_2\right) \tag{13.39}$$

The superposition of N photon particles in mode 1 with $|0>_2$ particles in mode 2 $|0>_2$ is represented by Eq. (13.35). Therefore the phase measurement ϕ changes from 0 to $\frac{\pi}{2}$. In this view, NOON states are essential in quantum sensing since they enable for precise phase measurements when applied to interferometry techniques. On the other hand, the NOON state without attenuation can be obtained by:

$$|\Psi_{\alpha}> = \frac{1}{\sqrt{2^N}}(|10> + |01>)^{\otimes N} \tag{13.40}$$

Even in the absence of decoherence caused by slick-covered water, Eq. (13.36) demonstrates that separable states cannot be used to implement the conventional quantum constraint. In terms of phase change, Eq. (13.36) can be written as:

$$\lim_{\alpha_1 \to 1} \lim_{\alpha_2 \to 1} \partial\phi = \frac{1}{\sqrt{N}} \tag{13.41}$$

The Heisenberg limit, or beating of the shot-noise limit, is now referred to as "super-sensitivity" and is based on Eqs. (13.36 and 13.37). The best performance is anticipated for the lowest uncertainty condition when the photon number uncertainty is the greatest. However, in this case, the quantitative uncertainty is spread out over the dual-photon paths A and B (Ou, 2007). If all N photons are in mode A (and none in B) or if all N photons are in mode B (and none in A), where the periods in both A and B would be terminated, then the situation in NOON states is fully unclear. In this view, the NOON state does not become much more confusing than that!

In this understanding, the mathematical description of the two complex SAR hologram interferometry images $H(x, y, x, z, t)$ is formulated as (Marghany, 2019a; 2019b):

$$H(x, y, x, z, t) = \varepsilon_0 c\left[S_{01}^2\cos^2(kz - \omega_1 t) + \frac{S_{02}^2}{B_r^2}\cos^2(kr - \omega_2 t)\right.$$
$$\left. + \frac{2S_{01}S_{02}}{B_r}\cos^2(kz - \omega_1 t)\cos(kr - \omega_2 t)\cos(\vartheta_{12})\right] \tag{13.42}$$

Eq. (13.38) says that the two complex radar images S_1 and S_2 are implemented to construct a 3-D hologram radar image. This hologram radar image involves the two signal wavenumber k, the baseline B_r between both acquired SAR images over the time of acquisition t, and the radial frequency ω. Additionally, Eq. (13.38) has a portion that only uses the radar complex signals and a term for interferometry that uses the cosine of the angle of 12 between the two vector amplitudes (which may be a function of position x,y,z). In this understanding, the estimation of phase maps of various holographic radar images can therefore be accomplished as follows:

$$\phi_{12}(x, y, z) = arctan$$
$$\left(\frac{\sin(\phi_2(x, y, z)) \otimes \cos(\phi_1(x, y, z)) - \cos\phi_2(x, y, z)) \otimes \sin(\phi_1(x, y, z))}{\cos(\phi_2(x, y, z)) \otimes \cos(\phi_1(x, y, z)) \oplus \sin(\phi_2(x, y, z)) \otimes \sin(\phi_1(x, y, z))}\right), \tag{13.43}$$

$$\phi_{32}(x, y, z) = arctan$$
$$\left(\frac{\sin(\phi_3(x, y, z)) \otimes \cos(\phi_2(x, y, z)) - \cos\phi_3(x, y, z)) \otimes \sin(\phi_2(x, y, z))}{\cos(\phi_3(x, y, z)) \otimes \cos(\phi_2(x, y, z)) \oplus \sin(\phi_3(x, y, z)) \otimes \sin(\phi_2(x, y, z))}\right), \tag{13.44}$$

$$\phi_{123}(x, y, z) = arctan$$
$$\left(\frac{\sin(\phi_{12}(x, y, z)) \otimes \cos(\phi_{23}(x, y, z)) - \cos\phi_{12}(x, y, z)) \otimes \sin(\phi_{23}(x, y, z))}{\cos(\phi_{12}(x, y, z)) \otimes \cos(\phi_{23}(x, y, z)) \oplus \sin(\phi_{12}(x, y, z)) \otimes \sin(\phi_{23}(x, y, z))}\right). \tag{13.45}$$

In this view, formulas $13.29 - 13.41$ say that $\phi_{123}(x, y, z)$ varies from 0 to 2π without 2π cutoffs, and then the image of the wrapped phase can be created. Consequently, reconstruction of the 3-D surface precisely using $\phi_{123}(x, y, z)$ cannot consign outstanding significances owing to $\phi_{123}(x, y, z)$ and does not expose 2π discontinuities since SAR long-wavelength often escalations in low quantity precision as the nature of its coherence signal. In other words, the phase $\phi_1(x, y)$ has the wavelength ratio of three complex SAR images; there are $\frac{S_{12}(\lambda)}{S_1(\lambda)}$ discontinuities with an identical layout $S_{12}(\lambda)$. In this circumstance, inside every $S_{12}(\lambda)$ there are discontinuities of $\frac{S_{123}(\lambda)}{S_{12}(\lambda)}$ on $\phi_{12}(x, y, z)$. In other words, the conjugate of three complex SAR images $\frac{S_{123}(\lambda)}{S_{12}(\lambda)} \otimes \frac{S_{12}(\lambda)}{S_1(\lambda)}$ forms discontinuities

$\phi_1(x,y,z)$. Overall, the hologram interferometry image can be mathematically expressed as:

$$H_k(x,y,z) = \alpha(x,y,z) + \beta(x,y,z)\sin(\phi(x,y,z) + 0.5(\pi k)) + G(x,y,z). \quad (13.46)$$

Eqs. (13.13 or 13.42) demonstrates the occurrence of white Gaussian noise $G(x,y,z)$ since Eqs. (13.36 − 13.38) contain degradation in the computing of the absolute phased (Ou, 2007; Marghany, 2021a). The main question arises is now: how can obtain 4-D phase unwrapping from 2-D phase unwrapping to construct 4-D visualization of chaotic shear-wave turbulence flow in hologram interferometry SAR data?

13.9 Marghany 4D quantized hologram interferometry algorithm

Marghany (2021b) has delivered a novel algorithm for 4-D phase unwrapping for 4-D object reconstruction. Following Marghany (2021b), the phase unwrapping technique is needed to estimate the 3-D object reconstruction in the complex SAR data, and the absolute phase cannot be used for this. For this purpose, the 4-D phase unwrapping can be stated as:

$$\Psi_3(x,y,z,t) = \phi(x,y,z,t) + 2\pi\left[\text{int}\frac{\phi_{12}(x,y,z,t)}{2\pi} \otimes \left(\frac{S_1(\lambda)}{S_2(\lambda)}\right) \otimes \left(\frac{S_2(\lambda)}{S_3(\lambda)}\right) + \text{int}\frac{\phi_{23}(x,y,z,t)}{2\pi} \otimes \left(\frac{S_2(\lambda)}{S_3(\lambda)}\right) \otimes \left(\frac{S_2(\lambda)}{S_3(\lambda)}\right) + G(\sigma)\right] \quad (13.47)$$

The innovation in Eq. (13.47) is based on the exchanging of the absolute phase estimates from the three complex SAR images, each with a unique acquisition time. In other words, the white Gaussian noise as a function of standard deviation σ is also taken into account in the innovative equation of the 4-D phase unwrapping (Marghany, 2021b). In addition to reconstructing phase unwrapping in 4-D, Eq. (13.47) seeks to eliminate a discontinuity. Let the wrapped phase ϕ_w be expressed in this case as the Laplacian of the real phase, which is derived by:

$$\nabla^2\phi = \cos\Psi_w \nabla^2(\sin\Psi_w) - \sin\Psi_w \nabla^2(\cos\Psi_w) \quad (13.48)$$

here ∇^2 is the Laplace operator, and by using an inverse of ∇^{-2}, the real phase ϕ' can be approximated in 4-D by assembling quantum Fourier-domain forward and ∇^{-2} as follows (Marghany, 2022):

$$\phi'(i,j,k,t)_Q = QFT^{-1} \frac{[QFT[\cos\phi_w(QFT^{-1}[|p^2+q^2+r^2+s^2> QFT(\sin\phi_w)]}{|p^2+q^2+r^2+s^2>} - QFT^{-1} \frac{[QFT[\sin\phi_w(QFT^{-1}[|p^2+q^2+r^2+s^2> QFT(\cos\phi_w)]}{|p^2+q^2+r^2+s^2>} \quad (13.49)$$

Eq. (13.49) reveals that QFT and QFT^{-1} are quantum Fourier inverse and forward cosine or sine transforms, respectively, in 4-D of time-domain coordinates (i,j,k) and the Quantum Fourier-domain coordinates. Eq. (13.13) comprises three sorts of operations: quantum forward and inverse cosine transforms, trigonometric operations, and the masking expression. Contrary to the conventional discrete Fourier transform, which requires $O(n^2)$ gates, the discrete Fourier transform on 2^n amplitudes can be implemented as a quantum circuit using only $O(n^2)$ Hadamard gates and controlled phase shift gates, where n is the number of qubits. Furthermore, the exact QFT algorithms only need *O (nlogn)* gates to perform a calculation (Marghany, 2021b; Marghany, 2022).

The specific equation for 4-D phase unwrapping based on the 3-D wave height distribution and sea-level variation would be given by:

$$\begin{pmatrix}\partial\Phi_1\\ \partial\Phi_2\\ \partial\Phi_3\\ \partial\Phi_4\end{pmatrix} = \frac{2\pi}{\lambda} \begin{pmatrix} (H_s,\partial\zeta,U_r)_{1i} & (H_s,\partial\zeta,U_r)_{1j} & (H_s,\partial\zeta,U_r)_{1k} & (H_s,\partial\zeta,U_r)_{1t}\\ (H_s,\partial\zeta,U_r)_{2i} & (H_s,\partial\zeta,U_r)_{2j} & (H_s,\partial\zeta,U_r)_{2k} & (H_s,\partial\zeta,U_r)_{2t}\\ (H_s,\partial\zeta,U_r)_{3i} & (H_s,\partial\zeta,U_r)_{3j} & (H_s,\partial\zeta,U_r)_{3k} & (H_s,\partial\zeta,U_r)_{3t}\\ (H_s,\partial\zeta,U_r)_{4i} & (H_s,\partial\zeta,U_r)_{4j} & (H_s,\partial\zeta,U_r)_{4k} & (H_s,\partial\zeta,U_r)_{4t} \end{pmatrix} \otimes \begin{pmatrix} W^x_{i,j,k,t}\\ W^y_{i,j,k,t}\\ W^z_{i,j,k,t}\\ W^t_{i,j,k,t}\end{pmatrix} \quad (13.50)$$

In this context, W stands for the user-defined weights, Cartesian coordinates *I, j, k*, and *t* for the 4-D vector representation in the hyperspace holographic interferometry. Consistent with Wnuk et al. (2019), the spatial variance of shear-wave turbulence flows significantly over time in this example of a 3D volume of a hologram image that is to be stored in 4D. Eq. (13.50) shows how the entanglement of the wave function can be used to determine the areas with the highest and lowest entanglement levels based on the wave function of nonunwrapped areas. In other words, the wave function exhibits two unwrapping phase states, lowest-entanglement |0> and highest-entanglement |1>, and is a superb estimator of the quality of phase unwrapping images generated from InSAR data (Marghany, 2019b). This novel formula, called Marghany 4-D quantized hologram interferometry, was developed by Marghany and exploited in recent publications (Marghany, 2021a, 2022).

In a four-dimensional space, each voxel possesses 80 neighbors, providing a robust criterion derived from the sum of 40 measurements. The construction of a 4-D hyperspace or hypercube from physical sea surface parameters, obtained through 4-D phase unwrapping, involves the implementation of a Hamming graph formula. This formula is executed as a quantum walk, represented by the expressions:

$$G_{Hc} \oplus H_m \otimes |\phi_{G_{H_c}} \otimes \phi_{H_m}(t)> = |\phi_{G_{H_c}}(t) \otimes |\phi_{H_m}(t)> \tag{13.51}$$

In the realm of quantum phase unwrapping walks, denoted as $\phi_{G_{H_c}}$ and ϕ_{H_m}, initiated from vertices v_{m_1} and v_{m_2}, respectively, Eq. (13.46) emerges as a pivotal component in the creation of hypercube ATI interferometry utilizing Hamming graphs. Broadly speaking, the Hamming graph serves as a potent tool for characterizing quantum uniform mixing. In the realm of quantum mechanics, the natural method to amalgamate dual systems is through the tensor product ⊗. In this context, the Cartesian graph artifact assumes a parallel role in facilitating quantum walks during the phase unwrapping process (Marghany, 2019b). Therefore the ultimate expression for the quantum walk, aiming for optimal exploration with a radius of 2r, can be elegantly presented through the pulsating sequence of the unitary operators $U_m^{''(+)}$:

$$\left|\phi^{(e)}{}_{G_{H_c}} \otimes \phi_{H_m}(2r)> = \left(U_m^{''(+)} \sum_{\alpha,\vec{q}} \right|\left|\alpha, \vec{q} \oplus \vec{e_\alpha} > |\alpha, \vec{q} > \left|C^{''(+)}\right)^r \right|\phi^{(e)}{}_{G_{H_c}}(t) > \otimes \phi_{H_m}(t)\right| \tag{13.52}$$

In this context, $C^{''(+)}$ represents the coin operator, exerting its influence on the combined Hilbert space $H^C \otimes H^V$; the $\sum_{\alpha,\vec{q}} ||\alpha, \vec{q} \oplus \vec{e_\alpha} > |\alpha, \vec{q} >|$ signifies a quantum walk propagator navigating through the hypercube vertices v_{m_1} and v_{m_2} in the direction α along the vertex edge $\vec{e}$, and $\oplus$ denotes the bitwise addition modulo 2 operator; and $\vec{q}$ is the Hamming weight of an integer, denoting the count of 1 in its binary string (Marghany, 2021c).

13.10 Conducting measurements in the natural environment

In the collection of surface current data, the Aquadopp 2 MHz current meter, produced by Nortek AS in Norway, was employed. This device, utilizing Doppler-based technology, operates independently to measure surface currents at the designated location. Designed with integrated memory and an internal battery pack, it can be programmed to internally record and store data for autonomous deployment (Fig. 13.7). The purpose of this sophisticated meter equipment is to measure sea surface conditions with precision as well as reliability. It can store a large amount of data thanks to its electronic memory, preventing the loss of important information. Its in-built battery pack also guarantees continuous functioning, which makes it a flexible instrument for long-term marine pattern monitoring and research.

The deployment of the Aquadopp 2 MHz current meter took place in the coastal waters of Teluk Kemang, Port Dickson, Malaysia, on May 6, 2017, (Fig. 13.8) and in Kelantan on October 7, 2020 (Fig. 13.9). Surface current data were systematically measured at 2-hour intervals during both phases. The retrieval of the current meter occurred following the conclusion of the TanDEM mission.

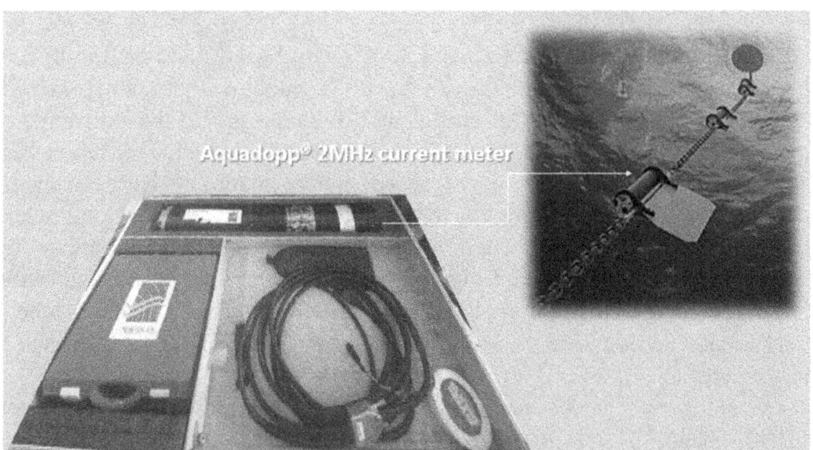

FIGURE 13.7 Deployment of Aquadopp 2 MHz current meter across water column.

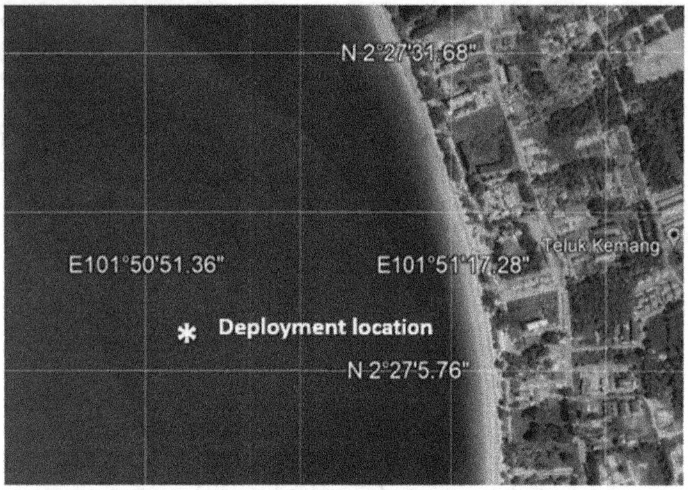

FIGURE 13.8 **Geographical positioning of deployment along the coastal waters of Teluk Kemang, Port Dickson, Malaysia.**

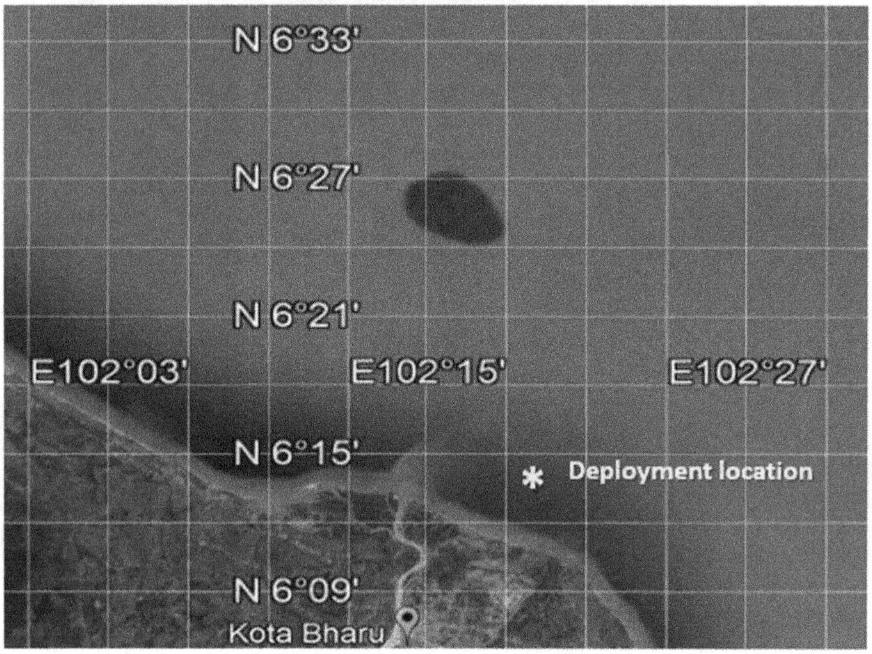

FIGURE 13.9 **Deployment location in the coastal waters of Kota Bharu, Kelantan, Malaysia.**

The researcher meticulously executed in situ current measurements coinciding with the TanDEM-X satellite overpasses. The majority of these deployments transpired during the late night and early morning hours, Fig. 13.10.

13.11 TanDEM-X satellite data

In the realm of space-borne synthetic aperture radar (SAR), along-track interferometry (ATI) emerges as a pioneering technique with diverse applications, spanning marine and terrestrial domains. It proves instrumental in tasks such as traffic monitoring, ship detection, and surface current mapping. When juxtaposed with single-satellite ATI configurations involving TerraSAR-X or RADARSAT-2, the interferometric imaging capabilities within the TanDEM-X (TerraSAR-X add-on for digital elevation measurements) satellite formation exhibit notable advancements. In this view, the instruments aboard both satellites represent cutting-edge technology in the form of advanced high-resolution X-band synthetic aperture radars (SARs), utilizing active-phased array technology. These SARs boast operational versatility in Spotlight, Stripmap, and ScanSAR modes, coupled with complete polarization capabilities. The instruments operate at a central frequency of 9.65 GHz, offering a selectable SAR chirp bandwidth of up to 300 MHz. The active-phased array antenna, with an expansive aperture size of 4.8 m × 0.7 m, stands securely affixed to the spacecraft body. Featuring 12 panels equipped with 32 waveguide subarrays for both Horizontal (H) and Vertical

(V) polarizations, this design facilitates agile beam pointing and adaptable beam shaping. These include extended and multiple ATI baselines, leading to heightened sensitivities for ground motion measurements and enhanced ambiguity resolution. However, unraveling motion information from TanDEM-X data through ATI encounters challenges, chiefly stemming from the amalgamation of across-track (XTI) and ATI baseline components (G. Krieger et al., 2006; Krieger et al., 2007; Suchandt & Runge, 2012).

FIGURE 13.10 Deployment of current meter by the author during TanDEM-X satellite overpasses.

From the aforementioned perspective, the inaugural set of TanDEM-X (TerraSAR-X add-on for Digital Elevation Measurements) satellite data was procured on May 6, 2017. The initial image was captured precisely at 7:27:17 a.m., followed by the second one at 19:20:06 p.m. Fig. 13.11. These data were obtained in spotlight mode, featuring X-band and dual polarization (HH and VV). It is imperative to note that these data are formatted as single-look complex data.

Subsequently, the second set of data was acquired on October 2, 2022, at 23:01:51 p.m. and the second image on October 7, 2022 at 11:22:35 a.m. This dataset was collected in Widescan mode, showcasing X-band and HH polarization (Fig. 13.12).

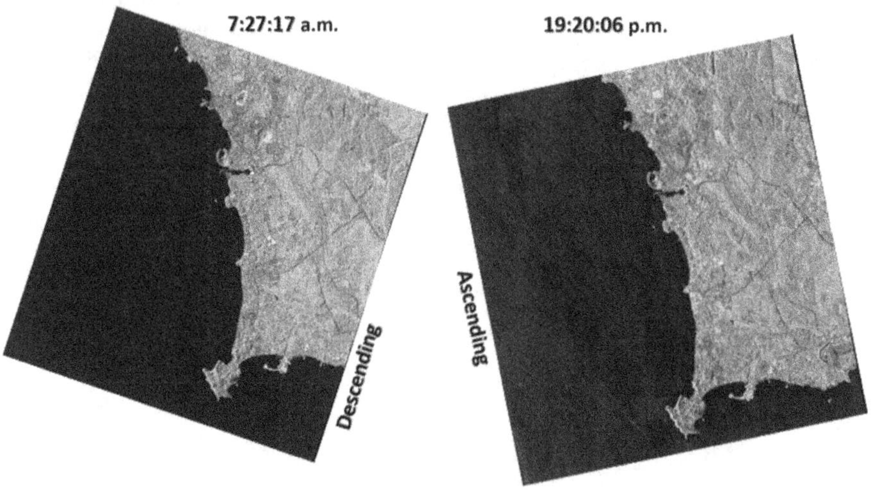

FIGURE 13.11 TanDEM-X satellite imaging at 7:27:17 a.m. and 7:20:06 p.m. on May 6, 2017.

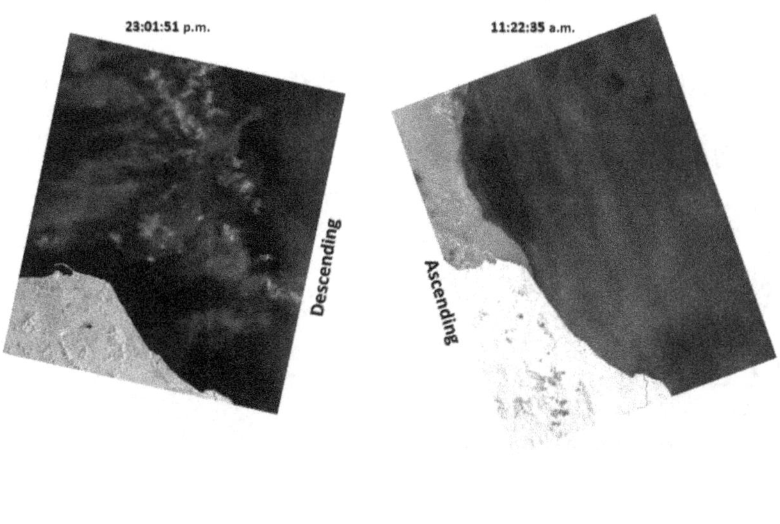

FIGURE 13.12 TanDEM-X satellite imaging at 23:01:51 p.m. and 11:22:35 a.m. on October 2 and 3, 2022; respectively.

To implement along-track interferometry, the SAR data must include both descending and ascending orbits. As observed, the SAR data utilized in this chapter comprise both descending and ascending acquisitions. In this context, when the satellite moves from the north pole to the south pole, it collects descending SAR data, offering a perspective from the north. This configuration is effective for observing areas in the northern hemisphere. Conversely, ascending SAR data are obtained as the satellite travels from the south pole to the north pole, providing a southward perspective. This proves beneficial for monitoring regions in the southern hemisphere (Fig. 13.13).

13.12 4-D shear-wave chaotic turbulence through hologram interferometry on TanDEM-X satellite

The interferogram phase, as determined through direct ATI phase measurement, exhibits a notable presence of noise, manifesting in a fuzzy pattern within the range of -0.9 to +0.9 degrees (Fig. 13.14). This phenomenon is attributed to the low coherence, registering below 0.5, prevalent along coastal waters. In stark contrast, regions dominated by land showcase high coherence nearing 1. Similar to the Spotlight mode, the Wide ScanSAR displays the lowest coherence, accompanied by an undefined interferogram phase (Fig. 13.15). This characteristic arises from heightened levels of noise and more pronounced fuzzy patterns compared with the Spotlight mode.

The cornerstone of both cross-track and along-track interferometry lies in phase unwrapping. Employing the Hamming weight graph, a 4-D phase unwrapping is executed, with excellent identification of the uwrapping phase cycle within 2-D (Fig. 13.16), 3-D (Fig. 13.17), and 4-D hologram phase unwrappings (Fig. 13.18). The clarity of hologram phase interferometry is notably enhanced as we progress from 2D to 3D and 4D. In both Spotlight mode and WideScanSAR mode, the achieved hologram phase interferometry cycles are sharper in 3D and 4D compared with 2D. The most robust chaotic shear waves in both modes are identified and traced more effectively in 3D and 4D, corresponding to both Spotlight mode and WideScanSAR mode.

Undoubtedly, the utilization of 4-D hologram phase unwrapping, incorporating the Hamming weight and a quantum walk algorithm, surpasses the brilliance observed in both 2-D and 3-D hologram phase unwrapping. The primary advantage of 4-D phase unwrapping lies in its preservation of the flawless phase cycle—maintaining the coherence across the entire cycle of the phase along-track interferometry patterns. Additionally, the implementation

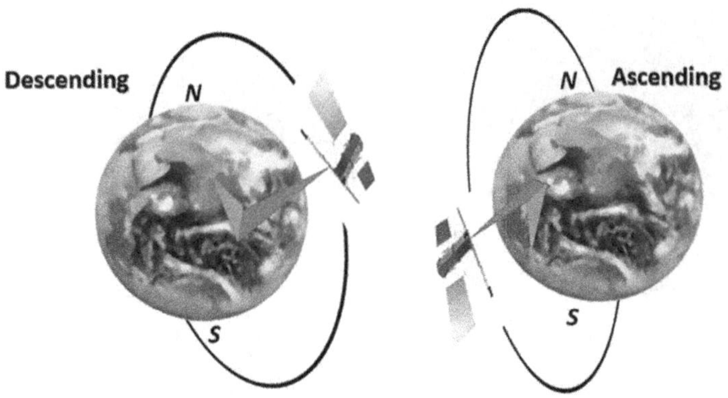

FIGURE 13.13 Observational perspectives: descending and ascending Orbits in SAR data acquisition . SAR, Synthetic aperture radar.

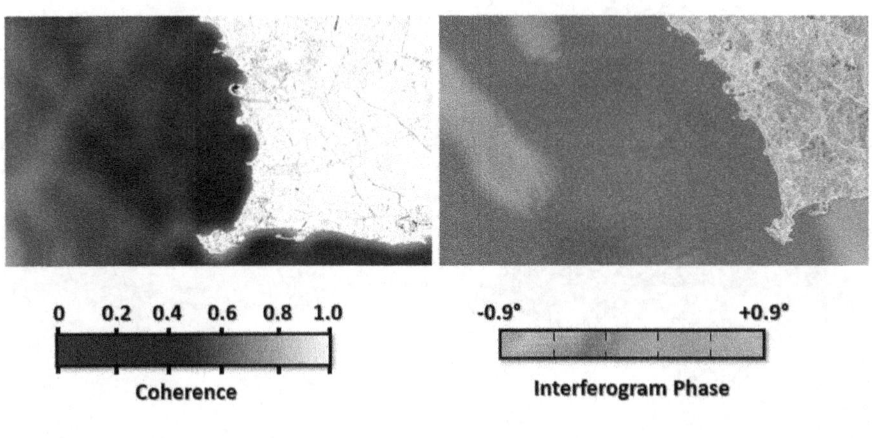

FIGURE 13.14 Interferogram phase retrieval through direct ATI phase estimation in spotlight mode . *ATI*, Along-track interferometry.

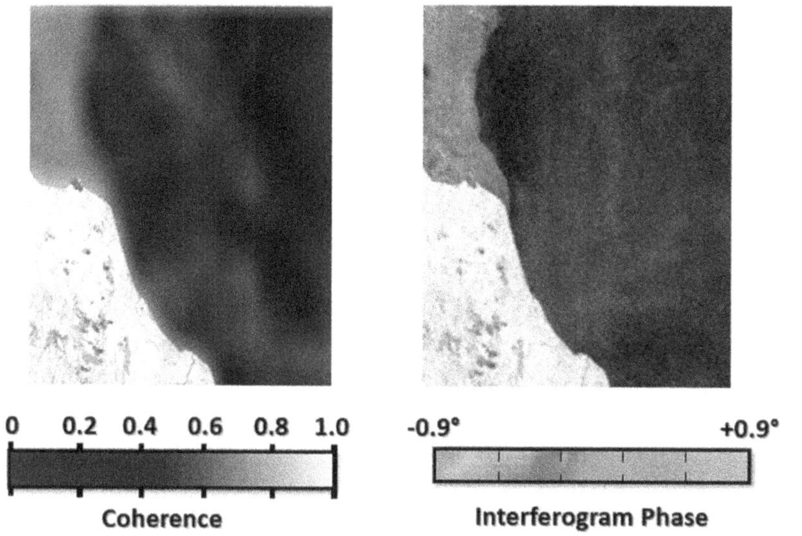

FIGURE 13.15 Interferogram phase retrieval through direct ATI phase estimation in wide scanSAR mode. *ATI*, Along-track interferometry.

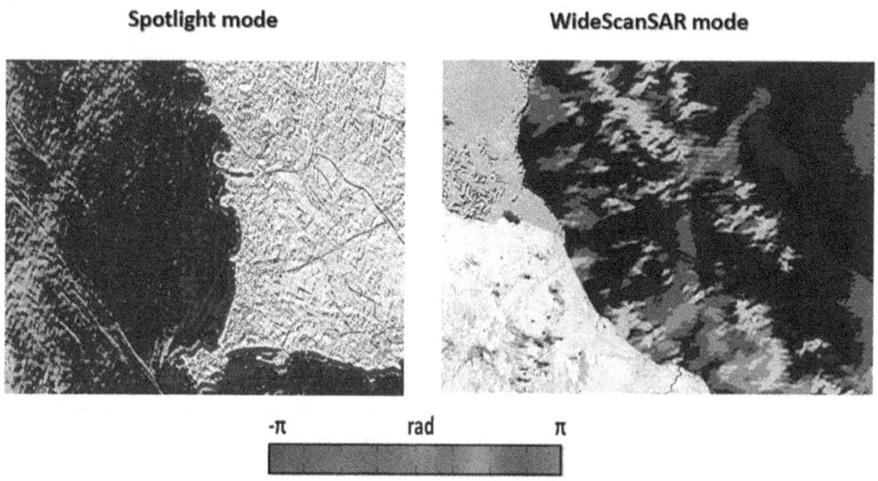

FIGURE 13.16 Two-dimensional hologram phase unwrapping.

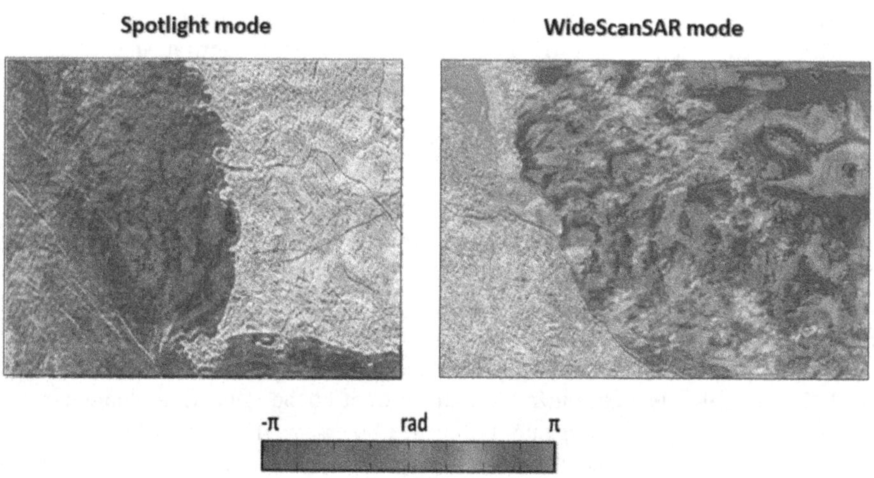

FIGURE 13.17 Three-dimensional hologram phase unwrapping.

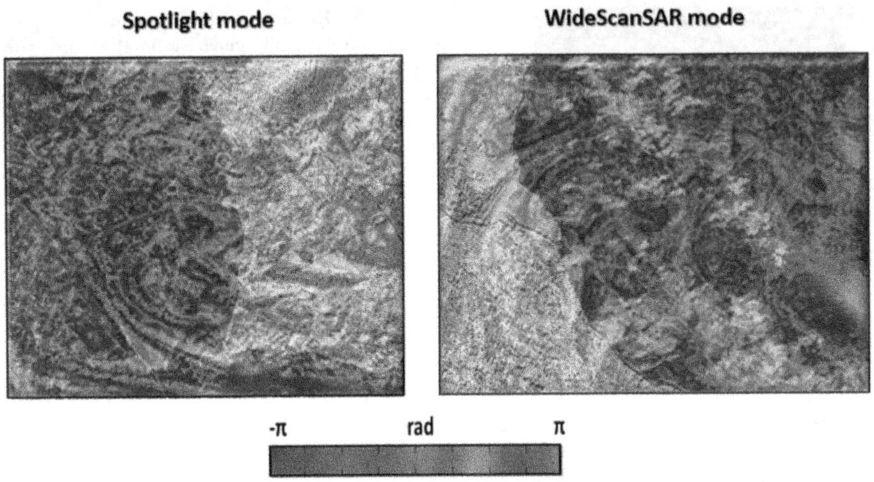

FIGURE 13.18 Four-dimensional hologram phase unwrapping.

of Hamming weight algorithms has effectively minimized errors in the interferogram cycle, particularly in regions with low coherence along water zones and coastlines, where high decorrelation is prevalent, notably in WideScanSAR mode.

In the comparative analysis between 3-D (Fig. 13.19) and 4-D visualizations in both Spotlight and WideScanSAR modes, respectively, it becomes evident that the identification of shear-wave chaotic turbulence flow is significantly enhanced in the 4-D hologram representation. The 4-D hologram interferometry provides richer insights into the characteristics of shear waves, including the generation of eddies and vorticities, particularly noticeable in the WideScanSAR mode with a maximum speed of 0.75 m/s (). Notably, during the month of October, which marks the transitional monsoon period between the southwest and northeast monsoons, previous eddies have been discussed. In contrast, May represents the southwest monsoon period, where the current predominantly slows down to 0.35 m/s along Port Dickson and due to low tide of less than 0.5 m as observed during in situ measurements. Additionally, shear-wave turbulence is observed along the coastal waters of Port Dickson, attributed to the shallowness of the region.

Conversely, the most robust shear waves are observed along the coastal waters of Kota Bharu, driven by strong current flows originating from the Gulf of Thailand. These flows generate eddies and well-defined quantized Marghany bound fronts due to the cold current flow along the coastal waters of Pattani, Thailand, toward the coastal waters of Kota Bharu, Kelantan, Malaysia, with a speed of 0.75 m/s. This shear-wave-induced turbulence results in a notable increase in sea level, reaching up to 1.2 meters.

In practice, the automated identification of nonlinear shear-wave chaotic patterns within the surrounding environment aligns seamlessly with the concept of “|NOON>”. In this context, |NOON> is employed to define a dual-state boundary for shear-wave chaotic turbulence and wave height variations within the 4-D image. Consequently, the fourth dimension vividly illustrates the presence of chaotic shear waves, along with eddy turbulence flows along coastal waters. The utilization of quantum hologram interferometry is regarded as a deterministic algorithm, carefully selected to locally regulate triangulation between two distinct pixels in TanDEM-X data. This process precisely renders a complete four-dimensional high-resolution image (Fig. 13.20), capturing intricate details of shear-wave chaotic turbulence flow. The Hamming graph introduces a wave − current interaction scenario that transcends the perspectives of both 2-D and 3-D in TanDEM-X data. This substantiates the effectiveness of quantum hologram interferometry based on the Hamming graph in accurately measuring movements reflected in phase shift fluctuations resulting from dynamic changes in the Doppler shift. Thus the development of a quantum interferometry based on the Hamming graph holds potential for unraveling specular reflection intricacies due to Doppler frequency shifts, while also facilitating the nonlinearities inherent in shear-wave chaotic imaging in synthetic aperture radar (SAR) and the nonlinear dynamics of shear-wave forming chaotic turbulence flow.

The 4-D simulation of shear-wave chaotic flows in both modes exhibits exceptional precision through the reconstruction of the 4-D hologram phase unwrapping using the Hamming graph algorithm grounded in quantum processing (Mills, 2004; Moghaddam et al., 2012). These outcomes provide a precise geometric morphology of both wave height propagation and sea surface current boundaries, delineated by the interplay between current movements and wave height transformations in coastal waters. The most distinct quantized Marghany bound front state is clearly discerned in 4-D hologram interferometry, as illustrated in Fig. 13.21.

Indeed, the application of 4-D hologram interferometry has conclusively illustrated and verified the presence of the quantized Marghany front bound state within the quantized Marghany turbulence boundary. This compelling evidence indicates that shear waves, recognized as an additional

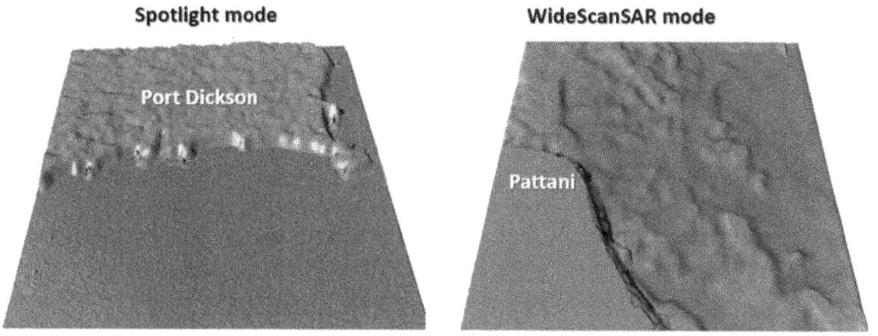

FIGURE 13.19 3-D shear-wave chaotic reconstruction in spotlight and WideScanSAR modes, respectively.

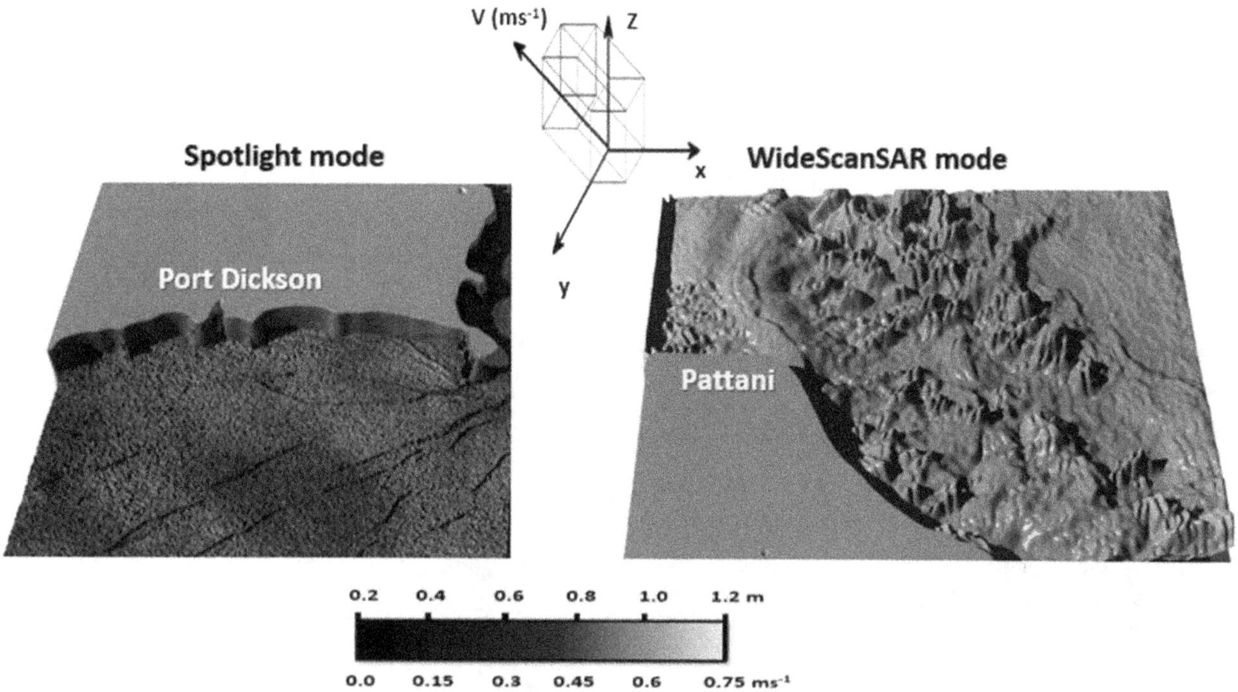

FIGURE 13.20 Simulation of shear-wave chaotic reconstruction in spotlight and WideScanSAR modes using 4-D hologram interferometry.

FIGURE 13.21 4-D hologram interferometry for tracing the quantized Marghany bound front state in the turbulence boundary of quantized Marghany.

turbulence feature, play a role in the formation and evolution of both the quantized Marghany front bound state, contributing to the establishment of the quantized Marghany turbulence boundary along the western boundary (Drazin & Riley, 2006; Karimova, Lavrova, et al., 2012; Marghany, 2013a; 2013b; 2019a).

13.13 Quantum relativity: four-dimensional sea surface reconstruction in TanDEM data

As we endure closer into the study of shear-wave chaotic turbulence, we must emphasize the complexity of the chaotic flow that is ingrained in ocean dynamics. This turbulent motion is characterized by an ensemble of nonlinear flow patterns involving vorticity, spirals, fronts, and eddies, among numerous additional phenomena. The combination of these several nonlinear behaviors leads to the formation of chaotic swirling motion in the complex structure of the ocean.

An intuitive grasp emerges as we continue to simulate this chaotic nonlinear flow: there are dynamics at work that go beyond simple spatial concerns. To fully understand these disordered movements, we need to go beyond space and include time in our investigation. This enlargement establishes the foundation for a theory of chaotic flow existing and changing inside the large four-dimensional space − time surface.

Imagine the tumultuous swirl of eddies, the complex maneuver of vorticity, the elegant spirals, and the spectacular creation of fronts—all of which are playing out not only in space but also constantly changing over the course of time. This chaotic spinning presents a complex and engrossing scenario of the turbulent responses of the ocean through the interaction of nonlinear components that shape the fluid dynamics inside a multidimensional space − time tapestry. This formulation expands beyond the conventional bounds of turbulent flow theory, inviting scientists to investigate the complex fabric of chaotic turbulence from a cohesive and multifaceted perspective.

In Einstein's groundbreaking theory of relativity, the fabric of the universe is described as four-dimensional, with the fourth dimension corresponding to time intricately woven into the three spatial dimensions. This concept of 4-D space − time is rooted in the principles of Lorentz transformations, a fundamental aspect of relativistic physics. In this context, the shear-wave chaotic features captured in synthetic aperture radar (SAR) data undergo a transformation from one inertial reference frame [e.g., (x, y, z)] to another [e.g., (p, q, r, s)].

The incorporation of 4-D quantum interferometry introduces a novel approach to comprehensively depict ocean dynamic features within SAR images. This approach leverages sophisticated algorithms such as Hamming graph, quantum walk search, and quantum hypercube to unveil a comprehensive scenario of quantum shear-wave chaotic flows. Essentially, 4-D quantum interferometry serves as a promising technique for the transformation of 2-D and 3-D images into a richer 4-D representation (Genderen & Marghany, 2014; Marghany, 2013a; 2013b).

This work aligns with the insights presented by Marghany (2021a), and Figs. 13.22 and 13.23 visually

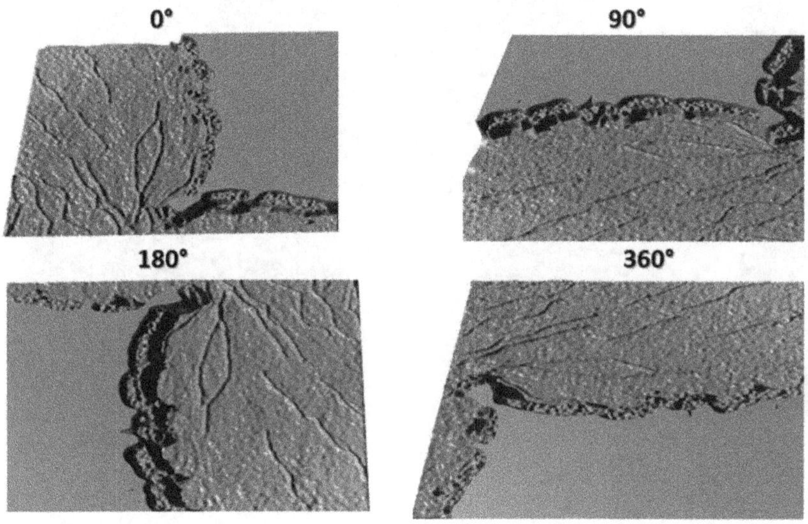

FIGURE 13.22 Various perspectives in 4-D quantum hologram interferometry simulation of chaotic shear-wave flow in spotlight mode.

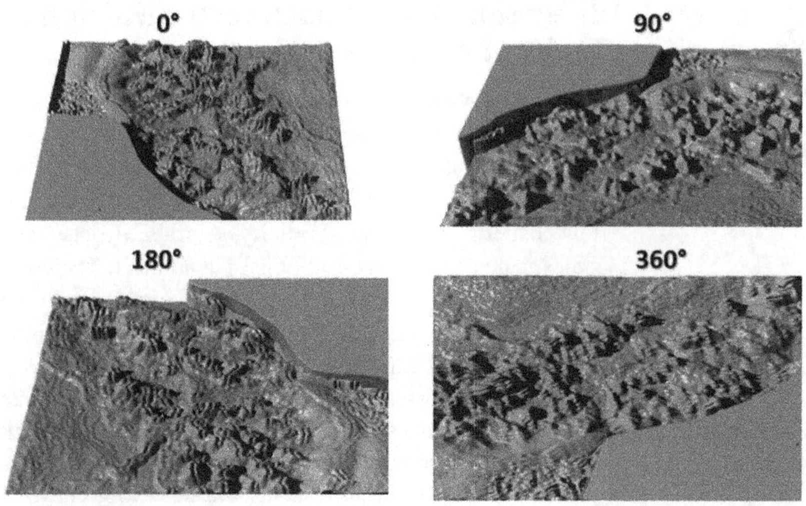

FIGURE 13.23 Various perspectives in 4-D quantum hologram interferometry simulation of chaotic shear-wave flow in WideScanSAR mode.

affirm the efficacy of this approach. Different viewing angles reveal distinct features in 4-D, with 0-degree angles highlighting wave height, ship footprints, and current boundaries through scattered bright patches. At 90- and 360-degree viewing angles, the intricate interplay between waves and currents becomes evident, showcasing turbulent movements along coastal waters with sharply defined crests at 180-degree viewing angles. This phenomenon is aptly showcased in both the Spotlight and WideScanSAR modes.

The utilization of duality in this context serves to provide a conceptual framework for understanding shear-wave chaotic turbulence flows in both modes. By considering the flows from a dual perspective, where they extend on one side of the hypercube and exhibit topological features on the other side, the analysis aims to capture the complexity and interplay of shear-wave chaotic features. This dual viewpoint allows for a nuanced exploration of the characteristics and behavior of shear-wave chaotic turbulence, offering insights into its dynamic nature and composition.

In the realm of quantum hologram interferometry applied to TanDEM-X data, a novel perspective emerges when viewing the chaotic turbulence through the lens of the hypercube. The hypercube, a higher-dimensional analog of a cube, offers a unique vantage point to dissect the intricate features of shear-wave chaotic turbulence flows. This perspective is particularly illuminating when considering the duality inherent in these turbulent flows.

Transitioning to the hypercube's viewpoint, the visualization of edge waves becomes a key focal point, as depicted in Fig. 13.24, and the chaotic shear-wave generates various water turbulence flows such as vorticity, eddies, and quantized Marghany front bound state as well demonstrated in WideScanSAR mode (Fig. 13.25). In the exploration of Lyapunov spectra within the dynamics of shear-wave chaotic turbulence, a notable divergence becomes apparent when comparing Spotlight mode and Wide ScanSAR mode. The distinction is particularly pronounced in the highest Lyapunov spectrum, which attains a value of 0.9 in the context of Wide ScanSAR mode. This numerical discrepancy signifies a heightened intensity of shear-wave chaotic turbulence within the domain of Wide ScanSAR mode as opposed to the conditions observed in Spotlight mode. The numerical values of Lyapunov spectra serve as quantitative indicators of the dynamic complexity inherent in shear-wave chaotic turbulence, shedding light on the distinct characteristics and intensities present in the two observation modes.

This phenomenon remains consistent across diverse viewing angles, with each facet of the hypercube unveiling

FIGURE 13.24 Exploring the Lyapunov spectrum of 4-D shear-wave chaotic turbulence in spotlight mode using the hypercube framework.

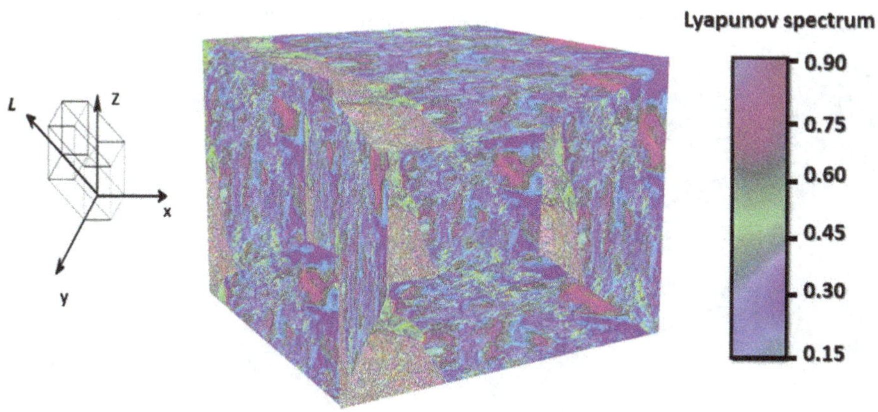

FIGURE 13.25 Exploring the Lyapunov spectrum of 4-D shear-wave chaotic turbulence in WideScanSAR mode using the hypercube framework.

a distinct facet of wave – current interaction. The TanDEM-X data, in particular, showcase the evolution of Lyapunov spectrum of 4-D quantized Marghany front bound states with highest value of 0.9 and its spectrum dissipation within Lyapunov spectrum value of 0.15 (Fig. 13.26). It also shows the shear-wave chaotic in various directions, including upward, downward, rightward, and leftward movements. Therefore the fourth dimension, akin to the structure of tesseracts, plays a pivotal role in this narrative. It extends and reconstructs the intricate patterns of wave – current interaction, providing a holistic understanding of ocean dynamics in a four-dimensional space – time continuum. The culmination of these dynamics gives rise to a cubic mass within the 4-D framework, comprising identical yet nontrivial tesseracts.

In essence, this 4-D exploration offers a profound insight into the quantum hologram interferometry's capacity to unravel the complexities of shear-wave chaotic turbulence. The establishment of tesseracts within the global space signifies a harmonious interplay of dimensions, encapsulating the multidimensional nature of oceanic turbulence with precision and depth.

The fundamental query that emerges is whether a 4-D quantum hologram can reconcile disparate periods and geographical locations with substantial temporal gaps. In this viewpoint, the 4-D quantum 3D image exhibits its ability to accommodate unmistakable periods, as outlined in Fig. 13.27. Indeed, the two distinct scenarios of shear-wave chaos demonstrate their integration into the 4-D quantum hologram. The initial overpasses are constructed in the first phase of time, and the subsequent scenarios of different times and geographical locations are constructed in diverse temporal coordinates.

This peculiarity can be explained from the angle of relativity hypothesis. Einstein's theory of relativity posits that time is not an absolute entity but is interconnected with space in a four-dimensional continuum known as spacetime. In this spacetime framework, events are not defined solely by their spatial coordinates but also by their temporal placement. Therefore the 4-D quantum hologram, encapsulating both time and space dimensions, aligns with the relativistic concept of spacetime, allowing for the compilation and representation of diverse scenarios across different temporal and spatial coordinates. The capacity of the 4-D quantum visualization to consistently coordinate unique shear-wave turbulent occasions is a demonstration of its adherence to the standards of relativity hypothesis.

FIGURE 13.26 4-D quantum hologram interferometry of Lyapunov spectrum of quantized Marghany front bound state 4.

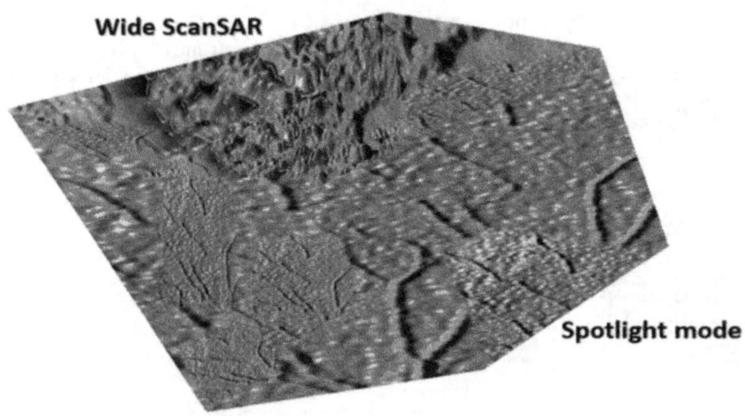

FIGURE 13.27 Unveiling shear-wave chaos across different time and different space. Insights from 4-D quantum hologram visualization.

Absolutely, the depiction of the initial stages of shear-wave occurrences in distinct temporal and geographical contexts, along with their anticipated developments, within the realm of 4-D quantum hologram interferometry is indeed significant.

This peculiarity fills in as a convincing exhibition and, as it were, validates the idea of quantum snare. While quantum entrapment is usually connected with the investigation of electron conduct, the utilization of this idea to the elements of maritime frameworks across different areas and time spans is a momentous expansion. The ensnarement guideline recommends that particles can be corresponded so that the condition of one molecule quickly impacts the condition of another, no matter what the distance between them.

In the context of the ocean dynamic systems, the 4-D quantum hologram's ability to encapsulate and display the interconnected evolution of shear-wave behaviors in different spatial and temporal domains aligns with the essence of quantum entanglement.

This extends how we might interpret quantum peculiarities past the tiny domain, representing its pertinence and appropriateness to naturally visible frameworks such as sea elements. The 4-D quantum 3D image, going about as a visual portrayal of caught shear-wave occasions, highlights the interconnected idea of these events across assorted areas and time spans, exhibiting the more extensive ramifications of quantum standards in figuring out complex normal peculiarities.

This chapter establishes that these 4D images unveil the Lyapunov spectrum derived from 4D quantum hologram interferometry. The Lyapunov spectrum, recognized as evidence of quantum entanglement theory, serves as a key indicator of the complex and dynamic nature of shear-wave chaotic flows. The outcomes demonstrate the impact of 4D shear-wave chaotic flows in generating phenomena such as eddies, vorticity, and a quantized Marghany turbulence boundary layer. This suggests that the quantum chaotic nonlinear system in ocean dynamics contributes to the formation of a quantized Marghany turbulence boundary layer.

In essence, the integration of quantum principles with the 4D phase unwrapping of TerraSAR-X satellite data is articulated as Marghany's 4D quantized phase unwrapping algorithm. This algorithm stands as a promising and high-quality technique for the 4D reconstruction of shear-wave chaotic turbulence flows and other features within complex SAR data, encompassing various modes of TerraSAR-X satellite data.

References

Bhattacharya, A., & Mukherjee, K. (2017). Review on InSAR based displacement monitoring of Indian Himalayas: Issues, challenges and possible advanced alternatives. *Geocarto International*, *32*(3), 298–321. Available from https://doi.org/10.1080/10106049.2016.1140820, http://www.tandfonline.com/toc/tgei20/current.

Drazin, P. G., & Riley, N. (2006). *The Navier-Stokes equations: A classification of flows and exact solutions*. Cambridge University Press.

Dritschel, D. G. (1988). Contour surgery: A topological reconnection scheme for extended integrations using contour dynamics. *Journal of Computational Physics*, *77*(1), 240–266. Available from https://doi.org/10.1016/0021-9991(88)90165-9.

Genderen, J. V., & Marghany, M. (2014). A three-dimensional sorting reliability algorithm for coastline deformation monitoring, using interferometric data. *IOP Conference Series: Earth and Environmental Science*, *18*(1), 012116. Available from https://doi.org/10.1088/1755-1315/18/1/012116.

Imel, D.D.A., Hensley, S., Pollard, B., Chapin, E., & Rodriguez, E. (2002). AIRSAR along-track interferometry. In chez 4th European Conference on Synthetic Aperture Radar (Vol. 117).

Johannessen, J. A., Shuchman, R. A., Digranes, G., Lyzenga, D. R., Wackerman, C., Johannessen, O. M., & Vachon, P. W. (1996). Coastal ocean fronts and eddies imaged with ERS 1 synthetic aperture radar. *Journal of Geophysical Research: Oceans*, *101*(C3), 6651–6667. Available from https://doi.org/10.1029/95jc02962.

Karimova, S. (2012). Spiral eddies in the Baltic, Black and Caspian seas as seen by satellite radar data. *Advances in Space Research*, *50*(8), 1107–1124. Available from https://doi.org/10.1016/j.asr.2011.100.027, http://www.journals.elsevier.com/advances-in-space-research/.

Karimova, S. S., Lavrova, O. Y., & Solov'ev, D. M. (2012). Observation of eddy structures in the Baltic Sea with the use of radiolocation and radiometric satellite data. *Izvestiya - Atmospheric and Ocean Physics*, *48*(9), 1006–1013. Available from https://doi.org/10.1134/S0001433812090071, http://www.maik.rssi.ru/journals/physatm.htm.

Krieger G. Moreira A. Fiedler H. Hajnsek I. Eineder M. Zink M. Werner M. 2006 22006/02 European Space Agency, (Special Publication) ESA SP 03796566 610 Germany. Tandem-x: A satellite formation for high resolution SAR interferometry.

Krieger, G., Moreira, A., Fiedler, H., Hajnsek, I., Werner, M., Younis, M., & Zink, M. (2007). TanDEM-X: A satellite formation for high-resolution SAR interferometry. *IEEE Transactions on Geoscience and Remote Sensing*, *45*(11), 3317–3341. Available from https://doi.org/10.1109/tgrs.2007.900693.

Li, S., Xu, W., & Li, Z. (2022). Review of the SBAS InSAR time-series algorithms, applications, and challenges. *Geodesy and Geodynamics*, *13*(2), 114–126. Available from https://doi.org/10.1016/j.geog.2021.090.007, http://www.keaipublishing.com/en/journals/geodesy-and-geodynamics/.

Lyzenga, D., Wackerman, C. (1997). Detection and classification of ocean eddies using ERS-1 and aircraft SAR images. European Space Agency, (Special Publication) ESA SP. (414), 1267–1271.

Macchiarulo, V., Milillo, P., Blenkinsopp, C., Reale, C., & Giardina, G. (2023). Multi-temporal InSAR for transport infrastructure monitoring: Recent trends and challenges. *Proceedings of the Institution of Civil Engineers - Bridge Engineering*, *176*(2), 92–117. Available from https://doi.org/10.1680/jbren.21.00039.

Marghany, M. (2013a). DInSAR technique for three-dimensional coastal spit simulation from RADARSAT-1 fine mode data. *Acta Geophysica*, *61*(2), 478–493. Available from https://doi.org/10.2478/s11600-012-0061-5.

Marghany M. 2013b 8 2013/08 Lecture Notes in Computer Science (including subseries Lecture Notes in Artificial Intelligence and

Lecture Notes in Bioinformatics) 10.1007/978-3-642-39643-4_43 16113349 2 599–610, Malaysia. Three dimensional coastline deformation from insar envisat satellite data, 7972.

Marghany, M. (2019a). *Automatic detection algorithms of oil spill in radar images*. CRC Press.

Marghany, M. (2019b). *Synthetic aperture radar imaging mechanism for oil spills* (pp. 1–322). Elsevier. Available from http://www.sciencedirect.com/science/book/9780128181119, https://doi.org/10.1016/C2018-0-02617-1.

Marghany, M. (2021a). *Four-dimensional Covid-19 simulation in slums using hologram interferometry of Sentinel-1A—Satellite Urban Book Series* (pp. 167–185). Indonesia: Springer Science and Business Media Deutschland GmbH. Available from http://springer.com/series/14773, https://doi.org/10.1007/978-3-030-71945-6_10.

Marghany, M. (2021b). *Nonlinear ocean dynamics: Synthetic aperture radar*. Elsevier.

Marghany, M. (2021c). *Advanced algorithms for mineral and hydrocarbon exploration using synthetic aperture radar*. Malaysia: Elsevier. Available from https://www.sciencedirect.com/book/9780128217962, https://doi.org/10.1016/C2019-0-04105-2.

Marghany, M. (2022). Remote sensing and image processing in mineralogy. CRC Press.

Mills, P. (2004). *Following the vapour trail: A study of chaotic mixing of water vapour in the upper troposphere*.

Moghaddam, F. F., Moghaddam, R. F., & Cheriet, M. (2012). *Curved space optimization: A random search based on general relativity theory*.

Oseen, C. W. (1911). *Uber die Stoke'sche Formel und über eine verwandte Aufgabe in der Hydrodynamik: Mitteilung, 2*.

Ott, E. (2002). *Chaos in dynamical systems*. Cambridge University Press.

Ou, Z. Y. J. (2007). *Multi-photon quantum interference* (pp. 1–268). Springer US, United States. Available from http://www.springerlink.com/openurl.asp?genre = book&isbn = 978-0-387-25532-3, https://doi.org/10.1007/978-0-387-25554-5.

Pattanayak, A. K. (2001). Characterizing the metastable balance between chaos and diffusion. *Physica D: Nonlinear Phenomena, 148*(1-2), 1–19. Available from https://doi.org/10.1016/S0167-2789(00)00186-X.

Paul, E. L., Atiemo-Obeng, V. A., & Kresta, S. M. (2004). *Handbook of industrial mixing* (pp. 1–143). Wiley-Blackwell.

Pinel, V., Hooper, A., De la Cruz-Reyna, S., Reyes-Davila, G., Doin, M. P., & Bascou, P. (2011). The challenging retrieval of the displacement field from InSAR data for andesitic stratovolcanoes: Case study of Popocatepetl and Colima Volcano, Mexico. *Journal of Volcanology and Geothermal Research, 200*(1-2), 49–61. Available from https://doi.org/10.1016/j.jvolgeores.2010.120.002.

Rankine, W. J. (1872). *A manual of applied mechanics*. Charles Griffin and Company.

Robinson, A. C., & Saffman, P. G. (1984). Stability and structure of stretched vortices. *Studies in Applied Mathematics, 70*(2), 163–181. Available from https://doi.org/10.1002/sapm1984702163UnitedStates, http://onlinelibrary.wiley.com/journal/10.1111/(ISSN)1467-9590.

Romeiser, R., Johannessen, J., Chapron, B., Collard, F., Kudryavtsev, V., Runge, H., Suchandt, S. (2010). Direct surface current field imaging from space by along-track InSAR and conventional SAR Oceanography from Space: Revisited. Springer Netherlands, United States, 73–91, Available from: http://www.springerlink.com/openurl.asp?genre = book&isbn = 978-90-481-8680-8. https://doi.org/10.1007/978-90-481-8681-5_5.

Rott, N. (1958). On the viscous core of a line vortex. *ZAMP Zeitschrift für Angewandte Mathematik und Physik, 9*(5-6), 543–553. Available from https://doi.org/10.1007/BF02424773.

Suchandt S. Runge H. 2012 1 2012/01 Proceedings of the European Conference on Synthetic Aperture Radar, EUSAR 9783800734047 392-395. Institute of Electrical and Electronics Engineers Inc. Germany. Along-Track interferometry using tandem-x: First results from marine and land applications, http://conference.vde.com/eusar/history/Pages/EUSARHistory.aspx.

Sullivan, R. D. (1959). A two-cell vortex solution of the Navier-Stokes equations. *Journal of the Aerospace Sciences, 26*(11), 767–768. Available from https://doi.org/10.2514/8.8303.

Waugh, D. W., & Plumb, R. A. (1994). Contour advection with surgery: a technique for investigating finescale structure in tracer transport. *Journal of the Atmospheric Sciences, 51*(4), 530–540. Available from https://doi.org/10.1175/1520-0469(1994)051 < 0530:CAWSAT > 2.0.CO;2.

Wnuk, K., Walton, G., & Zhou, W. (2019). Four-dimensional filtering of InSAR persistent scatterers elucidates subsidence induced by tunnel excavation in the Sri Lankan highlands. *Journal of Applied Remote Sensing, 13*(3). Available from https://doi.org/10.1117/1.JRS.13.034508, http://www.spie.org/x3636.xml.

Woodhouse, I. H. (2017). *Introduction to microwave remote sensing* (pp. 1–400). United Kingdom: CRC Press. Available from https://www.routledge.com/Introduction-to-Microwave-Remote-Sensing/Woodhouse/p/book/9780415271233, https://doi.org/10.1201/9781315272573.

Index

Note: Page numbers followed by "*f*" and "*t*" refer to figures and tables, respectively.

A

B

F

N

R

T

U

V

W

X

Y

Z

www.ingramcontent.com/pod-product-compliance
Lightning Source LLC
Chambersburg PA
CBHW080749090425
24824CB00036B/2213